国外优秀数学著作
原 版 系 列

非线性系统及其绝妙的数学结构

Nonlinear Systems and Their Remarkable Mathematical Structures(Volume 1)

（第1卷）

（英文）

●［墨］诺伯特·欧拉（Norbert Euler） 主编

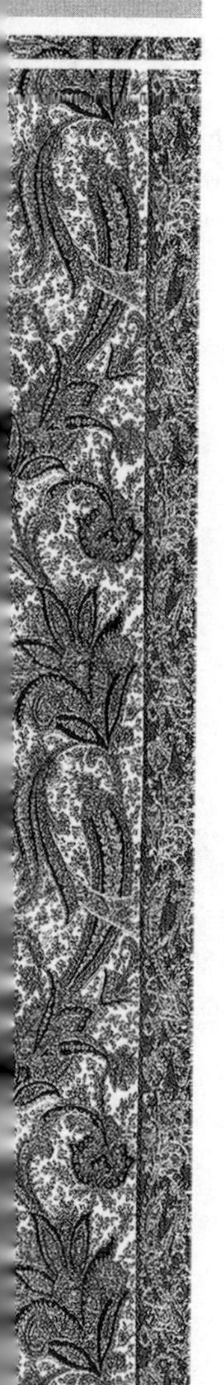

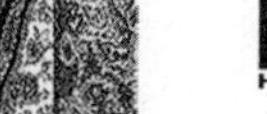

哈爾濱工業大學出版社
HARBIN NSTITUTE OF TECHNOLOGY PRESS

黑版贸审字 08-2020-199 号

Nonlinear Systems and Their Remarkable Mathematical Structures：Volume 1/by Norbert Euler/ISBN：978-1-1386-0100-0

图书在版编目（CIP）数据

非线性系统及其绝妙的数学结构. 第 1 卷 = Nonlinear Systems and Their Remarkable Mathematical Structures：Volume 1：英文/（墨）诺伯特・欧拉（Norbert Euler）主编. —哈尔滨：哈尔滨工业大学出版社，2024. 10
ISBN 978 - 7 - 5767 - 1283 - 4

Ⅰ. ①非… Ⅱ. ①诺 … Ⅲ. ①非线性控制系统-英文 Ⅳ. ①O231. 2

中国国家版本馆 CIP 数据核字（2024）第 050314 号

FEIXIANXING XITONG JI QI JUEMIAO DE SHUXUE JIEGOU. DI 1 JUAN

策划编辑 刘培杰 杜莹雪
责任编辑 张嘉芮 李兰静
封面设计 孙茵艾
出版发行 哈尔滨工业大学出版社
社 址 哈尔滨市南岗区复华四道街 10 号 邮编 150006
传 真 0451 - 86414749
网 址 http://hitpress. hit. edu. cn
印 刷 哈尔滨市工大节能印刷厂
开 本 787 mm×1 092 mm 1/16 印张 38. 5 字数 668 千字
版 次 2024 年 10 月第 1 版 2024 年 10 月第 1 次印刷
书 号 ISBN 978 - 7 - 5767 - 1283 - 4
定 价 88. 00 元

（如因印装质量问题影响阅读，我社负责调换）

Contents

Preface .... **ix**

The Authors .... **xi**

A1. **Systems of nonlinearly-coupled differential equations solvable by algebraic operations** **1**

F Calogero

1. Introduction 1
2. The main idea and some key identities 1
3. Two examples of systems of nonlinearly-coupled ODEs solvable by algebraic operations 5
4. A differential algorithm to evaluate all the zeros of a generic polynomial of arbitrary degree 9
5. Extensions 11

A2. **Integrable nonlinear PDEs on the half-line** **15**

A S Fokas and B Pelloni

1. Introduction 15
2. Transforms and Riemann-Hilbert problems 20
3. The structure of integrable PDEs: Lax pair formulation 24
4. An integral transform for nonlinear boundary value problems 26
5. Further considerations 37

A3. **Detecting discrete integrability: the singularity approach** **44**

B Grammaticos, A Ramani, R Willox and T Mase

1. Introduction 44
2. Singularity confinement 46
3. The full-deautonomisation approach 53
4. Halburd's exact calculation of the degree growth 57
5. Singularities and spaces of initial conditions 64

A4. **Elementary introduction to discrete soliton equations** **74**

J Hietarinta

1. Introduction 74
2. Basic set-up for lattice equations 74
3. Symmetries and hierarchies 77
4. Lax pairs 80
5. Continuum limits 82
6. Discretizing a continuous equation 85
7. Integrability test 91
8. Summary 92

A5. **New results on integrability of the Kahan-Hirota-Kimura discretizations** **94**

Yu B Suris and M Petrera

1. Introduction 94
2. General properties of the Kahan-Hirota-Kimura discretization 96
3. Novel observations and results 96
4. The general Clebsch flow 99
5. The first Clebsch flow 104
6. The Kirchhoff case 113
7. Lagrange top 116
8. Concluding remarks 117

B1. **Dynamical systems and isospectral matrices defined in terms of the zeros of orthogonal or otherwise special polynomial** **122**

O Bihun

1. Introduction 122
2. Zeros of generalized hypergeometric polynomial with two parameters and zeros of Jacobi polynomials 127
3. Zeros of generalized hypergeometric polynomials 133
4. Zeros of generalized basic hypergeometric polynomials 136
5. Zeros of Wilson and Racah polynomials 141
6. Zeros of Askey-Wilson and q-Racah polynomials 147

7. Discussion and Outlook 154

B2. Singularity methods for meromorphic solutions of differential equations 159

R Conte, T W Ng and C F Wu

1. Introduction 159
2. A simple pedagogical example 163
3. Lessons from this pedagogical example 167
4. Another characterization of elliptic solutions: the subequation method 169
5. An alternative to the Hermite decomposition 172
6. The important case of amplitude equations 173
7. Nondegenerate elliptic solutions 179
8. Degenerate elliptic solutions 180
9. Current challenges and open problems 182

B3. Pfeiffer-Sato solutions of Buhl's problem and a Lagrange-d'Alembert principle for heavenly equations 187

O E Hentosh, Ya A Prykarpatsky, D Blackmore and A Prykarpatski

1. Introduction 187
2. Lax–Sato compatible systems of vector field equations 190
3. Heavenly equations: Lie-algebraic integrability scheme 195
4. Integrable heavenly dispersionless equations: Examples 199
5. Lie-algebraic structures and heavenly dispersionless systems 202
6. Linearization covering method and its applications 207
7. Contact geometry linearization covering scheme 215
8. Integrable heavenly superflows: Their Lie-algebraic structure 217
9. Integrability and the Lagrange–d'Alembert principle 222

B4. Superposition formulae for nonlinear integrable equations in bilinear form 233

X B Hu

1. Introduction 233
2. Bianchi theorem of permutability and superposition formula of the KdV equation 235
3. Superposition formulae for a variety of soliton equations with examples 237

4. Superposition formulae for rational solutions 247
5. Superposition formulae for some other particular solutions 252

B5. **Matrix solutions for equations of the AKNS system** **257**

C Schiebold

1. Introduction 257
2. An operator approach to integrable systems 259
3. The nc AKNS system 263
4. Solution formulas for the AKNS system 265
5. Projection techniques revisited 268
6. Matrix- and vector-AKNS systems 269
7. Reduction 272
8. The finite-dimensional case 273
9. Solitons, strongly bound solitons (breathers), degeneracies 278
10. Multiple pole solutions 282
11. Solitons of matrix- and vector-equations 287

B6. **Algebraic traveling waves for the generalized KdV-Burgers equation and the Kuramoto-Sivashinsky equation** **295**

C Valls

1. Introduction and statement of the main results 295
2. Proof of Theorem 2 and some preliminary results 299
3. Proof of Theorem 3 with $n = 1$ 301
4. Proof of Theorem 3 with $n = 2$ 307
5. Final comments 313

C1. **Nonlocal invariance of the multipotentialisations of the Kupershmidt equation and its higher-order hierarchies** **317**

M Euler and N Euler

1. Introduction: symmetry-integrable equations and multipotentialisations 317
2. The multipotentialisation of the Kupershmidt equation 327
3. Invariance of the Kupershmidt equation and its chain of potentialisations 333

4. The hierarchies 338
5. Concluding remarks 342
Appendix A: A list of recursion operators 343
Appendix B: An equation that does not potentialise 348

C2. **Geometry of normal forms for dynamical systems 352**

G Gaeta

1. Introduction 352
2. Normal forms 354
3. Normal forms and symmetry 356
4. Michel theory 358
5. Unfolding of normal forms 360
6. Normal forms in the presence of symmetry 364
7. Normal forms and classical Lie groups 365
8. Finite normal forms 368
9. Gradient property 369
10. Spontaneous linearization 370
11. Discussion and conclusions 371
Appendix A: The normal forms construction 373
Appendix B: Examples of unfolding 376
Appendix C: Hopf and Hamiltonian Hopf bifurcations 379
Appendix D: Symmetry and convergence for normal forms 381

C3. **Computing symmetries and recursion operators of evolutionary super-systems using the SsTools environment 390**

A V Kiselev, A O Krutov and T Wolf

1. Notation and definitions 391
2. Symmetries 393
3. Recursions 396
4. Nonlocalities 399

C4. **Symmetries of Itô stochastic differential equations and their applications 408**

R Kozlov

1. Introduction 408
2. Illustrating example 409

3. Itô SDEs and Lie point symmetries 410
4. Properties of symmetries of Itô SDEs 413
5. Symmetry applications 420

C5. **Statistical symmetries of turbulence 437**

M Oberlack, M Wacławczyk and V Grebenev

1. Foreword 437
2. Stochastic behavior and symmetries of differential equations - an introduction 438
3. Statistics of the Navier-Stokes equations and its symmetries 444
4. Summary and outlook 464

D1. **Integral transforms and ordinary differential equations of infinite order 468**

A Chavez, H Prado and E G Reyes

1. Introduction 468
2. Differential operators of infinite order in mathematics and physics 470
3. Mathematical theory for nonlocal equations 473
4. The operator $f(\partial_t) : L^p(\mathbb{R}_+) \longrightarrow H^q(\mathbb{C}_+)$ 478
5. The initial value problem 483
6. From the Laplace to the Borel transform 488
7. Linear zeta-nonlocal field equations 492
8. Future work 495

D2. **On the rôle of nonlinearity in geostrophic ocean flows on a sphere 500**

A Constantin and R S Johnson

1. Introduction 500
2. Preliminaries 501
3. Governing equations 501
4. Geostrophy and the f- and β-plane approximations 503
5. Geostrophy in spherical coordinates 510

6. Discussion 516

D3. **Review of results on a system of type many predators - one prey** **520**

A V Osipov and G Söderbacka

1. Introduction 520
2. Lotka-Volterra equations 521
3. Rosenzweig-McArthur equations 523
4. Mathematical tools 524
5. Systems with more predators 528
6. Modified standard system 537

D4. **Ermakov-type systems in nonlinear physics and continuum mechanics** **541**

C Rogers and W K Schief

1. Overview 541
2. A rotating shallow water system. Ermakov-Ray-Reid reduction 544
3. Hamiltonian Ermakov-Ray-Reid reduction in magneto-gasdynamics. The pulsrodon 548
4. Hamiltonian Ermakov-Ray-Reid systems. Parametrisation and integration 553
5. Multi-component Ermakov systems. Genesis in $\boldsymbol{N}$-layer hydrodynamics 555
6. Multi-component Ermakov and many-body system connections 559
7. Multi-component Ermakov-Painlevé systems 562

Subject Index **577**

编辑手记 **583**

Preface

The aim of this book is to provide a comprehensive account of the state of the art on the mathematical description of nonlinear systems. The book consists of 20 invited contributions written by leading experts in different aspects of nonlinear systems that include ordinary and partial differential equations, difference equations and q-difference equations, discrete or lattice equations, non-commutative and matrix equations, stochastic equations, as well as supersymmetric equations. The contents is divided into four main parts, namely **Part A**: *Integrable Systems*, **Part B**: *Solution Methods and Solution Structures*, **Part C**: *Symmetry Methods of Nonlinear Systems*, and **Part D**: *Nonlinear Systems in Applications*. Below we give a short description of each contribution.

Part A consists of five contributions, numbered A1 to A5. In this part the authors mainly address the fundamental question of how to detect integrable systems. In **A1** the author *F Calogero* describes a somewhat novel technique to identify nonlinear systems of differential equations solvable by algebraic operations. In **A2** the authors *A S Fokas and B Pelloni* give a comprehensive review of the so-called Fokas Method to solve initial-boundary value problems applied to nonlinear partial differential equations on the half-line. In **A3** the authors *B Grammaticos, A Ramani, R Willox and T Mase* describe how to detect discrete integrability by the singularity approach; the singularities of which arise for second order rational mappings. In **A4** the author *J Hietarinta* provides an elementary introduction to discrete or lattice soliton equations, where a multidimensional setting is discussed in some detail. In the last contribution of this section, namely **A5**, the authors *Yu B Suris and M Petrera* discuss new results of the Hahan-Hirota-Kimura discretizations and provide, amongst other things, the integrals of motion.

Part B makes up the largest part of the book and consists of six contributions, numbered B1 to B6. Here the authors describe different methods by which to obtain explicit solutions of nonlinear systems and/or describe the solution structures of the systems. In **B1** the author *O Bihun* discusses a general method to construct isospectral matrices that are defined in terms of the zeros of certain polynomials. It is shown, amongst other things, how this leads to solvable nonlinear first-order ordinary differential systems. In **B2** the authors *R Conte, T W Ng and C F Wu* provide a tutorial introduction to methods that were recently developed by the authors in order to find all meromorphic particular solutions of nonintegrable, autonomous, algebraic ordinary differential equations of any order. Some examples in physics are given. In **B3** the authors *O E Hentosh , Ya A Prykarpatsky, D Blackmore and A Prykarpatski* give a review to the Buhl compatible vector field equation problem by emphasizing its Pfeiffer and Lax–Sato type solutions. They furthermore analyze the related Lie-algebraic structures and integrability of the so-called heavenly equations. An interesting related Lagrange–d'Alembert principle is also discussed in this contribution, as well as other related aspects. In **B4** the author *X B Hu* is concerned with superposition formulae and Bianchi identities that are related to bilinear Bäcklund transformations for nonlinear integrable equations in bilinear form. The author utilizes this to generate soliton solutions, rational solutions as well as some other special solutions to some nonlinear integrable equations. Many examples are provided. In **B5** the author *C Schiebold* constructs $m \times n$-matrix valued solutions for the AKNS system. A complete asymptotic description is given for multiple pole solutions, including wave packets of weakly bound breathers. The collision of vector solitons is also studied. In **B6** the author *C Valls* uses a new method, recently introduced by A Gasull and H Giacomini, to characterize all traveling wave solutions of the Generalized Korteweg-de-Vries-Burgers equation as well as for the Kuramoto-Sivashinsky equation.

In **Part C** the authors concentrate on symmetry methods for nonlinear systems. It consists of five contributions, numbered C1 to C5. In **C1** the authors *M Euler and N Euler* describe the multipotentialization process by which it is possible to construct nonlocal invariance of certain nonlinear symmetry-integrable evolution equations in $1+1$ dimensions. A complete account of this process for the fifth-order Kupershmidt equation and its hierarchies is given. In **C2** the author *G Gaeta* discusses several aspects of the geometry of vector fields in (Poincaré-Dulac) normal form. This relies substantially on Michel theory. The case, common in physics, of systems enjoying an *a priori* symmetry is also discussed in some detail. In **C3** the authors *A V Kiselev, A O Krutov and T Wolf* provide an informal discussion of the step-by-step computation of nonlocal recursions for symmetry algebras of nonlinear coupled boson-fermion $N=1$ supersymmetric systems by using the SsTools environment. In **C4** the author *R Kozlov* discusses Lie point symmetries of Itô stochastic differential equations, which correspond to Lie group transformations of the independent variable (time) and dependent variables that preserve the differential form of the equations and the properties of Brownian motion. In **C5** the authors *M Oberlack, M Wacławczyk and V Grebenev* describe the so-called statistical symmetries of turbulence and discuss their importance in understanding the statistics of turbulence such as intermittency and non-gaussianity.

Part D provides some of the applications of nonlinear systems. This part consists of four contributions, numbered D1 to D4. In **D1** the authors *A Chávez, H Prado and E G Reyes* review their work on integral transforms for ordinary differential equations of infinite order and apply the theory to equations defined with the help of the Riemann zeta function which are of interest in modern theoretical physics. In **D2** the authors *A Constantin and R S Johnson* discuss the rôle of nonlinearity in geostrophic ocean flows on a sphere and point out some of its challenges. In **D3** the authors *A V Osipov and G Söderbacka* give a review of their results on a class of systems of the type n predators - one prey. They also discuss, amongst other things, the extinction and different simple and complicated coexistence of the predators. In **D4** the authors *C Rogers and W K Schief* discuss Ermakov-type systems in nonlinear physics and continuum mechanics. They describe applications in rotating shallow water theory, magnetogasdynamics, multi-layer hydrodynamics and many-body theory. The authors also review their recent work on hybrid Ermakov-Painlevé systems.

Norbert Euler (Luleå, 3 July 2018)

The Authors

1. Oksana Bihun, *University of Colorado, Colorado Springs, USA* [**B1**]
2. Denis Blackmore, *New Jersey Institute of Technology, USA* [**B3**]
3. Francesco Calogero, *University of Rome, La Sapienza, Italy* [**A1**]
4. Alan Chávez, *Universidad Nacional de Trujillo, Perú* [**D1**]
5. Adrian Constantin, *Universität Wien, Austria* [**D2**]
6. Robert Conte, *École normale supérieure de Cachan, CNRS, Université Paris-Saclay, France* [**B2**]
7. Marianna Euler, *Luleå University of Technology, Sweden* [**C1**]
8. Norbert Euler, *Luleå University of Technology, Sweden* [**C1**]
9. Thanasis Fokas, *University of Cambridge, UK* [**A2**]
10. Giuseppe Gaeta, *Università degli Studi di Milano, Italy* [**C2**]
11. Basil Grammaticos, *Université Paris VII & XI, France* [**A3**]
12. Vladimir Grebenev, *Institute of Computational Technologies, Russian Academy of Sciences, Novosibirsk, Russia* [**C5**]
13. Oksana E Hentosh, *Institute for Applied Problems of Mechanics and Mathematics at the NAS, Lviv, Ukraine* [**B3**]
14. Jarmo Hietarinta, *University of Turku, Turku, Finland* [**A4**]
15. Xing-biao Hu, *LSEC, Institute of Computational Mathematics Chinese Academy of Sciences, China* [**B4**]
16. Robin S Johnson, *Newcaste University, UK* [**D2**]
17. Arthemy V Kiselev, *University of Groningen, The Netherlands* [**C3**]
18. Roman Kozlov, *Norwegian School of Economics, Bergen, Norway* [**C4**]
19. Andrey O Krutov, *Independent University of Moscow, Russia* [**C3**]
20. Takafumi Mase, *University of Tokyo, Japan* [**A3**]
21. Tuen Wai Ng, *The University of Hong Kong, China* [**B2**]
22. Martin Oberlack, *Technische Universität Darmstadt, Germany* [**C5**]
23. Alexandr V Osipov, *University of Saint Peterburg, Russia* [**D3**]

24. Beatrice Pelloni, *Heriot-Watt University, United Kingdom* [**A2**]

25. Matteo Petrera, *Technische Universität Berlin, Germany* [**A5**]

26. Humberto Prado, *Universidad de Santiago de Chile, Chile* [**D1**]

27. Anatolij Prykarpatski, *Polytechnical University, Krakow, Poland* [**B3**]

28. Yarema A Prykarpatsky, *University of Agriculture Krakow, Poland* [**B3**]

29. Alfred Ramani, *Université Paris VII & XI, France* [**A3**]

30. Enrique G Reyes, *Universidad de Santiago de Chile, Chile* [**D1**]

31. Colin Rogers, *University of New South Wales, Australia* [**D4**]

32. Cornelia Schiebold, *Mid Sweden University, Sweden* [**B5**]

33. Wolfgang Schief, *University of New South Wales, Australia* [**D4**]

34. Gunnar Söderbacka, *Åbo Akademi, Finland* [**D3**]

35. Yuri B Suris, *Technische Universität Berlin, Germany* [**A5**]

36. Claudia Valls, *Técnico Universidade de Lisboa, Portugal* [**B6**]

37. Marta Wacławczyk, *University of Warsaw, Poland* [**C5**]

38. Ralph Willox, *University of Tokyo, Japan* [**A3**]

39. Thomas Wolf, *Brock University, Canada* [**C3**]

40. Chengfa Wu, *Shenzhen University, China* [**B2**]

A1. Systems of nonlinearly-coupled differential equations solvable by algebraic operations

Francesco Calogero

Physics Department, University of Rome "La Sapienza"
Istituto Nazionale di Fisica Nucleare, Sezione di Roma 1, Italy

Abstract

In this paper an overview is provided of an approach to identify systems of nonlinearly-coupled Ordinary Differential Equations solvable by *algebraic* operations— also including models interpretable as many-body problems with Newtonian equations of motion ("accelerations equal forces"). This technique— which was introduced several decades ago—has had an important evolution recently. This development entails that the universe of solvable models thereby generated is quite vast; it also includes dynamical systems which feature non-trivially rather simple time evolutions, for instance *isochronous* or *asymptotically isochronous.* In this paper this development is tersely described and representative examples of the solvable models thereby obtained are exhibited and tersely discussed. A differential algorithm to evaluate all the zeros of a generic polynomial of arbitrary degree is also reported. Extensions of this approach to systems of nonlinearly-coupled Partial Differential Equations and to evolutions in "discrete time" are tersely mentioned.

1 Introduction

Four decades ago the idea was introduced to identify nonlinearly-coupled systems of Ordinary Differential Equations (ODEs) amenable to exact treatments by investigating the motion of zeros and poles of special solutions of solvable partial differential equations [1]. This idea led to many developments; see for instance the relevant chapters in the two books [2] [3] and references therein. Recently a new twist of this idea has emerged [4]; this breakthrough, and some of its developments (see for instance [5] and the last section of the present paper), have vastly expanded the identification of dynamical systems solvable by algebraic operations: see [6] and references therein. In this paper we provide a terse overview of these findings.

2 The main idea and some key identities

Let $p_N(z;t)$ be a generic t-dependent monic polynomial of arbitrary degree N in the (*complex*) variable z:

$$p_N(z;t) = z^N + \sum_{m=1}^{N} \left[y_m(t) \ z^{N-m} \right] = \prod_{n=1}^{N} \left[z - x_n(t) \right] . \tag{2.1}$$

Notation 2-1. Above and hereafter N is an arbitrary positive integer ($N \geq 2$); t is a *real* parameter (generally interpretable as *time* in the context of more physical applications such as the identification and investigation of many-body problems characterized by Newtonian equations of motion, "accelerations equal forces"; or, as a dimensionless parameter in purely mathematical contexts); the N (generally *complex*) numbers $y_m(t)$ are the *coefficients* of the polynomial $p_N(z;t)$ and the N (generally *complex*) numbers $x_n(t)$ are the *zeros* of this polynomial. Indices such as n, m, ℓ always run (unless otherwise indicated) over *all positive integers* from 1 to N. The N coefficients y_m are in the following, occasionally, collectively identified as the N components of the N-vector $\vec{y}$; and the N zeros x_n as the N elements of the unordered set $\tilde{x}$. Note that here we omitted to indicate the t-dependence of these quantities; this we shall often do in the following, whenever such omissions are unlikely to cause misunderstandings. Finally, let us mention that in the following superimposed dots denote time differentiations, for instance $\dot{x}_n \equiv dx_n(t)/dt$. ■

The two equalities in (2.1) indicate that the polynomial $p_N(z;t)$ can be defined either by assigning its N coefficients $y_m(t)$ or its N zeros $x_n(t)$; and they imply that the N coefficients $y_m(t)$ can be expressed in terms of the N zeros $x_n(t)$ via the well-known formulas

$$y_m = (-1)^m \; \sigma_m(\tilde{x}) \qquad (2.2a)$$

where

$$\sigma_m(\tilde{x}) = \sum_{1 \leq \ell_1 < \ell_2 \cdots \ell_m \leq N} (x_{\ell_1} x_{\ell_2} \cdots x_{\ell_m}) \qquad (2.2b)$$

is the symmetrical sum of degree m of the N components x_n of the unordered set $\tilde{x}$; so that, in particular,

$$\sigma_1(\tilde{x}) = \sum_{n=1}^{N} (x_n) \; , \quad \sigma_N(\tilde{x}) = \prod_{n=1}^{N} (x_n) \; . \qquad (2.2c)$$

Likewise the t-derivative $\dot{y}_m(t)$ can be *explicitly* expressed in terms of $\tilde{x}(t)$ and $\dot{\tilde{x}}(t)$ by t-differentiating these formulas.

On the other hand the zeros x_n—while being uniquely determined, up to permutations, whenever the N coefficients y_m are themselves assigned—cannot be generally expressed in terms of the coefficients by *explicit* formulas, unless $N \leq 4$. Nevertheless, for all values of N the fact that this transition from the N coefficients $y_m(t)$ to the zeros $x_n(t)$ is an *algebraic* operation has important implications: for instance if the time evolution of the N coefficients $y_m(t)$ is *isochronous* with period T—i. e. $y_m(t+T) = y_m(t)$ for all (or almost all) assignments of initial data—then clearly the evolution of the corresponding zeros is also *isochronous*, $x_n\left(t+\tilde{T}\right) = x_n(t)$; albeit with a period $\tilde{T}$ which is generally an *integer* multiple of the period T, $\tilde{T} = \nu T$, with the integer ν in the range $1 \leq \nu \leq \nu_{\text{Max}}(N)$ (for a clear discussion of this question, including bounds and estimates of the maximal value $\nu_{\text{Max}}(N)$, see [7]).

And analogous results obtain of course in the case of *asymptotic isochrony* (for this notion see for instance [8]).

It is as well evident that if the time evolution of the N coefficients $y_m(t)$ is *Hamiltonian*, the time evolution of the N coefficients $x_n(t)$ is also *Hamiltonian*—although it might not be possible to write *explicitly* the corresponding Hamiltonian nor the canonical momenta conjugated to the canonical coordinates $x_n(t)$; and likewise if the time evolution of the N coefficients $y_m(t)$ is *Hamiltonian* and *integrable*, the corresponding time evolution of the N coefficients $x_n(t)$ is also *Hamiltonian* and *integrable*.

The main idea to identify systems of N nonlinearly-coupled ODEs *solvable by algebraic operations* is to consider *solvable* time evolutions of the N coefficients $y_m(t)$ of a t-dependent polynomial and to then focus on the corresponding evolution of the N zeros $x_n(t)$ of that polynomial (see (2.1)); by taking advantage of the fact that the transition from the N zeros x_n to the N coefficients y_m of a polynomial is given by *explicit* formulas, and the inverse transition from the N coefficients y_m to the N zeros x_n of a polynomial by *algebraic* operations (see above). The solvable character of the t-evolution of the N coefficients $y_m(t)$ might be simple enough to yield an *explicit* solution, as would for instance be the case for a system of N *linear autonomous decoupled* ODEs; or the solution might itself be achieved by *algebraic* operations—such as, say, the evaluation of the N eigenvalues of an explicitly known $N \otimes N$ matrix—as it would be the case for a system of N *linearly-coupled autonomous* ODEs. Note that, even when the time evolution of the N coefficients $y_m(t)$ is characterized by *linear* ODEs, the corresponding evolution of the N zeros is generally quite *nonlinear*.

Clearly to implement this idea one needs to find an *explicit* way to transform the evolution ODEs satisfied by the N coefficients $y_m(t)$ into the corresponding ODEs satisfied by the N zeros $x_n(t)$. Indeed, a simple way to do so provided the foundation for the results reported in [1] and subsequently largely exploited to identify and investigate algebraically solvable models characterized by nonlinearly-coupled systems of ODEs; see for instance [2] [3] and references therein. But it was restricted to the case in which the system of ODEs satisfied by the N coefficients $y_m(t)$ had an essentially *linear* character.

Recently a—remarkably simple—way to bypass this important restriction was noted [4]. It is provided by the following key identities:

$$\dot{x}_n = \frac{-\sum_{m=1}^{N}\left[\dot{y}_m\,(x_n)^{N-m}\right]}{\prod_{\ell=1,\ell\neq n}^{N}(x_n - x_\ell)} \tag{2.3a}$$

$$= \frac{-\left.\frac{\partial p_N(z;t)}{\partial t}\right|_{z=x_n}}{\prod_{\ell=1,\ell\neq n}^{N}(x_n - x_\ell)}\,, \tag{2.3b}$$

$$\ddot{x}_n = \sum_{\ell=1,\ \ell\neq n}^{N} \left(\frac{2\dot{x}_n\dot{x}_\ell}{x_n - x_\ell}\right) - \frac{\sum_{m=1}^{N}\left[\ddot{y}_m\ (x_n)^{N-m}\right]}{\prod_{\ell=1,\ell\neq n}^{N}(x_n - x_\ell)} \tag{2.4a}$$

$$= \sum_{\ell=1,\ \ell\neq n}^{N} \left(\frac{2\dot{x}_n\dot{x}_\ell}{x_n - x_\ell}\right) - \frac{\left.\frac{\partial^2 p_N(z;t)}{\partial t^2}\right|_{z=x_n}}{\prod_{\ell=1,\ell\neq n}^{N}(x_n - x_\ell)}. \tag{2.4b}$$

Remark 2-1. The proofs of these 2 identities are so simple that we do not feel the need to detail them here: they obtain by (partially) t-differentiating, once respectively twice, the 2 expressions in the right-hand side of (2.1), and then setting $z = x_n(t)$. Readers who like to see these proofs are referred to the literature [4] [5] [6], where they will also find the generalization of these identities to t-differentiations of higher order than 2.

And let us report here, for future reference, the following additional relations obviously implied by (2.1):

$$\sum_{m=1}^{N}\left[y_m\ (x_n)^{N-m}\right] = -(x_n)^N\ . \ \blacksquare \tag{2.5}$$

Clearly these identities open the way to the explicit transformation of systems of ODEs satisfied by the N coefficients $y_m(t)$ to systems of ODEs satisfied by the N zeros $x_n(t)$. For instance, let the N coefficients $y_m(t)$ satisfy the system of first-order ODEs

$$\dot{\vec{y}} = \vec{f}(\vec{y};t)\ ,\quad \dot{y}_m(t) = f_m(\vec{y}(t);t)\ , \tag{2.6a}$$

with N assigned functions $f_m(\vec{y};t)$ (the N components of the N-vector $\vec{f}(\vec{y};t)$). Then clearly the identity (2.3) implies that the N zeros $x_n(t)$ satisfy the following system of ODEs:

$$\dot{x}_n = -\left[\prod_{\ell=1,\ \ell\neq n}^{N}(x_n - x_\ell)\right]^{-1}\sum_{m=1}^{N}\left[f_m(\vec{y};t)\ (x_n)^{N-m}\right]\ . \tag{2.6b}$$

In this system (2.6b) of ODEs for the N dependent variables $x_n(t)$ the N components $y_m(t)$ of the N-vector $\vec{y}(t)$ are of course supposed to be replaced by their *explicit* expressions (2.2) in terms of the N zeros $x_\ell(t)$.

It is plain how the same procedure allows to transform a system of *second-order* ODEs satisfied by the $y_m(t)$'s into a system of *second-order* ODEs satisfied by the $x_n(t)$'s—via the identities (2.3) and (2.4), as well as (2.2) and (if need be) their time-derivatives.

Let us conclude this section by spelling out—obvious as this might be—the *solvable* character via algebraic operations of a system of nonlinearly coupled ODEs such as (2.6b). We assume of course that the input—in addition to this explicit system of ODEs—is the assignment of the N initial data $x_n(0)$. The *first step* is then to compute—via the explicit formulas (2.2) (at $t = 0$)—the explicit values of the N initial values $y_m(0)$. The *second step* is to evaluate the quantities $y_m(t)$ by solving the system of ODEs (2.6a), starting from the N values $y_m(0)$ evaluated in the *first step*; note that we are assuming that this second step can be performed by *algebraic* operations (or possibly even *explicitly*). The *third step* is to evaluate the N quantities $x_n(t)$ by finding the N zeros of the polynomial, see (2.1), the N coefficients $y_m(t)$ of which have been computed in the *second step*; note that this is indeed an *algebraic* operation (which may be explicitly performed for $N \leq 4$; although the relevant formulas are complicated enough to be hardly useful for $N > 2$).

Remark 2-2. An important aspect of the procedure we just described needs to be emphasized. Because in the last step the quantities $x_n(t)$ are identified as the N zeros of a (known) t-dependent polynomial, see (2.1), they get defined as an *unordered* set $\tilde{x}(t)$ of N (generally *complex*) numbers. Hence the information is lost of which one of these N numbers $x_n(t)$ is, say, the solution $x_1(t)$ which has evolved—generally *continuously* over time—from the specific initial datum $x_1(0)$. In certain cases, this might be irrelevant: for instance, if the quantities $x_n(t)$ are the coordinates of N *indistinguishable* point particles moving in the *complex* plane (or, equivalently, in the *real* Cartesian plane); then only the evolution of the *unordered* set $\tilde{x}(t)$ is relevant. Otherwise the identification of each of the $x_n(t)$'s as they evolve over time can only be obtained by following their (generally *continuous*) evolution over time; or, in a more practical context, by breaking up that evolution—from the *initial* ($t = 0$) time to the *actual* time t—into subintervals: say, from τ_ℓ to $\tau_{\ell+1}$ with $\tau_\ell < \tau_{\ell+1}$, $\tau_1 = 0$, $\tau_L = t$, with appropriate assignments, of the *integer* L and of the L *positive* numbers τ_ℓ, adequate to guarantee that the identification of the quantities $x_n(t)$ obtain by an argument of *contiguity* (of $x_n(\tau_\ell)$ to $x_n(\tau_{\ell+1})$), *unambiguously* approximating *continuity*. ■

Let us however emphasize that—in spite of the limitation described in this **Remark 2-2**—certain *general* properties (say, *isochrony* or *integrability*; as discussed above) of the dynamical systems identified via the technique described above—see for instance (2.6b)—might be sufficient to single them out as interesting models, worthy of study in theoretical or applicative contexts.

3 Two examples of systems of nonlinearly-coupled ODEs solvable by algebraic operations

Since in this paper we are mainly interested in illustrating the main idea of our approach to identify systems of nonlinearly-coupled ODEs solvable by algebraic operations, we restrict our presentation in this section to the exhibition of just *two*, very simple, examples of such systems, characterized by equations of motion of *first*, respectively of *second*, order. The interested reader may find many more examples—less elementary than those displayed below—in the book [6] and in the papers referred to there.

Notation 3-1. In the following examples—trying to economize words—we report first the ODEs satisfied by the N coefficients $y_m(t)$ and their solutions, then the corresponding systems of nonlinearly-coupled ODEs satisfied by the N zeros $x_n(t)$—which are the models of interest—and finally we tersely provide some comments. Hereafter the parameters ω_m and α_m (possibly decorated with additional symbols, see below) are *a priori arbitrary real* numbers; occasionally we will assume that the parameters ω_m are *rational* numbers (up to an arbitrary common *real nonvanishing* parameter ω),

$$\omega_m = \left(\frac{p_m}{q_m}\right) \omega , \tag{3.7}$$

with the q_m's *positive integers* and the p_m's *nonvanishing integers.* $\mathbf{i}$ is throughout the *imaginary unit,* $\mathbf{i}^2 = -1$. ■

3.1 A solvable system of first-order nonlinearly-coupled ODEs

$$\dot{y}_m = (\mathbf{i}\omega_m - \alpha_m)\, y_m , \tag{3.8a}$$

$$y_m(t) = y_m(0) \ \exp\left[(\mathbf{i}\omega_m - \alpha_m)\, t\right] . \tag{3.8b}$$

$$\dot{x}_n = \left[\prod_{\ell=1,\ell\neq n}^{N} (x_n - x_\ell)\right]^{-1} \sum_{m=1}^{N} \left[(-1)^{m+1} (\mathbf{i}\omega_m - \alpha_m) \ \sigma_m(\tilde{x}) \ (x_n)^{N-m}\right] . \tag{3.9}$$

The solution of the equations of motions (3.9) is of course provided by the N zeros $x_n(t)$ of the polynomial $p_N(z;t)$ the N coefficients $y_m(t)$ of which are given by the explicit formulas (3.8b). Hence their time evolution is *isochronous*, respectively *asymptotically isochronous*, if *all* the parameters ω_m are *rational numbers,* and *all* the parameters α_m *vanish,* respectively *none* of the parameters α_m is *negative,* at least *one* of them is *positive* and at least *one* of them *vanishes*: so that the polynomial $p_N(z;t)$ with coefficients (3.8b) is itself *isochronous*, $p_N(z;t+T) = p_N(z;t)$ with

$$T = \left(\frac{2\pi}{\omega}\right) Q , \quad Q = \max_{m=1,2,\ldots,N} [q_m] , \tag{3.10}$$

respectively it features this property only *asymptotically,* i. e. up to corrections which vanish (exponentially) as $t \to \infty$.

In the special case with $N = 2$ the system of ODEs (3.9) reads as follows:

$$\begin{aligned} \dot{x}_1 &= -(x_1 - x_2)^{-1} \left[(\alpha_1 - \mathbf{i}\omega_1)\, x_1 + (\alpha - \mathbf{i}\omega)\, x_2\right] x_1 , \\ \dot{x}_2 &= (x_1 - x_2)^{-1} \left[(\alpha_1 - \mathbf{i}\omega_1)\, x_2 + (\alpha - \mathbf{i}\omega)\, x_1\right] x_2 , \\ \alpha &\equiv \alpha_1 - \alpha_2 , \quad \omega \equiv \omega_1 - \omega_2 . \end{aligned} \tag{3.11}$$

In the special case with $N = 3$ the system of ODEs (3.9) reads as follows:

$$\begin{aligned} \dot{x}_n &= \left[(x_n - x_{n+1})(x_n - x_{n+2})\right]^{-1} \Big[(\mathbf{i}\omega_1 - \alpha_1)(x_1 + x_2 + x_3)(x_n)^2 \\ &\quad - (\mathbf{i}\omega_2 - \alpha_2)(x_1 x_2 + x_2 x_3 + x_3 x_1)\, x_n + (\mathbf{i}\omega_3 - \alpha_3)\, x_1 x_2 x_3\Big] , \\ n &= 1, 2, 3 \mod(3) . \end{aligned} \tag{3.12}$$

3.2 A solvable system of second-order nonlinearly-coupled ODEs: the goldfish

$$\ddot{y}_m = \left[\mathbf{i}\left(\omega_m^{(+)} + \omega_m^{(-)}\right) - \left(\alpha_m^{(+)} + \alpha_m^{(-)}\right)\right]\dot{y}_m$$
$$+\left[\alpha_m^{(+)}\alpha_m^{(-)} - \omega_m^{(+)}\omega_m^{(-)} - \mathbf{i}\left(\alpha_m^{(+)}\omega_m^{(-)} + \alpha_m^{(-)}\omega_m^{(+)}\right)\right] y_m \,, \tag{3.13a}$$

$$y_m(t) = a_m^{(+)}\exp\left(\mathbf{i}\lambda_m^{(+)} t\right) + a_m^{(-)}\exp\left(\mathbf{i}\lambda_m^{(-)} t\right) ,$$
$$a_m^{(\pm)} = \frac{\dot{y}_m(0) - \lambda^{(\mp)} y_m(0)}{\lambda_m^{(+)} - \lambda_m^{(-)}} \,, \quad \lambda_m^{(\pm)} = \mathbf{i}\omega_m^{(\pm)} - \alpha_m^{(\pm)} \,. \tag{3.13b}$$

$$\ddot{x}_n = \sum_{\ell=1,\ \ell\neq n}^{N}\left(\frac{2\dot{x}_n\dot{x}_\ell}{x_n - x_\ell}\right) - \left[\prod_{\ell=1,\ell\neq n}^{N}(x_n - x_\ell)\right]^{-1} \cdot$$
$$\cdot\sum_{m=1}^{N}\left[\left\{\left[\mathbf{i}\left(\omega_m^{(+)} + \omega_m^{(-)}\right) - \left(\alpha_m^{(+)} + \alpha_m^{(-)}\right)\right]\dot{y}_m\right.\right.$$
$$\left.\left.+\left[\omega_m^{(+)}\omega_m^{(-)} - \alpha_m^{(+)}\alpha_m^{(-)} + \mathbf{i}\left(\omega_m^{(+)}\alpha_m^{(-)} + \omega_m^{(-)}\alpha_m^{(+)}\right)\right] y_m\right\}(x_n)^{N-m}\right] \tag{3.14}$$

In these formulas $\omega_m^{(\pm)}$ and $\alpha_m^{(\pm)}$ are *real* parameters (analogous to those introduced above without the additional upper decoration); and of course in the equations of motion (3.14) the quantities $y_m(t)$ (respectively $\dot{y}_m(t)$) are supposed to be replaced by their expressions (2.2) (respectively their time-derivatives) in terms of the dependent variables $x_n(t)$ (and their time derivatives).

Note that these equations can be interpreted as the Newtonian equations of motions ("accelerations equal forces") of N unit mass point-particles moving in the *complex* x-plane (or equivalently in the *real* Cartesian plane with coordinates $\mathrm{Re}(x)$ and $\mathrm{Im}(x)$).

We leave to the interested reader the analysis—analogous to that reported in Section 3.1—of the cases in which these dynamical systems are *isochronous* or *asymptotically isochronous.*

Note that, in the special case in which the parameters $\omega_m^{(\pm)}$ and $\alpha_m^{(\pm)}$ are restricted to satisfy the conditions

$$\omega_m^{(+)} + \omega_m^{(-)} = \Omega, \quad \alpha_m^{(+)} + \alpha_m^{(-)} = \alpha \,, \tag{3.15a}$$

$$\alpha_m^{(+)}\alpha_m^{(-)} - \omega_m^{(+)}\omega_m^{(-)} = A \,, \quad \alpha_m^{(+)}\omega_m^{(-)} + \alpha_m^{(-)}\omega_m^{(+)} = B \,, \tag{3.15b}$$

with Ω, α, A, B *real* parameters—independent of the parameter m—the equations of motion (3.14) get, via the identities (2.4) and (2.5), drastically simplified, reading

$$\ddot{x}_n = \sum_{\ell=1,\ \ell\neq n}^{N}\left(\frac{2\dot{x}_n\dot{x}_\ell}{x_n - x_\ell}\right) + (\mathbf{i}\Omega - \alpha)\,\dot{x}_n + (A - \mathbf{i}B)\left[\prod_{\ell=1,\ell\neq n}^{N}(x_n - x_\ell)\right]^{-1}(x_n)^N \,.$$

(3.15c)

In the even more special case with $\omega_m^{(+)} = \Omega$, $\omega_m^{(-)} = \alpha_m^{(+)} = \alpha_m^{(-)} = 0$ implying $\alpha = A = B = 0$ these equations of motion simplify even more, reading

$$\ddot{x}_n = \mathrm{i}\Omega\dot{x}_n + \sum_{\ell=1,\ \ell\neq n}^{N} \left(\frac{2\dot{x}_n\dot{x}_\ell}{x_n - x_\ell} \right) ; \tag{3.16}$$

while the corresponding equations for the $y_m(t)$'s read

$$\ddot{y}_m = \mathrm{i}\Omega\dot{y}_m \ , \tag{3.17a}$$

so that the solution of their initial-values problem reads simply

$$y_m(t) = y_m(0) + \dot{y}_m(0) \left[\frac{\exp(\mathrm{i}\Omega t) - 1}{\mathrm{i}\Omega} \right] . \tag{3.17b}$$

It is then easily seen that the solution of the initial-values problem for the system of nonlinearly-coupled Newtonian ODEs (3.16) is given by the following neat prescription: the values of the N coordinates $x_n(t)$ are the N solutions of the following neat equation in the independent variable z:

$$\sum_{n=1}^{N} \left[\frac{\dot{x}_n(0)}{z - x_n(0)} \right] = \frac{\mathrm{i}\Omega}{\exp(\mathrm{i}\Omega t) - 1} . \tag{3.18}$$

Note that this is equivalent to the statement that the N coordinates $x_n(t)$ are the N zeros of the polynomial equation of degree N in z that obtains by multiplying this formula by the polynomial $\prod_{\ell=1}^{N} [z - x_\ell(0)]$.

Remark 3.2-1. The system of Newtonian equations of motion (3.16) features several remarkable properties: it is invariant under an *arbitrary* rescaling of the time variable ($t \Rightarrow at$ with a an *arbitrary real* constant), under an *arbitrary* translation of the coordinates ($x_n(t) \Rightarrow x_n(t) + b$ with b an *arbitrary complex* constant), and under an *arbitrary* rescaling of the coordinates ($x_n(t) \Rightarrow c\, x_n(t)$ with c an *arbitrary complex* constant). Moreover—as indicated by the invariance just mentioned, with the assignment $c = \exp(\mathrm{i}\theta)$—these equations of motion (3.16) are invariant under *rotations* in the *complex* x-plane, so that they can be reformulated in *covariant* 2-vector form via the transition from the *complex* x-plane to the *real* Cartesian plane that obtains by introducing the *real* 2-vectors $\vec{r}_n(t) \equiv (\mathrm{Re}[x_n(t)], \mathrm{Im}[x_n(t)])$; it is indeed easily seen that they then read as follows,

$$\ddot{\vec{r}}_n = \Omega\ \hat{z} \wedge \dot{\vec{r}}_n + 2 \sum_{\ell=1, \ell\neq n}^{N} \left[\frac{\dot{\vec{r}}_n \left(\dot{\vec{r}}_\ell \cdot \vec{r}_{n\ell} \right) + \dot{\vec{r}}_\ell \left(\dot{\vec{r}}_n \cdot \vec{r}_{n\ell} \right) - \vec{r}_{n\ell} \left(\dot{\vec{r}}_n \cdot \dot{\vec{r}}_\ell \right)}{(r_{n\ell})^2} \right] , \tag{3.19}$$

where $\hat{z}$ denotes the unit-vector orthogonal to the Cartesian plane, the symbol $\wedge$ denotes the standard (3-dimensional) vector-product (so that, if $\vec{r}$ is a 2-vector in the Cartesian plane, $\hat{z} \wedge \vec{r}$ is that same 2-vector rotated counterclockwise by $\pi/2$), the dot sandwiched between two 2-vectors denotes the standard scalar product, $\vec{r}_{n\ell} \equiv \vec{r}_n - \vec{r}_\ell$ and $(r_{n\ell})^2 \equiv \vec{r}_{n\ell} \cdot \vec{r}_{n\ell}$.

Because of this neatness—and the simplicity of its *isochronous* behavior, as also demonstrated by its solution, see (3.18)—the name "goldfish" was suggested when this model was originally identified [9]; a name that seems to have caught on, see for instance [7] (and several other references reported in [3] and [6]). ■

Remark 3.2-2. The very simplest case of the goldfish model obtains when $\Omega = 0$, so that its equations of motion read

$$\ddot{x}_n = \sum_{\ell=1,\ \ell\neq n}^{N} \left(\frac{2\dot{x}_n \dot{x}_\ell}{x_n - x_\ell} \right) , \tag{3.20a}$$

or, equivalently,

$$\ddot{\vec{r}}_n = 2 \sum_{\ell=1, \ell\neq n}^{N} \left[\frac{\dot{\vec{r}}_n \left(\dot{\vec{r}}_\ell \cdot \vec{r}_{n\ell} \right) + \dot{\vec{r}}_\ell \left(\dot{\vec{r}}_n \cdot \vec{r}_{n\ell} \right) - \vec{r}_{n\ell} \left(\dot{\vec{r}}_n \cdot \dot{\vec{r}}_\ell \right)}{(r_{n\ell})^2} \right] . \tag{3.20b}$$

In this case, of course, its behavior is *not isochronous*; but it is also rather remarkable. The interested reader will find a detailed treatment in the book [2], see in particular its Subsection 4.2.4, the title of which provides a glimpse of its contents: "The simplest model: explicit solution (the game of musical chairs), Hamiltonian structure". ■

4 A differential algorithm to evaluate all the zeros of a generic polynomial of arbitrary degree

In this section we report a recent development [10] [11] [6] concerning a very old and classical problem, that of computing the N zeros of a generic (monic) polynomial $P_N(z)$ of arbitrary degree N in its argument z. This result is reported here because of its close connection with the findings discussed in this paper, see in particular Section 2.[1]

There are different variants of this result; in particular one based on a system of *first-order* ODEs [10] [6] and one based on a system of *second-order* ODEs [11] [6]. Here we report only the second version, which we consider more elegant. It is detailed by the following

[1] I like to thank Prof. Peter Olver for pointing out to me (at the SPT2018 Workshop, June 2018) that the differential algorithm to evaluate all the zeros of a generic polynomial described in this Section 4 may be considered a special case of the more general and more sophisticated "homotopy continuation technique" introduced over a decade ago to investigate the zeros of systems of polynomials: see for instance A. J. Sommese and C. W. Wampler, *Numerical Solution of Systems of Polynomials Arising in Engineering and Science*, World Scientific, Singapore, 2005, and the literature referred to there.

Proposition 4-1. Consider the following monic polynomial $P_N(z)$ defined in terms of its N coefficients c_m and featuring the N zeros z_n:

$$P_N(z) = z^N + \sum_{m=1}^{N} \left(c_m\, z^{N-m}\right) = \prod_{n=1}^{N} (z - z_n) \ . \tag{4.21}$$

Introduce the system of N nonlinear differential equations

$$\ddot{x}_n = \sum_{\ell=1,\ \ell\neq n}^{N} \left(\frac{2\,\dot{x}_n\,\dot{x}_\ell}{x_n - x_\ell}\right) , \tag{4.22}$$

with the independent variable t playing the mathematical role of "dimensionless time". And complement this system of ODEs with the following assignment of *initial* data: (i) the N *initial* values $x_n(0)$ of the N coordinates $x_n(t)$ are *arbitrary* (except for the restriction—motivated by eq. (4.23), see below—that they be *all* different among themselves, $x_n(0) \neq x_\ell(0)$ if $n \neq \ell$); (ii) the N *initial* values $\dot{x}_n(0)$ of the time-derivatives $\dot{x}_n(t)$ of the coordinates $x_n(t)$ are given, in terms of the N initial values $x_n(0)$ and of the polynomial $P_N(z)$ evaluated at the N initial values $x_n(0)$, by the following formula:

$$\dot{x}_n(0) = \frac{-P_N[x_n(0)]}{\prod_{\ell=1,\ \ell\neq n}^{N} [x_n(0) - x_\ell(0)]} \ . \tag{4.23}$$

Then integrate the system of second-order ODEs from $t = 0$ to $t = 1$; there thus obtain the N zeros of the polynomial $P_N(z)$:

$$P_N[x_n(1)] = 0\ , \quad \tilde{z} = \tilde{x}(1)\ . \ \blacksquare \tag{4.24}$$

The proof of this result is an elementary application of the two identities (2.3) and (2.4) to the specific t-dependent polynomial

$$p_N(z;t) = (1-t)\ \prod_{n=1}^{N} [z - x_n(0)] + t\ P_N(z)\ , \tag{4.25}$$

which clearly interpolates *linearly* from the polynomial featuring the N arbitrary zeros $x_n(0)$ (at $t = 0$) to the polynomial $P_N(z)$ at $t = 1$. It is indeed an elementary task to verify that the N zeros $x_n(t)$ of this polynomial $p_N(z;t)$ feature the following properties: (i) they are evidently consistent with the (arbitrary) assignment $x_n(0)$ at $t = 0$, and with the formula (4.24) at $t = 1$; (ii) the t-derivative of this polynomial $p_N(z;t)$, evaluated at $t = 0$, is consistent (via (2.3)) with the assignment (4.23); (iii) the *linear* dependence on t of this polynomial $p_N(z;t)$ implies that its second t-derivative vanishes identically, hence, via the identity (2.4b), that its zeros $x_n(t)$ evolve in time according to the simple system of ODEs (4.22). Q. E. D.

Remark 4-1. The alert reader will have noted that the system of ODEs (4.22) coincides with the simplest version of the *goldfish* model, see **Remark 3.2-2**. ■

Remark 4-2. Let us emphasize the remarkable flexibility of this algorithm implied by the possibility to assign *arbitrarily* the N initial values $x_n(0)$. An additional flexibility is provided by the possibility to rescale the independent variable t, namely to replace t with a convenient monotonic function $\tau(t)$. The exploration of these possibilities in the context of numerical analysis is an open task, as well as the possibility to use this approach to investigate certain polynomial properties; for an instance of applicative relevance see [12]. ■

Remark 4-3. The generalization of this approach to the investigation of the *infinite* set of zeros of *entire* functions is another interesting development [13]. ■

5 Extensions

Various extensions of the approach described in this paper have been investigated: (i) to the vistas opened by the possibility to *iterate* the procedure described above (to identify *algebraically solvable* systems of *nonlinearly-coupled* ODEs via the transition from the time evolution of the N coefficients of a time-dependent polynomial to that of its N zeros), and to thereby generate *infinite hierarchies* of such *algebraically solvable* systems [14]; (ii) to treat systems of PDEs rather than systems of ODEs [15]; (iii) to evolutions in *discrete* time rather than *continuous* time [16]. All these developments are reviewed in the book [6], where the interested reader will also find additional references.

Finally, let us report—without detailing their (quite easy) proofs, see [17] and [18]—some amusing twists (featuring a potential *Diophantine* connotation, see below) of the findings reported in Section 4 and in [16].

Proposition 5-1. Consider the system of N nonlinearly coupled *first-order* ODEs

$$\dot{x}_n(t) = \frac{\prod_{m=1}^{N}\left[x_n(t) - x_m(0)\right] - \prod_{m=1}^{N}\left[x_n(t) - f_m\right]}{\prod_{m=1,\, m\neq n}^{N}\left[x_n(t) - x_m(t)\right]}\,, \tag{5.1a}$$

where the N initial (possibly *complex*) values $x_n(0)$ of the dependent variables $x_n(t)$ are assigned *arbitrarily* (but all different among themselves—$x_n(0) \neq x_m(0)$ if $n \neq m$—to avoid that the denominator in the right-hand side of (5.1a) vanish at $t = 0$); and the N numbers f_m are assigned *arbitrarily.*

Then the values $x_n(1)$ of the dependent variables $x_n(t)$ at $t = 1$ coincide—up to permutations—with the N numbers f_m,

$$\tilde{x}(1) = \tilde{f}\,. \quad ■ \tag{5.1b}$$

Proposition 5-2. Consider the system of N *second-order discrete-time* evolution equations (with the *integers* ℓ now playing the role of *discrete* time)

$$2\prod_{m=1}^{N}\left[z_n(\ell+2) - z_m(\ell+1)\right] - \prod_{m=1}^{N}\left[z_n(\ell+2) - z_m(\ell)\right] = 0\,; \tag{5.2a}$$

note that this formula provides the unordered set $\tilde{z}(\ell+2)$, the elements of which are the N values $z_n(\ell+2)$, as the N zeros of the polynomial of degree N in z defined in terms of the two unordered sets $\tilde{z}(\ell)$ and $\tilde{z}(\ell+1)$ as follows:

$$P_N\left(z;\tilde{z}(\ell),\tilde{z}(\ell+1)\right)=2\prod_{m=1}^{N}\left[z-z_m(\ell+1)\right]-\prod_{m=1}^{N}\left[z-z_m(\ell)\right] . \tag{5.2b}$$

Let this system of second-order discrete-time evolution equations, (5.2a), be complemented by the following assignments of the two unordered sets $\tilde{z}(0)$ respectively $\tilde{z}(1)$ of $2N$ *initial* data $z_n(0)$ respectively $z_n(1)$: (i) the N data $z_n(0)$ are assigned *arbitrarily;* (ii) the N data $z_n(1)$—rather than being as well *arbitrarily* assigned—are defined as follows in terms of the *arbitrary positive integer* parameter L, the unordered set $\tilde{z}(0)$, and the unordered set $\tilde{f}$ the N elements of which are N *arbitrarily* assigned numbers f_m—by the N algebraic equations

$$\prod_{m=1}^{N}\left[z_n(1)-z_m(0)\right]+\frac{(-1)^N}{L-1}\prod_{m=1}^{N}\left[z_n(1)-f_m\right]=0 ; \tag{5.3a}$$

hence these N data $z_n(1)$ are the N roots of the polynomial $p_N^{(1)}\left(z;\tilde{z}(0),\tilde{f};L\right)$, of degree N in z, defined as follows in terms of the two unordered sets $\tilde{z}(0)$ and $\tilde{f}$:

$$p_N^{(1)}\left(z;\tilde{z}(0),\tilde{f};L\right)=\prod_{m=1}^{N}\left[z-z_m(0)\right]+\frac{(-1)^N}{L-1}\prod_{m=1}^{N}\left[z-f_m\right] . \tag{5.3b}$$

Then, at $\ell=L$, the solution $\tilde{z}(L)$ of the system of second-order discrete-time evolution equations (5.2a) coincides with the unordered set $\tilde{f}$:

$$\tilde{z}(L)=\tilde{f} . \quad \blacksquare \tag{5.4}$$

The *Diophantine* connotation of these findings obtains thanks to the possibility to assign the N, *a priori arbitrary*, numbers f_m to take *integer* values, for instance to coincide with the N *integers* from 1 to N. Note that—especially with such assignments—these findings might then be useful to test the accuracy of numerical routines to integrate systems of ODEs (see (5.1a)), respectively to find the zeros of polynomials of degree N (a task required in order to solve numerically, step by step, the system of *discrete-time* evolution equations (5.2)); by comparing the outcome of the numerical routines with the exact results predicted by **Proposition 5-1** respectively **Proposition 5-2**.

Let us end by mentioning the generalization of the approach discussed in this paper that emerges if the formula (2.1) is replaced by

$$p_N(z;t)=z^N+\sum_{m=1}^{N}\left[y_m(t)\ P_{N-m}(z)\right]=\prod_{n=1}^{N}\left[z-x_n(t)\right] , \tag{5.5}$$

where the N monic polynomials $P_{N-m}(z)$, of degree $N-m$, are arbitrarily assigned. It is then easily seen that identities analogous to (2.3) and (2.4) still hold, reading

$$\dot{x}_n = \frac{-\sum_{m=1}^{N}\left[\dot{y}_m\, P_{N-m}(x_n)\right]}{\prod_{\ell=1,\ell\neq n}^{N}(x_n - x_\ell)} \tag{5.6a}$$

$$= \frac{-\left.\frac{\partial p_N(z;t)}{\partial t}\right|_{z=x_n}}{\prod_{\ell=1,\ell\neq n}^{N}(x_n - x_\ell)}, \tag{5.6b}$$

$$\ddot{x}_n = \sum_{\ell=1,\ \ell\neq n}^{N}\left(\frac{2\dot{x}_n\dot{x}_\ell}{x_n - x_\ell}\right) - \frac{\sum_{m=1}^{N}\left[\ddot{y}_m\, P_{N-m}(x_n)\right]}{\prod_{\ell=1,\ell\neq n}^{N}(x_n - x_\ell)} \tag{5.6c}$$

$$= \sum_{\ell=1,\ \ell\neq n}^{N}\left(\frac{2\dot{x}_n\dot{x}_\ell}{x_n - x_\ell}\right) - \frac{\left.\frac{\partial^2 p_N(z;t)}{\partial t^2}\right|_{z=x_n}}{\prod_{\ell=1,\ell\neq n}^{N}(x_n - x_\ell)}. \tag{5.6d}$$

However the *ansatz* (5.5) does not generally entail simple explicit expressions—such as (2.2)—of the N quantities y_m in terms of the N quantities x_n. Relatively explicit expressions of the N quantities y_m in terms of the N quantities x_n can of course be obtained if the polynomials $P_{N-m}(z)$ were part of an *orthogonal* set, but the resulting treatment would still be somewhat more complicated; it would on the other hand allow the identification and investigation of new models solvable by algebraic operations, including examples featuring additional free parameters and yielding interesting properties—such as *isochrony*—which might hardly be guessed by just looking at their equations of motion.

References

[1] F. Calogero, Motion of Poles and Zeros of Special Solutions of Nonlinear and Linear Partial Differential Equations, and Related "Solvable" Many Body Problems, *Nuovo Cimento* **43B**, 177-241 (1978).

[2] F. Calogero, *Classical many-body problems amenable to exact treatments*, Lecture Notes in Physics Monograph **m66**, Springer, 2001 (749 pages).

[3] F. Calogero, *Isochronous systems*, Oxford University Press, 2008 (264 pages); a marginally updated paperback edition has been published by OUP in September 2012.

[4] F. Calogero, New solvable variants of the goldfish many-body problem, *Studies Appl. Math.* **137** (1), 123-139 (2016); DOI: 10.1111/sapm.12096.

[5] M. Bruschi and F. Calogero, A convenient expression of the time-derivative $z_n^{(k)}(t)$ of arbitrary order k of the zero $z_n(t)$ of a time-dependent polynomial $p_N(z;t)$ of arbitrary degree N in z, and solvable dynamical systems, *J. Nonlinear Math. Phys.* **23**, 474-485 (2016).

[6] F. Calogero, *Zeros of Polynomials and Solvable Nonlinear Evolution Equations*, Cambridge University Press, Cambridge, U. K., 2018.

[7] D. Gómez-Ullate and M. Sommacal, Periods of the goldfish many-body problem, *J. Nonlinear Math. Phys.***12** Suppl. **1**, 351-362 (2005).

[8] F. Calogero and D. Gómez-Ullate, Asymptotically isochronous systems, *J. Nonlinear Math. Phys.* **15**, 410-426 (2008).

[9] F. Calogero, The "neatest" many-body problem amenable to exact treatments (a "goldfish"?), *Physica D* **152-153**, 78-84 (2001).

[10] F. Calogero, Nonlinear differential algorithm to compute all the zeros of a generic polynomial, *J. Math. Phys.* **57**, 083508 (4 pages) (2016), http://dx.doi.org/10.1063/1.4960821; Comment on Nonlinear differential algorithm to compute all the zeros of a generic polynomial [*J. Math. Phys.* 57, 083508 (2016)], *J. Math. Phys.* **57**, 104101 (4 pages) (2016).

[11] F. Calogero, Novel differential algorithm to evaluate all the zeros of any generic polynomial, *J. Nonlinear Math. Phys.* **24**, 469-472 (2017). DOI: 10.1080/14029251.2017.1375685.

[12] M. K. Salehani, Identification of generic stable dynamical systems taking a nonlinear differential approach, *Discrete Contin. Dyn. Syst. Ser. B* **23** (10), 4541-4555, (2018). doi: 10.3934/dcdsb.2018175.

[13] F. Calogero, Zeros of entire functions and related systems of infinitely many nonlinearly-coupled evolution equations, *Theor. Math. Phys.* **196** (2), 1111-1128 (2018).

[14] O. Bihun and F. Calogero, Generations of monic polynomials such that the coefficients of each polynomial of the next generation coincide with the zeros of a polynomial of the current generation, and new solvable many-body problems, *Lett. Math. Phys.* **106** (7), 1011-1031 (2016); DOI: 10.1007/s11005-016-0836-8. arXiv: 1510.05017 [math-ph].

[15] F. Calogero, New C-integrable and S-integrable systems of nonlinear partial differential equation, *J. Nonlinear Math. Phys.* **24** (1), 142-148 (2017).

[16] O. Bihun and F. Calogero, Generations of solvable discrete-time dynamical systems, *J. Math. Phys.* **58**, 052701 (21 pages) (2017); doi: 10.1063/1.4982959.

[17] F. Calogero, Finite and infinite systems of nonlinearly coupled ordinary differential equations the solutions of which feature remarkable Diophantine findings, *J. Nonlinear Math. Phys.* **25**, 434-442 (2018).

[18] F. Calogero, Solvable nonlinear discrete-time evolutions and Diophantine findings, *J. Nonlinear Math. Phys.* **25**, 1-3 (2018).

A2. Integrable nonlinear PDEs on the half-line

A S Fokas [a] *and B Pelloni* [b]

[a] *Department of Applied Mathematics and Theoretical Physics, University of Cambridge, Cambridge CB3 0WA, United Kingdom*

[b] *Department of Mathematics, Heriot-Watt University, Edinburgh EH14 4AS, United Kingdom*

1 Introduction

The Inverse Scattering Transform, pioneered at the end of the 1960's by Kruskal et al. [50] and consolidated throughout the 1970's by the work of many others [5, 6, 55, 59], is considered one of the most important advances in the study of nonlinear evolution PDEs. This transform can be used for analysing integrable nonlinear evolution PDEs in one space dimension, and is essentially a nonlinear version of the Fourier transform in one variable. It can be applied to systems with the property that the nonlinearity is exactly balanced by other effects, such as dispersive effects. This implies that, in many important respects, the behaviour of the solutions of the system is highly regular. For example, when posed on the infinite line, these systems admit localised solutions, referred to as solitons, that interact elastically - the interaction does not affect the amplitude and speed of the solutions. More importantly still, localised initial conditions with sufficient energy will eventually evolve into a train of solitons, followed by a dispersive tail.

In addition to nonlinear evolution PDEs in one space variable, there exists a variety of other types of integrable systems, including PDEs in two space variables [39, 9], ODEs such as the classical Painlevé equations [49] and singular integral nonlinear equations such as the Benjamin-Ono equation [41]. The defining property of integrable systems is the existence of a Lax pair, namely the fact that the given equation can be written as the compatibility of two linear eigenvalue equations involving the eigenvalue $\lambda \in \mathbb{C}$ and the eigenfunction Ψ, which may be scalar or matrix-valued. The eigenvalue λ is sometimes referred to as the *spectral parameter*; in analogy with the Fourier transform, λ plays the role of the independent variable in Fourier space.

The Unified Transform or Fokas Method

The study of boundary value problems for integrable PDEs in the last fifteen years has motivated the development of a new powerful method in mathematical physics, usually referred to as *Unified Transform* or *Fokas Method.* This approach combines the main insights of the Inverse Scattering Transform with elements of the theory of Riemann-Hilbert problems, and uses essentially complex analytic properties to eliminate unknown boundary values from the solution representation. This method, proposed by Fokas and extensively developed by the authors and others, has produced a substantial body of results and, unexpectedly, has led to a new perspective and results on the theory of boundary value problems also for linear PDEs in two independent variables. For an introduction to this theory and a summary of the main results, see [30] (see also [19] for a review of linear boundary value problems and [38] for recent applications and advances).

The Unified Transform has its origin in two fundamental observations. These observations, seemingly pertaining to separate fields, yield the formulation in terms of a *Riemann-Hilbert problem*, namely to the problem of reconstructing a function of a complex variable λ, analytic everywhere off a given contour, from its known jump condition across the contour (a clear, non-technical review of Riemann-Hilbert problems in this context can be found in [52]. Riemann-Hilbert problems appear in a disparate variety of problems in mathematics and mathematical physics [21].

The two observations are the following:

1. The Lax pair formulation provides a highly nontrivial generalisation of the concept of *separation of variables*, valid for both linear and integrable nonlinear PDEs [40].

2. Integral transforms such as the Fourier transform can be derived via the spectral analysis of an ODE in the complex plane [32].

The origin of the Fokas Method is the realisation that the existence of a Lax pair makes it possible to solve both ODEs of this pair *simultaneously*. Just as the analysis of a single ODE yields a transform associated with one given independent variable, this simultaneous analysis yields a transform associated with both independent variables. Moreover, since this transform is associated with the particular Lax pair, it is also custom-made for the given PDE. Combining this idea with the two observations above, it is possible to construct algorithmically a formal integral representation of the solution of a given boundary value problem for an integrable PDE - this construction is the basis of the Unified Transform approach. It is important to note that this representation generally involves contours in the complex plane. Actually, for the solution of a linear PDE, the solution representation takes the form of an integral along such a complex contour.

The representation derived in this way has explicit dependence on the independent variables of the PDE, hence it is straightforward to show that it indeed provides a solution of the PDE. For evolution problems, one verifies easily that at $t = 0$ the solution representation satisfies the given initial condition.

However, this representation involves *all* the boundary values of the solution, while in general some of the boundary values cannot be prescribed independently as boundary conditions, and are therefore unknown. Indeed, the main difficulty in the Fokas Method is the elimination of the unknown boundary values from the representation.

However, already at this point, a significant advantage of the unified transform becomes clear: since it involves *explicit dependence* on the independent variables of the problem, asymptotic and qualitative information on the solution can be derived directly, even before the contribution of the unknown boundary values is eliminated.

Elimination of the Unknown Boundary Values

To give a concrete example and illustrate the difficulty of dealing with unknown boundary values, consider the simplest third order *linear* PDE, namely

$$\partial_t q(x,t) + \partial_{xxx} q(x,t) = 0, \qquad x \in I, \quad 0 < t < T, \qquad q(x,0) = q_0(x) \quad (1.1)$$

where I denotes a finite or infinite interval in $\mathbb{R}$, the given initial condition $q_0(x)$ is assumed sufficiently smooth and decaying if I is infinite, and T is a given positive constant.

When $I = \mathbb{R}$, hence the PDE is posed on the full line, assuming sufficiently rapid decay of the solution as $x \to \pm\infty$, one obtains a Cauchy initial value problem for (1.1). The standard approach for solving this problem involves taking the Fourier transform of this equation to obtain a first order ODE that can be solved explicitly:

$$q_t + \partial_x^3 q = 0 \ \ \textbf{PDE} \overset{\textbf{FT}}{\to} \textbf{ODE} \quad \frac{\partial \hat{q}(\lambda,t)}{\partial t} + (i\lambda)^3 \hat{q}(\lambda,t) = 0, \qquad (1.2)$$

where

$$\hat{q}(\lambda,t) = \int_{-\infty}^{\infty} \mathrm{e}^{-i\lambda x} q(x,t)dx.$$

Solving the ODE and inverting, one finds

$$\hat{q}(\lambda,t) = \mathrm{e}^{-(i\lambda)^3 t}\hat{q}(\lambda,0) \qquad \to \quad q(x,t) = \frac{1}{2\pi}\int_{-\infty}^{\infty} \mathrm{e}^{i\lambda x + i\lambda^3 t}\hat{q}(\lambda,0)d\lambda.$$

Consider now the case that $I = [0,\infty)$, assuming that the solution decays as $x \to \infty$, and that the boundary condition $q(0,t) = f_0(t)$ is prescribed. Following the same steps as for $I = \mathbb{R}$, namely taking the Fourier transform of $q(x,t)$ on the half line, solving the resulting ODE, which now contains three boundary values at $x = 0$, and evaluating the inverse transform of the result, one finds the following integral representation of the solution of (1.1) on the half-line:

$$q(x,t) = \frac{1}{2\pi}\int_{-\infty}^{\infty} \mathrm{e}^{i\lambda x + i\lambda^3 t}\left[\int_0^{\infty} \mathrm{e}^{-i\lambda y} q_0(y)dy\right] d\lambda + \qquad (1.3)$$

$$+ \ \frac{1}{2\pi}\int_{-\infty}^{\infty} \mathrm{e}^{i\lambda x + i\lambda^3 t}\left[\int_0^{t} \mathrm{e}^{-i\lambda^3 s}\left(q_{xx}(0,s) + i\lambda q_x(0,s) - \lambda^2 f_0(s)\right)ds\right] d\lambda.$$

The first term in this representation is the contribution of the initial condition. This would have been the only term present when solving the Cauchy value problem for decaying data, with the integration in y extending over $\mathbb{R}$. In this term, the x and t dependence is explicit through the exponential term.

The second term involves the boundary values of the solution at $x = 0$, but two of the boundary values involved in the integrand are not directly available. This is the generic case: to guarantee that the boundary value problem admits a unique solution, only one condition can be prescribed at this boundary, and therefore in general two boundary values are unknown [36, 37].

This difficulty can be overcome for certain PDEs of second order in x by using the sine or cosine transform. However, this is not possible for PDEs involving third order derivatives. In addition, even using the sine or cosine transform, the t-dependence of the resulting expression is not explicit, as t appears in the integration interval as well as in the exponential part of the integrand. Last but not least, the integral expressions derived by using the sine or cosine transform are *not uniformly convergent when* $x \to 0$. For example, in the case of the heat equation with $q(0,t) = f_0(t)$, the sine transform yields

$$q(x,t) = \frac{2}{\pi}\int_0^\infty \sin(\lambda x)\mathrm{e}^{-\lambda^2 t}\left[\int_0^\infty \sin(\lambda y)q_0(y)dy - \int_0^T \mathrm{e}^{\lambda^2 s} f_0(s)ds\right]d\lambda. \tag{1.4}$$

Letting $x = 0$ in the above expression and assuming that we can take the limit $x \to 0$ inside the integral, we find that $q(0,t) = 0$, not $q(0,t) = f_0(t)$. Thus we *cannot* take the limit $x \to 0$ inside the integral, i.e. the above integral is not uniformly convergent.

For the case of equation (1.1), where there does *not* exist a classical transform, the Unified Transform yields the following representation for the solution of the problem with $q(0,t) = f_0(t)$ as the prescribed boundary condition:

$$q(x,t) = \frac{1}{2\pi}\int_{-\infty}^\infty \mathrm{e}^{i\lambda x + i\lambda^3 t}\left[\int_0^\infty \mathrm{e}^{-i\lambda y}q_0(y)dy\right]d\lambda + \tag{1.5}$$

$$+ \frac{1}{2\pi}\int_{\partial D^+} \mathrm{e}^{i\lambda x + i\lambda^3 t}\left[\int_0^T \mathrm{e}^{-i\lambda^3 s}\left(q_{xx}(0,s) + i\lambda q_x(0,s) - \lambda^2 f_0(s)\right)ds\right]d\lambda,$$

where ∂D^+ is the contour in $\mathbb{C}^+$ defined by $Re(i\lambda^3) = 0$, see Figure 1.

The formulation (1.5) has fully explicit x and t dependence, and it is uniformly convergent at the boundary $x = 0$, a fact that plays an important role also in devising efficient numerical schemes [24, 25]. Moreover, by considering the transforms of the boundary values (with respect to t) as functions of the *complex* variable λ, it is possible to eliminate all unknown contributions. This elimination procedure is most naturally understood once the representation derived is of the form (1.5). Indeed, this elimination cannot be obtained by confining the spectral parameter λ to be real, hence restricting attention to real transforms.

As an example, the solution of the particular boundary value problem for the PDE (1.1) obtained when $q(0,t) = f_0(t)$ is the prescribed boundary condition for

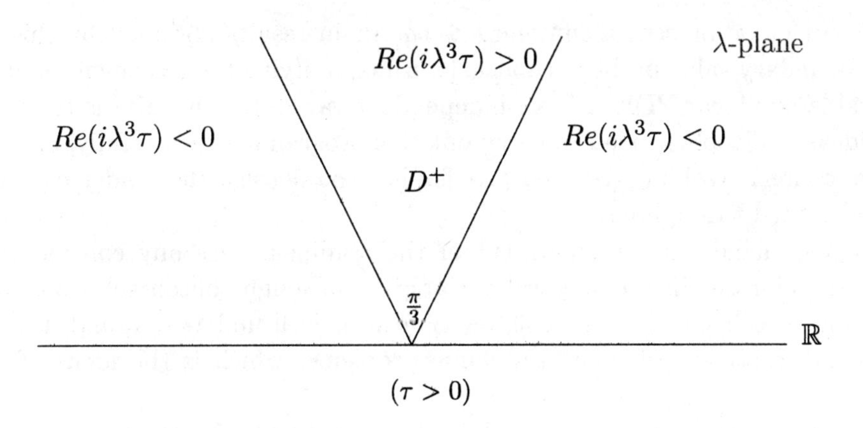

Figure 1. The domain D^+

$0 < t < T$, is given by

$$q(x,t) = \frac{1}{2\pi}\int_{\mathbb{R}} e^{i\lambda x + i\lambda^3 t}\hat{q}_0(\lambda)d\lambda \tag{1.6}$$
$$+\frac{1}{2\pi}\int_{\partial D^+} e^{i\lambda x + i\lambda^3 t}\left[(\theta-1)\hat{q}_0(\theta^2\lambda) + (1-\theta^2)\hat{q}_0(\theta\lambda) + 3\lambda^2\tilde{f}_0(\lambda)\right]d\lambda,$$

with

$$\hat{q}_0(\lambda) = \int_0^\infty e^{-i\lambda y}q_0(y)dy, \quad \lambda \in \mathbb{C}^-, \quad \tilde{f}_0(\lambda) = \int_0^T e^{-i\lambda^3 s}f(s)ds, \qquad \theta = e^{2\pi i/3}. \tag{1.7}$$

In this linear example, the expression (1.5) can also be derived without using a Lax pair formulation: it can be obtained from (1.3) by simply considering analyticity properties with respect to the variable λ, extended from $\mathbb{R}$ to $\mathbb{C}$, and deforming the contour of integration. However, this deformation is not possible in the nonlinear case. The general methodology to obtain the representation (1.5) is based instead on using the Lax pair formulation and on deriving a Riemann-Hilbert problem. This general approach can indeed be extended to the case of nonlinear integrable evolution equations in one space variable, e.g. to the famous KdV or mKdV equation, posed on the domain $0 < x < \infty, \ 0 < t < T$:

$$(KdV) \qquad q_t + q_{xxx} + q_x + 6qq_x = 0, \tag{1.8}$$
$$(mKdV) \qquad q_t + q_{xxx} + \nu 6q^2 q_x = 0, \qquad \nu = \pm 1. \tag{1.9}$$

However, in this case, the elimination of the unknown boundary values can be carried out as effectively as in the linear case only for special boundary conditions, called *linearisable* [27, 29]. For general boundary conditions, the characterisation of the unknown boundary values is more complicated [42, 43].

In this article, we present a summary of the main results obtained by this approach for boundary value problems on the half-line, unifying the treatment of linear and integrable nonlinear PDEs in two independent variables. Our aim is to review the main ideas and tools needed to carry out this programme successfully, and discuss its limitations, avoiding detailed proofs. For the details, the reader can refer to the papers cited throughout.

We will not include in the treatment of the nonlinear case any consideration of isolated singularities in the spectral variable λ, although soliton solutions arise precisely from these singularities. This mechanism is well understood and it is not specific to the treatment of boundary value problems, which is the focus of this paper [1, 4].

We note that problems periodic in x have been analysed with algebro-geometric techniques [12, 23, 54]. However, these techniques can be used only to construct particular solutions, namely x-periodic versions of soliton solutions. Although it has been established a long time ago that such problems are linearisable [48], the implementation of the unified transform to this class of problems remains open.

2 Transforms and Riemann-Hilbert problems

2.1 Fourier inversion via a Riemann-Hilbert problem

Let $q(x) \in \mathcal{S}(\mathbb{R})$ be a given, arbitrary function, and consider the following auxiliary ODE:

$$\mu_x - i\lambda\mu = q(x), \quad \mu = \mu(x,\lambda), \quad x \in \mathbb{R},\ \lambda \in \mathbb{C}. \tag{2.1}$$

Following the approach of [32], we seek a solution $\mu(x,\lambda)$ bounded as $\lambda \to \infty$, $\forall\lambda \in \mathbb{C}$. It is straightforward to find two solutions bounded in $\mathbb{C}^+$ and $\mathbb{C}^-$ respectively:

$$\mu_+(x,\lambda) = \int_{-\infty}^{x} e^{i\lambda(x-y)} q(y)dy, \quad \text{bounded for } \operatorname{Im}(\lambda) \geq 0,$$

$$\mu_-(x,\lambda) = -\int_{x}^{\infty} e^{i\lambda(x-y)} q(y)dy \quad \text{bounded for } \operatorname{Im}(\lambda) \leq 0.$$

In addition, the functions $\mu_\pm$ have the following properties (see Figure 2):

- $\mu_\pm$ is analytic in $\mathbb{C}^\pm$
- $\mu_\pm = O\left(\frac{1}{\lambda}\right)$ as $\lambda \to \infty$ in $\mathbb{C}^\pm$,
- For $\lambda \in \mathbb{R}$, where both functions $\mu^\pm$ are defined, the difference is

$$(\mu_+ - \mu_-)(x,\lambda) = e^{i\lambda x}\hat{q}(\lambda) \quad \lambda \in \mathbb{R} \tag{2.2}$$

where

$$\hat{q}(\lambda) = \int_{-\infty}^{\infty} e^{-i\lambda y} q(y) dy, \quad \lambda \in \mathbb{R}. \tag{2.3}$$

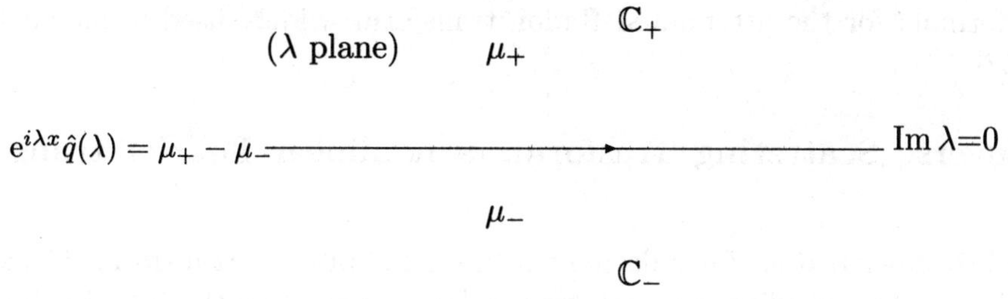

Figure 2. A Riemann-Hilbert problem on the real line

Hence, given $q(x)$, the transform $\hat{q}(\lambda)$ is obtained as the jump of the sectionally analytic solution μ such that $\mu = \mu^+$ for $\lambda \in \mathbb{C}^+$ and $\mu = \mu^-$ for $\lambda \in \mathbb{C}^-$.

Conversely, suppose $\hat{q}(\lambda)$ is given. Then the data above determine uniquely a scalar Riemann-Hilbert problem for μ, with jump on the real line given by the right hand side of (2.2). This Riemann-Hilbert problem can be solved explicitly by the so-called *Plemelj formula* [2]. This yields

$$\mu(x, \lambda) = \frac{1}{2\pi i} \int_{-\infty}^{\infty} \frac{[\mu_+ - \mu_-](x, \zeta)}{\zeta - \lambda} d\zeta = \frac{1}{2\pi i} \int_{-\infty}^{\infty} \frac{e^{i\zeta x}\hat{q}(\zeta)}{\zeta - \lambda} d\zeta. \tag{2.4}$$

Hence using

$$q(x) = \mu_x - i\lambda\mu,$$

one finds

$$q(x) = \frac{1}{2\pi} \int_{-\infty}^{\infty} e^{i\zeta x} \hat{q}(\zeta) d\zeta, \; x \in \mathbb{R}. \tag{2.5}$$

In summary: *the first order ODE (2.1) gives rise to the pair of equations (2.3)-(2.5), i.e. it encodes the Fourier transform.*

Note that the transform $\hat{q}(\lambda)$ can also be defined as

$$\hat{q}(\lambda) = \lim_{x \to -\infty} \left[-e^{-i\lambda x} \mu_-(x, \lambda) \right], \tag{2.6}$$

and that the expression (2.5) for $q(x)$ in terms of $\hat{q}(\lambda)$ can also be derived starting from the asymptotic information

$$\mu^{\pm} \sim \frac{iq(x)}{\lambda}, \quad |\lambda| \to \infty \Longrightarrow q(x) = -i \lim_{\lambda \to \infty} \lambda \mu(x, \lambda). \tag{2.7}$$

These latter formulae have a direct analogue in the nonlinear case presented in the next section. Indeed, the steps outlined here for the derivation of the Fourier transform can be used to give a derivation of other linear and nonlinear integral transforms, compute the associated inversion formula and prove rigorously its validity. This approach provides a new proof of the inversion formula for many classical transforms such as Mellin, Abel and Radon transforms, as well as the first proof of the inversion formula for the attenuated Radon transform, widely used in medical imaging [31, 57].

2.2 The Inverse Scattering Trasform: a nonlinear Fourier transform

The approach just described for formulating and inverting integral transforms has a nonlinear analogue. This nonlinear transform can be used to solve the initial value problem for integrable nonlinear evolution PDEs, in a way analogous to the use of Fourier transform in the linear case.

The starting point is a *matrix -valued* ODE, rather than the *scalar* ODE (2.1): define the matrix Q in terms of the given arbitrary function $q(x) \in \mathcal{S}(\mathbb{R})$ (although much less regularity is required, see [7, 9]) by

$$Q(x) = \begin{pmatrix} 0 & q(x) \\ \pm\bar{q}(x) & 0 \end{pmatrix}, \tag{2.8}$$

where the bar denotes complex conjugation, and consider the ODE

$$M_x + i\lambda[\sigma_3, M] = QM \quad x \in \mathbb{R},\ \lambda \in \mathbb{C}; \qquad M(x,\lambda)\ a\ 2\times 2\ matrix, \tag{2.9}$$

where

$$[\sigma_3, M] = \sigma_3 M - M\sigma_3, \quad \sigma_3 = diag(1,-1). \tag{2.10}$$

One seeks a solution $M(x,\lambda)$ of this ODE well-defined for all $\lambda \in \mathbb{C}$. As in the linear case, one considers two particular solutions: a solution $\mu_1(x,\lambda)$ which has the first column $\mu_1^{(1)}$ bounded for $\lambda \in \mathbb{C}^-$ and the second column $\mu_1^{(2)}$ bounded for $\lambda \in \mathbb{C}^+$ (we denote this concisely as $\lambda \in (\mathbb{C}^-, \mathbb{C}^+)$); and a a solution $\mu_2(x,\lambda)$ whose columns have the opposite boundedness. Explicitly, μ_1, and μ_2 are defined as the solutions of the linear integral equations

$$\mu_1(x,\lambda) = I - \int_x^\infty \mathrm{e}^{-i\lambda(x-y)\widehat{\sigma_3}} Q(y)\mu_1(y,\lambda)dy, \qquad \lambda \in (\mathbb{C}^-, \mathbb{C}^+),$$

$$\mu_2(x,\lambda) = I + \int_{-\infty}^x \mathrm{e}^{-i\lambda(x-y)\widehat{\sigma_3}} Q(y)\mu_2(y,\lambda)dy, \qquad \lambda \in (\mathbb{C}^+, \mathbb{C}^-).$$

The notation used above means that the action of $\exp(x\widehat{\sigma_3})$ on a 2×2 matrix A is given by

$$\mathrm{e}^{-x\widehat{\sigma_3}} A = \begin{pmatrix} a_{11} & \mathrm{e}^{-2x} a_{12} \\ \mathrm{e}^{2x} a_{21} & a_{22} \end{pmatrix}. \tag{2.11}$$

For $\lambda \in \mathbb{R}$, where both μ_1 and μ_2 are defined, these solutions are related by

$$\mu_1(x,\lambda) = \mu_2(x\lambda)e^{-ix\lambda\widehat{\sigma_3}}s(\lambda), \quad \lambda \in \mathbb{R}, \tag{2.12}$$

where, in analogy with (2.6), one has

$$s(\lambda) = \lim_{x\to-\infty}[e^{ix\lambda\widehat{\sigma_3}}\mu_1(x,\lambda)] = \begin{pmatrix} \overline{a(\lambda)} & b(\lambda) \\ \overline{b(\lambda)} & a(\lambda) \end{pmatrix}, \quad \lambda \in \mathbb{R}.$$

By re-arranging the jump matrix $s(\lambda)$ in (2.12), one finds that the two matrix-valued function $M^{\pm}$, defined as $M^{-} = (\mu_1^{(1)}, \mu_2^{(2)})$ and $M^{+} = (\mu_2^{(1)}, \mu_1^{(2)})$ and analytic in $\mathbb{C}^{\pm}$ respectively, satisfy a jump condition across $\mathbb{R}$ of the form

$$M^{-}(x,\lambda) = M^{+}(x,\lambda)e^{i\lambda x\widehat{\sigma_3}}J(\lambda), \qquad \lambda \in \mathbb{R}, \tag{2.13}$$

as well as asymptotic conditions

$$M^{\pm} = I + O\left(\frac{1}{\lambda}\right), \quad as\ |\lambda| \to \infty.$$

The jump $J(\lambda)$, defined for $\lambda \in \mathbb{R}$ by

$$J(\lambda) = \begin{pmatrix} 1 & -\gamma(\lambda) \\ \pm\overline{\gamma(\lambda)} & 1 + |\gamma(\lambda)|^2 \end{pmatrix}, \quad \lambda \in \mathbb{R}; \qquad \gamma(\lambda) = \frac{b(\lambda)}{\overline{a(\lambda)}}, \tag{2.14}$$

is now matrix-valued. The entries of the 2×2 matrix $J(\lambda)$ are defined (for $a(\lambda) \neq 0$) in terms of $\gamma(\lambda) = \frac{b(\lambda)}{\overline{a(\lambda)}}$, where the *spectral functions* $a(\lambda)$ and $b(\lambda)$ are the entries of $s(\lambda)$ as defined in (2.12).

Hence, given the matrix $Q(x)$, the relation (2.12) defines the spectral functions $a(\lambda)$, $b(\lambda)$, and then (2.14) yields the transform $J(\lambda)$.

Conversely, given $J(\lambda)$, the jump equation (2.13) defines a Riemann-Hilbert problem, with a jump on the real line, for $M(x,\lambda)$. Note that the jump matrix $J(\lambda)$ and the function $M(x,\lambda)$ are well-defined only modulo possible isolated singularities arising at the zeros of $a(\lambda)$ in the complex plane. While it is generally an open problem to characterise rigorously the properties of these singularities from the given initial datum, it is often assumed in the context of the inverse scattering transform that there are only finitely many, and then by giving residue conditions at the associated poles one can uniquely specify the associated Riemann-Hilbert problem. For the purpose of this review, we assume that there are no such zeros (see also remark 4.2).

The difference with the linear case is that in this case the jump condition is multiplicative. The multiplicative, non-commutative structure of the Riemann-Hilbert problem is a manifestation of the nonlinearity of the equation, and it implies that the solution does not have an explicit expression analogous to the one in (2.4). However, it is a classical result that the solution $M(x,\lambda)$ of the Riemann-Hilbert problem can be characterised through the solution of a linear integral equation, and its unique solvability can be rigorously proved, appealing to the symmetries of $s(\lambda)$ which are a consequence of the form of the matrix $Q(x)$ [9, 10, 11, 22].

From the expression for $M(x, \lambda)$ one must derive an expression for $q(x)$ and thus formulate the associated inverse transform. Recall that, in the linear case, $\mu \sim iq/\lambda$ as $\lambda \to \infty$. Similarly, from the expression for $M(x, \lambda)$, one can determine the arbitrary function q by computing the λ asymptotic of the $(1,2)$ element of the matrix $M(x, \lambda)$, as done in (2.7), and obtain

$$q(x) = 2i \lim_{|\lambda| \to \infty} (\lambda M_{12}(x, \lambda)).$$

The above procedure gives rise to a nonlinear version of the Fourier transform, known as the *Inverse Scattering Transform (IST)*. The jump matrix $J(\lambda)$ is the analogue of the direct Fourier transform. Indeed, the direct and inverse transform diagram can be summarised as follows:

Direct transform - obtain $J(\lambda)$ from $q(x)$ via solving an ODE for $M(x, t, \lambda)$:

$$q \longrightarrow Q = \begin{pmatrix} 0 & q(x) \\ \pm\bar{q}(x) & 0 \end{pmatrix} \longrightarrow$$

$$\longrightarrow \quad M : \; M_x + i\lambda[\sigma_3, M] = QM \longrightarrow e^{i\lambda x \widehat{\sigma_3}} J = (M^+)^{-1} M^-.$$

Inverse transform - obtain $q(x)$ from $J(\lambda)$ via solving a Riemann-Hilbert problem for $M(x, t, \lambda)$:

$$J \longrightarrow M : \; M^+ = e^{i\lambda x \widehat{\sigma_3}} J M^-, \; x \in \mathbb{R}, \; M = I + O\left(\frac{1}{\lambda}\right) \longrightarrow \quad q = 2i \lim_{\lambda \to \infty} (\lambda M_{12}).$$

3 The structure of integrable PDEs: Lax pair formulation

A useful Lax pair for several important evolution PDEs takes the general form

$$\begin{cases} M_x + if_1(\lambda)[\sigma_3, M] = Q(x, t, \lambda)M, \\ M_t + if_2(\lambda)[\sigma_3, M] = \tilde{Q}(x, t, \lambda)M, \end{cases} \tag{3.1}$$

where σ_3 and the commutator $[\cdot, \cdot]$ are defined in (2.8), (2.10).

The particular form of the functions $f_i(\lambda)$, $Q(x, t, \lambda)$, $\tilde{Q}(x, t, \lambda)$ depends on the specific PDE. Throughout this review, we will consider two of the most important integrable evolution PDEs, arising as models in mathematical physics, namely the nonlinear Schrödinger (NLS) and modified Korteweg-deVries (mKdV) equations. The case of this second- and third-order evolution equation illustrate the general approach for integrable evolution PDEs in one space variable. For these equations, the Lax pair is given by the following choices:

NLS

$$iq_t + q_{xx} - 2\nu q_x |q|^2 = 0, \qquad \nu = \pm 1; \tag{3.2}$$

$$\textit{Lax pair}: \quad f_1(\lambda) = \lambda, \quad f_2(\lambda) = 2\lambda^2,$$

$$Q = \begin{pmatrix} 0 & q(x) \\ \nu\bar{q}(x) & 0 \end{pmatrix}, \qquad \tilde{Q} = 2\lambda Q - iQ_x\sigma_3 \pm |q|^2\sigma_3;$$

mKdV

$$q_t + q_{xxx} - 6\nu q^2 q_x = 0, \qquad \nu = \pm 1; \tag{3.3}$$

Lax pair :
$$f_1(\lambda) = \lambda, \quad f_2(\lambda) = 4\lambda^3,$$
$$Q = \begin{pmatrix} 0 & q(x,t) \\ \nu q(x,t) & 0 \end{pmatrix},$$
$$\tilde{Q} = 2Q^3 - Q_{xx} - 2i\lambda[Q^2 + Q_x]\sigma_3 + 4\lambda^2 Q.$$

There is no general methodology for finding a Lax pair associated with a given nonlinear PDE, although the dressing method of Shabat and Zakharov can be used to derive an integrable PDE by starting from a Lax pair for a given linear PDE [30, 59].

3.1 The solution of the Cauchy problem on the line using the Inverse Scattering Transform

The Lax pair formulation of the NLS equation is given explicitly by

$$\begin{cases} M_x + i\lambda[\sigma_3, M] = QM \\ M_t + 2i\lambda^2[\sigma_3, M] = (2\lambda Q - iQ_x\sigma_3 - i\nu|q|^2\sigma_3)M \end{cases} \qquad \lambda \in \mathbb{C}, \tag{3.4}$$

where $M = M(x,t,\lambda)$, implies that $q(x,t)$ solves the PDE (3.2) if and only if for all $\lambda \in \mathbb{C}$ there exists an invertible matrix-valued function $M(x,t,\lambda)$ solving (3.4). The NLS is obtained by imposing the compatibility condition $M_{xt} = M_{tx}$.

The first ODE in this pair is precisely the ODE associated with the nonlinear Fourier transform derived earlier, where now $q(x,t)$ replaces $q(x)$. The second part of the Lax pair is used to determine the time evolution of $J(\lambda,t)$, which now depends on t. Given $q(x,0) = q_0(x) \in \mathcal{S}(\mathbb{R})$, one constructs $M(x,0,\lambda)$ and then $J_0(\lambda) = J(\lambda,0)$. The t-part of the Lax pair implies $J(\lambda,t) = e^{-i\lambda^2 t\hat{\sigma}_3}$.

Conversely, given $J(\lambda,t)$ the function $q(x,t)$ can be represented in terms of the elements of the matrix $M(x,t,\lambda)$ as

$$q(x,t) = 2i \lim_{|\lambda|\to\infty} (\lambda M_{12}(x,t,\lambda)),$$

where the function $M(x,t,\lambda)$, sectionally analytic for $\lambda \in \mathbb{C}$, is the solution of the Riemann-Hilbert problem determined by

$$M^- = M^+ e^{(i\lambda x - i\lambda^2 t)\widehat{\sigma_3}} J_0(\lambda),\ \lambda \in \mathbb{R}, \qquad M = I + O\left(\frac{1}{\lambda}\right)\ as\ |\lambda| \to \infty,$$

with the jump matrix $J_0(\lambda)$ given by

$$J_0(\lambda) = \begin{pmatrix} 1 & -\gamma(\lambda) \\ \pm\overline{\gamma(\lambda)} & 1 + |\gamma(\lambda)|^2 \end{pmatrix}, \quad \lambda \in \mathbb{R}; \qquad \gamma(\lambda) = \frac{\overline{b(\lambda)}}{a(\lambda)}, \tag{3.5}$$

and the spectral functions $a(\lambda)$, $b(\lambda)$ are defined by in terms of the given initial condition $q_0(x)$ as

$$(b(\lambda), a(\lambda)) = \lim_{x\to-\infty}(m(x,\lambda)_1, m(x,\lambda)_2), \qquad m(x,\lambda) = \begin{pmatrix} M_{21}(x,0,\lambda) \\ M_{22}(x,0,\lambda) \end{pmatrix}$$

(to simplify the exposition, here and in the sequel it is assumed that $a(\lambda)$ has no zeros).

The vector $m(x,\lambda)$, as the second column vector of the solution of the first ODE in the Lax pair at $t = 0$, satisfies

$$m_x(x,\lambda) + 2i\lambda \begin{pmatrix} 1 & 0 \\ 0 & 0 \end{pmatrix} m(x,\lambda) = \begin{pmatrix} 0 & q_0(x) \\ \pm\bar{q}_0 & 0 \end{pmatrix} m(x,\lambda),$$
$$\lim_{x\to\infty} m(x,\lambda) = \begin{pmatrix} 0 \\ 1 \end{pmatrix}.$$

The expressions above make it apparent that the spectral functions $a(\lambda)$, $b(\lambda)$ depend only on the initial condition $q_0(x)$, and that the time dependence only enters (explicitly) through the exponential term.

4 An integral transform for nonlinear boundary value problems

Consider the defocusing NLS equation (3.2) with $\nu = 1$ in the variables $x \in \mathbb{R}$ and $t \in \mathbb{R}$, posed in the domain

$$\Omega = \{(x,t) : 0 < x < \infty,\ 0 < t < T\} \subset \mathbb{R}^2. \tag{4.1}$$

Examples of such equations are the NLS, KdV and mKdV equations given by (3.2), (1.8), (1.9) respectively, as well as many other important equations of mathematical physics. In analogy with the case of the full line in section 3.1, the role of the Fourier transform of the initial condition $q_0(x)$ is played by the spectral data defined in terms of the initial information. These spectral data are the pair of functions $a(\lambda)$, $b(\lambda)$ such that

$$\begin{pmatrix} b(\lambda) \\ a(\lambda) \end{pmatrix} = \begin{pmatrix} M_{21}(x,0,\lambda) \\ M_{22}(x,0,\lambda) \end{pmatrix},$$

where $M(x,0,\lambda)$ is the solution of the x part of the Lax pair, i.e. the first ODE in (3.4), evaluated at $t = 0$. It is useful to express more transparently the relation between $q_0(x)$ and $a(\lambda), b(\lambda)$: let

$$\Psi(x,\lambda) = M(x,0,\lambda).$$

The function $\Psi(x,\lambda)$ has symmetries inherited from the matrices Q, $\tilde{Q}$, so it is enough to consider one of its columns. For example, the second column of $\Psi(x,\lambda)$

satisfies, for $0 < x < \infty$, the following ODE:

$$\partial_x \begin{pmatrix} \Psi_{21}(x,\lambda) \\ \Psi_{22}(x,\lambda) \end{pmatrix} + 2i\lambda \begin{pmatrix} 1 & 0 \\ 0 & 0 \end{pmatrix} \psi(x,\lambda) = \begin{pmatrix} 0 & q_0(x) \\ \overline{q_0}(x) & 0 \end{pmatrix} \begin{pmatrix} \Psi_{21}(x,\lambda) \\ \Psi_{22}(x,\lambda) \end{pmatrix},$$

$$\lim_{x\to\infty} \begin{pmatrix} \Psi_{21}(x,\lambda) \\ \Psi_{22}(x,\lambda) \end{pmatrix} = \begin{pmatrix} 0 \\ 1 \end{pmatrix}, \qquad \lambda \in \mathbb{C}^+. \tag{4.2}$$

The solution of these ODEs is equivalent to a linear Volterra integral equation, hence well defined.

Similarly, the boundary conditions $q(0,t), q_x(0,t)$ determine a pair of functions $A(\lambda)$, $B(\lambda)$ such that

$$\begin{pmatrix} B(\lambda) \\ A(\lambda) \end{pmatrix} = \begin{pmatrix} e^{2i\lambda^2 T} M_{21}(0,T,\lambda) \\ M_{22}(0,T,\lambda) \end{pmatrix},$$

where $M(0,t,\lambda)$ is the solution of the t part of the Lax pair, i.e. the second ODE in (3.4), evaluated at $x = 0$. It is desirable to express the relation between boundary values and these spectral functions, and to conform with the usual notation, set

$$\Phi(t,\lambda) = M(0,t,\lambda).$$

Again, symmetry considerations imply that it is enough to consider one column of this function. For example, the second column of $\Phi(t,\lambda)$ satisfies, for $0 < t < T$ and $\lambda \in \mathbb{C}$, the following ODE:

$$\partial_t \begin{pmatrix} \Phi_{21}(t,\lambda) \\ \Phi_{22}(t,\lambda) \end{pmatrix} + 4i\lambda^2 \begin{pmatrix} 1 & 0 \\ 0 & 0 \end{pmatrix} \begin{pmatrix} \Phi_{21}(t,\lambda) \\ \Phi_{22}(t,\lambda) \end{pmatrix} =$$

$$\begin{pmatrix} -i|q(0,t)|^2 & -iq_x(0,t) + 2\lambda q(0,t) \\ -i\overline{q_x}(0,t) + 2\lambda\overline{q}(0,t) & i|q(0,t)|^2 \end{pmatrix} \begin{pmatrix} \Phi_{21}(t,\lambda) \\ \Phi_{22}(t,\lambda) \end{pmatrix},$$

$$\begin{pmatrix} \Phi_{21}(0,\lambda) \\ \Phi_{22}(0,\lambda) \end{pmatrix} = \begin{pmatrix} 0 \\ 1 \end{pmatrix}, \qquad \lambda \in \mathbb{C}, \tag{4.3}$$

which again is equivalent to a linear Volterra integral equation, hence well defined.

Note that since in general only one boundary condition involving the two boundary values $q(0,t)$ and $q_x(0,t)$ can be prescribed, the boundary spectral functions $A(\lambda)$, $B(\lambda)$ are not fully characterised by the above ODE.

Nevertheless the solution $q(x,t)$ has a formal representation in terms of the solution $M(x,t,\lambda)$ of a Riemann-Hilbert problem defined on the real and imaginary axes, whose jump is defined in terms of the spectral functions $a(\lambda)$, $b(\lambda)$, $A(\lambda)$ and $B(\lambda)$. The function $q(x,t)$ exists uniquely, has explicit x and t dependence, and it represents a solution of the PDE satisfying the initial condition. In general, it will not satisfy prescribed boundary conditions. However, if the full set of boundary values is assumed *a priori* to satisfy the additional constraint given by the *global relation*, then the function $q(x,t)$ satisfies these boundary values.

For a general nonlinear integrable PDE the main question is how to determine the unknown boundary values in terms of the given initial and boundary conditions. Given $q_0(x) \in \mathcal{S}(\mathbb{R}^+)$ and a subset of set of the boundary values $\{f_k(t) = \partial_k^x q(0,t)\}_{k=0}^{n-1}$, the main problem becomes to show that the global relation characterises the remaining unknown boundary values. This involves *the analysis of the global relation in the complex λ plane to determine a representation depending only on the prescribed initial and boundary conditions.*

As discussed below, this step is fully successful in the nonlinear case *only* for certain special types of boundary conditions, called *linearisable boundary condition.*

For generic boundary conditions, the characterisation of the unknown boundary values via the global relation is itself a nonlinear problem, and it reduces to solving a nonlinear system of equations [29].

4.1 The integral representation of the solution

In this section we summarise the steps needed to derive the main formal statement regarding integrable evolution PDE in the two independent variables $(x,t) \in \Omega$, where Ω is given by (4.1). Rather than specifying a set of boundary conditions, one assumes a-priori that the initial condition and the full set of boundary values satisfy the global relation. See [42] for details.

For several of the most physically relevant PDEs in this class, the Lax pair takes the form (3.1). In this form, $M(x,t,\lambda)$ is a 2×2 matrix-valued function, while $f_1(\lambda)$, $f_2(\lambda)$ are given analytic (usually polynomial) functions of λ, encoding the dispersion relation of the PDE.

The Lax pair (3.1) can be written in terms of a differential form $W(x,t,\lambda)$ as

$$d[\mathrm{e}^{(if_1(\lambda)x+if_2(\lambda)t)\widehat{\sigma_3}}M(x,t,\lambda)] = \mathrm{e}^{(if_1(\lambda)x+f_2(\lambda)t)\widehat{\sigma_3}}W(x,t,\lambda), \tag{4.4}$$

where the meaning of the notation $e^{\widehat{\sigma_3}}$ is given in (2.11) and

$$W(x,t,\lambda) = \left[Q(x,t,\lambda)dx + \tilde{Q}(x,t,\lambda)dt\right] M(x,t,\lambda). \tag{4.5}$$

4.1.1 The direct problem

The first step is constructing *simultaneous* solutions of the two ODEs in the Lax pair, bounded and analytic as functions of λ in simply connected regions of the complex plane whose union is the whole plane. These basic eigenfunctions are given by

$$M_j(x,t,\lambda) = I + \int_{(x_j,t_j)}^{(x,t)} \mathrm{e}^{(-if_1(\lambda)(x-\xi)-if_2(\lambda)(t-\tau))\widehat{\sigma_3}}W_j(\xi,\tau,\lambda), \quad (x,t),(x_j,t_j) \in \Omega. \tag{4.6}$$

In order to define a solution $M(x,t,\lambda)$ defined and analytic everywhere except on a contour, it is sufficient to consider the points (x_j,t_j) as the vertices of the unbounded polygon Ω, namely

$$(x_1,t_1) = (0,T), \quad (x_2,t_2) = (0,0), \quad (x_3,t_3) = (\infty,t),$$

Thus one obtains three *sectionally analytic* basic eigenfunctions, M_1, M_2 and M_3. Their definition is independently of the path of integration,and the column vectors are bounded and analytic in certain domains. On the common boundary of these domains, these eigenfunctions satisfy the following jump conditions:

$$M_3(x,t,\lambda)=M_2(x,t,\lambda)\mathrm{e}^{(-if_1(\lambda)x-if_2(\lambda)t)\widehat{\sigma_3}}s(\lambda),\quad \lambda\in(\mathbb{C}^-,\mathbb{C}^+),\tag{4.7}$$
$$M_1(x,t,\lambda)=M_2(x,t,\lambda)\mathrm{e}^{(-if_1(\lambda)x-if_2(\lambda)t)\widehat{\sigma_3}}S(\lambda),\quad \lambda\in(D_2,D_3),\tag{4.8}$$

where $\lambda\in(D,\tilde{D})$ means the matrix identity is valid for the first column in the domain D and for the second column in the domain D,

$$D_1=\{\lambda: Imf_1(\lambda)>0\cup Imf_2(\lambda)>0\},\quad D_2=\{\lambda: Imf_1(\lambda)>0\cup Imf_2(\lambda)<0\},$$
$$D_3=\{\lambda: Imf_1(\lambda)<0\cup Imf_2(\lambda)>0\},\quad D_4=\{\lambda: Imf_1(\lambda)<0\cup Imf_2(\lambda)<0\},\tag{4.9}$$

and

$$s(\lambda)=M_3(0,0,\lambda);\qquad S(\lambda)=[\mathrm{e}^{if_2(\lambda)T\hat{\sigma}_3}M_2(0,T,\lambda)]^{-1}.\tag{4.10}$$

4.1.2 The spectral functions

Letting

$$\Psi(x,\lambda)=M_3(x,0,\lambda),\ \lambda\in(\mathbb{C}^-,\mathbb{C}^+),\qquad \Phi(t,\lambda)=M_2(0,t,\lambda),\ \lambda\in\mathbb{C},\tag{4.11}$$

one can write the matrices in (4.10) as

$$s(\lambda)=\Psi(0,\lambda),\qquad \lambda\in(\mathbb{C}^-,\mathbb{C}^+),\tag{4.12}$$
$$S(\lambda)=\left[\mathrm{e}^{if_2(\lambda)T\widehat{\sigma_3}}\Phi(T,\lambda)\right]^{-1},\qquad \lambda\in\mathbb{C}.\tag{4.13}$$

Since they solve the two ODEs in the Lax pair, these functions are the solutions of the following linear Volterra integral equations:

$$\Psi(x,\lambda)=I-\int_x^\infty \mathrm{e}^{-if_1(\lambda)(x-\xi)\widehat{\sigma_3}}Q(\xi,0,\lambda)\Psi(\xi,\lambda)d\xi,\quad x\in(0,\infty);\ \lambda\in(\mathbb{C}^-,\mathbb{C}^+),\tag{4.14}$$

$$\Phi(t,\lambda)=I+\int_0^t \mathrm{e}^{-if_2(\lambda)(t-\tau)\widehat{\sigma_3}}\tilde{Q}(0,\tau,\lambda)\Phi(\tau,\lambda)d\tau,\quad t\in(0,T);\ \lambda\in\mathbb{C},\tag{4.15}$$

which are respectively equivalent to ODE (4.2) and to the following analogue of (4.3):

$$\partial_t\begin{pmatrix}\Phi_{21}(t,\lambda)\\ \Phi_{22}(t,\lambda)\end{pmatrix}+2if_2(\lambda)\begin{pmatrix}1&0\\0&0\end{pmatrix}\begin{pmatrix}\Phi_{21}(t,\lambda)\\ \Phi_{22}(t,\lambda)\end{pmatrix}=\tilde{Q}(0,t,\lambda)\begin{pmatrix}\Phi_{21}(t,\lambda)\\ \Phi_{22}(t,\lambda)\end{pmatrix},$$
$$\begin{pmatrix}\Phi_{21}(0,\lambda)\\ \Phi_{22}(0,\lambda)\end{pmatrix}=\begin{pmatrix}0\\1\end{pmatrix},\qquad \lambda\in\mathbb{C}.\tag{4.16}$$

In particular, the spectral functions satisfy the following:

- $s(\lambda)$ is defined by the values of the solution at $t=0$ - the initial condition;
- $S(\lambda)$ is defined by the values of the solution at $x=0$ - the boundary values.

The matrices Q, $\tilde{Q}$ for the integrable PDE considered have symmetry properties that imply that $s(\lambda)$, $S(\lambda)$ can be written as

$$s(\lambda)=\begin{pmatrix} \overline{a(\bar{\lambda})} & b(\lambda) \\ \pm\overline{b(\bar{\lambda})} & a(\lambda) \end{pmatrix}, \qquad S(\lambda)=\begin{pmatrix} \overline{A(\bar{\lambda})} & B(\lambda) \\ \pm\overline{B(\bar{\lambda})} & A(\lambda) \end{pmatrix}. \tag{4.17}$$

Hence the jump matrices depend on the four distinct functions of the spectral parameter λ defined by (4.17).

4.1.3 The global relation

To obtain an additional relation involving the spectral functions $s(\lambda)$ and $S(\lambda)$, one observes that the function Φ defined by (4.11) and the function $M_3(0,t,\lambda)$ satisfy the same differential equation, namely the t-part of the Lax pair at $x=0$. Hence they must be related, and the explicit identity that this induces yields the global relation:

$$S^{-1}(\lambda)s(\lambda)=\mathrm{e}^{if_2(\lambda)T\widehat{\sigma_3}}M_3(0,T,\lambda), \qquad \lambda\in(\mathbb{C}^-,\mathbb{C}^+). \tag{4.18}$$

This relation can be written explicitly in terms of the components of the matrices $s(\lambda)$ and $S(\lambda)$, namely the spectral functions. The $(1,2)$ element of the relation then yields

$$a(\lambda)B(\lambda)-A(\lambda)b(\lambda)=\mathrm{e}^{2if_2(\lambda)T}C(\lambda), \quad \lambda\in\mathbb{C}^+, \tag{4.19}$$

where $C(\lambda)$ is a function analytic in $\mathbb{C}^+$, defined in terms of the solution at the final time $t=T$, and such that $C(\lambda)=O\left(\frac{1}{\lambda}\right)$ as $\lambda\to\infty$. This function does not contribute to the final solution representation.

4.1.4 The inverse problem

Assume that a suitable initial condition $q(x,0)=q_0(x)$ is specified, and that the functions $a(\lambda)$, $b(\lambda)$ are defined in terms of $q_0(x)$ as solutions of the ODE (4.2).

Assume also that there exists a set of boundary values $q(0,t), q_x(0,t)$ (and $q_{xx}(0,t)$ for mKdV) such that the spectral functions $A(\lambda)$, $B(\lambda)$ given as solutions of ODE (4.16), together with $(a(\lambda),b(\lambda))$, satisfy the global relation (4.19).

Equations (4.7)-(4.8) can be rewritten in the form

$$M_-(x,t,\lambda)=M_+(x,t,\lambda)J(t,\lambda), \quad \lambda\in\mathcal{L} \tag{4.20}$$

where $\mathcal{L}=\mathcal{L}_1\cup..\cup\mathcal{L}_4$ with

$$\mathcal{L}_k=D_k\cap D_{k+1},\ k=1,2,3, \quad \mathcal{L}_4=D_4\cap D_1. \tag{4.21}$$

The matrix J can be computed explicitly in terms of the entries of the spectral functions $s(\lambda)$, $S(\lambda)$ of (4.17), and has the form $J = e^{(if_1(\lambda)x+if_2(\lambda)t)\hat{\sigma}_3}\tilde{J}$ with

$$\tilde{J}(\lambda) = J_k(\lambda), \quad \lambda \in \mathcal{L}_k, \quad k = 1,..,4, \tag{4.22}$$

$$J_1 = \begin{pmatrix} 1 & 0 \\ \Gamma(\lambda) & 1 \end{pmatrix}; \quad J_3 = \begin{pmatrix} 1 & -\nu\overline{\Gamma(\bar{\lambda})} \\ 0 & 1 \end{pmatrix};$$

$$J_4 = \begin{pmatrix} 1 & -\gamma(\lambda) \\ \nu\overline{\gamma(\bar{\lambda})} & 1-\nu|\gamma(\lambda)|^2 \end{pmatrix}, \quad J_2 = J_3 J_4^{-1} J_1.$$

where

$$\gamma(\lambda) = \frac{b(\lambda)}{\overline{a(\bar{\lambda})}}, \quad \Gamma(\lambda) = \frac{\nu\frac{\overline{B(\bar{\lambda})}}{A(\bar{\lambda})}}{a(\lambda)[a(\lambda) - \nu b(\lambda)\frac{\overline{B(\bar{\lambda})}}{\overline{A(\bar{\lambda})}}]}. \tag{4.23}$$

For the case of NLS, the contours $\mathcal{L}_1, .., \mathcal{L}_4$ are the four semiaxes issuing from $\lambda = 0$, so M is well defined everywhere except the real and imaginary axes. For the case of mKdV, M is well defined off the real axis and the rays that make angles of $\pi/3$, $2\pi/3$, $4\pi/3$ and $5\pi/3$ with the positive real axis.

Assuming in addition that $M = I + O\left(\frac{1}{\lambda}\right)$, as $|\lambda| \to \infty$, the above Riemann-Hilbert problem admits a unique solution $M(x, t, \lambda)$. From this solution, the solution $q(x, t)$ of the PDE is determined through the relation

$$q(x,t) = \pm 2i \lim_{\lambda\to\infty} (\lambda M(x,t,\lambda))_{12} \quad (+: NLS, \quad -: mKdV). \tag{4.24}$$

The main theorem stated below summarises these results, as well as formalising the fact that the function $q(x, t)$ given by (4.24) also satifies $q(x, 0) = q_0(x)$ and the given set of boundary values.

4.1.5 The main theorem

The theorem below was first proven in the context of the NLS equation, [33]. This result justifies the construction of the solution outlined above for any integrable PDE with a Lax pair of the form (3.1) and satisfying symmetry conditions, though to avoid technicalities we restrict attention to PDE with spatial derivatives up to third order, and really have in mind the two examples of the NLS and mKdV equations. We state the main result for the case of these two equations, without giving its proof, that can be found in [30].

The result justifies the main intuition behind the Unified Transform approach: combining the main idea of the Inverse Scattering Transform of starting with the Lax pair formulation with the idea of solving boundary value problem by the simultaneous solution of both ODEs in the Lax pair.

Theorem 1. *Let a function $q_0(x) \in \mathcal{S}(\mathbb{R}^+)$ be given, and define $a(\lambda)$, $b(\lambda)$ through the solution of ODE (4.2), where Q is given by (2.8), and assume additionally that $a(\lambda)$ has no zeros.*

Assume that a set of functions $f_0(t)$, $f_1(t)$, $(f_2(t))$ are given in such a way that the spectral functions $A(\lambda)$, $B(\lambda)$ defined in terms of the boundary values $\partial_x^k q(0,t) = f_k(t)$, $k = 0, 1, (2)$, through the solution of ODE (4.16), with $\tilde{Q}$ as in (3.2) or (3.3), satisfy the global relation (4.19).

Define $M(x,t,\lambda)$ as the solution the Riemann-Hilbert problem defined on the contour $\mathcal{L}$ of (4.21), with jump matrix J given by (4.22) in terms of the spectral functions $\{a(\lambda), b(\lambda), A(\lambda), B(\lambda)\}$, and such that $M(x,t,\lambda) = I + O\left(\frac{1}{\lambda}\right)$.

This Riemann-Hilbert problem admits a unique solution $M(x,t,\lambda)$. In addition the function $q(x,t)$ defined in terms of of M by (4.24) satisfies either the NLS equation or mKdV equations, as well as

$$q(x,0) = q_0(x); \quad q(0,t) = f_0(t),\ q_x(0,t) = f_1(t),\ (q_{xx}(0,t) = f_2(t)\ for\, mKdV).$$

Remark 4.1. *The proof that $q(x,t)$ solves the given nonlinear PDE uses the standard arguments of the dressing method. The proof that $q(x,0) = q_0(x)$ is based on the fact that the Riemann-Hilbert problem satisfied by $M(x,0,\lambda)$ is equivalent to the Riemann-Hilbert problem defined by $s(\lambda)$, namely the Riemann-Hilbert problem which characterises $q_0(x)$. The proof that $\partial_x^k q(0,t)$, $k = 0, ..., n-1$ are the boundary values of the solution makes crucial use of the global relation [33]. Indeed, the Riemann-Hilbert problem satisfied by $M(0,t,\lambda)$ is equivalent to the Riemann-Hilbert problem defined by $S(\lambda)$, which characterises the boundary values, if and only if the spectral functions satisfy this global relation, hence this relation is a* necessary and sufficient condition *for the existence of a solution.*

Remark 4.2. *To simplify the exposition and stress the points that are of specific relevance for the case of boundary value problems, we have assumed the spectral functions have no isolated singularities. However, any zeros in the denominator of the expressions given in (4.23), namely any zero of the function $a(\lambda)$ or $d(\lambda) = a(\lambda) - \nu b(\lambda)\frac{\overline{B(\bar{\lambda})}}{\overline{A(\bar{\lambda})}}$, will result in such a singularity. The singularities generated by the zeros of $a(\lambda)$ are well understood, in particular it is known that the residue conditions at these singularities, for initial as well as for boundary value problems, describe the soliton part of the solution [1]. The question of whether $d(\lambda)$ has zeros remains open.*

4.2 Linearisable boundary conditions

The most difficult step in solving a boundary vlue problem by any method, including the Fokas method, is the characterisation of the two spectral functions $A(\lambda)$, $B(\lambda)$ in terms of the given initial and boundary data, i.e. the characterisation of the unknown boundary values. For certain boundary conditions, called linearisable, this can be achieved simply using algebraic manipulations analogous to what can be done for linear evolution PDEs. It should be noticed that all boundary conditions that are linearisable for integrable evolution PDEs had been known before: for the second order case (NLS) these are conditions that allow the problem to be solved essentially by restriction of a problem posed on the full line, see the review in [13]; for the sine-Gordon equation, they had been found in an important work of

Sklyanin based on physical considerations,[61, 60]; and, for KdV, they are conditions for which the Bäcklund transformation can be linearised, [51]. Nevertheless, as I explain below, in the approach described here these conditions all follow from the requirement that global invariant transformation of the spectral functions can be defined.

The strategy of evaluating the global relation at all transformations that leave $f_2(\lambda)$ invariant, which is fully successful in the linear case, fails in general. This failure is due to the fact that the two relevant spectral functions, namely $A(\lambda)$ and $B(\lambda)$, involve not only $e^{if_2(\lambda)t}$ but also the components of the function $\Phi(t, \lambda)$ given by (4.11). These components in general are not invariant under transformations that leave $f_2(\lambda)$ invariant.

Linearisable boundary conditions are precisely the conditions such that the components of $\Phi(t, \lambda)$ admit this additional invariance property. A more precise statement is given in the following proposition, see [30].

Proposition 1. *Suppose that the t part of the Lax pair of an integrable nonlinear PDE is characterised by the scalar function $f_2(\lambda)$ and by the 2×2 matrix-valued function $\tilde{Q}(x, t, \lambda)$ given in (2.8). Let $\theta(\lambda)$ be the transformations of complex (λ)-plane which leave $f_2(\lambda)$ invariant.*

Define $U(t, \lambda)$ by

$$U(t, \lambda) = if_2(\lambda)\sigma_3 - \tilde{Q}(0, t, \lambda). \tag{4.25}$$

If it is possible to define a matrix-valued function $N(\lambda)$, in terms only of the prescribed boundary conditions, such that

$$U(t, \theta(\lambda))N(\lambda) = N(\lambda)U(t, \lambda) \tag{4.26}$$

then the boundary spectral function $A(\lambda)$, $B(\lambda)$ defined in (4.17) possess explicit symmetry properties of the form

$$A(\theta(\lambda)) = L_1(A(\lambda), B(\lambda)), \qquad B(\theta(\lambda)) = L_2(A(\lambda), B(\lambda))$$

where L_1, L_2 are linear functions of $A(\lambda)$, $B(\lambda)$, $\overline{A(\bar{\lambda})}$, $\overline{B(\bar{\lambda})}$ with coefficients depending only on the entries of the matrix $N(\lambda)$.

When the condition of this proposition is satisfied, the functions $A(\lambda)$, $B(\lambda)$ can be computed as effectively as in the linear case in terms of $a(\lambda)$, $b(\lambda)$ and the prescribed boundary conditions.

It follows from this proposition that a *necessary* condition for the existence of linearisable boundary conditions is that the determinant of the matrix $U(t, \lambda)$ defined by (4.25) is a function of λ only through $f_2(\lambda)$. However, this condition is *not sufficient.* In particular, since the function U depends on the particular choice of Lax pair, it follows that different Lax pairs allow one to uncover different linearisable conditions. An explicit example is given by the case of the sine-Gordon equation, derived following this strategy in [28].

Particular conditions that are linearisable are listed below:

NLS In this case, there are three linearisable boundary conditions satisfying the necessary condition on the determinant of $U(t, \lambda)$ with $\tilde{Q}$ defined by (2.8):

$$(a)\ q(0,t) = 0; \quad (b)\ q_x(0,t) = 0; \quad (c)\ q_x(0,t) - \chi q(0,t) = 0,\ \chi \in \mathbb{R}^+; \quad (4.27)$$

KdV- This refers to the *KdV equation with dominant surface tension*, hence with a negative sign in front of the third derivative term:

$$q_t + q_x - q_{xxx} + 6qq_x = 0. \quad (4.28)$$

In this case, $N = 2$ so two boundary conditions must be prescribed at $x = 0$.

$$(a)\ q(0,t) = \chi,\ q_{xx}(0,t) = \chi + 3\chi^2, \quad \chi \in \mathbb{R}; \quad (4.29)$$

4.3 General boundary conditions

For general boundary conditions, not necessarily linearisable, the invariance analysis of the global relation is not sufficient to characterise the solution of the problem. However, for general boundary conditions that *decay for large t*, the representation obtained through the Unified Transform yields useful asymptotic information even without the explicit characterisation of the spectral functions.

In addition, two different approaches for analysing the *generalised Dirichlet to Neumann map* for the case of the NLS equation, i.e. to express $q_x(0,t)$ in terms of the given boundary condition $f(t)$ and initial condition $q_0(x)$, and presented in [42, 43].

For non-decaying boundary conditions, the computation of the large t behaviour of the solution and of its boundary values requires new ideas. The most significant example of this situation is the case of a *time-periodic* given boundary condition, an important condition in practice. For example, the KdV equation with given zero initial condition $q(x,0) = 0$ and a periodic boundary condition such as $q(0,t) = a\sin(\phi t)$, $0 < \phi < \pi$, corresponds to the very realistic situation of shallow water waves in a tank, excited by a periodic wavemaker. The linear case of this model is studied in [14].

The first results on the analysis of a periodic boundary value problem of this type, for the NLS equation, were obtained in [15, 16, 17] for the particular case that $f(t) = ae^{i\phi t}$. More recently, using the general approach of this review, coupled with perturbation techniques, significant progress has been achieved for the physically significant case of the NLS, and mKdV equations given the boundary condition $f(t) = a\sin t$, $a \in \mathbb{R}$, using a perturbation scheme. However this perturbative approach becomes cumbersome and only a few terms in the perturbative expansion can be computed.

The most recent result, presented in [44, 45], uses a new perturbative approach to compute the asymptotic behaviour of $q_x(0,t)$ given a periodic Dirichlet datum. By carrying out the analysis directly in the large t limit, the algorithm for characterising the perturbative expansion of $q_x(0,t)$ is greatly simplified, and used to prove that this function is asymptotically periodic of the same period as the given datum.

4.3.1 Nonlinearisable problems for NLS on the half-line

Recall that the unknown boundary values enter the solution representation through the spectral functions $A(\lambda)$, $B(\lambda)$ given by (4.17) as particular value of the function Φ solution of (4.15). The functions $A(\lambda)$ and $B(\lambda)$ are in general characterised by a system of integral equations. Indeed, for the particular example of NLS, this system is given explicitly as follows

$$A(\lambda) = \overline{\varphi_2(T, \bar{\lambda})}, \quad B(\lambda) = -e^{4i\lambda^2 T}\varphi_1(T, \lambda),$$

where $\varphi_1(t, \lambda)$, $\varphi_2(t, \lambda)$ are solutions of

$$\varphi_1(t, \lambda) = \int_0^t e^{4i\lambda^2(s-t)}[\nu|f_0(t)|^2\varphi_1 + (2\lambda f_0(t) + if_1(t))\varphi_2](s, \lambda)ds \tag{4.30}$$

$$\varphi_2(t, \lambda) = 1 \pm \int_0^t [(2\lambda\bar{f}_0(t) - i\bar{f}_1(t))\varphi_1 + i|f_0(t)|^2\varphi_2](s, \lambda)ds \qquad 0 < t < T, \ \ \lambda \in \mathbb{C}.$$

where we use the notation

$$f_0(t) = q(0, t), \qquad f_1(t) = q_x(0, t).$$

By substituting into the equations above the expression for f_0 and f_1 given below in equations (4.31)-(4.32), it becomes apparent that this system is itself nonlinear.

Indeed the following result summarises the situation for the general (non-homogeneous) Dirichlet or Neumann case, directly in terms of boundary functions in physical variables.

Proposition 2. *Let $T < \infty$. Consider the NLS equation on the positive half-line*

$$iq_t + q_{xx} - 2\nu q|q|^2 = 0, \quad x \in \mathbb{R}^+, 0 < t < T.$$

Let $q_0(x) \in \mathcal{S}(\mathbb{R}^+)$ be a given function, and consider one of the following two cases of boundary conditions (BC):

(a) Dirichlet BC

Let $q(0, t) = f_0(t)$ be given, smooth and compatible with $q_0(x)$ for $t = 0$.

(b) Neumann BC

Let $q_x(0, t) = f_1(t)$ be given, smooth and compatible with $q_0(x)$ for $t = 0$.

Suppose that the spectral function $a(\lambda)$ defined in terms of $q_0(x)$ has a finite set of simple zeros $\{\lambda_j\}$, none on which is on the real or imaginary axis.

Let

$$\chi_j(t, \lambda) = \varphi_j(t, \lambda) - \varphi_j(t, -\lambda); \ \tilde{\chi}_j = \varphi_j(t, \lambda) + \varphi_j(t, -\lambda), \ j = 1, 2; \ \ 0 < t < T, \ \ \lambda \in \mathbb{C}.$$

Then

(a) For $f_0(t)$ given,

$$
\begin{aligned}
f_1(t) &= \frac{2}{\pi i}\int\limits_{\partial D_3}(\lambda\chi_1(t,\lambda)+if_0(t))d\lambda+\frac{2f_0(t)}{\pi}\int\limits_{\partial D_3}\chi_2(t,\lambda)d\lambda \qquad (4.31)\\
&- \frac{4}{\pi i}\int\limits_{\partial D_3}\lambda \mathrm{e}^{-4i\lambda^2 t}\frac{b(-\lambda)}{a(-\lambda)}\overline{\varphi_2(t,-\bar{\lambda})}d\lambda+8\sum_{\lambda_j\in D_1}\lambda_j \mathrm{e}^{-4i\lambda_j^2 t}\frac{b(\lambda_j)}{\dot{a}(\lambda_j)}\overline{\varphi_2(t,\bar{\lambda}_j)},
\end{aligned}
$$

(b) For $f_1(t)$ given,

$$
\begin{aligned}
f_0(t) &= \frac{1}{\pi}\int\limits_{\partial D_3}\tilde{\chi}_1(t,\lambda)d\lambda+\frac{2}{\pi}\int\limits_{\partial D_3}\mathrm{e}^{-4i\lambda^2 t}\frac{b(-\lambda)}{a(-\lambda)}\overline{\varphi_2(t,-\bar{\lambda})}d\lambda\\
&+ 4i\sum_{\lambda_j\in D_1}\lambda_j \mathrm{e}^{-4i\lambda_j^2 t}\frac{b(\lambda_j)}{\dot{a}(\lambda_j)}\overline{\varphi_2(t,\bar{\lambda}_j)}. \qquad (4.32)
\end{aligned}
$$

It remains to show that the resulting nonlinear systems for φ_1, φ_2, obtained after substituting expressions (4.31) or (4.32) in the system (4.30), provide *effective* characterisation of the spectral functions $A(\lambda)$, $B(\lambda)$.

Here, effective means that:

(a) the linear limit yields the effective solution of the linearised boundary value problem, i.e. a representation involving only the known boundary data;

(b) for sufficiently small boundary data, the characterisation yields a perturbative scheme in which all terms can be computed uniquely via a well defined recursive scheme.

The general approach of [42] uses three ingredients.

1. The large λ asymptotics of the matrix-valued function $\Phi(t,\lambda)$ solution of (4.15), which defines $A(\lambda)$, $B(\lambda)$.

2. The global relation and the equations obtained under the transformations that fix the linearised dispersion relation

3. A perturbative scheme to show that the methodology is effective.

The analysis of large λ asymptotics to obtain additional conditions on the spectral functions was first employed in [20] for the case of the NLS equation, but it is only the combination of all three ingredients above that yields a general result that generalises also, for example, to equations such as mKdV involving third order derivatives .

4.3.2 Periodic boundary conditions

As already mentioned, even before the explicit characterisation of the unknown boundary data, the representation of the solution obtained via the Riemann-Hilbert approach described allows one to obtain precise asymptotic information on the behaviour of the solution for large times. However, this is only possible when the solution, and the given boundary conditions, decays when $t \to \infty$. For other cases, such as the case of given boundary conditions periodic in time, it is very hard to extract asymptotic information from the representation formulas directly.

After the pioneering results of [17], recent progress on this problem has been reported in [44, 45]. In this paper, Fokas and Lenells use the ideas of the perturbative approach described in the previous section to show that, given a t-periodic Dirichlet boundary condition, for either the NLS or the modified KdV equation, the coefficient in the perturbative expansion of the Neumann datum, to any given order, is periodic in the limit as $t \to \infty$. This result is shown by proving the following two properties:

(a) The perturbative approach allows one to show that given a t-periodic Dirichlet datum the Neumann datum also becomes periodic as $t \to \infty$. This analysis is based on analyticity considerations. It is carried out directly in the asymptotic limit, and this simplifies the procedure to the extent that it is possibly to compute the coefficient of the perturbative expansion to all order.

(b) Assuming that the Neumann datum is periodic the coefficients of the Fourier series of this periodic function can be characterised uniquely. Since this step is constructive, using the fact that the there exists a unique solution one can justify the a-priori periodicity assumption.

The idea of the construction is as follows. Starting from the expression (4.31), assuming a zero initial condition and additionally *assuming*

$$\varphi_1 = a\varphi_{11} + a^2\varphi_{12} + O(a^3), \quad a \to 0, \tag{4.33}$$

$$\varphi_2 = 1 + a\varphi_{21} + a^2\varphi_{22} + O(a^3), \quad a \to 0, \tag{4.34}$$

$$f_0(t) = af_{01}(t), \quad f_1(t) = af_{11}(t) + a^2 f_{12} + a^3 f_{13} + O(a^4), \quad a \to 0. \tag{4.35}$$

one can construct a recursive scheme for the coefficients of $\varphi_1(t,\lambda)$, $\varphi_2(t,\lambda)$ in terms of those of $f_0(t)$ and $f_1(t)$. If the latter two are arbitrary functions, then all one knows is that $\varphi_1(t,\lambda)$, $\varphi_2(t,\lambda)$ are entire functions of λ bounded as $\lambda \to \infty$ in $D_2 \cup D_4$. However, if $f_0(t)$ and $f_1(t)$ are the boundary values of a solution of the boundary value problem, with $q_0(x) = 0$, then this boundedness must also hold in D_1. Using this additional analyticity constraint it is possible and indeed rather straightforward to compute the coefficients in the expansion of $f_1(t)$ in terms of the given $f_0(t)$.

5 Further considerations

in this review we have concentrated on the implementation of the Unified Transform to integrable nonlinear evolution PDEs on the half-line. However, this method has much wider applicability.

Evolution PDEs on a Finite Interval

It was first shown in [58] that even for linear evolution PDEs involving a third order derivative, it is *not* possible to express the solution in terms of an infinite series. The implementation of the Fokas Method for a variety of linear and nonlinear boundary value problems can be found in [30, 38].

Well-posedness of non-integrable PDEs

In order to establish well posedness of nonlinear PDEs which are not integrable, one usually considers an associate linear forced problem, where the nonlinearity is treated as the forcing term, and then one uses a contraction mapping argument. Taking into consideration that the Fokas Method yields the explicit construction of the solution of any linear forced evolution PDE on the half-line, it is natural to expect that in can provide an efficient way of proving well-posedness. This plan has been carried out in [47, 46], where the Unified Method is placed within a Sobolev space framework.

Nonlinear Elliptic PDEs

The Fokas Method yields new integral representations for the solution of linear elliptic PDEs in polygonal domains, which in the case of simple domains, can be used to obtain the analytical solution of several problems which apparently cannot be solved by the standard methods [62, 53]. Some of these results have been extended to nonlinear integrable PDEs like the elliptic sine Gordon [34].

The importance of the Global Relation

Although the so called global relation is only one of the ingredients of the Fokas method, still this relation has had important analytical and numerical implications: first,it has led to novel analytical formulations of a variety of important physical problems from water waves [3, 35, 8] to three-dimensional layer scattering [56]. Second,it has led to the development of a new numerical techniques for the Laplace, modified Helmholtz, Helmholtz and biharmonic equations, as well as for Elliptic PDEs with variable coefficients, see [18] and the exhaustive reference list therein.

It should be emphasised that the Fokas method has a significant pedagogical advantage: both the numerical calculation of the analytical solutions obtained for linear evolution PDEs, as well as the implementation of the numerical technique to linear elliptic PDEs, are straightforward, so that even undergraduate students can implement them using MATLAB.

Acknowledgements

A.S. Fokas was supported by EPSRC in the form of a senior fellowship. B. Pelloni was supported by EPSRC grant F15R10379.

References

[1] M. J. Ablowitz and P.A. Clarkson. *Solitons, nonlinear evolution equations and inverse scattering.* Cambridge University Press, 1991.

[2] M. J. Ablowitz and A. S. Fokas. *Complex Variables.* Cambridge University Press, 1997.

[3] M. J. Ablowitz, A. S. Fokas, and Z. H. Musslimani. On a new non-local formulation of water waves. *J. Fluid Mech.*, 562:313–343, 2006.

[4] Mark J Ablowitz and Harvey Segur. *Solitons and the inverse scattering transform*, volume 4. SIAM, 1981.

[5] M.J. Ablowitz, D.J. Kaup, A.C. Newell, and H. Segur. Method for solving the sine-Gordon equation. *Physical Review Letters*, 30(25):1262–1264, 1973.

[6] M.J. Ablowitz, D.J. Kaup, A.C. Newell, and H. Segur. The inverse scattering transform-Fourier analysis for nonlinear problems. *Studies in Applied Mathematics*, 53:249–315, 1974.

[7] T. Aktosun. Inverse scattering transform and the theory of solitons. In *Encyclopedia of Complexity and Systems Science*, pages 4960–4971. Springer, 2009.

[8] A.C.L. Ashton and A. S. Fokas. A nonlocal formulation of rotational water waves. *Journal of Fluid Mechanics*, 689:129-48, 2011.

[9] R. Beals and R.R. Coifman. Scattering and inverse scattering for first order systems. *Comm. Pure Appl. Math.*, 37:39–90, 1984.

[10] R. Beals and R.R. Coifman. Scattering and inverse scattering for first order systems: Ii. *Inv. Prob.*, 3:577–593, 1987.

[11] R. Beals and R.R. Coifman. Linear spectral problems, non-linear equations and the $\bar{\partial}$-method. *Inv. Prob.*, 5:87–130, 1989.

[12] ED Belokolos, AI Bobenko, VZ Enol'Skii, AR Its, and VB Matveev. *Algebro-geometric approach to nonlinear integrable equations*, volume 1994. Springer-Verlag Berlin, 1994.

[13] G. Biondini, A.S. Fokas, and D. Shepelsky. Comparison of two approaches to the initial-boundary value problem for the nonlinear Schrödinger equation on the half line with Robin boundary conditions. In A.S. Fokas and B. Pelloni, editors, *Unified Transform for Boundary Value Problems: Applications and Advances.* SIAM, 2014.

[14] J. L. Bona and A. S. Fokas. Initial-boundary-value problems for linear and integrable nonlinear dispersive partial differential equations. *Nonlinearity*, 21:195–203, 2008.

[15] A. Boutet de Monvel and V. Kotlyarov. Generation of asymptotic solitons of the nonlinear Schrödinger equation by boundary data. *J. Math. Phys.*, 44:3185–3215, 2003.

[16] A. Boutet de Monvel and V. Kotlyarov. Characteristic properties of the scattering data for the mKdV equation on the half-line. *Comm. Math. Phys.*, 253(1):51–79, 2005.

[17] A. Boutet de Monvel and V. Kotlyarov. The focusing nonlinear Schrödinger equation on the quarter plane with time-periodic boundary condition: a Riemann-Hilbert approach. *J. Inst. Math. Jussieu*, 6:579–611, 2007.

[18] M.J. Colbrook. Extending the Fokas method: curvilinear polygons and variable coefficient PDEs. (submitted), 2018.

[19] B Deconinck, T. Trogdon, and V. Vasan. The method of Fokas for solving linear partial differential equations. *SIAM Reviews*, 56:159–186, 2014.

[20] A. Degasperis, S. Manakov, and P. Santini. On the initial-boundary value problems for soliton equations. *JETP Letters*, 74:481–485, 2001.

[21] P. Deift. Universality for mathematical and physical systems. *arXiv preprint math-ph/0603038*, 2006.

[22] P. Deift and E. Trubowitz. Inverse scattering on the line. *Communications on Pure and Applied Mathematics*, 32:121–251, 1979.

[23] B. A. Dubrovin, V. B. Matveev, and S. P. Novikov. Non-linear equations of Korteweg-deVries type, finite-zone linear operators, and abelian varieties. *Russian mathematical surveys*, 31(1):59, 1976.

[24] N. Flyer and A. S. Fokas. A hybrid analytical numerical method for solving evolution partial differential equations. I. the half-line. *Proc. R. Soc. Lond. Ser. A Math. Phys. Eng. Sci.*, 464(2095):1823–1849, 2008.

[25] N. Flyer and B. Fornberg. A numerical implementation of Fokas boundary integral approach: Laplace's equation on a polygonal domain. *Proc. R. Soc. Lond. A*, 467, 2011.

[26] A. S. Fokas. A unified transform method for solving linear and certain nonlinear PDEs. *Proc. R. Soc. Lond. Ser. A Math. Phys. Eng. Sci.*, 453:1411–1443, 1997.

[27] A. S. Fokas. Integrable nonlinear evolution equations on the half-line. *Comm. Math. Phys.*, 230(1):1–39, 2002.

[28] A. S. Fokas. Linearizable initial boundary value problems for the sine-Gordon equation on the half-line. *Nonlinearity*, 17(4):1521–1534, 2004.

[29] A. S. Fokas. The generalized Dirichlet-to-Neumann map for certain nonlinear evolution PDEs. *Comm. Pure Appl. Math.*, 58:639–670, 2005.

[30] A. S. Fokas. *Unified Transform for Boundary Value Problems.* CBMS-SIAM, 2008.

[31] A. S. Fokas and Novikov R. G. Discrete analogues of $\bar{\partial}$-equation and of Radon transform. *CR Acad Sci Paris Ser I Math*, 313:75, 1991.

[32] A. S. Fokas and I. M. Gelfand. Integrability of linear and nonlinear evolution equations and the associated nonlinear fourier transforms. *Lett. Math. Phys.*, 32:189–210, 1994.

[33] A. S. Fokas, A. R. Its, and L. Y. Sung. The nonlinear Schrödinger equation on the half-line. *Nonlinearity*, 18(4):1771–1822, 2005.

[34] A. S. Fokas, J. Lenells, and B. Pelloni. Explicit solutions for the elliptic sine-Gordon equation in a semistrip. *J Nonlin Science*, 23:241–282, 2013.

[35] A. S. Fokas and A. Nachbin. Water waves over a variable bottom: a non-local formulation and conformal mappings. (submitted), 2011.

[36] A. S. Fokas and B. Pelloni. Integral transforms, spectral representations and the d-bar problem. *Proc. R. Soc. Lond. Ser. A Math. Phys. Eng. Sci.*, 456:805–833, 2000.

[37] A. S. Fokas and B. Pelloni. Two-point boundary value problems for linear evolution equations. *Math. Proc. Cambridge Philos. Soc.*, 131:521–543, 2001.

[38] A. S. Fokas and B. Pelloni. *Unified Transform for Boundary Value Problems: Applications and Advances.* CBMS-SIAM, 2014.

[39] A. S. Fokas and L. Y. Sung. Generalised Fourier transforms, their nonlinearisation and the imaging of the brain. *Notices Amer. Math. Soc.*, 52(10):1178–1192, 2005.

[40] AS Fokas. Lax pairs: a novel type of separability. *Inverse Problems*, 25:1–44, 2009.

[41] AS Fokas and MJ Ablowitz. The inverse scattering transform for the Benjamin-Ono equationa pivot to multidimensional problems. *Studies in Applied Mathematics*, 68(1):1–10, 1983.

[42] A.S. Fokas and J. Lenells. The unified method: I. nonlinearizable problems on the half-line. *Journal of Physics A: Mathematical and Theoretical*, 45(19):195201, 2012.

[43] A.S. Fokas and J. Lenells. The unified method: II. NLS on the half-line with t-periodic boundary conditions. *Journal of Physics A: Mathematical and Theoretical*, 45(19):195202, 2012.

[44] A.S. Fokas and J. Lenells. The nonlinear Schrdinger equation with t-periodic data: I. Exact results. arXiv:1412.0304, 2014.

[45] A.S. Fokas and J. Lenells. The nonlinear Schrdinger equation with t-periodic data: II. Perturbative results. arXiv:1412.0306, 2014.

[46] A.S. Fokas, A Himonas, and D. Mantzavinos. The nonlinear schrödinger equation on the half-line. *Transactions of the American Mathematical Society*, 369(1):681–709, 2017.

[47] A.S. Fokas, A.Himonas, and D. Mantzavinos. The Korteweg–de Vries equation on the half-line. *Nonlinearity*, 29(2):489, 2016.

[48] A.S. Fokas and A.R. Its, The nonlinear Schrödinger equation on the interval, *Phys. A: Math*, 37:6091, 2004.

[49] A.S. Fokas, A.R. Its, A.I. Kapaev, and V.I. Novokshenov. *Painlevé transcendents: the Riemann-Hilbert approach.* Number 128. American Mathematical Soc., 2006.

[50] C. S. Gardner, J.M. Greene, M. D. Kruskal, and R.M. Miura. Method for solving the Korteweg-deVries equation. *Physical Review Letters*, 19(19):1095–1097, 1967.

[51] I.T. Habibulin. Backlund transformations and integrable initial-boundary value problems. *Mathematical Notes*, 49:418–423, 1991.

[52] A. R. Its. The Riemann-Hilbert problem and integrable systems. *Notices Amer. Math. Soc.*, 50(11):1389–1400, 2003.

[53] K. Kalimeris. *Initial and boundary value problems in two and three dimensions.* PhD thesis, University of Cambridge, 2010.

[54] I. M. Krichever. Methods of algebraic geometry in the theory of non-linear equations. *Russian Mathematical Surveys*, 32(6):185–213, 1977.

[55] P.D. Lax. Integrals of nonlinear equations of evolution and solitary waves. *Communications on pure and applied mathematics*, 21(5):467–490, 1968.

[56] David P Nicholls. A high-order perturbation of surfaces (hops) approach to fokas integral equations: three-dimensional layered media scattering. *Quart. Appl. Math.*, 2016.

[57] R. G. Novikov. An inversion formula for the attenuated x-ray transformation. *Arkiv för matematik*, 40:145–167, 2002.

[58] B. Pelloni. The spectral representation of two-point boundary-value problems for third-order linear evolution partial differential equations. *Proc. R. Soc. Lond. Ser. A Math. Phys. Eng. Sci.*, 461:2965–2984, 2005.

[59] A Shabat and V Zakharov. Exact theory of two-dimensional self-focusing and one-dimensional self-modulation of waves in nonlinear media. *Soviet Physics JETP*, 34:62–69, 1972.

[60] E. K. Sklyanin. Boundary conditions for integrable quantum systems. *Journal of Physics A: Mathematical and General*, 21:2375, 1988.

[61] E.K. Sklyanin. Boundary conditions for integrable equations. *Funct. Anal. Appl.*, 21:86–87, 1987.

[62] E. A. Spence. *Boundary Value Problems for Linear Elliptic PDEs*. Phd, University of Cambridge, 2009.

A3. Detecting discrete integrability: the singularity approach

B Grammaticos [a], *A Ramani* [a], *R Willox* [b] *and T Mase* [b]

[a] *IMNC, Université Paris VII & XI, CNRS, UMR 8165, Bât. 440, 91406 Orsay, France*

[b] *Graduate School of Mathematical Sciences, the University of Tokyo, 3-8-1 Komaba, Meguro-ku, 153-8914 Tokyo, Japan*

Abstract

We describe the various types of singularities that can arise for second order rational mappings and we discuss the historical and present-day practical role the singularity confinement property plays as an integrability detector. In particular, we show how singularity analysis can be used to calculate explicitly the dynamical degree for such mappings. [1]

1 Introduction

The history of integrability, in its broadest sense, goes back to the beginnings of differential calculus in the 17th century and to the quest for solutions to differential equations. Mathematical modelling of physical problems led the founding fathers of calculus, Leibniz, Newton and the Bernoullis, to the study of differential equations and their solutions. The domain blossomed over the next two centuries through the contributions of many great mathematicians like Euler, Lagrange, Gauss, to name but a few. What can be considered as the real beginning of the modern era of integrability though, is the pioneering work of Kovalevskaya on the heavy top, spinning around a fixed point. Kovalevskaya noticed that for all cases where a solution to the equations of motion was known, this solution was given in terms of meromorphic functions and in particular elliptic functions. She was thus led to investigate the existence of other cases with meromorphic solutions and actually discovered one previously unknown case, which was subsequently dubbed the Kovalevski top in her honour.

Meromorphicity of solutions lay also at the origin of Painlevé's approach to the construction of new functions through the solution of nonlinear differential equations. Painlevé first studied the first-order case showing that the only equation without movable (i.e. initial-condition dependent) critical (i.e. multivaluedness inducing) singularities was the Riccati equation. The Riccati equation, however, can be transformed to a linear equation and therefore does not introduce any new functions, since linear equations are regarded as solvable, essentially, in terms of 'known' functions. Painlevé therefore went on to study second-order equations, requiring that their solutions be meromorphic in the independent variable apart from

[1] **This work is dedicated to the memory of professor K.M. Tamizhmani, collaborator and beloved friend**

possible fixed critical singularities, which can easily be taken care of. This intuition was brilliant since it turned out that several nontrivial such examples do exist and that new functions can be introduced by some of the equations he derived. These new functions are since known as the Painlevé transcendents. But even more important than the discovery of the *Painlevé equations* these functions satisfy, was the realisation that the absence of movable critical singularities (a property which later came to be known as the *Painlevé property*) could provide a powerful integrability criterion for differential systems [28].

Curiously, the interest in integrable differential systems waned over the next half century, a situation which only changed with the advent of electronic computers and the possibilities for numerical simulation these offered. While studying the famous Fermi-Pasta-Ulam (FPU) model, describing particles interacting on a lattice, Kruskal and Zabusky [46] considered its continuum limit and found that it led to a partial differential equation, introduced at the end of the 19th century by Korteweg and de Vries (known today as the KdV equation), which possesses an explicit solitary wave solution. Kruskal and Zabusky observed through simulations that the evolution of an initial profile led to its separation into several solitary waves that interact elastically. They chose the name of 'soliton' for these special waves. Analytical studies soon followed, showing that the KdV equation possesses infinitely many conservation quantitites and, as was shown by Hirota, allows for an arbitrary number of solitons. The complete solution of KdV was subsequently obtained through the use of methods of quantum mechanical inverse scattering. One important result of these early studies was the observation that Painlevé equations often appear as reductions of integrable evolution equations. This led, on the one hand, to the integration of the Painlevé equations through inverse scattering methods and, on the other hand, to the formulation of the Ablowitz-Ramani-Segur conjecture [1]. The ARS conjecture reaffirmed the Painlevé property as an integrability criterion, positing that "every ordinary differential equation which arises as a reduction of a partial differential equation integrable by inverse scattering techniques, is of Painlevé type".

While its continuum limit is integrable, the same is not true for the FPU model itself. An important question therefore was whether one could find a nonlinear lattice that is completely integrable. Toda [40] showed that this is indeed possible, by introducing a lattice with exponential interactions between nearest neighbours (the system that now bears his name). The Toda lattice is however a semi-discrete system since it describes the positions of particles on a lattice as continuous functions in time. But what about fully discrete systems? In fact, while studying the possible discretisations of the logistic equation (which is a simplified, constant coefficient, Riccati equation), Skellam [38] and Morishita [22] obtained a discrete form that has the fundamental property of the continuous one: it can be transformed into a linear equation. This discovery, however, remained mainly unnoticed, which was also the case for the groundbreaking work of Hirota who, in the 1970s, singlehandedly produced the integrable, fully discrete, forms of a host of famous integrable partial differential equations [14].

Fortunately though, this lack of interest was short-lived and at the end of the

80's several findings finally brought about the discrete integrability epoch. One important observation was that Baxter's solution to the Yang-Baxter equations is associated to the Euler-Chasles correspondence [42]. This result was cast in integrable systems parlance by Quispel, Roberts and Thompson (QRT) who defined a family of integrable second-degree mappings [25] solvable in terms of elliptic functions. Around the same time, work on string theory led to the discovery of integrable non-autonomous recursion relations which turned out (upon derivation of their continuum limits) to be discrete analogues of the Painlevé equations [29]. All this made integrability specialists turn to the until then largely unexplored discrete domain, developing techniques that were, at times, in perfect parallel to those for continuous systems, but on other occasions specific to the discrete case. One important question was "What is the discrete analogue of the Painlevé property?" or, to put it in a less pretentious way, "Is there an easy-to-use discrete integrability criterion?". Our answer to this question, known under the name of *singularity confinement property* [9] will be the subject of this chapter.

2 Singularity confinement

Let us start with an example of what may happen when one iterates an integrable mapping. For this we shall use the McMillan mapping [17], which we shall consider over $\mathbb{P}^1 = \mathbb{C} \cup \{\infty\}$ (and where μ is an arbitrary non-zero complex number):

$$x_{n+1} + x_{n-1} = \frac{2\mu x_n}{1 - x_n^2}. \tag{2.1}$$

Suppose that, due to a special choice of initial condition, at some iteration step, x_{n+1} takes precisely the value 1, while x_n is generic. This leads to values $x_{n+2} = \infty$, $x_{n+3} = -1$ and an indeterminate result for $x_{n+4} : \infty - \infty$. The way to lift this indeterminacy is through continuity with respect to the initial conditions, by introducing a small quantity ϵ and assuming that $x_{n+1} = 1 + \epsilon$. If one performs the above iteration for this initial condition, taking the limit for $\epsilon \to 0$ we then find that $x_{n+4} = -x_n$. In other words, by lifting the indeterminacy that arose at step $n + 4$, we in fact recovered the memory of the initial condition that was lost at the level of x_{n+2}. The loss of a degree of freedom, here the memory of the value of x_n, constitutes a *singularity* for the mapping. Note that this loss of memory of the initial condition in fact means that the inverse mapping is not defined at this point, which is another possible definition of a singularity. It should be stressed that a degree of freedom lost in a singularity can only be recovered if, when iterating beyond the singularity, an indeterminacy appears that can be lifted in the way we just described. This is what we call *confinement* of the singularity.

The behaviour of the McMillan mapping can, in fact, be easily understood once one realises that (2.1) is just the addition formula for elliptic sines. Indeed, if we consider a discretisation of the continuous variable $t_n = n\delta + t_0$, we have $x_n = x(t_n)$, $x_{n\pm1} = x(t_n \pm \delta)$ and the solution of (2.1) is simply $x_n = k \,\mathrm{sn}\,\delta\, \mathrm{sn}(t_n)$ (where k is the modulus of the elliptic sine) with $\mu = \mathrm{cn}\delta\, \mathrm{dn}\delta$. Using the addition formulae we can easily verify that if x_n diverges then $x_{n\mp 1} = \pm 1$ and $x_{n+2} = -x_{n-2}$. Thus x_{n-1} has precisely the value that guarantees the divergence of x_n and for x_{n+1} the value that compensates for this divergence. Moreover, the memory of the value of x_{n-2} survives past the singularity and is recovered at x_{n+2}.

The McMillan mapping is not the only one that can be solved in terms of elliptic functions. As is pointed out in the introduction, a whole family of such mappings, first proposed by Quispel, Roberts and Thompson, does exist [25]. It has the form

$$x_{n+1} = \frac{f_1(x_n) - x_{n-1} f_2(x_n)}{f_2(x_n) - x_{n-1} f_3(x_n)}, \tag{2.2}$$

where the f_i are specific quartic polynomials. The QRT mapping possesses an invariant, which can be expressed in the form of the Euler-Chasles correspondence [42]

$$\alpha x_n^2 x_{n-1}^2 + \beta x_n x_{n-1}(x_n + x_{n-1}) + \gamma(x_n^2 + x_{n-1}^2) + \epsilon x_n x_{n-1} + \zeta(x_n + x_{n-1}) + \mu = 0, \tag{2.3}$$

and can be solved in terms of elliptic functions. By the same argument as for the McMillan mapping, one can therefore conclude that all mappings of the QRT family possess the singularity confinement property (unless the invariant (2.3) is a rational curve, in which case the mapping is linearisable and, as we shall see, the situation becomes more complicated).

Singularity confinement was first discovered, not on mappings, but while studying a fully discrete version of the KdV equation [9], the lattice KdV equation [14]. In this chapter however, we choose to focus on the case of mappings and we shall not delve into the properties of integrable lattice equations, which are an even more complicated topic. Suffice it to say that it is the observation that all discrete systems integrable by inverse scattering techniques we studied, in fact possess the singularity confinement property, which led us to propose the latter as a discrete integrability criterion. We shall come back to this point, but let us first discuss the various types of singularities which may appear in second order rational mappings. In the following we shall always work over the compactified complex plane, i.e. $x_n \in \mathbb{P}^1$, for all n.

Let us perform the singularity analysis (for $a \in \mathbb{C}, a \neq 0$) of the QRT mapping

$$x_{n+1} x_{n-1} = a\left(1 - \frac{1}{x_n}\right). \tag{2.4}$$

One should first, of course, detect the singularities of the mapping. Remember that a singularity occurs at step n whenever the value of x_{n+1} is independent of x_{n-1} (chosen generically). For (2.4) this can clearly only happen when $x_n = 1$ (yielding $x_{n+1} = 0$ for generic x_{n-1}) or when $x_n = 0$ (yielding $x_{n+1} = \infty$). Iterating (2.4) for a generic value $x_0 = f$ and for $x_1 = 1 + \epsilon$, and taking the limit $\epsilon \to 0$ of all ensuing iterates, it is straightforward to check that one obtains the successive values $0, \infty, \infty, 0$ and 1, and that the indeterminacy $(1-1)/0$ that arises at x_7 is lifted, yielding $x_7 = f$, after which the iteration proceeds normally. Hence we have recovered the information on the initial condition x_0 that was lost in the singularity at x_2 and we conclude that the mapping possesses a *confined* singularity pattern $\{1, 0, \infty, \infty, 0, 1\}$. However, the second singularity for mapping (2.4), resulting from $x_n = 0$, is of a very different type. Starting from $x_0 = f$ (generic) and $x_1 = \epsilon$, we obtain (at the limit $\epsilon \to 0$) the sequence of values $f, 0, \infty, \infty, 0, 1/f, \infty, af, 0, \infty, \infty, 0, 1/(af), \infty, a^2 f, 0, \cdots$, from which it becomes clear that the pattern $\{0, \infty, \infty, 0, f', \infty, f''\}$ of length seven keeps repeating indefinitely. Moreover, it is easy to verify that if we iterate the mapping backwards, we again find the same succession of values, which means that the basic pattern keeps repeating for all n. We call such a singularity pattern *cyclic*. As will be briefly explained in Section 5, cyclic patterns are perfectly compatible with the integrable character of a given mapping such as (2.4).

As a second example, let us consider the nonintegrable mapping

$$x_{n+1} x_{n-1} = 1 - \frac{1}{x_n^2}, \tag{2.5}$$

which has singularities at $x_n = \pm 1$ and $x_n = 0$. The first two singularities both behave in exactly the same way: they yield an *unconfined* singularity pattern $\{\pm 1, 0, \infty^2, \infty, 0^2, \infty^3, \infty^2, 0^3, \infty^4, \cdots\}$ which continues indefinitely. (The meaning of the exponents of ∞ and 0 is the following: had we introduced a small quantity ϵ by assuming that $x_n = \pm 1 + \epsilon$, we would have found that x_{n+2} is of order $1/\epsilon^2$, x_{n+4} of order ϵ^2 and so on). Note that iterating (2.5) backwards from $x_n = \pm 1$ and some generic $x_{n-1} = f$, we do not encounter any singularities. This is what sets apart the above unconfined patterns from the singularity pattern obtained for the singularity at $x_n = 0$. Iterating (2.5) forwards from $x_{n-1} = f$ (generic) and $x_n = 0$, we obtain a sequence of values similar to that for the unconfined patterns: $\infty^2, \infty, 0^2, \infty^3, \infty^2, \cdots$. However, iterating backwards from these initial conditions we find the sequence $\infty, 1/f, 0, \infty^2, \infty, 0^2, \infty^3, \infty^2, \cdots$, showing that in this case the inverse mapping also leads to an unconfined singularity. We thus find what we call an *anticonfined* singularity pattern

$$\{\cdots, \infty^4, 0^3, \infty^2, \infty^3, 0^2, \infty, \infty^2, 0, \frac{1}{f}, \infty, f, 0, \infty^2, \infty, 0^2, \infty^3, \infty^2, 0^3, \infty^4, \cdots\} \tag{2.6}$$

in which singularities extend indefinitely, both ways, from a finite set of regular values.

For non-linearisable mappings, the appearance of an unconfined pattern indicates its nonintegrability. Anticonfined patterns on the other hand come in different varieties (as will become clear at the end of this Section), some compatible with integrability, some with linearisability and others indicating nonintegrability.

Deautonomising integrable mappings The main application of the singularity confinement criterion has been the so-called *deautonomisation* [26] procedure. This procedure consists in deriving integrable, non-autonomous, extensions of integrable autonomous mappings by assuming that the parameters of the latter are functions of the independent variable, the precise form of which is obtained by applying the confinement criterion. We can illustrate this in the case of the mapping (2.4), which we rewrite as

$$x_{n+1}x_{n-1} = a_n\left(1 - \frac{1}{x_n}\right). \tag{2.7}$$

We require that the confined singularity pattern be the same as in the autonomous case and obtain for the function $a(n)$ the constraint

$$a_{n+5}a_{n+2} = a_{n+4}a_{n+3}, \tag{2.8}$$

the solution of which is $\log a_n = \alpha n+\beta+\gamma(-1)^n$. This non-autonomous form of (2.4) was first derived in [18] where we showed that it is a q-discrete form of the Painlevé II equation. The cyclic singularity pattern for (2.1) carries over to the non-autonomous case as well. Starting from the same initial conditions $x_0 = f$ and $x_1 = 0$ we find, using (2.8), the succession of values $f, 0, \infty, \infty, 0, 1/f, \infty, a_6 f, 0, \infty, \infty, 0, 1/(a_6 f), \cdots$, i.e. the pattern $\{0, \infty, \infty, 0, f', \infty, f''\}$ still repeats indefinitely. (The fact that the cyclic pattern remains cyclic after deautonomisation is not a general feature: in many cases a cyclic pattern becomes a genuinely confined one when deautonomised, see e.g. [45]).

The deautonomisation procedure has been instrumental in deriving the non-autonomous forms of most discrete Painlevé equations known to date. In fact, it is by this very method that the first q-discrete Painlevé equations were obtained. Moreover, the structure of the singularity patterns can provide an indication as to where to look for more integrable systems. When studying equations associated to the affine Weyl group $\mathrm{E}_8^{(1)}$ (in the Sakai classification [37]) we observed that the two previously known equations, obtained by two of the present authors in collaboration with Ohta [23], had singularity patterns of length 7 and 3, and 5 and 5 respectively. Thus we surmised that mappings with singularity patterns of lengths 8 and 2, and 6 and 4 should also exist. This turned out to be indeed the case, allowing us to complement the list of the $\mathrm{E}_8^{(1)}$-related discrete Painlevé equations [27]. Also, when deautonomising QRT mappings, the cyclic patterns of the original autonomous mapping yield important information on the geometric structure of the discrete Painlevé equations one obtains (as explained in [45]). The algebro-geometric underpinnings of the deautonomisation of QRT mappings have been developed in [5].

Singularities and degree growth A most interesting aspect of the singularity structure of rational mappings is that it is intimately related to the degree growth of their iterates. Quoting Veselov, we remind the reader here that "integrability has an essential correlation with the weak growth of certain characteristics". The *dynamical degree* of a rational mapping is a measure of this growth. It is obtained from the degrees d_n of the iterates of some initial condition and is defined as $\lambda = \lim_{n\to\infty} d_n^{1/n}$. Note that $\lambda \geq 1$, and that integrable mappings have a dynamical degree equal to 1, while a dynamical degree greater than 1 indicates nonintegrability. (One often encounters an alternative measure of growth, dubbed algebraic entropy [3]:

it is simply the logarithm of the dynamical degree). In order to illustrate how the singularity structure is linked to the degree growth we consider the McMillan mapping (2.1). We take initial conditions $x_0 = r$ and $x_1 = p/q$, and after iterating we find

$$x_2 = \frac{r(p^2 - q^2) + 2\mu pq}{q^2 - p^2}, \qquad x_3 = \frac{P_5}{q\,P_2^+ P_2^-}, \qquad x_4 = \frac{(q^2 - p^2)^2\,P_8}{(q^2 - p^2)^2\,P_4^+ P_4^-}, \tag{2.9}$$

where $P_n, P_n^\pm$ are irreducible polynomials in p, q, r of degree n in p, q. We remark that if $q = \pm p$, we have $P_2^\pm\big|_{q^2=p^2} = \pm 2\mu p^2$, $P_5\big|_{q^2=p^2} = 4\mu^2 p^5$ (and $P_4^\pm\big|_{q^2=p^2} = \pm 4\mu^2 p^3 q$, $P_8\big|_{q^2=p^2} = 16\mu^4 p^8 r$), from which we precisely obtain the confined singularity patterns for (2.1): $\{\pm 1, \infty, \mp 1\}$. Note that we have written these iterates without enforcing any simplifications, and that they therefore show that the very first cancellation happens in the iterate x_4, which is exactly where the indeterminacy occurs that allows the singularities to confine. Clearly, in x_4, the degree will drop by four because of the cancellation of $(q^2 - p^2)^2$. Iterating further we find

$$x_5 = \frac{(q^2 - p^2)^4\,\left(P_2^+ P_2^-\right)^2\,P_{13}}{q(q^2 - p^2)^4\,\left(P_2^+ P_2^-\right)^2\,P_6^+ P_6^-}, \tag{2.10}$$

in which we now have an impressive cancellation (of a polynomial of degree 16). These massive simplifications, which start appearing at the confinement step and which become increasingly important as one iterates further, have as a result that the degree growth of this mapping is polynomial, rather than exponential (and the mapping thus has a dynamical degree equal to 1). Indeed, computing the degree in p, q of the successive iterates we find the sequence of values, $0, 1, 2, 5, 8, 13, 18, 25, 32, 41, 50, 61, 72, 85, 98, 113, 128\cdots$. The degree growth can in fact be rigourously shown to be quadratic: $d_n = \left(n^2 + \psi_2(n)\right)/2$, where $\psi_2(n)$ is a periodic function, with period two, defined by $\psi_2(0) = 0$ and $\psi_2(1) = 1$ (cf. (4.2) in Section 4).

The question of late confinement As explained above, when deautonomising an integrable mapping using singularity confinement, standard practice is to take the patterns obtained for the original autonomous mapping and require that the singularity be confined at the same step as in the autonomous case. Consider for instance the mapping

$$x_{n+1} + x_{n-1} = \frac{a_n}{x_n} + \frac{1}{x_n^2}. \tag{2.11}$$

In the autonomous case its (confined) singularity pattern is $\{0, \infty^2, 0\}$, which means that if we start from $x_n = f$ (generic) and $x_{n+1} = 0$, we recover the information on f at x_{n+4} (in fact, $x_{n+4} \equiv f$). Repeating the same procedure when a is a function of n we find that for x_{n+4} to depend on f, the function a_n must obey the constraint

$$a_{n+3} - 2a_{n+2} + a_{n+1} = 0, \tag{2.12}$$

the solution of which is $a_n = \alpha n + \beta$. This is the well-known deautonomisation of (2.11) leading to a discrete form of the Painlevé I equation. However, one could also ask what happens if one does not impose the constraint (2.12). In that case the singularity does not confine at the fourth step but continues along the pattern $\{0, \infty^2, 0, \infty, 0, \infty^2, 0, \infty, 0, \infty^2, 0, \cdots\}$. Examining these iterates carefully, it turns out that it is possible for the singularity to be confined at the eighth iterate, provided one imposes the constraint

$$a_{n+7} - 2a_{n+6} + a_{n+5} - a_{n+4} + a_{n+3} - 2a_{n+2} + a_{n+1} = 0. \tag{2.13}$$

Looking for the solutions of (2.13) in the form $a_n = \lambda^n$ we obtain the characteristic equation $(\lambda^2 - \lambda + 1)(\lambda^4 - \lambda^3 - \lambda^2 - \lambda + 1) = 0$ which leads to solutions for a_n that contain no secular terms. Similar results are obtained if one postpones the confinement even further. Note that, as a rule, such *late* confinements always lead to nonintegrable systems [13, 20].

The case of (2.11) is not exceptional. All (at least to the authors' knowledge) integrable mappings, when deautonomised, offer the possibility for late confinement. Thus the rule of thumb when deautonomising with singularity confinement should be to confine at the first occasion, lest a late confinement induce nonintegrability. (A word of caution is necessary here. It may happen in some exceptional cases that a possibility for confinement appears before that which is considered as the timely one. It turns out that in all such cases, such an *early* confinement leads to a trivial mapping, usually a periodic one).

Nonintegrable mappings with confined singularities The above example shows that at least for late confining mappings, the singularity confinement criterion is not sufficient to ensure integrability, but one could wonder if this a problem that is particular to non-autonomous systems. It turns out that this is not the case. The best known example of a confining but nonintegrable, autonomous mapping is the Hietarinta-Viallet (H-V) mapping [12]

$$x_{n+1} + x_{n-1} = x_n + \frac{1}{x_n^2}, \tag{2.14}$$

the confined singularity pattern of which is $\{0, \infty^2, \infty^2, 0\}$. The H-V mapping has been extensively studied and its nonintegrability has been rigorously established [39], e.g.: its dynamical degree is $(3+\sqrt{5})/2$. The mapping proposed by Hone [15], who attributes it to Viallet,

$$x_{n+1}x_{n-1} = x_n + \frac{1}{x_n}, \tag{2.15}$$

is another example. Its confined singularity patterns are $\{\pm i, 0, \infty, \infty^2, \infty, 0, \mp i\}$ and its dynamical degree is greater than 1. Its exact value was obtained in [43]: $(1+\sqrt{17})/4 + \sqrt{(1+\sqrt{17})/8} \approx 2.0810$. A third example, proposed by Mimura and collaborators [21], is the mapping

$$x_{n+1}x_{n-1} = \frac{x_n^4 - 1}{x_n^4 + 1}, \tag{2.16}$$

the singularity patterns of which are, $\{\pm 1, 0 \mp 1\}$, $\{\pm i, 0 \mp i\}$, $\{\pm r, 0 \mp r\}$ and $\{\pm ir, 0 \mp ir\}$, where r is the square root of i, and which has dynamical degree $2+\sqrt{3}$.

It should be clear that infinitely many such examples exist. Moreover, they are not limited to second-order mappings: higher order ones with similar properties do exist and the same is true for lattice equations [16, 44].

As we saw in the preceding paragraphs the confinement of a singularity is intimately linked to cancellations in the iterates of the mapping, which lower its degree growth. For integrable mappings this results in polynomial growth, but while for nonintegrable mappings with confined singularities such cancellations still occur, these turn out to be insufficient and the growth of the mapping remains exponential.

Linearisable mappings and confinement Though it might not be a sufficient integrability criterion, the fact that all discrete systems integrable by inverse scattering techniques that we have studied do possess confined singularities, is a strong indication for the necessary character of the confinement property. However, if one relaxes the notion of 'integrability' so as to include systems that can be integrated by different means, this conclusion has to be adjusted. For example, it is a well-known fact that continuous systems that can be considered to be integrable because they are *linearisable*, do not necessarily possess the Painlevé property [34]. The same turns out to be true for discrete systems.

As has been thoroughly documented, linearisable second-order mappings belong to one of three classes. The first is that of *projective* mappings [30]. These have the form

$$x_{n+1}x_nx_{n-1} + ax_nx_{n-1} + bx_{n-1} + c = 0, \tag{2.17}$$

and can be linearised through a Cole-Hopf transformation $x_n = w_{n+1}/w_n$ to the linear equation $w_{n+2} + a_nw_{n+1} + b_nw_n + c_nw_{n-1} = 0$. Projective mappings do have the confinement property, the confined pattern being simply $\{0, \infty\}$.

The second class of linearisable mappings is that of so-called *Gambier* mappings [8] which consist of two coupled homographic mappings in cascade. One example is $x_{n+1} = y_n + c/x_n$, $y_n = a + 1/y_{n-1}$ which can be rewritten as

$$x_{n+1} = a + \frac{c}{x_n} + \frac{x_{n-1}}{x_nx_{n-1} - c}. \tag{2.18}$$

Equation (2.18) has one confined singularity, with pattern $\{0, \infty\}$, but $x_n = \infty$ gives rise to an unconfined one: iterating from this point onwards the memory of x_{n-1} is irretrievably lost. Most Gambier mappings possess unconfined singularities. However, in some cases, when the mapping is sufficiently rich, it is possible to turn these unconfined singularities into confined ones by imposing appropriate constraints on the parameters in the mapping.

Another interesting Gambier mapping is

$$x_{n+1} = a + x_n - \frac{1}{x_n} + \frac{1}{x_{n-1}}, \tag{2.19}$$

obtained by the composition of $x_{n+1} = y_n - 1/x_n$ and $y_n = a + y_{n-1}$. This mapping has the confined singularity pattern $\{0, \infty\}$, but also an anticonfined one: $\cdots, 0, 0, 0, x, \infty, \infty, \infty, \cdots$ (remember that in an anticonfined pattern, singularities extend both ways from a finite set, here a singleton, of regular values).

The third type of linearisable mapping is known under the moniker of *third kind.* These were first discovered in [31] where we have given a general framework for their linearisation. An example of such a mapping is

$$\frac{2z}{x_n + x_{n+1}} + \frac{2z}{x_n + x_{n-1}} = 1 + \frac{2z}{x_n}. \tag{2.20}$$

It has the confined singularity pattern $\{0, 0\}$, but also the anticonfined pattern

$$\{\cdots, \infty^2, \infty, \infty, x, x', x'', \infty, \infty, \infty^2, \infty^2, \infty^3, \infty^3, \cdots\}. \tag{2.21}$$

Notice the linearly growing exponents in the singularities in (2.21), which are the signature of third-kind mappings. The question of whether there exist third-kind autonomous mappings with unconfined singularities remains open. Deautonomising (2.20), based on the pattern $\{0, 0\}$, we obtain

$$\frac{z_n + z_{n+1}}{x_n + x_{n+1}} + \frac{z_n + z_{n-1}}{x_n + x_{n-1}} = 1 + \frac{z_{n+1} + z_{n-1}}{x_n}, \tag{2.22}$$

where z_n is a free function of the independent variable. The linearisation of (2.22) was presented in [32]. Note that although the deautonomisation preserves the confined pattern, the anticonfined one turns into the unconfined pattern $\{\infty, \infty, \infty, \dots\}$, in which all infinities have exponent 1.

3 The full-deautonomisation approach

All this leads to the question: why is deautonomisation, based on singularity confinement, so successful ? If we leave aside the special case of linearisable mappings the answer is rather obvious: when deautonomising one starts from an autonomous integrable mapping, i.e. a system where the degree growth is slow, associated to given confined and cyclic singularity patterns. As the deautonomisation process preserves the singularity patterns, with the only change that some cyclic patterns may become confined, the simplifications in the iterates that occur in the autonomous case carry over to the non-autonomous one, thus guaranteeing slow growth and the integrability of the deautonomised system. However, the deautonomisation approach is even more successful than it would appear at first sight. In fact, when studying the phenomenon of late confinement, we made the remarkable discovery that the confinement conditions on the parameters in the deautonomised mapping, such as (2.12) or (2.13), actually contain information on the value of the dynamical degree for the resulting non-autonomous mapping. For example, the largest root of the characteristic equation for (2.12) is clearly 1, the value of the dynamical degree for the associated Painlevé I equation, but more strikingly, the largest root for the characteristic equation for (2.13) is approximately 1.72208, which fits exactly with the dynamical degree for the corresponding nonintegrable mapping. The deeper algebro-geometric reasons for this remarkable phenomenon will be briefly touched upon in Section 5. Here, we shall first explain how, based on these findings, one can extract the value of the dynamical degree for a given mapping, from a 'sufficiently general' deautonomisation of that mapping. As will become clear in the following, this approach, dubbed *full-deautonomisation* [36], in fact redeems singularity confinement as an integrability criterion.

Deautonomising nonintegrable mappings Let us start from mapping (2.15) which has the confined singularity patterns $\{\pm i, 0, \infty, \infty^2, \infty, 0, \mp i\}$, and let us deautonomise it as

$$x_{n+1}x_{n-1} = x_n - \frac{q_n}{x_n}. \tag{3.1}$$

When we require the non-autonomous mapping to possess the same singularity patterns as the autonomous one, we obtain the constraint

$$q_{n+7}q_{n+1} = q_{n+6}^2 q_{n+2}^2, \tag{3.2}$$

the largest root of the characteristic equation of which, $(\lambda^2-\lambda+1)(\lambda^4-\lambda^3-2\lambda^2-\lambda+1) = 0$, takes precisely the value of the dynamical degree $(1+\sqrt{17})/4+\sqrt{(1+\sqrt{17})/8}$ given in the previous Section.

The mapping (2.16), can be deautonomiseed as

$$x_{n+1}x_{n-1} = \frac{x_n^4 - q_n^4}{x_n^4 + 1}, \tag{3.3}$$

in which form it has the same singularity patterns as in the autonomous case if

$$q_{n+1}q_{n-1} = q_n^4. \tag{3.4}$$

The characteristic equation for this constraint, $\lambda^2 - 4\lambda + 1 = 0$, has $2 + \sqrt{3}$ as its largest solution, which is precisely the value of the dynamical degree for the mapping.

Full deautonomisation Simple deautonomisations such as those for the two previous mappings however do not, in general, suffice to obtain the exact value of the dynamical degree and often a more sophisticated approach is required. The H-V mapping (2.14) is a case in point. Let us first try the simple deautonomisation,

$$x_{n+1} + x_{n-1} = x_n + \frac{q_n}{x_n^2}, \tag{3.5}$$

requiring that the singularity pattern remain the same as for the autonomous case: $\{0, \infty^2, \infty^2, 0\}$. Unfortunately, the resulting confinement constraint is just $q_{n+3} = q_n$ and q_n therefore exhibits no growth. However, (3.5) is not the only possible deautonomisation of the H-V mapping. In particular, one can add a term inversely proportional to x_n and still conserve the singularity pattern of the autonomous mapping:

$$x_{n+1} + x_{n-1} = x_n + \frac{f_n}{x_n} + \frac{1}{x_n^2}. \tag{3.6}$$

The confinement constraint is now

$$f_{n+3} - 2f_{n+2} - 2f_{n+1} + f_n = 0, \tag{3.7}$$

with characteristic equation $(\lambda + 1)(\lambda^2 - 3\lambda + 1) = 0$, the largest root of which is precisely the dynamical degree of the H-V mapping: $(3 + \sqrt{5})/2$.

We can now define what we mean by 'full-deautonomisation'. Ordinarily, when deautonomising a mapping we just assume that its coefficients depend on the independent variable and we fix this dependence by requiring that the singularity patterns of the non-autonomous mapping be the same as that of the autonomous mapping. Full-deautonomisation carries this one step further. Namely, we extend the mapping by adding terms which, though initially absent, do not, when present, modify the singularity patterns and we deautonomise these terms as well.

Let us clarify this approach in a second example,

$$x_{n+1} + x_{n-1} = \frac{1}{x_n^4}, \tag{3.8}$$

the confined singularity pattern of which is $\{0, \infty^4, 0\}$. Considering all possible terms that leave this singularity pattern unchanged, we find that the extension which leads to an interesting result is, again, through adding a term inversely proportional to x_n:

$$x_{n+1} + x_{n-1} = \frac{a_n}{x_n} + \frac{b_n}{x_n^4}. \tag{3.9}$$

The confinement constraints in this case are $b_{n+1} = b_{n-1}$ (which shows that a simple deautonomisation of (3.8) does not lead to a helpful result) and

$$a_{n+1} - 4a_n + a_{n-1} = 0. \tag{3.10}$$

This last constraint gives rise to the characteristic equation $\lambda^2 - 4\lambda + 1 = 0$, the largest root of which is $2 + \sqrt{3}$, coinciding with the dynamical degree of (3.8).

Lest one get the impression that the full-deautonomisation approach is only applicable to nonintegrable mappings, let us give an example of its use in the case of an integrable system. Let us consider the integrable mapping

$$x_{n+1} + x_{n-1} = \frac{1}{x_n^2}, \tag{3.11}$$

with confined singularity pattern $\{0, \infty^2, 0\}$, which we can deautonomise simply by replacing the numerator of the right-hand side by a function b_n. Imposing the same singularity pattern we find the constraint $b_{n+1} = b_{n-1}$, which is a trivial non-autonomous extension of (3.11) since the freedom introduced by the period-2 coefficient b_n can be absorbed in a proper gauge of x_n. In the spirit of full-deautonomisation however, it is possible to extend (3.11) by adding a term proportional to $1/x$ on the right-hand side, while still preserving the singularity pattern. We then obtain the mapping

$$x_{n+1} + x_{n-1} = \frac{a_n}{x_n} + \frac{1}{x_n^2}, \tag{3.12}$$

which is of course mapping (2.11) of Section 2, where it was shown that a_n must satisfy the constraint $a_{n+1} - 2a_n + a_{n-1} = 0$. The corresponding characteristic equation has a double root at $\lambda = 1$, which will be seen in Section 5 to be the hallmark of an integrable system.

Late confinement revisited In Section 2 we introduced the notion of late confinement and showed that it leads to nonintegrable systems. In particular, when examining mapping (2.11), we found for its first late confinement the constraint (2.13) with characteristic equation

$$(\lambda^2 - \lambda + 1)(\lambda^4 - \lambda^3 - \lambda^2 - \lambda + 1) = 0. \tag{3.13}$$

We remarked there that the deautonomisation associated with this late confinement does not give rise to a discrete Painlevé equation because the characteristic equation leads to solutions without secular terms. We can now carry this argument one step further since, in the light of the full-deautonomisation approach, we expect the dynamical degree of the late-confined mapping subject to (2.13) to be given, in fact, by the largest root of (3.13). The latter turns out to take the value $(1+\sqrt{13}+\sqrt{2}\sqrt{\sqrt{13}-1})/4$, which is approximately 1.7221. In order to verify that this is indeed the value of the dynamical degree of the mapping we can compute its degree growth for a coefficient a_n that satisfies $a_{n+2} = a_{n+1} - a_n$ (which automatically satisfies the constraint (2.13), as is clear from the factor $(\lambda^2-\lambda+1)$ in (3.13)). This yields the sequence of degrees $0, 1, 2, 5, 9, 17, 30, 54, 94, 164, 283, 489, 843, 1454, 2505, 4316\cdots$, the last entries in which grow as powers of 1.723, in good agreement with the expected value for the dynamical degree.

Another interesting question we can now address is what happens when one postpones the confinement indefinitely. Clearly, such a situation corresponds to an unconfined singularity. Let us start by considering a confinement which is delayed k times for mapping (2.11). The singularity pattern in that case consists of k blocks $\{0, \infty^2, 0, \infty\}$ terminated by a $\{0, \infty^2, 0\}$ block, and the characteristic polynomial for the ensuing confinement constraint is

$$P_k(\lambda) = \left(\frac{\lambda^{5k}-1}{\lambda-1}\right)\lambda^3(\lambda P_0(\lambda) - 1) + P_0(\lambda), \tag{3.14}$$

where $P_0(\lambda) = \lambda^2 - 2\lambda + 1$ (the characteristic polynomial of the integrable case). Since for $k \geq 1$ (3.14) always has a root greater than 1, we can easily take the limit $k \to \infty$ and we find that the nonconfining version of (2.11) (in which a_n does not satisfy condition (2.12) or any of the late ones), has a dynamical degree given by the largest root of the equation $\lambda P_0(\lambda)-1 \equiv \lambda^3-2\lambda^2+\lambda-1=0$, which is approximately $\lambda = 1.7549$. Iterating mapping (2.11) for arbitrary a_n, we obtain the sequence of degrees $0, 1, 2, 5, 9, 17, 30, 54, 95, 168, 295, 519, 911, 1600, 2808, 4929 \cdots$, from which we find that the degree grows approximately as a power of 1.755, which fits quite well with the value we obtained above. Note that the above degree sequences start to differ at x_{n+8}, from which point onwards extra cancellations appear for the late-confined mapping.

The property of late confinement is of course not limited to integrable mappings and one can apply the above approach to confining nonintegrable mappings, using the full-deautonomisation procedure, to obtain their dynamical degree. Several such examples can be found in [10].

4 Halburd's exact calculation of the degree growth

In the previous Section we showed that in the full-deautonomisation approach, a simple singularity analysis allows us to decide whether a given mapping is integrable or not and, in the latter case, that it even yields an exact value for the dynamical degree of the mapping. However, applying the full deautonomisation method is not always easy. Although the process of finding and studying all possible extensions of a mapping with confined singularities that do not modify the singularity patterns can be simplified through experience and intuition, in some cases (in particular for higher-order systems) there are substantial difficulties. Thankfully, a simpler method appeared in the ingenious work of Halburd [11].

Halburd's method starts from the basic fact that the degree of a rational function, $f_n(z)$ say, is equal to the number of preimages of some arbitrary value w for that function, i.e., the number of solutions in z, counted with the appropriate multiplicity, of $f_n(z) = w$ (in $\mathbb{P}^1$). This is, in fact, the same notion as Arnold's complexity [2]. The innovative feature in Halburd's approach is that this computation of the degree is performed on the n-th iterate of a rational mapping, not just for any arbitrary value of w, but for values that appear in the singularity patterns of the mapping. Starting from Halburd's approach, we presented in [35] a simpler method – dubbed 'express' – which yields the exact value of the dynamical degree of a rational mapping based on its singularity patterns, though not the degree. This information however suffices to decide on the integrability of the mapping, and this in a very efficient way.

Halburd's method in a nutshell In order to calculate the degree of a given mapping we shall view the iterates of the mapping, starting from initial conditions $x_0, x_1 \in \mathbb{P}^1$, as rational functions $f_n(z)$ in $x_1 = z$ (in which x_0 simply appears as a generic constant). Furthermore, as mentioned above, the degree d_n of each iterate will be calculated as the number of solutions, $d_n(w)$, of the equation $f_n(z) = w$, counted with multiplicities, for certain values of $w \in \mathbb{P}^1$. Needless to say that $d_0 = 0, d_1 = 1$ and that $d_n = d_n(w)$ for any choice of w. Let us show how this calculation works for the McMillan mapping (2.1),

$$x_{n+1} + x_{n-1} = \frac{2\mu x_n}{1 - x_n^2}, \tag{4.1}$$

which has exactly two confined singularity patterns: $\{1, \infty, -1\}$ and $\{-1, \infty, 1\}$.

We are interested in knowing how many times, at iteration step $n > 1$, a particular value of w, such as 1 or -1, can appear as the value of $f_n(z)$ for a special choice of z. Let us denote the number of appearances of 1 at step n by U_n and those of -1 by M_n. Now, a value 1 can appear either 'spontaneously' by an accidental choice of an appropriate z, or it arises 'automatically' two steps after a value -1 appeared for some z, as is clear from the singularity pattern $\{-1, \infty, 1\}$. The same is true, mutatis mutandis, for the value -1. Hence we have that $d_n(1) = U_n + M_{n-2}$ and $d_n(-1) = M_n + U_{n-2}$. What about the number of possible appearances of ∞? Such a value clearly appears automatically one step after a 1 or -1, but can it arise in other circumstances? It is easy to check that the only other possibility for an ∞ to appear is two steps after another ∞, i.e., as part of a 'cyclic pattern' $\{\infty, f\}$. Note that this pattern does not contain any singularities and is therefore, strictly speaking, not a singularity pattern at all. However, it does allows us to conclude that in addition to those generated by values ± 1, an additional ∞ automatically appears one step out of two. Hence, $d_n(\infty) = U_{n-1} + M_{n-1} + \psi_2(n)$, where

$$\psi_2(n) = \frac{1 - (-1)^n}{2}. \tag{4.2}$$

Furthermore, $M_0 = U_0 = 0, M_1 = U_1 = 1$ and, clearly, $M_n = U_n$ for all $n \geq 0$. Hence, from $d_n(1) = d_n(-1) = d_n(\infty)$ we find

$$U_n - 2U_{n-1} + U_{n-2} = \psi_2(n), \tag{4.3}$$

which has

$$U_n = \alpha n + \beta + \frac{n^2}{4} - \frac{(-1)^n}{8}, \tag{4.4}$$

as its general solution. The initial conditions for U_n determine the constants, $\alpha = 1/2, \beta = 1/8$, and the degree d_n of the n-th iterate is then obtained, for example, from $d_n = d_n(1)$:

$$d_n = \frac{n^2 + \psi_2(n)}{2}, \tag{4.5}$$

which fits exactly with the sequence of degrees calculated in Section 2, i.e.: $0, 1, 2, 5, 8, 13, 18, 25, 32, 41, 50, 61, 72, 85, 98, 113, 128, \cdots$.

The express method The *express* method is based on the observation that in order to deduce that the degree of the iterates of a mapping does not grow exponentially, it is in fact not necessary at all to solve recursion relations such as (4.3) exactly. Since, for the McMillan mapping, it is readily established that the value ∞

for an iterate $f_n(z)$ can only occur either because it is induced by a confined singularity pattern, or because it arises in a cyclic pattern, it is clear that the fact that the homogeneous part of equation (4.3) does not give rise to exponential growth, is sufficient to conclude that the general solution to the full equation cannot exhibit exponential growth. Indeed, the characteristic equation for the homogeneous part of (4.3)

$$\lambda^2 - 2\lambda + 1 = 0, \tag{4.6}$$

lacking any roots greater than 1, and the contribution in the righthand side of (4.3) due to the occurrences of ∞ outside the confined singularity patterns being bounded (in fact, periodic) in n, it is clear that the growth of the degree for this mapping can only be polynomial and never exponential. Hence, its dynamical degree is exactly 1 and the mapping is integrable.

This is the crux of the express method: if singularity analysis shows that a given mapping only has confined singularity patterns and cyclic patterns (we will come to the slightly more complicated case of anticonfined patterns in a moment), then it suffices to study the homogeneous parts of the recursion relations that hold for the occurrences of the values that make up the confined singularity patterns. If the resulting characteristic equations do not possess any roots greater than 1, then the mapping at hand is an integrable one. However, if the characteristic equations do possess roots greater than 1, then this is an indication of nonintegrability, as we shall see in a moment. Let us first give two more integrable examples.

In Section 2 we saw that the mapping (2.11) (a_n satisfying $a_{n+1}-2a_n+a_{n-1}=0$)

$$x_{n+1} + x_{n-1} = \frac{a_n}{x_n} + \frac{1}{x_n^2}, \tag{4.7}$$

has the (unique) confined singularity pattern $\{0, \infty^2, 0\}$. Denoting the number of appearances of the values 0 and ∞ in the iterates of this mapping by Z_n and I_n respectively, it is clear that $d_n(0) = Z_n + Z_{n-2}$. Moreover, it is easy to establish that in this case as well, an ∞ can only appear either due to the confined pattern or in the cyclic pattern $\{\infty, f\}$, the contribution of which we shall neglect and we write $d_n(\infty) \simeq 2Z_{n-1}$ (where the factor 2 is due to the multiplicty of ∞ in the confined pattern and where $\simeq$ signifies equality up a bounded function of n). From $d_n(0) = d_n(\infty)$ we then obtain the equation

$$Z_n + Z_{n-2} - 2Z_{n-1} \simeq 0, \tag{4.8}$$

the characteristic equation for which, again, has no roots greater than 1, and we conclude that the mapping is integrable.

In Section 2 we also saw that mapping (2.4) has a confined singularity pattern $\{1, 0, \infty, \infty, 0, 1\}$ and a cyclic pattern $\{0, \infty, \infty, 0, f', \infty, f''\}$. Denoting the number of appearances of a value 1 in the iterates of the mapping by U_n and neglecting the cyclic occurrences of the values 0 and ∞, we can therefore write (from $d_n(1) = d_n(0) = d_n(\infty)$)

$$U_n + U_{n-5} \simeq U_{n-1} + U_{n-4} \simeq U_{n-2} + U_{n-3}, \tag{4.9}$$

the characteristic equations

$$(\lambda-1)(\lambda^4-1)=0,\quad (\lambda^2-1)(\lambda^3-1)=0,\quad (\lambda-1)(\lambda^2-1)=0, \tag{4.10}$$

for which clearly do not possess any roots that are greater than 1, in accordance with the integrable character of the mapping.

The case of nonintegrable mappings Up to now we have only given examples of integrable mappings, but the express method works even better for nonintegrable mappings since it actually yields the exact value of the dynamical degree. A particularly interesting example is that of the Hone-Viallet mapping (2.15) we already encountered in Section 2,

$$x_{n+1}x_{n-1}=x_n+\frac{1}{x_n}, \tag{4.11}$$

which has the confined singularity patterns $\{\pm i,0,\infty,\infty^2,\infty,0,\mp i\}$. This mapping also has a long cyclic pattern involving the values 0 and ∞ but not $\pm i$: $\{f,0,\infty,\infty^2,\infty,0,f',\infty,\infty\}$. (Note that the confined singularity patterns were not correctly identified in [15, 10]).

Denoting by P_n (M_n) the number of appearances of $+i$ ($-i$) at the nth iterate, and neglecting the contributions from the cyclic pattern, we find from the various possible expressions for the degrees ($d_n(i)=d_n(-i)=d_n(0)=d_n(\infty)$) that

$$\begin{aligned}P_n+M_{n-6}=M_n+P_{n-6}&\simeq(P_{n-1}+M_{n-1})+(P_{n-5}+M_{n-5})\\&\simeq(P_{n-2}+M_{n-2})+2(P_{n-3}+M_{n-3})+(P_{n-4}+M_{n-4}).\end{aligned} \tag{4.12}$$

Now, since P_n and M_n clearly play the same role, we can take $M_n=P_n$ and we find

$$(\lambda^2-\lambda+1)Q(\lambda)=0,\qquad (\lambda^2+\lambda+1)Q(\lambda)=0,\qquad Q(\lambda)=0, \tag{4.13}$$

as characteristic equations for (4.12), where $Q(\lambda)=(\lambda^4-\lambda^3-2\lambda^2-\lambda+1)$. The solution to the general recursion relations that can be established for the degree d_n of the iterates of this mapping, can therefore be described in terms of the roots of the polynomial $Q(\lambda)$, among which $(1+\sqrt{17})/4+\sqrt{(1+\sqrt{17})/8}$ is the largest one. In [43] it has been shown that this largest root of $Q(\lambda)$, in fact, gives the dynamical degree for this mapping.

Another interesting nonintegrable example is that of a late confined one. As such we choose the first late confinement of (2.11) corresponding to the pattern $\{0,\infty^2,0,\infty,0,\infty^2,0\}$. Denoting by Z_n the number of appearances of the value 0 at some iterate n and neglecting all cyclic patterns that might arise, we find from $d_n(0)=d_n(\infty)$ the expression

$$Z_n+Z_{n-2}+Z_{n-4}+Z_{n-6}\simeq 2Z_{n-1}+Z_{n-3}+2Z_{n-5}, \tag{4.14}$$

the characteristic equation for which is:

$$(\lambda^2-\lambda+1)(\lambda^4-\lambda^3-\lambda^2-\lambda+1)=0. \tag{4.15}$$

This equation is identical to (3.13) and its largest root, $(1+\sqrt{13}+\sqrt{2}\sqrt{\sqrt{13}-1})/4$, therefore coincides with the value of the dynamical degree obtained by the full-deautonomisation approach.

Many more interesting examples can be found in [35], where it is also explained that if the confined singularity patterns are too short, a straightforward application of the express method might not lead to useful conclusions, and it is shown how one can deal with such a situation.

Non-confining mappings The fact that the express method can be applied to cases with late confinement raises the possibility that the approach might also be applicable to mappings with unconfined singularities. This turns out to be indeed the case if the unconfined singularity patterns, although extending indefinitely, are made up of blocks that keep repeating. To illustrate this we consider the mapping

$$x_{n+1} + x_{n-1} = x_n + \frac{1}{x_n^3}, \tag{4.16}$$

which has the unconfined singularity $\{0, \infty^3, \infty^3, 0, \infty^3, \infty^3, 0, \cdots\}$. Denoting by Z_n the number of appearances of the value 0, we have for the preimages of 0 and ∞,

$$d_n(0) = Z_n + Z_{n-3} + Z_{n-6} + \cdots = \sum_{k=0}^{\infty} Z_{n-3k}, \quad d_n(\infty) = 3\sum_{k=0}^{\infty}(Z_{n-3k-1} + Z_{n-3k-2}), \tag{4.17}$$

where $Z_m = 0$ whenever $m \leq 0$ because of the initial conditions we imposed for the iteration, and all sums in these expressions are therefore finite ones. As a result we obtain the relation

$$\sum_{k=0}^{\infty} Z_{n-3k} - 3\sum_{k=0}^{\infty}(Z_{n-3k-1} + Z_{n-3k-2}) \simeq 0, \tag{4.18}$$

and its corresponding characteristic equation by setting $Z_n = \lambda^n$. Assuming now that this characteristic equation has a root with modulus greater than 1, we can take the limit $n \to \infty$ to obtain

$$\frac{1}{1-\frac{1}{\lambda^3}} = \left(\frac{1}{\lambda} + \frac{1}{\lambda^2}\right)\frac{3}{1-\frac{1}{\lambda^3}}, \tag{4.19}$$

which yields,

$$\lambda^2 - 3\lambda - 3 = 0. \tag{4.20}$$

The largest root of (4.20) is $\lambda = (3+\sqrt{21})/2$, approximately equal to 3.791. This value of the dynamical degree is in perfect agreement with the one obtained by the direct calculation of the sequence of degrees: 0, 1, 4, 15, 58, 220, 834, 3163, $\cdots$.

Linearisable mappings Linearisable mappings are special in the sense that they are integrable without necessarily having confined singularities. Still, their dynamical degree is always equal to 1 and the degree of their iterates grows linearly with n. Applying the express method to linearisable mappings is straightforward. Let us consider the mapping

$$\frac{x_{n+1}+x_n}{x_{n-1}+x_n}=\frac{1-x_n}{1+x_n}, \tag{4.21}$$

which can be linearized as $y_n = y_{n-1}+1$, $x_{n+1}=(1-x_n y_n)/(y_n-1)$ [6]. The mapping has two singularity patterns, a confined one, $\{1,-1\}$, and an unconfined one, $\{-1,\infty,\infty,\infty,\dots\}$, and it is easily checked that, starting from $x_0=r, x_1=p/q$, the growth of the degree (in p,q) of its iterates is linear: 0, 1, 2, 3, 4, 5, 6, 7, 8, 9, ...

We denote by U_n and M_n the number of appearances of the values $+1$ and -1, respectively, at the nth iterate of the mapping. From the preimages of the values $+1,-1$ and ∞ we find that

$$U_n = M_n + U_{n-1} \simeq M_{n-1}+M_{n-2}+M_{n-3}+\cdots=\sum_{k=1}^{\infty} M_{n-k}, \tag{4.22}$$

with $U_m = M_m = 0$ if $m \le 0$, in which we substitute $U_n = U_0\lambda^n$ and $M_n = M_0\lambda^n$ in order to obtain a characteristic equation for which we assume again that it has a root greater than 1 so that we can resum the geometric series that arises at the limit $n\to\infty$. We obtain finally

$$M_0 = U_0\left(1-\frac{1}{\lambda}\right) \quad \text{and} \quad U_0 = M_0\left(\frac{1}{1-1/\lambda}-1\right), \tag{4.23}$$

and find $\lambda=1$ as the only possible solution. This would mean that the root $\lambda>1$ we assumed cannot exist. However, as by definition the dynamical degree cannot be less than 1, it then follows that the dynamical degree of (4.21) must be equal to 1, as expected.

Of course, not all linearisable mappings have unconfined singularities. For instance, the mapping (linearised in [33])

$$x_{n+1}x_{n-1}=x_n^2-1, \tag{4.24}$$

has two confined singularity patterns $\{\pm1,0,\mp1\}$ and an anticonfined one

$$\cdots,\infty^4,\infty^3,\infty^2,\infty,f,0,f',\infty,\infty^2,\infty^3,\infty^4,\cdots. \tag{4.25}$$

The degree of its iterates (for $x_0=r, x_1=p/q$) grows linearly as $0,1,2,4,6,8,10,\dots$ and it belongs to the class that we call 'linearisable of the third kind'. From these singularity patterns it is immediately clear that, if we denote by U_n, M_n the number of appearances of the values $+1,-1$ in the iteration, we have that $d_n(1)=U_n+M_{n-2}$, $d_n(-1)=M_n+U_{n-2}$ and $d_n(0)=U_{n-1}+M_{n-1}+\delta_{n1}$ (where the Kronecker δ indicates the single appearance of a 0 in the anticonfining pattern). Taking $M_n=U_n$

(since +1 and -1 obviously play the same role) and neglecting – just as we did for the cyclic patterns – the contribution of the anticonfined pattern, we have from $d_n(1) = d_n(0)$ that

$$U_n + U_{n-2} \simeq 2U_{n-1}. \tag{4.26}$$

As the characteristic equation for this relation does not have any roots greater than 1, we find, as expected, that the dynamical degree for this mapping is 1.

The effect of anticonfinement The previous example suggests that there might exist cases where one cannot just brush aside the anticonfined singularities. The mapping introduced by Tsuda and collaborators [41] is a case in point:

$$x_{n+1} = x_{n-1}\left(x_n - \frac{1}{x_n}\right). \tag{4.27}$$

This nonintegrable mapping has two confined singularity patterns, $\{\pm 1, 0, \infty, \mp 1\}$, as well as an anticonfined one

$$\{\cdots, 0^8, 0^5, 0^3, 0^2, 0, 0, f, 0, \infty, f', \infty, \infty, \infty^2, \infty^3, \infty^5, \infty^8 \cdots\}, \tag{4.28}$$

in which the exponents form a Fibonacci sequence. Due to this exponential growth of the number of occurrences of the value ∞ in the iterates of the mapping, one concludes that its dynamical degree must be at least equal to the golden mean $\varphi = (1+\sqrt{5})/2 \approx 1.6180$. Direct calculation of the degree, 0, 1, 2, 4, 8, 14, 24, 40, 66, 108, 176, 286, 464, 752, 1218, 1972, 3192..., indicates that the growth rate indeed converges to the golden mean.

Denoting by U_n the number of spontaneous occurrences of the value 1 in the iteration (and taking into account that +1 and −1 play the same role) we find that $d_n(\pm 1) = U_n + U_{n-3}$, $d_n(0) = 2U_{n-1} + \delta_{n1}$ where the δ_{n1} is due to the single appearance of 0, after a generic value, in the anticonfined pattern. We thus find the equation

$$U_n + U_{n-3} = 2U_{n-1} + \delta_{n1}, \tag{4.29}$$

and the dynamical degree of the mapping, given by the largest root of the characteristic equation for (4.29), is precisely the golden mean already obtained above.

The interesting point here is that, had we tried to compute the degree of the mapping from the number of appearances of the value ∞, we would have found $d_n(\infty) = 2U_{n-2} + (f_{n-3} + \delta_{n2}) + f_n$ (due to the two generic values for f that are followed by an infinite sequence of infinities with Fibonacci exponents in the anticonfined pattern) as an alternative right-hand side for (4.29). Here, f_n is defined by $f_{n+1} = f_n + f_{n-1}$ for $n \geq 1$ with $f_1 = 1$ and $f_n = 0$ for $n \leq 0$, and this type of contribution to the equations cannot simply be discarded in an express-type treatment. In general, when one has an anticonfined singularity pattern with exponential growth in its exponents, it may happen that this growth coincides with the dynamical degree of the mapping, but there are cases where it only offers a lower bound for the dynamical degree. However, as we have seen above, anticonfined singularities with non-exponential (or even no) growth do exist for some mappings, and the corresponding patterns can be neglected, just as the cyclic ones, when calculating the dynamical degree with the express method.

5 Singularities and spaces of initial conditions

In this Section we shall briefly explain the algebro-geometric background of the singularity confinement property. More precisely, we will show in an example that for a second order (bi-) rational mapping that only possesses confined and cyclic singularities, its points of indeterminacy can be resolved by means of a finite number of blow-ups, i.e.: that the mapping is birationally equivalent to a family of isomorphisms between rational surfaces (in the general, nonautonomous, case), obtained by blowing up $\mathbb{P}^1 \times \mathbb{P}^1$ a finite number of times. In the process, it will also become clear why the deautonomisation and, especially, the full-deautonomisation procedures work in the way they do.

As an example we choose the nonautonomous mapping (2.7), for $a_n \neq 0$, which we rewrite as the birational mapping $\varphi_n\colon \mathbb{P}^1 \times \mathbb{P}^1 \dashrightarrow \mathbb{P}^1 \times \mathbb{P}^1$:

$$(x_n, y_n) \mapsto (x_{n+1}, y_{n+1}) = \left(y_n, \frac{a_n(y_n - 1)}{x_n y_n}\right). \tag{5.1}$$

As usual, we cover $\mathbb{P}^1 \times \mathbb{P}^1$ with four copies of $\mathbb{C}^2$, as $\mathbb{P}^1 \times \mathbb{P}^1 = (x_n, y_n) \cup (x_n, t_n) \cup (s_n, y_n) \cup (s_n, t_n)$, where $s_n := 1/x_n$ and $t_n := 1/y_n$. From the definition (5.1) it is clear that the mapping is indeterminate at the points $(x_n, y_n) = (\infty, 0)$ and $(x_n, y_n) = (0, 1)$ (the so-called points of indeterminacy of φ_n), and that there exist exactly two curves in $\mathbb{P}^1 \times \mathbb{P}^1$ that contract to a point under its action, the curves $\{y_n = 1\}$ and $\{y_n = 0\}$:

$$\{y_n = 1\} \to (x_{n+1}, y_{n+1}) = (1, 0), \qquad \{y_n = 0\} \to (x_{n+1}, y_{n+1}) = (0, \infty). \tag{5.2}$$

Note that the images of these two curves are nothing but the points of indeterminacy for the inverse mapping φ_n^{-1}. These are the singularities of the mapping (5.1). Moreover, it is easily checked that, under the action of φ_n, both contracted curves end up in the point of indeterminacy $(\infty, 0)$ through the chains

$$\{y_n = 1\} \to (1, 0) \to (0, \infty) \to (\infty, \infty) \to (\infty, 0), \tag{5.3}$$

and

$$\{y_n = 0\} \to (0, \infty) \to (\infty, \infty) \to (\infty, 0), \tag{5.4}$$

which clearly correspond to the first parts of the confined singularity pattern $\{1, 0, \infty, \infty, 0, 1\}$ and the cyclic pattern $\{0, \infty, \infty, 0, f', \infty, f''\}$ for (2.7), respectively. Recall that in the singularity confinement approach the indeterminacy was lifted using a continuity argument in the initial conditions. Here we shall show that the same aim can be achieved by 8 successive blow-ups of $\mathbb{P}^1 \times \mathbb{P}^1$. For this purpose, let us define the points $P_n\colon (x_n, y_n) = (1, 0)$, $Q_n\colon (x_n, y_n) = (0, \infty)$, $R_n\colon (x_n, y_n) = (0, 1)$ and $S_n\colon (x_n, y_n) = (\infty, 0)$, for which it should be remarked that we have $\varphi_{n+1} \circ \varphi_n(Q_n) = S_{n+2}$ for all n and for any choice of a_n. Hence, it is clear that we will have achieved confinement, along the lines of the pattern $\{1, 0, \infty, \infty, 0, 1\}$ for mapping (2.7), if we can establish that $\varphi_{n+4} \circ \varphi_{n+3} \circ \varphi_{n+2} \circ \varphi_{n+1}(P_{n+1}) = R_{n+5}$.

In general, we shall say that the singularity for mapping (5.1) that arises at P_{n+1}, due to the collapse of the curve $\{y_n = 1\}$, is confined if there exists some value k for which we have that

$$\varphi_{n+k+1} \circ \cdots \circ \varphi_{n+2} \circ \varphi_{n+1}(P_{n+1}) = R_{n+k+2}. \tag{5.5}$$

Standard confinement Let us first see how this can be achieved for the standard confinement (or what we called, in Section 2, the 'timely' confinement) which arises at $k = 3$. More precisely, we want to show that

$$\varphi_{n+4} \circ \varphi_{n+3} \circ \varphi_{n+2} \circ \varphi_{n+1}(P_{n+1}) = R_{n+5} \quad \Leftrightarrow \quad \frac{a_{n+1}a_{n+4}}{a_{n+2}a_{n+3}} = 1. \tag{5.6}$$

This condition on the parameters a_n, which is automatically satisfied if a_n is independent of n (and the confinement is therefore indeed timely, i.e. the same as in the autonomous case) is of course nothing but the condition (2.8) we obtained from the deautonomisation approach in Section 2. In order to prove condition (5.6) we have to blow up $\mathbb{P}^1 \times \mathbb{P}^1$, for general n, at the 8 points,

$$\Big(x_n, y_n\Big) = (0, 1), \quad (s_n, y_n) = (0, 0), \quad (x_n, t_n) = (0, 0), \quad (x_n, y_n) = (1, 0) \tag{5.7a}$$

$$\left(\frac{x_n}{t_n}, t_n\right) = (-a_{n-1}, 0), \quad (s_n, t_n) = (0, 0), \quad \left(s_n, \frac{t_n}{s_n}\right) = \left(0, -\frac{a_{n-2}}{a_{n-1}}\right), \tag{5.7b}$$

$$\text{and} \quad \left(\frac{s_n}{y_n}, y_n\right) = \left(-\frac{a_{n-3}}{a_{n-2}a_{n-1}}, 0\right), \tag{5.7c}$$

as described in Figure 1 (below). The rational surface X_n (at each value of n) obtained from these blow-ups of $\mathbb{P}^1 \times \mathbb{P}^1$ corresponds to the case $\ell = 0$ of the surface depicted in Figure 2 (below).

It can be verified, subject to the condition on the parameters given in (5.6), that φ_n is now defined on the whole of X_n (for every n) and that it acts as an isomorphism from X_n to X_{n+1} (and as an automorphism on $X = X_n$ in the autonomous case). The surface X_n is called the space of initial conditions for the mapping, a notion which was first introduced by Okamoto in his study of the differential Painlevé equations [24].

It is easy to check that the curves $D_1, \ldots, D_7$ and C_1, C_2, C_3 on X_n, as defined in Figure 2, move in the following fashion under the action of the mapping:

$$D_1 \to D_2 \to \cdots \to D_7 \to D_1 \tag{5.8a}$$

$$\{y = 1\} \to C_1 \to C_2 \to \cdots \to C_5 \to \{x = 1\}. \tag{5.8b}$$

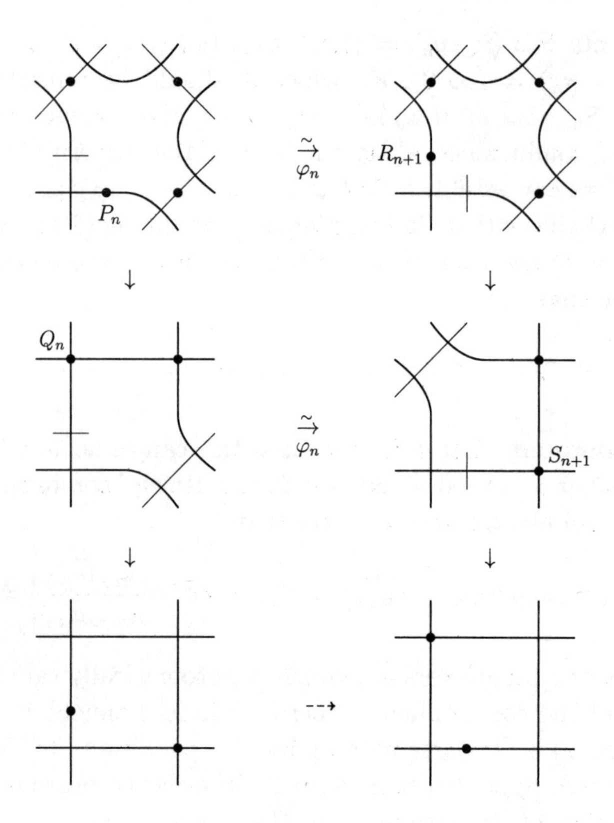

Figure 1. Pictorial representation of the 8 blow-ups of $\mathbb{P}^1 \times \mathbb{P}^1$ that are required to lift the indeterminacy for mapping (5.1) under condition (5.6).

The motion of the D curves clearly corresponds to the cyclic pattern $\{0, \infty, \infty, 0, f, \infty, f'\}$, and that of the C curves to the confined pattern $\{1, 0, \infty, \infty, 0, 1\}$ for (2.7). As the curves $D_1, \ldots, D_7, C_1, C_2, C_3$ form a basis for the Picard lattice (of rank 10) for X_n, (5.8a,5.8b) suffices to define the automorphism φ_* that is induced by φ_n on this Picard lattice, for all n. In particular, φ_* can be represented as the linear map $\begin{bmatrix} U & * \\ O & \Phi \end{bmatrix}$, where U is the permutation matrix of size 7 that corresponds to the motion of the D curves, and $\Phi = \begin{bmatrix} 0 & 0 & -1 \\ 1 & 0 & 1 \\ 0 & 1 & 1 \end{bmatrix}$. Here O stands for a size 3×7 null matrix and we have omitted the entries in the upper right-hand corner of the matrix for φ_* as these are irrelevant when it comes to deciding whether the mapping is integrable or not. The reason for this is that, according to [39] and [7], the dynamical degree of the mapping actually coincides with the largest eigenvalue for φ_*, which is obviously decided by the submatrices U and Φ. Since U is a permutation matrix and is thus unitary, its eigenvalues all have modulus 1. Moreover, from the form of the submatrix Φ (which is a companion matrix) it is clear that its eigenvalues are $1, 1$ and -1. Hence, all eigenvalues of the linear action φ_* have modulus 1 and the dynamical

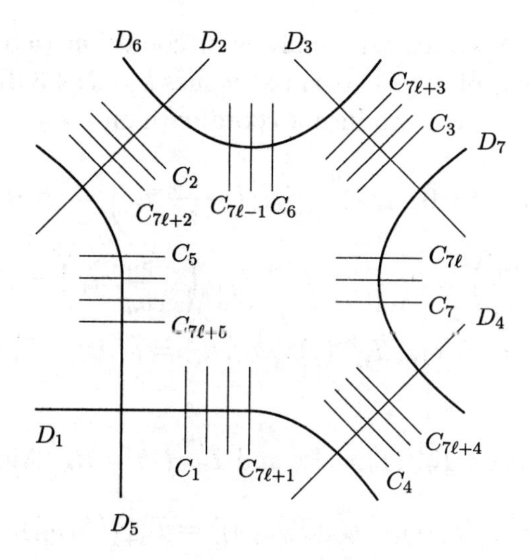

Figure 2. Pictorial representation of the rational surface obtained for mapping (5.1) by blowing up $\mathbb{P}^1 \times \mathbb{P}^1$ at $7\ell + 8$ points. The surface for the standard confinement – which is the same as for the autonomous mapping – corresponds to $\ell = 0$ (in which case curves C_6 and C_7 are absent).

degree of the mapping, under the constraint $a_n a_{n+3} = a_{n+1} a_{n+2}$, is necessarily equal to 1 as well and the mapping is therefore integrable. Note that this constraint can be written as $\left[A_{n+1}\ A_{n+2}\ A_{n+3}\right] = \left[A_n\ A_{n+1}\ A_{n+2}\right] \cdot \Phi$, where $A_n = \log a_n$. The statement that the submatrix Φ is free of eigenvalues with modulus greater than 1 is therefore equivalent to saying that the characteristic equation for the confinement constraint on the parameters a_n in the mapping does not have any roots with modulus greater than 1. This is the fundamental observation that underlies the full-deautonomisation approach, as explained in Section 2.

Another important observation is that the minimal polynomial for the submatrix Φ contains a factor $(\lambda - 1)^2$, and that the parameters which satisfy the confinement condition therefore have secular dependence on n. This is, in fact, a general feature: the non-permutation part of the linear action φ_* obtained by regularising a confining second order mapping by blow-up, can only have a double root 1 in its minimal polynomial if the mapping is integrable.

A last remark concerns the type of discrete Painlevé equation, in Sakai's classification [37], that we obtain from (5.6). As is easily established from Figure 2, the intersection diagram of the D curves that make up the cyclic pattern for this mapping (i.e. the diagram obtained by associating a vertex with each curve D_i and an edge between two vertices with each intersection of curves) is nothing but the Dynkin diagram for $\mathrm{A}_6^{(1)}$, which is exactly the type of surface in Sakai's clasification associated with this particular discrete Painlevé equation. Hence, it is clear that the cyclic singularity pattern for the original autonomous mapping (2.4) determines the surface type of its (standard) deautonomisation (see [45] for more details).

Late confinement A systematic analysis of condition (5.5) shows that all late confinements for the mapping (5.1) occur at values $k = 7\ell+3$, for some non-negative integer ℓ. To see this, let us introduce a coordinate on each D_i by defining:

$$T_n^{(1)}(\alpha)\colon (s_n, y_n) = (\alpha, 0), \quad T_n^{(2)}(\beta)\colon \left(-\frac{x_n}{t_n}, t_n\right) = (\beta, 0), \tag{5.9a}$$

$$T_n^{(3)}(\gamma)\colon \left(s_n, -\frac{t_n}{s_n}\right) = (0, \gamma), \quad T_n^{(4)}(\delta)\colon \left(-\frac{s_n}{y_n}, y_n\right) = (\delta, 0), \tag{5.9b}$$

$$T_n^{(5)}(\epsilon)\colon (x_n, y_n) = (0, \epsilon), \quad T_n^{(6)}(\zeta)\colon (x_n, t_n) = (\zeta, 0), \quad T_n^{(7)}(\eta)\colon (s_n, t_n) = (\eta, 0) \tag{5.9c}$$

Note that in this notation $T_n^{(1)}(1) = P_n$ and $T_n^{(5)}(1) = R_n$. Moreover, we have that

$$\varphi_n(T_n^{(1)}(\alpha)) = T_{n+1}^{(2)}(a_n\alpha), \quad \varphi_n(T_n^{(2)}(\beta)) = T_{n+1}^{(3)}(\beta/a_n), \tag{5.10a}$$

$$\varphi_n(T_n^{(3)}(\gamma)) = T_{n+1}^{(4)}(\gamma/a_n), \quad \varphi_n(T_n^{(4)}(\delta)) = T_{n+1}^{(5)}(a_n\delta), \tag{5.10b}$$

$$\varphi_n(T_n^{(5)}(\epsilon)) = T_{n+1}^{(6)}(\epsilon), \quad \varphi_n(T_n^{(6)}(\zeta)) = T_{n+1}^{(7)}(\zeta/a_n), \quad \varphi_n(T_n^{(7)}(\eta)) = T_{n+1}^{(1)}(\eta). \tag{5.10c}$$

Now, since

$$\varphi_{n+7\ell+3}\circ\cdots\circ\varphi_{n+1}\circ\varphi_n(P_n) = T_n^{(5)}\left(\frac{a_{n+7\ell}a_{n+7\ell+3}}{a_{n+7\ell+1}a_{n+7\ell+2}}\prod_{m=0}^{\ell-1}\frac{a_{n+7m}a_{n+7m+3}}{a_{n+7m+1}a_{n+7m+2}a_{n+7m+5}}\right) \tag{5.11}$$

we find that confinement can only happen through condition (5.5) and this if and only if $k = 7\ell + 3$, and

$$\frac{a_{n+7\ell+1}a_{n+7\ell+4}}{a_{n+7\ell+2}a_{n+7\ell+3}}\prod_{m=0}^{\ell-1}\frac{a_{n+7m+1}a_{n+7m+4}}{a_{n+7m+2}a_{n+7m+3}a_{n+7m+6}} = 1. \tag{5.12}$$

Note that when $\ell \geq 1$, this condition is not satisfied for a_n which are independent of n. For such late confinements, blowing up $\mathbb{P}^1 \times \mathbb{P}^1$ at the 8 points (5.7a–5.7c) as well as at the 7ℓ points (5.9a–5.9c) that appear in the chain (5.10a–5.10c) before confinement happens, we obtain the rational surface depicted in Figure 2 for general ℓ. On this surface, the D curves still form a cycle of length 7

$$D_1 \to D_2 \to \cdots \to D_7 \to D_1, \tag{5.13}$$

(which proves that in this case the cyclic pattern not only survives the standard deautonomisation but also any late one), but the confined pattern now becomes

$$\{y = 1\} \to C_1 \to C_2 \to \cdots \to C_{7\ell+5} \to \{x = 1\}, \tag{5.14}$$

or $\{1, 0, \infty, \infty, 0, \alpha, \infty, \alpha', \ldots, 0, \infty, \infty, 0, 1\}$ in the language of mapping (2.7) (where the entries α, α' etc. take specific values that depend on a_n but not on the initial condition).

The Picard lattice for this surface now has rank $7\ell+10$ and we can use the curves $D_1, \dots, D_7, C_1, \dots, C_{7\ell+3}$ as a basis. We then find that the linear action φ_* induced by the mapping (5.1) on this Picard lattice takes the form $\varphi_* \sim \begin{bmatrix} U & * \\ O & \Phi \end{bmatrix}$, where U is the same as in the standard confinement case and Φ is again a companion matrix, now of size $7\ell+3$, the last column of which is ${}^t[-1\,1\,1\,-1\,0\,1\,0 \cdots -1\,1\,1]$. The minimal polynomial of Φ is therefore

$$\lambda^{7\ell+3} - \lambda^{7\ell+2} - \lambda^{7\ell+1} + \lambda^{7\ell} - (\lambda^5 - \lambda^3 + \lambda^2 + \lambda - 1) \sum_{m=0}^{\ell-1} \lambda^{7m}, \tag{5.15}$$

which always has a real root greater than 1 unless $\ell = 0$. Hence, when $\ell \geq 1$, the linear action φ_* has an eigenvalue greater than 1 and we conclude that the mapping in the late confinement case is always nonintegrable.

Late confinement and full-deautonomisation Just as for the standard confinement, it is easy to check that the general confinement constraint (5.12) can be written as

$$\begin{bmatrix} A_{n+2} & \cdots & A_{n+7\ell+4} \end{bmatrix} = \begin{bmatrix} A_{n+1} & \cdots & A_{n+7\ell+3} \end{bmatrix} \cdot \Phi, \tag{5.16}$$

where $A_n = \log a_n$. Now, since the evolution of the coefficient a_n is written in terms of the submatrix Φ, the growth rate of $A_n = \log a_n$ is determined by the largest root of φ_*, which explains why the full-deautonomisation method works in the case of general late confinements as well. Note that while the full-deautonomisation method has a clear algebro-geometric justification only in the case of second-order mappings, several encouraging results do suggest however that it might be valid for higher order mappings as well [44].

Spaces of initial conditions and degree growth The fact that autonomous second order rational mappings that only have confined or cyclic singularities, in fact, possess a space of initial conditions (in the sense explained above) was first recognized by Takenawa [39]. The conditions under which a nonautonomous mapping can be said to enjoy the same property are set out in detail in [19].

In [7] it is shown that if an autonomous second order mapping has a space of initial conditions, then its degree growth must be either bounded, quadratic or exponential. Moreover, in the case of exponential growth, the value of the dynamical degree is strongly restricted: it can only be a reciprocal quadratic integer greater than 1 or a Salem number (i.e., a real algebraic integer greater than 1, such that its reciprocal is a conjugate and all (but at least one) of the other conjugates lie on the unit circle). Note that this result implies, in particular, that if a mapping has linear degree growth, it cannot possess a space of initial conditions. A similar result has been shown to hold for nonautonomous mappings as well [19].

If a mapping does not have a space of initial conditions, there is no general theory in the nonautonomous case. In the autonomous case however, it is known that such a mapping can only have linear or exponential degree growth. Moreover, again in the autonomous case, it is sometimes possible to verify whether or not a mapping has a space of initial conditions, only by studying the value of its dynamical degree [4]. This kind of classification can therefore also be of help when checking the integrability of an equation. For example, if an autonomous equation does not have a space of initial conditions and its degree growth is not linear, then one can immediately conclude that the mapping is nonintegrable.

Acknowledgements

RW and TM would like to acknowledge support from the Japan Society for the Promotion of Science (JSPS), through JSPS grants number 18K03355 and 18K13438, respectively.

References

[1] Ablowitz M J, Ramani A, Segur H, Nonlinear evolution equations and ordinary differential equations of Painlevé type, *Lett. Nuovo Cimento* **23** (1978) 333–338.

[2] Arnold V I, Dynamics of complexity of intersections, *Bol. Soc. Bras. Mat.* **21** (1990) 1–10.

[3] Bellon M P, Viallet C M, Algebraic Entropy, *Commun. Math. Phys.* **204** (1999) 425–437.

[4] Blanc J, Cantat S, *J. Amer. Math. Soc.* **29** (2016) 415–471.

[5] Carstea A S, Dzhamay A, Takenawa T, Fiber-dependent deautonomization of integrable 2D mappings and discrete Painlevé equations, *J. Phys. A: Math. Theor.* **50** (2017) 405202.

[6] Carstea A S, Ramani A, Grammaticos B, Deautonomising integrable non-QRT mappings, *J. Phys. A* **42** (2009) 485207.

[7] Diller J, Favre C, Dynamics of bimeromorphic maps of surfaces. *Amer. J. Math.* **123** (2001) 1135–1169.

[8] Grammaticos B, Ramani A, Lafortune S, The Gambier Mapping revisited, *Physica A* **253** (1998) 260–270.

[9] Grammaticos B, Ramani A, Papageorgiou V, Do integrable mappings have the Painlevé, property?, *Phys. Rev. Lett.* **67** (1991) 1825–1828.

[10] Grammaticos B, Ramani A, Willox R, Mase T, Satsuma J, Singularity confinement and full-deautonomisation: A discrete integrability criterion, *Physica D* **313** (2015) 11–25.

[11] Halburd R, Elementary exact calculations of degree growth and entropy for discrete equations, *Proc. R. Soc. A* **473** (2017) 20160831.

[12] Hietarinta J, Viallet C, Singularity confinement and chaos in discrete systems, *Phys. Rev. Lett.* **81** (1998) 325–328.

[13] Hietarinta J, Viallet C, Discrete Painlevé I and singularity confinement in projective space. *Chaos, Solitons Fractals* **11** (2000) 29–32.

[14] Hirota R, Discrete analogue of a generalized Toda equation, *J. Phys. Soc. Japan* **50** (1981) 3785–3791.

[15] Hone A N W, Laurent polynomials and superintegrable maps, *SIGMA* **3** (2007) 022.

[16] Kanki M, Mase T, Tokihiro T, Singularity confinement and chaos in two-dimensional discrete systems, *J. Phys. A: Math. Theor.* **49** (2016) 23LT01.

[17] McMillan E M, A problem in the stability of periodic systems, *Topics in Modern Physics: A Tribute to E.U. Condon*, eds. W.E. Brittin and H. Odabasi, Colorado Associated Univ. Press, Boulder 1971, p.219–244.

[18] Kruskal M D, Tamizhmani K M, Grammaticos B, Ramani A, Asymmetric discrete Painlevé equations, *Reg. Chaot. Dyn.* **5** (2000) 273–280.

[19] Mase T, Studies on spaces of initial conditions for nonautonomous mappings of the plane, *Journal of integrable Systems* **3** (2018) 1–47 (xyy010).

[20] Mase T, Willox R, Grammaticos B, Ramani A, Deautonomisation by singularity confinement: an algebro-geometric justification, Proc. Royal Soc. A **471** (2015) 20140956.

[21] Mimura N, Isojima S, Murata M, Satsuma J, Ramani A, Grammaticos B, Do ultradiscrete systems with parity variables satisfy the singularity confinement criterion?, *J. Math. Phys.* **53** (2012) 023510.

[22] Morishita M, The fitting of the logistic equation to the rate of increase of population density, *Res. Popul. Ecol.* **VII** (1965) 52–55.

[23] Ohta Y, Ramani A, Grammaticos B, Elliptic discrete Painlevé equations, *J. Phys. A* **35** (2002) L653–L659.

[24] Okamoto K, Sur les feuilletages associés aux équations du second ordre à points critiques fixes de P. Painlevé – Espace des conditions initiales, *Japanese Journal of Mathematics. New series* **5** (1979) 1–79.

[25] Quispel Q R W, Roberts J A G, Thompson C J, Integrable mappings and soliton equations II, *Physica D* **34** (1989) 183–192.

[26] Ramani A, Grammaticos B, Discrete Painlevé equations: coalescences, limits and degeneracies, *Physica A* **228** (1996) 160–171.

[27] Ramani A, Grammaticos B, Discrete Painlevé equations associated with the affine Weyl group E8, *J. Phys. A* **48** (2015) 355204.

[28] Ramani A, Grammaticos B, Bountis T, The Painlevé property and singularity analysis of integrable and non-integrable systems, *Physics Reports* **180** (1989) 159-245.

[29] Ramani A, Grammaticos B, Hietarinta J, Discrete versions of the Painlevé equations, *Phys. Rev. Lett.* **67** (1991) 1829–1832.

[30] Ramani A, Grammaticos B, Karra G, Linearizable mappings, *Physica A***180** (1992) 115–127.

[31] Ramani A, Grammaticos B, Ohta Y, Discrete integrable systems from continuous Painlevé equations through limiting procedures, *Nonlinearity* **13** (2000) 1073–1085.

[32] Ramani A, Grammaticos B, Satsuma J, On the explicit integration of a class of linearisable mappings, *J. Phys. A* **45** (2012) 365202.

[33] Ramani A, Grammaticos B, Satsuma J, Mimura N, Linearisable QRT mappings, *J. Phys. A* **44** (2011) 425201.

[34] Ramani A, Grammaticos B, Tremblay S, Integrable systems without the Painlevé property, *J. Phys. A* **33** (2000) 3045–3052.

[35] Ramani A, Grammaticos B, Willox R, Mase T, Calculating algebraic entropies: an express method, *J. Phys. A: Math. Theor.* **50** (2017) 185203.

[36] Ramani A, Grammaticos B, Willox R, Mase T, Kanki M, The redemption of singularity confinement, *J. Phys. A: Math. Theor.* **48** (2015) 11FT02.

[37] Sakai H, Rational surfaces associated with affine root systems and geometry of the Painlevé equations, *Commun. Math. Phys.* **220** (2001) 165–229.

[38] Skellam J G, random dispersal in theoretical populations, *Biometrika* **38** (1951) 196–218.

[39] Takenawa T, A geometric approach to singularity confinement and algebraic entropy, *J. Phys. A: Math. Gen.* **34** (2001) L95–L102.

[40] Toda M, *Theory of Nonlinear Lattices*, Springer Series in Solid State Physics **20**, Springer Verlag Berlin 1981.

[41] Tsuda T, Ramani A, Grammaticos B, Takenawa T, A class of integrable and nonintegrable mappings and their dynamics, *Lett. Math. Phys.* **82** (2007) 39–49.

[42] Veselov A P, What is an integrable mapping?, *What is Integrability?*, ed. V.E. Zakharov, Springer Verlag Berlin 1991, p.251–272.

[43] Viallet C M, Algebraic dynamics and algebraic entropy, *Int. J. Geom. Meth. in Mod. Phys.* **5** (2008) 1373–1391.

[44] Willox R, Mase T, Ramani A, Grammaticos B, Full-deautonomisation of a lattice equation, *J. Phys. A: Math. Theor.* **49** (2016) 28LT01.

[45] Willox R, Ramani A, Grammaticos B, A systematic method for constructing discrete Painlevé equations in the degeneration cascade of the E_8 group, *J. Math. Phys.* **58** (2017) 123504.

[46] Zabusky N J, Kruskal M D, Interaction of solitons in a collisionless plasma and the recurrence of initial states, *Phys. Rev. Lett.* **15** (1965) 240–243.

A4. Elementary introduction to discrete soliton equations

Jarmo Hietarinta

Department of Physics and Astronomy
University of Turku
FI-20014, Turku, Finland

Abstract

We will give a short introduction to discrete or lattice soliton equations, with the particular example of the Korteweg-de Vries as illustration. We will discuss briefly how Bäcklund transformations lead to equations that can be interpreted as discrete equations on a $\mathbb{Z}^2$ lattice. Hierarchies of equations and commuting flows are shown to be related to multidimensionality in the lattice context, and multidimensional consistency is one of the necessary conditions for integrability. The multidimensional setting also allows one to construct a Lax pair and a Bäcklund transformation, which in turn leads to a method of constructing soliton solutions. The relationship between continuous and discrete equations is discussed from two directions: taking the continuum limit of a discrete equation and discretizing a continuous equation following the method of Hirota.

1 Introduction

We are all familiar with the integrable (continuous) soliton equations that have been studied intensively since their resurrection in the late 1960's (see e.g. [1, 5, 6]). Many interesting and useful properties are associated with such systems, such as symmetries, the infinite number of conserved quantities, elastic scattering of solitons, and solvability using various methods such as the Inverse Scattering Transform and Hirota's bilinear method. All of these nice properties follow from some important underlying mathematical structure, which has been elaborated upon in many studies (e.g. by Mikio Sato and his collaborators in Kyoto, see e.g. [13]).

With such a beautiful continuous theory of soliton equations, what is the point of a discrete soliton theory? One might say that we need to discretize PDEs in order to do computations with them, or that there cannot be smooth continuity beyond the Planck scale where quantum aspects take over. But the reason proposed here is that discrete soliton equations should be studied because their mathematical properties are, if possible, even more beautiful than those of the continuum equations.

2 Basic set-up for lattice equations

When one mentions "lattice equations" perhaps the first thing that comes to mind is the ubiquitous Toda lattice equation given by:

$$\ddot{x}_i(t) = e^{-(x_i(t)-x_{i-1}(t))} - e^{-(x_i(t)-x_{i-1}(t))}, \quad \forall i.$$

Here x_i is the position of the particle having the name i and the equation gives the time evolution $x_i(t)$. Since time is still continuous we would call this a semi-discrete equation.

2.1 Equations on Cartesian lattice

Here we are considering fully discrete soliton equations and therefore the continuous $u(x,t)$ will be replaced by $u_{n,m}$; i.e., both the space and time coordinates are discretized. The most common discrete two-dimensional space is the Cartesian 2D lattice with dependent variables located at the vertices of the lattice, see Figure 1. Other lattices can also be considered, as well as variables not on the vertices but on the links between the vertices.

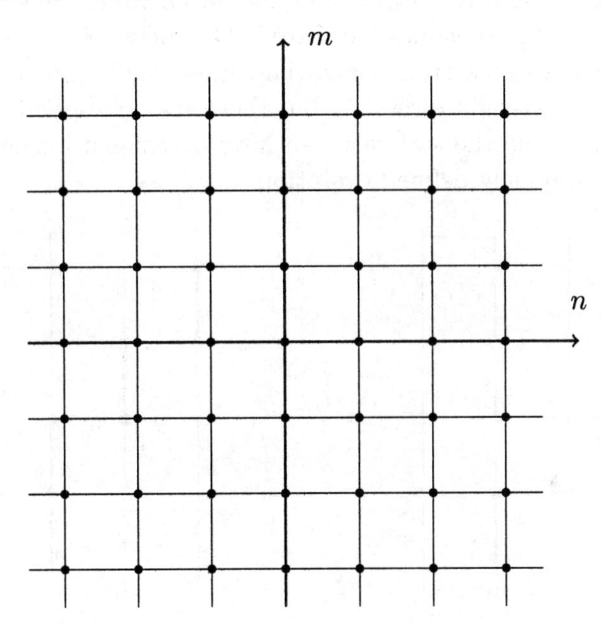

Figure 1. The Cartesian lattice; $u_{n,m}$ are located at lattice points.

The soliton equations are typically evolution equations and therefore we must discuss what kind of evolution we can have on the lattice. The simplest equation is the one relating the corners of a lattice square or quadrilateral. This involves four corners and if we give values at three corners we should be able to compute the value on the fourth corner, see Figure 2. For this to be possible we must require that the equation is *affine linear* in all the corner variables. As examples of such equations we have the lattice potential KdV equation (lpKdV)

$$(u_{n,m} - u_{n+1,m+1})(u_{n,m+1} - u_{n+1,m}) = p^2 - q^2, \quad \forall n, m, \tag{2.1}$$

and the lattice potential modified KdV equation (lpmKdV)

$$p\,(u_{n,m}\,u_{n+1,m} - u_{n,m+1}\,u_{n+1,m+1}) = q\,(u_{n,m}\,u_{n,m+1} - u_{n+1,m}\,u_{n+1,m+1}). \tag{2.2}$$

For these examples we can clearly compute, within each quadrilateral, the value at any corner once the other three corner values are given. This is the local situation.

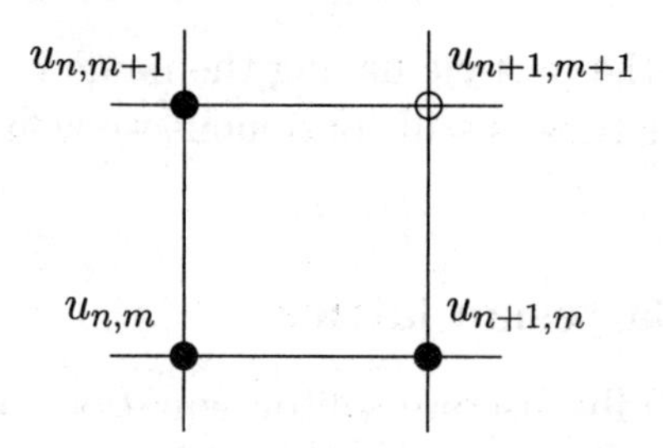

Figure 2. Equation on a quadrilateral: If values at three corners are given, one should be able to compute the value at the fourth corner.

For a global picture it is necessary to define initial values on some curve so that one can proceed to compute values "forward". One possibility is to give the values on a corner; another is to use staircase initial values; see Figure 3. In the figure the evolution is to the upper-right direction, but there are corresponding initial settings for other directions. Also the staircase can have occasional longer or higher stairs as long as we have uniquely defined evolution.

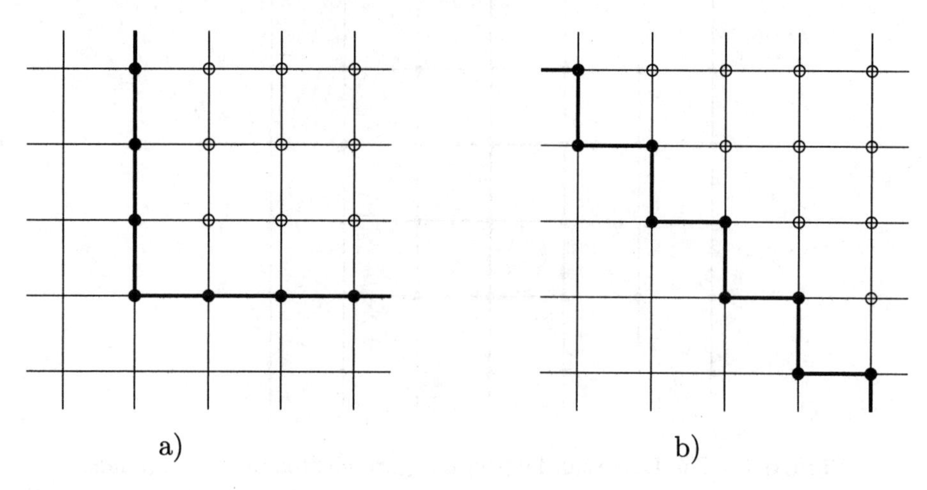

Figure 3. a): The Cartesian lattice with corner initial values (black disks) given; one can then compute the values at open circles in the upper right quadrant. b) The same with staircase initial values.

2.2 Discrete structure within continuous soliton equations

The examples above (2.1) and (2.2) did not come from thin air. If we use (2.1) with $n = 0, m = 0$ and solve for $u_{1,1}$ we get

$$u_{1,1} = u_{0,0} + \frac{p^2 - q^2}{u_{1,0} - u_{0,1}}. \tag{2.3}$$

This may look familiar. Indeed, in 1973 Wahlquist and Estabrook discussed [16] Bäcklund transformation (BT) for KdV solitons and found (translating notation to the present case) that if $u_{0,0}$ is a "seed" solution and $u_{1,0}$ is obtained from it by a BT with parameter p, and similarly $u_{0,1}$ with parameter q, then there is a *superposition principle*: If one applies BT with q on $u_{1,0}$ or with p on $u_{0,1}$ then the results can be the same (i.e., the BTs commute) and the unique result is given *algebraically* according to formula (2.3).

Similar results were derived even before, within the theory of surfaces. In his studies Bäcklund derived [3] the equation

$$\theta_{uv} - 2\sin(\theta/2),$$

that is now called the sine-Gordon equation and subsequently Bianchi derived [4] the permutability theorem for the BTs (c.f. (2.3)):

$$\theta_{12} = \theta + 4\arctan\left[\frac{\beta_2+\beta_1}{\beta_2-\beta_1}\tan\left(\frac{\theta_2-\theta_1}{4}\right)\right].$$

If we take tan on both sides and write the result in terms of $u := \exp(i\theta/2)$ we get

$$\beta_1(u\,u_1 - u_2\,u_{12}) = \beta_2(u\,u_2 - u_1\,u_{12}).$$

This can again be elevated to an abstract equation on the $\mathbb{Z}^2$ lattice, namely to lpmKdV as given in (2.2). Here we interpret subscripts as giving directions of steps in the lattice.

One can study various properties of abstract lattice equations, but if they have a connection to continuous soliton equations as noted above, some of the results may have concrete applications for them.

3 Symmetries and hierarchies

3.1 In the continuum

One of the essential concepts of integrable soliton equations is that they do not appear isolated but in hierarchies. For example for the KdV equation we have the hierarchy of equations

$$\begin{aligned} u_{t_1} &= \partial_x u && (3.1a)\\ u_{t_3} &= \tfrac{1}{4}\partial_x[u_{xx} + 3\,u^2] && (3.1b)\\ u_{t_5} &= \tfrac{1}{16}\partial_x[u_{4x} + 10\,u_{xx}\,u + 5\,u_x^2 + 10\,u^3] && (3.1c)\\ &\;\vdots \end{aligned}$$

Thus in the KdV case we have one space variable x and multiple time variables t_j, and the flows corresponding to the different times commute. Furthermore, if we assign weight 2 for u, weight 1 for ∂_x and weight j for ∂_{t_j} then all equations are weight homogeneous. There are elegant explanations for why the equations fit together so nicely, e.g. by the Sato theory [13].

3.2 Discrete multidimensionality

For the present discrete case we would also like to have hierarchical and multidimensional structure. To begin with, our (2.1) is fully symmetric between the n and m coordinates of the $\mathbb{Z}^2$ lattice, and therefore as we introduce higher dimensionality we would like to keep this symmetry. Thus we introduce a third dimension and the corresponding lattice index k by $u_{n,m} \to u_{n,m,k}$ and rewrite (2.1) as

$$(u_{n,m,k} - u_{n+1,m+1,k})(u_{n,m+1,k} - u_{n+1,m,k}) = p^2 - q^2, \quad \forall n, m, k. \tag{3.2}$$

This means that we have the same equation on all planes labeled by k. When we look at the situation from this point of view it is natural to propose [14] that we should equally well have equations in which m labels the plane while n, k label the corners of the quadrilateral:

$$(u_{n,m,k} - u_{n+1,m,k+1})(u_{n,m,k+1} - u_{n+1,m,k}) = p^2 - r^2, \quad \forall n, m, k. \tag{3.3}$$

Here we have also replaced q with r, which is the lattice constant for the k direction. Finally we can have a similar equation in the m, k plane

$$(u_{n,m,k} - u_{n,m+1,k+1})(u_{n,m,k+1} - u_{n,m+1,k}) = q^2 - r^2, \quad \forall n, m, k. \tag{3.4}$$

As the subscript notation starts to get lengthy it is common in the literature to introduce various kinds of shorthand notations. We sometimes use the notation in which shift in the n-direction is indicated by a tilde, in the m-direction by a hat and in the k-direction by a bar:

$$u_{n,m,k} = u, \quad u_{n+1,m,k} = \widetilde{u}, \quad u_{n,m+1,k} = \widehat{u}, \quad u_{n,m,k+1} = \overline{u},$$
$$u_{n+1,m+1,k} = \widehat{\widetilde{u}}, \quad u_{n+1,m,k+1} = \overline{\widetilde{u}}, \quad u_{n,m+1,k+1} = \overline{\widehat{u}}, \quad u_{n+1,m+1,k+1} = \overline{\widehat{\widetilde{u}}}.$$

Then our equations on the three planes read

$$\begin{aligned} Q_{12}(u, \widetilde{u}, \widehat{u}, \widehat{\widetilde{u}}; p, q) &:= (u - \widehat{\widetilde{u}})(\widehat{u} - \widetilde{u}) - p^2 + q^2 = 0, & (3.5a)\\ Q_{23}(u, \widehat{u}, \overline{u}, \overline{\widehat{u}}; q, r) &:= (u - \overline{\widehat{u}})(\overline{u} - \widehat{u}) - q^2 + r^2 = 0, & (3.5b)\\ Q_{31}(u, \overline{u}, \widetilde{u}, \overline{\widetilde{u}}; r, p) &:= (u - \overline{\widetilde{u}})(\widetilde{u} - \overline{u}) - r^2 + p^2 = 0, & (3.5c)\end{aligned}$$

when written using cyclic changes: tilde $\to$ hat $\to$ bar, $p \to q \to r$.

3.3 Commuting discrete flows

In the continuum case we know that we cannot introduce arbitrary flows in the different time directions because they would not be compatible; i.e., they would not commute. We have already discussed evolution on the lattice (see Figure 3) and when we assign equations on different planes, the evolution they generate must also satisfy some compatibility conditions. Let us look at this locally. Assuming a common corner (n, m, k) in the $\mathbb{Z}^3$ lattice we should have a situation as in Figure 2 in each of the three planes intersecting in that corner. If we keep just the elementary plaquettes we get Figure 4.

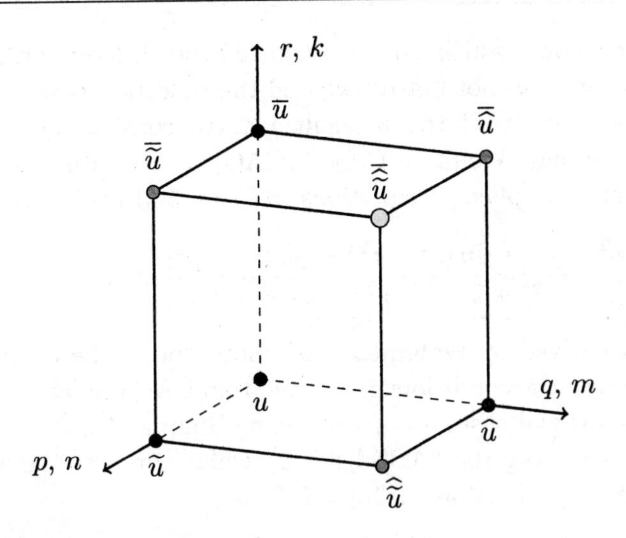

Figure 4. The consistency cube. Evolution can start on each plane from the corner with values at the black disks given. The values at gray circles can then be computed uniquely, but the value at the open circle may be ambiguous as it can be computed in three different ways.

In terms of equations the situation is as follows: At the various sides of the cube we have the corresponding equations:

$$\text{bottom:}\quad Q_{12}(u,\widetilde{u},\widehat{u},\widehat{\widetilde{u}};p,q)=0. \qquad \text{top:}\quad Q_{12}(\overline{u},\overline{\widetilde{u}},\overline{\widehat{u}},\overline{\widehat{\widetilde{u}}};p,q)=0, \tag{3.6a}$$

$$\text{back:}\quad Q_{23}(u,\widehat{u},\overline{u},\overline{\widehat{u}};q,r)=0, \quad \text{front:}\quad Q_{23}(\widetilde{u},\widehat{\widetilde{u}},\overline{\widetilde{u}},\overline{\widehat{\widetilde{u}}};q,r)=0, \tag{3.6b}$$

$$\text{left:}\quad Q_{31}(u,\overline{u},\widetilde{u},\overline{\widetilde{u}};r,p)=0, \quad \text{right:}\quad Q_{31}(\widehat{u},\overline{\widehat{u}},\widehat{\widetilde{u}},\overline{\widehat{\widetilde{u}}};r,p)=0. \tag{3.6c}$$

Here we get from the LHS to the RHS by applying on the dependent variables a shift in the direction not yet appearing on the LHS while keeping the equation itself unchanged. We would use this for any candidate equations which are uniform on parallel planes; for lpKdV we have (3.5).

From a corner we can start evolution and for the configuration of Figure 4 with $u,\widetilde{u},\widehat{u},\overline{u}$ as initial values we can compute using LHS equations the values of $\widehat{\widetilde{u}}$, $\overline{\widetilde{u}}$, $\overline{\widehat{u}}$. After this we can compute $\overline{\widehat{\widetilde{u}}}$ from each of the three RHS equations and the result should be the same. In the language of the commuting flows we have three different order of flows

- first, independently, (LHS Q_{12} to get $\widehat{\widetilde{u}}$, and LHS Q_{23} to get $\overline{\widehat{u}}$), after that RHS Q_{31} to get $\overline{\widehat{\widetilde{u}}}$.
- first, independently, (LHS Q_{23} to get $\overline{\widehat{u}}$, and LHS Q_{31} to get $\overline{\widetilde{u}}$), after that RHS Q_{12} to get $\overline{\widehat{\widetilde{u}}}$.
- first, independently, (LHS Q_{31} to get $\overline{\widetilde{u}}$, and LHS Q_{12} to get $\widehat{\widetilde{u}}$), after that RHS Q_{23} to get $\overline{\widehat{\widetilde{u}}}$.

Thus we have three flows, which can be arranged in six different orders, but since the order in the first pair does not matter we find the condition that the three possibilities listed above should give the same result, i.e., two consistency conditions. This is also called "Consistency-Around-a-Cube" (CAC) or Multidimensional consistency (MDC). When this is applied to equations (3.5) we find that in each case

$$\overline{\widehat{\widetilde{u}}} = -\frac{\widetilde{u}\widehat{u}\,(p^2-q^2)+\widehat{u}\overline{u}\,(q^2-r^2)+\overline{u}\widetilde{u}\,(r^2-p^2)}{\widetilde{u}\,(q^2-r^2)+\widehat{u}\,(r^2-p^2)+\overline{u}\,(p^2-q^2)}.$$

This was already derived by Wahlquist and Estabrook in the context of BT [16].

In the general case the conditions following from CAC are fairly complicated, but under suitable additional assumptions one can obtain a classification of equations, the most interesting being the "ABS list" [2], which contains the above mentioned lpKdV as "H1" and lpmKdV as "H3($\delta = 0$)".

4 Lax pairs

4.1 Constructing the Lax pair from CAC

In the discrete case we can use the equations on the consistency cube to generate a Lax pair by taking the bar-variables as the auxiliary linear variables.

Let us take the back and left equations of (3.6) and solve for $\overline{\widehat{u}}$ and $\overline{\widetilde{u}}$. In the case of lpKdV we get

$$\overline{\widehat{u}} = \frac{u(\overline{u}-\widehat{u})-q^2+r^2}{\overline{u}-\widehat{u}}, \tag{4.1a}$$

$$\overline{\widetilde{u}} = \frac{u(\overline{u}-\widetilde{u})-p^2+r^2}{\overline{u}-\widetilde{u}}. \tag{4.1b}$$

Now introducing

$$\overline{u} = \frac{f}{g}$$

we can write (4.1) as

$$\frac{\widehat{f}}{\widehat{g}} = \frac{u\,f-(u\widehat{u}+q^2-r^2)\,g}{f-\widehat{u}\,g}, \tag{4.2a}$$

$$\frac{\widetilde{f}}{\widetilde{g}} = \frac{u\,f-(u\widetilde{u}+p^2-r^2)\,g}{f-\widetilde{u}\,g}. \tag{4.2b}$$

This can be written as a matrix equation:

$$\widehat{\Phi} = \mathcal{M}\Phi, \quad \widetilde{\Phi} = \mathcal{L}\Phi, \quad \Phi := \begin{pmatrix} f \\ g \end{pmatrix},$$

where

$$\mathcal{M} := \mu(u,\widehat{u})\begin{pmatrix} u & -(u\widehat{u}+q^2-r^2) \\ 1 & -\widehat{u} \end{pmatrix}, \quad \mathcal{L} := \lambda(u,\widetilde{u})\begin{pmatrix} u & -(u\widetilde{u}+p^2-r^2) \\ 1 & -\widetilde{u} \end{pmatrix}. \tag{4.3}$$

(Here $\mu(u,\widehat{u})$ and $\lambda(u,\widetilde{u})$ are separation factors; one way to fix them is to require that $\det\mathcal{M}=1$, $\det\mathcal{L}=1$.) The compatibility condition arises from

$$(\widehat{\Phi})^{\sim}=(\widetilde{\Phi})^{\wedge}$$

which implies

$$\widetilde{\mathcal{M}}\mathcal{L}=\widehat{\mathcal{L}}\mathcal{M}. \tag{4.4}$$

Applying the matrices given in (4.3) (with $\mu=\lambda=1$) to this equation yields

$$\widetilde{\mathcal{M}}\mathcal{L}-\widehat{\mathcal{L}}\mathcal{M}=\left[(u-\widehat{\widetilde{u}})(\widehat{u}-\widetilde{u})-p^2+q^2\right]\begin{pmatrix}-1 & \widehat{u}+\widetilde{u}\\ 0 & 1\end{pmatrix},$$

and thus the Lax pair does generate the equation. However, it should be noted that there can also be "fake Lax pairs"; that is, even if an equation has the CAC property its Lax pair as constructed above might not generate the equation (for example if equation (4.4) is satisfied automatically).

4.2 Bäcklund transformation for constructing soliton solutions

The Lax pair and the Bäcklund transformation are different ways of interpreting the six equations (3.6): For a Lax pair we used the back and left equations and wrote them in matrix form. For BT we assume that $u_{n,m,1}$ solves the top equation and then we use the back and left equations to construct a solution to the bottom equation (which has the same form as the top equation). Since we have the extra lattice parameter r at our disposal, the solution to the bottom equation should be more general.

The starting point in this construction is a "seed" solution of the top equation. Usually this is just the null solution, but now we observe that $u_{n,m,k}\equiv 0$ is not a solution of (3.2) and the first problem is to find a suitable seed solution. One finds easily that

$$u_{n,m,k}=\pm p\,n\pm q\,m+c\,k$$

is a simple linear solution. With this in mind let us change dependent variables by

$$u_{n,m,k}=v_{n,m,k}-pn-qm-rk \tag{4.5}$$

after which the bottom, back and left equations can be written, respectively, as

$$(v_{n+1,m,k}-v_{n,m+1,k}-p+q)(v_{n+1,m+1,k}-v_{n,m,k}-p-q)=(p^2-q^2), \tag{4.6a}$$

$$(v_{n,m+1,k}-v_{n,m,k+1}-q+r)(v_{n,m+1,k+1}-v_{n,m,k}-q-r)=(q^2-r^2), \tag{4.6b}$$

$$(v_{n,m,k+1}-v_{n+1,m,k}-r+p)(v_{n+1,m,k+1}-v_{n,m,k}-r-p)=(r^2-p^2). \tag{4.6c}$$

We now use these equations for $k=0$, take $v_{n,m,1}=0$, $\forall n,m$, which solves the top equation, and solve for $\nu_{n,m}:=v_{n,m,0}$. We find

$$\nu_{n,m+1}=\frac{(q-r)\,\nu_{n,m}}{\nu_{n,m}+q+r}, \tag{4.7}$$

$$\nu_{n+1,m}=\frac{(p-r)\,\nu_{n,m}}{\nu_{n,m}+p+r}. \tag{4.8}$$

Again we would like to use matrix notation to write these results, and for that purpose we introduce

$$\psi_{n,m} = \begin{pmatrix} a_{n,m} \\ b_{n,m} \end{pmatrix}, \quad M := \mu \begin{pmatrix} q-r & 0 \\ 1 & q+r \end{pmatrix}, \quad L := \lambda \begin{pmatrix} p-r & 0 \\ 1 & p+r \end{pmatrix},$$

so that the equations to solve are

$$\psi_{n,m+1} = M\,\psi_{n,m}, \quad \psi_{n+1,m} = L\,\psi_{n,m}.$$

Since M, L are commuting constant matrices and

$$M^m = \begin{pmatrix} F^m & 0 \\ (1-F^m)/(2r) & 1 \end{pmatrix}, \quad L^n = \begin{pmatrix} G^n & 0 \\ (1-G^n)/(2r) & 1 \end{pmatrix}$$

where $F := (q-r)/(q+r)$, $G := (p-r)/(p+r)$, we find

$$\psi_{n,m} = M^m\,L^n\,\psi_{0,0}.$$

Putting everything together yields the result

$$\nu_{n,m} = \frac{a_{n,m}}{b_{n,m}} = 2r\frac{\rho_{n,m}}{1-\rho_{n,m}}, \quad \text{where} \quad \rho_{n,m} = \left(\frac{q-r}{q+r}\right)^m \left(\frac{p-r}{p+r}\right)^n \frac{v_{0,0}}{2r+v_{0,0}}. \tag{4.9}$$

5 Continuum limits

When we compare discrete and continuous spaces we will match the origins and then for a generic point we have $(x, y) = (\epsilon n, \delta m)$, where ϵ and δ measure the lattice distances. For a quad equation there are two ways to take a continuum limit as illustrated in Figure 5: In the top figure (straight limit) the square is squeezed in the m-direction; in the bottom figure (skew limit) it is first rotated 45° and then squeezed.

5.1 Skew limit

Here we will only consider the skew limit. For that purpose we rotate the coordinates by $(n, m) \to (n+m-1, m-n)$; furthermore let us denote $n+m = n'$, $m-n = m'$ and then equation (4.6a) reads

$$(v_{n',m'-1} - v_{n',m'+1} - p + q)(v_{n'+1,m} - v_{n'-1,m'} - p - q) = p^2 - q^2. \tag{5.1}$$

Since we take the limit in the m' direction we set

$$v_{n'+\nu,m'+\mu} = V_{n'+\mu}(t + \delta\mu), \tag{5.2}$$

where δ is the lattice distance in the m' direction. Thus we will take

$$m' \to \infty, \quad \delta \to 0, \quad \text{while} \quad m'\delta = t \quad \text{stays fixed.} \tag{5.3}$$

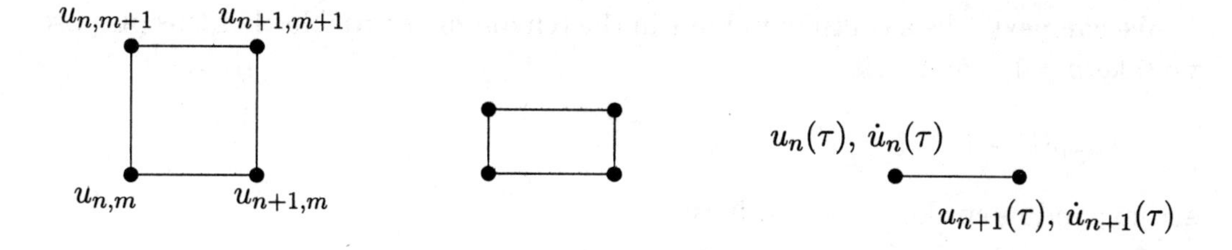

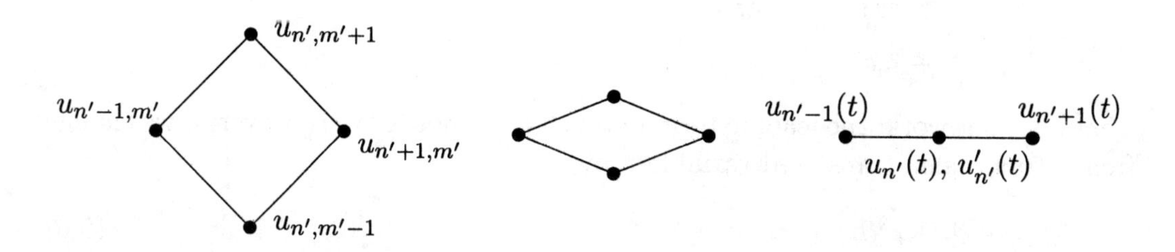

Figure 5. There are two ways to squeeze the quadrilateral to obtain continuum limits: Top, the straight limit; bottom, the skew limit; corresponding derivatives are $\dot{u}$ and u'.

We still have the question of how δ and p, q are related. Some help can be obtained from the form of ρ in (4.9). We know that the soliton solutions are constructed using exponential functions and ρ can be interpreted as a discrete form of the exponential, due to the well known limit formula

$$\lim_{n\to\infty}\left(1+\frac{x}{n}\right)^n = e^x. \tag{5.4}$$

In our case we have to consider the combination with $m-n$ power, i.e.

$$\left(\frac{q-r}{q+r}\cdot\frac{p+r}{p-r}\right)^{(m-n)} = \left(1+\frac{2r(q-p)}{(p-r)(q+r)}\right)^{m'}.$$

Since r is a soliton parameter it will stay finite and nonzero and therefore we take $q-p=\delta\to 0$. Substituting $q=p+\delta$ and using (5.3) we get

$$\left(1+\frac{2r\delta}{(p-r)(p+\delta+r)}\right)^{m'} = \left(1+\frac{t}{m'}\frac{2r}{(p-r)(p+t/m'+r)}\right)^{m'} \to \exp\left(\frac{2rt}{p^2-r^2}\right).$$

Thus the limit works and produces a reasonable plane wave factor. We then proceed to insert $q=p+\delta$ and (5.2) into (5.1) and expand in δ. This yields

$$\partial_t V_n(t) = 1 - \frac{2p}{V_{n-1}(t) - V_{n+1}(t) + 2p} \tag{5.5}$$

where we have dropped the primes in n. This equation is therefore the skew semi-discrete limit of the (translated) lpKdV equation given in (4.6a). It is a bona-fide integrable equation, having a Lax pair, etc.

We can next take a continuum limit in the remaining n variable. For this purpose we take $p = 1/\epsilon$ and write

$$V_{n+\nu}(t) = U(t, \xi + \nu\epsilon)$$

and expand in epsilon. The result is

$$\begin{aligned}\partial_t U &= \epsilon^2 U_x \\ &\quad +\epsilon^4 \tfrac{1}{6}[U_{xxx} + 6U_x^2] \\ &\quad + \text{h.o.}\end{aligned}$$

That does not work; the leading term in this limit is not KdV. However, if we change from t to a scaled-translated variable τ by

$$\partial_t = \epsilon^2 \partial_x + \epsilon^4 \partial_\tau \tag{5.6}$$

then the leading term ϵ^4 yields

$$U_\tau = \tfrac{1}{6}[U_{xxx} + 6U_x^2], \tag{5.7}$$

which is a potential form of the KdV equation (pKdV). The need for some sort of new "squeezed" variables as in (5.6) is obvious: the starting discrete equation is very symmetric while the continuum target equation is asymmetric, with x playing a different role in comparison to the t_i.

5.2 Double limit

On the basis of the above we could try to take a limit in n, m directions simultaneously, but at the same time we should somehow introduce suitable scaling. Thus we try

$$v_{n,m} = V(x + \epsilon(na_1 + mb_1), t + \epsilon^3(na_3 + mb_3)), \tag{5.8}$$

where we have chosen the powers of ϵ following the expected relative scaling of x and t_3. Inserting this with $p = \alpha/\epsilon$, $q = \beta/\epsilon$ into (4.6a) and expanding in ϵ we find that if we choose

$$a_j = \frac{2^j}{j\alpha^j}, \quad b_j = \frac{2^j}{j\beta^j}, \quad \alpha^2 \neq \beta^2, \tag{5.9}$$

we get, as the leading term, the pKdV equation in the form

$$V_t = \tfrac{1}{4}[V_{xxx} + 3V_x^2]. \tag{5.10}$$

But there is more: If we look at the next order in ϵ we find an x derivative of the above equation, and then at the next order, after using (5.10) to eliminate time derivatives, the expression

$$V_{xxxxx} + 10\, V_{xxx}\, V_x + 5\, V_{xx}^2 + 10 V_x^3,$$

which appears in the square brackets on the RHS of the fifth order KdV (3.1c). Thus it seems that the discrete equation contains inside it the whole hierarchy of continuum equations! In order to explore this further, let us use multiple time variables as follows:

$$v_{n,m} = V\left(x + \epsilon(na_1 + mb_1), t_3 + \epsilon^3(na_3 + mb_3), t_5 + \epsilon^5(na_5 + mb_5), \cdots\right)$$

When we expand (4.6a) using this multi-time expression with parameters (5.9) and in the results eliminate lower order times using lower order equations, and change $V_x = u$, we get the sequence of higher order members of the KdV hierarchy, some of which were given in (3.1).

The above observations can be made into precise statements using more refined mathematics, for example by using the Sato theory. In that formalism infinite number of time variables are used at the outset and one can find a simple correspondence between the discrete and continuum hierarchies. The main observation to take away from this is that the nicely symmetric and simple looking discrete formalism is in effect as rich as the corresponding continuum theory. And this statement holds also for the more general equations such as KP.

6 Discretizing a continuous equation

One approach to discrete equations is to take a known continuous integrable equation and try to construct a discrete version with as many integrability properties as possible. One important object that we would like to preserve is the class of solutions, perhaps in the sense that the discrete solutions approach the continuous ones in a smooth fashion.

6.1 A simple 1D example

Consider the nonlinear ODE (Verhulst's population model)

$$\dot{x}(t) = \alpha\, x(t)(1 - x(t)). \tag{6.1}$$

We would like to discretize this so that the solutions of the discrete version follow closely the continuous solution, which can be derived easily:

$$x(t) = \frac{1}{1 + e^{\alpha\,(t - t_0)}}. \tag{6.2}$$

How should this equation be discretized? A naive discretization would be to replace the derivative by a forward difference:

$$h^{-1}(x(t+h) - x(t)) = \alpha\, x(t)(1 - x(t)). \tag{6.3}$$

This is the logistic equation which is well known to lead to chaotic behavior for most values of the parameter α, while the solution (6.2) is always smooth. We need a different discretization.

In order to proceed we note that equation (6.1) can be linearized:

$$x(t) = \frac{1}{1+y(t)} \quad \Rightarrow \quad \dot{y}(t) + \alpha\, y(t) = 0. \tag{6.4}$$

For the linear equation the naive discretization works: The solution to the continuous y equation (6.4) is given by

$$y(t) = \exp[-\alpha(t-t_0)] \tag{6.5}$$

while the solution to the discretized version of (6.4)

$$h^{-1}(y(t+h) - y(t)) + \alpha\, y(t) = 0$$

is given by

$$y(t+nh) = (1-\alpha\, h)^{n+(t-t_0)/h}. \tag{6.6}$$

This solution approximates the solution (6.5), due to the limit formula (5.4):

$$(1-\alpha h)^{(t-t_0)\frac{1}{h}} \to e^{-\alpha(t-t_0)} \quad \text{as} \quad h \to 0^+.$$

Let us denote $y(t+nh) = y_n$, $(t-t_0)/h = -n_0$ and reverse the steps above. We find

$$y_n := (1-\alpha\, h)^{n-n_0} \quad \text{solves} \quad y_{n+1} = (1-\alpha\, h) y_n$$

and since $x = 1/(1+y)$,

$$x_n := \frac{1}{1+(1-\alpha\, h)^{n-n_0}}, \tag{6.7a}$$

solves

$$x_{n+1} = \frac{x_n}{1-\alpha\, h + \alpha\, h\, x_n}. \tag{6.7b}$$

The solution for x_n (6.7a) is a good approximation of (6.2) but the equation it solves (6.7b) is not at all like the one obtained by naive discretization (6.3).

6.2 Hirota's method of discretization

For PDE's the situation is much more complicated. This is the case in particular because we do not know all solutions, or rather, the general solution is too complicated to work with. One approach is to make sure that at least the soliton solutions carry over from continuous to discrete. For this we follow R. Hirota, who in a series of papers in 1977 discretized many soliton equations while preserving their N-soliton solutions [9, 10, 11]. The culmination of this work was the "DAGTE" equation [12] from which many other soliton equations follow.

6.2.1 Bilinear form of continuous KdV

Hirota's method is based on a transformation of the dependent variable so that in terms of the new dependent variable the soliton solutions are simply polynomials of exponentials with linear exponents. Instead of the standard form of the KdV equation,

$$u_t + 6\,u\,u_x + u_{xxx} = 0, \tag{6.8}$$

it is better to introduce the variable v by $u = v_x$, and integrate (6.8) into the pKdV equation

$$v_t + 3v_x^2 + v_{xxx} = 0. \tag{6.9}$$

The new dependent variable f is defined by

$$v = 2\partial_x \log(f), \quad \text{or} \quad u = 2\partial_x^2 \log(f), \tag{6.10}$$

and when this is used in (6.9) we get a fourth order equation

$$D_x(D_t + D_x^3)\, f \cdot f = 0, \tag{6.11}$$

which is written in terms of Hirota's bilinear derivatives, defined by

$$D_x^n\, D_t^m\, f \cdot g = (\partial_x - \partial_{x'})^n (\partial_t - \partial_{t'})^m\, f(x,t)\, g(x',t')\big|_{x'=x,\, y'=y}\,.$$

6.2.2 Discretizing KdV

In order to continue we need a discrete version of the bilinear derivative. For usual derivatives we have

$$e^{ax} f(x) = f(x+a)$$

and therefore we have, for example,

$$\begin{aligned} e^{aD_x}\, f\cdot g &= f(x+a)\,g(x-a), \\ \sinh(aD_x) f\cdot g &= \tfrac{1}{2}[f(x+a)g(x-a) - f(x-a)g(x+a)]. \end{aligned}$$

Since

$$\sinh(aD_x) = aD_x + \text{ higher order terms in } aD_x$$

it seems reasonable to use discretization rules like $D \to \sinh(aD)$. The precise replacement to (6.11) proposed by Hirota was (ref [9], equation (2.3))

$$\sinh[\tfrac{1}{4}(D_x + \delta D_t)]\,\{2\delta^{-1}\sinh[\tfrac{1}{2}\delta D_t] + 2\sinh[\tfrac{1}{2}D_x]\}\; f(x,t)\cdot f(x,t) = 0, \tag{6.12}$$

which can also be written as

$$\begin{aligned} &\{\cosh[\tfrac{1}{4}\delta D_t + \tfrac{3}{4}D_x] + \delta^{-1}\cosh[\tfrac{3}{4}\delta D_t + \tfrac{1}{4}D_x] \\ &\qquad - (1+\delta^{-1})\cosh[\tfrac{1}{4}\delta D_t - \tfrac{1}{4}D_x]\}\; f(x,t)\cdot f(x,t) = 0. \end{aligned}$$

In order to write this as shifts we note that

$$\cosh(\alpha D_x + \beta D_t)\, f(x,t) \cdot f(x,t) = f(x+\alpha, t+\beta) f(x-\alpha, t-\beta)$$

and if we convert shifts to discrete subscript notation

$$f(x + \tfrac{1}{4}\nu, t + \delta\tfrac{1}{4}\mu) = f_{n+\frac{1}{4}\nu, m+\frac{1}{4}\mu},$$

we can write (6.12) as

$$f_{n+\frac{3}{4},m+\frac{1}{4}} f_{n-\frac{3}{4},m-\frac{1}{4}} + \delta^{-1} f_{n+\frac{1}{4},m+\frac{3}{4}} f_{n-\frac{1}{4},m-\frac{3}{4}} \\ - (1+\delta^{-1}) f_{n-\frac{1}{4},m+\frac{1}{4}} f_{n+\frac{1}{4},m-\frac{1}{4}} = 0. \qquad (6.13)$$

This does not sit at the points of the $\mathbb{Z}^2$ lattice but if we make a 45° rotation and a shift according to

$$(n+\nu, m+\mu) \mapsto (n+m+\nu+\mu, n-m+\nu-\mu+1) = (n'+\nu+\mu, m'+\nu-\mu)$$

we get

$$f_{n'+1,m'+1} f_{n'-1,m'} + \delta^{-1} f_{n'+1,m'} f_{n'-1,m'+1} - (1+\delta^{-1}) f_{n',m'} f_{n',m'+1} = 0. \quad (6.14)$$

The dependent variables are now on the points of the $\mathbb{Z}^2$ lattice, but the equation connects points on two quadrilaterals. (This is typical for Hirota bilinear equations; in fact the only one-component equation that can exist on a single quadrilateral is trivial.)

Equations (6.13) and (6.14) have the main properties essential in Hirota's approach to constructing soliton solutions: a) $f_{n,m} \equiv 1$ is a solution, and b) in each product the sum of indices is the same. This last property implies gauge invariance: if $f_{n,m}$ is a solution, so is $f'_{n,m} := A^n B^m f_{n,m}$ for any constants A, B.

6.2.3 Soliton solutions

In Hirota's approach soliton solutions are constructed perturbatively:

Background solution: The bilinear form (6.14) obviously has $f_{n,m} \equiv 1$ as the vacuum or background solution.

One-soliton solution: The ansatz for the one-soliton solution of (6.14) is

$$f_{n,m} = 1 + cA(p,k_1)^n B(q,k_1)^m, \qquad (6.15)$$

where k_1 is the parameter of the soliton. We have also noted possible dependence on lattice parameters: the *plane wave factor* A may depend on p because it is associated with the n direction; similarly B may depend on q. The constant c is constant only in n, m but may depend on p, q, k_1. This ansatz leads to the *dispersion relation*

$$A(p,k_1) - B(q,k_1) = \delta\,[1 - A(p,k_1)B(q,k_1)],$$

and evidently δ should also depend on p, q. This relation is resolved by

$$A(p,k_1) = \frac{p-k_1}{p+k_1}, \quad B(q,k_1) = \frac{q-k_1}{q+k_1}, \quad \delta(p,q) = \frac{p-q}{p+q}. \qquad (6.16)$$

Note the beautiful symmetry which even encompasses the parameter δ.

Two-soliton solution: Following Hirota's perturbative approach the 2SS ansatz is

$$\begin{aligned} f_{n,m} &= 1 + c_1 A(p,k_1)^n B(q,k_1)^m + c_2 A(p,k_2)^n B(q,k_2)^m \\ &\quad + \mathcal{A}(k_1,k_2) c_1 c_2 A(p,k_1)^n B(q,k_1)^m A(p,k_2)^n B(q,k_2)^m. \end{aligned} \tag{6.17}$$

This form is dictated by the condition that when solitons are far apart they look like 1SS. There is a new parameter $\mathcal{A}(k_1, k_2)$ called the *phase factor.* When this ansatz is substituted into (6.14) with (6.16), we find that it is a solution, provided that the phase factor is given by

$$\mathcal{A}(k_1,k_2) = \frac{(k_1-k_2)^2}{(k_1+k_2)^2}.$$

This is exactly the same as for the continuous KdV equation.

***N*-soliton solution in determinant form:** We could follow this perturbative route and construct an ansatz for 3SS, with only k_3 as a new parameter, and verify that it is a solution. But we can do better and construct a general determinant formula for the N-soliton solution. For that purpose let us define

$$\psi_{n,m}(j,l) := \rho_j^+ \,(p+k_j)^n\,(q+k_j)^m\,k_j^l + \rho_j^- \,(p-k_j)^n\,(q-k_j)^m\,(-k_j)^l.$$

It is easy to verify that $\psi_{n,m}(1,0)$ is gauge equivalent to $f_{n,m}$ of (6.15) for $c = \rho_1^-/\rho_1^+$. With ψ given we write the 2SS as

$$f_{n,m}^{[2ss]} = \begin{vmatrix} \psi_{n,m}(1,0) & \psi_{n,m}(1,1) \\ \psi_{n,m}(2,0) & \psi_{n,m}(2,1) \end{vmatrix},$$

and this is gauge equivalent to (6.17) if we connect parameters c_j, ρ_j^+, ρ_j^- by

$$\frac{\rho_1^-}{\rho_1^+} = c_1 \frac{k_2-k_1}{k_1+k_2}, \quad \frac{\rho_2^-}{\rho_1^+} = c_2 \frac{k_1-k_2}{k_1+k_2}.$$

The N-soliton solution is given by the natural extension

$$f_{n,m}^{[Nss]} = \begin{vmatrix} \psi_{n,m}(1,0) & \psi_{n,m}(1,1) & \cdots & \psi_{n,m}(1,N-1) \\ \psi_{n,m}(2,0) & \psi_{n,m}(2,1) & \cdots & \psi_{n,m}(2,N-1) \\ \vdots & \vdots & \ddots & \vdots \\ \psi_{n,m}(N,0) & \psi_{n,m}(N,1) & \cdots & \psi_{n,m}(N,N-1) \end{vmatrix}. \tag{6.18}$$

That this is a solution of (6.14) for δ as given above can be shown by determinantal manipulations as in [8] (cf. Equations (2.20), (5.17b) with bar $\to$ tilde, and (5.18)).

6.2.4 From bilinear to nonlinear

Recall that the change of dependent variables from continuous nonlinear to continuous bilinear by (6.10) involved derivatives, which are easy. The reverse operation involves integration and is more involved, especially for the discrete case.

We start with (6.13) and shift it by $(n,m) \to (n-\frac{1}{4}, m+\frac{1}{4})$ and by $(n,m) \to (n+\frac{1}{4}, m-\frac{1}{4})$ and get

$$\begin{aligned} f_{n+\frac{1}{2},m+\frac{1}{2}} f_{n-1,m} + \delta^{-1} f_{n,m+1} f_{n-\frac{1}{2},m-\frac{1}{2}} - (1+\delta^{-1}) f_{n-\frac{1}{2},m+\frac{1}{2}} f_{n,m} &= 0, \\ f_{n+1,m} f_{n-\frac{1}{2},m-\frac{1}{2}} + \delta^{-1} f_{n+\frac{1}{2},m+\frac{1}{2}} f_{n,m-1} - (1+\delta^{-1}) f_{n,m} f_{n+\frac{1}{2},m-\frac{1}{2}} &= 0, \end{aligned}$$

respectively. Multiplying the first equation by $f_{n+\frac{1}{2},m-\frac{1}{2}}$ the second by $f_{n-\frac{1}{2},m+\frac{1}{2}}$ and subtracting them yields a four term equation and after multiplying it by $f_{n,m}/(f_{n+\frac{1}{2},m+\frac{1}{2}} f_{n+\frac{1}{2},m-\frac{1}{2}} f_{n-\frac{1}{2},m+\frac{1}{2}} f_{n-\frac{1}{2},m-\frac{1}{2}})$ we can write the result as

$$\begin{aligned} &\frac{f_{n-1,m} f_{n,m}}{f_{n-\frac{1}{2},m-\frac{1}{2}} f_{n-\frac{1}{2},m+\frac{1}{2}}} - \frac{f_{n+1,m} f_{n,m}}{f_{n+\frac{1}{2},m+\frac{1}{2}} f_{n+\frac{1}{2},m-\frac{1}{2}}} \\ &\quad = \frac{1}{\delta}\left(\frac{f_{n,m+1} f_{n,m}}{f_{n+\frac{1}{2},m+\frac{1}{2}} f_{n-\frac{1}{2},m+\frac{1}{2}}} - \frac{f_{n,m-1} f_{n,m}}{f_{n-\frac{1}{2},m-\frac{1}{2}} f_{n+\frac{1}{2},m-\frac{1}{2}}} \right). \end{aligned}$$

Now introduce the quantity

$$W := \frac{f_{n+\frac{1}{2},m} f_{n-\frac{1}{2},m}}{f_{n,m+\frac{1}{2}} f_{n,m-\frac{1}{2}}}, \tag{6.20}$$

in terms of which the above equation can be written as

$$W_{n-\frac{1}{2},m} - W_{n+\frac{1}{2},m} = \frac{1}{\delta}\left(\frac{1}{W_{n,m+\frac{1}{2}}} - \frac{1}{W_{n,m-\frac{1}{2}}} \right). \tag{6.21}$$

In order to get a convenient form we still make a change in the discrete variables by

$$(n+\nu, m+\mu) \mapsto (n+m+(\nu+\mu+\tfrac{1}{2}), n-m+(\nu-\mu+\tfrac{1}{2})),$$

after which equation (6.21), when written in terms of $n' := n+m$, $m' := n-m$, reads

$$W_{n',m'} - W_{n'+1,m'+1} = \frac{1}{\delta}\left(\frac{1}{W_{n'+1,m'}} - \frac{1}{W_{n',m'+1}} \right). \tag{6.22}$$

This now has the standard quad form as it depends on the corner variables of the quadrilateral as in Figure 2.

If we apply the double continuous limit (5.8) on the relation (6.20) we get

$$\lim_{\epsilon \to 0} \frac{1}{\epsilon^2}(W-1) = \partial_x^2 \log(f(x,t)).$$

Comparing this to (6.10) we see that the equation for $W-1$ should be taken as the discrete KdV equation.

6.2.5 Relation to the lpKdV version of KdV

Equation (6.22) was obtained from the potential KdV equation (6.9) by discretizing its bilinear form (6.11) as (6.14). The discrete bilinear form was then nonlinearized into (6.22). But this equation is different from the lpKdV equation (2.1) which we announced in the beginning as being the discrete form of KdV. The reason for the difference is that lpKdV (2.1) is discrete *potential* KdV while (6.22) is discrete KdV or lattice KdV (lKdV).

The explicit relation is obtained as follows: In (2.1) let us introduce new variables as follows:

$$W_{n,m} := u_{n,m+1} - u_{n+1,m}, \quad Z_{n,m} := u_{n,m} - u_{n+1,m+1}, \tag{6.23}$$

after which (2.1) can be written as

$$W_{n,m}\, Z_{n,m} = p^2 - q^2. \tag{6.24}$$

According to the definitions (6.23) W, Z are related by

$$W_{n,m} - W_{n+1,m+1} = Z_{n,m+1} - Z_{n+1,m} \tag{6.25}$$

and if we solve for $Z_{n,m}$ from (6.24) and substitute it into (6.25) we get

$$W_{n,m} - W_{n+1,m+1} = (p^2 - q^2)\left(\frac{1}{W_{n,m+1}} - \frac{1}{W_{n+1,m}}\right), \tag{6.26}$$

which is (6.22) up to the constant coefficient. The reason for calling this the lattice KdV and (2.1) the lattice *potential* KdV is seen from the relation $W_{n,m} := u_{n,m+1} - u_{n+1,m}$, which is analogous to $u = v_x$ for the continuous case.

7 Integrability test

As usual we are mainly concerned with integrable equations, but when faced with a new equation how can we tell whether it is potentially integrable? A method that only requires direct computation would be desirable.

In the continuous case we have the Painlevé property and for 1D discrete equations the singularity confinement (SC) idea has been proposed as an analogous property. SC has turned out to be a very useful concept as an indicator, although it is only a necessary condition.

For discrete equations there is also the concept of "algebraic entropy" which states that the complexity of the iterates should not grow exponentially, but only polynomially. Complexity is here measured by the degree of an iterate with respect to the initial values. This is computationally demanding but can be automatized.

As an example consider the corner initial values as in Figure 3 a) and a quadratic equation (such as (2.1)). If we define $u_{n,0} = x_n$, $u_{0,m} = y_n$ ($y_0 = x_0$) we find that generically $u_{1,1}$ is quadratic over linear in the initial values, $u_{1,2}$ and $u_{2,1}$ are cubic over quadratic, and $u_{2,2}$ is order six over order five. However, for the particular case of (2.1), which is integrable, the numerator and denominator of $u_{2,2}$ have a common factor which can be canceled and the degrees are just order five over order four. Such cancellations continue for higher orders. According to [15], for (2.1) the degree of the numerator is $d_{n,m} = nm + 1$; that is the growth is polynomial. For a generic quadratic equation the asymptotic growth of degrees is exponential.

8 Summary

In this brief introduction to the discrete or lattice soliton equations we have looked at some of their important features, and as an example we have given explicit details for the Korteweg-de Vries equation. We have discussed how the permutation property of Bäcklund transformations can be interpreted as discrete equations on a $\mathbb{Z}^2$ lattice and how lattice evolutions are typically defined. The fundamental idea of hierarchy of equations is in the lattice setting provided by multidimensionality. The multidimensional consistency condition was discussed in detail, along with its consequence of providing Lax pair and Bäcklund transformation, which was used to construct a one-soliton solution. We have also discussed continuum limits and discretization, in particular Hirota's discretization and the ensuing soliton solutions.

For further introductory material we refer the reader to the book [7].

Acknowledgements

I would like to thank Da-jun Zhang for comments on the manuscript.

References

[1] Ablowitz M J and Clarkson P A. *Solitons, nonlinear evolution equations and inverse scattering*, volume 149 of *London Mathematical Society Lecture Note Series.* Cambridge University Press, Cambridge, 1991.

[2] Adler V, Bobenko A, and Suris Yu. Classification of integrable equations on quad-graphs. The consistency approach. *Comm. Math. Phys.*, 233(3):513–543, 2003.

[3] Bäcklund A V, Om ytor med konstant negative krökning, *Lund Univ. Årsskrift*, **19** (1883), 1-48.

[4] Bianchi L, Sulla trasformatzione di Bäcklund per le superficie pseudoferiche, *Rend. Lincei,* **5** (1892), 3-12.

[5] Drazin P G and Johnson R S. *Solitons: an introduction.* Cambridge Texts in Applied Mathematics. Cambridge University Press, Cambridge, 1989.

[6] Faddeev L and Takhtajan L. *Hamiltonian methods in the theory of solitons.* Classics in Mathematics. Springer, Berlin, English edition, 2007.

[7] Hietarinta J, Joshi N, Nijhoff F, *Discrete Systems and Integrability*, Cambridge University Press, Cambridge, 2016.

[8] Hietarinta J and Zhang D-j. Soliton solutions for ABS lattice equations. II. Casoratians and bilinearization. *J. Phys. A*, 42(40):404006, 30, 2009.

[9] Hirota R. Nonlinear partial difference equations. I. A difference analogue of the Korteweg-de Vries equation. *J. Phys. Soc. Japan*, 43(4):1424–1433, 1977.

[10] Hirota R. Nonlinear partial difference equations. II. Discrete-time Toda equation. *J. Phys. Soc. Japan*, 43(6):2074–2078, 1977.

[11] Hirota R. Nonlinear partial difference equations. III. Discrete sine-Gordon equation. *J. Phys. Soc. Japan*, 43(6):2079–2086, 1977.

[12] Hirota R. Discrete analogue of a generalized Toda equation. *J. Phys. Soc. Japan*, 50(11):3785–3791, 1981.

[13] Miwa T, Jimbo M, and Date E. *Solitons: Differential equations, symmetries and infinite-dimensional algebras*, volume 135 of *Cambridge Tracts in Mathematics*. Cambridge University Press, Cambridge, 2000.

[14] Nijhoff F, Ramani A, Grammaticos B, and Ohta Y. On discrete Painlevé equations associated with the lattice KdV systems and the Painlevé VI equation. *Stud. Appl. Math.*, 106(3):261–314, 2001.

[15] Tremblay S, Grammaticos B, and Ramani A. Integrable lattice equations and their growth properties. *Phys. Lett. A*, 278(6):319–324, 2001.

[16] Wahlquist H, and Estabrook F, Bäcklund transformation for solutions of the Korteweg-de Vries equation, *Phys. Rev. Lett.*, **31** (1973), 1386–1390.

A5. New results on integrability of the Kahan-Hirota-Kimura discretizations

Matteo Petrera [a] *and Yuri B Suris* [b]

Institut für Mathematik, MA 7-1, Technische Universität Berlin,
Str. des 17. Juni 136, 10623 Berlin, Germany.
[a] *E-mail: petrera@math.tu-berlin.de,* [b] *E-mail: suris@math.tu-berlin.de*

Abstract

R. Hirota and K. Kimura discovered integrable discretizations of the Euler and the Lagrange tops, given by birational maps. Their method is an application, in the integrable context, of a general discretization scheme introduced by W. Kahan for arbitrary vector fields with a quadratic dependence on phase variables. We report several novel observations regarding integrability of the Kahan-Hirota-Kimura discretization. For several of the most complicated cases for which integrability is known (Clebsch system, Kirchhoff system, and Lagrange top),

- we give nice compact formulas for some of the more complicated integrals of motion and for the density of the invariant measure, and
- we establish the existence of higher order Wronskian Hirota-Kimura bases, generating the full set of integrals of motion.

While the first set of results admits nice algebraic proofs, the second one relies on computer algebra.

1 Introduction

The Kahan-Hirota-Kimura discretization method was introduced in the geometric integration literature by Kahan in the unpublished notes [10] as a method applicable to any system of ordinary differential equations with a quadratic vector field:

$$\dot{x} = f(x) = Q(x) + Bx + c, \tag{1.1}$$

where each component of $Q : \mathbb{R}^n \to \mathbb{R}^n$ is a quadratic form, while $B \in \mathrm{Mat}_{n\times n}(\mathbb{R})$ and $c \in \mathbb{R}^n$. Kahan's discretization (with stepsize 2ϵ) reads as

$$\frac{\widetilde{x} - x}{2\epsilon} = Q(x, \widetilde{x}) + \frac{1}{2}B(x + \widetilde{x}) + c, \tag{1.2}$$

where

$$Q(x, \widetilde{x}) = \frac{1}{2}\big(Q(x + \widetilde{x}) - Q(x) - Q(\widetilde{x})\big)$$

is the symmetric bilinear form corresponding to the quadratic form Q. We say that the expression on the right-hand side of (1.2) is the *polarization* of the expression on the right-hand side of (1.1). Equation (1.2) is *linear* with respect to $\widetilde{x}$ and therefore defines a *rational* map $\widetilde{x} = \Phi_f(x, \epsilon)$. Clearly, this map approximates the time 2ϵ shift along the solutions of the original differential system. Since equation (1.2) remains invariant under the interchange $x \leftrightarrow \widetilde{x}$ with the simultaneous sign inversion $\epsilon \mapsto -\epsilon$, one has the *reversibility* property

$$\Phi_f^{-1}(x, \epsilon) = \Phi_f(x, -\epsilon). \tag{1.3}$$

In particular, the map Φ_f is *birational.*

Kahan applied this discretization scheme to the famous Lotka-Volterra system and showed that in this case it possesses a very remarkable non-spiralling property. This property was explained by Sanz-Serna [18] by demonstrating that in this case the numerical method preserves an invariant Poisson structure of the original system.

The next intriguing appearance of this discretization was in two papers by Hirota and Kimura who (being apparently unaware of the work by Kahan) applied it to two famous *integrable* system of classical mechanics, the Euler top and the Lagrange top [8, 11]. Surprisingly, the discretization scheme produced in both cases *integrable* maps.

In [14, 12, 13], see also [9, 15, 16, 17], the authors undertook an extensive study of the properties of the Kahan's method when applied to integrable systems (we proposed to use in the integrable context the term "Hirota-Kimura method"). It was demonstrated that, in an amazing number of cases, the method preserves integrability in the sense that the map $\Phi_f(x, \epsilon)$ possesses as many independent integrals of motion as the original system $\dot{x} = f(x)$.

Further remarkable geometric properties of the Kahan's method were discovered by Celledoni, McLachlan, Owren and Quispel in [4], see also [5, 6]. These properties are unrelated to integrability. They demonstrated that for an arbitrary Hamiltonian vector field with a constant Poisson tensor and a cubic Hamilton function, the map $\Phi_f(x, \epsilon)$ possesses a rational integral of motion and an invariant measure with a polynomial density.

The goal of the present paper is to communicate several novel observations regarding integrability of the Kahan's method. These observations hold for several of the most complicated cases for which integrability of the Kahan-Hirota-Kimura discretization is established (Clebsch system, *so*(4) Euler top, Kirchhoff system, and Lagrange top). However, some of our new findings here are verifiable by hand calculations and do not require heavy computer algebra computations. This refers to nice compact formulas for some of the more complicated integrals of motion and for the density of the invariant measure. See Theorem 1 and Observation 2 in Section 3. We give these results for all of the above mentioned systems, but provide detailed proofs for the first Clebsch flow only (see Section 5). Another set of results still relies on the computer algebra. This refers to the so called higher order Wronskian Hirota-Kimura bases. See Observation 3 in Section 3. We expect that understanding of the latter phenomenon could be crucial for the whole integrability picture of the Kahan-Hirota-Kimura discretizations, but to this moment the origin of this phenomenon remains obscure.

2 General properties of the Kahan-Hirota-Kimura discretization

The explicit form of the map Φ_f defined by (1.2) is

$$\widetilde{x} = \Phi_f(x, \epsilon) = x + 2\epsilon(I - \epsilon f'(x))^{-1} f(x), \tag{2.1}$$

where $f'(x)$ denotes the Jacobi matrix of $f(x)$. As a consequence, each component x_i of x is a rational function,

$$x_i = \frac{p_i(x; \epsilon)}{\Delta(x; \epsilon)}, \tag{2.2}$$

with a common denominator

$$\Delta(x; \epsilon) = \det(I - \epsilon f'(x)). \tag{2.3}$$

Clearly, the degrees of all polynomials p_i and Δ are equal to n.

One has the following expression for the Jacobi matrix of the map Φ_f:

$$d\Phi_f(x) = \frac{\partial \widetilde{x}}{\partial x} = \big(I - \epsilon f'(x)\big)^{-1}\big(I + \epsilon f'(\widetilde{x})\big), \tag{2.4}$$

so that

$$\det\big(d\Phi_f(x)\big) = \frac{\Delta(\widetilde{x}; -\epsilon)}{\Delta(x; \epsilon)}. \tag{2.5}$$

For our investigations here, the notion of a Hirota-Kimura basis will be relevant. It was introduced and studied in some detail in [12]. We recall here the main facts.

For a given birational map $\Phi : \mathbb{R}^n \to \mathbb{R}^n$, a set of functions $(\varphi_1, \ldots, \varphi_m)$, linearly independent over $\mathbb{R}$, is called a *Hirota-Kimura basis* (HK basis), if for every $x \in \mathbb{R}^n$ there exists a non-vanishing vector of coefficients $c = (c_1, \ldots, c_m)$ such that

$$c_1\varphi_1(\Phi^i(x)) + \ldots + c_m\varphi_m(\Phi^i(x)) = 0 \quad \text{for all} \quad i \in \mathbb{Z}.$$

For a given $x \in \mathbb{R}^n$, the vector space consisting of all $c \in \mathbb{R}^m$ with this property, say $K(x)$, is called the *null-space* of the basis $(\varphi_1, \ldots, \varphi_m)$ at the point x.

The notion of HK-bases is closely related to the notion of integrals, even if they cannot be immediately translated into one another. For instance, if $\dim K(x) = 1$, and $K(x)$ is spanned by the vector $(c_1(x), \ldots, c_m(x))$, then the quotients $c_i(x) : c_j(x)$ are integrals of the map Φ.

3 Novel observations and results

The most complicated cases where integrability of the Kahan-Hirota-Kimura discretization was established in [12, 13] are 6-dimensional systems including the Clebsch and the Kirchhoff cases of the motion of a rigid body in an ideal fluid (the first one being linearly isomorphic to the $so(4)$-Euler top), and the Lagrange top. A common feature of these systems is that they are Hamiltonian with respect to linear Lie-Poisson brackets, and completely integrable in the Liouville-Arnold sense (possess four independent integrals in involution, two of them being the Casimir functions of the Lie-Poisson bracket). The discretizations turn out to possess four integrals of motion. These integrals are very complicated. However, for each of these discretizations, one simple *quadratic-fractional integral* was found. Integrals which are not simple tend to be extremely complex. Nevertheless, we find a compact representation for some of those complex integrals.

Theorem 1. *a) Suppose that there exists a symmetric bilinear expression* $\widehat{P}(x,\widetilde{x};\epsilon^2)$ *such that for* $\widetilde{x}=\Phi_f(x;\epsilon)$ *we have*

$$\widehat{P}(x,\widetilde{x};\epsilon^2)=\frac{p(x;\epsilon^2)}{\Delta(x;\epsilon)}. \tag{3.1}$$

Then the map $\Phi_f(x;\epsilon)$ *has an invariant measure*

$$\frac{dx_1\wedge\ldots\wedge dx_n}{p(x;\epsilon^2)}. \tag{3.2}$$

b) Suppose that there exists another symmetric bilinear expression $\widehat{Q}(x,\widetilde{x};\epsilon^2)$ *such that for for* $\widetilde{x}=\Phi_f(x;\epsilon)$ *we have*

$$\widehat{Q}(x,\widetilde{x};\epsilon^2)=\frac{q(x;\epsilon^2)}{\Delta(x;\epsilon)}. \tag{3.3}$$

Then the map $\Phi_f(x;\epsilon)$ *has an integral of motion*

$$J(x;\epsilon^2)=\frac{\widehat{P}(x,\widetilde{x};\epsilon^2)}{\widehat{Q}(x,\widetilde{x};\epsilon^2)}. \tag{3.4}$$

Proof. Changing in (3.1) ϵ to $-\epsilon$, we see that

$$\widehat{P}(x,\underset{\sim}{x};\epsilon^2)=\frac{p(x;\epsilon^2)}{\Delta(x;-\epsilon)}. \tag{3.5}$$

Applying the map $x\mapsto\widetilde{x}$ and taking into account symmetry of $\widehat{P}$, we find:

$$\widehat{P}(x,\widetilde{x};\epsilon^2)=\frac{p(\widetilde{x};\epsilon^2)}{\Delta(\widetilde{x};-\epsilon)}.$$

Comparing this with (3.1), we arrive at

$$\frac{\Delta(\widetilde{x};-\epsilon)}{\Delta(x;\epsilon)}=\frac{p(\widetilde{x};\epsilon^2)}{p(x;\epsilon^2)}.$$

According to (2.5), this is equivalent to the first statement of the theorem. The second one follows from (3.5). Indeed, we see that

$$\frac{\widehat{P}(x,\underset{\sim}{x};\epsilon^2)}{\widehat{Q}(x,\underset{\sim}{x};\epsilon^2)}=\frac{p(x;\epsilon^2)}{q(x;\epsilon^2)}=\frac{\widehat{P}(x,\widetilde{x};\epsilon^2)}{\widehat{Q}(x,\widetilde{x};\epsilon^2)}.$$

This finishes the proof. ∎

In general, it is not clear how to find bilinear expressions with the property described in Theorem 1. However, the following observation mysteriously holds true in a large number of cases.

Observation 2. *Suppose that the Kahan-Hirota-Kimura discretizatrion of $\dot{x} = f(x)$ possesses a* quadratic-fractional integral *of the form*

$$I(x, \epsilon) = \frac{P(x; \epsilon^2)}{Q(x; \epsilon^2)}, \tag{3.6}$$

where P and Q as functions of x are polynomials of degree ≤ 2, while as functions of ϵ they are polynomials of ϵ^2. In many cases, the polarizations $\widehat{P}$,$\widehat{Q}$ of the polynomials P,Q, with ϵ^2 replaced by $-\epsilon^2$, satisfy the conditions of Theorem 1. In particular, to the quadratic-fractional integral (3.6)*, there corresponds a further* bilinear-fractional *integral*

$$J(x, \epsilon) = \frac{\widehat{P}(x, \widetilde{x}; -\epsilon^2)}{\widehat{Q}(x, \widetilde{x}; -\epsilon^2)}, \tag{3.7}$$

while either of the numerators of $\widehat{P}(x, \widetilde{x}; -\epsilon^2)$, $\widehat{Q}(x, \widetilde{x}; -\epsilon^2)$ serves as a densitiy of an invariant measure.

The second observation is related to linear *Wronskian relations* with constant coefficients, which turn out to exist for all systems we consider in the present paper. These are relations of the type

$$\sum_{(i,j) \in \mathcal{J}} \gamma_{ij} w_{ij}(x) = 0, \quad \text{where} \quad w_{ij}(x) = \dot{x}_i x_j - x_i \dot{x}_j, \tag{3.8}$$

satisfied on solutions of $\dot{x} = f(x)$. Here $\mathcal{J} \subset \{1, 2, \ldots, n\}^2$ is some set, and γ_{ij} for $(i, j) \in \mathcal{J}$ are certain constants. Note that the existence of such a relation can be formulated in a different way, by saying that the Wronskians $w_{ij}(x) = \dot{x}_i x_j - x_i \dot{x}_j = x_j f_i(x) - x_i f_j(x)$ with $(i, j) \in \mathcal{J}$ build a HK basis for the continuous time system $\dot{x} = f(x)$.

Observation 3. *In many cases, to a Wronskian relation* (3.8) *satisfied on solutions of $\dot{x} = f(x)$, the discrete Wronskians*

$$W_{ij}^{(\ell)}(x) = x_i^{(\ell)} x_j - x_i x_j^{(\ell)}, \quad (i, j) \in \mathcal{J}, \tag{3.9}$$

of all orders $\ell \geq 1$, form a HK basis for the map Φ_f. Here we use the notation $x_i^{(\ell)} = x_i \circ \Phi_f^{\ell}(x; \epsilon)$ for the components of the iterates of x under the map Φ_f.

4 The general Clebsch flow

The motion of a heavy top and the motion of a rigid body in an ideal fluid can be described by the so called *Kirchhoff equations*

$$\begin{cases} \dot{m} = m \times \dfrac{\partial H}{\partial m} + p \times \dfrac{\partial H}{\partial p}, \\ \dot{p} = p \times \dfrac{\partial H}{\partial m}, \end{cases} \tag{4.1}$$

where $H = H(m,p)$ is a quadratic polynomial in $m = (m_1, m_2, m_3) \in \mathbb{R}^3$ and $p = (p_1, p_2, p_3) \in \mathbb{R}^3$; here $\times$ denotes vector product in $\mathbb{R}^3$. The physical meaning of m is the total angular momentum, whereas p represents the total linear momentum of the system. System (4.1) is Hamiltonian with the Hamilton function $H(m,p)$, with respect to the Lie-Poisson bracket of $e(3)^*$:

$$\{m_i, m_j\} = m_k, \qquad \{m_i, p_j\} = p_k, \tag{4.2}$$

where (i,j,k) is a cyclic permutation of (1,2,3) (all other pairwise Poisson brackets of the coordinate functions are obtained from these by the skew-symmetry, or otherwise vanish). The functions

$$K_1 = p_1^2 + p_2^2 + p_3^2, \quad K_2 = m_1p_1 + m_2p_2 + m_3p_3$$

are Casimirs of the bracket (4.2), thus integrals of motion of an arbitrary Hamiltonian system (4.1). The complete integrability of Kirchhoff equations is guaranteed by the existence of a fourth integral of motion, functionally independent of H, K_1, K_2 and in Poisson involution with the Hamiltonian H.

Consider a homogeneous quadratic Hamiltonian

$$H = \frac{1}{2}\langle m, Am\rangle + \frac{1}{2}\langle p, Bp\rangle, \tag{4.3}$$

where $A = \operatorname{diag}(a_1, a_2, a_3)$ and $B = \operatorname{diag}(b_1, b_2, b_3)$. The corresponding Kirchhoff equations read

$$\begin{cases} \dot{m} = m \times Am + p \times Bp, \\ \dot{p} = p \times Am, \end{cases} \tag{4.4}$$

or, in components,

$$\begin{cases} \dot{m}_1 = (a_3 - a_2)m_2m_3 + (b_3 - b_2)p_2p_3, \\ \dot{m}_2 = (a_1 - a_3)m_3m_1 + (b_1 - b_3)p_3p_1, \\ \dot{m}_3 = (a_2 - a_1)m_1m_2 + (b_2 - b_1)p_1p_2. \\ \dot{p}_1 = a_3m_3p_2 - a_2m_2p_3, \\ \dot{p}_2 = a_1m_1p_3 - a_3m_3p_1, \\ \dot{p}_3 = a_2m_2p_1 - a_1m_1p_2. \end{cases} \tag{4.5}$$

An additional (fourth) integral of motion exists, if the following *Clebsch condition* is satisfied:

$$\frac{b_1 - b_2}{a_3} + \frac{b_2 - b_3}{a_1} + \frac{b_3 - b_1}{a_2} = 0. \tag{4.6}$$

Note that condition (4.6) is equivalent to the existence of ω_1, ω_2, ω_3 such that

$$\omega_1 - \omega_2 = \frac{b_1 - b_2}{a_3}, \quad \omega_2 - \omega_3 = \frac{b_2 - b_3}{a_1}, \quad \omega_3 - \omega_1 = \frac{b_3 - b_1}{a_2}. \tag{4.7}$$

Another way to express condition (4.6) is to say that the following three expressions have the same value:

$$\frac{b_1 - b_2}{a_3(a_1 - a_2)} = \frac{b_2 - b_3}{a_1(a_2 - a_3)} = \frac{b_3 - b_1}{a_2(a_3 - a_1)} = \frac{1}{\beta}. \tag{4.8}$$

With this notation, and taking into account (4.7), we easily derive that there exists a constant α such that

$$a_i = \alpha + \beta\omega_i, \quad b_i = \alpha\omega_i - \beta\omega_j\omega_k. \tag{4.9}$$

Thus, under condition (4.6), the flow (4.5) is a linear combination of two flows:

- the *first Clebsch flow* corresponding to $\alpha = 1$ and $\beta = 0$, so that

$$a_1 = a_2 = a_3 = 1, \quad b_1 = \omega_1, \quad b_2 = \omega_2, \quad b_3 = \omega_3, \tag{4.10}$$

  with the Hamilton function $\frac{1}{2}H_1$, where

$$H_1 = m_1^2 + m_2^2 + m_3^2 + \omega_1 p_1^2 + \omega_2 p_2^2 + \omega_3 p_3^2, \tag{4.11}$$

- and the *second Clebsch flow* corresponding to $\alpha = 0$ and $\beta = 1$, so that

$$\begin{cases} a_1 = \omega_1, & a_2 = \omega_2, & a_3 = \omega_3, \\ b_1 = -\omega_2\omega_3, & b_2 = -\omega_3\omega_1, & b_3 = -\omega_1\omega_2. \end{cases} \tag{4.12}$$

  with the Hamilton function $\frac{1}{2}H_2$, where

$$H_2 = \omega_1 m_1^2 + \omega_2 m_2^2 + \omega_3 m_3^2 - \omega_2\omega_3 p_1^2 - \omega_3\omega_1 p_2^2 - \omega_1\omega_2 p_3^2. \tag{4.13}$$

The key statement is:

Theorem 4. *Hamilton functions* (4.11) *and* (4.13) *are in involution, so that the first and the second Clebsch flows commute. Thus, under condition* (4.6)*, the flow* (4.5) *is completely integrable.*

As a matter of fact, Clebsch condition (4.6) admits a different characterization.

Theorem 5. *Condition* (4.6) *is equivalent to the existence of a Wronskian relation*

$$A_1(\dot{m}_1 p_1 - m_1 \dot{p}_1) + A_2(\dot{m}_2 p_2 - m_2 \dot{p}_2) + A_3(\dot{m}_3 p_3 - m_3 \dot{p}_3) = 0 \tag{4.14}$$

with constant coefficients A_i*, satisfied on solutions of* (4.5)*. In this case, coefficients* A_i *are given, up to a common factor, by*

$$A_1 = \frac{1}{a_2} + \frac{1}{a_3} - \frac{1}{a_1}, \quad A_2 = \frac{1}{a_3} + \frac{1}{a_1} - \frac{1}{a_2}, \quad A_3 = \frac{1}{a_1} + \frac{1}{a_2} - \frac{1}{a_3}. \tag{4.15}$$

Proof. Wronskian relation (4.14) is satisfied by virtue of equations of motion (4.5), if and only if coefficients A_i satisfy

$$\begin{cases} A_1(a_3 - a_2) + A_2 a_3 - A_3 a_2 = 0, \\ -A_1 a_3 + A_2(a_1 - a_3) + A_3 a_1 = 0, \\ A_1 a_2 - A_2 a_1 + A_3(a_2 - a_1) = 0, \end{cases} \tag{4.16}$$

and

$$A_1(b_3 - b_2) + A_2(b_1 - b_3) + A_3(b_2 - b_1) = 0. \tag{4.17}$$

System (4.16) is equivalent to

$$(A_1 + A_2)a_3 = (A_3 + A_1)a_2 = (A_2 + A_3)a_1,$$

so that, up to an inessential common factor, we find:

$$A_1 + A_2 = \frac{2}{a_3}, \quad A_2 + A_3 = \frac{2}{a_1}, \quad A_3 + A_1 = \frac{2}{a_2}, \tag{4.18}$$

leading to (4.15). With these values of A_i, equation (4.17) is equivalent to Clebsch condition (4.6):

$$\begin{aligned} & A_1(b_3 - b_2) + A_2(b_1 - b_3) + A_3(b_2 - b_1) \\ &= b_1(A_2 - A_3) + b_2(A_3 - A_1) + b_3(A_1 - A_2) \\ &= b_1\left(\frac{1}{a_3} - \frac{1}{a_2}\right) + b_2\left(\frac{1}{a_1} - \frac{1}{a_3}\right) + b_3\left(\frac{1}{a_2} - \frac{1}{a_1}\right) \\ &= \frac{b_1 - b_2}{a_3} + \frac{b_2 - b_3}{a_1} + \frac{b_3 - b_1}{a_2} = 0, \end{aligned}$$

which concludes the proof. ∎

In what follows, we assume that Clebsch condition (4.8) is satisfied with $\beta \neq 0$, and we define the constants A_i as in (4.15).

Applying the Kahan-Hirota-Kimura scheme to the general flow (4.5) of Clebsch system, we arrive at the following discretization:

$$\begin{cases} \widetilde{m}_1 - m_1 = \epsilon(a_3 - a_2)(\widetilde{m}_2 m_3 + m_2 \widetilde{m}_3) + \epsilon(b_3 - b_2)(\widetilde{p}_2 p_3 + p_2 \widetilde{p}_3), \\ \widetilde{m}_2 - m_2 = \epsilon(a_1 - a_3)(\widetilde{m}_3 m_1 + m_3 \widetilde{m}_1) + \epsilon(b_1 - b_3)(\widetilde{p}_3 p_1 + p_3 \widetilde{p}_1), \\ \widetilde{m}_3 - m_3 = \epsilon(a_2 - a_1)(\widetilde{m}_1 m_2 + m_1 \widetilde{m}_2) + \epsilon(b_2 - b_1)(\widetilde{p}_1 p_2 + p_1 \widetilde{p}_2), \\ \widetilde{p}_1 - p_1 = \epsilon a_3(\widetilde{m}_3 p_2 + m_3 \widetilde{p}_2) - \epsilon a_2(\widetilde{m}_2 p_3 + m_2 \widetilde{p}_3), \\ \widetilde{p}_2 - p_2 = \epsilon a_1(\widetilde{m}_1 p_3 + m_1 \widetilde{p}_3) - \epsilon a_3(\widetilde{m}_3 p_1 + m_3 \widetilde{p}_1), \\ \widetilde{p}_3 - p_3 = \epsilon a_2(\widetilde{m}_2 p_1 + m_2 \widetilde{p}_1) - \epsilon a_1(\widetilde{m}_1 p_2 + m_1 \widetilde{p}_2). \end{cases} \tag{4.19}$$

Linear system (4.19) defines a birational map $\Phi_f : \mathbb{R}^6 \to \mathbb{R}^6$, $(m,p) \mapsto (\widetilde{m},\widetilde{p})$.

A simple (quadratic-fractional) integral of Φ_f and a first order Wronskian HK basis for this map were found in our previous work.

Theorem 6. (Quadratic-fractional integral, [12]) *The function*

$$I_0(m,p,\epsilon) = \frac{A_1 a_2 a_3 g_1 + A_2 a_3 a_1 g_2 + A_3 a_1 a_2 g_3}{1+\epsilon^2 \dfrac{a_1 a_2 a_3}{\beta}(g_1+g_2+g_3)}, \tag{4.20}$$

where

$$g_i(m,p) = p_i^2 + \frac{\beta a_i}{a_j a_k} m_i^2, \tag{4.21}$$

is an integral of motion of the map Φ_f. The set

$$\Psi_0 = (g_1, g_2, g_3, 1) \tag{4.22}$$

is a HK basis for the map Φ_f, with a one-dimensional null space

$$K_{\Psi_0}(m,p) = [c_1 a_2 a_3 : c_2 a_3 a_1 : c_3 a_1 a_2 : -c_0], \tag{4.23}$$

where

$$\begin{aligned}
c_1 &= A_1 + \epsilon^2 A_3 a_1 (b_1 - b_2) g_2 + \epsilon^2 A_2 a_1 (b_1 - b_3) g_3, && (4.24a)\\
c_2 &= A_2 + \epsilon^2 A_1 a_2 (b_2 - b_3) g_3 + \epsilon^2 A_3 a_2 (b_2 - b_1) g_1, && (4.24b)\\
c_3 &= A_3 + \epsilon^2 A_2 a_3 (b_3 - b_1) g_1 + \epsilon^2 A_1 a_3 (b_3 - b_2) g_2, && (4.24c)\\
c_0 &= A_1 a_2 a_3 g_1 + A_2 a_3 a_1 g_2 + A_3 a_1 a_2 g_3. && (4.24d)
\end{aligned}$$

Theorem 7. (Discrete Wronskians HK basis, [13]) *Functions $W_i^{(1)}(m,p) = \widetilde{m}_i p_i - m_i \widetilde{p}_i$, $i = 1,2,3$, form a HK basis for the map Φ_f, with a one-dimensional null space spanned by $[c_1 : c_2 : c_3]$, with the functions c_i given in* (4.24a)–(4.24c). *In other words, on orbits of the map Φ_f there holds:*

$$\sum_{i=1}^{3} c_i (\widetilde{m}_i p_i - m_i \widetilde{p}_i) = 0. \tag{4.25}$$

Novel results, exemplifying Observations 2 and 3, are as follows.

Theorem 8. (Bilinear-fractional integral) *The function*

$$J_0(m,p,\epsilon) = \frac{A_1 a_2 a_3 G_1 + A_2 a_3 a_1 G_2 + A_3 a_1 a_2 G_3}{1-\epsilon^2 \dfrac{a_1 a_2 a_3}{\beta}(G_1+G_2+G_3)}, \tag{4.26}$$

where

$$G_i(m,p) = p_i \widetilde{p}_i + \frac{\beta a_i}{a_j a_k} m_i \widetilde{m}_i, \tag{4.27}$$

is an integral of motion of the map Φ_f. *The set*

$$\Psi_1 = (G_1, G_2, G_3, 1) \tag{4.28}$$

is a HK basis for the map Φ_f, *with a one-dimensional null space*

$$K_{\Psi_1}(m,p) = [C_1 a_2 a_3 : C_2 a_3 a_1 : C_3 a_1 a_2 : -C_0], \tag{4.29}$$

where

$$\begin{aligned}
C_1 &= A_1 - \epsilon^2 A_3 a_1 (b_1 - b_2) G_2 - \epsilon^2 A_2 a_1 (b_1 - b_3) G_3, && \text{(4.30a)}\\
C_2 &= A_2 - \epsilon^2 A_1 a_2 (b_2 - b_3) G_3 - \epsilon^2 A_3 a_2 (b_2 - b_1) G_1, && \text{(4.30b)}\\
C_3 &= A_3 - \epsilon^2 A_2 a_3 (b_3 - b_1) G_1 - \epsilon^2 A_1 a_3 (b_3 - b_2) G_2, && \text{(4.30c)}\\
C_0 &= A_1 a_2 a_3 G_1 + A_2 a_3 a_1 G_2 + A_3 a_1 a_2 G_3. && \text{(4.30d)}
\end{aligned}$$

Actually, the numerator and the denominator of the fraction (4.26) satisfy conditions of Theorem 1.

Corollary 9. (Density of an invariant measure) *The map* $\Phi_f(x;\epsilon)$ *has an invariant measure*

$$\frac{dm_1 \wedge dm_2 \wedge dm_3 \wedge dp_1 \wedge dp_2 \wedge dp_3}{\phi(m,p;\epsilon)},$$

where $\phi(m,p;\epsilon)$ *can be taken as the numerator of either of the functions* C_i, $i = 0, 1, 2, 3$.

Theorem 10. (Second order Wronskians HK basis) *The functions*

$$W_i^{(2)}(m,p) = \tilde{\tilde{m}}_i p_i - m_i \tilde{\tilde{p}}_i, \quad i = 1, 2, 3,$$

form a HK basis for the map Φ_f, *with a one-dimensional null space spanned by* $[C_1 : C_2 : C_3]$, *with the functions* C_i *given in* (4.30a)–(4.30c). *In other words, on orbits of the map* Φ_f *there holds:*

$$\sum_{i=1}^{3} C_i(\tilde{\tilde{m}}_i p_i - m_i \tilde{\tilde{p}}_i) = 0. \tag{4.31}$$

Theorem 11. (Higher order Wronskians HK bases)

1. *The functions* $W_i^{(3)}(m,p) = \tilde{\tilde{\tilde{m}}}_i p_i - m_i \tilde{\tilde{\tilde{p}}}_i$, $i = 1, 2, 3$, *form a HK basis for the map* Φ_f, *with a one-dimensional null space. On orbits of the map* Φ_f *there holds*

$$\sum_{i=1}^{3} D_i(\tilde{\tilde{\tilde{m}}}_i p_i - m_i \tilde{\tilde{\tilde{p}}}_i) = 0.$$

   The two integrals of motion $J_1 = D_1/D_3$ *and* $J_2 = D_2/D_3$ *are functionally independent.*

2. *The functions* $W_i^{(4)}(m,p) = \widetilde{\widetilde{\widetilde{m}}}_i p_i - m_i \widetilde{\widetilde{\widetilde{p}}}_i$, $i = 1,2,3$, *form a HK basis for the map* Φ_f, *with a one-dimensional null space. On orbits of the map* Φ_f *there holds*

$$\sum_{i=1}^{3} E_i(\widetilde{\widetilde{\widetilde{m}}}_i p_i - m_i \widetilde{\widetilde{\widetilde{p}}}_i) = 0.$$

The two integrals of motion $J_3 = E_1/E_3$ *and* $J_4 = E_2/E_3$ *are functionally independent.*

3. *Among the integrals of motion* $I_0, J_0, \ldots, J_4$ *four are functionally independent. In particular, each of the sets* $\{I_0, J_0, J_1, J_2\}$, $\{I_0, J_0, J_3, J_4\}$, $\{J_1, J_2, J_3, J_4\}$ *consists of four independent integrals of motion.*

5 The first Clebsch flow

We collect here the formulas for the first Clebsch flow, since this will be our main example. Equations of motion:

$$\begin{cases} \dot{m}_1 = (\omega_3 - \omega_2)p_2p_3, \\ \dot{m}_2 = (\omega_1 - \omega_3)p_3p_1, \\ \dot{m}_3 = (\omega_2 - \omega_1)p_1p_2, \\ \dot{p}_1 = m_3p_2 - m_2p_3, \\ \dot{p}_2 = m_1p_3 - m_3p_1, \\ \dot{p}_3 = m_2p_1 - m_1p_2. \end{cases} \tag{5.1}$$

The Wronskian relation satisfied on solutions of (5.1) is:

$$(\dot{m}_1p_1 - m_1\dot{p}_1) + (\dot{m}_2p_2 - m_2\dot{p}_2) + (\dot{m}_3p_3 - m_3\dot{p}_3) = 0. \tag{5.2}$$

Applying the Kahan-Hirota-Kimura scheme to the first Clebsch flow (5.1), we arrive at the following discretization:

$$\begin{cases} \widetilde{m}_1 - m_1 = \epsilon(\omega_3 - \omega_2)(\widetilde{p}_2p_3 + p_2\widetilde{p}_3), \\ \widetilde{m}_2 - m_2 = \epsilon(\omega_1 - \omega_3)(\widetilde{p}_3p_1 + p_3\widetilde{p}_1), \\ \widetilde{m}_3 - m_3 = \epsilon(\omega_2 - \omega_1)(\widetilde{p}_1p_2 + p_1\widetilde{p}_2), \\ \widetilde{p}_1 - p_1 = \epsilon(\widetilde{m}_3p_2 + m_3\widetilde{p}_2) - \epsilon(\widetilde{m}_2p_3 + m_2\widetilde{p}_3), \\ \widetilde{p}_2 - p_2 = \epsilon(\widetilde{m}_1p_3 + m_1\widetilde{p}_3) - \epsilon(\widetilde{m}_3p_1 + m_3\widetilde{p}_1), \\ \widetilde{p}_3 - p_3 = \epsilon(\widetilde{m}_2p_1 + m_2\widetilde{p}_1) - \epsilon(\widetilde{m}_1p_2 + m_1\widetilde{p}_2). \end{cases} \tag{5.3}$$

Linear system (5.3) defines a birational map $\Phi_f : \mathbb{R}^6 \to \mathbb{R}^6$, $(m,p) \mapsto (\widetilde{m}, \widetilde{p})$.

Theorem 12. (Quadratic-fractional integral, [12]) *The function*

$$I_0(m,p,\epsilon) = \frac{p_1^2 + p_2^2 + p_3^2}{1 - \epsilon^2(\omega_1p_1^2 + \omega_2p_2^2 + \omega_3p_3^2)} \tag{5.4}$$

is an integral of motion of the map $\Phi_f : (m, p) \mapsto (\widetilde{m}, \widetilde{p})$. *The set*

$$\Psi_0 = (p_1^2, p_2^2, p_3^2, 1) \tag{5.5}$$

is a HK-basis for the map Φ_f, *with a one-dimensional null-space*

$$K_{\Psi_0}(m, p) = [c_1 : c_2 : c_3 : -c_0], \tag{5.6}$$

where

$$\begin{aligned} c_1 &= 1 + \epsilon^2(\omega_1 - \omega_2)p_2^2 + \epsilon^2(\omega_1 - \omega_3)p_3^2, && (5.7a)\\ c_2 &= 1 + \epsilon^2(\omega_2 - \omega_1)p_1^2 + \epsilon^2(\omega_2 - \omega_3)p_3^2, && (5.7b)\\ c_3 &= 1 + \epsilon^2(\omega_3 - \omega_1)p_1^2 + \epsilon^2(\omega_3 - \omega_2)p_2^2, && (5.7c)\\ c_0 &= p_1^2 + p_2^2 + p_3^2. && (5.7d) \end{aligned}$$

Equivalently:

$$\begin{aligned} K_{\Psi_0}(m, p) &= [\,1 + \epsilon^2\omega_1 I_0 : 1 + \epsilon^2\omega_2 I_0 : 1 + \epsilon^2\omega_3 I_0 : -I_0\,] && (5.8a)\\ &= \left[\frac{1}{I_0} + \epsilon^2\omega_1 : \frac{1}{I_0} + \epsilon^2\omega_2 : \frac{1}{I_0} + \epsilon^2\omega_3 : -1\right]. && (5.8b) \end{aligned}$$

Theorem 13. (Discrete Wronskians HK basis, [13]) *The functions* $W_i^{(1)}(m, p) = \widetilde{m}_i p_i - m_i \widetilde{p}_i$, $i = 1, 2, 3$, *form a HK basis for the map* Φ_f, *with a one-dimensional null space spanned by* $[c_1 : c_2 : c_3]$, *with the functions* c_i *given in* (5.7a)–(5.7c). *In other words, on orbits of the map* Φ_f *there holds:*

$$\sum_{i=1}^{3} c_i(\widetilde{m}_i p_i - m_i \widetilde{p}_i) = 0. \tag{5.9}$$

Novel results, illustrating Observations 2 and 3, are as follows.

Theorem 14. (Bilinear-fractional integral) *The function*

$$J_0(m, p, \epsilon) = \frac{p_1\widetilde{p}_1 + p_2\widetilde{p}_2 + p_3\widetilde{p}_3}{1 + \epsilon^2(\omega_1 p_1\widetilde{p}_1 + \omega_2 p_2\widetilde{p}_2 + \omega_3 p_3\widetilde{p}_3)} \tag{5.10}$$

is an integral of motion of the map Φ_f. *The set*

$$\Psi_1 = (p_1\widetilde{p}_1, p_2\widetilde{p}_2, p_3\widetilde{p}_3, 1) \tag{5.11}$$

is a HK-basis for the map Φ_f, *with a one-dimensional null space*

$$K_{\Psi_1}(m, p) = [C_1 : C_2 : C_3 : -C_0], \tag{5.12}$$

where

$$\begin{aligned} C_1 &= 1 + \epsilon^2(\omega_2 - \omega_1)p_2\widetilde{p}_2 + \epsilon^2(\omega_3 - \omega_1)p_3\widetilde{p}_3, && (5.13a)\\ C_2 &= 1 + \epsilon^2(\omega_1 - \omega_2)p_1\widetilde{p}_1 + \epsilon^2(\omega_3 - \omega_2)p_3\widetilde{p}_3, && (5.13b)\\ C_3 &= 1 + \epsilon^2(\omega_1 - \omega_3)p_1\widetilde{p}_1 + \epsilon^2(\omega_2 - \omega_3)p_2\widetilde{p}_2, && (5.13c)\\ C_0 &= p_1\widetilde{p}_1 + p_2\widetilde{p}_2 + p_3\widetilde{p}_3. && (5.13d) \end{aligned}$$

Equivalently:

$$K_{\Psi_1}(m,p) = [\,1-\epsilon^2\omega_1 J_0 : 1-\epsilon^2\omega_2 J_0 : 1-\epsilon^2\omega_3 J_0 : -J_0\,] \tag{5.14a}$$

$$= \left[\frac{1}{J_0}-\epsilon^2\omega_1 : \frac{1}{J_0}-\epsilon^2\omega_2 : \frac{1}{J_0}-\epsilon^2\omega_3 : -1\right]. \tag{5.14b}$$

Remark 1. One can show that the numerators of the expressions C_i, $i=0,\dots,3$, are irreducible polynomials of m,p depending on ϵ^2 rather than on ϵ; thus they satisfy Observation 2.

Corollary 15. (Density of an invariant measure) *The map $\Phi_f(x;\epsilon)$ has an invariant measure*

$$\frac{dm_1\wedge dm_2\wedge dm_3\wedge dp_1\wedge dp_2\wedge dp_3}{\phi(m,p;\epsilon)},$$

where for $\phi(m,p;\epsilon)$ one can take the numerator of the function $p_1\widetilde{p}_1+p_2\widetilde{p}_2+p_3\widetilde{p}_3$, or the numerator of the function $1+\epsilon^2(\omega_1p_1\widetilde{p}_1+\omega_2p_2\widetilde{p}_2+\omega_3p_3\widetilde{p}_3)$ (the quotient of both densities is an integral of motion J_0).

Theorem 16. (Second order Wronskians HK basis) *The functions*

$$W_i^{(2)}(m,p)=\widetilde{\widetilde{m}}_ip_i-m_i\widetilde{\widetilde{p}}_i,\quad i=1,2,3,$$

form a HK basis for the map Φ_f, with a one-dimensional null space spanned by $[C_1:C_2:C_3]$, with the functions C_i given in (5.13a)–(5.13c). *In other words, on orbits of the map Φ_f there holds:*

$$\sum_{i=1}^{3}C_i(\widetilde{\widetilde{m}}_ip_i-m_i\widetilde{\widetilde{p}}_i)=0. \tag{5.15}$$

Finally, there holds a theorem which reads literally as Theorem 11 on higher order Wronskians HK bases.

We now turn to the proofs of the above results.

Proof of Theorem 12. We show that the function I_0 in (5.4) is an integral of motion, i.e., that

$$\frac{p_1^2+p_2^2+p_3^2}{1-\epsilon^2(\omega_1p_1^2+\omega_2p_2^2+\omega_3p_3^2)}=\frac{\widetilde{p}_1^2+\widetilde{p}_2^2+\widetilde{p}_3^2}{1-\epsilon^2(\omega_1\widetilde{p}_1^2+\omega_2\widetilde{p}_2^2+\omega_3\widetilde{p}_3^2)}.$$

This is equivalent to

$$\widetilde{p}_1^2-p_1^2+\widetilde{p}_2^2-p_2^2+\widetilde{p}_3^2-p_3^2$$

$$=\epsilon^2\left[(\omega_2-\omega_1)(\widetilde{p}_1^2p_2^2-\widetilde{p}_2^2p_1^2)+(\omega_3-\omega_2)(\widetilde{p}_2^2p_3^2-\widetilde{p}_3^2p_2^2)+(\omega_1-\omega_3)(\widetilde{p}_3^2p_1^2-\widetilde{p}_1^2p_3^2)\right].$$

On the left-hand side of this equation we replace $\widetilde{p}_i-p_i$ through the expressions from the last three equations of motion (5.3); on the right-hand side we replace

$\epsilon(\omega_k-\omega_j)(\widetilde{p}_j p_k+p_j\widetilde{p}_k)$ by $\widetilde{m}_i-m_i$, according to the first three equations of motion (5.3). This brings the equation we want to prove into the form

$$\begin{aligned}
&(\widetilde{p}_1+p_1)(\widetilde{m}_3p_2+m_3\widetilde{p}_2-\widetilde{m}_2p_3-m_2\widetilde{p}_3)+\\
&(\widetilde{p}_2+p_2)(\widetilde{m}_1p_3+m_1\widetilde{p}_3-\widetilde{m}_3p_1-m_3\widetilde{p}_1)+\\
&(\widetilde{p}_3+p_3)(\widetilde{m}_2p_1+m_2\widetilde{p}_1-\widetilde{m}_1p_2-m_1\widetilde{p}_2)\;=\\
&\quad=(\widetilde{p}_1p_2-p_1\widetilde{p}_2)(\widetilde{m}_3-m_3)+(\widetilde{p}_2p_3-p_2\widetilde{p}_3)(\widetilde{m}_1-m_1)+(\widetilde{p}_3p_1-p_3\widetilde{p}_1)(\widetilde{m}_2-m_2).
\end{aligned}$$

But this is an algebraic identity in twelve variables $m_k,p_k,\widetilde{m}_k,\widetilde{p}_k$. ■

Proof of Theorem 14. We show that the function J_0 in (5.10) is an integral of motion, i.e., that

$$\frac{p_1\underset{\sim}{p}_1+p_2\underset{\sim}{p}_2+p_3\underset{\sim}{p}_3}{1+\epsilon^2(\omega_1p_1\underset{\sim}{p}_1+\omega_2p_2\underset{\sim}{p}_2+\omega_3p_3\underset{\sim}{p}_3)}=\frac{\widetilde{p}_1p_1+\widetilde{p}_2p_2+\widetilde{p}_3p_3}{1+\epsilon^2(\omega_1\widetilde{p}_1p_1+\omega_2\widetilde{p}_2p_2+\omega_3\widetilde{p}_3p_3)}.$$

This is equivalent to

$$\begin{aligned}
&p_1(\widetilde{p}_1-\underset{\sim}{p}_1)+p_2(\widetilde{p}_2-\underset{\sim}{p}_2)+p_3(\widetilde{p}_3-\underset{\sim}{p}_3)\\
&=\;\epsilon^2\left[(\omega_1-\omega_2)(\widetilde{p}_1p_1p_2\underset{\sim}{p}_2-\widetilde{p}_2p_2p_1\underset{\sim}{p}_1)+(\omega_2-\omega_3)(\widetilde{p}_2p_2p_3\underset{\sim}{p}_3-\widetilde{p}_3p_3p_2\underset{\sim}{p}_2)\right.\\
&\qquad\left.+(\omega_3-\omega_1)(\widetilde{p}_3p_3p_1\underset{\sim}{p}_1-\widetilde{p}_1p_1p_3\underset{\sim}{p}_3)\right].
\end{aligned}$$

On the left-hand side of this equation we use

$$2p_i(\widetilde{p}_i-\underset{\sim}{p}_i)=(\widetilde{p}_i+p_i)(p_i-\underset{\sim}{p}_i)+(\widetilde{p}_i-p_i)(p_i+\underset{\sim}{p}_i),$$

and then replace $\widetilde{p}_i-p_i$ and $p_i-\underset{\sim}{p}_i$ through the expressions from the last three equations of motion (5.3). On the right-hand side we use

$$2(\widetilde{p}_jp_jp_k\underset{\sim}{p}_k-\widetilde{p}_kp_kp_j\underset{\sim}{p}_j)=(\widetilde{p}_jp_k+\widetilde{p}_kp_j)(p_j\underset{\sim}{p}_k-p_k\underset{\sim}{p}_j)+(\widetilde{p}_jp_k-\widetilde{p}_kp_j)(p_j\underset{\sim}{p}_k+p_k\underset{\sim}{p}_j),$$

and then replace $\epsilon(\omega_k-\omega_j)(\widetilde{p}_jp_k+p_j\widetilde{p}_k)$ by $\widetilde{m}_i-m_i$, and $\epsilon(\omega_k-\omega_j)(p_j\underset{\sim}{p}_k+p_k\underset{\sim}{p}_j)$ by $m_i-\underset{\sim}{m}_i$, according to the first three equations of motion (5.3). This brings the equation we want to prove into the form

$$\begin{aligned}
&(p_1+\underset{\sim}{p}_1)(\widetilde{m}_3p_2+m_3\widetilde{p}_2-\widetilde{m}_2p_3-m_2\widetilde{p}_3)\\
&+(\widetilde{p}_1+p_1)(m_3\underset{\sim}{p}_2+\underset{\sim}{m}_3p_2-m_2\underset{\sim}{p}_3-\underset{\sim}{m}_2p_3)\\
&+(p_2+\underset{\sim}{p}_2)(\widetilde{m}_1p_3+m_1\widetilde{p}_3-\widetilde{m}_3p_1-m_3\widetilde{p}_1)\\
&+(\widetilde{p}_2+p_2)(m_1\underset{\sim}{p}_3+\underset{\sim}{m}_1p_3-m_3\underset{\sim}{p}_1-\underset{\sim}{m}_3p_1)\\
&+(p_3+\underset{\sim}{p}_3)(\widetilde{m}_2p_1+m_2\widetilde{p}_1-\widetilde{m}_1p_2-m_1\widetilde{p}_2)\\
&+(\widetilde{p}_3+p_3)(m_2\underset{\sim}{p}_1+\underset{\sim}{m}_2p_1-m_1\underset{\sim}{p}_2-\underset{\sim}{m}_1p_2)\\
&\quad=(p_2\underset{\sim}{p}_1-\underset{\sim}{p}_2p_1)(\widetilde{m}_3-m_3)+(\widetilde{p}_2p_1-p_2\widetilde{p}_1)(m_3-\underset{\sim}{m}_3)\\
&\quad\;+(p_3\underset{\sim}{p}_2-\underset{\sim}{p}_3p_2)(\widetilde{m}_1-m_1)+(\widetilde{p}_3p_2-p_3\widetilde{p}_2)(m_1-\underset{\sim}{m}_1)\\
&\quad\;+(p_1\underset{\sim}{p}_3-\underset{\sim}{p}_1p_3)(\widetilde{m}_2-m_2)+(\widetilde{p}_1p_3-p_1\widetilde{p}_3)(m_2-\underset{\sim}{m}_2).
\end{aligned}$$

On the left-hand side there are many cancellations, so that we get:

$$
\begin{aligned}
&p_1(m_3\widetilde{p}_2 - m_2\widetilde{p}_3) + p_1(m_3\underset{\sim}{p}_2 - m_2\underset{\sim}{p}_3) + \underset{\sim}{p}_1(\widetilde{m}_3 p_2 - \widetilde{m}_2 p_3) + \widetilde{p}_1(\underset{\sim}{m}_3 p_2 - \underset{\sim}{m}_2 p_3)\\
&+p_2(m_1\widetilde{p}_3 - m_3\widetilde{p}_1) + p_2(m_1\underset{\sim}{p}_3 - m_3\underset{\sim}{p}_1) + \underset{\sim}{p}_2(\widetilde{m}_1 p_3 - \widetilde{m}_3 p_1) + \widetilde{p}_2(\underset{\sim}{m}_1 p_3 - \underset{\sim}{m}_3 p_1)\\
&+p_3(m_2\widetilde{p}_1 - m_1\widetilde{p}_2) + p_3(m_2\underset{\sim}{p}_1 - m_1\underset{\sim}{p}_2) + \underset{\sim}{p}_3(\widetilde{m}_2 p_1 - \widetilde{m}_1 p_2) + \widetilde{p}_3(\underset{\sim}{m}_2 p_1 - \underset{\sim}{m}_1 p_2)\\
&\quad = (p_2\underset{\sim}{p}_1 - \underset{\sim}{p}_2 p_1)(\widetilde{m}_3 - m_3) + (\widetilde{p}_2 p_1 - p_2\widetilde{p}_1)(m_3 - \underset{\sim}{m}_3)\\
&\qquad +(p_3\underset{\sim}{p}_2 - \underset{\sim}{p}_3 p_2)(\widetilde{m}_1 - m_1) + (\widetilde{p}_3 p_2 - p_3\widetilde{p}_2)(m_1 - \underset{\sim}{m}_1)\\
&\qquad +(p_1\underset{\sim}{p}_3 - \underset{\sim}{p}_1 p_3)(\widetilde{m}_2 - m_2) + (\widetilde{p}_1 p_3 - p_1\widetilde{p}_3)(m_2 - \underset{\sim}{m}_2).
\end{aligned}
$$

But this is an algebraic identity in the variables $m_k, p_k, \widetilde{m}_k, \widetilde{p}_k, \underset{\sim}{m}_k, \underset{\sim}{p}_k$. ∎

Proof of Theorem 13 is based on the following four identities which hold on orbits of the map Φ_f:

$$\sum_{i=1}^{3} c_i\widetilde{m}_i p_i = \sum_{i=1}^{3} C_i m_i p_i, \tag{5.16a}$$

$$\sum_{i=1}^{3} c_i m_i\widetilde{p}_i = \sum_{i=1}^{3} C_i m_i p_i, \tag{5.16b}$$

$$\sum_{i=1}^{3} \widetilde{c}_i m_i\widetilde{p}_i = \sum_{i=1}^{3} C_i\widetilde{m}_i\widetilde{p}_i, \tag{5.16c}$$

$$\sum_{i=1}^{3} \widetilde{c}_i\widetilde{m}_i p_i = \sum_{i=1}^{3} C_i\widetilde{m}_i\widetilde{p}_i. \tag{5.16d}$$

Indeed, from these relations there follows immediately:

$$\sum_{i=1}^{3} c_i(\widetilde{m}_i p_i - m_i\widetilde{p}_i) = \sum_{i=1}^{3} \widetilde{c}_i(\widetilde{m}_i p_i - m_i\widetilde{p}_i) = 0,$$

which proves the theorem. ∎

Proof of formula (5.16a). Using the first three equations of motion (5.3), we compute:

$$
\begin{aligned}
&\sum_{i=1}^{3}(\widetilde{m}_i - m_i)p_i\\
&= \epsilon(\omega_3-\omega_2)(p_1p_2\widetilde{p}_3 + p_1p_3\widetilde{p}_2) + \epsilon(\omega_1-\omega_3)(p_2p_3\widetilde{p}_1 + p_2p_1\widetilde{p}_3)\\
&\quad +\epsilon(\omega_2-\omega_1)(p_3p_1\widetilde{p}_2 + p_3p_2\widetilde{p}_1)\\
&= \epsilon(\omega_1-\omega_2)p_1p_2\widetilde{p}_3 + \epsilon(\omega_2-\omega_3)p_2p_3\widetilde{p}_1 + \epsilon(\omega_3-\omega_1)p_3p_1\widetilde{p}_2\\
&= \epsilon(\omega_1-\omega_2)p_1p_2(\widetilde{p}_3 - p_3) + \epsilon(\omega_2-\omega_3)p_2p_3(\widetilde{p}_1 - p_1) + \epsilon(\omega_3-\omega_1)p_3p_1(\widetilde{p}_2 - p_2).
\end{aligned}
$$

Using the last three equations of motion (5.3), we continue the computation:

$$\begin{aligned}
&= \epsilon^2(\omega_1-\omega_2)p_1p_2\big((\widetilde{m}_2p_1+m_2\widetilde{p}_1)-(\widetilde{m}_1p_2+m_1\widetilde{p}_2)\big)\\
&\quad+\epsilon^2(\omega_2-\omega_3)p_2p_3\big((\widetilde{m}_3p_2+m_3\widetilde{p}_2)-(\widetilde{m}_2p_3+m_2\widetilde{p}_3)\big)\\
&\quad+\epsilon^2(\omega_3-\omega_1)p_3p_1\big((\widetilde{m}_1p_3+m_1\widetilde{p}_3)-(\widetilde{m}_3p_1+m_3\widetilde{p}_1)\big)
\end{aligned}$$

$$\begin{aligned}
&= \big(\epsilon^2(\omega_2-\omega_1)p_2^2+\epsilon^2(\omega_3-\omega_1)p_3^2\big)\widetilde{m}_1p_1\\
&\quad+\big(\epsilon^2(\omega_3-\omega_2)p_3^2+\epsilon^2(\omega_1-\omega_2)p_1^2\big)\widetilde{m}_2p_2\\
&\quad+\big(\epsilon^2(\omega_1-\omega_3)p_1^2+\epsilon^2(\omega_2-\omega_3)p_2^2\big)\widetilde{m}_3p_3\\
&\quad+\big(\epsilon^2(\omega_2-\omega_1)\widetilde{p}_2p_2+\epsilon^2(\omega_3-\omega_1)\widetilde{p}_3p_3\big)m_1p_1\\
&\quad+\big(\epsilon^2(\omega_3-\omega_2)\widetilde{p}_3p_3+\epsilon^2(\omega_1-\omega_2)\widetilde{p}_1p_1\big)m_2p_2\\
&\quad+\big(\epsilon^2(\omega_1-\omega_3)\widetilde{p}_1p_1+\epsilon^2(\omega_2-\omega_3)\widetilde{p}_2p_2\big)m_3p_3.
\end{aligned}$$

This proves (5.16a). We remark that this relation is the discrete time analog of $\sum_{i=1}^3 \dot{m}_ip_i=0$; the crucial point in this analogy is that $\sum_{i=1}^3(\widetilde{m}_i-m_i)p_i$ turns out to be of order ϵ^2. ∎

Proof of formula (5.16b). Using the first three equations of motion (5.3), we compute:

$$\begin{aligned}
&\sum_{i=1}^3 m_i(\widetilde{p}_i-p_i)\\
&= \epsilon m_1\big((\widetilde{m}_3p_2+m_3\widetilde{p}_2)-(\widetilde{m}_2p_3+m_2\widetilde{p}_3)\big)\\
&\quad+\epsilon m_2\big((\widetilde{m}_1p_3+m_1\widetilde{p}_3)-(\widetilde{m}_3p_1+m_3\widetilde{p}_1)\big)\\
&\quad+\epsilon m_3\big((\widetilde{m}_2p_1+m_2\widetilde{p}_1)-(\widetilde{m}_1p_2+m_1\widetilde{p}_2)\big)
\end{aligned}$$

$$= \epsilon(m_1p_2-m_2p_1)\widetilde{m}_3+\epsilon(m_2p_3-m_3p_2)\widetilde{m}_1+\epsilon(m_3p_1-m_1p_3)\widetilde{m}_2$$

$$\begin{aligned}
&= \epsilon(m_1p_2-m_2p_1)(\widetilde{m}_3-m_3)+\epsilon(m_2p_3-m_3p_2)(\widetilde{m}_1-m_1)\\
&\quad+\epsilon(m_3p_1-m_1p_3)(\widetilde{m}_2-m_2).
\end{aligned}$$

Using the last three equations of motion (5.3), we continue:

$$\begin{aligned}
&= \epsilon^2(\omega_2-\omega_1)(m_1p_2-m_2p_1)(p_1\widetilde{p}_2+p_2\widetilde{p}_1)\\
&\quad+\epsilon^2(\omega_3-\omega_2)(m_2p_3-m_3p_2)(p_2\widetilde{p}_3+p_3\widetilde{p}_2)\\
&\quad+\epsilon^2(\omega_1-\omega_3)(m_3p_1-m_1p_3)(p_3\widetilde{p}_1+p_1\widetilde{p}_3)
\end{aligned}$$

$$\begin{aligned}
&= \big(\epsilon^2(\omega_2-\omega_1)p_2^2+\epsilon^2(\omega_3-\omega_1)p_3^2\big)m_1\widetilde{p}_1\\
&\quad+\big(\epsilon^2(\omega_3-\omega_2)p_3^2+\epsilon^2(\omega_1-\omega_2)p_1^2\big)m_2\widetilde{p}_2\\
&\quad+\big(\epsilon^2(\omega_1-\omega_3)p_1^2+\epsilon^2(\omega_2-\omega_3)p_2^2\big)m_3\widetilde{p}_3\\
&\quad+\big(\epsilon^2(\omega_2-\omega_1)\widetilde{p}_2p_2+\epsilon^2(\omega_3-\omega_1)\widetilde{p}_3p_3\big)m_1p_1\\
&\quad+\big(\epsilon^2(\omega_3-\omega_2)\widetilde{p}_3p_3+\epsilon^2(\omega_1-\omega_2)\widetilde{p}_1p_1\big)m_2p_2\\
&\quad+\big(\epsilon^2(\omega_1-\omega_3)\widetilde{p}_1p_1+\epsilon^2(\omega_2-\omega_3)\widetilde{p}_2p_2\big)m_3p_3.
\end{aligned}$$

This proves (5.16b). Again, this relation is the discrete time analog of $\sum_{i=1}^{3} m_i\dot{p}_i = 0$, since $\sum_{i=1}^{3} m_i(\widetilde{p}_i - p_i)$ turns out to be of order ϵ^2. ■

Proof of formula (5.16c). We start by using the first three equations of motion (5.3):

$$\begin{aligned}
&\sum_{i=1}^{3}(\widetilde{m}_i - m_i)\widetilde{p}_i \\
&\quad= \epsilon(\omega_3-\omega_2)(\widetilde{p}_1\widetilde{p}_3p_2+\widetilde{p}_1\widetilde{p}_2p_3)+\epsilon(\omega_1-\omega_3)(\widetilde{p}_2\widetilde{p}_1p_3+\widetilde{p}_2\widetilde{p}_3p_1) \\
&\qquad+\epsilon(\omega_2-\omega_1)(\widetilde{p}_3\widetilde{p}_2p_1+\widetilde{p}_3\widetilde{p}_1p_2) \\
&\quad= \epsilon(\omega_1-\omega_2)\widetilde{p}_1\widetilde{p}_2p_3+\epsilon(\omega_2-\omega_3)\widetilde{p}_2\widetilde{p}_3p_1+\epsilon(\omega_3-\omega_1)\widetilde{p}_3\widetilde{p}_1p_2 \\
&\quad= \epsilon(\omega_2-\omega_1)\widetilde{p}_1\widetilde{p}_2(\widetilde{p}_3-p_3)+\epsilon(\omega_3-\omega_2)\widetilde{p}_2\widetilde{p}_3(\widetilde{p}_1-p_1)+\epsilon(\omega_1-\omega_3)\widetilde{p}_3\widetilde{p}_1(\widetilde{p}_2-p_2),
\end{aligned}$$

and continue by using the last three equations of motion (5.3):

$$\begin{aligned}
&= \epsilon^2(\omega_2-\omega_1)\widetilde{p}_1\widetilde{p}_2\big((\widetilde{m}_2p_1+m_2\widetilde{p}_1)-(\widetilde{m}_1p_2+m_1\widetilde{p}_2)\big) \\
&\quad+\epsilon^2(\omega_3-\omega_2)\widetilde{p}_2\widetilde{p}_3\big((\widetilde{m}_3p_2+m_3\widetilde{p}_2)-(\widetilde{m}_2p_3+m_2\widetilde{p}_3)\big) \\
&\quad+\epsilon^2(\omega_1-\omega_3)\widetilde{p}_3\widetilde{p}_1\big((\widetilde{m}_1p_3+m_1\widetilde{p}_3)-(\widetilde{m}_3p_1+m_3\widetilde{p}_1)\big) \\
&= \big(\epsilon^2(\omega_1-\omega_2)\widetilde{p}_2^2+\epsilon^2(\omega_1-\omega_3)\widetilde{p}_3^2\big)m_1\widetilde{p}_1 \\
&\quad+\big(\epsilon^2(\omega_2-\omega_3)\widetilde{p}_3^2+\epsilon^2(\omega_2-\omega_1)\widetilde{p}_1^2\big)m_2\widetilde{p}_2 \\
&\quad+\big(\epsilon^2(\omega_3-\omega_1)\widetilde{p}_1^2+\epsilon^2(\omega_3-\omega_2)\widetilde{p}_2^2\big)m_3\widetilde{p}_3 \\
&\quad+\big(\epsilon^2(\omega_1-\omega_2)\widetilde{p}_2p_2+\epsilon^2(\omega_1-\omega_3)\widetilde{p}_3p_3\big)\widetilde{m}_1\widetilde{p}_1 \\
&\quad+\big(\epsilon^2(\omega_2-\omega_3)\widetilde{p}_3p_3+\epsilon^2(\omega_2-\omega_1)\widetilde{p}_1p_1\big)\widetilde{m}_2\widetilde{p}_2 \\
&\quad+\big(\epsilon^2(\omega_3-\omega_1)\widetilde{p}_1p_1+\epsilon^2(\omega_3-\omega_2)\widetilde{p}_2p_2\big)\widetilde{m}_3\widetilde{p}_3.
\end{aligned}$$

This proves (5.16c). Again, this relation is the discrete time analog of $\sum_{i=1}^{3}\dot{m}_ip_i = 0$. ■

Proof of formula (5.16d). We compute:

$$\begin{aligned}
&\sum_{i=1}^{3}\widetilde{m}_i(\widetilde{p}_i-p_i) \\
&\quad= \epsilon\widetilde{m}_1\big((\widetilde{m}_3p_2+m_3\widetilde{p}_2)-(\widetilde{m}_2p_3+m_2\widetilde{p}_3)\big) \\
&\qquad+\epsilon\widetilde{m}_2\big((\widetilde{m}_1p_3+m_1\widetilde{p}_3)-(\widetilde{m}_3p_1+m_3\widetilde{p}_1)\big) \\
&\qquad+\epsilon\widetilde{m}_3\big((\widetilde{m}_2p_1+m_2\widetilde{p}_1)-(\widetilde{m}_1p_2+m_1\widetilde{p}_2)\big) \\
&\quad= \epsilon(\widetilde{m}_1\widetilde{p}_2-\widetilde{m}_2\widetilde{p}_1)m_3+\epsilon(\widetilde{m}_2\widetilde{p}_3-\widetilde{m}_3\widetilde{p}_2)m_1+\epsilon(\widetilde{m}_3\widetilde{p}_1-\widetilde{m}_1\widetilde{p}_3)m_2
\end{aligned}$$

$$\begin{aligned}
&= \epsilon(\tilde{m}_2\tilde{p}_1 - \tilde{m}_1\tilde{p}_2)(\tilde{m}_3 - m_3) \\
&\quad +\epsilon(\tilde{m}_3\tilde{p}_2 - \tilde{m}_2\tilde{p}_3)(\tilde{m}_1 - m_1) \\
&\quad +\epsilon(\tilde{m}_1\tilde{p}_3 - \tilde{m}_3\tilde{p}_1)(\tilde{m}_2 - m_2) \\
&= \epsilon^2(\omega_2 - \omega_1)(\tilde{m}_2\tilde{p}_1 - \tilde{m}_1\tilde{p}_2)(p_1\tilde{p}_2 + p_2\tilde{p}_1) \\
&\quad +\epsilon^2(\omega_3 - \omega_2)(\tilde{m}_3\tilde{p}_2 - \tilde{m}_2\tilde{p}_3)(p_2\tilde{p}_3 + p_3\tilde{p}_2) \\
&\quad +\epsilon^2(\omega_1 - \omega_3)(\tilde{m}_1\tilde{p}_3 - \tilde{m}_3\tilde{p}_1)(p_3\tilde{p}_1 + p_1\tilde{p}_3) \\
&= \big(\epsilon^2(\omega_1 - \omega_2)\tilde{p}_2^2 + \epsilon^2(\omega_2 - \omega_3)\tilde{p}_3^2\big)\tilde{m}_1 p_1 \\
&\quad +\big(\epsilon^2(\omega_2 - \omega_3)\tilde{p}_3^2 + \epsilon^2(\omega_2 - \omega_1)\tilde{p}_1^2\big)\tilde{m}_2 p_2 \\
&\quad +\big(\epsilon^2(\omega_3 - \omega_1)\tilde{p}_1^2 + \epsilon^2(\omega_3 - \omega_2)\tilde{p}_2^2\big)\tilde{m}_3 p_3 \\
&\quad +\big(\epsilon^2(\omega_1 - \omega_2)\tilde{p}_2 p_2 + \epsilon^2(\omega_1 - \omega_3)\tilde{p}_3 p_3\big)\tilde{m}_1\tilde{p}_1 \\
&\quad +\big(\epsilon^2(\omega_2 - \omega_3)\tilde{p}_3 p_3 + \epsilon^2(\omega_2 - \omega_1)\tilde{p}_1 p_1\big)\tilde{m}_2\tilde{p}_2 \\
&\quad +\big(\epsilon^2(\omega_3 - \omega_1)\tilde{p}_1 p_1 + \epsilon^2(\omega_3 - \omega_2)\tilde{p}_2 p_2\big)\tilde{m}_3\tilde{p}_3.
\end{aligned}$$

This proves (5.16d) and provides us with another discrete time analog of the formula $\sum_{i=1}^3 m_i\dot{p}_i = 0$. ■

Corollary 17. *The function*

$$K(m,p) = \sum_{i=1}^{3} \frac{C_i}{C_0}\frac{m_i p_i}{c_0} \tag{5.17}$$

is an integral of motion of the map Φ_f.

Proof. Compare (5.16a) with (5.16d), taking into account that c_i/c_0 are integrals of motion, that is, $c_i/c_0 = \tilde{c}_i/\tilde{c}_0$. We arrive at

$$\sum_{i=1}^{3} C_i \frac{m_i p_i}{c_0} = \sum_{i=1}^{3} C_i \frac{\tilde{m}_i\tilde{p}_i}{\tilde{c}_0}.$$

Since C_i/C_0 are integrals of motion, that is $C_i/C_0 = \tilde{C}_i/\tilde{C}_0$, we see that

$$\sum_{i=1}^{3} \frac{C_i}{C_0}\frac{m_i p_i}{c_0} = \sum_{i=1}^{3} \frac{\tilde{C}_i}{\tilde{C}_0}\frac{\tilde{m}_i\tilde{p}_i}{\tilde{c}_0}.$$

This proves the statement. ■

Proof of Theorem 16. We start with

$$\begin{aligned}
&\sum_{i=1}^{3}(\tilde{\tilde{m}}_i - \tilde{m}_i)p_i + \sum_{i=1}^{3}(\tilde{m}_i - m_i)\tilde{\tilde{p}}_i \\
&= \epsilon(\omega_3 - \omega_2)(\tilde{\tilde{p}}_2\tilde{p}_3 + \tilde{p}_2\tilde{\tilde{p}}_3)p_1 + \epsilon(\omega_1 - \omega_3)(\tilde{\tilde{p}}_3\tilde{p}_1 + \tilde{p}_3\tilde{\tilde{p}}_1)p_2 \\
&\quad +\epsilon(\omega_2 - \omega_1)(\tilde{\tilde{p}}_1\tilde{p}_2 + \tilde{p}_1\tilde{\tilde{p}}_2)p_3 \\
&\quad +\epsilon(\omega_3 - \omega_2)(\tilde{p}_2 p_3 + p_2\tilde{p}_3)\tilde{\tilde{p}}_1 + \epsilon(\omega_1 - \omega_3)(\tilde{p}_3 p_1 + p_3\tilde{p}_1)\tilde{\tilde{p}}_2 \\
&\quad +\epsilon(\omega_2 - \omega_1)(\tilde{p}_1 p_2 + p_1\tilde{p}_2)\tilde{\tilde{p}}_3
\end{aligned}$$

$$\begin{aligned}
&= \epsilon(\omega_3-\omega_1)\tilde{\tilde{p}}_1\tilde{p}_2p_3+\epsilon(\omega_1-\omega_2)\tilde{\tilde{p}}_1\tilde{p}_3p_2+\epsilon(\omega_1-\omega_2)\tilde{\tilde{p}}_2\tilde{p}_3p_1\\
&\quad+\epsilon(\omega_2-\omega_3)\tilde{\tilde{p}}_2\tilde{p}_1p_3+\epsilon(\omega_2-\omega_3)\tilde{\tilde{p}}_3\tilde{p}_1p_2+\epsilon(\omega_3-\omega_1)\tilde{\tilde{p}}_3\tilde{p}_2p_1
\end{aligned}$$

$$\begin{aligned}
&= \epsilon(\omega_3-\omega_2)\tilde{\tilde{p}}_1p_2p_3+\epsilon(\omega_1-\omega_3)\tilde{\tilde{p}}_2p_3p_1+\epsilon(\omega_2-\omega_1)\tilde{\tilde{p}}_3p_1p_2\\
&\quad+\epsilon(\omega_3-\omega_1)\tilde{\tilde{p}}_1(\tilde{p}_2-p_2)p_3+\epsilon(\omega_1-\omega_2)\tilde{\tilde{p}}_1(\tilde{p}_3-p_3)p_2\\
&\quad+\epsilon(\omega_1-\omega_2)\tilde{\tilde{p}}_2(\tilde{p}_3-p_3)p_1+\epsilon(\omega_2-\omega_3)\tilde{\tilde{p}}_2(\tilde{p}_1-p_1)p_3\\
&\quad+\epsilon(\omega_2-\omega_3)\tilde{\tilde{p}}_3(\tilde{p}_1-p_1)p_2+\epsilon(\omega_3-\omega_1)\tilde{\tilde{p}}_3(\tilde{p}_2-p_2)p_1.
\end{aligned}$$

This is equal to

$$\begin{aligned}
&= \epsilon(\omega_3-\omega_2)(\tilde{\tilde{p}}_1-\tilde{p}_1)p_2p_3+\epsilon(\omega_1-\omega_3)(\tilde{\tilde{p}}_2-\tilde{p}_2)p_3p_1+\epsilon(\omega_2-\omega_1)(\tilde{\tilde{p}}_3-\tilde{p}_3)p_1p_2\\
&\quad+\epsilon(\omega_3-\omega_2)\tilde{p}_1p_2p_3+\epsilon(\omega_1-\omega_3)\tilde{p}_2p_3p_1+\epsilon(\omega_2-\omega_1)\tilde{p}_3p_1p_2\\
&\quad+\epsilon(\omega_3-\omega_1)\tilde{\tilde{p}}_1(\tilde{p}_2-p_2)p_3+\epsilon(\omega_1-\omega_2)\tilde{\tilde{p}}_1(\tilde{p}_3-p_3)p_2+\epsilon(\omega_1-\omega_2)\tilde{\tilde{p}}_2(\tilde{p}_3-p_3)p_1\\
&\quad+\epsilon(\omega_2-\omega_3)\tilde{\tilde{p}}_2(\tilde{p}_1-p_1)p_3+\epsilon(\omega_2-\omega_3)\tilde{\tilde{p}}_3(\tilde{p}_1-p_1)p_2+\epsilon(\omega_3-\omega_1)\tilde{\tilde{p}}_3(\tilde{p}_2-p_2)p_1.
\end{aligned}$$

Here the second line is equal to

$$-\sum_{i=1}^{3}(\tilde{m}_i-m_i)p_i,$$

so that, upon use of equations of motion, we find:

$$\begin{aligned}
&\sum_{i=1}^{3}(\tilde{\tilde{m}}_ip_i-m_i\tilde{\tilde{p}}_i)-\sum_{i=1}^{3}m_ip_i+\sum_{i=1}^{3}\tilde{m}_i\tilde{p}_i\\
&= \epsilon^2(\omega_3-\omega_2)\big((\tilde{\tilde{m}}_3\tilde{p}_2+\tilde{m}_3\tilde{\tilde{p}}_2)-(\tilde{\tilde{m}}_2\tilde{p}_3+\tilde{m}_2\tilde{\tilde{p}}_3)\big)p_2p_3\\
&\quad+\epsilon^2(\omega_1-\omega_3)\big((\tilde{\tilde{m}}_1\tilde{p}_3+\tilde{m}_1\tilde{\tilde{p}}_3)-(\tilde{\tilde{m}}_3\tilde{p}_1+\tilde{m}_3\tilde{\tilde{p}}_1)\big)p_3p_1\\
&\quad+\epsilon^2(\omega_2-\omega_1)\big((\tilde{\tilde{m}}_2\tilde{p}_1+\tilde{m}_2\tilde{\tilde{p}}_1)-(\tilde{\tilde{m}}_1\tilde{p}_2+\tilde{m}_1\tilde{\tilde{p}}_2)\big)p_1p_2\\
&\quad+\epsilon^2(\omega_3-\omega_1)\tilde{\tilde{p}}_1\big((\tilde{m}_1p_3+m_1\tilde{p}_3)-(\tilde{m}_3p_1+m_3\tilde{p}_1)\big)p_3\\
&\quad+\epsilon^2(\omega_1-\omega_2)\tilde{\tilde{p}}_1\big((\tilde{m}_2p_1+m_2\tilde{p}_1)-(\tilde{m}_1p_2+m_1\tilde{p}_2)\big)p_2\\
&\quad+\epsilon^2(\omega_1-\omega_2)\tilde{\tilde{p}}_2\big((\tilde{m}_2p_1+m_2\tilde{p}_1)-(\tilde{m}_1p_2+m_1\tilde{p}_2)\big)p_1\\
&\quad+\epsilon^2(\omega_2-\omega_3)\tilde{\tilde{p}}_2\big((\tilde{m}_3p_2+m_3\tilde{p}_2)-(\tilde{m}_2p_3+m_2\tilde{p}_3)\big)p_3\\
&\quad+\epsilon^2(\omega_2-\omega_3)\tilde{\tilde{p}}_3\big((\tilde{m}_3p_2+m_3\tilde{p}_2)-(\tilde{m}_2p_3+m_2\tilde{p}_3)\big)p_2\\
&\quad+\epsilon^2(\omega_3-\omega_1)\tilde{\tilde{p}}_3\big((\tilde{m}_1p_3+m_1\tilde{p}_3)-(\tilde{m}_3p_1+m_3\tilde{p}_1)\big)p_1.
\end{aligned}$$

(So, the strategy of transformations is: leave one variable with double tilde, one with single tilde and two without tilde.) Collecting terms, we have:

$$
\begin{aligned}
&\sum_{i=1}^{3}(\tilde{\tilde{m}}_i p_i - m_i\tilde{\tilde{p}}_i) - \sum_{i=1}^{3} m_i p_i + \sum_{i=1}^{3} \tilde{\tilde{m}}_i\tilde{\tilde{p}}_i \\
&= \tilde{\tilde{m}}_1 p_1\Big(\epsilon^2(\omega_1-\omega_3)\tilde{p}_3 p_3 + \epsilon^2(\omega_1-\omega_2)\tilde{p}_2 p_2\Big) \\
&\quad + \tilde{\tilde{m}}_2 p_2\Big(\epsilon^2(\omega_2-\omega_3)\tilde{p}_3 p_3 + \epsilon^2(\omega_2-\omega_1)\tilde{p}_1 p_1\Big) \\
&\quad + \tilde{\tilde{m}}_3 p_3\Big(\epsilon^2(\omega_3-\omega_2)\tilde{p}_2 p_2 + \epsilon^2(\omega_3-\omega_1)\tilde{p}_1 p_1\Big) \\
&\quad + m_1\tilde{\tilde{p}}_1\Big(\epsilon^2(\omega_3-\omega_1)\tilde{p}_3 p_3 + \epsilon^2(\omega_2-\omega_1)\tilde{p}_2 p_2\Big) \\
&\quad + m_2\tilde{\tilde{p}}_2\Big(\epsilon^2(\omega_1-\omega_2)\tilde{p}_1 p_1 + \epsilon^2(\omega_3-\omega_2)\tilde{p}_3 p_3\Big) \\
&\quad + m_3\tilde{\tilde{p}}_3\Big(\epsilon^2(\omega_2-\omega_3)\tilde{p}_2 p_2 + \epsilon^2(\omega_1-\omega_3)\tilde{p}_1 p_1\Big) \\
&\quad + m_1 p_1\Big(\epsilon^2(\omega_2-\omega_1)\tilde{\tilde{p}}_2\tilde{p}_2 + \epsilon^2(\omega_3-\omega_1)\tilde{\tilde{p}}_3\tilde{p}_3\Big) \\
&\quad + m_2 p_2\Big(\epsilon^2(\omega_1-\omega_2)\tilde{\tilde{p}}_1\tilde{p}_1 + \epsilon^2(\omega_3-\omega_2)\tilde{\tilde{p}}_3\tilde{p}_3\Big) \\
&\quad + m_3 p_3\Big(\epsilon^2(\omega_1-\omega_3)\tilde{\tilde{p}}_1\tilde{p}_1 + \epsilon^2(\omega_2-\omega_3)\tilde{\tilde{p}}_2\tilde{p}_2\Big) \\
&\quad + \tilde{\tilde{m}}_1\tilde{\tilde{p}}_1\Big(\epsilon^2(\omega_2-\omega_1)p_2^2 + \epsilon^2(\omega_3-\omega_1)p_3^2\Big) \\
&\quad + \tilde{\tilde{m}}_2\tilde{\tilde{p}}_2\Big(\epsilon^2(\omega_1-\omega_2)p_1^2 + \epsilon^2(\omega_3-\omega_2)p_3^2\Big) \\
&\quad + \tilde{\tilde{m}}_3\tilde{\tilde{p}}_3\Big(\epsilon^2(\omega_2-\omega_3)p_2^2 + \epsilon^2(\omega_1-\omega_3)p_1^2\Big).
\end{aligned}
$$

This can be put as

$$\sum_{i=1}^{3} C_i(\tilde{\tilde{m}}_i p_i - m_i\tilde{\tilde{p}}_i) = \sum_{i=1}^{3}\tilde{C}_i m_i p_i - \sum_{i=1}^{3} c_i\tilde{\tilde{m}}_i\tilde{\tilde{p}}_i.$$

From (5.17), (5.16c), and from relations $\tilde{C}_i/\tilde{C}_0 = C_i/C_0$ and $c_i/c_0 = \tilde{c}_i/\tilde{c}_0$, we derive:

$$\sum_{i=1}^{3}\tilde{C}_i m_i p_i = \tilde{C}_0\cdot c_0 K(m,p), \quad \sum_{i=1}^{3} c_i\tilde{\tilde{m}}_i\tilde{\tilde{p}}_i = c_0\cdot\tilde{C}_0 K(\tilde{\tilde{m}},\tilde{\tilde{p}}).$$

By Corollary 17, $K(m,p) = K(\tilde{\tilde{m}},\tilde{\tilde{p}})$. This finishes the proof. ■

As for Theorem 11, at present we only have a proof based on symbolic computations by Maple, even for the first Clebsch flow.

6 The Kirchhoff case

The *Kirchhoff case* of the motion of the rigid body in an ideal fluid corresponds to the following values of the parameters in (4.3), (4.5):

$$a_1 = a_2, \quad b_1 = b_2. \tag{6.1}$$

Equations of motion read:

$$\begin{cases} \dot{m}_1 = (a_3 - a_1)m_2m_3 + (b_3 - b_1)p_2p_3, \\ \dot{m}_2 = (a_1 - a_3)m_1m_3 + (b_1 - b_3)p_1p_3, \\ \dot{m}_3 = 0, \\ \dot{p}_1 = a_3p_2m_3 - a_1p_3m_2, \\ \dot{p}_2 = a_1p_3m_1 - a_3p_1m_3, \\ \dot{p}_3 = a_1(p_1m_2 - p_2m_1). \end{cases} \tag{6.2}$$

Thus, m_3 is an obvious fourth integral, due to the rotational symmetry of the system. It is easy to see that for (6.1) the Clebsch condition (4.6) is satisfied, as well. Thus, formally the Kirchhoff case is the particular case of the Clebsch case. One can choose the parameters ω_i in (4.7) as

$$\omega_1 = \omega_2 = \frac{b_1}{a_1}, \quad \omega_3 = \frac{b_3}{a_1}. \tag{6.3}$$

Correspondingly, integral H_1 becomes proportional to

$$a_1H_1 = a_1(m_1^2 + m_2^2 + m_3^2) + b_1(p_1^2 + p_2^2) + b_3p_3^2.$$

Taking into account that the Hamilton function H is given by

$$2H = a_1(m_1^2 + m_2^2) + a_3m_3^2 + b_1(p_1^2 + p_2^2) + b_3p_3^2,$$

we see that the fourth integral H_1 can be replaced just by m_3^2.

The Wronskian relation satisfied on solutions of (6.2) is:

$$(\dot{m}_1p_1 - m_1\dot{p}_1) + (\dot{m}_2p_2 - m_2\dot{p}_2) + \left(\frac{2a_3}{a_1} - 1\right)(\dot{m}_3p_3 - m_3\dot{p}_3) = 0. \tag{6.4}$$

Applying the Kahan-Hirota-Kimura scheme to the Kirchhoff system (6.2), we arrive at the following discretization:

$$\begin{cases} \widetilde{m}_1 - m_1 = \epsilon(a_3 - a_1)(\widetilde{m}_2m_3 + m_2\widetilde{m}_3) + \epsilon(b_3 - b_1)(\widetilde{p}_2p_3 + p_2\widetilde{p}_3), \\ \widetilde{m}_2 - m_2 = \epsilon(a_1 - a_3)(\widetilde{m}_3m_1 + m_3\widetilde{m}_1) + \epsilon(b_1 - b_3)(\widetilde{p}_3p_1 + p_3\widetilde{p}_1), \\ \widetilde{m}_3 - m_3 = 0, \\ \widetilde{p}_1 - p_1 = \epsilon a_3(\widetilde{m}_3p_2 + m_3\widetilde{p}_2) - \epsilon a_1(\widetilde{m}_2p_3 + m_2\widetilde{p}_3), \\ \widetilde{p}_2 - p_2 = \epsilon a_1(\widetilde{m}_1p_3 + m_1\widetilde{p}_3) - \epsilon a_3(\widetilde{m}_3p_1 + m_3\widetilde{p}_1), \\ \widetilde{p}_3 - p_3 = \epsilon a_1(\widetilde{m}_2p_1 + m_2\widetilde{p}_1 - \widetilde{m}_1p_2 - m_1\widetilde{p}_2). \end{cases} \tag{6.5}$$

As usual, linear system (6.5) defines a birational map $\Phi_f : \mathbb{R}^6 \to \mathbb{R}^6$, $(m, p) \mapsto (\widetilde{m}, \widetilde{p})$.

Theorem 18. (Quadratic-fractional integral, [13]) *The function*

$$I_0(m, p; \epsilon) = \frac{c_3(m, p; \epsilon)}{c_1(m, p; \epsilon)}, \tag{6.6}$$

where

$$c_1 = 1+\epsilon^2 a_3(a_1-a_3)m_3^2+\epsilon^2 a_1(b_1-b_3)p_3^2, \quad (6.7a)$$

$$c_3 = \frac{2a_3}{a_1}-1+\epsilon^2 a_1(a_3-a_1)(m_1^2+m_2^2)+\epsilon^2 a_3(b_3-b_1)(p_1^2+p_2^2), \quad (6.7b)$$

is an integral of motion of the map Φ_f.

Theorem 19. (Discrete Wronskians HK basis, [13]) *Functions* $W_i^{(1)}(m,p)=\widetilde{m}_i p_i - m_i\widetilde{p}_i$, $i-1,2,3$, *form a HK basis for the map* Φ_f *with a one-dimensional null space spanned by* $[c_1:c_1:c_3]=[1:1:I_0]$, *where* I_0 *is the integral of* Φ_f *given by* (6.6).

Novel results, illustrating Observations 2 and 3, are as follows.

Theorem 20. (Bilinear-fractional integral) *The function*

$$J_0(m,p;\epsilon)=\frac{C_3(m,p;\epsilon)}{C_1(m,p;\epsilon)}, \quad (6.8)$$

where

$$C_1 = 1-\epsilon^2 a_3(a_1-a_3)m_3^2-\epsilon^2 a_1(b_1-b_3)p_3\widetilde{p}_3, \quad (6.9a)$$

$$C_3 = \frac{2a_3}{a_1}-1-\epsilon^2 a_1(a_3-a_1)(m_1\widetilde{m}_1+m_2\widetilde{m}_2) -\epsilon^2 a_3(b_3-b_1)(p_1\widetilde{p}_1+p_2\widetilde{p}_2), \quad (6.9b)$$

is an integral of motion of the map Φ_f.

Corollary 21. (Density of an invariant measure) *The map* $\Phi_f(x;\epsilon)$ *has an invariant measure*

$$\frac{dm_1\wedge dm_2\wedge dm_3\wedge dp_1\wedge dp_2\wedge dp_3}{\phi(m,p;\epsilon)},$$

where for $\phi(m,p;\epsilon)$ *one can take the numerator of either of the functions* C_1, C_3.

Theorem 22. (Second order Wronskians HK basis) *Functions*

$$W_i^{(2)}(m,p)=\widetilde{\widetilde{m}}_i p_i - m_i\widetilde{\widetilde{p}}_i, \quad i=1,2,3,$$

form a HK basis for the map Φ_f, *with a one-dimensional null space spanned by* $[C_1:C_1:C_3]=[1:1:J_0]$, *with the function* J_0 *given in* (6.8).

Theorem 23. (Third order Wronskians HK basis) *Functions* $W_i^{(3)}(m,p)=\widetilde{\widetilde{\widetilde{m}}}_i p_i - m_i\widetilde{\widetilde{\widetilde{p}}}_i$, $i=1,2,3$, *form a HK basis for the map* Φ_f, *with a one-dimensional null space. On orbits of the map* Φ_f *there holds*

$$(\widetilde{\widetilde{\widetilde{m}}}_1 p_1 - m_1\widetilde{\widetilde{\widetilde{p}}}_1)+(\widetilde{\widetilde{\widetilde{m}}}_2 p_2 - m_2\widetilde{\widetilde{\widetilde{p}}}_2)+J_1(\widetilde{\widetilde{\widetilde{m}}}_3 p_3 - m_3\widetilde{\widetilde{\widetilde{p}}}_3)=0,$$

where the function $J_1(m,p;\epsilon)$ *is an integral of motion. The four integrals of motion* $\{I_0,J_0,J_1,m_3\}$ *are functionally independent .*

7 Lagrange top

The Hamilton function of the Lagrange top is $H = \frac{1}{2}H_1$, where

$$H_1 = m_1^2 + m_2^2 + \alpha m_3^2 + 2\gamma p_3. \tag{7.1}$$

Unlike the Clebsch and the Kirchhoff cases, this function is not homogeneous. Equations of motion of Lagrange top read

$$\begin{cases} \dot{m}_1 = (\alpha-1)m_2m_3 + \gamma p_2, \\ \dot{m}_2 = (1-\alpha)m_1m_3 - \gamma p_1, \\ \dot{m}_3 = 0, \\ \dot{p}_1 = \alpha p_2m_3 - p_3m_2, \\ \dot{p}_2 = p_3m_1 - \alpha p_1m_3, \\ \dot{p}_3 = p_1m_2 - p_2m_1. \end{cases} \tag{7.2}$$

So, like in the Kirchhoff case, m_3 is an obvious fourth integral, due to the rotational symmetry of the system.

The Wronskian relation satisfied on solutions of (7.2) is:

$$(\dot{m}_1p_1 - m_1\dot{p}_1) + (\dot{m}_2p_2 - m_2\dot{p}_2) + (2\alpha-1)(\dot{m}_3p_3 - m_3\dot{p}_3) = 0. \tag{7.3}$$

Applying the Kahan-Hirota-Kimura scheme to the vector field f from (7.2), we arrive at a birational map $\Phi_f : \mathbb{R}^6 \to \mathbb{R}^6$, $(m,p) \mapsto (\widetilde{m},\widetilde{p})$, defined by the following linear system:

$$\begin{cases} \widetilde{m}_1 - m_1 = \epsilon(\alpha-1)(\widetilde{m}_2m_3 + m_2\widetilde{m}_3) + \epsilon\gamma(p_2+\widetilde{p}_2), \\ \widetilde{m}_2 - m_2 = \epsilon(1-\alpha)(\widetilde{m}_3m_1 + m_3\widetilde{m}_1) - \epsilon\gamma(p_1+\widetilde{p}_1), \\ \widetilde{m}_3 - m_3 = 0, \\ \widetilde{p}_1 - p_1 = \epsilon\alpha(p_2\widetilde{m}_3 + \widetilde{p}_2m_3) - \epsilon(p_3\widetilde{m}_2 + \widetilde{p}_3m_2), \\ \widetilde{p}_2 - p_2 = \epsilon(p_3\widetilde{m}_1 + \widetilde{p}_3m_1) - \epsilon\alpha(p_1\widetilde{m}_3 + \widetilde{p}_1m_3), \\ \widetilde{p}_3 - p_3 = \epsilon(p_1\widetilde{m}_2 + \widetilde{p}_1m_2 - p_2\widetilde{m}_1 - \widetilde{p}_2m_1). \end{cases} \tag{7.4}$$

Theorem 24. (Quadratic-fractional integral, [13])

$$I_0(m,p;\epsilon) = \frac{r(m,p;\epsilon)}{s(m,p;\epsilon)}, \tag{7.5}$$

where

$$s = 1 + \epsilon^2\alpha(1-\alpha)m_3^2 - \epsilon^2\gamma p_3, \tag{7.6a}$$

$$r = (2\alpha-1) + \epsilon^2(\alpha-1)(m_1^2+m_2^2) + \frac{\epsilon^2\gamma}{m_3}(m_1p_1+m_2p_2), \tag{7.6b}$$

is an integral of motion of the map Φ_f.

Theorem 25. (Discrete Wronskians HK basis, [13]) *Functions* $W_i^{(1)}(m,p) = \widetilde{m}_i p_i - m_i \widetilde{p}_i$, $i = 1,2,3$, *form a HK basis for the map* Φ_f *with a one-dimensional null space spanned by* $[1 : 1 : I_0]$, *where* I_0 *is the integral of* Φ_f *given by* (7.5).

Novel results, supporting Observations 2 and 3, are as follows.

Theorem 26. (Bilinear-fractional integral) *The function*

$$J_0(m,p;\epsilon) = \frac{R(m,p;\epsilon)}{S(m,p;\epsilon)}, \tag{7.7}$$

where

$$S = 1 - \epsilon^2\alpha(1-\alpha)m_3^2 + \tfrac{1}{2}\epsilon^2\gamma(p_3 + \widetilde{p}_3), \tag{7.8a}$$

$$R = (2\alpha - 1) - \epsilon^2(\alpha - 1)(m_1\widetilde{m}_1 + m_2\widetilde{m}_2) - \frac{\epsilon^2\gamma}{2m_3}(\widetilde{m}_1 p_1 + m_1\widetilde{p}_1 + \widetilde{m}_2 p_2 + m_2\widetilde{p}_2), \tag{7.8b}$$

is an integral of motion of the map Φ_f.

Corollary 27. (Density of an invariant measure) *The map* $\Phi_f(x;\epsilon)$ *has an invariant measure*

$$\frac{dm_1 \wedge dm_2 \wedge dm_3 \wedge dp_1 \wedge dp_2 \wedge dp_3}{\phi(m,p;\epsilon)},$$

where $\phi(m,p;\epsilon)$ *can be taken as the numerator of either of the functions* R, S.

Theorem 28. (Second order Wronskians HK basis) *Functions*

$$W_i^{(2)}(m,p) = \widetilde{\widetilde{m}}_i p_i - m_i \widetilde{\widetilde{p}}_i, \quad i = 1,2,3,$$

form a HK basis for the map Φ_f, *with a one-dimensional null space spanned by* $[1 : 1 : J_0]$, *with the function* J_0 *given in* (7.7).

There holds a theorem which reads literally as Theorem 23 on the third order Wronskians HK basis.

8 Concluding remarks

We would like to remark that the existence of quadratic-fractional integrals of the Kahan discretizations is a rather common phenomenon which is not even related to integrability. In this connection, we refer to the recent paper [7], where the following result is established.

Theorem 29. *Let two components of a quadratic vector field be of the form*

$$\begin{cases} \dot{x}_1 = \ell(x)\dfrac{\partial H}{\partial x_2} = \ell(x)(bx_1 + cx_2), \\[2ex] \dot{x}_2 = -\ell(x)\dfrac{\partial H}{\partial x_1} = -\ell(x)(ax_1 + bx_2), \end{cases} \tag{8.1}$$

where $\ell(x)$ is an affine function on $\mathbb{R}^n$, and

$$H(x_1,x_2)=\frac{1}{2}(ax_1^2+2bx_1x_2+cx_2^2) \tag{8.2}$$

is a quadratic form of x_1, x_2. The Kahan discretization of this vector field, with the first two equations of motion

$$\begin{cases} \widetilde{x}_1-x_1=\epsilon\ell(x)(b\widetilde{x}_1+c\widetilde{x}_2)+\epsilon\ell(\widetilde{x})(bx_1+cx_2), \\ \widetilde{x}_2-x_2=-\epsilon\ell(x)(a\widetilde{x}_1+b\widetilde{x}_2)-\epsilon\ell(\widetilde{x})(ax_1+bx_2), \end{cases} \tag{8.3}$$

admits a quadratic-fractional integral of motion

$$F(x,\epsilon)=\frac{ax_1^2+2bx_1x_2+cx_2^2}{1+\epsilon^2(ac-b^2)\ell^2(x)}. \tag{8.4}$$

Actually, their result holds true for any (not necessarily homogeneous) quadratic polynomial $H(x_1,x_2)$, but this generalization easily follows by a shift of variables. This result does not depend on equations of motion for x_k with $3\le k\le n$, and therefore it is unrelated to integrability. It turns out that the procedure described in Observation 2 (polarization of the quadratic polynomials in the numerator and in the denominator, accompanied by the change $\epsilon^2\to-\epsilon^2$) works for the whole class of vector fields described in Theorem 29, but, amazingly, it does not lead to new integrals. We have the following result.

Theorem 30. *On orbits of the map (8.3), we have: $\widehat{F}(x,\epsilon)=F(x,\epsilon)$, where*

$$\widehat{F}(x,\epsilon)=\frac{ax_1\widetilde{x}_1+b(x_1\widetilde{x}_2+\widetilde{x}_1x_2)+cx_2\widetilde{x}_2}{1-\epsilon^2(ac-b^2)\ell(x)\ell(\widetilde{x})}. \tag{8.5}$$

Proof. Relation

$$\frac{ax_1\widetilde{x}_1+b(x_1\widetilde{x}_2+x_2\widetilde{x}_1)+cx_2\widetilde{x}_2}{1-\epsilon^2(ac-b^2)\ell(\widetilde{x})\ell(x)}=\frac{ax_1^2+2bx_1x_2+cx_2^2}{1+\epsilon^2(ac-b^2)\ell^2(x)}$$

is equivalent to

$$\begin{aligned} & ax_1(\widetilde{x}_1-x_1)+bx_1(\widetilde{x}_2-x_2)+bx_2(\widetilde{x}_1-x_1)+cx_2(\widetilde{x}_2-x_2) \\ & =-\epsilon^2(ac-b^2)\ell(x)\Big(ax_1\big(\widetilde{x}_1\ell(x)+x_1\ell(\widetilde{x})\big)+bx_1\big(\widetilde{x}_2\ell(x)+x_2\ell(\widetilde{x})\big) \\ & \quad +bx_2\big(\widetilde{x}_1\ell(x)+x_1\ell(\widetilde{x})\big)+cx_2\big(\widetilde{x}_2\ell(x)+x_2\ell(\widetilde{x})\big)\Big). \end{aligned}$$

We transform the left-hand side, using equations of motion (8.3):

$$\begin{aligned} & (ax_1+bx_2)(\widetilde{x}_1-x_1)+(bx_1+cx_2)(\widetilde{x}_2-x_2) \\ & =\epsilon(ax_1+bx_2)\Big((b\widetilde{x}_1+c\widetilde{x}_2)\ell(x)+(bx_1+cx_2)\ell(\widetilde{x})\Big) \\ & \quad -\epsilon(bx_1+cx_2)\Big((a\widetilde{x}_1+b\widetilde{x}_2)\ell(x)+(ax_1+bx_2)\ell(\widetilde{x})\Big) \\ & =\epsilon(ac-b^2)(x_1\widetilde{x}_2-x_2\widetilde{x}_1)\ell(x). \end{aligned}$$

A similar transformation of the right-hand side leads to:

$$-\epsilon^2(ac-b^2)\ell(x)\Big(x_1\big((a\widetilde{x}_1+b\widetilde{x}_2)\ell(x)+(ax_1+bx_2)\ell(\widetilde{x})\big)$$
$$+x_2\big((b\widetilde{x}_1+c\widetilde{x}_2)\ell(x)+(bx_1+cx_2)\ell(\widetilde{x})\big)\Big)$$
$$=\epsilon(ac-b^2)\ell(x)\Big(x_1(\widetilde{x}_2-x_2)-x_2(\widetilde{x}_1-x_1)\Big)$$
$$=\epsilon(ac-b^2)\ell(x)(x_1\widetilde{x}_2-x_2\widetilde{x}_1).$$

This proves the theorem. ■

The latter result makes the applicability of Observation 2 all the more intriguing. In the majority of cases when the Kahan discretization possesses a quadratic-fractional integral of motion $F(x,\epsilon)$, the polarization of the latter (i.e., the function $\widehat{F}(x,\epsilon)$ obtained by the polarization of the numerator and of the denominator of $F(x,\epsilon)$, accompanied by the change $\epsilon^2\to-\epsilon^2$) turns out to be an integral, as well. Further examples of this observation are delivered by integrable systems with the Lax representation $\dot{L}=[L^2,A]$ in the following situations:

- L is a 3×3 or a 4×4 skew-symmetric matrix and A is a constant diagonal matrix (Euler top on the algebras $so(3)$ and $so(4)$);
- L is a symmetric 3×3 matrix and A is a constant skew-symmetric 3×3 matrix [2, 3];
- L is a 3×3 matrix with vanishing diagonal and A is a constant diagonal matrix (3-wave system, see [13]).

The only counterexample we are aware of, is given by the system $\dot{L}=[L^2,A]$ with a general 3×3 matrix L and a constant diagonal matrix A [1]. For this system, polarization applied to quadratic-fractional integrals of the Kahan-Hirota-Kimura discretization (there are two independent such integrals) does not lead to integrals of motion.

Clarifying all the mysterious observations related to the Kahan-Hirota-Kimura discretization remains an intriguing and a rewarding task.

Acknowledgements

This research is supported by the DFG Collaborative Research Center TRR 109 “Discretization in Geometry and Dynamics”.

References

[1] K. Aleshkin, A. Izosimov. Euler equations on the general Lie group, cubic curves, and inscribed hexagons, *Enseign. Math.* **62** (2016), 143–170.

[2] A.M. Bloch, A. Iserles. On an isospectral Lie-Poisson system and its Lie algebra, *Foundations Comput. Math.* **6** (2006), 121–144.

[3] A.M. Bloch, V. Brinzanescu, A. Iserles, J.E. Marsden, T. Ratiu. A class of integrable flows on the space of symmetric matrices, *Commun. Math. Phys.* **290** (2009), 399–435.

[4] E. Celledoni, R.I. McLachlan, B. Owren, G.R.W. Quispel. Geometric properties of Kahan's method, *J. Phys. A* **46** (2013), 025201, 12 pp.

[5] E. Celledoni, R.I. McLachlan, D.I. McLaren, B. Owren, G.R.W. Quispel. Integrability properties of Kahan's method, *J. Phys. A* **47** (2014), 365202, 20 pp.

[6] E. Celledoni, R.I. McLachlan, D.I. McLaren, B. Owren, G.R.W. Quispel. Discretization of polynomial vector fields by polarization, *Proc. R. Soc. A* **471** (2015), 20150390, 10 pp.

[7] E. Celledoni, D.I. McLaren, B. Owren, G.R.W. Quispel. Geometric and integrability properties of Kahan's method, `arXiv: 1805.08382 [math.NA]`, 8 pp.

[8] R. Hirota, K. Kimura, Discretization of the Euler top, *J. Phys. Soc. Japan* **69**, Nr. 3 (2000), 627–630.

[9] A.N.W. Hone, M. Petrera, Three-dimensional discrete systems of Hirota-Kimura type and deformed Lie-Poisson algebras, *J. Geom. Mech.* **1**, Nr. 1 (2009), 55–85.

[10] W. Kahan, *Unconventional numerical methods for trajectory calculations*, Unpublished lecture notes, 1993.

[11] K. Kimura, R. Hirota, Discretization of the Lagrange top, *J. Phys. Soc. Japan* **69**, Nr. 10 (2000), 3193–3199.

[12] M. Petrera, A. Pfadler, Yu.B. Suris, On integrability of Hirota-Kimura-type discretizations: experimental study of the discrete Clebsch system, *Exp. Math.* **18**, Nr. 2 (2009), 223–247.

[13] M. Petrera, A. Pfadler, Yu.B. Suris, On integrability of Hirota-Kimura-type discretizations, *Reg. Chaotic Dyn.* **16**, Nr. 3/4 (2011), 245–289.

[14] M. Petrera, Yu.B. Suris, On the Hamiltonian structure of Hirota-Kimura discretization of the Euler top, *Math. Nachr.* **283**, Nr. 11 (2010), 1654–1663.

[15] M. Petrera, Yu.B. Suris, S. Kovalevskaya system, its generalization and discretization, *Frontiers of Math. in China* **8**, Nr. 11 (2013), 1047–1065.

[16] M. Petrera, Yu.B. Suris, Spherical geometry and integrable systems, *Geometriae Dedicata* **169**, Nr. 1 (2014), 83–98.

[17] M. Petrera, Yu.B. Suris, A construction of a large family of commuting pairs of integrable symplectic birational 4-dimensional maps, *Proc. Royal Soc. A* **473** (2017), 20160535, 16 pp.

[18] J.M. Sanz-Serna. An unconventional symplectic integrator of W. Kahan, *Appl. Numer. Math.* **16** (1994), 245–250.

B1. Solvable dynamical systems and isospectral matrices defined in terms of the zeros of orthogonal or otherwise special polynomials

Oksana Bihun

Department of Mathematics, University of Colorado, Colorado Springs, 1420 Austin Bluffs Pkwy, Colorado Springs 80918, USA

Abstract

Several recently discovered properties of multiple families of special polynomials (some orthogonal and some not) that satisfy certain differential, difference or q-difference equations are reviewed. A general method of construction of isospectral matrices defined in terms of the zeros of such polynomials is discussed. The method involves reduction of certain partial differential, differential difference and differential q-difference equations to systems of ordinary differential equations and their subsequent linearization about the zeros of polynomials in question. Via this process, solvable (in terms of algebraic operations) nonlinear first order systems of ordinary differential equations are constructed.

1 Introduction

Orthogonal or otherwise special polynomials play an important role in mathematical physics, for example, in construction of exact solutions of quantum mechanical systems [36]. They are indispensable in numerical analysis, in part due to the best approximation properties of their linear combinations in certain L^2 spaces and the efficiency of the spectral methods for solving differential equations based on these types of approximation [37, 28]. Zeros of special polynomials are significant in many ways, for example, as equilibria of important solvable N-body problems [22] or as the nodes in numerical quadrature formulas that yield a higher degree of exactness compared to other nodes [42, 25, 34]. The zeros of a polynomial P_N from a family $\{P_n\}_{n=0}^{\infty}$ orthogonal with respect to a positive measure have remarkable properties; in particular, they are distinct and real, remain in the support of the measure, and interlace with the zeros of P_{N+1} [42].

The interest in algebraic properties of the zeros of special polynomials dates back to 1885, when Stieltjes established algebraic relations satisfied by the zeros of classical orthogonal polynomials (Jacobi, Laguerre and Hermite) [39, 40, 41]. For example, the zeros ζ_n of the Hermite polynomial H_N satisfy

$$\sum_{j=1, j\neq n}^{N} \frac{1}{\zeta_j - \zeta_n} = -\zeta_n, \quad n = 1, 2, \ldots, N.$$

The last family of algebraic relations has an electrostatic interpretation: it is the extremum condition of an energy functional for the system of N unit charges placed on the real line, interacting pairwise according to a repulsive logarithmic potential in the harmonic field [39, 40, 41, 42, 43]. Other electrostatic models for zeros of polynomials are surveyed in [33]; see also [29, 30].

Since the groundbreaking work of Stieltjes, the literature on algebraic relations satisfied by the zeros of orthogonal or otherwise special polynomials is abundant; see for example [3, 1, 2, 19, 20, 38, 4, 8, 9, 10, 11, 7, 16].

In this chapter we describe a general approach to construction of isospectral matrices defined in terms of the zeros of special polynomials. The spectrum of these matrices is independent of the zeros and possibly some of the parameters of the polynomials, thus the use of the term "isospectral" to describe them. The upshot is that given a polynomial P_N of degree N with the zeros $\boldsymbol{\zeta} = (\zeta_1, \ldots, \zeta_N)$, construction of an $N \times N$ matrix $L \equiv L(\boldsymbol{\zeta})$ with the eigenvalues λ_j independent of the zeros $\boldsymbol{\zeta}$ gives rise to multiple algebraic identities satisfied by these zeros, the identities $\operatorname{Trace} L = \sum_{j=1}^{N} \lambda_j$ and $\det L = \prod_{j=1}^{N} \lambda_j$ among them. This approach relies on the differential, difference or q-difference equations satisfied by the polynomial P_N and does not use the (possible) orthogonality properties of the family of polynomials that contains P_N. Therefore, the results for orthogonal polynomials stated here remain true even if the parameters of these polynomials are outside of the orthogonality range.

The method goes back to the ideas of Stieltjes and Szegö, which exploit the nonlinear relation between time-dependent coefficients $\boldsymbol{c}(t) = (c_1(t), \ldots, c_N(t))$ and zeros $\boldsymbol{z}(t) = (z_1(t), \ldots, z_N(t))$ of a monic polynomial

$$\psi_N(z,t) = z^N + \sum_{j=1}^{N} c_j(t) z^{N-j} = \prod_{j=1}^{N} [z - z_j(t)]. \tag{1.1}$$

Note the abuse of notation as z is a complex variable in the polynomial $\psi_N(z,t)$, while $\boldsymbol{z}(t)$ is a t-dependent vector of its zeros. Numerous solvable, in terms of algebraic operations, N-body problems have been constructed by assuming that $\psi_N(z,t)$ satisfies a linear partial differential equation (PDE) [22]. By recasting the PDE as a linear, hence solvable, system for the coefficients $\boldsymbol{c}(t)$ of the polynomial $\psi_N(z,t)$ and, on the other hand, a nonlinear system for the zeros $\boldsymbol{z}(t)$ of the same polynomial, one concludes that the nonlinear system is solvable. Indeed, solutions $\boldsymbol{z}(t)$ of the nonlinear system can be recovered via the algebraic operation of finding the zeros of the t-dependent polynomial $\psi_N(z,t)$ with the coefficients $\boldsymbol{c}(t)$ that solve the corresponding linear system. If the solution $\boldsymbol{c}(t)$ of the linear system is isochronous, i.e. periodic with the period independent of the initial conditions, then the solution $\boldsymbol{z}(t)$ of the corresponding nonlinear system is isochronous. This observation is instrumental in construction of many solvable isochronous systems with remarkable properties [23]. Similar constructions of solvable nonlinear dynamical systems have been carried out by comparing the evolution of time-dependent nodes and (generalized or standard) Lagrange basis coefficients of a time-dependent polynomial that satisfies a linear PDE [22].

Recently, convenient formulas that relate the derivatives of the coefficients $\boldsymbol{c}(t)$ with the derivatives of the zeros $\boldsymbol{z}(t)$ of the polynomial $\psi_N(z,t)$ have been discovered [24, 17]. This has expanded the possibilities for construction of solvable nonlinear dynamical systems [12, 13]; for example, see [13] for a new solvable N-body problem that may be interpreted as a hybrid between the Calogero-Moser [18, 35] and the goldfish [21] systems. Using the approach outlined in these recent developments, one may assume that the coefficients $\boldsymbol{c}(t)$ of the time-dependent monic polynomial $\psi_N(z,t)$ satisfy a solvable nonlinear dynamical system (as opposed to a linear one, a constraint of the method described above). Then, as before, the corresponding nonlinear system satisfied by the zeros $\boldsymbol{z}(t)$ is solvable. Upon iteration of this procedure, one may in fact construct infinite hierarchies of solvable dynamical systems [14, 15], a remarkable find given that integrable or solvable dynamical systems are rare (Poincaré showed that the integrability of Hamiltonian systems is not a generic property).

In this chapter we will review construction of several solvable (in terms of algebraic operations) first order systems of nonlinear differential equations of the form

$$\dot{\boldsymbol{z}}(t) = F\big(\boldsymbol{z}(t)\big), \tag{1.2}$$

where $\boldsymbol{z}(t) = \big(z_1(t), \ldots, z_N(t)\big)$ and the "dot" denotes differentiation with respect to the time-variable t. The aim is to obtain isospectral matrices defined in terms of the zeros of certain special polynomials. To this end, an equation

$$\psi_t = \mathcal{A}\psi \tag{1.3}$$

that admits a polynomial solution $\psi_N(z,t)$ of the form (1.1) is constructed, with $\mathcal{A}$ being a linear differential, difference or q-difference operator. An additional requirement is that the last equation (1.3) has a t-independent solution $\psi(z,t) = P_N(z)$ with $P_N(z)$ being the N-th member of a polynomial family of interest. This ensures that every vector $\boldsymbol{\zeta} = \big(\zeta_1, \ldots, \zeta_N\big)$ of the zeros of $P_N(z)$ is an equilibrium of the nonlinear system (1.2) (there are $N!$ of such vectors if the zeros ζ_n are distinct). The polynomial $\psi_N(z,t)$ solves the equation $\psi_t = \mathcal{A}\psi$ if and only if its coefficients $\boldsymbol{c}(t) = \big(c_1(t), \ldots, c_N(t)\big)$ solve a linear system

$$\dot{\boldsymbol{c}}(t) = A\boldsymbol{c}(t) \tag{1.4}$$

with the $N \times N$ coefficient matrix A, if and only if its zeros $\boldsymbol{z}(t)$ solve system (1.2) with an appropriate right-hand side $F\big(\boldsymbol{z}(t)\big)$. A conclusion is that the last system (1.2) is solvable since its solutions $\boldsymbol{z}(t)$ are the zeros of the polynomial $\psi_N(z,t)$ given by (1.1) with the coefficients $\boldsymbol{c}(t)$ that solve system (1.4). If the $N \times N$ coefficient matrix of the linear system is upper or lower diagonal, its eigenvalues $\{\lambda_j\}_{j=1}^N$ are easily found. One may then argue that the eigenvalues of the coefficient matrix $L = L(\boldsymbol{\zeta})$ of the linearization of system (1.2) about the equilibrium $\boldsymbol{\zeta}$ are the same and moreover do not depend on some of the parameters of the polynomial $P_N(z)$. The main result is that the matrix $L(\boldsymbol{\zeta})$ defined in terms of the zeros $\boldsymbol{\zeta}$ of $P_N(z)$ is isospectral.

A crucial step in this method is construction of a linear operator $\mathcal{A}$ such that equation (1.3) admits polynomial solutions and possesses a t-independent equilibrium solution $P_N(z)$. To accomplish this task, the corresponding differential, difference or q-difference equation satisfied by the special polynomial P_N is utilized.

The Askey and the q-Askey schemes contain polynomials orthogonal with respect to a positive measure supported on the real line, which satisfy certain second order linear differential, difference or q-difference equations [31]. All these polynomials satisfy certain standard (yet remarkable) properties: three term recurrence relations, Rodrigues-type formulas, formulas that represent their derivatives as linear combinations of polynomials from the same type of families with shifted parameter(s), explicit formulas for the generating functions and other. Moreover, by taking certain limits with respect to parameters of the polynomials positioned in higher levels of the Askey or the q-Askey scheme hierarchy, one obtains polynomials in the lower levels. The Wilson and the Racah polynomials are at the top of the Askey scheme, while the Askey-Wilson and the q-Racah polynomials are at the top of the q-Askey scheme, thus our interest in proving algebraic identities for the zeros of these polynomials. All the polynomials in the Askey scheme can be written as special cases of the generalized hypergeometric function, while all the polynomials in the q-Askey scheme can be written as particular cases of the generalized basic hypergeometric function. The last two functions play a very important role in mathematics.

The generalized hypergeometric function can be defined as an infinite series $\sum_{j=0}^{\infty} A_j z^j$ such that for each j, the ratio $\frac{A_{j+1}}{A_j}$ is a rational function of j. The following is a standard notation for the generalized hypergeometric function:

$$\begin{aligned} &{}_{p+1}F_q\left(\alpha_0,\alpha_1,\dots,\alpha_p;\beta_1,\dots,\beta_q;z\right)\\ &\equiv {}_{p+1}F_q\left(\begin{matrix}\alpha_0,\alpha_1,\dots,\alpha_p\\ \beta_1,\dots,\beta_q\end{matrix}\,\middle|\,z\right)\\ &=\sum_{j=0}^{\infty}\left[\frac{(\alpha_0)_j\ (\alpha_1)_j\cdots(\alpha_p)_j\ z^j}{j!\ (\beta_1)_j\cdots(\beta_q)_j}\right], \end{aligned} \tag{1.5}$$

where $(\alpha)_j$ is the Pochhammer symbol defined by

$$(\alpha)_0=1;\quad (\alpha)_j=\alpha\ (\alpha+1)\cdots(\alpha+j-1)=\frac{\Gamma(\alpha+j)}{\Gamma(\alpha)}\quad \text{for}\quad j=1,2,3,\dots$$

and Γ is the gamma function. Here, $\{\alpha_j\}_{j=0}^{p}$ and $\{\beta_k\}_{k=1}^{q}$ are complex numbers such that, after appropriate cancellations, the denominators in the series in (1.5) do not vanish (for example, each β_k is not zero and is not a negative integer). Note that if one of the parameters $\alpha_j=-N$ is a negative integer, then series (1.5) becomes a finite sum and the corresponding generalized hypergeometric function is a polynomial of degree at most N in z. In particular, we define the N-th generalized basic hypergeometric polynomial by

$$\begin{aligned} P_N\left(\alpha_1,\dots,\alpha_p;\beta_1,\dots,\beta_q;z\right)&=\sum_{m=0}^{N}\left[\frac{(-N)_m\,(\alpha_1)_m\cdots(\alpha_p)_m\ z^{N-m}}{m!\ (\beta_1)_m\cdots(\beta_q)_m}\right]\\ &=z^N\ {}_{p+1}F_q\left(-N,\alpha_1,\dots,\alpha_p;\beta_1,\dots,\beta_q;1/z\right). \end{aligned} \tag{1.6}$$

In the case where series (1.5) is not a polynomial, its radius of convergence ρ can be determined by the ratio test: $\rho=+\infty$ if $p<q$, $\rho=1$ if $p=q$ and $\rho=0$ if $p>q$, see [32, 27]. Of particular interest is the linear differential equation satisfied by the generalized basic hypergeometric function (1.5):

$$\left[\mathcal{D}\prod_{j=1}^{q}(\mathcal{D}+\beta_j-1)-z\prod_{j=0}^{p}(\mathcal{D}+\alpha_j)\right]u(z)=0, \tag{1.7}$$

where $\mathcal{D}=z\,d/dz$, see [5]. The generalized hypergeometric polynomial P_N, on the other hand, satisfies the following differential equation:

$$\left[\mathcal{D}_N\prod_{j=1}^{q}(\beta_j-1-\mathcal{D}_N)-\frac{d}{dz}\prod_{j=1}^{p}(\alpha_j-\mathcal{D}_N)\right]u(z)=0, \tag{1.8}$$

where

$$\mathcal{D}_N=z\,d/dz-N; \tag{1.9}$$

see Section 3 of [8] for a proof.

The generalized basic hypergeometric, or q-hypergeometric, function with the basis $q\neq 1$ can be defined as an infinite series $\sum_{j=0}^{\infty}A_jz^j$ such that for each j, the ratio $\frac{A_{j+1}}{A_j}$ is a rational function of q^j. A standard notation for the generalized basic hypergeometric function is the following:

$$\begin{aligned}
&{}_{r+1}\phi_s\left(\alpha_0,\alpha_1,\ldots,\alpha_r;\beta_1,\ldots,\beta_s;q;z\right)\\
&\equiv {}_{r+1}\phi_s\left(\begin{array}{c}\alpha_0,\alpha_1,\ldots,\alpha_r\\ \beta_1,\ldots,\beta_s\end{array}\middle|\, q;z\right)\\
&=\sum_{m=0}^{\infty}\left\{\frac{(\alpha_0;q)_m\,(\alpha_1;q)_m\cdots(\alpha_r;q)_m}{(q;q)_m\quad(\beta_1;q)_m\cdots(\beta_s;q)_m}\left[(-1)^m\,q^{m(m-1)/2}\right]^{s-r}z^m\right\},
\end{aligned} \tag{1.10}$$

where $\{\alpha_j\}_{j=0}^{r}$ and $\{\beta_k\}_{k=1}^{s}$ are complex parameters such that, after appropriate cancellations, the denominators in the terms of the above series do not vanish and

$$(\gamma;q)_0=1,\ (\gamma;q)_m=(1-\gamma)(1-\gamma q)\cdots(1-\gamma q^{m-1})\text{ for }m=1,2,3,\ldots$$

is the q-Pochhammer symbol.

The generalized basic hypergeometric function is a q-analogue of the generalized hypergeometric function in the following sense [32]:

$$\begin{aligned}
&\lim_{q\to 1}{}_{r+1}\phi_s\left(\begin{array}{c}q^{\alpha_0},q^{\alpha_1},\ldots,q^{\alpha_r}\\ q^{\beta_1},\ldots,q^{\beta_s}\end{array}\middle|\, q;(q-1)^{s-r}z\right)\\
&={}_{r+1}F_s\left(\begin{array}{c}\alpha_0,\alpha_1,\ldots,\alpha_r\\ \beta_1,\ldots,\beta_s\end{array}\middle|\, z\right).
\end{aligned} \tag{1.11}$$

In general, the radius of convergence ρ of the series (1.10) is $\rho=\infty$ if $r<s$, $\rho=1$ if $r=s$, and $\rho=0$ if $r>s$, see [32]. However, for some choices of the parameters, the series reduces to a finite sum, producing a polynomial. Indeed, if $m>N$, $(q^{-N};q)_m=(1-q^{-N})\cdots(1-q^{-N}q^N)\cdots(1-q^{-N}q^{m-1})=0$, hence ${}_{r+1}\phi_s\left(q^{-N},\alpha_1,\ldots,\alpha_r;\beta_1,\ldots,\beta_s;q;z\right)$ is a polynomial of degree at most N. Motivated by this observation, define the N-th generalized basic hypergeometric polynomial by [10]

$$\begin{aligned}
&P_N\left(\alpha_1,\ldots,\alpha_r;\beta_1,\ldots,\beta_s;q;z\right)\\
&=\sum_{m=0}^{N}\left[\frac{\left(q^{-N};q\right)_m\left(\alpha_1;q\right)_m\cdots\left(\alpha_r;q\right)_m}{\left(q;q\right)_m\;\left(\beta_1;q\right)_m\cdots\left(\beta_s;q\right)_m}\left[(-1)^m\;q^{\frac{m(m-1)}{2}}\right]^{s-r}z^m\right]\\
&={}_{r+1}\phi_s\left(q^{-N},\alpha_1,\ldots,\alpha_r;\beta_1,\ldots,\beta_s;q;z\right).
\end{aligned}\tag{1.12}$$

The generalized basic hypergeometric function (1.10) and the generalized basic hypergeometric polynomial (1.12) satisfy certain q-difference equations. To formulate these equations, let us introduce some q-difference operators.

Recall that for $q\neq1$, the q-derivative operator $\mathcal{D}_q$ is defined by [32]

$$\mathcal{D}_q f(z)=\begin{cases}\frac{f(z)-f(qz)}{z-qz} & \text{if } z\neq0,\\ f'(0) & \text{if } z=0.\end{cases}\tag{1.13}$$

This operator generalizes the differentiation operator d/dz: If a function $f(z)$ is differentiable at z, then $\lim\limits_{q\to1}\mathcal{D}_q f(z)=d/dz\;f(z)$. For the purposes of our study, it is convenient to use the q-difference operators δ_q and Δ_γ defined by

$$\delta_q f(z)=f(qz),\tag{1.14}$$

$$\Delta_\gamma f(z)=\left(\gamma\delta_q-1\right)f(z)=\gamma f(qz)-f(z).\tag{1.15}$$

Note that because $\Delta_\gamma z^m=\left(\gamma q^m-1\right)z^m$ for every nonnegative integer m, the action of the operator Δ_γ on a polynomial does not raise its degree.

With the above notation, we can state the q-difference equation satisfied by the generalized basic hypergeometric function ${}_{r+1}\phi_s\left(\alpha_0,\alpha_1,\ldots,\alpha_r;\beta_1,\ldots,\beta_s;z\right)$ (see Exercise 1.31 on page 27 of [6]):

$$\Delta_1\left[\prod_{k=1}^{s}\left(\Delta_{\beta_k/q}\right)\right]u(z)-z\,\Delta_{q^{-N}}\left[\prod_{j=1}^{r}\left(\Delta_{\alpha_j}\right)\right]u(zq^{s-r})=0.\tag{1.16}$$

Therefore, the generalized basic hypergeometric polynomial (1.12) satisfies the following q-difference equation [10]:

$$\left[\Delta_1\prod_{k=1}^{s}\left(\Delta_{\beta_k/q}\right)-z\,\Delta_{q^{-N}}\prod_{j=1}^{r}\left(\Delta_{\alpha_j}\right)\left(\delta_q\right)^{s-r}\right]u(z)=0.\tag{1.17}$$

2 Zeros of generalized hypergeometric polynomial with two parameters and zeros of Jacobi polynomials

In this section we illustrate the general method of construction of an isospectral matrix defined in terms of the zeros of a special polynomial $P_N(z)$ for the simple example where

$$P_N(z) \equiv P_N(\alpha_1;\beta_1;z) = \sum_{m=0}^{N} \left[\frac{(-N)_m (\alpha_1)_m \ z^{N-m}}{m! \ (\beta_1)_m} \right] \tag{2.1}$$

is the generalized basic hypergeometric polynomial (1.6) with two parameters α_1 and β_1, see [8]. By taking $p = q = 1$ in (1.8), we conclude that this polynomial satisfies the differential equation

$$\mathcal{A}u(z) = 0, \tag{2.2}$$

where $\mathcal{A}$ is the linear differential operator

$$\mathcal{A} = \sum_{j=1}^{2} b_j (\mathcal{D}_N)^j - \frac{d}{dz} \sum_{j=0}^{1} a_j (\mathcal{D}_N)^j \tag{2.3}$$

with $\mathcal{D}_N = z\frac{d}{dz} - N$ or $\mathcal{D}_N = z\frac{\partial}{\partial z} - N$ depending on the context, see (1.9), and

$$a_0 = \alpha_1, \ a_1 = -1, \ b_1 = \beta_1 - 1 \text{ and } \ b_2 = -1. \tag{2.4}$$

To construct an $N \times N$ isospectral matrix defined in terms of the zeros $\boldsymbol{\zeta} = (\zeta_1, \ldots, \zeta_N)$ of the polynomial $P_N(z)$, consider the PDE

$$\frac{\partial}{\partial t}\psi(z,t) = -\mathcal{A}\psi(z,t), \tag{2.5}$$

which has a time-independent solution $\psi(z,t) = P_N(z)$. Equation (2.5) admits time-dependent monic polynomial solutions of the form (1.1). Indeed, upon substitution of the anzatz $\psi_N(z,t) = z^N + \sum_{j=1}^{N} c_j(t) z^{N-j}$ into (2.5), we obtain the following linear system of differential equations for the coefficients $\boldsymbol{c}(t) = (c_1(t), \ldots, c_N(t))$:

$$\dot{c}_m = m(\beta_1 - 1 + m)\, c_m + (N + 1 - m)(\alpha_1 - 1 + m)\, c_{m-1}, \tag{2.6}$$

where $m = 1, \ldots, N$, $c_0 = 1$ and $c_j = 0$ for all integer j outside of the interval $[0, N]$. The last system can be recast in the form

$$\dot{\boldsymbol{c}}(t) = A\boldsymbol{c}(t) + \boldsymbol{h}, \tag{2.7}$$

where A is a lower diagonal $N \times N$ matrix with the eigenvalues

$$\lambda_m = m(\beta_1 - 1 + m) \tag{2.8}$$

that do not depend on the parameter α_1 of the polynomial $P_N(z)$ and $\boldsymbol{h}$ in the N-vector $\boldsymbol{h} = (N\alpha_1, 0, \ldots, 0)^T$. Assuming that these eigenvalues are all different among themselves, a general solution of linear system (2.7) can be written as

$$\boldsymbol{c}(t) = \sum_{j=1}^{N} \eta_m e^{\lambda_m t} \boldsymbol{v}_m + \boldsymbol{c}_p, \tag{2.9}$$

where η_m are N arbitrary constants, each $\boldsymbol{v}_m$ is an eigenvector of the matrix A that corresponds to the eigenvalue λ_m and $\boldsymbol{c}_p = -A^{-1}h$ is a particular solution of (2.7) (note a misprint in (34) of [8]).

Let us now substitute the representation $\psi_N(z,t) = \prod_{m=1}^{N}[z - z_m(t)]$ of the time-dependent polynomial (1.1) in terms of its zeros $\boldsymbol{z}(t) = (z_1(t), \ldots, z_N(t))$ into PDE (2.5). In terms of computations, this last substitution is somewhat less trivial, given that our aim is to obtain a first order nonlinear system of ODEs for the zeros $\boldsymbol{z}(t)$ of the form (1.2). Before we discuss a convenient method for construction of system (1.2) satisfied by the zeros $\boldsymbol{z}(t)$ of the polynomial $\psi_N(z,t)$ satisfying PDE (2.5), let us outline a general strategy for construction of an $N \times N$ isospectral matrix $L(\boldsymbol{\zeta})$ defined in terms of the zeros $\boldsymbol{\zeta}$ of the polynomial $P_N(z)$.

Because $P_N(z)$ is a time-independent polynomial solution of PDE (2.5), the vector $\boldsymbol{\zeta}$ of its zeros is an equilibrium of system (1.2); that is, $F(\boldsymbol{\zeta}) = 0$. It is therefore natural to linearize this last system about its equilibrium $\boldsymbol{\zeta}$ to obtain the system

$$\dot{x}(t) = Lx(t), \tag{2.10}$$

where the $N \times N$ matrix $L \equiv L(\boldsymbol{\zeta}) = DF(\boldsymbol{\zeta})$ is the Jacobian matrix of the function F in the right-hand side of (1.2), evaluated at $\boldsymbol{\zeta}$. The next step is to compare the eigenvalues of the matrix $L(\boldsymbol{\zeta})$ with the eigenvalues $\{\lambda_m\}_{m=1}^{N}$ of the matrix of coefficients A of system (2.7) to conclude that these eigenvalues must be equal. A corollary of the last statement is that the $N \times N$ matrix $L(\boldsymbol{\zeta})$ defined in terms of the zeros $\boldsymbol{\zeta}$ of the polynomial $P_N(z)$ has the N eigenvalues $\lambda_m = m(\beta_1 - 1 + m)$, see (2.8), which depend neither on the zeros $\boldsymbol{\zeta}$ nor on the parameter α_1 of the polynomial $P_N(z)$ and, moreover, are rational if the parameter β_1 is rational, a diophantine property.

To construct system (1.2), we use the following notation, see [23]. For a monic polynomial

$$\phi(z) = \prod_{n=1}^{N}(z - z_n) \tag{2.11}$$

with the vector of zeros $(z_1, \ldots, z_N)$ and a linear differential operator $\mathcal{B}$ with polynomial coefficients, acting in the variable z, we write

$$\mathcal{B}\phi\,(z) \longleftrightarrow f_n\,(z_1, \ldots, z_N) \tag{2.12}$$

to denote the identity

$$\mathcal{B}\phi\,(z) = \phi(z)\sum_{n=1}^{N}\left[(z - z_n)^{-1}\ f_n\,(z_1, \ldots, z_N)\right]; \tag{2.13}$$

note a misprint in (48b) of [8]. For example, by logarithmic differentiation of (2.11), one obtains

$$\left(\frac{d}{dz}\right)\phi\,(z) \longleftrightarrow 1, \tag{2.14}$$

and, from the last identity, by using $z/(z-z_n) = 1 + z_n/(z-z_n)$, one obtains

$$\mathcal{D}_N\phi(z) \longleftrightarrow z_n, \tag{2.15}$$

where $\mathcal{D}_N$ is given by (1.9), see (A.4) and (A.6a) in [23].

Let us find $f_n^{\mathcal{A}}(z_1,\ldots,z_N)$ such that

$$\mathcal{A}\phi(z) \longleftrightarrow f_n^{\mathcal{A}}(z_1,\ldots,z_N). \tag{2.16}$$

A computation of $f_n^{\mathcal{A}}(z_1,\ldots,z_N)$ will accomplish the goal of construction of system (1.2). Indeed, for the polynomial $\psi_N(z,t) = \prod_{n=1}^N [z - z_n(t)]$,

$$-\mathcal{A}\psi_N = -\psi_N \sum_{n=1}^N [z - z_n(t)]^{-1} f_n^{\mathcal{A}}(z_1,\ldots,z_N)$$

and

$$\partial\psi_N/\partial t = -\psi_N \sum_{n=1}^N [z - z_n(t)]^{-1} \dot{z}_n;$$

hence $\psi_N(z,t)$ solves PDE (2.5) if and only if its zeros $\boldsymbol{z}(t) = (z_1(t),\ldots,z_N(t))$ satisfy the differential equations

$$\dot{z}_n(t) = f_n^{\mathcal{A}}(z_1(t),\ldots,z_N(t)), \tag{2.17}$$

where $n = 1,2,\ldots,N$. System (2.17) is an explicit form of system (1.2).

Motivated by formula (2.3) for the differential operator $\mathcal{A}$, we introduce the expressions $f_n^{(j)}(z_1,\ldots,z_N)$ and $g_n^{(j)}(z_1,\ldots,z_N)$ to denote

$$(\mathcal{D}_N)^j\,\phi(z) \longleftrightarrow f_n^{(j)}(z_1,\ldots,z_N) \tag{2.18}$$

and

$$\frac{d}{dz}(\mathcal{D}_N)^j\,\phi(z) \longleftrightarrow g_n^{(j)}(z_1,\ldots,z_N) \tag{2.19}$$

so that

$$\begin{aligned} f_n^{\mathcal{A}}(z_1,\ldots,z_N) &= b_1 f_n^1(z_1,\ldots,z_N) + b_2 f_n^2(z_1,\ldots,z_N) \\ &+ a_0 g_n^0(z_1,\ldots,z_N) + a_1 g_n^1(z_1,\ldots,z_N). \end{aligned} \tag{2.20}$$

By (2.15) and (2.14),

$$f_n^{(1)}(z_1,\ldots,z_N) = z_n \text{ and } g_n^{(0)}(z_1,\ldots,z_N) = 1. \tag{2.21}$$

To compute $f_n^{(2)}(z_1,\ldots,z_N)$, let us apply the differential operator $(\mathcal{D}_N)^2$ to the polynomial $\phi(z)$ defined by (2.11):

$$\begin{aligned} (\mathcal{D}_N)^2\phi &= \mathcal{D}_N\left[\phi\sum_{n=1}^N (z-z_n)^{-1} z_n\right] = (\mathcal{D}_N\phi)\sum_{n=1}^N \frac{z_n}{z-z_n} + \phi\sum_{n=1}^N z\frac{d}{dz}\left(\frac{z_n}{z-z_n}\right) \\ &= \phi\left[\sum_{n,m=1}^N \frac{z_n z_m}{(z-z_n)(z-z_m)} - \sum_{n=1}^N \frac{(z-z_n)z_n}{(z-z_n)^2} - \sum_{n=1}^N \frac{(z_n)^2}{(z-z_n)^2}\right] \\ &= \phi\left[\sum_{n,m=1,n\neq m}^N \frac{z_n z_m}{(z-z_n)(z-z_m)} - \sum_{n=1}^N \frac{z_n}{z-z_n}\right] \end{aligned}$$

$$= \phi \left\{ \sum_{n,m=1,n\neq m}^{N} \frac{z_n z_m}{z_n - z_m} \left[\frac{1}{z - z_n} - \frac{1}{z - z_m} \right] - \sum_{n=1}^{N} \frac{z_n}{z - z_n} \right\}$$

$$= \phi \sum_{n=1}^{N} (z - z_n)^{-1} \left[\sum_{m=1,m\neq n}^{N} \frac{2 z_n z_m}{z_n - z_m} - z_n \right]. \tag{2.22}$$

Therefore,

$$f_n^{(2)}(z_1, \ldots, z_N) = \sum_{m=1,m\neq n}^{N} \frac{2 z_n z_m}{z_n - z_m} - z_n. \tag{2.23}$$

To compute $g_n^{(2)}(z_1, \ldots, z_N)$, consider

$$\frac{d}{dz}(\mathcal{D}_N \phi) = \frac{d}{dz} \left[\phi \sum_{n=1}^{N} (z - z_n)^{-1} z_n \right] = \left(\frac{d\phi}{dz} \right) \sum_{n=1}^{N} \frac{z_n}{z - z_n} - \phi \sum_{n=1}^{N} \frac{z_n}{(z - z_n)^2}$$

$$= \phi \left[\sum_{n,m=1,n\neq m}^{N} \frac{z_n}{(z - z_n)(z - z_m)} \right]$$

$$= \phi \left\{ \sum_{n,m=1,n\neq m}^{N} \frac{z_n}{z_n - z_m} \left[\frac{1}{z - z_n} - \frac{1}{z - z_m} \right] \right\}$$

$$= \phi \sum_{n=1}^{N} (z - z_n)^{-1} \left[\sum_{m=1,m\neq n}^{N} \frac{z_n + z_m}{z_n - z_m} \right]. \tag{2.24}$$

Therefore,

$$g_n^{(1)}(z_1, \ldots, z_N) = \sum_{m=1,m\neq n}^{N} \frac{z_n + z_m}{z_n - z_m}. \tag{2.25}$$

Using (2.20), (2.4), (2.21), (2.23) and (2.25), we find

$$f_n^{\mathcal{A}}(z_1, \ldots, z_N) = -\alpha_1 + \beta_1 z_n + \sum_{m=1,m\neq n}^{N} \frac{z_n + z_m - 2 z_n z_m}{z_n - z_m}$$

$$= N - 1 - \alpha_1 + \beta_1 z_n + 2(1 - z_n) \sum_{m=1,m\neq n}^{N} \frac{z_m}{z_n - z_m}. \tag{2.26}$$

We proved the following Lemma.

Lemma 1. *The time-dependent monic polynomial* (1.1) *satisfies PDE* (2.5) *if and only if its coefficients* $\mathbf{c}(t)$ *satisfy the linear system* (2.6) *if and only if its zeros* $\mathbf{z}(t)$ *satisfy the nonlinear system*

$$\dot{z}_n(t) = N - 1 - \alpha_1 + \beta_1 z_n(t) + 2[1 - z_n(t)] \sum_{m=1,m\neq n}^{N} \frac{z_m(t)}{z_n(t) - z_m(t)}. \tag{2.27}$$

In particular, system (2.27) *is solvable in terms of algebraic operations.*

Because $P_N(z)$ given by (2.1) is a time-independent solution of PDE (2.5), the vector of its zeros $\boldsymbol{\zeta} = (\zeta_1, \ldots, \zeta_N)$ is an equilibrium of system (2.27). Therefore, the following corollary holds.

Corollary 1. *If the (possibly complex) zeros $\zeta_1, \ldots, \zeta_N$ of the generalized basic hypergeometric polynomial $P_N(z) = P_N(\alpha_1, \beta_1; z)$ defined by (2.1) are all different among themselves, they satisfy the following family of algebraic identities [8]:*

$$2(\zeta_n - 1) \sum_{m=1, m\neq n}^{N} \frac{\zeta_m}{\zeta_n - \zeta_m} = N - 1 - \alpha_1 + \beta_1 \zeta_n, \tag{2.28}$$

where $n = 1, 2, \ldots, N$.

Let us now linearize system (2.27) about its equilibrium $\boldsymbol{\zeta} = (\zeta_1, \ldots, \zeta_N)$ to obtain linear system (2.10). The $N \times N$ matrix of coefficients $L \equiv L(\boldsymbol{\zeta})$ of system (2.10) is given componentwise by

$$L_{nm} = \frac{\partial}{\partial z_m} f_n^A(z_1, \ldots, z_N) \Big|_{z_j = \zeta_j,\ j=1,\ldots,N,} \tag{2.29}$$

where $f_n^A(z_1, \ldots, z_N)$ are given by (2.26). Its eigenvalues coincide with the eigenvalues (2.8) of the matrix of coefficients A of system (2.6), (2.7) (see the proof of the next theorem); therefore, the following Theorem holds.

Theorem 1. *Let $\boldsymbol{\zeta} = (\zeta_1, \ldots, \zeta_N)$ be a vector of the zeros of the generalized basic hypergeometric polynomial $P_N(z) = P_N(\alpha_1, \beta_1; z)$ defined by (2.1). If ζ_n are all different among themselves, then the $N \times N$ matrix $L \equiv L(\boldsymbol{\zeta})$ defined componentwise by [8]*

$$\begin{aligned} L_{nm} \equiv\ & L_{nm}(\alpha_1, \beta_1; N; \boldsymbol{\zeta}) = \delta_{nm} \left\{ \beta_1 + 2 \sum_{\ell=1; \ell\neq n}^{N} \left[\frac{\zeta_\ell\ (\zeta_\ell - 1)}{(\zeta_n - \zeta_\ell)^2} \right] \right\} \\ & -2\ (1 - \delta_{nm})\ \frac{\zeta_n\ (\zeta_n - 1)}{(\zeta_n - \zeta_m)^2} \end{aligned} \tag{2.30}$$

has the eigenvalues $\lambda_m \equiv \lambda_m(\beta_1; N) = m(\beta_1 - 1 - m)$, where $m = 1, 2, \ldots, N$.

Note that the eigenvalues λ_m of the last matrix L do not depend on the parameter α_1 and are moreover rational if β_1 is rational, a diophantine property.

Proof. Formulas (2.30) follow from (2.29). Let $\{\tilde{\lambda}_m\}_{m=1}^N$ be the eigenvalues of the matrix L. In case these eigenvalues are all different among themselves, a general solution of system (2.10) is given by

$$\boldsymbol{x}(t) = \sum_{j=1}^{N} \tilde{\eta}_m e^{\tilde{\lambda}_m t} \tilde{\boldsymbol{v}}_m, \tag{2.31}$$

where $\tilde{\eta}_m$ are N arbitrary constants and each $\tilde{\boldsymbol{v}}_m$ is an eigenvector of the matrix L that corresponds to the eigenvalue $\tilde{\lambda}_m$. But the behavior of the solution of the linearization (2.10) of system (2.27) in the vicinity of its equilibrium $\boldsymbol{\zeta}$ cannot differ from its general behavior. The solutions $\boldsymbol{z}(t)$ of system (2.27) are the zeros of the monic time-dependent polynomial (1.1) such that the behavior of its coefficients $\boldsymbol{c}(t)$ is characterized by the N exponential functions $e^{\lambda_m t}$, see (2.9). Therefore, the set of eigenvalues $\tilde{\lambda}_m$ of the matrix $L(\boldsymbol{\zeta})$ coincides with the set of eigenvalues λ_m, see (2.8), of the matrix of coefficients A of system (2.6), (2.7). ■

The Jacobi polynomials $P_N^{(\alpha,\beta)}(x)$ can be expressed in terms of the generalized hypergeometric polynomial $P_N(\alpha_1;\beta_1;z)$ defined by (2.1); see (1.8.1) in [31]:

$$P_N^{(\alpha,\beta)}(x) = \frac{(\alpha+1)_N}{N!}\left(\frac{1-x}{2}\right)^N P_N\left(N+\alpha+\beta+1;\alpha+1;\frac{2}{1-x}\right). \quad (2.32)$$

Therefore, the results of Corollary 1 and Theorem 1 can be recast in terms of the zeros $\boldsymbol{x} = (x_1, \ldots, x_N)$ of the Jacobi polynomial $P_N^{(\alpha,\beta)}(x)$. In this setting, Corollary 1 reduces to eqs. (5.2a) respectively (5.2b) of [3], while Theorem 1 produces results discovered more recently [8], which we state below.

Theorem 2. *If $\boldsymbol{x} = (x_1, \ldots, x_N)$ is a vector of distinct zeros of the Jacobi polynomial $P_N^{(\alpha,\beta)}(x)$, then the $N \times N$ matrix $L(\boldsymbol{x})$ defined componentwise by [8]*

$$L_{nm} \equiv L_{nm}(\alpha,\beta;N;\boldsymbol{x}) = \delta_{nm}\left\{\alpha+1+\sum_{\ell=1,\ \ell\neq n}^{N}\left[\frac{(1+x_\ell)\ (1-x_n)^2}{(x_n-x_\ell)^2}\right]\right\}$$
$$-(1-\delta_{nm})\left[\frac{(1+x_n)\ (1-x_m)^2}{(x_n-x_m)^2}\right] \quad (2.33)$$

has the N eigenvalues

$$\lambda_m \equiv \lambda_m(\alpha;N) = m\ (m+\alpha), \quad m = 1, \ldots, N. \quad (2.34)$$

Note the isospectral property of the matrix $L(\alpha,\beta;N;\boldsymbol{x})$: its elements depend on the zeros $\boldsymbol{x} \equiv \boldsymbol{x}(\alpha,\beta)$ and, via these zeros, on the two parameters α and β, while its eigenvalues depend only on the parameter α. Moreover, if α is rational, then the eigenvalues λ_m are rational, a diophantine property.

Let us also note that in the last Theorem, the parameters α, β need not be in the real range $(-1, +\infty)$ that ensures the orthogonality of the Jacobi polynomial family. However, it is assumed that the complex parameters α, β are such that the (possibly complex) zeros x_n of $P_N^{(\alpha,\beta)}(x)$ are all different among themselves. This is true, for example, for the quasiorthogonal case where $\alpha > -1$ and $-2 < \beta < -1$; see [26] and references therein.

3 Zeros of generalized hypergeometric polynomials

Using an approach similar to that described in Section 2, it is possible to prove algebraic identities and to construct isospectral matrices defined in terms of the zeros of the generalized hypergeometric polynomial (1.6). Before stating these results, let us introduce a few new definitions.

Given p complex parameters $\alpha_1, \ldots, \alpha_p$ and q complex parameters $\beta_1, \ldots, \beta_q$ of the generalized hypergeometric polynomial $P_N(\alpha_1, \ldots, \alpha_p; \beta_1, \ldots, \beta_q; z)$, let the complex numbers $a_0, \ldots, a_p$ and $b_1, \ldots, b_{q+1}$ be defined as the unique set of coefficients that makes the following polynomial equations true:

$$\prod_{j=1}^{p} (\alpha_j - x) = \sum_{j=0}^{p} a_j \, x^j, \tag{3.1}$$

$$x \prod_{k=1}^{q} (\beta_k - 1 - x) = \sum_{k=1}^{q+1} \left(b_k \, x^k \right). \tag{3.2}$$

Then

$$\begin{aligned} a_0 &= \prod_{j=1}^{p} (\alpha_j), \quad a_1 = -\sum_{k=1}^{p} \left[\prod_{j=1;\ j \neq k}^{p} (\alpha_j) \right], \\ a_2 &= \frac{1}{2} \sum_{\ell,k=1;\ell \neq k}^{p} \left[\prod_{j=1;\ j \neq \ell,k}^{p} (\alpha_j) \right], \ldots, \quad a_p = (-1)^p \end{aligned} \tag{3.3}$$

and

$$\begin{aligned} b_1 &= \prod_{k=1}^{q} (\beta_k - 1), \quad b_2 = -\sum_{j=1}^{q} \left[\prod_{k=1,\ k \neq j}^{q} (\beta_k - 1) \right], \\ b_3 &= \frac{1}{2} \sum_{\ell,j=1;\ell \neq j}^{q} \left[\prod_{k=1,\ k \neq \ell,j}^{q} (\beta_k - 1) \right], \ldots, b_{q+1} = (-1)^q. \end{aligned} \tag{3.4}$$

Given a vector $\boldsymbol{\zeta} = (\zeta_1, \ldots, \zeta_N)$ in $\mathbb{C}^N$ and an integer n such that $1 \leq n \leq N$, define the functions $f_n^{(j)}(\boldsymbol{\zeta})$ and $g_n^{(j)}(\boldsymbol{\zeta})$ as follows:

$$\begin{aligned} f_n^{(j+1)}(\boldsymbol{\zeta}) &= -f_n^{(j)}(\boldsymbol{\zeta}) + \sum_{\ell=1;\ \ell \neq n}^{N} \left[\frac{\zeta_n \, f_\ell^{(j)}(\boldsymbol{\zeta}) + \zeta_\ell \, f_n^{(j)}(\boldsymbol{\zeta})}{\zeta_n - \zeta_\ell} \right], \\ f_n^{(1)}(\boldsymbol{\zeta}) &= \zeta_n, \end{aligned} \tag{3.5}$$

and

$$\begin{aligned} g_n^{(j)}(\boldsymbol{\zeta}) &= \sum_{\ell=1;\ \ell \neq n}^{N} \left[\frac{f_n^{(j)}(\boldsymbol{\zeta}) + f_\ell^{(j)}(\boldsymbol{\zeta})}{\zeta_n - \zeta_\ell} \right], \\ g_n^{(0)}(\boldsymbol{\zeta}) &= 1, \end{aligned} \tag{3.6}$$

where $j = 1, 2, \ldots$. This definition is consistent with the notation (2.18), (2.19) of the previous section.

Using the notation

$$\sigma_n^{(r,\rho)}(\boldsymbol{\zeta}) = \sum_{\ell=1;\ell\neq n}^{N} \frac{(\zeta_\ell)^r}{(\zeta_n - \zeta_\ell)^\rho}, \tag{3.7}$$

one may express the above functions, for small values of j, as follows:

$$f_n^{(2)}(\boldsymbol{\zeta}) = \zeta_n \left[-1 + 2\,\sigma_n^{(1,1)}(\boldsymbol{\zeta})\right],$$

$$f_n^{(3)}(\boldsymbol{\zeta}) = \zeta_n \left\{1 - 6\,\sigma_n^{(1,1)}(\boldsymbol{\zeta}) - 3\,\sigma_n^{(2,2)}(\boldsymbol{\zeta}) + 3\left[\sigma_n^{(1,1)}(\boldsymbol{\zeta})\right]^2\right\},$$

$$f_n^{(4)}(\boldsymbol{\zeta}) = \zeta_n \left\{-1 + 14\,\sigma_n^{(1,1)}(\boldsymbol{\zeta}) + 18\,\sigma_n^{(2,2)}(\boldsymbol{\zeta}) + 8\,\sigma_n^{(3,3)}(\boldsymbol{\zeta}) \right.$$
$$\left. -18\left[\sigma_n^{(1,1)}(\boldsymbol{\zeta})\right]^2 - 12\,\sigma_n^{(1,1)}(\boldsymbol{\zeta})\,\sigma_n^{(2,2)}(\boldsymbol{\zeta}) + 4\left[\sigma_n^{(1,1)}(\boldsymbol{\zeta})\right]^3\right\},$$

and

$$g_n^{(1)}(\boldsymbol{\zeta}) = N - 1 + 2\,\sigma_n^{(1,1)}(\boldsymbol{\zeta}),$$

$$g_n^{(2)}(\boldsymbol{\zeta}) = 1 - N + 2\,(N-3)\,\sigma_n^{(1,1)}(\boldsymbol{\zeta}) - 3\,\sigma_n^{(2,2)}(\boldsymbol{\zeta}) + 3\left[\sigma_n^{(1,1)}(\boldsymbol{\zeta})\right]^2,$$

$$g_n^{(3)}(\boldsymbol{\zeta}) = N - 1 - 2\,(3N-7)\,\sigma_n^{(1,1)}(\boldsymbol{\zeta}) - \frac{9}{4}\,(N-7)\,\sigma_n^{(2,2)}(\boldsymbol{\zeta}) + 6\,\sigma_n^{(3,3)}(\boldsymbol{\zeta})$$
$$+\frac{9}{4}\,(N-7)\left[\sigma_n^{(1,1)}(\boldsymbol{\zeta})\right]^2 - 9\,\sigma_n^{(1,1)}(\boldsymbol{\zeta})\,\sigma_n^{(2,2)}(\boldsymbol{\zeta}) + 3\left[\sigma_n^{(1,1)}(\boldsymbol{\zeta})\right]^3.$$

Proposition 1. *Let $\boldsymbol{\zeta} = (\zeta_1, \ldots, \zeta_N)$ be a vector of the zeros of the generalized hypergeometric polynomial $P_N(\alpha_1, \ldots, \alpha_p; \beta_1, \ldots, \beta_q; z)$ defined by (1.6). If these zeros are all different among themselves, then they satisfy the following system of N algebraic equations [8]:*

$$\sum_{k=1}^{q+1}\left[b_k\; f_n^{(k)}(\boldsymbol{\zeta})\right] - \sum_{j=0}^{p}\left[a_j\; g_n^{(j)}(\boldsymbol{\zeta})\right] = 0, \quad n = 1, \ldots, N, \tag{3.8}$$

where the coefficients b_k and a_j are defined by (3.1), (3.2), while the functions $f_n^{(j)}(\boldsymbol{\zeta})$ and $g_n^{(j)}(\boldsymbol{\zeta})$ are defined by (3.5), (3.6).

Note that in the last proposition, the functions $f_n^{(j)}(\boldsymbol{\zeta})$ and $g_n^{(j)}(\boldsymbol{\zeta})$ are universal in the sense that they do not depend on the generalized hypergeometric polynomial under consideration.

Let us now introduce the following functions of the variable $\boldsymbol{\zeta} = (\zeta_1, \ldots, \zeta_N)$:

$$f_{n,m}^{(j)}(\boldsymbol{\zeta}) = \frac{\partial\, f_n^{(j)}}{\partial \zeta_m}(\boldsymbol{\zeta}), \quad g_{n,m}^{(j)}(\boldsymbol{\zeta}) = \frac{\partial\, g_n^{(j)}}{\partial\, \zeta_m}(\boldsymbol{\zeta}). \tag{3.9}$$

Explicit expressions for $f_{n,m}^{(j)}(\boldsymbol{\zeta})$ and $g_{n,m}^{(j)}(\boldsymbol{\zeta})$ for small values of j are reported in [8].

Proposition 2. *Let $\boldsymbol{\zeta} = (\zeta_1, \ldots, \zeta_N)$ be a vector of the zeros of the generalized hypergeometric polynomial $P_N(\alpha_1, \ldots, \alpha_p; \beta_1, \ldots, \beta_q; z)$ defined by (1.6). If these zeros are all different among themselves, then the $N \times N$ matrix $L(\boldsymbol{\zeta})$ defined componentwise by [8]*

$$L_{nm} \equiv L_{nm}(\alpha_1, \ldots, \alpha_p; \beta_1, \ldots, \beta_q; N; \boldsymbol{\zeta})$$
$$= \sum_{k=1}^{q+1} \left[b_k \, f_{n,m}^{(k)}(\boldsymbol{\zeta}) \right] - \sum_{j=1}^{p} \left[a_j \, g_{n,m}^{(j)}(\boldsymbol{\zeta}) \right], \tag{3.10}$$

where the coefficients a_j and b_k are defined by (3.1) and (3.2), while the functions $f_{n,m}^{(k)}(\boldsymbol{\zeta})$ and $g_{n,m}^{(j)}(\boldsymbol{\zeta})$ are defined by (3.9)*, has the N eigenvalues*

$$\lambda_m \equiv \lambda_m(\beta_1, \ldots, \beta_q; N) = m \prod_{k=1}^{q} (\beta_k - 1 + m), \quad m = 1, \ldots, N. \tag{3.11}$$

Note that in the last theorem, the functions $f_{n,m}^{(j)}(\boldsymbol{\zeta})$ and $g_{n,m}^{(j)}(\boldsymbol{\zeta})$ are universal: they do not depend on the generalized hypergeometric polynomial under consideration.

The $N \times N$ matrix L defined in the last theorem depends on the zeros $\boldsymbol{\zeta} = (\zeta_1, \ldots, \zeta_N)$ of the generalized hypergeometric polynomial $P_N(\alpha_1, \ldots, \alpha_p; \beta_1, \ldots, \beta_q; z)$ defined by (1.6) and, via these zeros, on the $p+q$ parameters $(\alpha_1, \ldots, \alpha_p; \beta_1, \ldots, \beta_q)$. This matrix is isospectral because its eigenvalues λ_m given by (3.11) depend only on the parameters $\beta_1, \ldots, \beta_q$. Moreover, these eigenvalues are rational if the parameters $\beta_1, \ldots, \beta_q$ are rational, a diophantine property.

An immediate generalization of Proposition 2 stems from the following observation. If in the generalized hypergeometric polynomial

$$Q_N(z) \equiv P_N(\alpha_1, \ldots, \alpha_p, \ldots, \alpha_{p+r}; \beta_1, \ldots, \beta_q, \ldots, \beta_{q+r}; z)$$

the last r parameters of the α and β type are equal, that is, $\alpha_{p+j} = \beta_{q+j}$ for all $j = 1, \ldots, r$, then the polynomial $Q_N(z)$ reduces to the generalized hypergeometric polynomial $P_N(z) \equiv P_N(\alpha_1, \ldots, \alpha_p; \beta_1, \ldots, \beta_q; z)$. In this setting, one may apply Proposition 2 to the polynomial $Q_N(z)$ to obtain an isospectral matrix defined in terms of the zeros of the polynomial $P_N(z) = Q_N(z)$ that is different from the matrix L defined by (3.10). For example, by considering the polynomial $P_N(\alpha_1, \alpha_2; \beta_1, \beta_2; z)$ with $\alpha_2 = \beta_2$, one can find an isospectral matrix different from (2.30) defined in terms of the zeros of the polynomial $P_N(\alpha_1; \beta_1; z)$, see Sect. 2.4 of [8].

4 Zeros of generalized basic hypergeometric polynomials

In this section, while summarizing the results of [10], we extend the method introduced in Section 2 to construct an isospectral matrix defined in terms of the zeros of generalized basic hypergeometric polynomial $P_N(z) \equiv P_N(\alpha_1, \ldots, \alpha_r; \beta_1, \ldots, \beta_s; q; z)$ defined in (1.12). To this end, let us consider the differential q-difference equation (DqDE)

$$\frac{\partial}{\partial t}\psi(z,t) = \mathcal{A}\psi(z,t), \tag{4.1a}$$

where $\mathcal{A}$ is the linear q-difference operator

$$\mathcal{A} = z^{-1}\left[\Delta_1 \prod_{k=1}^{s}(\Delta_{\beta_k/q}) - z\,\Delta_{q^{-N}}\prod_{j=1}^{r}(\Delta_{\alpha_j})(\delta_q)^{s-r}\right] \tag{4.1b}$$

and note that the generalized basic hypergeometric polynomial $P_N(z)$ is a t-independent solution of the last DqDE, see (1.17). The operator $\mathcal{A}$ equals the operator in the left-hand side of the q-difference equation (1.17), times a factor of z^{-1}. The presence of this factor ensures existence of monic polynomial solutions $\psi_N(z,t)$, given by (1.1), of DqDE (4.1). More precisely, because $\frac{\partial}{\partial t}\psi_N(z,t)$ is a polynomial (in z) of degree at most $N-1$, for DqDE (4.1) to have a polynomial solution $\psi_N(z,t)$, $\mathcal{A}\psi_N(z,t)$ must be a polynomial of degree at most $N-1$. Let us confirm that this is indeed the case.

Recall that the operator Δ_γ does not raise degrees of polynomials when acting on them; see the remark following display (1.15). Also, the operator Δ_1 annihilates functions independent of z, while the operator $\Delta_{q^{-N}}$ annihilates z^N. Therefore, the expression

$$z\mathcal{A}\psi_N(z,t) = \left[\Delta_1 \prod_{k=1}^{s}(\Delta_{\beta_k/q}) - z\,\Delta_{q^{-N}}\prod_{j=1}^{r}(\Delta_{\alpha_j})(\delta_q)^{s-r}\right]\psi_N(z,t)$$

is a polynomial in z of degree at most N with zero constant term, rendering $\mathcal{A}\psi_N(z,t)$ to be a polynomial in z of degree at most $N-1$. Upon substitution of the anzatz $\psi_N(z,t) = z^N + \sum_{n=1}^{N} c_n(t) z^{N-n}$ into equation (4.1), we find that $\psi_N(z,t)$ solves the equation if and only if the coefficients $\boldsymbol{c}(t) = \left(c_1(t), \ldots, c_N(t)\right)$ satisfy the following linear system of ODEs:

$$\begin{aligned}\dot{c}_m(t) = &-\left[q^{(s-r)(N-m)}\left(q^{-m}-1\right)\prod_{j=1}^{r}\left(\alpha_j\, q^{N-m}-1\right)\right]c_m(t),\\ &+\left[\left(q^{N-m+1}-1\right)\prod_{k=1}^{s}\left(\beta_k\, q^{N-m}-1\right)\right]c_{m-1}(t)\\ &m = 1, 2, \ldots, N, \quad c_0 = 1.\end{aligned} \tag{4.2}$$

In matrix form, the last system reads

$$\dot{\boldsymbol{c}}(t) = A\boldsymbol{c}(t) + \boldsymbol{h}, \tag{4.3}$$

where A is a lower diagonal $N \times N$ matrix with the N eigenvalues

$$\lambda_m = -q^{(s-r)(N-m)}\left(q^{-m}-1\right)\prod_{j=1}^{r}\left(\alpha_j\, q^{N-m}-1\right), \quad m=1,2,\ldots,N \tag{4.4}$$

and $\boldsymbol{h}$ is the N-vector $\boldsymbol{h} = \left((q^N-1)\prod_{k=1}^{s}(\beta_k q^{N-1}-1), 0, \ldots, 0\right)^T$.

On the other hand, by a substitution of $\psi_N(z,t) = \prod_{n=1}^{N}[z - z_n(t)]$ into (4.1), we obtain the following nonlinear system of ODEs for the zeros $\boldsymbol{z}(t)$:

$$\begin{aligned}
\dot{z}_n = (-1)^{s+1}\Bigg\{ & (q-1) f_n(1,\boldsymbol{z}) \\
& + \sum_{k=1}^{s} b_k \frac{(-1)^k}{q^k}\Big[(q^{k+1}-1) f_n(k+1,\boldsymbol{z}) - (q^k-1) f_n(k,\boldsymbol{z})\Big]\Bigg\} \\
& + (-1)^r z_n \Bigg\{ q^{-N}(q^{s-r+1}-1) f_n(s-r+1,\boldsymbol{z}) - (q^{s-r}-1) f_n(s-r,\boldsymbol{z}) \\
& + \sum_{j=1}^{r} a_j (-1)^j \Big[q^{-N}(q^{j+s+1-r}-1) f_n(j+s+1-r,\boldsymbol{z}) \\
& - (q^{j+s-r}-1) f_n(j+s-r,\boldsymbol{z})\Big]\Bigg\},
\end{aligned} \tag{4.5}$$

where

$$f_n(p,\boldsymbol{z}) = f_n(p, z_1, \ldots, z_N) = \prod_{\ell=1,\ell\neq n}^{N}\left(\frac{q^p\, z_n - z_\ell}{z_n - z_\ell}\right), \quad n=1,2,\ldots,N \tag{4.6}$$

and the complex coefficients $\{a_j\}_{j=1}^{r}$ and $\{b_k\}_{k=1}^{s}$ are defined as the unique set of complex numbers that satisfy the following polynomial identities:

$$\prod_{j=1}^{r}(1+\alpha_j\, x) = 1 + \sum_{j=1}^{r} a_j\, x^j, \tag{4.7}$$

$$\prod_{k=1}^{s}(1+\beta_k\, x) = 1 + \sum_{k=1}^{s} b_k\, x^k. \tag{4.8}$$

We thus proved the following Lemma.

Lemma 2. *A time-dependent monic polynomial* (1.1) *with distinct zeros* $z_m(t)$ *satisfies DqDE* (4.1) *if and only if its coefficients* $\mathbf{c}(t)$ *satisfy linear system* (4.2) *if and only if its zeros* $\boldsymbol{z}(t) = \left(z_1(t), \ldots, z_N(t)\right)$ *satisfy nonlinear system* (4.5). *In particular, system* (4.5) *is solvable in terms of algebraic operations.*

Because the generalized basic hypergeometric polynomial $P_N(z)$ defined by (1.12) is a t-independent solution of DqDE (4.1), every vector $\boldsymbol{\zeta} = (\zeta_1, \ldots, \zeta_N)$ of the zeros of $P_N(z)$ is an equilibrium of system (4.5), provided that the zeros are distinct. We thus obtain the following algebraic identities satisfied by the zeros of $P_N(z)$.

Proposition 3. *Let* $\boldsymbol{\zeta} = (\zeta_1, \ldots, \zeta_N)$ *be a vector of zeros of the generalized basic hypergeometric polynomial* $P_N(\alpha_1, \ldots, \alpha_r; \beta_1, \ldots, \beta_s; q; z)$ *defined by* (1.12). *If these zeros are all different among themselves, they satisfy the following set of N algebraic equations [10]:*

$$
\begin{aligned}
&-\prod_{m=1}^{N}(\zeta_n\, q - \zeta_m) + \sum_{k=1}^{s}(-q)^{-k} b_k \left[\prod_{m=1}^{N}\left(\zeta_n\, q^k - \zeta_m\right) - \prod_{m=1}^{N}\left(\zeta_n\, q^{k+1} - \zeta_m\right)\right] \\
&-(-1)^{r-s}\, \zeta_n \left[\prod_{m=1}^{N}\left(\zeta_n\, q^{s-r} - \zeta_m\right) - q^{-N}\prod_{m=1}^{N}\left(\zeta_n\, q^{s-r+1} - \zeta_m\right)\right] \\
&-(-1)^{r-s}\, \zeta_n \left\{\sum_{j=1,\ j\neq r-s}^{r}\left[(-1)^j a_j \prod_{m=1}^{N}\left(\zeta_n\, q^{s-r+j} - \zeta_m\right)\right]\right. \\
&\left. -q^{-N} \sum_{j=1,\ j\neq r-s-1}^{r}\left[(-1)^j a_j \prod_{m=1}^{N}\left(\zeta_n\, q^{s-r+j+1} - \zeta_m\right)\right]\right\} = 0, \\
&n = 1, 2, .., N,
\end{aligned}
\tag{4.9}
$$

where the sets of coefficients $\{a_j\}_{j=1}^{r}$ *and* $\{b_k\}_{k=1}^{s}$ *are defined by identities* (4.7) *and* (4.8) *in terms of the parameters* $\alpha_1, \ldots, \alpha_r$ *and* $\beta_1, \ldots, \beta_s$, *respectively.*

The linearization of system (4.5) about its equilibrium $\boldsymbol{\zeta} = (\zeta_1, \ldots, \zeta_N)$, a vector of zeros of the polynomial (1.12), yields the following system of ODEs:

$$
\dot{\boldsymbol{x}}(t) = L\boldsymbol{x}(t), \tag{4.10}
$$

where the $N \times N$ matrix $L \equiv L(\boldsymbol{\zeta})$ is given by (4.11). Using an argument similar to the one employed in the proof of Theorem 1, one may conclude that the set of eigenvalues of L coincides with the set of eigenvalues of the matrix of coefficients A in the linear system (4.3). Hence the following proposition holds.

Proposition 4. *Let* $\boldsymbol{\zeta} = (\zeta_1, \ldots, \zeta_N)$ *be a vector of the zeros of the generalized basic hypergeometric polynomial* $P_N(\alpha_1, \ldots, \alpha_r; \beta_1, \ldots, \beta_s; q; z)$ *defined by* (1.12). *If these zeros are all different among themselves, then the* $N \times N$ *matrix* $L(\boldsymbol{\zeta})$

defined componentwise by [10]

$$
\begin{aligned}
&L_{nn} \equiv L_{nn}\left(\alpha_1, \ldots, \alpha_r; \beta_1, \ldots, \beta_s; q; N; \zeta\right) \\
&= (-1)^s \Bigg\{ (q-1)^2 g_n(1, \zeta) \\
&+ \sum_{k=1}^{s} b_k \frac{(-1)^k}{q^k} \left[(q^{k+1}-1)^2 g_n(k+1, \zeta) - (q^k - 1)^2 g_n(k, \zeta) \right] \Bigg\} \\
&+ (-1)^{r+1} \zeta_n \Bigg\{ q^{-N} (q^{s-r+1} - 1)^2 g_n(s-r+1, \zeta) - (q^{s-r}-1)^2 g_n(s-r, \zeta) \\
&+ \sum_{j=1}^{r} a_j (-1)^j \left[q^{-N} (q^{j+s+1-r} - 1)^2 g_n(j+s+1-r, \zeta) \right. \\
&\left. - (q^{j+s-r} - 1)^2 g_n(j+s-r, \zeta) \right] \Bigg\} \\
&+ (-1)^r \Bigg\{ q^{-N} (q^{s-r+1} - 1) f_n(s-r+1, \zeta) - (q^{s-r} - 1) f_n(s-r, \zeta) \\
&+ \sum_{j=1}^{r} a_j (-1)^j \Big[q^{-N} (q^{j+s+1-r} - 1) f_n(j+s+1-r, \zeta) \\
&- (q^{j+s-r} - 1) f_n(j+s-r, \zeta) \Big] \Bigg\}, \\
&n = 1, 2, \ldots, N,
\end{aligned}
\tag{4.11a}
$$

$$
\begin{aligned}
&L_{nm} \equiv L_{nm}\left(\alpha_1, \ldots, \alpha_r; \beta_1, \ldots, \beta_s; q; N; \zeta\right) \\
&= (-1)^{s+1} \frac{\zeta_n}{(\zeta_n - \zeta_m)^2} \Bigg\{ (q-1)^2 f_{nm}(1, \zeta) \\
&+ \sum_{k=1}^{s} b_k \frac{(-1)^k}{q^k} \left[(q^{k+1}-1)^2 f_{nm}(k+1, \zeta) - (q^k - 1)^2 f_{nm}(k, \zeta) \right] \Bigg\} \\
&+ (-1)^r \frac{\zeta_n^2}{(\zeta_n - \zeta_m)^2} \Bigg\{ q^{-N} (q^{s-r+1} - 1)^2 f_{nm}(s-r+1, \zeta) \\
&- (q^{s-r} - 1)^2 f_{nm}(s-r, \zeta) \\
&+ \sum_{j=1}^{r} a_j (-1)^j \left[q^{-N} (q^{j+s+1-r} - 1)^2 f_{nm}(j+s+1-r, \zeta) \right. \\
&\left. - (q^{j+s-r} - 1)^2 f_{nm}(j+s-r, \zeta) \right] \Bigg\}, \\
&n, m = 1, 2, \ldots, N, \quad n \neq m,
\end{aligned}
\tag{4.11b}
$$

where

$$f_{nm}(p,\zeta) = \prod_{\ell=1,\ell\neq n,m}^{N} \left(\frac{q^p\, \zeta_n - \zeta_\ell}{\zeta_n - \zeta_\ell} \right), \tag{4.12}$$

$$g_n(p,\zeta) = \sum_{k=1,k\neq n}^{N} \left[f_{nk}(p,\zeta)\, \frac{\zeta_k}{(\zeta_n - \zeta_k)^2} \right], \tag{4.13}$$

and the coefficients a_j and b_k are defined by (4.7) and (4.8), has the N eigenvalues

$$\lambda_n \equiv \lambda_n(\alpha_1,\ldots,\alpha_r;q;N) = -q^{(s-r)(N-n)}\left(q^{-n}-1\right)\prod_{j=1}^{r}\left(\alpha_j\, q^{N-n}-1\right),$$

$$n = 1,2,\ldots,N. \tag{4.14}$$

Note that the eigenvalues λ_n of the last matrix L do not depend on the parameters $(\beta_1,\ldots,\beta_s)$ and are moreover rational if the parameters $(\alpha_1,\ldots,\alpha_r;q)$ are rational, a diophantine property.

5 Zeros of Wilson and Racah polynomials

Wilson and Racah polynomials are at the top of the Askey scheme of orthogonal polynomials [31]. They are defined in terms of the generalized hypergeometric function (1.5), yet are not particular cases of the generalized hypergeometric polynomial (1.6). In this section we summarize the results of [9] and provide isospectral matrices defined in terms of the zeros of these polynomials.

5.1 Wilson Polynomials

The N-th degree Wilson polynomial $W_N(z;a,b,c,d)$ with $z = x^2$ is defined by

$$W_N(z) \equiv W_N(z;a,b,c,d) = (a+b)_N\ (a+c)_N\ (a+d)_N$$
$$\cdot {}_4F_3\left(\begin{matrix} -N,\ N+a+b+c+d-1,\ a+\mathbf{i}x,\ a-\mathbf{i}x \\ a+b,\ a+c,\ a+d \end{matrix} \middle| 1 \right), \tag{5.1a}$$

see [31], or equivalently, by

$$W_N(z;a,b,c,d) = (a+b)_N\ (a+c)_N\ (a+d)_N$$
$$\cdot \sum_{k=0}^{N} \left[\frac{(-N)_k\ (N+a+b+c+d-1)_k\ [a;z]_k}{k!\ (a+b)_k\,(a+c)_k\,(a+d)_k} \right], \tag{5.1b}$$

see [9], where a,b,c,d are complex parameters such that, after appropriate cancellations, the denominators in (5.1b) do not vanish, $\mathbf{i}$ is the imaginary unit and the modified Pochhammer symbol

$$\left[a;x^2\right]_k = (a+\mathbf{i}x)_k\ (a-\mathbf{i}x)_k\,,$$
$$[a;z]_0 = 1,$$
$$[a;z]_k = \left(a^2+z\right)\left[(a+1)^2+z\right]\cdots\left[(a+k-1)^2+z\right],$$
$$k = 1,2,3,\ldots \tag{5.2}$$

Because $[a;z]_k$ is a polynomial of degree k in z, the Wilson polynomial $W_N(z;a,b,c,d)$ is indeed a polynomial of degree N in $z=x^2$, and of degree $2N$ in x.

The polynomial

$$w_{2N}(x) \equiv w_{2N}(x;a,b,c,d) = W_N(x^2;a,b,c,d) \tag{5.3}$$

satisfies the following difference equation [31]:

$$\Big[B(-x)\left(1-\delta_{\mathbf{i}}^{(+)}\right) + B(x)\left(1-\delta_{\mathbf{i}}^{(-)}\right) \\ +N(N+a+b+c+d-1)\Big]w_{2N}(x) = 0, \tag{5.4}$$

where $B(x)$ is defined by

$$B(x) \equiv B(x;a,b,c,d) = \frac{(a+\mathbf{i}\,x)(b+\mathbf{i}\,x)(c+\mathbf{i}\,x)(d+\mathbf{i}\,x)}{2\,\mathbf{i}\,x\,(2\,\mathbf{i}\,x+1)}, \tag{5.5}$$

and the operator $\delta_\gamma^{(\pm)}$ is defined by

$$\delta_\gamma^{(\pm)} f(x) = f(x \pm \gamma). \tag{5.6}$$

Consider the following differential difference equation (DDE):

$$\frac{\partial \psi(x,t)}{\partial t} = \mathcal{A}\psi(x,t), \tag{5.7a}$$

where $\mathcal{A}$ is the following difference operator:

$$\mathcal{A} = \mathbf{i}\Big[B(-x)\,(1-\delta_{\mathbf{i}}^{(+)}) + B(x)\,(1-\delta_{\mathbf{i}}^{(-)}) + N(N+a+b+c+d-1)\Big]. \tag{5.7b}$$

The last DDE has solutions that are t-dependent monic polynomials of degree N. Indeed, let $p_\ell(z) \equiv p_\ell(z;a,b,c,d)$ be the monic version of the Wilson polynomial $W_\ell(z)$ of degree ℓ defined by (5.1) with $N=\ell$, and let

$$\psi_{2N}(x,t) = p_N(x^2) + \sum_{m=1}^{N}\left[c_m(t)\; p_{N-m}(x^2)\right] = \prod_{n=1}^{N}\left[x^2 - x_n^2(t)\right] \tag{5.8}$$

be a symmetric monic polynomial in x of degree $2N$, expressed in terms of its coefficients $c_m(t)$ in the basis $\left\{p_{N-m}(x^2)\right\}_{m=1}^{N}$ and its zeros $\{\pm x_n(t)\}_{n=1}^{N}$. A substitution of $\psi_{2N}(x,t)$ into (5.7) yields the following decoupled linear system of ODEs for the coefficients $c_m(t)$:

$$\dot{c}_m(t) = \mathbf{i}\,m\,(2N-m+\alpha_1-1)\,c_m(t) \tag{5.9}$$

and the following nonlinear system of ODEs for the zeros $x_n(t)$:

$$2x_n\dot{x}_n = -\mathbf{i}\left\{\frac{D(x_n)}{2\,\mathbf{i}\,x_n}\left[\prod_{m=1,\;m\neq n}^{N}\left(\frac{x_n^2-x_m^2-1-2\,\mathbf{i}\,x_n}{x_n^2-x_m^2}\right)\right] \\ +\Big[(x_s \to (-x_s))\Big]\right\}, \tag{5.10}$$

where

$$D\left(x_n\right) = \alpha_4 + \mathbf{i}\,\alpha_3\, x_n - \alpha_2\, x_n^2 - \mathbf{i}\,\alpha_1\, x_n^3 + x_n^4, \tag{5.11a}$$

$$\begin{aligned}
\alpha_1 &= a + b + c + d,\\
\alpha_2 &= ab + ac + ad + bc + bd + cd,\\
\alpha_3 &= bcd + acd + abd + abc,\\
\alpha_4 &= abcd
\end{aligned} \tag{5.11b}$$

and the symbol $+\Big[\left(x_s \rightarrow (-x_s)\right)\Big]$ denotes the addition of everything that comes before it (within the curly brackets), with the replacement of x_s with $(-x_s)$ for all $s = 1, 2, \ldots, N$. Because system (5.9) is explicitly solvable, we conclude that DDE (5.7) has monic symmetric polynomial solutions of the form (5.8) and that the nonlinear system (5.10) is solvable in terms of algebraic operations (indeed its solution is a vector function of zeros $x_n(t)$ of $\psi_{2N}(x,t)$).

Because the polynomial $w_{2N}(x) = W_N(x^2)$ defined by (5.3) is a t-independent solution of DDE (5.7), see (5.4), for every vector $\bar{\boldsymbol{z}} = (\bar{z}_1, \ldots, \bar{z}_N) = (\bar{x}_1^2, \ldots, \bar{x}_N^2)$ of the zeros of the Wilson polynomial $W_N(z)$ with distinct components satisfying $\bar{z}_n \neq 0$, the vector $\bar{\boldsymbol{x}} = (\bar{x}_1, \ldots, \bar{x}_N)$ is an equilibrium of system (5.10). By linearizing system (5.10) about the equilibrium $\bar{\boldsymbol{x}}$, we obtain the following linear system of ODEs:

$$\dot{\boldsymbol{\xi}}(t) = \mathbf{i}\, L\, \boldsymbol{\xi}(t), \tag{5.12}$$

where the $N \times N$ matrix $L \equiv L(\bar{\boldsymbol{x}})$ is given by (5.13). Using an argument similar to the one employed in the proof of Theorem 1, one may conclude that the set of eigenvalues of L coincides with the set of the eigenvalues of the (diagonal) matrix of coefficients A in the linear system (5.9). Hence the following proposition holds.

Proposition 5. *Suppose that $\bar{\boldsymbol{z}} = (\bar{z}_1, \ldots, \bar{z}_N) = (\bar{x}_1^2, \ldots, \bar{x}_N^2)$ is a vector of zeros of Wilson polynomial $W_N(z; a, b, c, d) \equiv W_N\left(x^2; a, b, c, d\right)$ of degree N in $z = x^2$ defined by (5.1). If these zeros are all different among themselves and such that $\bar{z}_n = \bar{x}_n^2 \neq 0$ for all $n = 1, \ldots, N$, then the $N \times N$ matrix L defined componentwise by [9]*

$$\begin{aligned}
L_{nn} &\equiv L_{nn}(a, b, c, d; N; \bar{\boldsymbol{x}})\\
&= (2\bar{x}_n)^{-2} \left\{ \left[\frac{2D\left(\bar{x}_n\right)}{\mathbf{i}\bar{x}_n} + \mathbf{i}D'\left(\bar{x}_n\right) \right] \prod_{\ell=1,\ \ell\neq n}^{N} \left(1 - \frac{1 + 2\mathbf{i}\bar{x}_n}{\bar{x}_n^2 - \bar{x}_\ell^2}\right) \right.\\
&\quad + 2D\left(\bar{x}_n\right) \sum_{m=1,\ m\neq n} \frac{\mathbf{i}\bar{x}_n - \left(\bar{x}_n^2 + \bar{x}_m^2\right)}{\left(\bar{x}_n^2 - \bar{x}_m^2\right)^2} \prod_{\ell=1,\ \ell\neq n,m}^{N} \left(1 - \frac{1 + 2\mathbf{i}\bar{x}_n}{\bar{x}_n^2 - \bar{x}_\ell^2}\right)\\
&\quad \left. + \left[\left(\bar{x}_s \rightarrow (-\bar{x}_s)\right)\right] \right\}, \quad n = 1, 2, \ldots, N,
\end{aligned} \tag{5.13a}$$

$$L_{nm} \equiv L_{nm}(a,b,c,d;N;\bar{\boldsymbol{x}})$$
$$= -\left(2\bar{x}_n\right)^{-2}\left\{2D\left(\bar{x}_n\right)\frac{\mathrm{i}\bar{x}_m\left(1+2\mathrm{i}\bar{x}_n\right)}{\left(\bar{x}_n^2-\bar{x}_m^2\right)^2}\prod_{\ell=1,\ \ell\neq n,m}^{N}\left(1-\frac{1+2\mathrm{i}\bar{x}_n}{\bar{x}_n^2-\bar{x}_\ell^2}\right)\right.$$
$$\left.+\left[\left(\bar{x}_s\to\left(-\bar{x}_s\right)\right)\right]\right\},\quad m,n=1,2,\ldots,N,\quad m\neq n, \tag{5.13b}$$

where $D\left(\bar{x}_n\right)$ is defined by (5.11) *and the symbol* $+\left[\left(\bar{x}_s\to\left(-\bar{x}_s\right)\right)\right]$ *denotes the addition of everything that comes before it (within the curly brackets), with the replacement of $\bar{x}_s$ with $(-\bar{x}_s)$ for all $s=1,2,\ldots,N$, has the N eigenvalues*

$$\lambda_m \equiv \lambda_m(a+b+c+d;N) = m\,(2N-m+a+b+c+d-1),$$
$$m=1,2,\ldots,N. \tag{5.14}$$

Note that the last matrix elements L_{nm} in (5.13) are functions of $\bar{z}_s=\bar{x}_s^2$ because all the terms in L_{nm} that are odd functions of $\bar{x}_s$ cancel due to the addition implied by the symbol $+\left[\left(\bar{x}_s\to\left(-\bar{x}_s\right)\right)\right]$. The eigenvalues λ_m of the last matrix L depend only on the sum $a+b+c+d$ of the parameters a,b,c,d as opposed to all the parameters a,b,c,d. These eigenvalues are moreover rational if this sum is rational, a diophantine property.

5.2 Racah Polynomials

The N-th degree Racah polynomial is defined by

$$R_N(\lambda(x);\alpha,\beta,\gamma,\delta)$$
$$={}_4F_3\left(\begin{array}{c}-N,\ N+\alpha+\beta+1,\ -x,\ x+\gamma+\delta+1\\ \alpha+1,\ \beta+\delta+1,\ \gamma+1\end{array}\middle|\,1\right), \tag{5.15a}$$

see [31], or, equivalently, by

$$R_N(\lambda(x);\alpha,\beta,\gamma,\delta)=\sum_{n=0}^{N}\frac{(-N)_n(N+\alpha+\beta+1)_n[\lambda(x)]_n}{n!(\alpha+1)_n(\beta+\delta+1)_n(\gamma+1)_n}, \tag{5.15b}$$

where $\lambda(x)=x(x+\gamma+\delta+1)$, $(a)_n$ is the Pochhammer symbol,

$$\begin{aligned}[\lambda(x)]_n &= (-x)_n(x+\gamma+\delta+1)_n\\ &= -\lambda(x)\left[-\lambda(x)+(\gamma+\delta+1)+1\right]\left[-\lambda(x)+2(\gamma+\delta+1)+2^2\right]\\ &\quad\cdots\left[-\lambda(x)+(n-1)(\gamma+\delta+1)+(n-1)^2\right],\end{aligned} \tag{5.15c}$$

and $\alpha,\beta,\gamma,\delta$ are complex parameters such that, after appropriate cancellations, the denominators in the finite sum (5.15b) do not vanish.

The standard definition of Racah polynomials imposes the restriction $\alpha+1=-M$ or $\beta+\delta+1=-M$ or $\gamma+1=-M$ on the complex parameters $\alpha,\beta,\gamma,\delta$, with M being a nonnegative integer, together with the inequality $0\le N\le M$. However, in this study, none of the last diophantine relations is required or assumed. This is because the only property of Racah polynomials used in the construction of the isospectral matrices provided below is that these polynomials satisfy difference equation (5.19), which is valid even if the diophantine restrictions mentioned above do not hold.

Consider the monic polynomial

$$q_{2N}(x) = \frac{(\alpha+1)_N(\beta+\delta+1)_N(\gamma+1)_N}{(N+\alpha+\beta+1)_N} R_N(\lambda(x);\alpha,\beta,\gamma,\delta) \tag{5.16}$$

of degree $2N$ in x and the related polynomial

$$\tilde{q}_{2N}(y) = q_{2N}(y-\theta), \tag{5.17}$$

where

$$y = x+\theta, \quad \theta = \frac{\gamma+\delta+1}{2}, \tag{5.18}$$

so that $\lambda(x) = x(x+2\theta) = y^2 - \theta^2$. This last polynomial satisfies the difference equation

$$\left[\tilde{D}(y)(\delta_1^{(+)}-1) + \tilde{D}(-y)(\delta_1^{(-)}-1) - N\,(N+\alpha+\beta+1)\right]\tilde{q}_{2N}(y) = 0, \tag{5.19a}$$

where

$$\begin{aligned}\tilde{D}(y) \;&=\; [32\,y\,(2y+1)]^{-1}\,(2y+\gamma+\delta+1)(2y+\gamma-\delta+1) \\ &\quad \cdot(2y+2\alpha-\gamma-\delta+1)(2y+2\beta-\gamma+\delta+1)\end{aligned} \tag{5.19b}$$

and $\delta_1^{(\pm)} f(y) = f(y\pm 1)$; see (5.6).

Let us now consider the DDE

$$\frac{\partial}{\partial t} f(y,t) = \mathcal{A} f(y,t), \tag{5.20a}$$

where

$$\mathcal{A} = \mathbf{i}\left[\tilde{D}(y)(\delta_1^{(+)}-1) + \tilde{D}(-y)(\delta_1^{(-)}-1) - N(N+\alpha+\beta+1)\right]. \tag{5.20b}$$

The t-dependent symmetric monic polynomial in the y variable

$$\tilde{\psi}_{2N}(y,t) = \tilde{q}_{2N}(y) + \sum_{m=1}^{N} c_m(t)\tilde{q}_{(2N-2m)}(y) = \prod_{m=1}^{N}\left[y^2 - y_m^2(t)\right] \tag{5.21}$$

solves the last DDE (5.20) if and only if its coefficients $c_m(t)$ (with respect to the basis $\{\tilde{q}_{(2N-2m)}(y)\}_{m=1}^{N}$) satisfy the decoupled linear system of ODEs

$$\dot{c}_m(t) = \mathbf{i}\, m\,(m-2N-\alpha-\beta-1)\, c_m(t) \tag{5.22}$$

if and only if the zeros $y_n(t)$ satisfy the nonlinear system of ODEs

$$2y_n\dot{y}_n = -\mathbf{i}\left\{\tilde{D}(y_n)(2y_n+1)\prod_{\ell=1,\ell\neq n}^{N}\left(1+\frac{1+2y_n}{y_n^2-y_\ell^2}\right) + [(y_s\to(-y_s))]\right\}, \tag{5.23}$$

where, again, the symbol $+[(y_s \to (-y_s))]$ denotes the addition of the expression preceding it within the curly brackets, with y_s replaced by $(-y_s)$ for all $s = 1, 2, \ldots, N$. Because system (5.22) is explicitly solvable, we conclude that DDE (5.20) has symmetric monic polynomial solutions of the form (5.21) and that the nonlinear system (5.23) is solvable in terms of algebraic operations (indeed its solution is a vector function of zeros $y_n(t)$ of $\tilde{\psi}_{2N}(y,t)$).

Because $\tilde{q}_{2N}(y) = (\alpha+1)_N(\beta+\delta+1)_N(\gamma+1)_N[(N+\alpha+\beta+1)_N]^{-1}R_N(\lambda(y-\theta); \alpha, \beta, \gamma, \delta)$ is a t-independent solution of DDE (5.20), every vector of its zeros $\bar{\boldsymbol{y}} = (\bar{y}_1, \ldots, \bar{y}_N)$ is an equilibrium of the nonlinear system (5.23). By linearizing the last system about the equilibrium $\bar{\boldsymbol{y}}$, we obtain the linear system

$$\dot{\boldsymbol{\eta}}(t) = \mathbf{i}\, L\, \boldsymbol{\eta}(t), \tag{5.24}$$

where the $N \times N$ matrix $L \equiv L(\bar{\boldsymbol{y}})$ is given by (5.25). Arguing as in the proof of Theorem 1, one may conclude that the set of eigenvalues of L coincides with the set of eigenvalues of the (diagonal) matrix of coefficients of system (5.22); thus the following proposition holds.

Proposition 6. *Suppose that $\bar{\boldsymbol{z}} = (\bar{z}_1, \ldots, \bar{z}_N) = (\bar{y}_1^2 - \theta^2, \ldots, \bar{y}_N^2 - \theta^2)$ is a vector of zeros of the Racah polynomial $R_N(z; \alpha, \beta, \gamma, \delta) \equiv R_N(y^2 - \theta^2; \alpha, \beta, \gamma, \delta)$ of degree N in $z = y^2 - \theta^2$ defined by (5.15), where $\theta = (\gamma + \delta + 1)/2$. If these zeros are all different among themselves and such that $\bar{z}_n + \theta^2 = \bar{y}_n^2 \neq 0$ for all $n = 1, \ldots, N$, then the $N \times N$ matrix L defined componentwise by [9]*

$$\begin{aligned} L_{nn} &\equiv L_{nn}(\alpha, \beta, \gamma, \delta; N; \bar{\boldsymbol{y}}) \\ &= \frac{1}{2}\Bigg\{ \Bigg[\left(\frac{\tilde{D}(\bar{y}_n)}{\bar{y}_n^2} - \frac{\tilde{D}'(\bar{y}_n)}{\bar{y}_n} \right)(1 + 2\bar{y}_n) - 2\frac{\tilde{D}(\bar{y}_n)}{\bar{y}_n} \Bigg] \prod_{\ell=1, \ell \neq n}^{N} \left(1 + \frac{1 + 2\bar{y}_n}{\bar{y}_n^2 - \bar{y}_\ell^2} \right) \\ &+ 2\frac{\tilde{D}(\bar{y}_n)}{\bar{y}_n}(1 + 2\bar{y}_n) \sum_{m=1, m \neq n}^{N} \left[\frac{\bar{y}_n^2 + \bar{y}_m^2 + \bar{y}_n}{(\bar{y}_n^2 - \bar{y}_m^2)^2} \prod_{\ell=1, \ell \neq n, m}^{N} \left(1 + \frac{1 + 2\bar{y}_n}{\bar{y}_n^2 - \bar{y}_\ell^2} \right) \right] \\ &+ [(\bar{y}_s \to (-\bar{y}_s))] \Bigg\}, \quad n = 1, 2, \ldots, N \end{aligned} \tag{5.25a}$$

and

$$\begin{aligned} L_{nm} &\equiv L_{nm}(\alpha, \beta, \gamma, \delta; N; \bar{\boldsymbol{y}}) \\ &= -\frac{1}{(\bar{y}_n^2 - \bar{y}_m^2)^2} \Bigg\{ \frac{\bar{y}_m \tilde{D}(\bar{y}_n)}{\bar{y}_n}(1 + 2\bar{y}_n)^2 \prod_{\ell=1, \ell \neq n, m}^{N} \left(1 + \frac{1 + 2\bar{y}_n}{\bar{y}_n^2 - \bar{y}_\ell^2} \right) \\ &+ \Big[(\bar{y}_s \to (-\bar{y}_s)) \Big] \Bigg\}, \quad n, m = 1, 2, \ldots, N, \quad n \neq m, \end{aligned} \tag{5.25b}$$

where $\tilde{D}(y)$ is defined by (5.19b) and again the symbol $+\Big[(\bar{y}_s \to (-\bar{y}_s)) \Big]$ denotes the addition of everything that comes before it (within the curly brackets), with $\bar{y}_s$ replaced by $(-\bar{y}_s)$ for all $s = 1, 2, \ldots, N$, has the N eigenvalues

$$\lambda_m \equiv \lambda_m(\alpha + \beta; N) = m(m - 2N - \alpha - \beta - 1), \quad m = 1, 2, \ldots, N. \tag{5.26}$$

Note that the last matrix elements L_{nm} are functions of $\bar{z}_s = \bar{y}_s^2 - \theta^2$ because the addition implied by the symbol $+\left[\left(\bar{y}_s \to (-\bar{y}_s)\right)\right]$ results in cancellation of all the terms odd in $\bar{y}_s$, $s = 1, 2, \ldots, N$. The eigenvalues λ_m of the last matrix L depend only on the sum $\alpha + \beta$ of the parameters α, β and do not depend on the parameters γ, δ. These eigenvalues are moreover rational if this sum is rational, a diophantine property.

6 Zeros of Askey-Wilson and q-Racah polynomials

Askey-Wilson and q-Racah polynomials are at the top of the q-Askey scheme of orthogonal polynomials [31]. They are defined in terms of the generalized basic hypergeometric function (1.10), yet are not particular cases of the generalized basic hypergeometric polynomial (1.12). In this section we provide isospectral matrices defined in terms of the zeros of these polynomials, which were constructed in [11].

6.1 Zeros of Askey-Wilson polynomials

The Askey-Wilson polynomial $p_N(a, b, c, d; q; x)$ with $x = \cos\theta$ is defined by

$$p_N(a, b, c, d; q; \cos\theta) = \frac{(ab, ac, ad; q)_N}{a^N}$$
$$\cdot {}_4\phi_3\left(\begin{array}{c} q^{-N},\ abcd\ q^{N-1},\ a\ \exp(\mathbf{i}\theta),\ a\ \exp(-\mathbf{i}\theta) \\ ab,\ ac,\ ad \end{array}\middle|\, q; q\right), \tag{6.1a}$$

see [31], or, equivalently by

$$p_N(a, b, c, d; q; x) = \frac{(ab, ac, ad; q)_N}{a^N} \cdot$$
$$\cdot \sum_{m=0}^{N}\left[\frac{q^m\left(q^{-N}; q\right)_m\ \left(abcd\ q^{N-1}; q\right)_m}{(q; q)_m\ (ab; q)_m\ (ac; q)_m\ (ad; q)_m}\ \{a; q; x\}_m\right], \tag{6.1b}$$

see [11], where the modified q-Pochhammer symbol

$$\{a; q; x\}_0 = 1\ ;$$
$$\{a; q; x\}_m = \left(1 + a^2 - 2ax\right)\left(1 + q^2a^2 - 2aqx\right)\cdots\left(1 + a^2q^{2(m-1)} - 2aq^{m-1}x\right),$$
$$m = 1, 2, 3, \ldots \tag{6.1c}$$

and the complex parameters a, b, c, d together with the base $q \neq 1$ are such that, after appropriate cancellations, the denominators in the sum (6.1b) do not vanish. Because $\{a; q; x\}_m$ is a polynomial of degree m in x, the Askey-Wilson polynomial $p_N(a, b, c, d; q; x)$ defined by (6.1) is indeed a polynomial of degree N in x. It is a q-analogue of the Wilson polynomial (5.1).

Consider the related rational function

$$Q_N(z) \equiv Q_N\left(a, b, c, d; q; z\right) = p_N\left(a, b, c, d; q; \frac{z^2 + 1}{2\ z}\right), \tag{6.2}$$

and note the relations

$$p_N(a,b,c,d;q;x) = Q_N\left(a,b,c,d;q;x+\sqrt{x^2-1}\right), \tag{6.3}$$

$$x=\cos\theta = \frac{z^2+1}{2\,z}, \quad z=e^{i\theta}=x+\sqrt{x^2-1}. \tag{6.4}$$

Due to definition (6.2), the choice of the square root $\sqrt{x^2-1}$ from among the two possibilities is irrelevant because in both cases $(z^2+1)/(2z)=x$.

The rational function $Q_N(z)$ satisfies the following q-difference equation:

$$\Big[\left(q^{-N}-1\right)\left(1-abcd\,q^{N-1}\right)+D(z)\left(1-\delta_q\right) \\ +D\left(z^{-1}\right)\left(1-\delta_{q^{-1}}\right)\Big]\,Q_N(z)=0, \tag{6.5}$$

where

$$D(z)\equiv D(a,b,c,d;q;z)=\frac{(1-az)\ (1-bz)\ (1-cz)\ (1-dz)}{(1-z^2)\ (1-qz^2)}, \tag{6.6}$$

$\delta_q f(z)=f(qz)$ and $\delta_{q^{-1}}f(z)=f(q^{-1}z)$, see (1.14).

Consider the differential q-difference equation (DqDE)

$$\frac{\partial\Psi(z,t)}{\partial t}=\mathcal{A}\,\Psi(z,t), \tag{6.7a}$$

where the q-difference operator $\mathcal{A}$ is defined by

$$\mathcal{A}=\left(q^{-N}-1\right)\left(1-abcd\,q^{N-1}\right)+D(z)\left(1-\delta_q\right)+D\left(z^{-1}\right)\left(1-\delta_{q^{-1}}\right). \tag{6.7b}$$

It is clear that the rational function (6.2) is a t-independent solution of DqDE (6.7). Moreover, this DqDE has a t-dependent rational solution of the form

$$\Psi_N(z,t)=\sum_{m=0}^{N}\left[c_m(t)\ Q_{N-m}(z)\right], \tag{6.8}$$

where each $Q_{N-m}(z)$ is a rational function defined by (6.2) with N replaced by $N-m$. Indeed, the rational function (6.8) solves DqDE (6.7) if and only if the coefficients $c_m(t)$ solve the decoupled linear system of ODEs

$$\dot{c}_m(t)=q^{-N}\ (1-q^m)\left(1-abcd\,q^{2N-1-m}\right)\ c_m(t)\ , \\ m=0,1,,2...,N, \tag{6.9}$$

which is, of course, explicitly solvable. Note that $c_0(t)=c_0$ is constant.

Let us now perform the following change of variables in DqDE (6.7):

$$\Psi(z,t)=\psi\left(\frac{z^2+1}{2\,z},t\right),\quad \psi(x;t)=\Psi\left(x+\sqrt{x^2-1},t\right) \tag{6.10}$$

to recast it as

$$\frac{\partial\psi(x,t)}{\partial t}=\tilde{A}\,\psi(x,t)\,, \tag{6.11a}$$

where

$$\tilde{A}=\left(q^{-N}-1\right)\left(1-abcd\,q^{N-1}\right)-D(z)\;S^{(+)}-D\left(z^{-1}\right)\;S^{(-)}$$
$$+D(z)+D\left(z^{-1}\right), \tag{6.11b}$$

$z=x+\sqrt{x^2-1}$ and

$$S^{(\sigma)}\;f(x)=f\left(q^{\sigma}x+\sigma\frac{1-q^2}{2qz}\right),\quad \sigma=\pm1. \tag{6.11c}$$

It is then clear that the polynomial

$$\psi_N(x,t)=\Psi_N\left(x+\sqrt{x^2-1},t\right)=\sum_{m=0}^{N}\left[c_m(t)\;p_{N-m}(x)\right]$$
$$=c_0\,p_N(x)+\sum_{m=1}^{N}\left[c_m(t)\;p_{N-m}(x)\right], \tag{6.12}$$

where $p_{N-m}(x)$ are Askey-Wilson polynomials, see (6.1), (6.8) and (6.3), solves DqDE (6.11).

Let us now represent the last polynomial $\psi_N(x,t)$ in terms of its zeros $x_n(t)$:

$$\psi_N(x,t)=C_N\prod_{n=1}^{N}\left[x-x_n(t)\right], \tag{6.13}$$

where C_N is the leading coefficient of $\psi_N(x,t)$, which does not depend on t because c_0 is constant; see (6.12) and the remark after display (6.9). Upon substitution of (6.13) into DqDE (6.7), one obtains the following nonlinear system of ODEs:

$$\dot{x}_n=\frac{(q-1)}{2q^N}\left\{G(z_n)\prod_{\ell=1,\ell\neq n}^{N}K(z_n,z_\ell)+G\left(\frac{1}{z_n}\right)\prod_{\ell=1,\ell\neq n}^{N}K\left(\frac{1}{z_n},\frac{1}{z_\ell}\right)\right\},$$
$$n=1,\ldots,N, \tag{6.14a}$$

where

$$G(z_n)=D(z_n)\left(qz_n-\frac{1}{z_n}\right), \tag{6.14b}$$

$$K(z_n,z_m)=\frac{(z_m-qz_n)\,(qz_nz_m-1)}{(z_m-z_n)\,(z_nz_m-1)}, \tag{6.14c}$$

and $z_s=\exp(\mathrm{i}\theta_s)$, $x_s=\cos\theta_s$, see (6.4). Note that the right-hand side of (6.14) is defined if $x_n\neq x_m$; it is an even function of θ_s, hence a function of x_s, $s=$

$1,\ldots,N$. System (6.14) is solvable because its solutions are vectors of the zeros of the polynomial (6.12) whose coefficients $c_m(t)$ are components of (explicitly known) solutions of the linear system (6.9).

Because the Askey-Wilson polynomial $p_N(x) \equiv p_N(a,b,c,d;q;x)$ is a t-independent solution of DqDE (6.11), every vector $\bar{\boldsymbol{x}} = (\bar{x}_1,\ldots,\bar{x}_N)$ of its zeros is an equilibrium of system (6.14), provided that these zeros are distinct. We may therefore linearize system (6.14) about its equilibrium $\bar{\boldsymbol{x}} = (\bar{x}_1,\ldots,\bar{x}_N)$ to obtain a linear system

$$\dot{\boldsymbol{\xi}}(t) = L\,\boldsymbol{\xi}(t), \tag{6.15}$$

where the $N\times N$ matrix $L \equiv L(\bar{\boldsymbol{x}})$ is defined by (6.16) and has the set of eigenvalues that coincides with the set of the eigenvalues of the (diagonal) coefficient matrix of the linear system (6.9) with $m = 1,\ldots,N$. Therefore, the following proposition holds.

Proposition 7. *Suppose that $\bar{\boldsymbol{x}} = (\bar{x}_1,\ldots,\bar{x}_N)$ is vector of zeros of the Askey-Wilson polynomial $p_N(a,b,c,d;q;x)$ of degree N in x defined by (6.1). Let $\bar{z}_n$ be related to $\bar{x}_n$ via $\bar{x}_n = \cos\bar{\theta}_n = (\bar{z}_n^2+1)/(2\bar{z}_n), \bar{z}_n = e^{i\bar{\theta}_n} = \bar{x}_n + \sqrt{\bar{x}_n^2-1}$. If the zeros $\bar{x}_n$ are all different among themselves, then the $N\times N$ matrix L defined componentwise by [11]*

$$\begin{aligned}
&L_{nn} \equiv L_{nn}(a,b,c,d,;q;N;\bar{\boldsymbol{x}})\\
&= \frac{(q-1)}{2q^N}\Bigg\{\Bigg[\frac{2\bar{z}_n^2}{\bar{z}_n^2-1}G(\bar{z}_n)\sum_{m=1,m\neq n}^{N}\Bigg(-\frac{q}{\bar{z}_m-q\bar{z}_n}+\frac{q\bar{z}_m}{q\bar{z}_n\bar{z}_m-1}\\
&+\frac{1}{\bar{z}_m-\bar{z}_n}-\frac{\bar{z}_m}{\bar{z}_n\bar{z}_m-1}\Bigg)+\frac{2\bar{z}_n^2G'(\bar{z}_n)}{\bar{z}_n^2-1}\Bigg]\prod_{\ell=1,\ell\neq n}^{N}K(\bar{z}_n,\bar{z}_\ell)\\
&+\left[\left(\bar{z}_s\to\frac{1}{\bar{z}_s}\right)\right]\Bigg\}
\end{aligned} \tag{6.16a}$$

and

$$\begin{aligned}
&L_{nm} \equiv L_{nm}(a,b,c,d,;q;N;\bar{\boldsymbol{x}})\\
&= \frac{(q-1)}{2q^N}\Bigg\{\frac{2\bar{z}_m^2}{\bar{z}_m^2-1}G(\bar{z}_n)\Bigg[\frac{1}{\bar{z}_m-q\bar{z}_n}+\frac{q\bar{z}_n}{q\bar{z}_n\bar{z}_m-1}\\
&-\frac{1}{\bar{z}_m-\bar{z}_n}-\frac{\bar{z}_n}{\bar{z}_n\bar{z}_m-1}\Bigg]\prod_{\ell=1,\ell\neq n}^{N}K(\bar{z}_n,\bar{z}_\ell)+\left[\left(\bar{z}_s\to\frac{1}{\bar{z}_s}\right)\right]\Bigg\}
\end{aligned} \tag{6.16b}$$

for $n,m = 1,...,N$ with $n\neq m$, where $G(\bar{z}_n)$, $K(\bar{z}_n,\bar{z}_m)$ are defined by (6.14b), (6.14c) and the symbol $+\left[\left(\bar{z}_s\to\frac{1}{\bar{z}_s}\right)\right]$ indicates addition of everything that comes before it, within the curly brackets, with $\bar{z}_s$ replaced by $\frac{1}{\bar{z}_s}$ for all $s = 1,\ldots,N$, has the N eigenvalues

$$\begin{aligned}
&\lambda_n \equiv \lambda_n(abcd;q;N) = q^{-N}\left(1-q^n\right)\left(1-abcd\;q^{2N-1-n}\right)\\
&= \left(q^{-N}+abcd\;q^{N-1}-q^{n-N}-abcd\;q^{N-1-n}\right),\\
&n = 1,2,...,N.
\end{aligned} \tag{6.17}$$

Note that after the substitution $\bar{x}_n = \cos\bar{\theta}_n$ and $\bar{z}_n = \exp(\mathrm{i}\bar{\theta}_n)$, the components L_{nm} of the matrix L in the last proposition become even functions of each $\bar{\theta}_s$; hence L_{nm} are functions of $\bar{x}_s$, $s = 1, \ldots, N$. The eigenvalues λ_n of the last matrix L depend only on the product $abcd$ and the basis q as opposed to all the parameters $(a, b, c, d; q)$ and are moreover rational if $abcd$ and q are rational, a diophantine property.

6.2 Zeros of q-Racah polynomials

The q-Racah polynomial $R_N(\alpha, \beta, \gamma, \delta; q; z)$ is defined by

$$R_N(\alpha,\beta,\gamma,\delta;q;z) = {}_4\phi_3\left(\begin{matrix} q^{-N},\ \alpha\beta q^{N+1},\ q^{-x},\ \gamma\delta q^{x+1} \\ \alpha q,\ \beta\delta q,\ \gamma q \end{matrix} \middle| q; q \right), \tag{6.18a}$$

where

$$z \equiv z(\gamma\delta; q; x) = q^{-x} + \gamma\delta q^{x+1}, \tag{6.18b}$$

see [31], or, equivalently by

$$R_N(\alpha,\beta,\gamma,\delta;q;z) = \\ = \sum_{m=0}^{N} \left[\frac{q^m \left(q^{-N};q\right)_m \left(\alpha\beta\, q^{N+1};q\right)_m \left(q^{-x};q\right)_m \left(\gamma\delta q^{x+1};q\right)_m}{(q;q)_m\, (\alpha q;q)_m\, (\beta\delta q;q)_m\, (\gamma q;q)_m} \right], \tag{6.18c}$$

see [11], where

$$\left(q^{-x};q\right)_m \left(\gamma\delta q^{x+1};q\right)_m = \prod_{s=0}^{m-1} \left(1 - zq^s + \gamma\delta q^{2s+1}\right). \tag{6.18d}$$

and the complex parameters $\alpha, \beta, \gamma, \delta$ together with the base $q \neq 1$ are such that, after appropriate cancellations, the denominators in the sum (6.18c) do not vanish. Because $(q^{-x};q)_m \left(\gamma\delta q^{x+1};q\right)_m$ is a polynomial of degree m in z, the q-Racah polynomial $R_N(\alpha, \beta, \gamma, \delta; q; z)$ defined by (6.18) is indeed a polynomial of degree N in z. It is a q-analogue of the Racah polynomial (5.15).

The standard definition of q-Racah polynomials imposes the restriction $\alpha q = q^{-M}$ or $\beta\delta q = q^{-M}$ or $\gamma q = q^{-N}$ on the complex parameters $\alpha, \beta, \gamma, \delta$, with M being a nonnegative integer, together with the inequality $0 \leq N \leq M$. However, in this study, none of the last diophantine relations is required or assumed. This is because the only property of the q-Racah polynomials used in the construction of the isospectral matrices provided below is that these polynomials satisfy difference equation (6.19), which is valid even if the diophantine restrictions mentioned above do not hold.

The q-Racah polynomial $R_N(z) \equiv R_N(\alpha, \beta, \gamma, \delta; q; z)$ satisfies the following q-difference equation:

$$B(z)\ R_N(z^{(+)}) - [B(z) + D(z)]\ R_N(z) + D(z)\ R_N(z^{(-)}) \\ = \left(q^{-N} - 1\right)\left(1 - \alpha\beta q^{N+1}\right)\ R_N(z), \tag{6.19a}$$

where

$$z^{(\pm)} = z\,(x \pm 1) = q^{\pm 1} z \pm \left(\frac{1-q^2}{2q}\right)\left[z - \sqrt{z^2 - 4\gamma\delta q}\right] \tag{6.19b}$$

and

$$\begin{aligned} B\,(z) = &\left\{\left[1 - \gamma\delta q Z^2\,(q;z)\right]\left[1 - \gamma\delta q^2 Z^2\,(q;z)\right]\right\}^{-1}\left[1 - \alpha q Z\,(q;z)\right] \\ &\cdot\left[1 - \beta\delta q Z\,(q;z)\right]\ \left[1 - \gamma q Z\,(q;z)\right]\ \left[1 - \gamma\delta q Z\,(q;z)\right], \end{aligned} \tag{6.19c}$$

$$\begin{aligned} D\,(z) = &\left\{\left[1 - \gamma\delta Z^2\,(\gamma\delta q;z)\right]\left[1 - \gamma\delta q Z^2\,(\gamma\delta q;z)\right]\right\}^{-1} q\left[1 - Z\,(\gamma\delta q;z)\right] \\ &\cdot\left[1 - \delta Z\,(\gamma\delta q;z)\right]\left[\beta - \gamma Z\,(\gamma\delta q;z)\right]\left[\alpha - \gamma\delta Z\,(\gamma\delta q;z)\right], \end{aligned} \tag{6.19d}$$

with

$$Z\,(\gamma\delta q;z) = q^x = \frac{z + \sqrt{z^2 - 4\gamma\delta q}}{2\gamma\delta q}\ . \tag{6.19e}$$

Note that the determination of the square root in (6.19b) and (6.19e) is irrelevant as long as it is consistent throughout.

Consider the differential q-difference equation (DqDE)

$$\frac{\partial\Psi\,(z,t)}{\partial t} = \mathcal{A}\,\Psi\,(z,t)\,, \tag{6.20a}$$

where the q-difference operator $\mathcal{A}$ is defined by

$$\mathcal{A} = \left(q^{-N} - 1\right)\left(1 - \alpha\beta q^{N+1}\right) + B\,(z)\left(1 - \Delta^{(+)}\right) + D\,(z)\left(1 - \Delta^{(-)}\right) \tag{6.20b}$$

with

$$\Delta^{(\pm)} f(z) = f\left(z^{(\pm)}\right), \tag{6.20c}$$

see (6.19b). From (6.19a) it is clear that the q-Racah polynomial (6.18a) is a t-independent solution of DqDE (6.20). Moreover, this DqDE has a t-dependent polynomial solution of the form

$$\Psi_N\,(z,t) = \sum_{m=0}^{N}\left[c_m\,(t)\ R_{N-m}\,(z)\right], \tag{6.21}$$

where each $R_{N-m}\,(z) \equiv R_{N-m}\,(\alpha,\beta,\gamma,\delta;q;z)$ is a q-Racah polynomial defined by (6.18) with N replaced by $N-m$. Indeed, polynomial (6.21) solves DqDE (6.20) if and only if the coefficients $c_m(t)$ solve the decoupled linear system of ODEs

$$\begin{aligned} &\dot{c}_m\,(t) = q^{-N}\ \left(1 - q^m\right)\left(1 - \alpha\beta\ q^{2N-m+1}\right)\ c_m\,(t)\,, \\ &m = 0, 1, , 2..., N, \end{aligned} \tag{6.22}$$

which is, of course, explicitly solvable. Note that $c_0(t) = c_0$ is constant.

Let us now represent the same polynomial $\Psi_N(z,t)$ in terms of its zeros $z_n(t)$:

$$\Psi_N(z,t) = C_N \prod_{n=1}^{N} [z - z_n(t)], \tag{6.23}$$

where C_N the leading coefficient of $\Psi_N(z,t)$, which does not depend on t because c_0 is constant, see (6.21) and the remark after display (6.22). Upon substitution of (6.23) into DqDE (6.20), one obtains the following nonlinear system of ODEs:

$$\begin{aligned} \dot{z}_n = {} & B(z_n) \left(z_n^{(+)} - z_n\right) \prod_{\ell=1,\ \ell\neq n}^{N} \left(\frac{z_n^{(+)} - z_\ell}{z_n - z_\ell}\right) \\ & + D(z_n) \left(z_n^{(-)} - z_n\right) \prod_{\ell=1,\ \ell\neq n}^{N} \left(\frac{z_n^{(-)} - z_\ell}{z_n - z_\ell}\right), \\ & n = 1, 2, ..., N, \end{aligned} \tag{6.24}$$

where $B(z) \equiv B(\alpha,\beta,\gamma,\delta;q;z)$ respectively $D(z) \equiv D(\alpha,\beta,\gamma,\delta;q;z)$ are defined by (6.19c) respectively (6.19d) with (6.19e), and $z_n^{(\pm)}$ are defined by (6.19b). System (6.24) is solvable because its solutions are vectors of the zeros of the polynomial (6.23) whose coefficients $c_m(t)$ are components of (explicitly known) solutions of the linear system (6.22).

Because the q-Racah polynomial $R_N(z) \equiv R_N(\alpha,\beta,\gamma,\delta;q;z)$ is a t-independent solution of DqDE (6.20), every vector $\bar{\boldsymbol{z}} = (\bar{z}_1,\ldots,\bar{z}_N)$ of its zeros is an equilibrium of system (6.24), provided that these zeros are distinct. We may therefore linearize system (6.24) about its equilibrium $\bar{\boldsymbol{z}} = (\bar{z}_1,\ldots,\bar{z}_N)$ to obtain a linear system

$$\dot{\boldsymbol{\xi}}(t) = L\,\boldsymbol{\xi}(t), \tag{6.25}$$

where the $N \times N$ matrix $L = L(\bar{\boldsymbol{z}})$ is defined by (6.26) and has the set of eigenvalues that coincides with the set of the eigenvalues of the (diagonal) coefficient matrix of the linear system (6.22) with $m = 1,\ldots,N$. Therefore, the following Proposition holds.

Proposition 8. *Let $\bar{\boldsymbol{z}} = (\bar{z}_1,\ldots,\bar{z}_N)$ be a vector of zeros of the q-Racah polynomial $R_N(\alpha,\beta,\gamma,\delta;q;z)$ of degree N in z defined by (6.18). Let $\bar{z}_n^{(\pm)}$ be related to $\bar{z}_n$ via $z_n^{(\pm)} = q^{\pm 1} z_n \pm \left(\frac{1-q^2}{2q}\right)\left[z_n - \sqrt{z_n^2 - 4\gamma\delta q}\right]$; see (6.19b). If the zeros $\bar{z}_n$ are all different among themselves, then the $N \times N$ matrix L defined componentwise by [11]*

$$L_{nn} \equiv L_{nn}(\alpha,\beta,\gamma,\delta;q;N;\bar{z}) = \Bigg\{ B'(\bar{z}_n)(\bar{z}_n^{(+)} - \bar{z}_n)$$

$$+B(\bar{z}_n)\Big[C^{(+)}(\bar{z}_n) - 1 + (\bar{z}_n^{(+)} - \bar{z}_n) \sum_{m=1,m\neq n}^{N} W^{(+)}(\bar{z}_n,\bar{z}_m)\Big]\Bigg\} \prod_{\ell=1,\ell\neq n}^{N} \frac{\bar{z}_n^{(+)} - \bar{z}_\ell}{\bar{z}_n - \bar{z}_\ell}$$

$$+\Bigg\{ D'(\bar{z}_n)(\bar{z}_n^{(-)} - \bar{z}_n)$$

$$+D(\bar{z}_n)\Big[C^{(-)}(\bar{z}_n) - 1 + (\bar{z}_n^{(-)} - \bar{z}_n) \sum_{m=1,m\neq n}^{N} W^{(-)}(\bar{z}_n,\bar{z}_m)\Big]\Bigg\} \prod_{\ell=1,\ell\neq n}^{N} \frac{\bar{z}_n^{(-)} - \bar{z}_\ell}{\bar{z}_n - \bar{z}_\ell},$$

$$n = 1,...,N, \tag{6.26a}$$

$$L_{nm} \equiv L_{nm}(\alpha,\beta,\gamma,\delta;q;N;\bar{z})$$
$$= B(\bar{z}_n)\left(\frac{\bar{z}_n^{(+)} - \bar{z}_n}{\bar{z}_n - \bar{z}_m}\right)^2 \prod_{\ell=1\,\ell\neq n,m}^{N} \frac{\bar{z}_n^{(+)} - \bar{z}_\ell}{\bar{z}_n - \bar{z}_\ell}$$
$$+D(\bar{z}_n)\left(\frac{\bar{z}_n^{(-)} - \bar{z}_n}{\bar{z}_n - \bar{z}_m}\right)^2 \prod_{\ell=1\,\ell\neq n,m}^{N} \frac{\bar{z}_n^{(-)} - \bar{z}_\ell}{\bar{z}_n - \bar{z}_\ell},$$
$$n,m = 1,...,N,\ n \neq m, \tag{6.26b}$$

where the functions $B(z)$ and $D(z)$ are defined by (6.19c), (6.19d), (6.19e),

$$C^{(\pm)}(\bar{z}_n) = \frac{d\bar{z}_n^{(+)}}{d\bar{z}_n} = q^{\pm 1} \pm \frac{1-q^2}{2q}\left(1 - \frac{\bar{z}_n}{\sqrt{\bar{z}_n^2 - 4\gamma\delta q}}\right),$$

$$W^{(\pm)}(\bar{z}_n,\bar{z}_m) = \frac{C^{(\pm)}(\bar{z}_n)(\bar{z}_n - \bar{z}_m) - \bar{z}_n^{(\pm)} + \bar{z}_m}{(\bar{z}_n - \bar{z}_m)(\bar{z}_n^{(\pm)} - \bar{z}_m)},$$

has the N eigenvalues

$$\lambda_n \equiv \lambda_n(\alpha\beta;q;N) = q^{-N}(1-q^m)\left(1 - \alpha\beta\, q^{2N-m+1}\right),$$
$$n = 1,2,...,N. \tag{6.27}$$

The eigenvalues λ_n of the last matrix L depend only on the product $\alpha\beta$ and the basis q as opposed to all the parameters $(\alpha,\beta,\gamma,\delta;q)$ and are moreover rational if $\alpha\beta$ and q are rational, a diophantine property.

7 Discussion and Outlook

We have reviewed some recent results on the properties of the zeros of several families of special polynomials: generalized hypergeometric, generalized basic hypergeometric, Wilson and Racah as well as Askey-Wilson and q-Racah polynomials. For each polynomial family $\{P_n(z)\}_{n=1}^{\infty}$ listed above, we have stated an $N \times N$ isospectral matrix L defined in terms of the zeros of the polynomial $P_N(z)$. The eigenvalues $\{\lambda_m\}_{m=1}^{N}$ of the matrix L are given by neat formulas that do not involve the zeros of the polynomial $P_N(z)$ and depend on fewer parameters than the polynomial $P_N(z)$. Moreover, these eigenvalues are rational in the case where some of the parameters of $P_N(z)$ are rational, a diophantine property. These isospectral matrices are obtained via construction of several solvable (in terms of algebraic operations) first order nonlinear systems of ODEs, see (2.27), (4.5), (5.10), (5.23), (6.14), (6.24). Properties of these solvable systems pose an interesting subject for future studies.

Because Wilson and Racah polynomials are at the top of the Askey scheme, while Askey-Wilson and q-Racah polynomials are at the top of the q-Askey scheme, it must be possible, by taking appropriate limits, to use the results of Propositions 5, 6, 7, 8 to obtain new isospectral matrices defined in terms of the zeros of all the other polynomials in the Askey and the q-Askey schemes. This is another topic for future investigation.

In each among the Propositions 2, 4, 5, 6, 7, 8, the N eigenvalues of the isospectral matrix L are given, while the corresponding eigenvectors and the transition matrix T that diagonalizes the matrix L via the similarity transformation $T^{-1}LT$ are not known. Finding similarity matrices for each isospectral matrix L in Propositions 2, 4, 5, 6, 7, 8 would provide a tool for construction of additional algebraic identities satisfied by the zeros of the appropriate polynomials. A method that compares the spectral and the pseudospectral matrix representations of linear differential operators has been employed in [4, 7, 16] to construct new and re-obtain known algebraic identities satisfied by classical, Krall and Sonin-Markov orthogonal polynomials. It would be interesting to adapt the last method to the cases of the generalized hypergeometric, generalized basic hypergeometric, Wilson and Racah as well as Askey-Wilson and q-Racah polynomials and to compare the results with those reviewed in this chapter as well as to utilize the new developments to study properties of the solvable nonlinear systems (2.27), (4.5), (5.10), (5.23), (6.14), (6.24).

Acknowledgements

Results reviewed in this chapter were obtained together with Francesco Calogero, in particular during several visits to the "La Sapienza" University of Rome, for which hospitality the author is grateful.

References

[1] Ahmed S, Novel properties of the zeros of Laguerre polynomials, *Lettere Nuovo Cimento* **22** (1978) 371–375.

[2] Ahmed S, A general technique to obtain nonlinear equations for the zeros of classical orthogonal polynomials, *Lettere Nuovo Cimento* **26** (1979) 285–288.

[3] Ahmed S, Bruschi M, Calogero F, Olshanetsky M A, Perelomov A M, Properties of the zeros of the classical polynomials and of Bessel functions, *Nuovo Cimento* **49B** (1979) 173-199.

[4] Alici H, Taşeli H, Unification of Stieltjes-Calogero type relations for the zeros of classical orthogonal polynomials, *Math. Meth. Appl. Sci.*, **38**(14) (2015) 3118–3129.

[5] Beals R, Szmigielski J, Meijer G-functions: a gentle introduction, *Notices of the AMS* **60**(7) (2013) 866-872.

[6] Gasper G, Rahman M, *Basic Hypergeometric Series*, Cambridge University Press, 1990.

[7] Bihun O, New properties of the zeros of Krall polynomials, *J. Nonlinear Math. Phys.* **24**(4) (2017) 495–515.

[8] Bihun O, Calogero F, Properties of the zeros of generalized hypergeometric polynomials, *J. Math. Analysis Appl.* **419**(2) (2014) 1076–1094.

[9] Bihun O, Calogero F, Properties of the zeros of the polynomials belonging to the Askey scheme, *Lett. Math. Phys.* **104**(12) (2014) 571–588.

[10] Bihun O, Calogero F, Properties of the zeros of generalized basic hypergeometric polynomials, *J. Math. Phys.*, **56** (2015) 112701, 1–15.

[11] Bihun O, Calogero F, Properties of the zeros of the polynomials belonging to the q-Askey scheme, *J. Math. Analysis Appl.* **433**(1) (2016) 525–542.

[12] Bihun O, Calogero F, Novel *solvable* many-body problems, *J. Nonlinear Math. Phys.* **23**(2) (2016) 190–212.

[13] Bihun O, Calogero F, A new solvable many-body problem of goldfish type, *J. Nonlinear Math. Phys.* **23**(1) (2016) 28–46.

[14] Bihun O, Calogero F, Generations of monic polynomials such that the coefficients of the polynomials of the next generation coincide with the zeros of the polynomials of the current generation, and new solvable many-body problems, *Lett. Math. Phys.* **106**(7) (2016) 1011–1031.

[15] Bihun O, Calogero F, Generations of solvable discrete-time dynamical systems, *J. Math. Phys.* **58** (2017) 052701, 21 pages.

[16] Bihun O, Mourning C, Generalized Pseudospectral Method and Zeros of Orthogonal Polynomials, *Adv. Math. Phys.* Vol. 2018 (2018), Article ID 4710754, 10 pages.

[17] Bruschi M, Calogero F, A convenient expression of the time-derivative , of arbitrary order k, of the zero $z_n(t)$ of a time dependent polynomial $p_N(z;t)$ of arbitrary degree N in z, and solvable dynamical systems, *J. Nonlinear Math. Phys.* **23**(4) 474–485.

[18] Calogero F, Solution of the one-dimensional N-body problem with quadratic and/or inversely quadratic pair potentials, *J. Math. Phys.* **12** (1971) 419-436; "Erratum", ibidem 37, 3646 (1996).

[19] Calogero F, On the zeros of Hermite polynomials, *Lettere Nuovo Cimento* **20** (1977) 489–490.

[20] Calogero F, On the zeros of the classical polynomials, *Lettere Nuovo Cimento* **19** (1977) 505–508.

[21] Calogero F, The "neatest" many-body problem amenable to exact treatments (a "goldfish"?), *Physica D* **152-153** (2001) 78-84.

[22] Calogero F, *Classical many-body problems amenable to exact treatments*, Lecture Notes in Physics Monographs **m66**, Springer, Heidelberg, 2001.

[23] Calogero F, *Isochronous systems*, Oxford University Press, 2012.

[24] Calogero F, New solvable variants of the goldfish many-body problem, *Studies Appl. Math.* **137**(1) (2016) 123–139.

[25] Chihara T, *An introduction to orthogonal polynomials*, Gordon and Breach, 1978.

[26] Driver K, Jordaan K, Zeros of quasi-orthogonal Jacobi polynomials, *SIGMA* **12** (2016) 042, 13 pages.

[27] *Higher Transcendental Functions*, Volumes 1, 2, Bateman Project, Erdélyi A, Editor, McGraw-Hill, New York, 1953.

[28] Funaro D, *Polynomial approximations of differential equations*, Springer-Verlag, Berlin, 1992.

[29] Ismail M E H, An electrostatic model for zeros of general orthogonal polynomials, *Pacific J. Math.* **193** (2000) 355–369.

[30] Ismail M E H, More on electrostatic models for zeros of orthogonal polynomials, *Numer. Funct. Anal. Opt.* **21** (2000) 191–204.

[31] Koekoek R, Swarttouw R F, *The Askey-scheme of hypergeometric orthogonal polynomials and its q-analogue*, Delft University of Technology, Faculty of Technical Mathematics and Informatics, Report no. 94-05 (1994), revised in Report no. 98-17, 1998, available online at http://homepage.tudelft.nl/11r49/askey/.

[32] Koekoek R, Lesky P A, Swarttouw R F, *Hypergeometric orthogonal polynomials and their q- analogues*, Springer, 2010.

[33] Marcellán F, Martínez-Finkelshtein A, Martínez-González P, Electrostatic models for zeros of polynomials: Old, new, and some open problems, *J. Computational Applied Math* **207**(2) (2007) 258–272.

[34] Mastroianni G, Milovanović G, *Interpolation processes: basic theory and applications*, Springer, 2008.

[35] Moser J, Three integrable Hamiltonian systems connected with isospectral deformations, *Adv. Math.* bf 16 (1975) 197-220.

[36] Nikiforov A F, Uvarov V B, *Special functions of mathematical physics*, Birkhäuser, 1988.

[37] Phillips G, *Interpolation and approximation by polynomials*, Canadian Mathematical Society, 2003.

[38] Sasaki R, Perturbations around the zeros of classical orthogonal polynomials, *J. Math. Phys.*, **56** (2015) 042106.

[39] Stieltjes T J, Sur quelques théorèmes d'algèbre, *Comptes Rendus de l'Académie des Sciences* **100** (1885) 439–440.

[40] Stieltjes T J, Sur les polynômes de Jacobi, *Comptes Rendus de l'Acadmie des Sciences* **100** (1885) 620–622.

[41] Stieltjes T J, Sur certains polynômes que vérifient uneéquation différentielle linéaire du second ordre et sur la théorie des fonctions de Lamé, *Acta Math.* **6** (1885) 321–326.

[42] Szegö G., *Orthogonal polynomials*, American Mathematical Society, 1939.

[43] Veselov A P, On Stieltjes relations, Painlevé-IV hierarchy and complex monodromy, *J. Physics A: Mathematical and General* **34** 2001 3511–3519.

B2. Singularity methods for meromorphic solutions of differential equations

Robert Conte [ab], *Tuen Wai Ng* [b] *and Cheng-Fa Wu* [c]

[a] *Centre de mathématiques et de leurs applications*
École normale supérieure de Cachan, CNRS, Université Paris-Saclay,
61, avenue du président Wilson, F–94235 Cachan Cedex, France.
Robert.Conte@cea.fr

[b] *Department of Mathematics, The University of Hong Kong, Pokfulam Road, Hong Kong.*
ntw@maths.hku.hk

[c] *Institute for Advanced Study, Shenzhen University, Nanshan District, Shenzhen, Guangdong, China 518060.*
cfwu@szu.edu.cn, wuchengfa9@126.com

Abstract

This contibution is mainly a tutorial introduction to methods recently developed in order to find all (as opposed to some) meromorphic particular solutions of given nonintegrable, autonomous, algebraic ordinary differential equations of any order. The examples are taken from physics and include Kuramoto-Sivashinsky and complex Ginzburg-Landau equations.

1 Introduction

Consider an N-th order algebraic ordinary differential equation (ODE) for $u(x)$ which may or may not possess the Painlevé property (PP), defined as the absence of movable critical singularities in the general solution, a singularity being called "critical" if multivaluedness takes place around it. This does not prevent the existence of particular solutions obeying a lower order ODE with the PP. Let us therefore address the problem to find *all* such solutions (i.e. without movable critical singularities) in closed form. We restrict here to autonomous ODEs, i.e. which do not depend explicitly on the independent variable $x \in \mathbb{C}$.

Moreover, in order to shorten the writing, instead of the mathematically correct adjectives "uniformizable" (synonym of "without movable critical singularities") and "nonuniformizable", we will use "singlevalued" and "multivalued". For instance, if the singularity $x = 0$ is fixed and the singularity $x = x_0$ is movable, the expression $u(x) = \sqrt{x}(x - x_0)^{-3}$ will be called here "singlevalued".

Let us denote $\overline{\mathbb{C}}$ the analytic plane, i.e. the complex plane $\mathbb{C}$ compactified by addition of the unique point at infinity.

Definition 1. A function $x \to u(x)$ from $\mathbb{C}$ to $\overline{\mathbb{C}}$ is called *meromorphic* on $\mathbb{C}$ if its only singularities in $\mathbb{C}$ are poles. Example: $1/x + \coth(x - \pi)$.

Two kinds of methods will be used: (i) those based on the structure of singularities of the solutions (Painlevé analysis [11]), (ii) those based on the growth of solutions near infinity (Nevanlinna theory [27]). We will not use here methods based on Lie symmetries.

These two methods present an important difference: Painlevé analysis always considers the general solution (or particular solutions, i.e. obtained from the general solution by assigning numerical values to the integration constants) and only assumes singlevaluedness, while Nevanlinna theory considers particular solutions (which may be the general solution) and always assumes meromorphy on $\mathbb{C}$. However, the synergy of these two methods is considerable, the main achievement being that, under mild assumptions, any solution meromorphic on $\mathbb{C}$ is necessarily elliptic or degenerate elliptic and hence it allows one to find *all* meromorphic solutions.

As compared with previous tutorial presentations of ours [10] [11] [12] [13] [40] this short paper insists on the following features.

1. The criterium allowing one to conclude that any solution which is meromorphic on $\mathbb{C}$ is necessarily elliptic or degenerate elliptic.

2. The advantages, when dealing with amplitude equations (like complex Ginzburg-Landau or even nonlinear Schrödinger), to consider the logarithmic derivative of the complex amplitude A instead of the natural physical variables $|A|^2$ (square modulus) and $\arg A$ (phase).

3. The Hermite decomposition of elliptic or degenerate elliptic functions as a finite sum of poles.

First, it is important to understand the consequence of the term "closed form". By the construction of the Painlevé school [11], the closed form of any singlevalued solution (whether particular or general) of an N-th order ODE is a finite expression built from

1. the general solution of linear ODEs of any order;

2. the general solution of irreducible nonlinear ODEs of order one (i.e. elliptic functions),

3. the general solution of irreducible nonlinear ODEs of order two (i.e. the six functions Pn of Painlevé), three (no such function), ..., up to N included.

As a consequence, since Hölder proved the nonexistence of an ODE obeyed by the Γ function, this function can never contribute to the solution of an autonomous ODE.

Let us now explain why there exist two privileged subsets of singlevalued functions, the elliptic functions and the meromorphic functions.

1.1 First order equations and elliptic functions

Irreducible first order ODEs are privileged in this search for closed form singlevalued solutions, because order one is the smallest nonlinear order.

As proven by Lazarus Fuchs, Poincaré and Painlevé [37, vol II §141], irreducible first order ODEs with the PP (autonomous and nonautonomous) have a general solution which is a rational function of the Weierstrass function $\wp(x, g_2, g_3)$ and its first derivative.

Definition 2. A function $x \to u(x)$ from $\mathbb{C}$ to $\overline{\mathbb{C}}$ is called *elliptic* if it is doubly periodic and meromorphic on $\mathbb{C}$. Example: $\wp'(x) + \wp(2x)$.

All the elliptic functions $f(x)$ (among them the twelve Jacobi functions) are rational functions of $\wp(x)$ and $\wp'(x)$,

$$f(x) = R(\wp(x), \wp'(x)). \tag{1.1}$$

This function $\wp$ of Weierstrass is defined as the general solution of the first order ODE

$$u'^2 = 4u^3 - g_2 u - g_3 = 4(u - e_1)(u - e_2)(u - e_3), \tag{1.2}$$

it is doubly periodic and its only singularities are a lattice of double poles. Elliptic functions $f(x)$ have two successive degeneracies,

- when one root e_j is double ($g_2^3 - 27g_3^2 = 0$), degeneracy to simply periodic functions (i.e. rational functions of one exponential e^{kx}) according to

$$\forall x, d: \ \wp(x, 3d^2, -d^3) = 2d - \frac{3d}{2} \coth^2 \sqrt{\frac{3d}{2}} x, \tag{1.3}$$

- when the root e_j is triple ($g_2 = g_3 = 0$), degeneracy to rational functions of x.

The reason why elliptic functions (one should say *the* elliptic function because they are all equivalent under birational transformations) are privileged is that they are the only functions defined by irreducible (i.e. not linearizable) first order algebraic ODEs. They are therefore the natural building blocks for representing a wide class of particular solutions of ODEs of higher order.

1.2 Second order equations and meromorphic functions

Irreducible second order first degree ODEs in the class

$$u'' = F(u', u, x), \tag{1.4}$$

with F rational in u' and u, analytic in x, have the property that all their movable singularities are poles [32]. Note that, if the dependence on u is algebraic, there exist equations with movable isolated essential singularities in their general solution.

Therefore, elliptic functions naturally occur at first order, and meromorphic functions at second order.

1.3 Insufficiency of the truncation methods

Let us now explain why the so-called "truncation methods" cannot achieve our goal (to find all singlevalued solutions).

The four types of expressions considered up to now obey the sequence of inclusions

$$\text{multivalued} \supset \text{singlevalued} \supset \text{meromorphic} \supset \text{elliptic}\ , \tag{1.5}$$

and this sequence admits an important group of invariance.

Definition 3. One calls *homographic (or Möbius) transformation* any transformation $u \to U$, $u = (aU + b)/(cU + d)$, $ad - bc = 1$, with a, b, c, d constant.

Homographic transformations are the unique bijections of the analytic plane $\overline{\mathbb{C}}$.

Each of the four above mentioned subsets is invariant under homographies (on the condition to consider holomorphic functions as particular meromorphic functions, so that for instance the transform U of $u = 1/x$ by $u = 1/U$ is meromorphic).

Therefore, any method claiming to possibly find all singlevalued solutions *must* be invariant under homographies.

Truncation methods[1] are the main class of methods able to find singlevalued particular solutions. Their common feature (see the summer school lecture notes [7]) is to assume the unknown solution to belong to a given class of expressions, for instance polynomials in $\tanh(kx)$ and $\operatorname{sech}(kx)$ [8]. They essentially rely on the pioneering work of Weiss, Tabor and Carnevale [39]. Although very successful to find some solutions, they miss by construction any solution outside the given class. For instance, if one considers the rational trigonometric function

$$u = \frac{\tanh(\xi - \xi_0)}{2 + \tanh^2(\xi - \xi_0)}, \tag{1.6}$$

and builds the first order ODE which it obeys (this is a common way to construct examples),

$$u'^2 + \left(12u^2 - \frac{3}{2}\right) u' + 36u^4 - \frac{17}{2}u^2 + \frac{1}{2} = 0, \tag{1.7}$$

then no assumption u polynomial in $\tanh(kx)$ can succeed to find its solution.

One immediately notices the reason for this failure: this is the noninvariance of these truncation methods under homographies.

It is therefore necessary to build a method invariant under homographies.

Up to now, we do not know of a method able to find all singlevalued solutions, but there do exist methods to possibly find all meromorphic solutions. This is the subject of the present chapter.

[1] "New expansion methods" which are not new at all are regularly published, see a critical account in Ref. [26].

2 A simple pedagogical example

Let us first outline the method on a very simple example.

The Kuramoto and Sivashinsky (KS) equation

$$v_t + \nu v_{xxxx} + b v_{xxx} + \mu v_{xx} + v v_x = 0,\ \ \nu \neq 0, \tag{2.1}$$

with (ν, b, μ) constants in $\mathbb{R}$, admits a travelling wave reduction defined as

$$v(x,t) = c + u(\xi),\ \xi = x - ct,\ \nu u''' + b u'' + \mu u' + \frac{u^2}{2} + A = 0, \tag{2.2}$$

in which A is an integration constant. It has a chaotic behavior [28], and it depends on two dimensionless parameters, $b^2/(\mu\nu)$ and $\nu A/\mu^3$.

It admits only eight meromorphic solutions, which are all elliptic or degenerate of elliptic,

1. one nondegenerate elliptic solution [19, 25] with codimension one,

$$\begin{cases} u = -60\nu\wp' - 15b\wp - \dfrac{b\mu}{4\nu}, \\ b^2 = 16\mu\nu,\ g_2 = \dfrac{\mu^2}{12\nu^2},\ g_3 = \dfrac{13\mu^3 + \nu A}{1080\nu^3}; \end{cases} \tag{2.3}$$

2. six degenerate elliptic solutions rational in one exponential [24] with codimension two,

$$\begin{aligned} u &= 120\nu\tau^3 - 15b\tau^2 + \left(-30\nu k^2 - \frac{15(b^2 - 16\mu\nu)}{4 \times 19\nu}\right)\tau \\ &\quad + \frac{5}{2}bk^2 - \frac{13b^3}{32 \times 19\nu^2} + \frac{7\mu b}{4 \times 19\nu},\ \ \tau = \frac{k}{2}\coth\frac{k}{2}(x - x_0), \end{aligned} \tag{2.4}$$

detailed in Table 1;

Table 1. The six trigonometric solutions of KS, Eq. (2.2). They all have the form (2.4). The last line is a degeneracy of the elliptic solution (2.3).

$b^2/(\mu\nu)$	$\nu A/\mu^3$	$\nu k^2/\mu$
0	$-4950/19^3,\ 450/19^3$	$11/19,\ -1/19$
144/47	$-1800/47^3$	1/47
256/73	$-4050/73^3$	1/73
16	$-18,\ -8$	$1,\ -1$

3. one rational solution with codimension three,

$$u = 120\nu(x - x_0)^{-3},\ b = \mu = A = 0. \tag{2.5}$$

Let us prove that no more meromorphic solutions exist.

2.1 Outline of the method

The successive steps of the method, to be detailed soon, are

1. Count the number of different Laurent series of u.

2. Apply a useful result of Nevanlinna theory, known as Clunie's lemma to show that u has infinitely many poles if u is nonrational.

   If in addition the number of Laurent series is finite (which follows from the non-existence of positive Fuchs indices), one can prove that any meromorphic solution is elliptic or degenerate elliptic.

3. If the solution is nondegenerate elliptic, compute its closed form, for instance by its Hermite decomposition.

4. If the solution is rational in one exponential e^{kx}, with k unknown, then its closed form is again provided by its Hermite decomposition.

5. If the solution is rational in x, then its closed form is provided by the classical partial fraction decomposition.

Before proceeding, let us mention the main contributions to the method, in chronological order: [30], [23], [18], [12], [13], [17]. We would also like to reiterate that this method (originated from Eremenko's work [18]) allows one to find *all* meromorphic solutions if they exist.

Let us now detail the successive steps of our pedagogical example.

2.2 Step 1. Counting the Laurent series

Taylor series are here excluded.

One applies the method of Kowalevski and Gambier [6] to look for all algebraic behaviours

$$u = u_0 \chi^p, \ u_0 \neq 0, \ \chi = x - x_0, \tag{2.6}$$

in which p is not a positive integer. This defines the unique dominant behaviour $p = -3, u_0 = 120\nu$, obtained by requiring the terms $\nu u'''$ and $u^2/2$ to both behave like the same power χ^q, here $q = -6$.

Next, one must compute the Fuchs indices [11] of the linearized equation near this triple pole. This is a classical computation, which can be detailed as

$$\lim_{\chi \to 0} \chi^{-j-q}(\nu \partial_x^3 + u_0 \chi^p)\chi^{j+p} \tag{2.7}$$

$$= \nu(j-3)(j-4)(j-5) + 120\nu = \nu(j+1)(j^2 - 13j + 60) \tag{2.8}$$

$$= \nu(j+1)\left(j - \frac{13 + i\sqrt{71}}{2}\right)\left(j - \frac{13 - i\sqrt{71}}{2}\right) = 0. \tag{2.9}$$

Apart from -1, no Fuchs index is a positive integer, therefore the Laurent series exists,

$$u = 120\nu\chi^{-3} - 15b\chi^{-2} - \frac{15(b^2 - 16\mu\nu)}{4 \times 19\nu}\chi^{-1} + \frac{13(4\mu\nu - b^2)b}{32 \times 19\nu^2} + O(\chi), \qquad (2.10)$$

where $\chi = x - x_0$ and it is unique (no arbitrary coefficient enters the series); consequently, for each set of arbitrary values of ν, b, μ, A, there exists at most one one-parameter singlevalued function solution of (2.2).

2.3 Step 2. Clunie's lemma

This is a very useful lemma in Nevanlinna theory to show that certain transcendental (i.e. nonrational) meromorphic function has infinitely many poles. This theory uses a specific vocabulary which we will not detail here, referring to the book [27]. Combined with the finiteness of the number of Laurent series, its main conclusion is that "meromorphic implies elliptic or degenerate elliptic".

Lemma 1 (Clunie's lemma). *Let $f(z)$ be a nonrational meromorphic solution of*

$$f^n P(z, f, f', ...) = Q(z, f, f', ...), \qquad (2.11)$$

where n is a nonzero positive integer, P and Q are polynomials in f and its derivatives with meromorphic coefficients $\{a_\lambda | \lambda \in I\}$, such that for all $\lambda \in I$, $m(r, a_\lambda) = S(r, f)$, where I is some known index set. If the total degree[2] *of Q as a polynomial in f and its derivatives is less than or equal to n, then*

$$m(r, P(z, f, f', ...)) = S(r, f). \qquad (2.12)$$

Equation (2.2) does fit (2.11), with $n = 1$, $P = u$, $Q = -2(\nu u''' + bu'' + \mu u' + A)$, and all its coefficients obey the smallness assumption of the lemma because they are constant. Hence, it follows from Clunie's lemma that $m(r, P) = m(r, u) = S(r, u)$. By the First Main Theorem of Nevanlinna theory ([27]), one can deduce that u must have infinitely many poles $\{z_n\}$ if u is a transcendental meromorphic solution of (2.2). Notice that if a meromorphic solution u has a pole at a, then $u(z + z_n - a)$ is also a meromorphic solution with a pole at a because the ODE (2.2) is *autonomous.* Then a simple demonstration [13], which essentially assumes the number of Laurent series to be finite (which is our case), proves that any meromorphic solution must be periodic and hence elliptic or degenerate elliptic.

Remark 1. Using this method, one can prove that for a large class of n th order autonomous algebraic ODEs, their meromorphic solutions must be elliptic or degenerate elliptic (see [31]). The method fails if there is a positive Fuchs index and hence one cannot conclude the finiteness of the number of Laurent series. In such case, it is possible to have meromorphic solutions which are neither elliptic nor degenerate elliptic (see Lemma 4.5 of [16]).

The remaining work is now to obtain all these solutions in closed form.

[2]Defined as the global degree in all derivatives $f^{(j)}, j \geq 0$.

2.4 Hermite decomposition

The classical decomposition of a rational function as a finite sum of poles has been extended by Hermite to rational functions of one exponential, nondegenerate elliptic functions and even to elliptic functions of the second and third kinds [20]. It fits our goal (to find closed forms) for two reasons: it is a finite expression and therefore closed form; it makes use of the singularity structure established in Step 1. Let us recall these results of Hermite, a nice account of which can be found in a review by Paul Appell [1].

Given a function $u(x)$ which is elliptic or degenerate elliptic, its partial fraction decomposition (Ref. [21] in the elliptic case, Ref. [20, pages 2, 321, 352, 365] in the trigonometric or rational case) is the sum of two parts: the principle part (sum of the principal parts at the poles) and the regular part (an entire function). If one requires both parts to be elliptic or degenerate elliptic, then the entire function can only be a constant (elliptic case), the sum of a polynomial of e^{kx} and a polynomial of e^{-kx} (rational of e^{kx} case), a polynomial of x (rational case), and the decomposition is unique,

$$\text{elliptic}: \ u = \left(\sum_{j=1}^{N}\sum_{q=0}^{n_j} c_{jq}\zeta^{(q)}(x-x_j)\right) + (\text{constant}), \ \sum_{j=1}^{N} c_{j0} = 0, \ 2 \le N, \tag{2.13}$$

$$\text{rat}(e^{kx}), k\neq 0: u = \left(\sum_{j=1}^{N}\sum_{q=0}^{n_j} c_{jq}\left(\frac{k}{2}\coth\frac{k}{2}(x-x_j)\right)^{(q)}\right) + \left(\sum_{m=M_1}^{M_2} d_m (e^{kx})^m\right), 0 \le N \tag{2.14}$$

$$\text{rat}(x): \ u = \left(\sum_{j=1}^{N}\sum_{q=0}^{n_j} c_{jq}(x-x_j)^{-q}\right) + \left(\sum_{m=0}^{M} d_m x^m\right), \ 0 \le N, \tag{2.15}$$

in which ζ is the function introduced by Weierstrass($\zeta' = -\wp$), N is the number of poles in a period parallelogram, x_j the affixes of the poles, n_j the order of the pole x_j, M a positive integer, M_1 and M_2 are integers, c_{jq}, d_m, k complex constants.

Note that each elementary unit (*élément simple*) has by convention one simple pole of residue unity at the origin.

The c_{jq} coefficients are in one-to-one correspondence with the coefficients of the principal part of the finitely many Laurent series established in Step 1.

2.5 Step 3. Nondegenerate elliptic solutions

In our example (2.2), since there is only one pole ($N = 1$), the constraint that the sum of residues of the Laurent series should vanish imposes $b^2 = 16\mu\nu$. The closed form of Hermite is then

$$u = 60\nu\zeta'' + 15b\zeta' + 0\zeta + \text{constant}, \ b^2 = 16\mu\nu, \tag{2.16}$$

and a straightforward computation yields (2.3).

2.6 Step 4. Solutions rational in one exponential

In the closed form (2.14), a simple computation (we will come back to it later) first shows that the entire part reduces to a constant, yielding the closed form,

$$u=60\nu\tau''+15b\tau'-\left(\frac{15(b^2-16\mu\nu)}{4\times 19\nu}+30\nu k^2\right)\tau+\text{constant},\tau=\frac{k}{2}\coth\frac{k}{2}(\xi-\xi_0) \quad (2.17)$$

and there only remains to compute k^2, the constant and the possible constraints on the fixed parameters ν, b, μ, A. The output is the six codimension-two solutions (2.4) and Table 1.

2.7 Step 5. Rational solutions

Finally, the decomposition (2.15) of rational solutions reduces to

$$u=120\nu\chi^{-3}-15b\chi^{-2}-\frac{15(b^2-16\mu\nu)}{4\times 19\nu}\chi^{-1}+\text{constant},\ \chi=x-x_0 \quad (2.18)$$

which yields the unique codimension-three solution (2.5).

3 Lessons from this pedagogical example

3.1 On truncations

1. This example (2.2) is too simple in the sense that truncation methods also succeed to achieve the full results. Indeed, if one assumes the truncation [19, 25]

$$u=c_0\wp'+c_1\wp+c_2,\ c_0\neq 0, \quad (3.1)$$

and eliminates the derivatives of $\wp$ with

$$\wp''=6\wp^2-\frac{g_2}{2},\ \wp'^2=4\wp^3-g_2\wp-g_3, \quad (3.2)$$

the left-hand side of (2.2) becomes an expression similar to (3.1), i.e. a polynomial in $\wp, \wp'$ of degree one in $\wp'$,

$$E(u)=E_{3,0}\wp^3+E_{1,1}\wp\wp'+E_{2,0}\wp^2+E_{0,1}\wp'+E_{1,0}\wp+E_{0,0}=0, \quad (3.3)$$

and the resolution of the six *determining equations* $E_{j,k}=0$ yields the unique solution (2.3).

Similarly, the trigonometric assumption

$$u=\sum_{j=0}^{-p}c_j\tau^{-j-p},\ c_0\neq 0,\ p=-3,\ q=-6, \quad (3.4)$$

if which τ is defined so as to have one simple pole of residue unity,

$$\tau' + \tau^2 + \frac{S}{2} = 0, \ S = -\frac{k^2}{2} = \text{ constant} \in \mathbb{C}, \tag{3.5}$$

generates the seven determining equations

$$E(u) = \sum_{j=0}^{-q} E_j \tau^{-j-q} = 0, \ \forall j : \ E_j = 0, \tag{3.6}$$

whose solutions are the six trigonometric solutions (case $S \neq 0$ and $\tau = \frac{k}{2} \coth \frac{k}{2}(x - x_0)$) and the rational solution (case $S = 0$ and $\tau = 1/(x - x_0)$).

2. *A contrario*, the example (1.7), built to escape the truncation methods, is easily integrated by the above procedure. It admits two Laurent series with equal residues ($\chi = x - x_0$),

$$u = \frac{1}{6\chi} + \varepsilon\frac{i\sqrt{2}}{24} + \frac{35}{144}\chi - \varepsilon\frac{3i\sqrt{2}}{64}\chi^2 + \dots, \ \varepsilon = \pm 1, \ \text{ Fuchs index } (-1), \tag{3.7}$$

it fits Clunie's lemma ($n = 3, P = u, Q = -(1/36)(u'^2 + 12u^2u' - (3/2)u' - (17/2)u^2 + 1/2)$), and its two-pole Hermite decomposition with a constant entire part is,

$$u = \frac{1}{6}\left(\frac{k}{2} \coth \frac{k}{2}(x - x_1) + \frac{k}{2} \coth \frac{k}{2}(x - x_2)\right) + c_0. \tag{3.8}$$

One then identifies the Laurent expansion of (3.8) near $x = x_1$ (resp. $x = x_2$) to the Laurent series (3.7) with $\varepsilon = 1$ (resp. $\varepsilon = -1$). This yields $c_0 = 0, k^2 = 4$, $\tanh^2((x_1 - x_2)/2) = -2$ or $-1/2$ (remember that tanh and coth only differ by a shift).

3.2 On positive integer Fuchs indices

Let us explain on an elementary example why positive integer Fuchs indices harm the procedure. Given the ODE with a meromorphic general solution [5],

$$u'' + 3uu' + u^3 = 0, \ u = \frac{1}{x - a} + \frac{1}{x - b}, \ a \text{ and } b \text{ arbitrary}, \tag{3.9}$$

the question is to integrate this ODE with the Hermite decomposition, in the rational case to simplify. Since it admits the two sets of Laurent series (χ denotes $x - x_0$, with x_0 movable),

$$\left\{ \begin{array}{l} u = \chi^{-1} + \sum_{j=0}^{+\infty} (-1)^j a_1^{j+1} \chi^j, \ a_1 \text{ arbitrary, Fuchs indices } (-1, 1), \\ u = 2\chi^{-1}, \text{ Fuchs indices } (-2, -1), \end{array} \right. \tag{3.10}$$

the number of admissible rational Hermite decompositions is undetermined,

$$\begin{cases} u = \left(\sum_{j=1}^{N} \frac{1}{x - x_j}\right), \ 1 \le N, \\ u = \left(\sum_{j=1}^{N} \frac{1}{x - x_j}\right) + \frac{2}{x}, \ 0 \le N, \end{cases} \tag{3.11}$$

which is unpleasant, although one of these decompositions does succeed. One way out is to look for a first integral of (3.9) associated to the positive Fuchs index 1. Such a first integral K does exist in the present case, it is easily obtained with the assumption (4.2) below with $m = 2$,

$$(u' + u^2)^2 + 4K(2u' + u^2) = 0, \ K = a_1^2, \tag{3.12}$$

it admits finitely many (two) distinct Laurent series,

$$u = \chi^{-1} + \sum_{j=0}^{+\infty} (-1)^j (\pm\sqrt{K})^{j+1} \chi^j, \text{ Fuchs index } (-1), \tag{3.13}$$

and none of them admits the suppressed positive Fuchs index. This defines two admissible Hermite decompositions,

$$u = \left(\sum_{j=1}^{N} \frac{1}{x - x_j}\right), \ N = 1, 2, \tag{3.14}$$

and one of them ($N = 2$) succeeds to represent the general solution.

Remark 2. Once reduced to its first integral (3.12), this example is similar to the previous one (two Laurent series only).

4 Another characterization of elliptic solutions: the subequation method

If one knows (or assumes) that the solutions of the given N-th order ODE are elliptic, the structure of its polar singularities allows one to characterize each such solution by a first order ODE, which will still remain to be integrated. This is the subequation method [30, 12], based on two classical theorems of Briot and Bouquet, which we first recall.

Definition 4. Given an elliptic function, its **elliptic order** is defined as the number of poles in a period parallelogram, counting multiplicity. It is equal to the number of zeros.

Theorem 1. *[3, theorem XVII p. 277]. Given two elliptic functions u, v with the same periods of respective elliptic orders m, n, they are linked by an algebraic equation*

$$F(u,v) \equiv \sum_{k=0}^{m} \sum_{j=0}^{n} a_{j,k} u^j v^k = 0, \tag{4.1}$$

with $\deg(F,u) = \mathrm{order}(v)$, $\deg(F,v) = \mathrm{order}(u)$. *If in particular v is the derivative of u, the first order ODE obeyed by u takes the precise form*

$$F(u,u') \equiv \sum_{k=0}^{m} \sum_{j=0}^{2m-2k} a_{j,k} u^j u'^k = 0, \ a_{0,m} \neq 0. \tag{4.2}$$

Theorem 2. *(Briot and Bouquet [37, vol II §139]). If a first order m-th degree autonomous ODE has a singlevalued general solution,*

- *it must have the form (4.2),*
- *its general solution is either elliptic (two periods) or rational in one exponential e^{kx} (one period) or rational in x (no period) (successive degeneracies $g_2^3 - 27g_3^2 = 0$, then $g_2 = 0$ in $\wp'^2 = 4\wp^3 - g_2\wp - g_3$).*

Remark 3. Equation (4.2) is invariant under an arbitrary homographic transformation having constant coefficients, this is another useful feature of elliptic equations.

The algorithm of the subequation method is the following.
Input: an N-th order ($N \geq 2$) any degree autonomous algebraic ODE

$$E(u, u', \ldots, u^{(N)}) = 0, \ ' = \frac{\mathrm{d}}{\mathrm{d}x}, \tag{4.3}$$

admitting at least one Laurent series

$$u = \chi^p \sum_{j=0}^{+\infty} u_j \chi^j, \ \chi = x - x_0, \ -p \in \mathbb{N}. \tag{4.4}$$

Output: all its elliptic or degenerate elliptic solutions in closed form.
Successive steps [30, 11]:

1. Enumerate the pole orders m_i of all distinct Laurent series (excluding Taylor series). This is Step 1 section 2.2. Deduce the list of elliptic orders (m,n) of (u,u'), with m equal to all possible partial sums of the m_i's. For each element (m,n) of the list, perform the next steps.

   Example 1: the ODE (2.2) admits only one series u with a triple pole, therefore (m,n) can only be $(3,4)$.

   Example 2: the ODE

$$E(u) \equiv \text{some differential consequence of } a^2 u'^2 - (u^2+b)^2 + c = 0, \tag{4.5}$$

admits at least two Laurent series ($m_i = 1, n_i = 2,\ i = 1,2$),

$$u = \pm a\chi^{-1} + \dots \tag{4.6}$$

therefore the list of elliptic orders (m,n) is $(1,2)$, $(2,4)$.

2. Compute J terms in the Laurent series, with J slightly greater than $(m+1)^2$.

3. Define the first order m-th degree subequation $F(u,u') = 0$ (it contains at most $(m+1)^2$ coefficients $a_{j,k}$),

$$F(u,u') \equiv \sum_{k=0}^{m} \sum_{j=0}^{2m-2k} a_{j,k} u^j u'^k = 0,\ a_{0,m} \neq 0. \tag{4.7}$$

4. For each Laurent series (4.4) whose elliptic order m_i contributes to the current sum m, require the series to obey $F(u,u') = 0$,

$$F \equiv \chi^{m(p-1)} \left(\sum_{j=0}^{J} F_j \chi^j + \mathcal{O}(\chi^{J+1}) \right),\ \forall j\ :\ F_j = 0. \tag{4.8}$$

   and solve this **linear overdetermined** system for $a_{j,k}$.

5. Integrate each resulting first order subequation $F(u,u') = 0$. This can be achieved either by the Hermite decomposition (section 2.4 above), or section 5 below.

Remark 4. The fourth step generates a *linear*, infinitely overdetermined, system of equations $F_j = 0$ for the unknown finite set of coefficients $a_{j,k}$. It is quite an easy task to solve such a system, and this is the key advantage of the present algorithm.

4.1 Tutorial examples

The travelling wave reduction of the Korteweg-de Vries (KdV) equation

$$u''' - \frac{6}{a} u u' = 0, \tag{4.9}$$

admits an infinite number of Laurent series (notation $\chi = x - x_0$),

$$u = 2a\chi^{-2} + U_4\chi^2 + U_6\chi^4 + \frac{U_4^2}{6a}\chi^6 + \dots,\ (U_4, U_6) \text{ arbitrary constants.} \tag{4.10}$$

In the infinite list $(m,n) = (2k, 3k)$, $k \in \mathbb{N}$, let us start with $k = 1$ and define (step 3)

$$F \equiv u'^2 + a_{0,1}u' + a_{1,1}uu' + a_{0,0} + a_{1,0}u + a_{2,0}u^2 + a_{3,0}u^3, a_{0,2} = 1. \tag{4.11}$$

This generates (step 4) the linear overdetermined system (4.8),

$$\begin{cases} F_0 \equiv 16a^2 a_{0,2} + 8a^3 a_{3,0} = 0, \\ F_1 \equiv -8a^2 a_{1,1} = 0, \\ F_2 \equiv 4a^2 a_{2,0} = 0, \\ F_3 \equiv -4a a_{0,1} = 0, \\ F_4 \equiv 2a a_{1,0} - 16a a_{0,2} U_4 + 12a^2 a_{3,0} U_4 = 0, \\ F_5 \equiv 0, \\ F_6 \equiv a_{0,0} + 4a a_{2,0} U_4 - 32a a_{0,2} U_6 + 12a^2 a_{3,0} U_6 = 0, \\ \dots \end{cases} \tag{4.12}$$

whose unique solution is

$$u'^2 - (2/a)u^3 + 20U_4 u + 56aU_6 = 0. \tag{4.13}$$

Therefore U_4 and U_6 are interpreted as the two first integrals of (4.9), and they are generated by the method.

Consider a second example which requires no computation,

$$E(u) \equiv a^2 u'^2 - (u^2 + b)^2 + c = 0, \tag{4.14}$$

whose Laurent series are (4.6), defining a list (m, n) made of two elements, $(1, 2)$ and $(2, 4)$. Step 4 generates the constraint $c = 0$ for $(m, n) = (1, 2)$ and no constraint for $(m, n) = (2, 4)$. Step 5 yields the general solution in each case, respectively a coth function ($c = 0$) and a Jacobi elliptic function ($c \neq 0$).

For the application to KS ODE (2.2), see [12].

5 An alternative to the Hermite decomposition

This section only applies to first order autonomous equations which have the Painlevé property, such as (4.2), and it represents its elliptic or degenerate elliptic solution with a closed form different from that of Hermite.

A classical 19th century result due to Picard [37, §33] states that, if an algebraic curve $F(u, v) = 0$ can be parametrized by functions u and v meromorphic on $\mathbb{C}$, then the genus can only be one or zero.

If the genus is one (nondegenerate elliptic case), there exists a birational transformation between (u, v) and $(\wp, \wp')$,

$$u = R_1(\wp, \wp'),\ v = R_2(\wp, \wp'),\ \wp = R_3(u, v),\ \wp' = R_4(u, v), \tag{5.1}$$

thus generating the representation (1.1) of the elliptic function u.

If the genus is zero (degenerate elliptic), the algebraic curve $F(u, v) = 0$ admits a proper rational parametric representation,

$$u = R_1(t),\ v = R_2(t), \tag{5.2}$$

and the condition $\mathrm{d}u/\mathrm{d}x = v$ yields for t either a rational function of one exponential e^{kx} or a rational function of x.

Both cases are implemented in the package **algcurves** written by Mark van Hoeij [22] in the computer algebra language Maple.

The genus one case is processed as

`with(algcurves);` load the package

`genus` $((4.13),u,u')$; check that genus is one

`Weierstrassform` $((4.13),u,u',\wp,\wp')$;

the last command yielding the four formulae (5.1), the first one being the desired answer.

In the genus zero case, the statement **parametrization** $((3.12),u,u',t)$; answers $(K=1)$,

$$u=\frac{3t^2+2t-1}{2t(t-1)},\ u'=\frac{du}{dx}=-\frac{1+2t^2+8t^3+5t^4}{4t^2(1-t)^2}. \tag{5.3}$$

Remark 5. The output (5.1) of **Weierstrassform** is returned *modulo* the addition formula of $\wp$ and thus may not be simplified enough. For instance, **Weierstrassform** $(u^4-u^3+u^2+u+4-u'^2,u,u',\wp,\wp'$,**Weierstrass**) answers

$$u=\frac{9\wp-75-18\wp'}{-143-6\wp+9\wp^2},\ g_2=-\frac{208}{3},\ g_3=-\frac{568}{27}, \tag{5.4}$$

while the simplest answer is u independent of $\wp'$ and homographic in $\wp$.

6 The important case of amplitude equations

Let us denote $A(x,t)$ a complex amplitude depending on the time t and on one space variable x. We consider two amplitude equations, i.e. partial differential equations obeyed by the complex amplitudes. In the following, p, q, r denote complex constants, and γ a real constant. These equations are: the one-dimensional cubic complex Gingburg-Landau equation (CGL3)

$$\text{(CGL3)}\ iA_t+pA_{xx}+q|A|^2A-i\gamma A=0,\ pq\gamma\,\mathrm{Im}(q/p)\neq 0, \tag{6.1}$$

the one-dimensional cubic-quintic complex Gingburg-Landau equation (CGL5)

$$\text{(CGL5)}\ iA_t+pA_{xx}+q|A|^2A+r|A|^4A-i\gamma A=0,\ pr\gamma\,\mathrm{Im}(r/p)\neq 0. \tag{6.2}$$

We will also use the real notation

$$\frac{q}{p}=d_r+id_i,\ \frac{r}{p}=e_r+ie_i,\ \frac{1}{p}=s_r-is_i. \tag{6.3}$$

These equations are generic equations for slowly varying amplitudes, with many physical applications (pattern formation, superconductivity, nonlinear optics, ...), see the reviews [2, 36]. In most physical applications, the quintic term $r|A|^4A$ is zero. Only when the cubic term fails to describe the required features (bifurcation, stability, etc) is the quintic term necessary. We will only consider the most interesting (and most challenging) physical situations, i.e. the so-called "complex case" in which q/p (CGL3) or r/p (CGL5) is not real. Moreover, we will discard the plane wave solutions

$$A = \text{constant } e^{-i\omega t + iKx}, \tag{6.4}$$

because they don't capture the nonlinearity.

Since these two PDEs are autonomous, they admit a traveling wave reduction, i.e. a reduction to an ordinary differential equation (ODE) in the independent variable $\xi = x - ct$, with c an arbitrary real constant.

Physicists prefer to represent the complex amplitudes by their modulus and phase, and therefore to define the traveling wave reduction of (6.2) by two real variables $M(\xi)$, $\varphi(\xi)$,

$$\begin{cases} A(x,t) = \sqrt{M(\xi)} e^{i(-\omega t + \varphi(\xi))}, \ \xi = x - ct, \ c \text{ and } \omega \in \mathbb{R}, \\ \\ \dfrac{M''}{2M} - \dfrac{M'^2}{4M^2} + i\varphi'' - \varphi'^2 + i\varphi' \dfrac{M'}{M} - i\dfrac{c}{2p}\dfrac{M'}{M} + \dfrac{c}{p}\varphi' + \dfrac{q}{p}M + \dfrac{r}{p}M^2 + \dfrac{\omega - i\gamma}{p} = 0. \end{cases} \tag{6.5}$$

From a mathematical point of view, the search for solutions becomes much simpler if one chooses a different representation for the complex amplitudes. One first represents the traveling wave reduction by one complex function a,

$$\begin{cases} A(x,t) = a(\xi)e^{-i\omega t}, \ \overline{A(x,t)} = \bar{a}(\xi)e^{i\omega t}, \ c \text{ and } \omega \in \mathbb{R}, \\ pa'' + qa^2\bar{a} + ra^3\bar{a}^2 - ica' + (\omega - i\gamma)a = 0 \text{ and complex conjugate (c.c.)} \end{cases} \tag{6.6}$$

Then, after introducing the logarithmic derivative,

$$(\log a)' = U, \ (\log \bar{a})' = -U + (\log M)', \tag{6.7}$$

the optimal system to be considered is the closed system made of,

$$p(U' + U^2) + qM + rM^2 - icU + \omega - i\gamma = 0, \tag{6.8}$$

and its complex conjugate,

$$\left[\bar{p}\left(\frac{\mathrm{d}^2}{\mathrm{d}\xi^2} - 2U\frac{\mathrm{d}}{\mathrm{d}\xi} - U' + U^2\right) + \bar{q}M + \bar{r}M^2 + ic\left(\frac{\mathrm{d}}{\mathrm{d}\xi} - U\right) + \omega + i\gamma\right] M = 0. \tag{6.9}$$

The further elimination of the modulus M between (6.8) and (6.9) defines a single third order ODE for the logarithmic derivative U. We will not write explicitly this equation, simply referring to it as (6.9U).

Remark 6. This single ODE (6.9U) is a differential consequence of the Riccati ODE defined by setting $M = 0$ in (6.8).

Remark 7. For CGL3 and for the pure quintic case $q = 0$ of CGL5, the degree of this third order ODE is one. For the cubic-quintic case of CGL5 ($q \neq 0$), because $M + q/(2r)$ as defined by (6.8) is the square root of a differential polynomial of U, this degree is two, therefore (6.9U) may admit a singular solution (defined by canceling an odd-multiplicity factor of the discriminant); this is indeed the case, but this singular solution, defined by $M = 0$, represents the plane wave solution (6.4), which we have discarded.

Definition 5. We call "meromorphic traveling wave" of CGL3/CGL5 any solution in which either M or φ' or U is a function of ξ meromorphic on $\mathbb{C}$.

Step 2 (section 2.3) of the method succeeds to prove that, for both CGL3 and CGL5 and for all values of the fixed parameters $p, q, r, \gamma, c, \omega$ of the reduced ODE, all meromorphic traveling waves M are elliptic or degenerate elliptic. However, the proof (see Refs. [30, 23] for CGL3 and Refs. [15] for CGL5) needs a lot of details and for this reason will not be reproduced in this short chapter. In particular, the three functions M, φ', U of ξ are birationally equivalent [15], therefore the meromorphy of anyone implies the meromorphy of the two others. As a consequence, all meromorphic traveling waves M of both CGL3 and CGL5 have been determined. Several results had been previously obtained [30, 23, 38] for finding such solutions, but they were incomplete.

Let us rather focus here on the following topics: (i) the advantages of the logarithmic derivative U over the physical variables square modulus M and phase φ'; (ii) the singularity structure (Step 1); (iii) the integration by the Hermite decomposition.

6.1 Advantages of the logarithmic derivative variable

There are two advantages. The first one is that all the poles of the logarithmic derivative variable are simple.

The second one is the following. Suppose one has found an admissible Hermite decomposition for U. Then, by construction, the corresponding complex amplitude A is recovered simply by one quadrature, see (6.7),

$$A = A_0 e^{-i\omega t} e^{\int (\text{entire function part})\, d\xi} \prod_{j \in J} (\sigma(\xi - \xi_j))^{c_{j0}}, \tag{6.10}$$

in which $\sigma(z)$ is the Weierstrass entire function or one of its degeneracies $\sinh(kz)/k$ or z.

As opposed to previous work using the physical variables M and φ', we therefore choose here to consider the logarithmic derivative U.

6.2 Laurent series of U of CGL3 and CGL5

As is well known, an elimination may create extraneous solutions, therefore Laurent series computed from the single equation (6.9U) may not all be admissible for the system (6.8)–(6.9). To be safe, one should proceed as follows.

Using the single equation (6.9U), one first computes all the polar behaviours of U. These poles are all simple, the residues U_0 being the three roots of

$$(\text{CGL3}): \ (U_0 - 1)\left[(U_0 + 2)(U_0 + 3)\bar{p}q - U_0(U_0 - 1)p\bar{q}\right] = 0, \tag{6.11}$$

$$(\text{CGL5}): \ (U_0 - 1)\left[(U_0 + 1)(U_0 + 2)\bar{p}r - U_0(U_0 - 1)p\bar{r}\right] = 0, \tag{6.12}$$

i.e. the three values

$$(\text{CGL3}): \ U_0 = 1, -1 + i\alpha_1, \ -1 + i\alpha_2, \tag{6.13}$$

$$(\text{CGL5}): \ U_0 = 1, -\frac{1}{2} + i\alpha_1, \ -\frac{1}{2} + i\alpha_2, \tag{6.14}$$

where α_1, α_2 are the two real roots defined by [4] [29],

$$(\text{CGL3}): \ \alpha_k^2 - 3\frac{d_r}{d_i}\alpha_k - 2 = 0, \ d_i \neq 0, \tag{6.15}$$

$$(\text{CGL5}): \ \alpha_k^2 - 2\frac{e_r}{e_i}\alpha_k - \frac{3}{4} = 0, \ e_i \neq 0. \tag{6.16}$$

Next, one computes the behaviour of M when $U \sim U_0/(\xi - \xi_0)$, by solving the algebraic first degree (CGL3 case) or second degree (CGL5 case) equation (6.8) for M, then by requiring M to also obey (6.9). The behaviour of $(\log \bar{a})'$ follows by (6.7), and the result is summarized in Table 2.

Table 2. Laurent series of U. For each polar behaviour of U, this table displays: the leading behaviours of M and $(\log \bar{a})'$, the Fuchs indices of the Laurent series of U, the number of distinct Laurent series of U of the considered line. The (unphysical) solution $M = 0$ of (6.9) is recovered by setting both arbitrary constants m_1, m_2 to zero.

	Poles of U	M	$(\log \bar{a})'$	Indices	Nb(U)	Detail
CGL3	$(-1+i\alpha_k)\xi^{-1}$	$\frac{3\alpha_k}{d_i}\xi^{-2}$	$(-1-i\alpha_k)\xi^{-1}$	$-1, \frac{7\pm\sqrt{1-24\alpha_k^2}}{2}$	2	
	$\xi^{-1}+\frac{ic}{2p}$	$m_1\xi + m_2\xi^2$	$\frac{m_2}{m_1} - \frac{ic}{2p}$	$-1, 3, 4$	und.	
CGL5 $q\neq 0$	$\left(-\frac{1}{2}+i\alpha_k\right)\xi^{-1}$	$\pm\sqrt{\frac{2\alpha_k}{e_i}}\xi^{-1}$	$\left(-\frac{1}{2}-i\alpha_k\right)\xi^{-1}$	$-1, \frac{5\pm\sqrt{1-32\alpha_k^2}}{2}$	4	(6.17)
	$\xi^{-1}+\frac{ic}{2p}$	$m_1\xi + m_2\xi^2$	$\frac{m_2}{m_1} - \frac{ic}{2p}$	$-1, 3, 4$	und.	(6.18)
CGL5 $q=0$	$\left(-\frac{1}{2}+i\alpha_k\right)\xi^{-1}$	$\pm\sqrt{\frac{2\alpha_k}{e_i}}\xi^{-1}$	$\left(-\frac{1}{2}-i\alpha_k\right)\xi^{-1}$	$-1, \frac{5\pm\sqrt{1-32\alpha_k^2}}{2}$	2	(6.17)
	$\xi^{-1}+\frac{ic}{2p}$	$m_1\xi + m_2\xi^2$	$\frac{m_2}{m_1} - \frac{ic}{2p}$	$-1, 4, 5$	und.	(6.18)

Remark 8. The table enumerating all the Laurent series of M is quite different from Table 2. For instance, whatever be q in CGL5, the poles of M with principal parts $\pm\sqrt{\frac{2\alpha_k}{e_i}}\xi^{-1}$ define four distinct Laurent series of M, and these are the only ones.

For each of the two families with irrational Fuchs indices, the number of Laurent series U is equal to the degree (one or two) of the third order algebraic ODE for U in its highest derivative U'''. For CGL3 and the case $q = 0$ of CGL5, this degree is one. For the case $q \neq 0$ of CGL5, this degree is two and the two Laurent series only differ by signs,

$$\begin{aligned}
(\text{CGL5})\ \forall q: \ U_0 &= -\frac{1}{2} + i\alpha_k, \ k = 1, 2, \\
U &= U_0\chi^{-1} + ic\frac{(U_0+1)(U_0+3)}{8p} - ic\frac{U_0^2 - 1}{8\bar{p}} \\
&\pm \sqrt{U_0(1-U_0)\frac{p}{r}}\left[-\frac{(U_0+1)(U_0+3)q}{8p} + \frac{(U_0-1)\bar{q}}{8\bar{p}}\right] + O(\chi).
\end{aligned} \tag{6.17}$$

For the residue unity of U in the CGL5 case, the dependence on m_1, m_2 (the two arbitrary coefficients) and q, $\bar{q}$ of the couple (U, M) is,

$$\begin{cases} U = \frac{1}{\xi} + \frac{ic}{2p} + .\xi - \frac{m_1 q}{4p}\xi^2 + (.m_1^2 + .m_2 q + .)\xi^3 + (.m_1 m_2 + .m_1 q + .m_2 q)\xi^4 + O(\xi^5), \\ M = m_1\xi + m_2\xi^2 + (.m_1 + .m_2)\xi^3 + (.m_1 + .m_2 + .m_1^2 q + .m_1^2 \bar{q})\xi^4 + O(\xi^5). \end{cases} \quad (6.18)$$

in which the dots (.) represent constants independent of m_1, m_2, q, $\bar{q}$. This allows one to compute the Fuchs indices of U in both subcases $q = \bar{q} = 0$ and $q\bar{q} \neq 0$ and to check that the expansion is free from movable logarithms.

To conclude, the set of poles in the Hermite decomposition of any solution U which is elliptic or degenerate elliptic is made of: an undetermined number of poles of residue unity, two poles (CGL3, resp. CGL5 $q = 0$) of residues $-1 + i\alpha_k$, resp. $-1/2 + i\alpha_k$, or four poles (CGL5 $q \neq 0$) of residues $-1/2 + i\alpha_k$.

6.3 Entire part of a toy ODE

In the method, one must also compute the regular parts of the partial fraction decompositions, Eqs. (2.14) and (2.15). Let us first present this computation on a toy ODE.

Let us consider (then forget) the rational function of e^{kx},

$$u \;= k \coth k(x - x_0)/2 + a e^{k(x-x_0)} + b e^{-2k(x-x_0)}, \quad (6.19)$$

build the first order autonomous ODE by elimination of x_0,

$$2u'^4 + \cdots - 4k^3 u^5 = 0, \quad (6.20)$$

then process it in order to compute the Laurent polynomial $\sum_{m=-2}^{1} d_m e^{mkx}$ (last two terms of (6.19)).

1. Assuming u to be a rational function of $X = e^{qx}$, one builds the ODE for for $U(X) = u(x)$,

$$u = U,\ X = e^{qx},\ E(U', U, X, q, k) \equiv 2q^4 X^4 \left(\frac{\mathrm{d}U}{\mathrm{d}X}\right)^4 + \cdots - 4k^3 U^5 = 0. \quad (6.21)$$

2. Look for an algebraic behaviour $U \sim cX^p$, as $X \to \infty$, with $cp \neq 0$,

$$X \to \infty,\ U \sim cX^p,\ E(U', U, X, q, k) \sim X^{5p} c^5 (pq - k)(pq + 2k)^2, \quad (6.22)$$

which yields the two solutions,

$$\begin{cases} pq = k, & u \sim ce^{kx}, \quad c \text{ undetermined}, \\ pq = -2k, & u \sim ce^{-2kx}, \quad c \text{ undetermined}, \end{cases} \quad (6.23)$$

and the final assumption for the Hermite decomposition of (6.20)

$$u = q \coth \frac{q}{2} x + \left(\sum_{m=-2}^{1} d_m (e^{kx})^m \right), \quad (6.24)$$

with q and the d_m's to be determined.

6.4 Entire part of CGL3 and CGL5

Let us show that it reduces to a constant for both CGL3 and CGL5.

Following previous section 6.3, one performs in (6.9U) the change of function

$$(U,\xi) \to (F,X): \; U=F, \; X=e^{k\xi}, \; E(F''',F'',F',F,X,k)=0, \tag{6.25}$$

and one looks for power law behaviours of $F(X)$ when $X\to\infty$. The result

$$\begin{aligned}
&X\to\infty, \; F\sim aX^m, \; m\neq 0,\\
&(\text{CGL3}) \; E(F''',F'',F',F,X,k)\sim a^4 p(\bar p q-\bar q p)X^{4m},\\
&(\text{CGL5}) \; E(F''',F'',F',F,X,k)\sim a^{12}2^6 p^4 r^3(\bar p r-\bar r p)^2 X^{12m},
\end{aligned} \tag{6.26}$$

yields no solution for mk (as opposed to the toy ODE, see Eq. (6.22)), therefore the regular part in the partial fraction decomposition of the solution U of (6.9U) reduces to a constant.

6.5 All admissible partial fraction decompositions

The information already obtained is (i) the list of distinct Laurent series, recalled in the end of section 6.2; (ii) the entire part of each decomposition, which is constant.

Since all the poles of U are simple (advantage of the logarithmic derivative), the list of admissible Hermite decompositions of U is identical to the list of residues, counting multiplicity. Let us prove that this list is finite.

Indeed, as proven in [15], the number of Laurent series of M is finite (and equal to 2 or 4), therefore the number of admissible Hermite decompositions of M is finite, and so is the number of solutions M in the elliptic or degenerate elliptic class.

Since U is a rational function of M, M', M'', M''' (see details for instance in [14]),

$$U=\frac{M'}{M}+\frac{cs_r}{2}+\frac{G'-2cs_iG}{2M^2(g_r-d_iM-e_iM^2)}, \tag{6.27}$$

$$G=\frac{1}{2}MM''-\frac{1}{4}M'^2-\frac{cs_i}{2}MM'+g_iM^2+d_rM^3+e_rM^4, \tag{6.28}$$

the number of solutions U in that class is also finite, and so is the number of their Hermite decompositions. In particular, one concludes that the number N_1 of poles of U with residue unity in any Hermite decomposition of U is finite, and we leave it to the reader to establish the upper bound of N_1.

For CGL3, restricting to $N_1\le 1$, the list contains six elements, characterized by their residues,

$$\left\{\begin{aligned}
&N=1: \; \{\text{residues}\}=\{-1+i\alpha_k\},\\
&\\
&N=2: \; \{\text{residues}\}=\{1,-1+i\alpha_k\},\\
&N=2: \; \{\text{residues}\}=\{-1+i\alpha_1,-1+i\alpha_2\},\\
&\\
&N=3: \; \{\text{residues}\}=\{1,-1+i\alpha_1,-1+i\alpha_2\}.
\end{aligned}\right. \tag{6.29}$$

The list for the pure quintic case $q = 0$ of CGL5 is the same, after replacing $-1 + i\alpha_k$ by $-1/2 + i\alpha_k$.

For the cubic-quintic case of CGL5 ($q \neq 0$), the multiplicity of each residue $-1/2 + i\alpha_k$ is two, thus defining a list made of twenty-four elements (restricting to $N_1 \leq 2$),

$$\left\{\begin{array}{l}
N = 1: \ \{\text{residues}\} = \{-1/2 + i\alpha_k\}, \\
\\
N = 2: \ \{\text{residues}\} = \{1, -1/2 + i\alpha_k\}, \\
N = 2: \ \{\text{residues}\} = \{-1/2 + i\alpha_1, -1/2 + i\alpha_2\}, \\
N = 2: \ \{\text{residues}\} = \{-1/2 + i\alpha_k, -1/2 + i\alpha_k\}, \\
\\
N = 3: \ \{\text{residues}\} = \{1, 1, -1/2 + i\alpha_k\}, \\
N = 3: \ \{\text{residues}\} = \{1, -1/2 + i\alpha_1, -1/2 + i\alpha_2\}, \\
N = 3: \ \{\text{residues}\} = \{1, -1/2 + i\alpha_k, -1/2 + i\alpha_k\}, \\
N = 3: \ \{\text{residues}\} = \{-1/2 + i\alpha_1, -1/2 + i\alpha_2, -1/2 + i\alpha_k\}, \\
\\
N = 4: \ \{\text{residues}\} = \{1, 1, -1/2 + i\alpha_1, -1/2 + i\alpha_2\}, \\
N = 4: \ \{\text{residues}\} = \{1, 1, -1/2 + i\alpha_k, -1/2 + i\alpha_k\}, \\
N = 4: \ \{\text{residues}\} = \{1, -1/2 + i\alpha_1, -1/2 + i\alpha_2, -1/2 + i\alpha_k\}, \\
N = 4: \ \{\text{residues}\} = \{-1/2 + i\alpha_1, -1/2 + i\alpha_2, -1/2 + i\alpha_1, -1/2 + i\alpha_2\}, \\
\\
N = 5: \ \{\text{residues}\} = \{1, 1, -1/2 + i\alpha_1, -1/2 + i\alpha_2, -1/2 + i\alpha_k\}, \\
N = 5: \ \{\text{residues}\} = \{1, -1/2 + i\alpha_1, -1/2 + i\alpha_2, -1/2 + i\alpha_1, -1/2 + i\alpha_2\}, \\
\\
N = 6: \ \{\text{residues}\} = \{1, 1, -1/2 + i\alpha_1, -1/2 + i\alpha_2, -1/2 + i\alpha_1, -1/2 + i\alpha_2\}.
\end{array}\right. \tag{6.30}$$

7 Nondegenerate elliptic solutions

7.1 Nondegenerate elliptic solutions of CGL3

Given the set of three residues defined in (6.13), with respective multiplicities (undetermined, zero or one, zero or one), the only realization of the constraint $\sum_{j=1}^{N} c_{j0} = 0$ in (2.13) is: two copies of the Laurent series with residue one, plus one copy of each of the two others, with the additional condition $\alpha_1 = -\alpha_2 = \sqrt{2}$ (i.e. $d_r = 0$). However, the zero sum condition further applied to various differential polynomials of U (which must also be elliptic) generates additional necessary conditions which prevent CGL3 to admit nondegenerate elliptic solutions.

This result was first established by Hone [23] by considering the variable $|A|^2$, but his argument is here simplified by the proof that each residue $-1 + i\alpha_k$ is associated to one and only one Laurent series.

7.2 Nondegenerate elliptic solutions of CGL5

Similarly, given the three residues (6.14), with respective multiplicities (unknown, zero or one or two, zero or one or two), there exist two sets whose weighted sum is zero, obtained by taking each residue once or twice with the additional condition $\alpha_1 = -\alpha_2 = \sqrt{2}$ (i.e. $e_r = 0$),

$$\text{first set, any } q: \ 1, \ \frac{-1+i\sqrt{3}}{2}, \ \frac{-1-i\sqrt{3}}{2}, \tag{7.1}$$

$$\text{second set, } q \neq 0: \ 1, \ 1, \ \frac{-1+i\sqrt{3}}{2}, \ \frac{-1+i\sqrt{3}}{2}, \ \frac{-1-i\sqrt{3}}{2}, \ \frac{-1-i\sqrt{3}}{2}. \tag{7.2}$$

For the first set, the zero sum condition further applied to various differential polynomials of U generates three more necessary conditions, equivalent to the conditions [14, Eq. (25)] generated by the consideration of M instead of U. The result is a unique elliptic solution,

$$\begin{cases} q = 0, \ \mathrm{Re}\left(\dfrac{r}{p}\right) = 0, \ \mathrm{Im}\left(\dfrac{\gamma + i\omega}{p}\right) = -\dfrac{1}{4}(cs_r)^2 - \dfrac{3}{16}(cs_i)^2, \\ \dfrac{\mathrm{d}}{\mathrm{d}\xi}\log\left(Ae^{i\omega t - i\frac{cs_r}{2}\xi}\right) = \dfrac{cs_i}{2} + \displaystyle\sum_{k=1}^{3} e^{k2i\pi/3}\left(\zeta(\xi - \xi_{j,k}^U, G_2, G_3) + \zeta(\xi_{j,k}^U, G_2, G_3)\right), \\ M = \dfrac{3^{1/4}}{\sqrt{-e_i}}\displaystyle\sum_{k=1}^{4} e^{k2i\pi/4}\left(\zeta(\xi - \xi_{j,k}^M, g_2, g_3) + \zeta(\xi_{j,k}^M, g_2, g_3)\right). \end{cases} \tag{7.3}$$

Detailed in [15, Eq. (61)], this codimension-four solution is an extrapolation of a previous result [38], and the relation $M = |A|^2$ defines a Landen transformation [15, Appendix] between $\zeta(*, g_2, g_3)$ and $\zeta(*, G_2, G_3)$.

Remark 9. Out of the three ζ functions in the expression (7.3) for U, two arise from the Laurent series (6.14) with irrational Fuchs indices and one from the Laurent series with positive integer indices. This is in contrast with the expression for M, made of four Laurent series with irrational Fuchs indices (see [15, Eq. (21)]), without any contribution from some Laurent series with positive integer indices.

As to the second set, it cannot define a nondegenerate elliptic solution since, as proven in [15], (7.3) is the unique such solution.

8 Degenerate elliptic solutions

8.1 Method of resolution

Processing the single third order ODE (6.9U) might result in too large expressions, therefore it is better to again take advantage of the logarithmic derivative, as we now illustrate on the example of trigonometric solutions of CGL3 made of one principal part.

Given the trigonometric one-pole assumption

$$U = c_a + U_0 \frac{k}{2} \coth \frac{k}{2}\xi, \ k \neq 0, \ U_0 \neq 1, \tag{8.1}$$

one quadrature yields

$$a = A_0 e^{c_a \xi}\left(\sinh \frac{k}{2}\xi\right)^{U_0}, \ A_0 = \text{ arbitrary complex constant}, \tag{8.2}$$

and, by complex conjugation (using the property $\mathrm{Re}(U_0) = -1$),

$$\bar{a} = \bar{A}_0 e^{\bar{c_a}\xi}\left(\sinh\frac{k}{2}\xi\right)^{-2-U_0},\quad M = m_0\, e^{(c_a+\bar{c_a})\xi}\,\frac{k^2}{4}\left(\coth^2\frac{k}{2}\xi - 1\right), \tag{8.3}$$

therefore the constant c_a must be purely imaginary.

The above values of U and M inserted in (6.8) and (6.9) generate two polynomials,

$$(6.8) = \sum_{j=0}^{2} E_j z^j,\ (6.9) = \sum_{j=0}^{4} E'_j z^j,\ z = \frac{k}{2}\coth\frac{k}{2}\xi. \tag{8.4}$$

Requiring that they identically vanish defines eight determining equations in the unknowns $(U_0,\, m_0,\, c_a,\, \omega,\, k^2,\, c)$, and the parameters $(p,\, \bar{p},\, q,\, \bar{q},\, \gamma)$,

$$\begin{cases} E_2 \equiv q m_0 + p U_0(U_0-1) = 0, \\ E_1 \equiv U_0(2pc_a - ic) = 0, \\ E_0 \equiv -4icc_a - 4i\gamma + 4\omega + 4pc_a^2 + (pU_0 - qm_0)k^2 = 0, \end{cases} \tag{8.5}$$

and

$$\begin{cases} E'_4 \equiv (\bar{q}m_0 + \bar{p}(U_0+2)(U_0+3))m_0 = 0, \\ E'_3 \equiv (U_0+2)(2\bar{p}c_a - ic)m_0 = 0, \\ E'_2 \equiv \left[-4icc_a + 4i\gamma + 4\omega + 4\bar{p}c_a^2 + (-\bar{p}(U_0+2)(U_0+4) - 2\bar{q}m_0)k^2\right] m_0 = 0, \\ E'_1 \equiv k^2 E'_3 = 0, \\ E'_0 \equiv k^4 E'_4 - k^2 E'_2 = 0. \end{cases} \tag{8.6}$$

Two kinds of resolutions can be performed.

The first one is to solve the three equations $E_j = 0$ (which are linearly independent) on the field $\mathbb{C}$ as a *linear* system of Cramer type (for instance in $m_0,\, c_a,\, \omega$), then to require that $m_0,\, ic_a,\, \omega,\, k^2$ and c be real. These reality conditions ensure by construction that the other five equations $E'_j = 0$ are satisfied.

The second one is to solve the eight equations $E_j = 0,\ E'_j = 0$ (which are no more independent) again on the field $\mathbb{C}$, as an overdetermined *linear* system for some set of variables, be they unknowns or parameters (this makes no difference, as argued in [9, Appendix A]). As usual, equations should be processed by decreasing values of the singularity degree j. Such a set is, for instance, $(m_0, \bar{q})$, then (c_a, c), then (ω, γ). After that, there is no need to enforce any reality condition since they are already taken into account.

Since the residue $U_0 = 1$ has been excluded in assumption (8.1), the solution is unique,

$$\begin{cases} m_0 = -U_0(U_0-1)\dfrac{p}{q},\ \bar{q} = \dfrac{(U_0+2)(U_0+3)}{U_0(U_0-1)}\dfrac{\bar{p}q}{p},\ c_a = i\dfrac{c}{2p},\ c(p-\bar{p}) = 0, \\ \omega = -\dfrac{k^2}{8}(pU_0^2 + \bar{p}(U_0+2)^2) - \dfrac{c^2}{4p},\ \gamma = -i\dfrac{k^2}{8}(pU_0^2 - \bar{p}(U_0+2)^2), \end{cases} \tag{8.7}$$

or equivalently in the physical variables,

$$\begin{cases} A = A_0 e^{-i\omega t + ic\xi/(2p)} \frac{k}{2} \left(\sinh \frac{k}{2}\xi\right)^{-1+i\alpha}, \ M = |A_0|^2 \frac{k^2}{4}\left(\coth^2 \frac{k}{2}\xi - 1\right), \\ |A_0|^2 = \frac{3\alpha}{d_i}, \ \frac{i\gamma - \omega}{p} = \left(\frac{c}{2p}\right)^2 + \frac{k^2}{4}(1-i\alpha)^2, \ \mathrm{Im}(c/p) = 0, \end{cases} \tag{8.8}$$

which splits in two solutions. The case $c = 0$ is the stationary pulse or solitary wave [33],

$$\begin{cases} A = A_0 e^{-i\omega t} (\cosh Kx)^{-1+i\alpha}, \ \frac{i\gamma - \omega}{p} = (1-i\alpha)^2 K^2, \ c = 0, \ \lim_{x\to -\infty} |A| = \lim_{x\to +\infty} |A| = 0, \\ c = 0, \ |A_0|^2 = \frac{3\alpha}{d_i}, \ \frac{\gamma}{2\alpha s_r + (\alpha^2 - 1)s_i} = \frac{\omega}{2\alpha s_i - (\alpha^2-1)s_r} = -\frac{k^2}{4(s_r^2 + s_i^2)}, \end{cases} \tag{8.9}$$

and the case $\mathrm{Im}(1/p) = 0$ is,

$$s_i = 0, \ |A_0|^2 = \frac{3\alpha}{d_i}, \ k^2 = -\frac{2s_r\gamma}{\alpha}, \ \omega = \frac{2(1-\alpha^2)\gamma - \alpha s_r c^2}{4\alpha}, \ c \text{ arbitrary.} \tag{8.10}$$

8.2 Results

By lack of space, we postpone the exhaustive list of results to a forthcoming paper, but, essentially, the only new recent result is the elliptic solution of the purely quintic CGL5 [15, Eq. (61)], recalled in Eq. (7.3).

9 Current challenges and open problems

The two examples of this chapter are not independent, and CGL3 admits a scaling limit [35] under which the variable $u = \arg A$ obeys the KS PDE (2.1), this is why in this section we concentrate on KS.

1. All meromorphic solutions of the third order ODE (2.2) have been found.

2. Since they all have a nonzero codimension, the problem remains open to find a closed form of the Laurent series (2.10) for generic values of (ν, b, μ, A). If it exists, the results of Eremenko prove that it is not meromorphic.

3. A numerical investigation by Padé approximants [41] of the singularities of the sum of the Laurent series (2.10) for generic values of (ν, b, μ, A) confirms (this is not a proof) the absence of any multivaluedness and displays a nearly doubly periodic pattern for the singularities, the unit cell being made of one triple pole and three simple zeroes (Fig. 1). Then singlevalued nonmeromorphic closed forms matching this description could involve the expression of Picard [34] $\wp((\omega/(i\pi)) \log(x - c_1) + c_2, g_2, g_3)$ (with 2ω a period of $\wp$), which admits the isolated movable noncritical essential singularities $x = c_1 + 2ni\pi$, n integer.

Acknowledgments. *This work was partially funded by the Hong Kong RGC grant 17301115. The first author is grateful to the Institute of mathematical research of HKU for financial support.*

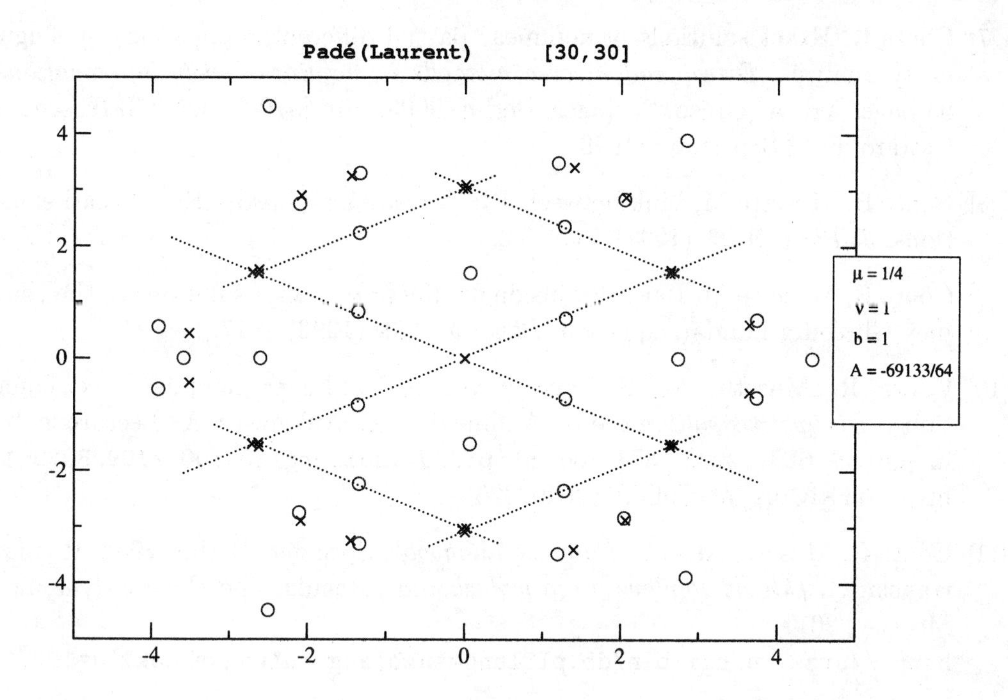

Figure 1. Singularities of the Padé approximant [L/M] of the unknown solution of KS ODE 2.2 (courtesy of Tony Yee Tat-leung). Crosses are poles, circles zeroes. The numerical values are $L = 30, M = 30, \nu = 1, b = 1, \mu = 1/4, A = -69133/64$.

References

[1] Appell P, Sur la décomposition d'une fonction méromorphe en éléments simples, *Mémorial des sciences mathématiques* **36** (1929) 1–37. Gauthier-Villars, Paris, 1929. http://gallica.bnf.fr/ark:/12148/bpt6k38984v

[2] Aranson I S, Kramer L, The world of the complex Ginzburg-Landau equation, *Rev. Math. Phys.* **74** (2002) 99–143. http://arXiv.org/abs/cond-mat/0106115

[3] Briot C, Bouquet J-C, *Théorie des fonctions elliptiques*, Mallet-Bachelier, Paris, 1859. `http://gallica.bnf.fr/document?O=N099571`

[4] Cariello F, Tabor M, Painlevé expansions for nonintegrable evolution equations, *Physica D* **39** (1989) 77–94.

[5] Conte R, Fordy A P, Pickering A, A perturbative Painlevé approach to nonlinear differential equations, *Physica D* **69** (1993) 33–58.

[6] Conte R, The Painlevé approach to nonlinear ordinary differential equations, *The Painlevé property, one century later*, 77–180, ed. R. Conte, CRM series in mathematical physics, Springer, New York, 1999. Solv-int/9710020.

[7] Conte R, Exact solutions of nonlinear partial differential equations by singularity analysis, *Direct and inverse methods in nonlinear evolution equations*, 83 pages, ed. A. Greco, Springer, Berlin, 2003. nlin.SI/0009024. CIME school, Cetraro, 5–12 September 1999.

[8] Conte R, Musette M, Link between solitary waves and projective Riccati equations, *J. Phys. A* **25** (1992) 5609–5623.

[9] Conte R, Musette M, Linearity inside nonlinearity: exact solutions to the complex Ginzburg-Landau equation, *Physica D* **69** (1993) 1–17.

[10] Conte R, Musette M, Solitary waves of nonlinear nonintegrable equations, *Dissipative solitons*, eds. Akhmediev A, Ankiewicz A, Lecture notes in physics **661** (2005) 373–406. `http://dx.doi.org/10.1007/10928028_15` http://arXiv.org/abs/nlin.PS/0407026

[11] Conte R, Musette M, *The Painlevé handbook*, Springer, Berlin, 2008. Russian translation *(Metod Penleve y ego prilozhenia)*, Regular and chaotic dynamics, Moscow, 2011.
`http://urss.ru/cgi-bin/db.pl?lang=en&blang=ru&page=Book&id=122718`

[12] Conte R, Musette M, Elliptic general analytic solutions, *Studies in applied mathematics* **123** (2009) 63–81. http://arxiv.org/abs/0903.2009 math.CA, math.DS

[13] Conte R, Ng T W, Meromorphic solutions of a third order nonlinear differential equation, *J. Math. Phys.* **51** (2010) 033518 (9 pp). http://arxiv.org/abs/1002.1209

[14] Conte R, Ng T W, Detection and construction of an elliptic solution to the complex cubic-quintic Ginzburg-Landau equation, *Teoreticheskaya i Matematicheskaya Fizika* **172** (2012) 224–235. *Theor. math. phys.* **172** (2012) 1073–1084. http://arxiv.org/abs/1204.3028

[15] Conte R, Ng T W, Meromorphic traveling wave solutions of the complex cubic-quintic Ginzburg-Landau equation, *Acta applicandae mathematicae* **122** (2012) 153–166. doi:10.1007/s10440-012-9734-y
http://arxiv.org/abs/1204.3032

[16] Conte R, Ng T W, Wu C F, Hayman's classical conjecture on some nonlinear second order algebraic ODEs, *Complex Variables and Elliptic Equations* **60** (2015) 1539–1552. doi.org/10.1080/17476933.2015.1033414

[17] Demina M V, Kudryashov N A, Explicit expressions for meromorphic solutions of autonomous nonlinear ordinary differential equations, *Commun. nonlinear sci. numer. simul.* **16** (2011) 1127–1134.

[18] Eremenko A E, Meromorphic traveling wave solutions of the Kuramoto-Sivashinsky equation, *J. of mathematical physics, analysis and geometry* **2** (2006) 278–286.

[19] Fournier J D, Spiegel E A, and Thual O, Meromorphic integrals of two nonintegrable systems, *Nonlinear dynamics*, 366–373, ed. Turchetti G, World Scientific, Singapore, 1989.

[20] Hermite C, *Cours d'analyse de l'École polytechnique*, Gauthier-Villars, Paris, 1873. `https://archive.org/details/coursdanalysedel01hermuoft`

[21] Hermite C, Remarques sur la décomposition en éléments simples des fonctions doublement périodiques, *Annales de la faculté des sciences de Toulouse* **II** (1888) C1–C12. *Œuvres d'Hermite*, vol IV, pp 262–273.

[22] van Hoeij M, package "algcurves", Maple V (1997). http://www.math.fsu.edu/~hoeij/algcurves.html

[23] Hone A N W, Non-existence of elliptic travelling wave solutions of the complex Ginzburg-Landau equation, *Physica D* **205** (2005) 292–306.

[24] Kudryashov N A, Exact soliton solutions of the generalized evolution equation of wave dynamics, *Prikladnaia Matematika i Mekhanika* **52** (1988) 465–470 [English : *Journal of applied mathematics and mechanics* **52** (1988) 361–365]

[25] Kudryashov N A, Exact solutions of a generalized equation of Ginzburg-Landau, *Matematicheskoye modelirovanie* **1** (1989) 151–158 [in Russian].

[26] Kudryashov N A, Loguinova N B, Be careful with the Exp-function method, *Commun. nonlinear sci. numer. simul.* **14** (2009) 1881–1890.

[27] Laine I, *Nevanlinna theory and complex differential equations*, de Gruyter, Berlin and New York, 1992.

[28] Manneville P, *Dissipative structures and weak turbulence*, Academic Press, Boston, 1990. French adaptation: *Structures dissipatives, chaos et turbulence*, Aléa-Saclay, Gif-sur-Yvette, 1991.

[29] Marcq P, Chaté H, and Conte R, Exact solutions of the one-dimensional quintic complex Ginzburg-Landau equation, *Physica D* **73** (1994) 305–317. http://arXiv.org/abs/patt-sol/9310004

[30] Musette M, Conte R, Analytic solitary waves of nonintegrable equations, *Physica D* **181** (2003) 70–79. http://arXiv.org/abs/nlin.PS/0302051

[31] Ng T W, Wu C F, Nonlinear Loewy factorizable algebraic ODEs and Hayman's conjecture, *Israel Journal of Mathematics* (to appear). https://arxiv.org/abs/1710.08593

[32] Painlevé P, Sur les équations différentielles du second ordre et d'ordre supérieur dont l'intégrale générale est uniforme, *Acta Math.* **25** (1902) 1–85.

[33] Pereira N R, Stenflo L, Nonlinear Schrödinger equation including growth and damping, *Phys. Fluids* **20** (1977) 1733–1743.

[34] Picard É, Remarques sur les équations différentielles, *Acta Math.* **17** (1893) 297–300.

[35] Pomeau Y, Manneville P, Stability and fluctuations of a spatially periodic flow, *J. Physique Lett.* **40** (1979) L609–L612.

[36] van Saarloos W, Front propagation into unstable states, *Physics reports* **386** (2003) 29–222.

[37] Valiron G, *Cours d'analyse mathématique*, I. – *Théorie des fonctions*, 522 pages, 2ième éd., Masson, Paris, 1948. II. – *Équations fonctionnelles. Applications*, 605 pages, 2ième éd., Masson, Paris, 1950.

[38] Vernov S Yu, Elliptic solutions of the quintic complex one-dimensional Ginzburg-Landau equation, *J. Phys. A* **40** (2007) 9833–9844. http://arXiv.org/abs/nlin.PS/0602060

[39] Weiss J, Tabor M, Carnevale G, The Painlevé property for partial differential equations, *J. Math. Phys.* **24** (1983) 522–526.

[40] Wu C F, Meromorphic solutions of complex differential equations, Ph D thesis, The University of Hong Kong (2014).

[41] Yee T L, Conte R, Musette M, Sur la "solution analytique générale" d'une équation différentielle chaotique du troisième ordre, 195–212, *From combinatorics to dynamical systems*, eds. Fauvet F, Mitschi C, IRMA lectures in mathematics and theoretical physics **3**, de Gruyter, Berlin, 2003. http://arXiv.org/abs/nlin.PS/0302056 Journées de calcul formel, Strasbourg, IRMA, 21–22 mars 2002.

B3. Pfeiffer–Sato solutions of Buhl's problem and a Lagrange–d'Alembert principle for heavenly equations

Oksana E Hentosh [a], *Yarema A Prykarpatsky* [b], *Denis Blackmore* [c] *and Anatolij Prykarpatski* [d]

[a] *Institute for Applied Problems of Mechanics and Mathematics at the NAS, Lviv, 79060 Ukraine*

[b] *Department of Applied Mathematics at the University of Agriculture Krakow, 30059, Poland*

[c] *Department of Mathematical Sciences at New Jersey Institute of Technology, University Heights, Newark, NJ 07102 USA*

[d] *Department of Physics, Mathematics and Computer Science Cracow University of Technology, Krakow, 31-155, Poland*

Dedicated to our colleague, teacher, friend, and brilliant mathematician Anatolij M. Samoilenko on his 80th Birthday Jubilee

Abstract

This review is devoted to the Buhl compatible vector field equation problem, emphasizing its Pfeiffer and Lax–Sato type solutions. We analyze the related Lie-algebraic structures and integrability of the heavenly equations. AKS-algebraic and related $\mathcal{R}$-structure schemes are used to study the corresponding co-adjoint actions. Their compatibility conditions are shown to coincide with the corresponding heavenly equations, all of which originate in this way and can be represented as a Lax compatibility condition. The infinite hierarchy of conservation laws for the heavenly equations is described along with its Casimir invariant connection and several examples are presented. An interesting related Lagrange–d'Alembert principle is also discussed. A generalization of the scheme, related to the loop Lie superalgebra of the Lie super group of superconformal diffeomorphisms of the $1|N$-dimensional supertorus, is used to construct Lax–Sato integrable supersymmetric analogs of the Mikhalev–Pavlov heavenly equation for every $N \in \mathbb{N}\backslash\{4;5\}$. Super-analogs of Liouville equations are constructed using superconformal maps.

1 Introduction

In his classical 1928 works [15, 16, 17], the French mathematician M.A. Buhl posed the problem of classifying all infinitesimal symmetries of a given linear vector field equation

$$A\psi = 0, \tag{1.1}$$

where $\psi \in C^2(\mathbb{R}^n;\mathbb{R})$, and

$$A := \sum_{j=\overline{1,n}} a_j(x)\frac{\partial}{\partial x_j} \tag{1.2}$$

is a vector field operator on $\mathbb{R}^n$ with smooth coefficients $a_j \in C^\infty(\mathbb{R}^n;\mathbb{R}), j=\overline{1,n}$. It is easy to show that this problem reduces [68] to describing all possible vector fields

$$A^{(k)} := \sum_{j=\overline{1,n}} a_j^{(k)}(x)\frac{\partial}{\partial x_j} \tag{1.3}$$

with $a_j^{(k)} \in C^\infty(\mathbb{R}^n;\mathbb{R}), j,k=\overline{1,n}$, satisfying the Lax-type commutator condition

$$[A, A^{(k)}] = 0 \tag{1.4}$$

for all $x \in \mathbb{R}^n$ and $k=\overline{1,n}$. The Buhl problem above was solved in 1931 by the Ukrainian mathematician M.G. Pfeiffer in [65, 66, 67, 68, 69, 70], wherein he constructed the set of independent vector fields (1.3) by making use of the invariants for (1.2) and the related solution set structure of the Jacobi–Mayer system that follows from (1.4). Some incomplete results were also obtained by Popovici [72].

The equivalent Buhl-type problem was reanalyzed by Takasaki & Takebe [92, 93] and by Bogdanov, Dryuma & Manakov [11] for a case when (1.2) depends analytically on a "spectral" parameter $\lambda \in \mathbb{C}$:

$$\tilde{A} := \frac{\partial}{\partial t} + \sum_{j=\overline{1,n}} a_j(t,x;\lambda)\frac{\partial}{\partial x_j} + a_0(t,x;\lambda)\frac{\partial}{\partial \lambda}. \tag{1.5}$$

Using Sato theory [81, 82], the above authors have shown that for some special kinds of (1.5) there exists an infinite hierarchy of the symmetry vector fields

$$\tilde{A}^{(k)} := \frac{\partial}{\partial \tau_k} + \sum_{j=\overline{1,n}} a_j^{(k)}(\tau,x;\lambda)\frac{\partial}{\partial x_j} + a_0^{(k)}(\tau,x;\lambda)\frac{\partial}{\partial \lambda}, \tag{1.6}$$

where $\tau = (t;\tau_1,\tau_2,...) \in \mathbb{R}^{\mathbb{Z}_+}, k \in \mathbb{Z}_+$, satisfying the Lax–Sato commutator conditions

$$[\tilde{A}, \tilde{A}^{(k)}] = 0 = [\tilde{A}^{(j)}, \tilde{A}^{(k)}] \tag{1.7}$$

for all $k, j \in \mathbb{Z}_+$. Moreover, in these cases, the conditions (1.7) proved to be equivalent to some (important for applications) heavenly dispersionless partial differential equations (PDEs).

Here we investigate Lax–Sato compatible systems, their related Lie-algebraic structures and integrability of an interesting class of nonlinear systems called heavenly equations, which were introduced by Plebański [71] and analyzed in [49, 50,

11, 60, 61, 62, 63, 83, 92, 93]. We use the Adler–Kostant–Symes (AKS)-algebraic and related $\mathcal{R}$-structure schemes [8, 7, 9, 94, 57, 80, 78] applied to the holomorphic Birkhoff factorized loop Lie algebra $\tilde{\mathcal{G}} := \widetilde{diff}(\mathbb{T}_{\mathbb{C}}^{1+n})$ of vector fields on the complexified torus $\mathbb{T}_{\mathbb{C}}^{1+n}, n \in \mathbb{Z}_{+}$. The orbits of their coadjoint actions on $\tilde{\mathcal{G}}^*$, related to the Lie–Poisson structures, are reanalyzed in detail. By constructing two commuting flows on $\tilde{\mathcal{G}}^*$, generated by a root element $\tilde{l} \in \tilde{\mathcal{G}}^*$ and Casimir invariants, we show their compatibility condition coincides with the corresponding heavenly equations.

As a by product of thc construction in [38, 77], we prove all heavenly equations have a similar origin and can be represented as a Lax compatibility condition for special vector fields on $\mathbb{T}_{\mathbb{C}}^{1+n}$. We analyze the structure of the infinite hierarchy of conservations laws, related to the heavenly equations, and show that their analytical structure connected with the Casimir invariants is generated by the Lie–Poisson structure on $\tilde{\mathcal{G}}^*$. Also we generalized the Lie-algebraic scheme of [38, 77] subject to the holomorphic loop Lie algebra $\widetilde{diff}(\mathbb{T}_{\mathbb{C}}^{1|N})$ of superconformal vector fields on $\mathbb{T}^{1|N}$, which is a Lie algebra of the Lie group of superconformal diffeomorphisms of the $1|N$-dimensional supertorus $\mathbb{T}^{1|N} \simeq \mathbb{T}^1 \times \Lambda_1^N$, where $\Lambda := \Lambda_0 \oplus \Lambda_1$ is an infinite-dimensional Grassmann algebra over $\mathbb{C}$, $\Lambda_0 \supset \mathbb{C}$. This is applied to constructing the Lax–Sato integrable superanalogs of the Mikhalev–Pavlov heavenly super-equation for every $N \in \mathbb{N} \setminus \{4; 5\}$. Using this scheme, we can construct a natural derivation of the Lax–Sato representations for an infinite hierarchy of heavenly equations, related to the canonical Lie–Poisson structure on the adjoint space $\tilde{\mathcal{G}}^*$. As a result of suitably chosen superconformal mappings in the space of variables $(z; \theta_1, \ldots, \theta_N) \in \mathbb{T}_{\mathbb{C}}^{1|N}$ the superanalogs of Liouville type equations are obtained. We also mention an interesting aspect of our integrability approach to the heavenly dynamical systems, closely related to their classical Lagrange–d'Alembert type mechanical interpretation.

There are few multi-dimensional integrable systems that have been analyzed in great detail. As was mentioned in [91], the heavenly equations comprise an important class of such integrable systems. This is due in part to the fact that some of them are obtained by a reduction of the Einstein equations with Euclidean (and neutral) signature for (anti-) self-dual gravity, which includes the theory of gravitational instantons. This and other examples of important applications of multi-dimensional integrable equations motivated us to study this class of equations. In particular, the main motivation was based both on the paper by Kulish [45] studying the super-conformal Korteweg–de-Vries equation as an integrable Hamiltonian flow on the adjoint space to the holomorphic loop Lie superalgebra of super-conformal vector fields on the circle, and on the insightful investigation by Mikhalev [55], which studied Hamiltonian structures on the adjoint space to the holomorphic loop Lie algebra of smooth vector fields on the circle. We were also impressed by the deep results [92, 93] of Takasaki & Takebe, who fully realized the vector field scheme of the Lax–Sato theory. Additionally, we were influenced by the results of Pavlov, Bogdanov, Dryuma, Konopelchenko and Manakov [13, 11, 12, 40], as well as those of Ferapontov and Moss [30], Błaszak, Szablikowski and Sergyeyev, Krasil'shchik [9, 85, 89, 90, 86, 87, 41] in which they devised new effective differential-geometric and analytical tools for studying an integrable degenerate multi-dimensional dispersionless heavenly type hierarchy of equations, the importance of which is still under-appreciated. Other Lie-algebraic approaches to constructing integrable heavenly equations are Ovsienko [60, 61] and Kruglikov & Morozov [44].

We shall present examples of the Lie-algebraic description of typical integrable heavenly equations including the Mikhalev–Pavlov type [55], the Shabat types [2], new Sergyeyev [90] 3-D types, the Hirota [25] and general [83, 84] types, and the Liouville type [12].

2 Lax–Sato compatible systems of vector field equations

Consider a simple vector field $X : \mathbb{R}\times\mathbb{T}^n \to T(\mathbb{R}\times\mathbb{T}^n)$ on the $(n+1)$-dimensional toroidal manifold $\mathbb{R}\times\mathbb{T}^n$ for arbitrary $n \in \mathbb{Z}_+$, which we write in the form

$$A = \frac{\partial}{\partial t} + < a(t,x), \frac{\partial}{\partial x} >, \tag{2.1}$$

where $(t,x) \in \mathbb{R}\times\mathbb{T}^n, a(t,x) \in \mathbb{E}^n, \frac{\partial}{\partial x} := (\frac{\partial}{\partial x_1}, \frac{\partial}{\partial x_2}, ..., \frac{\partial}{\partial x_n})^\intercal$ and $< \cdot, \cdot >$ is the usual scalar product on the Euclidean space $\mathbb{E}^n$. With (2.1), one can associate the equation

$$A\psi = 0 \tag{2.2}$$

for a $\psi \in C^2(\mathbb{R}\times\mathbb{T}^n;\mathbb{R})$, which we will call an "*invariant*" of the vector field.

Next, we study the existence and number of such functionally-independent invariants of (2.2). For this we pose the following Cauchy problem for (2.2): Find a function $\psi \in C^2(\mathbb{R}\times\mathbb{T}^n;\mathbb{R})$, such that $\psi(t,x)|_{t=t^{(0)}} = \bar{\psi}(x)$, $x \in \mathbb{R}^n$ for a given function $\bar{\psi} \in C^2(\mathbb{T}^n;\mathbb{R})$. For (2.2) there is a naturally related parametric vector field on the torus $\mathbb{T}^n$ in the form of the ordinary differential equation (ODE)

$$dx/dt = a(t,x), \tag{2.3}$$

with its corresponding Cauchy problem: find a function $x : \mathbb{R} \to \mathbb{T}^n$ satisfying

$$x(t)|_{t=t^{(0)}} = z \tag{2.4}$$

for an arbitrary constant vector $z \in \mathbb{T}^n$. Assuming that $a \in C^1(\mathbb{R}\times\mathbb{T}^n;\mathbb{R}^n)$, it follows from the Cauchy theorem [19] that there is a unique solution to (2.3) - (2.4) of the form $x = \Phi(t,z)$ with $\Phi \in C^1(\mathbb{R}\times\mathbb{T}^n;\mathbb{T}^n)$ such that the matrix $\partial\Phi(t,z)/\partial z$ is nondegenerate for all $t \in \mathbb{R}$ sufficiently close to $t^{(0)} \in \mathbb{R}$. Hence, the Implicit Function Theorem [1, 3, 19, 20] implies that there exists a map $\Psi : \mathbb{R}\times\mathbb{T}^n \to \mathbb{T}^n$, such that

$$\Psi(t,x) = z \tag{2.5}$$

for all z and t sufficiently near $t^{(0)}$. If $\Psi(t,x) = (\psi^{(1)}(t,x), \psi^{(2)}(t,x), ..., \psi^{(n)}(t,x))^\intercal$, $(t,x) \in \mathbb{R}\times\mathbb{T}^n$ is constructed, the arbitrariness of the parameter $z \in \mathbb{T}^n$ implies that all functions $\psi^{(j)} : \mathbb{R} \times \mathbb{T}^n \to \mathbb{R}$, $j = \overline{1,n}$, are functionally independent invariants of the vector field equation (2.2), i.e $A\psi^{(j)} = 0, j = \overline{1,n}$. Thus, (2.2) has exactly $n \in \mathbb{Z}_+$ functionally independent invariants, which make it possible, in particular, to solve the Cauchy problem posed above. Namely, let a map $\alpha : \mathbb{R}^n \to \mathbb{R}$ be chosen such that $\alpha(\Psi(t,x))|_{t=t^{(0)}} = \bar{\psi}(x)$ for all $x \in \mathbb{R}^n$ and a fixed $t^{(0)} \in \mathbb{R}$. Inasmuch as the composition $\alpha \circ \Psi : \mathbb{R} \times \mathbb{T}^n \to \mathbb{R}$ is, evidently, also an invariant for (2.2), it provides the solution to this Cauchy problem, which we can formulate as follows.

Lemma 1. *The linear equation (2.2), generated by the vector field (2.3) on the toroidal manifold $\mathbb{R} \times \mathbb{T}^n$, has exactly $n \in \mathbb{Z}_+$ functionally independent invariants.*

Consider now a differential form $\chi^{(n)} \in \Lambda^n(\mathbb{T}^n)$, generated by the vector of independent invariants (2.5), depending on the vector evolution parameter $t \in \mathbb{R}^n$:

$$\chi^{(n)} := d\psi^{(1)} \wedge d\psi^{(2)} \wedge ... \wedge d\psi^{(n)}, \tag{2.6}$$

where for any $\psi^{(k)} \in C^2(\mathbb{R}^n \times \mathbb{T}^n; \mathbb{R}), k = \overline{1,n}$, the differentials

$$d\psi^{(k)} = \sum_{j=\overline{1,n}} \frac{\partial \psi^{(k)}}{\partial t_j} dt_j + < \frac{\partial \psi^{(k)}}{\partial x}, dx > . \tag{2.7}$$

The vector $\Psi \in C^2(\mathbb{R}^n \times \mathbb{T}^n; \mathbb{R}^n)$ consists of invariants and satisfies for all $(t,x) \in \mathbb{R}^n \times \mathbb{T}^n$ a compatible system of the equations

$$\frac{\partial \psi^{(s)}}{\partial t_j} + \sum_{k=\overline{1,n}} a_j^{(k)}(t,x) \frac{\partial \psi^{(s)}}{\partial x_k} = 0, \tag{2.8}$$

where $a_j^{(s)} \in C^1(\mathbb{R}^n \times \mathbb{T}^n; \mathbb{R}), j, s = \overline{1,n}$, are given vector field coefficients. Owing to (2.8), one easily obtains from (2.6) that the following

$$\left|\frac{\partial \psi}{\partial x}\right|^{-1} d\psi^{(1)} \wedge d\psi^{(2)} \wedge \cdots \wedge d\psi^{(n)} = \left(dx_1 - \sum\nolimits_{j=1}^n a_j^{(1)}(t,x)dt_j\right) \wedge$$
$$\left(dx_2 - \sum\nolimits_{j=1}^n a_j^{(2)}(t,x)dt_j\right) \wedge \cdots \wedge \left(dx_n - \sum\nolimits_{j=1}^n a_j^{(n)}(t,x)dt_j\right) \tag{2.9}$$

hold on $\mathbb{T}^n$. Moreover, as follows from the Frobenius theorem [19, 34, 40], the Plücker type form (2.6), owing to the functional independence of $\Psi \in C^2(\mathbb{R}^n \times \mathbb{T}^n; \mathbb{R}^n)$ on the torus $\mathbb{T}^n$, makes it possible to construct [70] the Jacobi–Mayer type relationship (2.9), giving rise to the corresponding set of compatible vector field relationships

$$-\frac{\partial \Psi}{\partial t_s} + \sum_{j,k=\overline{1,n}} \left[\left(\frac{\partial \Psi}{\partial x}\right)^{-1}_{jk} \frac{\partial \psi^{(k)}}{\partial t_s}\right] \frac{\partial \Psi}{\partial x_j} = 0 \tag{2.10}$$

for any $s = \overline{1,n}$. This, as demonstrated by Pfeiffer [70], makes it possible to solve the Buhl problem and has interesting applications [11, 40] in the theory of completely integrable dynamical systems of heavenly type, which are considered in the next section.

2.1 Vector field hierarchies on the torus with "*spectral*" parameter and the Lax–Sato integrable heavenly dynamical systems.

Consider a naturally ordered infinite set of parametric vector fields (2.1) on the infinite-dimensional toroidal manifold $\mathbb{R}^{\mathbb{Z}_+} \times \mathbb{T}^n$ in the form

$$A^{(k)} = \frac{\partial}{\partial t_k} + < a^{(k)}\,(t, x; \lambda), \frac{\partial}{\partial x} > + a_0^{(k)}(t, x; \lambda)\frac{\partial}{\partial \lambda} := \frac{\partial}{\partial t_k} + \mathrm{A}^{(k)}, \qquad (2.11)$$

where $t_k \in \mathbb{R}, k \in \mathbb{Z}_+, (t, x; \lambda) \in (\mathbb{R}^{\mathbb{Z}_+} \times \mathbb{T}^n) \times \mathbb{C}$ are the evolution parameters; the dependence of smooth vectors $(a_0^{(k)}, a^{(k)})^\intercal \in \mathbb{E} \times \mathbb{E}^n$, $k \in \mathbb{Z}_+$, on the "*spectral*" parameter $\lambda \in \mathbb{C}$ is assumed to be holomorphic. Suppose now that the infinite hierarchy

$$A^{(k)}\psi = 0 \qquad (2.12)$$

for $k \in \mathbb{Z}_+$ has exactly $n + 1 \in \mathbb{Z}_+$ common functionally independent invariants $\psi^{(j)}(\lambda) \in C^2(\mathbb{R}^{\mathbb{Z}_+} \times \mathbb{T}^n; \mathbb{C}), j = \overline{0, n}$ on $\mathbb{T}^n$, suitably depending on the parameter $\lambda \in \mathbb{C}$. Then, it follows from standard results for ODEs depending analytically on $\lambda \in \mathbb{C}$ [19, 20] that one can assume that these invariants may be analytically continued in $\lambda \in \mathbb{C}$ both inside $\mathbb{D}^1_+ \subset \mathbb{C}$ of some disc $\mathbb{D}^1 \subset \mathbb{C}$ and subject to $\lambda^{-1} \in \mathbb{C}, |\lambda| \to \infty$, outside $\mathbb{D}^1_- \subset \mathbb{C}$ of this disc $\mathbb{D}^1 \subset \mathbb{C}$. Hence, as $|\lambda| \to \infty$ we have the expansions:

$$\begin{aligned} &\psi^{(0)}(\lambda) \sim \lambda + \sum_{k=0}^{\infty} \psi_k^{(0)}(t, x)\lambda^{-k}, \\ &\psi^{(1)}(\lambda) \sim \sum_{k=0}^{\infty} \tau_k^{(1)}(t, x)\psi_0(\lambda)^k + \sum_{k=0}^{\infty} \psi_k^{(1)}(t, x)\psi_0(\lambda)^{-k}, \cdots, \\ &\psi^{(n)}(\lambda) \sim \sum_{k=0}^{\infty} \tau_k^{(n)}(t, x)\psi_0(\lambda)^k + \sum_{k=0}^{\infty} \psi_k^{(n)}(t, x)\psi_0(\lambda)^{-k}, \end{aligned} \qquad (2.13)$$

where we took into account that $\psi^{(0)}(\lambda) \in C^2(\mathbb{R}^{\mathbb{Z}_+} \times \mathbb{T}^n; \mathbb{C})$, $\lambda \in \mathbb{C}$, is the basic invariant solution to (2.12), the functions $\tau_l^{(s)} \in C^2(\mathbb{R}^{\mathbb{Z}_+} \times \mathbb{T}^n; \mathbb{R})$for all $s = \overline{1, n}, l \in \mathbb{Z}_+$, are assumed to be independent and $\psi_k^{(j)} \in C^2(\mathbb{R}^{\mathbb{Z}_+} \times \mathbb{T}^n; \mathbb{R})$ for all $k \in \mathbb{N}, j = \overline{0, n}$, are arbitrary. Write (2.9) on $\mathbb{C} \times \mathbb{T}^n$ with $\mathrm{x} := (\lambda, x) \in \mathbb{C} \times \mathbb{T}^n$ as

$$\begin{aligned} &\left|\frac{\partial \psi}{\partial x}\right|^{-1} d\psi^{(1)} \wedge d\psi^{(2)} \wedge \cdots \wedge d\psi^{(n)} = \left(dx_0 - \sum\nolimits_{j=1}^{n} a_j^{(0)}(t, x)dt_j\right) \wedge \\ &\left(dx_1 - \sum\nolimits_{j=1}^{n} a_j^{(1)}(t, x)dt_j\right) \wedge \cdots \wedge \left(dx_n - \sum\nolimits_{j=1}^{n} a_j^{(n)}(t, x)dt_j\right) \end{aligned} \qquad (2.14)$$

where $|\frac{\partial \Psi}{\partial \mathrm{x}}|$ is the Jacobi determinant of $\Psi := (\psi^{(0)}, \psi^{(1)}, \psi^{(2)}, ..., \psi^{(n)})^\intercal \in C^2(\mathbb{C} \times (\mathbb{R}^{\mathbb{Z}_+} \times \mathbb{T}^n); \mathbb{C}^{n+1})$ on $\mathbb{C} \times \mathbb{T}^n$. Since this map depending on $\lambda \in \mathbb{C}$ can be analytically continued [20] inside $\mathbb{S}^1_+ \subset \mathbb{C}$ of the disc $\mathbb{D}^1 \subset \mathbb{C}$, subject to $\lambda^{-1} \in \mathbb{C}$ as $|\lambda| \to \infty$ outside $\mathbb{D}^1_- \subset \mathbb{C}$ of this disc $\mathbb{D}^1 \subset \mathbb{C}$, one readily obtains from

$$d\psi^{(j)} = < \frac{\partial \psi^{(j)}}{\partial \mathrm{x}}, d\mathrm{x} > + \sum_{k=0}^{\infty} \frac{\partial \psi^{(j)}}{\partial \tau_k^{(j)}} d\tau_k^{(j)} = 0 \qquad (2.15)$$

for all $j = \overline{1,n}$, and (2.14) on $\mathbb{C} \times \mathbb{T}^n$ of the variables $\mathrm{x} \in \mathbb{C} \times \mathbb{T}^n$, evolving analytically with respect to $\tau_k^{(j)} \in \mathbb{R}$, $j = \overline{1,n}, k \in \mathbb{Z}_+$, the Lax–Sato criterion:

$$\left(|\frac{\partial \Psi}{\partial \mathrm{x}}|^{-1} d\psi^{(0)} \wedge d\psi^{(1)} \wedge d\psi^{(2)} \wedge ... \wedge d\psi^{(n)} \right)_{-} = 0, \tag{2.16}$$

where $(...)_-$ denotes the asymptotic part of an expression in the bracket, depending on the parameter $\lambda^{-1} \in \mathbb{C}$ as $|\lambda| \to \infty$. Substitution of (2.15) into (2.16) yields

$$-\frac{\partial \Psi}{\partial \tau_k^{(j)}} = \left[\left(\frac{\partial \Psi}{\partial \mathrm{x}} \right)^{-1}_{0j} \psi^{(0)}(\lambda)^k \right]_+ \frac{\partial \Psi}{\partial \lambda} + \sum_{s=1}^{n} \left[\left(\frac{\partial \Psi}{\partial \mathrm{x}} \right)^{-1}_{sj} \psi^{(0)}(\lambda)^k \right]_+ \frac{\partial \Psi}{\partial x_s} \tag{2.17}$$

for all $k \in \mathbb{Z}_+, j = \overline{1,n}$. Now (2.17) comprise an infinite hierarchy of Lax–Sato compatible [92, 93] linear equations, where $(...)_+$ denotes the asymptotic part of an expression in the bracket, depending on nonnegative powers of $\lambda \in \mathbb{C}$. The functional parameters $\tau_k^{(j)} \in C^2(\mathbb{R}^{\mathbb{Z}_+} \times \mathbb{T}^n; \mathbb{R})$ for all $k \in \mathbb{Z}_+, j = \overline{1,n}$, are functionally independent owing to their *a priori* linear dependence on the evolution parameters $t_k \in \mathbb{R}$, $k \in \mathbb{Z}_+$. On the other hand, taking into account the form of the hierarchy (2.17), following [11], it is not hard to show that the corresponding vector fields

$$\mathrm{A}_k^{(j)} := \left[\left(\frac{\partial \Psi}{\partial \mathrm{x}} \right)^{-1}_{0j} \psi^{(0)}(\lambda)^k \right]_+ \frac{\partial}{\partial \lambda} + \sum_{s=1}^{n} \left[\left(\frac{\partial \Psi}{\partial \mathrm{x}} \right)^{-1}_{sj} \psi^{(0)}(\lambda)^k \right]_+ \frac{\partial}{\partial x_s} \tag{2.18}$$

on $\mathbb{C} \times \mathbb{T}^n$ satisfy for all $k, m \in \mathbb{Z}_+, j, l = \overline{1,n}$, the Lax–Sato compatibility conditions

$$\frac{\partial \mathrm{A}_m^{(l)}}{\partial \tau_k^{(j)}} - \frac{\partial \mathrm{A}_k^{(j)}}{\partial \tau_m^{(l)}} = [\mathrm{A}_k^{(j)}, \mathrm{A}_m^{(l)}], \tag{2.19}$$

which are equivalent to the independence of the functional parameters $\tau_k^{(j)} \in C^1(\mathbb{R}^{\mathbb{Z}_+} \times \mathbb{T}^n; \mathbb{R})$, $k \in \mathbb{Z}_+, j = \overline{1,n}$. As a corollary of the above analysis, one can show that the infinite hierarchy (2.11) is a linear combination of the basic vector fields (2.18) and also satisfies the Lax compatibility condition (2.19). Inasmuch as the coefficients of (2.18) are suitably smooth functions on the manifold $\mathbb{R}^{\mathbb{Z}_+} \times \mathbb{T}^n$, the compatibility conditions (2.19) yield the corresponding sets of differential-algebraic relationships on their coefficients, which have the common infinite set of invariants. Hence, they comprise an infinite hierarchy of completely integrable *heavenly* nonlinear dynamical systems on the corresponding multidimensional functional manifolds. Thus, the above can be considered as an introduction to a recently devised [11, 13, 21, 38, 92, 93] constructive algorithm for generating infinite hierarchies of completely integrable nonlinear heavenly dynamical systems on functional manifolds of arbitrary dimension. We stress here that the above algorithm for nonlinear multidimensional dynamical systems still does not make it possible to directly show they are Hamiltonian and construct other related mathematical structures. This problem is solved by other means, for example, via the analytical properties of the related loop diffeomorphism groups generated by (2.11).

2.2 Example: Representation for the Mikhalev–Pavlov heavenly equation.

The Mikhalev–Pavlov equation was first constructed in [55] and has the form

$$u_{xt} + u_{yy} = u_y u_{xx} - u_x u_{xy}, \tag{2.20}$$

where $u \in C^\infty(\mathbb{R}^2 \times \mathbb{T}^1; \mathbb{R})$ and $(t, y; x) \in \mathbb{R}^2 \times \mathbb{T}^1$. Assume now [11] that the two functions

$$\psi^{(0)} = \lambda, \quad \psi^{(1)} \sim \sum_{k=3}^{\infty} \lambda^k \tau_k - \lambda^2 t + \lambda y + x + \sum_{j=1}^{\infty} \psi_j^{(1)}(t, y, \tau; x)\, \lambda^{-j}, \tag{2.21}$$

where $\psi_1^{(1)}(t, y, \tau; x) = u$, $(t, y, \tau; x) \in \mathbb{R}^2 \times \mathbb{R}^\infty \times \mathbb{T}^1$, are invariants of (2.12) for an infinite set of constant parameters $\tau_k \in \mathbb{R}, k = \overline{3, \infty}$, as $\lambda \to \infty$. By applying to the invariants (2.21) the criterion (2.16) in the form

$$((\partial \psi^{(1)}/\partial x)^{-1} d\psi^{(1)})_- = 0, \tag{2.22}$$

one can easily obtain the following compatible linear vector field equations

$$\psi_t + (\lambda^2 + \lambda u_x - u_y)\, \psi_x = 0,\ \psi_y + (\lambda + u_x)\, \psi_x = 0, \cdots, \psi_{\tau_k} + P_k(u, \lambda)\psi_x = 0, \tag{2.23}$$

where $P_k(u; \lambda), k = \overline{3, \infty}$, are independent differential-algebraic polynomials in $u \in C^\infty(\mathbb{R}^2 \times \mathbb{R}^\infty \times \mathbb{T}^1)$ and algebraic polynomials in $\lambda \in \mathbb{C}$, calculated from the expressions (2.17). Moreover, the compatibility condition (2.19) for the first two equations of (2.23) yields the Mikhalev–Pavlov equation (2.20).

2.3 Example: The Dunajski metric equation

The Dunajski metric equations [23], with $(u, v) \in C^\infty(R^2 \times T^2; R^2)$, $(y, t; x_1, x_2) \in R^2 \times T^2$, are

$$\begin{aligned} &u_{x_1 t} + u_{y x_2} + u_{x_1 x_1} u_{x_2 x_2} - u_{x_1 x_2} - v = 0, \\ &v_{x_1 t} + v_{x_2 y} + u_{x_1 x_1} v_{x_2 x_2} - 2u_{x_1 x_2} v_{x_1 x_2} = 0, \end{aligned} \tag{2.24}$$

where $(u, v) \in C^\infty\left(\mathbb{R}^2 \times \mathbb{T}^2; \mathbb{R}^2\right)$ and $(y, t; x_1, x_2) \in \mathbb{R}^2 \times \mathbb{T}^2$.

Using an approach analogous to that employed for the Dunajski equations, we obtain a compatible hierarchy of the following Lax–Sato type vector field equations:

$$\begin{aligned} A^{(t_0)}\psi &:= \frac{\partial \psi}{\partial t} + \mathrm{A}^{(t_0)}\psi = 0, \mathrm{A}^{(t_0)} := u_{x_2 x_2}\frac{\partial}{\partial x_1} - (\lambda + u_{x_1 x_2})\frac{\partial}{\partial x_2} + v_{x_2}\frac{\partial}{\partial \lambda} = 0, \\ A^{(t_1)}\psi &:= \frac{\partial \psi}{\partial y} + \mathrm{A}^{(t_1)}\psi = 0, \mathrm{A}^{(t_1)} := (\lambda - u_{x_1 x_2})\frac{\partial}{\partial x_1} + u_{x_1 x_1}\frac{\partial}{\partial x_2} - v_{x_1}\frac{\partial}{\partial \lambda} = 0, \\ A^{(t_k^{(s)})}\psi &:= \frac{\partial \psi}{\partial \tau_k^s} + < P_k^s(u; \lambda), \frac{\partial \psi}{\partial \mathrm{x}} >= 0, \end{aligned} \tag{2.25}$$

where $P_k^s(u,v;\lambda)\in\mathbb{E}^3, s=\overline{1,2}, k\in\mathbb{N}\backslash\{1\}$, are independent vector-valued differential-algebraic polynomials [13] in $(u,v)\in C^\infty(\mathbb{R}^2\times\mathbb{R}^{\mathbb{Z}_+}\times\mathbb{T}^2;\mathbb{R}^2)$ and algebraic polynomials in $\lambda\in\mathbb{C}$, calculated from (2.17).

The above indicates (as shown in the sequel) that the Birkhoff [73] loop groups $\widetilde{Diff}(\mathbb{T}^n)$, $Diff_{hol}(\mathbb{C}\times\mathbb{T}^n)$ should play key roles in analyzing the Lie-algebraic nature of the integrable heavenly systems.

3 Heavenly equations: Lie-algebraic integrability scheme

Let $\tilde{G}_\pm := \widetilde{Diff}_\pm(\mathbb{T}^n_{\mathbb{C}})$, $n\in\mathbb{Z}_+$, be subgroups of the Birkhoff [73] loop group $\widetilde{Diff}(\mathbb{T}^n) := \{\mathbb{C}\supset\mathbb{S}^1\to Diff_{hol}(\mathbb{T}^n)\}$, holomorphically extended in the interior $\mathbb{D}^1_+\subset\mathbb{C}$ and in the exterior $\mathbb{D}^1_-\subset\mathbb{C}$ regions of the unit disc $\mathbb{D}^1\subset\mathbb{C}^1$, such that $\partial\mathbb{D}^1=\mathbb{S}^1$, and for any $g(\lambda)\in\tilde{G}_\pm$, either for $\lambda\in\mathbb{D}^1_-$ the value $g(\infty)=1\in Diff(\mathbb{T}^n)$ or for $\lambda\in\mathbb{D}^1_+$ $g(0)=1\in Diff(\mathbb{T}^n)$. The corresponding Lie subalgebras $\tilde{\mathcal{G}}_\pm := \widetilde{diff}_\pm(\mathbb{T}^n_{\mathbb{C}})$ of the loop subgroups $\tilde{G}_\pm$ are vector fields on $\mathbb{T}^n$ holomorphic, respectively, on $\mathbb{S}^1_\pm\subset\mathbb{C}^1$, where either for any $\tilde{a}(\lambda)\in\tilde{\mathcal{G}}_-$ $\tilde{a}(\infty)=0$, or for any $\tilde{a}(\lambda)\in\tilde{\mathcal{G}}_+$ $\tilde{a}(0)=0$. The split loop Lie algebra $\tilde{\mathcal{G}}=\tilde{\mathcal{G}}_+\oplus\tilde{\mathcal{G}}_-$ can be naturally identified with a dense subspace of the dual space $\tilde{\mathcal{G}}^*$ via the pairing

$$(\tilde{l},\tilde{a})_{s;q} := \underset{\lambda\in\mathbb{C}}{res}(\lambda^{-s}l(\lambda;x),a(\lambda;x))_{H^q}, \tag{3.1}$$

for some fixed $s\in\mathbb{Z}$ and $q\in\mathbb{Z}_+$. In [19, 77], we defined a loop vector field $\tilde{a}\in\Gamma(\tilde{T}(\mathbb{T}^n))$ and a loop differential 1-form $\tilde{l}\in\tilde{\Lambda}^1(\mathbb{T}^n)$ given as

$$\begin{aligned}\tilde{a} &= \sum_{j=1}^n a^{(j)}(\lambda;x)\frac{\partial}{\partial x_j}+a^{(0)}(\lambda;x)\frac{\partial}{\partial\lambda} := \left\langle a(\mathrm{x}),\frac{\partial}{\partial\mathrm{x}}\right\rangle, \\ \tilde{l} &= \sum_{j=1}^n l_j(\lambda;x)dx_j+l_0(\lambda;x)d\lambda := \langle l(\mathrm{x}),d\mathrm{x}\rangle,\end{aligned} \tag{3.2}$$

denoted $(\lambda;x):=\mathrm{x}\in\mathbb{T}^n_{\mathbb{C}}$, introduced the gradient operator $\frac{\partial}{\partial\mathrm{x}} := \left(\frac{\partial}{\partial\lambda};\frac{\partial}{\partial x_1},\frac{\partial}{\partial x_2},\ldots\right.$ $\left.\ldots,\frac{\partial}{\partial x_n}\right)^\intercal$ in $\mathbb{C}\times\mathbb{E}^n$ and chose the Sobolev type metric $(\cdot,\cdot)_{H^q}$ on the dense subspace $C^\infty(\mathbb{T}^n;\mathbb{R}^n)\subset H^q(\mathbb{T}^n;\mathbb{R}^n)$ for some $q\in\mathbb{Z}_+$ as

$$(l(x;\lambda),a(x;\lambda))_{H^q} := \sum_{j=1}^n\sum_{|\alpha|=0}^q\int_{\mathbb{T}^n}dx\left(\frac{\partial^{|\alpha|}l_j(x;\lambda)}{\partial x^\alpha}\frac{\partial^{|\alpha|}a^{(j)}(x;\lambda)}{\partial x^\alpha}\right), \tag{3.3}$$

where $\partial x^\alpha := \partial x_1^{\alpha_1}\partial x_2^{\alpha_2}\ldots\partial x_2^{\alpha_n}$, $|\alpha|=\sum_{j=1}^n\alpha_j$ for $\alpha\in\mathbb{Z}^n_+$, generalizing the metric, used in [56]. The standard Lie commutator of vector fields $\tilde{a},\tilde{b}\in\tilde{\mathcal{G}}$ is

$$[\tilde{a},\tilde{b}]=\tilde{a}\tilde{b}-\tilde{b}\tilde{a}=\left\langle\left\langle a(\mathrm{x}),\circ\frac{\partial}{\partial\mathrm{x}}\right\rangle b(\mathrm{x}),\frac{\partial}{\partial\mathrm{x}}\right\rangle-\left\langle\left\langle b(\mathrm{x}),\frac{\partial}{\partial\mathrm{x}}\right\rangle a(\mathrm{x}),\frac{\partial}{\partial\mathrm{x}}\right\rangle. \tag{3.4}$$

As the above Lie algebra $\tilde{\mathcal{G}}$ is canonically split into the direct sum $\tilde{\mathcal{G}} = \tilde{\mathcal{G}}_+ \oplus \tilde{\mathcal{G}}_-$, one can identify the dual spaces $\tilde{\mathcal{G}}_+^* \simeq \lambda^{s-1}\tilde{\mathcal{G}}_-$, $\quad \tilde{\mathcal{G}}_-^* \simeq \lambda^{s-1}\tilde{\mathcal{G}}_+$, where either for any $l(\lambda) \in \tilde{\mathcal{G}}_+^*$ one has the constraint $\lim_{\lambda\to\infty} \lambda^{-s}\tilde{l}(\lambda) = 0$ or for any $l(\lambda) \in \tilde{\mathcal{G}}_-^*$ one has the constraint $\lim_{\lambda\to 0} \lambda^{-s}\tilde{l}(\lambda) = 0$. By defining the projections

$$P_\pm \tilde{\mathcal{G}} := \tilde{\mathcal{G}}_\pm \subset \tilde{\mathcal{G}}, \tag{3.5}$$

one can construct an $\mathcal{R}$-structure [94, 80, 88] on $\tilde{\mathcal{G}}$ as the endomorphism $\mathcal{R} : \tilde{\mathcal{G}} \to \tilde{\mathcal{G}}$, where

$$\mathcal{R} := \ (P_+ - P_-)/2, \tag{3.6}$$

which allows to determine on the vector space $\tilde{\mathcal{G}}$ the new Lie algebra structure

$$[\tilde{a}, \tilde{b}]_{\mathcal{R}} := [\mathcal{R}\tilde{a}, \tilde{b}] + [\tilde{a}, \mathcal{R}\tilde{b}] \tag{3.7}$$

for any $\tilde{a}, \tilde{b} \in \tilde{\mathcal{G}}$, satisfying the standard Jacobi identity.

Let $\mathrm{D}(\tilde{\mathcal{G}}^*)$ denote the space of smooth functions on $\tilde{\mathcal{G}}^*$. Then for any $f, g \in \mathcal{D}(\tilde{\mathcal{G}}^*)$ one has the canonical [94, 31, 80, 76, 7] Lie–Poisson bracket

$$\{f, g\} := (\tilde{l}, [\nabla f(\tilde{l}), \nabla g(\tilde{l})]), \tag{3.8}$$

where $\tilde{l} \in \tilde{\mathcal{G}}^*$ is a seed element and $\nabla f(\tilde{l})$, $\nabla g(\tilde{l}) \in \tilde{\mathcal{G}}$ are the standard functional gradients at $\tilde{l} \in \tilde{\mathcal{G}}^*$ with respect to the metric (3.1). The space $\mathrm{I}(\tilde{\mathcal{G}}^*)$ of Casimir invariants related to (3.8) is the set $\mathrm{I}(\tilde{\mathcal{G}}^*) \subset \mathrm{D}(\tilde{\mathcal{G}}^*)$ of smooth independent functions $h \in \mathrm{D}(\tilde{\mathcal{G}}^*)$ for which

$$ad^*_{\nabla h(\tilde{l})} \tilde{l} = 0. \tag{3.9}$$

For any seed element

$$\tilde{l} =< l(\mathrm{x}), d\mathrm{x} > \tag{3.10}$$

the gradients

$$\nabla h(\tilde{l}) := \left\langle \nabla h(l), \frac{\partial}{\partial \mathrm{x}} \right\rangle \tag{3.11}$$

and the coadjoint action (3.9) can be recast with respect to the metric

$$(\tilde{l}, \tilde{a})_0 := \underset{\lambda\in\mathbb{C}}{res} (l(\lambda; x), a(\lambda; x))_{H^0} \tag{3.12}$$

in the case $q = 0 = s$ as

$$\left\langle \frac{\partial}{\partial \mathrm{x}}, \circ \nabla h^{(j)}(l) \right\rangle l + \left\langle l, (\frac{\partial}{\partial \mathrm{x}} \nabla h^{(j)}(l)) \right\rangle = 0 \tag{3.13}$$

for $j \in \mathbb{Z}_+$, where "$\circ$" denotes the composition of mappings. The second Poisson bracket of any two smooth functions $h^{(1)}, h^{(2)} \in \mathrm{I}(\tilde{\mathcal{G}}^*) \subset \mathrm{D}(\tilde{\mathcal{G}}^*)$,

$$\{h^{(1)}, h^{(2)}\}_{\mathcal{R}} := (\tilde{l}, [\nabla h^{(1)}, \nabla h^{(2)}]_{\mathcal{R}}) \tag{3.14}$$

on the space $\tilde{\mathcal{G}}^*$ vanishes, that is

$$\{h^{(1)}, h^{(2)}\}_{\mathcal{R}} = 0 \tag{3.15}$$

at any seed element $\tilde{l} \in \tilde{\mathcal{G}}^*$. Since the functions $h^{(1)}, h^{(2)} \in \mathrm{I}(\tilde{\mathcal{G}}^*)$, the following hold:

$$ad^*_{\nabla h^{(1)}(\tilde{l})}\tilde{l} = 0, \qquad ad^*_{\nabla h^{(2)}(\tilde{l})}\tilde{l} = 0, \tag{3.16}$$

which can be rewritten (as above in the case $q = 0 = p$) as

$$\begin{aligned}\left\langle \frac{\partial}{\partial \mathrm{x}}, \circ\nabla h^{(1)}(l)\right\rangle l + \left\langle l, (\frac{\partial}{\partial \mathrm{x}}\nabla h^{(1)}(l))\right\rangle &= \left\langle \nabla h^{(1)}(l), \frac{\partial}{\partial \mathrm{x}}\right\rangle l + \\ \left\langle (\frac{\partial}{\partial \mathrm{x}}, \nabla h^{(1)}(l))\right\rangle l + \left\langle l, (\frac{\partial}{\partial \mathrm{x}}\nabla h^{(1)}(l))\right\rangle &:= (\nabla h^{(1)}(\tilde{l}) + B_{\nabla h^{(1)}(\tilde{l})})l\end{aligned} \tag{3.17}$$

and similarly

$$\langle \tfrac{\partial}{\partial \mathrm{x}}, \circ\nabla h^{(2)}(l)\rangle l + \langle l, (\tfrac{\partial}{\partial \mathrm{x}}, \nabla h^{(2)}(l))\rangle := (\nabla h^{(2)}(\tilde{l}) + B_{\nabla h^{(2)}(\tilde{l})})l, \tag{3.18}$$

where the expressions

$$\nabla h^{(1)}(\tilde{l}) := \left\langle \nabla h^{(1)}(l), \frac{\partial}{\partial \mathrm{x}}\right\rangle, \quad \nabla h^{(2)}(\tilde{l}) := \left\langle \nabla h^{(2)}(l), \frac{\partial}{\partial \mathrm{x}}\right\rangle \tag{3.19}$$

are true vector fields on $\mathbb{T}^{1+n}_C$, yet the expressions

$$\begin{aligned}B_{\nabla h^{(1)}(\tilde{l})} &:= \left\langle \frac{\partial}{\partial \mathrm{x}}, \nabla h^{(1)}(l)\right\rangle \mathbf{1} + \left(\frac{\partial}{\partial \mathrm{x}}\nabla h^{(1)}(l)\right), \\ B_{\nabla h^{(2)}(\tilde{l})} &:= \left\langle \frac{\partial}{\partial \mathrm{x}}, \nabla h^{(2)}(l)\right\rangle \mathbf{1} + \left(\frac{\partial}{\partial \mathrm{x}}\nabla h^{(2)}(l)\right),\end{aligned} \tag{3.20}$$

are the usual matrix homomorphisms of the Euclidean space $\mathbb{E}^n$.

Now consider the following Hamiltonian flows on the space $\tilde{\mathcal{G}}^*$:

$$\partial\tilde{l}/\partial y := \{h^{(1)}, \tilde{l}\}_{\mathcal{R}} = -ad^*_{\nabla h^{(1)}(\tilde{l})_+}\tilde{l}, \partial\tilde{l}/\partial t := \{h^{(2)}, \tilde{l}\}_{\mathcal{R}} = -ad^*_{\nabla h^{(2)}(\tilde{l})_+}\tilde{l}, \tag{3.21}$$

where $h^{(1)}, h^{(2)} \in \mathrm{I}(\tilde{\mathcal{G}}^*)$ and $y, t \in \mathbb{R}$ are the corresponding evolution parameters. Since $h^{(1)}, h^{(2)} \in \mathrm{I}(\tilde{\mathcal{G}}^*)$ are Casimirs, the flows (3.21) commute. Thus, owing to (3.17), one can recast the flows (3.21) as (which satisfy the following lemma proved in [38])

$$\partial l/\partial t = -(\nabla h^{(y)}(\tilde{l}) + B_{\nabla h^{(y)}(\tilde{l})})l, \qquad \partial l/\partial y = -(\nabla h^{(y)}(\tilde{l}) + B_{\nabla h^{(y)}(\tilde{l})})l, \tag{3.22}$$

where

$$\nabla h^{(y)}(\tilde{l}) := \nabla h^{(1)}(\tilde{l})|_+, \qquad \nabla h^{(t)}(\tilde{l}) := \nabla h^{(2)}(\tilde{l})|_+. \tag{3.23}$$

Lemma 2. *The compatibility of the flows (3.22) is equivalent to the Lax–Sato relationship*

$$\frac{\partial}{\partial y}\nabla h^{(t)}(\tilde{l})-\frac{\partial}{\partial t}\nabla h^{(y)}(\tilde{l})+[\nabla h^{(y)}(\tilde{l}),\nabla h^{(t)}(\tilde{l})]=0, \tag{3.24}$$

which holds for all $y,t\in\mathbb{R}$ and arbitrary $\lambda\in\mathbb{C}$.

For the exact representatives of $h^{(1)},h^{(2)}\in \mathrm{I}(\tilde{\mathcal{G}}^*)$, it is necessary to solve (3.13), taking into account [20] that if the chosen element $\tilde{l}\in\tilde{\mathcal{G}}^*$ is singular as $|\lambda|\to\infty$, the related expansion

$$\nabla h^{(p)}(l)\sim\lambda^p\sum_{j\in\mathbb{Z}_+}\eta_j(l)\lambda^{-j}, \tag{3.25}$$

where the degree $p\in\mathbb{Z}_+$ can be taken as arbitrary. Upon substituting (3.25) into (3.13), one can recurrently find the coefficients $\nabla h^{(p)}(l)$, $p\in\mathbb{Z}_+$, and then construct the gradients of the Casimir functions $h^{(1)},h^{(2)}\in \mathrm{I}(\tilde{\mathcal{G}}^*)$ projected on $\tilde{\mathcal{G}}_+$ as

$$\nabla h^{(t)}(l):=\lambda^{p_t}\eta(l)|_+,\qquad \nabla h^{(y)}(l):=\lambda^{p_y}\eta(l)|_+ \tag{3.26}$$

for some positive integers $p_y,p_t\in\mathbb{Z}_+$.

Remark 1. *The expansion (3.25) is effective if a chosen seed element $\tilde{l}\in\tilde{\mathcal{G}}^*$ is singular as $|\lambda|\to\infty$. When it is singular as $|\lambda|\to 0$, the basic Lie subalgebras $\tilde{\mathcal{G}}_\pm:=\widetilde{diff_\pm(\mathbb{T}^n_{\mathbb{C}})}$* are given as

$$\tilde{\mathcal{G}}_+:=\{\sum_{j=\overline{1,\infty}}\lambda^j<a_j(x),\frac{\partial}{\partial \mathrm{x}}>:\ a_j\in C^\infty(\mathbb{T}^n;\mathbb{C}\times\mathbb{E}^n),j=\overline{1,\infty}\}; \tag{3.27}$$

$$\tilde{\mathcal{G}}_-:=\cup_{m\in\mathbb{Z}_+}\{\sum_{j=0}^{m}\lambda^{-j}<b_j(x),\frac{\partial}{\partial \mathrm{x}}>:\ b_j\in C^\infty(\mathbb{T}^n;\mathbb{C}\times\mathbb{E}^n),j=\overline{1,m}\},$$

subject to which the *expression (3.25) should be replaced by the expansion*

$$\nabla h^{(p)}(l)\sim\lambda^{-p}\sum_{j\in\mathbb{Z}_+}\eta_j(l)\lambda^j \tag{3.28}$$

for an arbitrary $p\in\mathbb{Z}_+$. Then, the projected Casimir function gradients are given by

$$\nabla h^{(y)}(l):=\nabla h^{(p_y)}(l)|_-,\qquad \nabla h^{(t)}(l):=\nabla h^{(p_t)}(l)|_- \tag{3.29}$$

for some $p_y,p_t\in\mathbb{Z}_+$ and the corresponding flows are, respectively, written as

$$\partial\tilde{l}/\partial t=-ad^*_{\nabla h^{(t)}(\tilde{l})}\tilde{l},\quad \partial\tilde{l}/\partial y=-ad^*_{\nabla h^{(y)}(\tilde{l})}\tilde{l}. \tag{3.30}$$

The above results, owing to Lemma 2, can be formulated as a main result:

Proposition 1. *Let a seed element* $\tilde{l} \in \tilde{\mathcal{G}}^*$ *and let* $h^{(1)}, h^{(2)} \in \mathrm{I}(\tilde{\mathcal{G}}^*)$ *be subject to* $\tilde{\mathcal{G}} = \widetilde{diff}(\mathbb{T}^n_{\mathbb{C}})$ *and its natural coadjoint action on* $\tilde{\mathcal{G}}^*$*. Then the following*

$$\partial\tilde{l}/\partial y = -ad^*_{\nabla h^{(y)}(\tilde{l})}\tilde{l}, \quad \partial\tilde{l}/\partial t = -ad^*_{\nabla h^{(t)}(\tilde{l})}\tilde{l} \tag{3.31}$$

are commuting Hamiltonian flows for all $y, t \in \mathbb{R}$ *and their compatibility condition is equivalent to*

$$(\partial/\partial t + \nabla h^{(t)}(\tilde{l}))\psi = 0, \qquad (\partial/\partial y + \nabla h^{(y)}(\tilde{l}))\psi = 0, \tag{3.32}$$

where $\psi \in C^2(\mathbb{R}^2 \times \mathbb{T}^n; \mathbb{C})$ *and* $\nabla h^{(y)}(\tilde{l}), \nabla h^{(t)}(\tilde{l}) \in \tilde{\mathcal{G}}_+$*, given by (3.23) and (3.26), satisfy the Lax–Sato relationship (3.24).*

The proposition above makes it possible to effectively describe the Bäcklund transformations between two solution sets of the dispersionless heavenly equations resulting from the Lax compatibility condition (3.24). Namely, let a diffeomorphism $\xi \in \widetilde{Diff}(\mathbb{T}^{1+n}_{\mathbb{C}})$, depending parametrically on $\lambda, \mu \in \mathbb{C}$ and evolution variables $(y,t) \in \mathbb{R}^2$, be such that a seed loop differential form $\tilde{l}(x;\lambda,\mu) \in \tilde{\mathcal{G}}^* \simeq \tilde{\Lambda}^1(\mathbb{T}^n)$ satisfies the invariance condition

$$\tilde{l}(\xi(x;\lambda,\mu);\lambda) = k\tilde{l}(x;\mu) \tag{3.33}$$

for some constant $k \in \mathbb{C}\backslash\{0\}$, any $x \in \mathbb{T}^n$ and an arbitrary $\lambda \in \mathbb{C}$. As $\tilde{l}(\xi(x;\lambda,\mu);\lambda) \in \tilde{\Lambda}^1(\mathbb{T}^{1+n}_{\mathbb{C}})$ simultaneously satisfies the system of compatible equations following from (3.31), the loop diffeomorphism $\xi \in \widetilde{Diff}(\mathbb{T}^{1+n}_{\mathbb{C}})$, found from (3.33), should satisfy

$$\frac{\partial}{\partial y}\xi = \nabla h^{(y)}(l), \quad \frac{\partial}{\partial t}\xi = \nabla h^{(t)}(l), \tag{3.34}$$

leading to the Bäcklund relationships for coefficients of $\tilde{l} \in \tilde{\mathcal{G}}^* \simeq \tilde{\Lambda}^1(\mathbb{T}^{1+n}_{\mathbb{C}})$.

4 Integrable heavenly dispersionless equations: Examples

We next present examples of heavenly dispersionless equations whose hidden Lie-algebraic structure is based on the loop Lie algebra splitting $\tilde{\mathcal{G}} = \widetilde{diff}(\mathbb{T}^n_{\mathbb{C}}) \simeq \widetilde{diff}(\mathbb{T}^n) := \widetilde{diff}_+(\mathbb{T}^n) \oplus \widetilde{diff}_-(\mathbb{T}^n)$, with degenerate loop subalgebras $\tilde{\mathcal{G}}_\pm := \widetilde{diff}_\pm(\mathbb{T}^n)$; that is

$$\tilde{\mathcal{G}}_- := \{\sum_{j=\overline{1,\infty}} \lambda^{-j} < a_j(x), \frac{\partial}{\partial x} > : \ a_j \in C^\infty(\mathbb{T}^n; \mathbb{E}^n), j = \overline{1,\infty}\} \tag{4.1}$$

$$\tilde{\mathcal{G}}_+ := \cup_{m\in\mathbb{Z}_+}\{\sum_{j=0}^{m} \lambda^j < b_j(x), \frac{\partial}{\partial x} > : \ b_j \in C^\infty(\mathbb{T}^n; \mathbb{E}^n), j = \overline{1,m}\}. \tag{4.2}$$

In this case the inverse $\mathbb{T}^1_{\mathbb{C}} \ni \lambda \to 1/\lambda \in \mathbb{T}^1_{\mathbb{C}}$ isomorphically maps the Lie subalgebras (4.1) and (4.2) into the dual subalgebras

$$\tilde{\mathcal{G}}_+ := \{ \sum_{j=\overline{1,\infty}} \lambda^j < a_j(x), \frac{\partial}{\partial x} > : \ a_j \in C^\infty(\mathbb{T}^n; \mathbb{E}^n), j = \overline{1,\infty}\} \tag{4.3}$$

$$\tilde{\mathcal{G}}_- := \cup_{m \in \mathbb{Z}_+} \{ \sum_{j=0}^{m} \lambda^{-j} < b_j(x), \frac{\partial}{\partial x} > : \ b_j \in C^\infty(\mathbb{T}^n; \mathbb{E}^n), j = \overline{1,m}\}, \tag{4.4}$$

making no difference between them.

4.1 Example: The Whitham heavenly equation.

Consider the [29, 64, 51, 53, 43] heavenly equation:

$$u_{ty} = u_x u_{xy} - u_y u_{xx}, \tag{4.5}$$

where $u \in C^2(\mathbb{R}^2 \times \mathbb{T}^1; \mathbb{R})$ and $(t, y; x) \in \mathbb{R}^2 \times \mathbb{T}^1$. To prove the Lax–Sato type integrability of (4.5), we consider a seed element $\tilde{l} \in \tilde{\mathcal{G}}^*$, defined as

$$\tilde{l} = (u_y^{-2}\lambda^{-1} + 2u_x + \lambda)dx, \tag{4.6}$$

where $\lambda \in \mathbb{C}$ is a complex parameter. The following asymptotic expressions are gradients of the Casimirs $h^{(1)}, h^{(2)} \in \mathrm{I}(\tilde{\mathcal{G}}^*)$, related to the holomorphic loop Lie algebra $\tilde{\mathcal{G}} = \widetilde{diff}(\mathbb{T}^1)$:

$$\nabla h^{(1)}(l) \sim [(u_x\lambda^{-1} - 1) + O(1/\lambda^2), \tag{4.7}$$

as $\lambda \to \infty$, and

$$\nabla h^{(2)}(l) \sim u_y + O(\lambda), \tag{4.8}$$

as $\lambda \to 0$. Using (4.7) and (4.8), one can construct [77] the commuting Hamiltonian flows

$$\frac{\partial}{\partial y}\tilde{l} = -ad^*_{\nabla h^{(y)}(\tilde{l})}\tilde{l}, \quad \frac{\partial}{\partial t}\tilde{l} = -ad^*_{\nabla h^{(t)}(\tilde{l})}\tilde{l} \tag{4.9}$$

with respect to the evolution parameters $y, t \in \mathbb{R}$, where

$$\nabla h^{(y)}(l) := \lambda \nabla h^{(1)}(l)|_+, \quad \nabla h^{(t)}(l) := \lambda^{-1}\nabla h^{(2)}(l)|_-. \tag{4.10}$$

Now, the following equations are easily obtained from (4.9) and (4.10)

$$u_{yt} = u_x u_{xy} - u_y u_{xx}, u_t = -u_y^{-2}/2 + 3u_x^2/2, u_{yy} = -u_y^3[(u_x u_y)_x + u_x u_{xy}], \tag{4.11}$$

where the projected gradients $\nabla h^{(y)}(\tilde{l})$, $\nabla h^{(t)}(\tilde{l}) \in \tilde{\mathcal{G}}$ equal to the loop vector fields

$$\nabla h^{(t)}(\tilde{l}) = (u_x - \lambda)\frac{\partial}{\partial x}, \quad \nabla h^{(y)}(\tilde{l}) = \frac{u_y}{\lambda}\frac{\partial}{\partial x}, \tag{4.12}$$

satisfying for evolution parameters y, t the Lax–Sato compatibility condition:

$$\frac{\partial}{\partial y}\nabla h^{(y)}(\tilde{l}) - \frac{\partial}{\partial t}\nabla h^{(y)}(\tilde{l}) + [\nabla h^{(t)}(\tilde{l}), \nabla h^{(y)}(\tilde{l})] = 0. \tag{4.13}$$

A simple consequence of this is that the first equation of (4.11) coincides with (4.5). Thus, this equation is completely integrable with respect to both evolution parameters.

Remark 2. It is worth observing that the third equation of (4.11) implies

$$\frac{\partial}{\partial y}(1/u_y) = \frac{\partial}{\partial x}(u_x u_y^2), \tag{4.14}$$

whose compatibility makes it possible to introduce a new function $v \in C^2(\mathbb{S}^1; \mathbb{R})$ satisfying

$$v_x = 1/u_y, \qquad v_y = u_x u_y^2, \tag{4.15}$$

which hold for all $(x, y) \in \mathbb{T}^1 \times \mathbb{R}$. Using (4.15) the seed element (4.6) can be rewritten as

$$\tilde{l} = (v_x^2 \lambda^{-1} + 2u_x + \lambda)dx, \tag{4.16}$$

and the vector fields (4.12) are rewritten as

$$\nabla h^{(t)}(\tilde{l}) = (u_x - \lambda)\frac{\partial}{\partial x}, \quad \nabla h^{(y)}(\tilde{l}) = \frac{1}{v_x \lambda}\frac{\partial}{\partial x}, \tag{4.17}$$

whose compatibility condition (4.13) gives rise to the system of heavenly integrable flows:

$$v_y = u_x v_x^{-2}, \ v_{xt} = u_x v_{xy} + u_{xx} v_x, \ u_y = 1/v_x, \ u_t = -v_x^2/2 + 3u_x^2/2, \tag{4.18}$$

compatible for arbitrary evolution parameters $y, t \in \mathbb{R}$.

4.2 The Hirota heavenly equation

The Hirota equation describing [25, 54, 96] 3-dimensional Veronese webs is

$$\alpha u_x u_{yt} + \beta u_y u_{xt} + \gamma u_t u_{xy} = 0 \tag{4.19}$$

for $u \in C^\infty(\mathbb{R}^2 \times \mathbb{T}^1; \mathbb{R})$ with respect to the evolution parameters $t, y \in \mathbb{R}$ and the spatial variable $x \in \mathbb{T}^1$, where α, β and $\gamma \in \mathbb{R}$ are real constants satisfying $\alpha + \beta + \gamma = 0$.

To demonstrate the Lax integrability of (4.19), we proceed as above, except we choose a seed vector field $\tilde{l} \in \tilde{\mathcal{G}}^* := \widetilde{diff}^*(\mathbb{T}^1)$ of the form

$$\tilde{l} = \left(\frac{u_x^2}{u_t^2(\lambda + \alpha)} - \frac{u_x^2(u_y^2 + u_t^2)}{2\alpha u_t^2 u_y^2} + \frac{u_x^2}{u_y^2(\lambda - \alpha)} \right) dx. \tag{4.20}$$

It is then easy to show that the compatibility of the resulting vector fields gives rise to (4.19), whose equivalent Lax–Sato vector field representation is

$$\frac{\partial \psi}{\partial t} - \frac{2\gamma u_t}{u_x(\lambda + \alpha)}\frac{\partial \psi}{\partial x} = 0, \quad \frac{\partial \psi}{\partial y} + \frac{2\beta u_y}{u_x(\lambda - \alpha)}\frac{\partial \psi}{\partial x} = 0, \tag{4.21}$$

satisfied for $\psi \in C^2(\mathbb{R}^2 \times \mathbb{T}^1; \mathbb{C})$ for all $(y, t; x) \in \mathbb{R}^2 \times \mathbb{T}^1$ and $\lambda \in \mathbb{C} \backslash \{\pm\alpha\}$.

4.3 The first reduced Shabat heavenly type equation

This equation [2] is

$$u_{yt} + u_t u_{xy} - u_{xt} u_y = 0 \tag{4.22}$$

for $u \in C^\infty(\mathbb{R}^2 \times \mathbb{T}^1; \mathbb{R})$, where $(y, t; x) \in \mathbb{R}^2 \times \mathbb{T}^1$. To prove Lax–Sato integrability of (4.22), we use the seed element $\tilde{l} \in \tilde{\mathcal{G}}^* := \widetilde{diff}^*(\mathbb{T}^1)$

$$\tilde{l} = \left(\frac{u_t^{-2}}{\lambda + 1} + \frac{u_t^2 - u_y^2}{u_y^2 u_t^2} + \frac{u_y^{-2}}{\lambda} \right) dx, \tag{4.23}$$

where $\lambda \in \mathbb{C} \backslash \{0, -1\}$. Then, our method yields the compatible Lax–Sato representation

$$\frac{\partial \psi}{\partial t} + \frac{u_t}{\lambda + 1} \frac{\partial \psi}{\partial x} = 0, \qquad \frac{\partial \psi}{\partial y} + \frac{u_y}{\lambda} \frac{\partial \psi}{\partial x} = 0, \tag{4.24}$$

satisfied for $\psi \in C^2(\mathbb{R}^2 \times \mathbb{T}^1; \mathbb{C})$, any $(t, y; x) \in \mathbb{R}^2 \times \mathbb{T}^1$ and all $\lambda \in \mathbb{C} \backslash \{0, -1\}$.

4.4 The second reduced Shabat heavenly equation

The second equation [2] is

$$u_{yy} - u_{xt} u_y + u_t u_{xy} = 0 \tag{4.25}$$

for $u \in C^\infty(\mathbb{R}^2 \times \mathbb{T}^1; \mathbb{R})$, where $(y, t; x) \in \mathbb{R}^2 \times \mathbb{T}^1$. Then for Lax–Sato integrability of (4.25), we take a seed element $\tilde{l} \in \tilde{\mathcal{G}}^* := \widetilde{diff}^*(\mathbb{T}^1)$ as

$$\tilde{l} = 2u_t \left(u_y^{-3} + (\lambda u_y^2)^{-1} \right) dx, \tag{4.26}$$

and ultimately obtain the compatible Lax–Sato vector field representation

$$\frac{\partial \psi}{\partial t} + (\lambda^{-1} u_t - \lambda^{-2} u_y) \frac{\partial \psi}{\partial x} = 0, \qquad \frac{\partial \psi}{\partial y} + \lambda^{-1} u_y \frac{\partial \psi}{\partial x} = 0 \tag{4.27}$$

for (4.22), satisfied for $\psi \in C^2(\mathbb{R}^2 \times \mathbb{T}^1; \mathbb{C})$, any $(t, y; x) \in \mathbb{R}^2 \times \mathbb{T}^1$ and all $\lambda \in \mathbb{C}$.

5 Lie-algebraic structures and heavenly dispersionless systems

We now present examples of heavenly dispersionless equations whose hidden Lie-algebraic structure is based on the general structural loop Lie algebra splitting $\tilde{\mathcal{G}} := \widetilde{diff}(\mathbb{T}^n_{\mathbb{C}}) = \widetilde{diff}_+(\mathbb{T}^n_{\mathbb{C}}) \oplus \widetilde{diff}_-(\mathbb{T}^n_{\mathbb{C}})$, and whose loop subalgebras $\tilde{\mathcal{G}}_\pm := \widetilde{diff}_\pm(\mathbb{T}^n_{\mathbb{C}})$ are nondegenerate, that is

$$\tilde{\mathcal{G}}_- : = \{ \sum_{j=\overline{1,\infty}} \lambda^{-j} < a_j(x), \frac{\partial}{\partial \mathrm{x}} > : \ a_j \in C^\infty(\mathbb{T}^n; \mathbb{C} \times \mathbb{E}^n) \}, \tag{5.1}$$

$$\tilde{\mathcal{G}}_+ := \cup_{m \in \mathbb{Z}_+} \{ \sum_{j=0}^{m} \lambda^j < b_j(x), \frac{\partial}{\partial \mathrm{x}} > : \ b_j \in C^\infty(\mathbb{T}^n; \mathbb{C} \times \mathbb{E}^n) \}, \tag{5.2}$$

for which $\tilde{\mathcal{G}}_-|_{\lambda\to\infty} = 0$. The second non-isomorphic splittings of $\tilde{\mathcal{G}} := \widetilde{diff}(\mathbb{T}^n_{\mathbb{C}})$ are

$$\tilde{\mathcal{G}}_+ := \{ \sum_{j=\overline{1,\infty}} \lambda^j < a_j(x), \frac{\partial}{\partial \mathrm{x}} > : \ a_j \in C^\infty(\mathbb{T}^n; \mathbb{C}\times\mathbb{E}^n)\}, \tag{5.3}$$

$$\tilde{\mathcal{G}}_- := \cup_{m\in\mathbb{Z}_+}\{ \sum_{j=0}^{m} \lambda^{-j} < b_j(x), \frac{\partial}{\partial \mathrm{x}} > : \ b_j \in C^\infty(\mathbb{T}^n; \mathbb{C}\times\mathbb{E}^n)\}, \tag{5.4}$$

for which $\tilde{\mathcal{G}}_+|_{\lambda\to 0} = 0$.

5.1 Example: The Einstein–Weyl metric equation

Define $\tilde{\mathcal{G}} := diff(\mathbb{T}^2_{\mathbb{C}})$, where for any $\tilde{a}(\lambda) \in \tilde{\mathcal{G}}_-$, $\tilde{a}(\infty) = 0$, and take the seed element $\tilde{l} \in \tilde{\mathcal{G}}^*$ to be $\tilde{l} = (u_x\lambda - 2u_xv_x - u_y)\,dx + \left(\lambda^2 - v_x\lambda + v_y + v_x^2\right)d\lambda$, where $(u,v) \in C^\infty(\mathbb{R}^2\times\mathbb{T}^1)$ and which generates with respect to the metric (3.12) (as before for $q = 0 = s$) the gradients of the Casimir invariants $h^{(1)}, h^{(2)} \in \mathrm{I}(\tilde{\mathcal{G}}^*)$ in the form

$$\begin{aligned} \nabla h^{(2)}(l) &\sim \lambda^2(0,1)^\intercal + (-u_x, v_x)^\intercal\lambda + (u_y, u - v_y)^\intercal + O(\lambda^{-1}), \\ \nabla h^{(1)}(l) &\sim \lambda(0,1)^\intercal + (-u_x, v_x)^\intercal + (u_y, -v_y)^\intercal\lambda^{-1} + O(\lambda^{-2}) \end{aligned} \tag{5.5}$$

as $|\lambda| \to \infty$ at $p_t = 2$, $p_y = 1$. For the gradients of $h^{(1)}, h^{(2)} \in \mathrm{I}(\tilde{\mathcal{G}}^*)$, determined by (3.13), one easily obtains the corresponding Hamiltonian vector field generators

$$\begin{aligned} \nabla h^{(t)}(\tilde{l}) &:= \left\langle \nabla h^{(2)}(l)_+, \frac{\partial}{\partial \mathrm{x}} \right\rangle = (\lambda^2 + \lambda v_x + u - v_y)\frac{\partial}{\partial x} + (-\lambda u_x + u_y)\frac{\partial}{\partial\lambda}, \\ \nabla h^{(y)}(\tilde{l}) &= \left\langle \nabla h^{(1)}(l)_+, \frac{\partial}{\partial \mathrm{x}} \right\rangle = (\lambda + v_x)\frac{\partial}{\partial x} - u_x\frac{\partial}{\partial\lambda}, \end{aligned} \tag{5.6}$$

satisfying the compatibility condition (3.24), which is equivalent to

$$u_{xt} + u_{yy} + (uu_x)_x + v_xu_{xy} - v_yu_{xx} = 0, v_{xt} + v_{yy} + uv_{xx} + v_xv_{xy} - v_yv_{xx} = 0, \tag{5.7}$$

describing general integrable Einstein–Weyl metric equations [26].

As is well known [49], the invariant reduction of (5.7) at $v = 0$ gives rise to the famous dispersionless Kadomtsev–Petviashvili equation

$$(u_t + uu_x)_x + u_{yy} = 0, \tag{5.8}$$

for which the reduced vector field representation (3.13) follows from (5.6) and is given by

$$\nabla h^{(t)}(\tilde{l}) = (\lambda^2 + u)\frac{\partial}{\partial x} + (-\lambda u_x + u_y)\frac{\partial}{\partial\lambda}, \ \ \nabla h^{(y)}(\tilde{l}) = \lambda\frac{\partial}{\partial x} - u_x\frac{\partial}{\partial\lambda}, \tag{5.9}$$

satisfying the compatibility condition (3.32), equivalent to (5.8). In particular, (3.13) and (5.9) imply

$$\frac{\partial\psi}{\partial t} + (\lambda^2 + u)\frac{\partial\psi}{\partial x} + (-\lambda u_x + u_y)\frac{\partial\psi}{\partial\lambda} = 0, \ \ \frac{\partial\psi}{\partial y} + \lambda\frac{\partial\psi}{\partial x} - u_x\frac{\partial\psi}{\partial\lambda} = 0, \tag{5.10}$$

satisfied for $\psi \in C^2(\mathbb{R}^2\times\mathbb{T}^2_{\mathbb{C}};\mathbb{C})$, any $y, t \in \mathbb{R}$ and all $(x,\lambda) \in \mathbb{T}^2_{\mathbb{C}}$.

5.2 Example: The modified Einstein–Weyl metric equation

This system is

$$
\begin{aligned}
u_{xt} &= u_{yy} + u_x u_y + u_x^2 w_x + u u_{xy} + u_{xy} w_x + u_{xx} a, \\
w_{xt} &= u w_{xy} + u_y w_x + w_x w_{xy} + a w_{xx} - a_y,
\end{aligned} \tag{5.11}
$$

where $(u, w) \in C^\infty(\mathbb{R}^2 \times \mathbb{T}^1), a_x := u_x w_x - w_{xy}$, and was recently derived in [86]. In this case we also use $\tilde{\mathcal{G}} = \widetilde{diff}(\mathbb{T}^2_{\mathbb{C}})$, where for any $\tilde{b}(\lambda) \in \tilde{\mathcal{G}}_-$, $\tilde{b}(\infty) = 0$, yet for a seed element $\tilde{l} \in \tilde{\mathcal{G}}^*$ we choose

$$
\begin{aligned}
\tilde{l} = {} & [\lambda^2 u_x + (2u_x w_x + u_y + 3uu_x)\,\lambda + 2u_x \partial_x^{-1} u_x w_x + 2u_x \partial_x^{-1} u_y + \\
& + 3u_x {w_x}^2 + 2u_y w_x + 6uu_x w_x + 2uu_y + 3u^2 u_x - 2au_x]dx + \\
& + [\lambda^2 + (w_x + 3u)\,\lambda + 2\partial_x^{-1} u_x w_x + 2\partial_x^{-1} u_y + {w_x}^2 + 3uw_x + 3u^2 - a]d\lambda.
\end{aligned} \tag{5.12}
$$

Then an approach completely analogous to that in the preceding section yields the compatible Lax–Sato system

$$
\begin{aligned}
& \frac{\partial \psi}{\partial y} + (-\lambda + w_x)\frac{\partial \psi}{\partial x} + u_x \lambda \frac{\partial \psi}{\partial \lambda} = 0, \\
& \frac{\partial \psi}{\partial t} + [-\lambda^2 + (w_x - u)\lambda + uw_x - a)\frac{\partial \psi}{\partial x} + (u_x \lambda^2 + (uu_x + u_y)\lambda]\frac{\partial \psi}{\partial \lambda} = 0,
\end{aligned} \tag{5.13}
$$

satisfied for $\psi \in C^2(\mathbb{R}^2 \times \mathbb{T}^2_{\mathbb{C}}; \mathbb{C})$, any $y, t \in \mathbb{R}$ and all $(\lambda; x) \in \mathbb{T}^2_{\mathbb{C}}$.

5.3 Example: The Kupershmidt hydrodynamic heavenly system

This mutually compatible hydrodynamic system [6, 46, 47, 87, 62] is given as

$$
\begin{aligned}
& 3v_y - 6uv_x + 6u_x v + 6uu_y - 6u^2 u_x - 2u_t = 0, \\
& -12v_x + 6u_y - 12uu_x = 0, \\
& 6uv_{xx} - 3v_{xy} - 6u_{xx} v - 6u_x u_y + 6u^2 u_{xx} - 6uu_{xy} + 12u{u_x}^2 + 2u_{xt} = 0, \\
& 6v_{xx} + 6uu_{xx} - 3u_{xy} + 6{u_x}^2 = 0
\end{aligned} \tag{5.14}
$$

for smooth functions $(u, v) \in C^\infty(\mathbb{R}^2 \times \mathbb{T}^1; \mathbb{R}^2)$ with respect to "*hidden*" evolution parameters $t, y \in \mathbb{R}$ and the spatial variable $x \in \mathbb{T}^1$. Its Lax–Sato integrability stems from a seed element $\tilde{l} \in \tilde{\mathcal{G}}^*$ of the form

$$
\tilde{l} = [\lambda(v_x + 2uu_x) + \lambda^2 u_x]dx + [(v + u^2) + 2\lambda u + \lambda^2]d\lambda, \tag{5.15}
$$

which leads to compatability equations equivalent to the Lax–Sato representation

$$
\begin{aligned}
\tfrac{\partial \psi}{\partial t} - 3(\lambda^2 + 2\lambda u + u^2 + v)\tfrac{\partial \psi}{\partial x} + 3\lambda(\lambda u_x + 2uu_x + v_x)\tfrac{\partial \psi}{\partial \lambda} &= 0, \\
\tfrac{\partial \psi}{\partial y} - 2(\lambda + u)\tfrac{\partial \psi}{\partial x} + 2\lambda u_x \tfrac{\partial \psi}{\partial \lambda} &= 0,
\end{aligned} \tag{5.16}
$$

satisfied for $\psi \in C^2(\mathbb{R}^2 \times \mathbb{T}^2_{\mathbb{C}}; \mathbb{C})$ for all $(y, t; \lambda, x) \in \mathbb{R}^2 \times \mathbb{T}^2_{\mathbb{C}}$. The result obtained can be formulated as follows:

Proposition 2. *The Kupershmidt system (5.14) is representable as commuting Hamiltonian flows on orbits of the coadjoint action of* $\tilde{\mathcal{G}} = \widetilde{diff}(\mathbb{T}^2_{\mathbb{C}})$ *and are equivalent to the Lax–Sato condition (5.16).*

5.4 Example: First Sergyeyev spatially $3D$-integrable heavenly system

A new spatially $3D$-integrable heavenly system, recently constructed in [90] using techniques from contact geometry [14, 48], is

$$\begin{aligned} u_t - v_y - vu_z - ru_x + uv_z + wv_x &= 0, \\ 2u_z - r_z + w_x + 2ww_z &= 0, \\ 2r_x - 3u_x - 2w_y - v_z + 2wu_z - 2ww_x + 2uw_z &= 0, \\ w_t - r_y + 2v_x - 4wu_x + wr_x - rw_x - vw_z + ur_z &= 0 \end{aligned} \tag{5.17}$$

with respect to two evolution parameters $t, y \in \mathbb{R}$ for $(u, v, w, r) \in C^\infty(\mathbb{R}^2 \times \mathbb{T}^2; \mathbb{R}^4)$. We set $\tilde{\mathcal{G}} := \widetilde{diff}(\mathbb{T}^3_{\mathbb{C}})$ and take the corresponding seed element $\tilde{l} \in \tilde{\mathcal{G}}^*$ as

$$\begin{aligned} \tilde{l} = &\left[3\partial_z^{-1}\left(3a_{xxx} + 6w_z a_{xx} + 14w_{xz}w_x + 6(2ww_x + r_x)w_z + 8ww_{xx} + \right.\right. \\ &\left. 6(w^2 + r)w_{xz} + 16w_x^2 + 6r_z w_x + 2v_{xz} + r_{xx} + 6wr_{xz}\right) + 6\lambda a_{xx} + \\ &\left. 12(3\lambda a + \lambda^2)a_x + 6\lambda r_x\right] dx + \left[27a_{xx} + 70w_z a_x + 6\left(5a^2 + 6\lambda a + \right.\right. \\ &\left.\left. 5r + 2\lambda^2\right) w_z + (64w + 6\lambda)w_z + 10v_z + 6(5w + \lambda)r_z + 9r_z\right] dz + \\ &6[7a_z + 9w^2 + 8\lambda w + 3r + 2\lambda^2]d\lambda, \end{aligned} \tag{5.18}$$

where $w_x = (r - w^2 - 2u)_z$ and $a \in C^\infty(\mathbb{R}^2 \times \mathbb{T}^2; \mathbb{R})$is such that $w = a_z, r - w^2 - 2u = a_x$for all $x, z \in \mathbb{T}^2$. Mimicking the approach in the preceding sections one obtains for (5.17) the representation

$$\begin{gathered} \frac{\partial\psi}{\partial t} + (2\lambda + w, u - \lambda^2)\frac{\partial\psi}{\partial x} + (u - \lambda^2)\frac{\partial\psi}{\partial z} + (w_z\lambda^2 + (u_z - w_x)\lambda - u_x)\frac{\partial\psi}{\partial\lambda} = 0, \\ \frac{\partial\psi}{\partial y} + (2w_z\lambda^3 + (-2w_x + r_z)\lambda^2 + (-r_x + v_z)\lambda - v_x)\frac{\partial\psi}{\partial x} + (3\lambda^2 + 4w\lambda + r)\frac{\partial\psi}{\partial z} + \\ +(2w_z\lambda^3 + (-2w_x + r_z)\lambda^2 + (-r_x + v_z)\lambda - v_x)\frac{\partial\psi}{\partial\lambda} = 0, \end{gathered} \tag{5.19}$$

satisfied for an invariant $\psi \in C^2(\mathbb{R}^2 \times \mathbb{T}^3_{\mathbb{C}}; \mathbb{C})$, any $(t, y; x, z) \in \mathbb{R}^2 \times \mathbb{T}^2$ and all $\lambda \in \mathbb{T}_{\mathbb{C}}$.

5.5 Example: Second Sergyeyev spatially $3D$-integrable heavenly system

The second spatially 3-D integrable heavenly type system, presented in [90], is given by two separate flows

$$\begin{aligned} u_t &= 2rv_x - 2vw_z + vr_x + wu_x, \\ v_t &= vw_x + wv_x, \\ w_y &= 2vr_z - 2ru_x + uw_z + rv_z, \\ r_y &= ur_z + ru_z \end{aligned} \tag{5.20}$$

with respect to two independent evolution parameters $t, y \in \mathbb{R}$ for four smooth functions $(u, v, w, r) \in C^\infty(\mathbb{R}^2 \times \mathbb{T}^2; \mathbb{R}^4)$. Things are a bit more complicated than in the above examples. Inasmuch the system (5.20) is not completely specified as evolution flows and is obviously invariant with respect to the involution mapping Symm$\{x \rightleftarrows z, t \rightleftarrows y; u \rightleftarrows w, r \rightleftarrows v\}$, we need to imbed this system into the above Lie-algebraic integrability scheme.

We start with the basic vector fields Lie algebra $\widetilde{diff}(\mathbb{T}^3_{\mathbb{C}})$ on the torus $\mathbb{T}^3_{\mathbb{C}}$ with coefficients from the differential-algebra $\mathbb{R}\{u,v,w,r|(x,z;\lambda)\}$ and consider its splitting

$$\tilde{\mathcal{G}} := \tilde{\mathcal{G}}_+ \oplus \tilde{\mathcal{G}}_- \tag{5.21}$$

with $\tilde{\mathcal{G}}_\pm := \widetilde{diff}(\mathbb{T}^3_{\mathbb{C}})|_\pm$. Then, we chose a seed element $l^* \in \tilde{\mathcal{G}}^*_+$ of the form

$$\begin{aligned} l^* = {} & \left[r^3\partial_z^{-1}(3w_z + 5r_x) + \lambda r^4\right] d\lambda + \{r^2\partial_z^{-1}\left(-2ar^{-4}r_x w_z + \right.\\ & 3br^{-3}w_z - rw_{zx} + r^{-3}aw_{xz} + wr_{xx} - ar^{-3}r_{xx} + r^{-1}wr_x^2 + \\ & r^{-4}ar_x^2 - r^{-3}a_x r_x + 2r^{-2}b_x\left.\right) + \lambda r^3\partial_z^{-1}\left[r_x r^{-1}(3w_z + 5r_x) + \right.\\ & (w_z + r_x)_x + r^{-4}ar_x\left.\right] + \lambda^2 r^3 r_x\} \, dx + \{(2r^{-1}aw_z - r^3 w_x + \\ & 2r^2 wr_x - r^{-1}ar_x + a_x + b) + \lambda r^3(w_z + r_x) + r^{-1}ar_z + \lambda r^3 r_z\} dz. \end{aligned} \tag{5.22}$$

Then, using an approach analogous to that employed in the above examples, we obtain the Lax–Sato compatible representation (5.20), namely

$$\begin{aligned} &\psi_t + [r_z\lambda^3 + (w_z - r_x)\lambda^2 - w_x\lambda]\psi_\lambda + (2\lambda r + w)\psi_x - r\lambda^2\psi_z = 0, \\ &\psi_y + (\lambda u_z - u_x + v_z - v_x\lambda^{-1})\psi_\lambda - v\lambda^{-2}\psi_x + (u + 2v\lambda^{-1})\psi_z = 0, \end{aligned} \tag{5.23}$$

satisfied for an invariant $\psi \in C^2(\mathbb{R}^2 \times \mathbb{T}^3_{\mathbb{C}}; \mathbb{C})$, any $(t,y;x,z) \in \mathbb{R}^2 \times \mathbb{T}^2$ and all $\lambda \in \mathbb{T}_{\mathbb{C}}$.

Remark 3. It should be mentioned here that a vector field in the construction for this example possesses no proto-Casimir functionals $h_j^{(y)} \in \mathrm{D}(\tilde{\mathcal{G}}^*), j = \overline{1,2}$, playing the key role of generating the corresponding gradient projections.

5.6 Example: A generalized Liouville type equation

In a study [12] of Grassmannians, differential forms and related integrable systems, the authors presented a Lax–Sato type representation for the well-known Liouville equation

$$\frac{\partial^2\varphi}{\partial y \partial t} = \exp\varphi, \tag{5.24}$$

expressed in the so called "laboratory" coordinates $y, t \in \mathbb{R}^2$ for a function $\varphi \in C^2(\mathbb{R}^2; \mathbb{R})$ and having different geometric interpretations. Their related formally obtained result is that the system

$$\partial\psi/\partial y + (\lambda^2 + v\lambda + 1)\partial\psi/\partial\lambda = 0, \qquad \partial\psi/\partial t - u\partial\psi/\partial\lambda = 0 \tag{5.25}$$

for $\psi \in C^2(\mathbb{R}^2;\mathbb{C})$ is compatible for all $y,t \in \mathbb{R}^2$, where $u,v \in C^2(\mathbb{R}^2;\mathbb{R})$ are functional coefficients and $\lambda \in \mathbb{C}$. Note that the transformation $u = 1/2\exp\varphi$ converts (5.25)to(5.24).

Since we are interested in the Lie-algebraic nature of (5.25) for (5.24), we define the Lie group $G := \widetilde{Diff}(\mathbb{T}^1_{\mathbb{C}})$, holomorphically extended in the interior $\mathbb{D}^1_+ \subset \mathbb{C}$ and in the exterior $\mathbb{D}^1_- \subset \mathbb{C}$ regions of the unit disc $\mathbb{D}^1 \subset \mathbb{C}$, such that for any $g(z) \in G|_{\mathbb{D}^1_-}$, $z \in \mathbb{D}^1_-$, $g(\infty) = 1 \in \widetilde{Diff}(\mathbb{T}^1_{\mathbb{C}})$, and study specially chosen coadjoint orbits, related to (5.25).

Then, starting with $\tilde{\mathcal{G}} = \tilde{\mathcal{G}}_+ \oplus \tilde{\mathcal{G}}_-$ and the corresponding Laurent series as described above, we can employ Adler–Kostant–Symes theory, to construct, as in preceding sections, commuting Hamiltonian flows and associated Casimirs that lead to the *a priori* compatible system of vector field equations

$$\frac{\partial\psi}{\partial y} + (z^2 + vz + w)\frac{\partial\psi}{\partial\lambda} = 0, \qquad \frac{\partial\psi}{\partial y} - u\frac{\partial\psi}{\partial\lambda} = 0 \tag{5.26}$$

for the corresponding $\psi \in C^2(\mathbb{R}^2 \times \mathbb{T}^1_{\mathbb{C}};\mathbb{C})$, giving rise to the system of heavenly equations:

$$v_t - 2u = 0, \quad u_y - uv + w_t = 0. \tag{5.27}$$

This can be, in particular, parameterized by means of the substitution $u := 1/2\exp\varphi$ as follows:

$$\varphi_{yt} = \exp\varphi - (2w_t\exp(-\varphi))_t. \tag{5.28}$$

Then, the specifications $w := const = 1$ or $w := -\frac{1}{2}\exp\varphi$ give rise to the Liouville equations

$$\varphi_{yt} = \exp\varphi, \quad \varphi_{yt} - \varphi_{tt} = \exp\varphi, \tag{5.29}$$

respectively, which, as is well known, possess standard Lax isospectral representations. The above analysis implies the following results:

Proposition 3. *The system (5.27) has a Lax–Sato compatible representation, whose Lie-algebraic structure is governed by the classical Adler-Kostant-Symes theory.*

Remark 4. Just as above, one can describe the Lie-algebraic structure for other generalized Liouville type heavenly equations in [12] for a higher order in $\lambda \in \mathbb{T}^1_{\mathbb{C}}$ systems.

6 Linearization covering method and its applications

Some three years ago Krasil'shchik [41] analyzed a Gibbons-Tsarev equation

$$z_{yy} + z_t z_{ty} - z_y z_{tt} + 1 = 0. \tag{6.1}$$

and its so called nonlinear first-order differential *covering*

$$\frac{\partial w}{\partial t} - \frac{1}{z_y + z_t w - w^2} = 0, \quad \frac{\partial w}{\partial y} + \frac{z_t - w}{u_y + z_t w - w^2} = 0, \tag{6.2}$$

which for any solution $z : \mathbb{R}^2 \to \mathbb{R}$ to (6.1) is compatible for all $(t, y) \in \mathbb{R}^2$. He showed that this approach makes it possible to determine for any smooth solution $w : \mathbb{R}^2 \to \mathbb{R}$ to (6.2) a suitable smooth function $\psi : \mathbb{R} \times \mathbb{R}^2 \to \mathbb{R}$, satisfying the corresponding Lax–Sato linear representation:

$$\frac{\partial \psi}{\partial t} + \frac{1}{z_y + z_t\lambda - \lambda^2}\frac{\partial \psi}{\partial \lambda} = 0, \quad \frac{\partial \psi}{\partial y} - \frac{z_t - \lambda}{z_y + z_t\lambda - \lambda^2}\frac{\partial \psi}{\partial \lambda} = 0, \tag{6.3}$$

which for any solution to (6.1) is also compatible for all $(t, y) \in \mathbb{R}^2$ and any parameter $\lambda \in \mathbb{R}$.

Krasil'shchik also posed the problem of providing a differential-geometric explanation of the linearization procedure for a given nonlinear differential-geometric relationship $J^1(\mathbb{R}^n \times \mathbb{R}^2; \mathbb{R})|_{\mathcal{E}}$ in the jet-manifold $J^1(\mathbb{R}^n \times \mathbb{R}^2; \mathbb{R})$, $n \in \mathbb{Z}_+$, realizing a compatible *covering* [4] for the corresponding differential equation $\mathcal{E}[x, \tau; u] = 0$, imbedded in an adjacent jet-manifold $J^k(\mathbb{R}^n \times \mathbb{R}^2; \mathbb{R}^m)$ for some $k, m \in \mathbb{Z}_+$. His explanation in [41] was so abstract it was difficult to find any new applications. Our work is aimed at shedding light on some important points of this linearization procedure in the framework of classical nonuniform vector field equations and presenting new and important applications. We consider the jet manifold $J^k(\mathbb{R}^n \times \mathbb{R}^2; \mathbb{R}^m)$ for some fixed $k, m \in \mathbb{Z}_+$ and a differential relationship [35] in a general form $\mathcal{E}[x, \tau; u] = 0$, satisfied for all $(x; \tau) \in \mathbb{R}^n \times \mathbb{R}^2$ and suitable smooth maps $u : \mathbb{R}^n \times \mathbb{R}^2 \to \mathbb{R}^m$.

As a new example, we take $n = 1 = m, k = 2$ and choose $\mathcal{E}[x; y, t; u] = 0$ to be

$$u_t u_{xy} - k_1 u_x u_{ty} - k_2 u_y u_{tx} = 0, \tag{6.4}$$

the so called ABC- or Zakharevich equation, first discussed in [96] if $k_1 + k_2 - 1 = 0$, where $u : \mathbb{R} \times \mathbb{R}^2 \to \mathbb{R}$ and $k_1, k_2 \in \mathbb{R}$ are arbitrary parameters, satisfying

$$k_1 + k_2 - 1 = 0 \ \vee k_1 + k_2 - 1 \neq 0. \tag{6.5}$$

The first case $k_1 + k_2 - 1 = 0$ was investigated in [25, 54, 96] and recently in [38], where its Lax–Sato linearization was found along with many other properties. For the second case $k_1 + k_2 - 1 \neq 0$ the following result was stated in equivalent form by Burovskiy, Ferapontov and Tsarev in [18, 42] and recently by Krasil'shchik, Sergyeyev & Morozov [42].

Proposition 4. *A dual to (6.4) covering system $J^1(\mathbb{R} \times \mathbb{R}^2; \mathbb{R})|_{\mathcal{E}}$ of relationships*

$$\begin{aligned} &\frac{\partial w}{\partial t} + \frac{u_t w}{u_x k_1(k_1 + k_2 - 1)}\frac{\partial w}{\partial x} - \frac{w(w + k_1 + k_2 - 1)u_{tx}}{u_x k_1} = 0, \\ &\frac{\partial w}{\partial y} + \frac{u_y w}{u_x k_1(w + k_1 + k_2 - 1)}\frac{\partial w}{\partial x} - \frac{w(k_1 + k_2 - 1)u_{yx}}{u_x k_1} = 0 \end{aligned} \tag{6.6}$$

on $J^1(\mathbb{R} \times \mathbb{R}^2; \mathbb{R})$ is compatible; that is, it holds for any its smooth solution $w : \mathbb{R} \times \mathbb{R}^2 \to \mathbb{R}$ at all points $(x; y, t) \in \mathbb{R} \times \mathbb{R}^2$ iff $u : \mathbb{R} \times \mathbb{R}^2 \to \mathbb{R}$ satisfies (6.4).

Moreover, this result was generalized in [42] to the following "*linearizing*" proposition.

Proposition 5. *A system $J^1_{lin}(\mathbb{R}^2 \times \mathbb{R}^2; \mathbb{R})|_{\mathcal{E}}$ of linear first order differential relationships*

$$\begin{gathered}
\frac{\partial \psi}{\partial t} + \frac{\lambda u_x^{\frac{k_2-1}{k_1}} u_t}{k_1(k_1+k_2-1)} \frac{\partial \psi}{\partial x} - \frac{\lambda^2 u_x^{\frac{k_2-k_1-1}{k_1}} (u_t u_{xx} - k_1 u_{xx} u_{xt})}{k_1^2} \frac{\partial \psi}{\partial \lambda} = 0, \\
\frac{\partial \psi}{\partial y} + \frac{k_2 \lambda u_x^{\frac{k_2-1}{k_1}} u_y}{k_1(\lambda u_x^{\frac{k_2-k_1-1}{k_1}} + k_1+k_2-1)} \frac{\partial \psi}{\partial x} - \frac{\lambda^2 k_2(k_1+k_2-1) u_x^{\frac{k_2-k_1-1}{k_1}} u_y u_{xx}}{k_1^2(\lambda u_x^{\frac{k_2-k_1-1}{k_1}} + k_1+k_2-1)} \frac{\partial \psi}{\partial \lambda} = 0
\end{gathered} \tag{6.7}$$

on the covering jet-manifold $J^1(\mathbb{R}^2 \times \mathbb{R}^2; \mathbb{R})$ is compatible; that is, it holds for any its smooth solution $\psi : \mathbb{R}^2 \times \mathbb{R}^2 \to \mathbb{R}$ at all points $(x, \lambda; y, t) \in \mathbb{R}^2 \times \mathbb{R}^2$ iff $u : \mathbb{R} \times \mathbb{R}^2 \to \mathbb{R}$ satisfies (6.4).

We proved a similar result, when $n = 1$, $m, k = 2$, for the Manakov–Santini equations

$$u_{tx} + u_{yy} + (u u_x)_x + v_x u_{xy} - v_y u_{xx} = 0, \; v_{xt} + v_{yy} + u v_{xx} + v_x u_{xy} - v_y v_{xx} = 0, \tag{6.8}$$

whose Lax–Sato integrability was studied in [24, 27, 49, 58, 13, 38]. The system (6.8) as a jet-submanifold $J^1(\mathbb{R} \times \mathbb{R}^2; \mathbb{R})|_{\mathcal{E}} \subset J^1(\mathbb{R} \times \mathbb{R}^2; \mathbb{R})$ allows the following representation

$$\begin{gathered}
\frac{\partial w}{\partial t} + (w^2 - w v_x + u - v_y) \frac{\partial w}{\partial x} + u_x w - u_y + v_{yy} + v_x (v_y - u)_x = 0, \\
\frac{\partial w}{\partial y} + w \frac{\partial w}{\partial x} - v_{xx} w + (u - v_y)_x = 0,
\end{gathered} \tag{6.9}$$

compatible on solutions to $\mathcal{E}[x, \tau; u] = 0$ (6.8) on $J^2(\mathbb{R} \times \mathbb{R}^2; \mathbb{R}^2)$. The existence of the compatible representation (6.9) implies the following result:

Proposition 6. *A covering system $J^1_{lin}(\mathbb{R}^2 \times \mathbb{R}^2; \mathbb{R})|_{\mathcal{E}}$ of relationships*

$$\frac{\partial \psi}{\partial t} + (\lambda^2 + \lambda v_x + u - v_y) \frac{\partial \psi}{\partial x} + (u_y - \lambda u_x) \frac{\partial \psi}{\partial \lambda} = 0, \; \frac{\partial \psi}{\partial y} + (v_x + \lambda) \frac{\partial \psi}{\partial x} - u_x \frac{\partial \psi}{\partial \lambda} = 0 \tag{6.10}$$

on $J^1(\mathbb{R}^2 \times \mathbb{R}^2; \mathbb{R})$is compatible; i.e., it holds for any its smooth solution $\psi : \mathbb{R}^2 \times \mathbb{R}^2 \to \mathbb{R}$ at all points $(x, \lambda; y, t) \in \mathbb{R}^2 \times \mathbb{R}^2$ iff $u : \mathbb{R} \times \mathbb{R}^2 \to \mathbb{R}$ satisfies (6.8).

From the point of view of the above propositions, the main focus of our present analysis, similar to that suggested in [41], is to recover the intrinsic mathematical structure responsible for the existence of the "*linearizing*" covering jet-manifold mappings

$$J^1_{lin}(\mathbb{R}^{n+1} \times \mathbb{R}^2; \mathbb{R})|_{\mathcal{E}} \quad \rightleftharpoons \quad J^1(\mathbb{R}^n \times \mathbb{R}^2; \mathbb{R})|_{\mathcal{E}} \tag{6.11}$$

for any dimension $n \in \mathbb{Z}_+$, compatible with $\mathcal{E}[x, \tau; u] = 0$, as was presented above in the form (6.6) and (6.7) for (6.4). Thus, for a given relationship $\mathcal{E}[x, \tau; u] = 0$ on $J^k(\mathbb{R}^n \times \mathbb{R}^2; \mathbb{R}^m)$ for some $k \in \mathbb{Z}_+$ one can formulate the following **problem:**

If there is a compatible system $J^1(\mathbb{R}^n \times \mathbb{R}^2; \mathbb{R}^m)|_{\mathcal{E}} \subset J^1(\mathbb{R}^n \times \mathbb{R}^2; \mathbb{R}^m)$ of quasi-linear first order differential relationships, how does one construct a linearizing first order differential system $J^1_{lin}(\mathbb{R}^{(1+n)+2}; \mathbb{R})|_{\mathcal{E}} \subset J^1(\mathbb{R}^{(1+n)+2}; \mathbb{R})$ in vector field form on the covering space $\mathbb{R}^{n+1} \times \mathbb{R}^2$, realizing the implications (6.11). This can be interpreted as the corresponding Lax–Sato representation [97], recently developed in [10, 11, 12, 27, 28, 92, 93], for the given $\mathcal{E}[x, \tau; u] = 0$ on $J^k(\mathbb{R}^n \times \mathbb{R}^2; \mathbb{R}^m)$.

As a dual approach to this linearization covering scheme, we also present the contact geometry linearization, suggested in [90] and slightly generalizing the well-known [87] Hamiltonian linearization covering method. As an example, we have the following result.

Proposition 7. *The [18] nonlinear singular manifold Toda differential relationship*

$$u_{xy} sh^2 u_t = u_x u_y u_{tt} \tag{6.12}$$

on the jet manifold $J^2(\mathbb{R}^2 \times \mathbb{R}^2; \mathbb{R})$ *allows the Lax–Sato type linearization covering*

$$\begin{gathered} \frac{\partial \psi}{\partial t} + \frac{(e^{-2u_t} - 1)}{2u_x} \frac{\partial \psi}{\partial x} - \left[\lambda \left(\frac{e^{-2u_t} - 1}{2u_x} \right)_x + \lambda^2 \left(\frac{e^{-2u_t} - 1}{2u_x} \right)_z \right] \frac{\partial \psi}{\partial \lambda} = 0, \\ \frac{\partial \psi}{\partial y} - \frac{u_y e^{-2u_t}}{u_x} \frac{\partial \psi}{\partial x} + \left[\lambda \left(\frac{u_y e^{-2u_t}}{u_x} \right)_x + \lambda^2 \left(\frac{u_y e^{-2u_t}}{u_x} \right)_z \right] \frac{\partial \psi}{\partial \lambda} = 0 \end{gathered} \tag{6.13}$$

for invariant $\psi \in C^2(\mathbb{R}^3 \times \mathbb{R}^2; \mathbb{R})$, *all* $(x, z, \lambda; \tau) \in \mathbb{R}^3 \times \mathbb{R}^2$ *and any smooth solution* $u : \mathbb{R}^2 \times \mathbb{R}^2 \rightarrow \mathbb{R}$ *to (6.12).*

6.1 The linearization covering scheme

A realization of the scheme (6.11) is based on invariants of specified vector fields on the extended base space $\mathbb{R}^{n+1} \times \mathbb{R}^2$, whose definition suitable for our needs is as follows: a smooth map $\psi : \mathbb{R}^{n+1} \times \mathbb{R}^2 \rightarrow \mathbb{R}$ subject to parameters $\tau \in \mathbb{R}^2$ is an invariant of a set of vector fields

$$X^{(k)} := \frac{\partial}{\partial \tau_k} + \sum_{j=\overline{1,n}} a_j^{(k)}(x, \lambda; \tau) \frac{\partial}{\partial x_j} + b^{(k)}(x, \lambda; \tau) \frac{\partial}{\partial \lambda} \tag{6.14}$$

on $\mathbb{R}^{n+1} \times \mathbb{R}^2$ with smooth coefficients $(a^{(k)}, b^{(k)}) : \mathbb{R}^{n+1} \times \mathbb{R}^2 \rightarrow \mathbb{E}^n \times \mathbb{R}$, $k = \overline{1,2}$, if

$$X^{(k)} \psi = 0 \tag{6.15}$$

holds for $k = \overline{1,2}$ and all $(x, \lambda; \tau) \in \mathbb{R}^{n+1} \times \mathbb{R}^2$. The system (6.15) is also representable as a jet-submanifold $J^1_{lin}(\mathbb{R}^{n+1} \times \mathbb{R}^2; \mathbb{R})|_{\mathcal{E}} \subset J^1(\mathbb{R}^{n+1} \times \mathbb{R}^2; \mathbb{R})$. It is also well known [19] that the flows

$$\frac{\partial x_j}{\partial \tau_k} = a_j^{(k)}(x, \lambda; \tau), \quad \frac{\partial \lambda}{\partial \tau_k} = b^{(k)}(x, \lambda; \tau) \tag{6.16}$$

are compatible for any $j = \overline{1,n}, k = \overline{1,2}$ and all $(x, \lambda; \tau) \in \mathbb{R}^{n+1} \times \mathbb{R}^2$. Since there is an invariant function $\psi : \mathbb{R}^{n+1} \times \mathbb{R}^2 \rightarrow \mathbb{R}$, which can be represented as

$\psi(x,\lambda;\tau) = w(x;\tau) - \lambda := 0$ for some smooth map $w : \mathbb{R}^n \times \mathbb{R}^2 \to \mathbb{R}$, its substitution into (6.15) yields *a priori* the compatible reduced system relationships

$$\frac{\partial w}{\partial \tau_k} + \sum_{j=\overline{1,n}} a_j^{(k)}(x,w;\tau)\frac{\partial w}{\partial x_j} - b^{(k)}(x,w;\tau) = 0 \tag{6.17}$$

for $k = \overline{1,2}$ on the jet-manifold $J^0(\mathbb{R}^n \times \mathbb{R}^2;\mathbb{R})$. Moreover, subject to (6.17) one can also observe [19, 34] that, modulo solutions to (6.16), the expression $w(x;\tau) = \psi(x,\lambda(\tau);\tau)+\lambda(\tau)$ for all $(x;\tau) \in \mathbb{R}^n \times \mathbb{R}^2$, where $\psi : \mathbb{R}^{n+1} \times \mathbb{R}^2 \to \mathbb{R}$ is a first integral of the flows (6.16). Thus, the reduction scheme above provides the algorithm

$$J^1_{lin}(\mathbb{R}^{n+1} \times \mathbb{R}^2;\mathbb{R})|_{\mathcal{E}} \ \rightarrow\ J^1(\mathbb{R}^n \times \mathbb{R}^2;\mathbb{R})|_{\mathcal{E}} \tag{6.18}$$

from the implications (6.11) formulated above. The corresponding *inverse* implication

$$J^1_{lin}(\mathbb{R}^{n+1} \times \mathbb{R}^2;\mathbb{R})|_{\mathcal{E}} \ \leftarrow\ J^1(\mathbb{R}^n \times \mathbb{R}^2;\mathbb{R})|_{\mathcal{E}} \tag{6.19}$$

can be algorithmically described as follows.

Consider a compatible system $J^1(\mathbb{R}^{n+2};\mathbb{R})|_{\mathcal{E}} \subset J^1(\mathbb{R}^{n+2};\mathbb{R})$ of relationships

$$\frac{\partial w}{\partial \tau_k} + \sum_{j=\overline{1,n}} a_j^{(k)}(x,w;\tau)\frac{\partial w}{\partial x_j} - b^{(k)}(x,w;\tau) = 0 \tag{6.20}$$

with smooth coefficients $(a^{(k)}, b^{(k)}) : \mathbb{R}^{n+1} \times \mathbb{R}^2 \to \mathbb{E}^n \times \mathbb{R}$, $k = \overline{1,2}$. As the first step it is necessary to check *whether the adjacent system of vector field flows*

$$\frac{\partial x_j}{\partial \tau_k} = a_j^{(k)}(x,w;\tau) \tag{6.21}$$

on $\mathbb{R}^{n+1}$ modulo the flows (6.20) for all $j = \overline{1,n}$ and $k = \overline{1,2}$ *is also compatible.* If the answer is "*yes*", it means [19] that any solution to (6.20) as a complex function $w : \mathbb{R}^n \times \mathbb{R}^2 \to \mathbb{R}$ is representable as $w(x;\tau) - \lambda = \alpha(\psi(x,\lambda;\tau))$ for any $\lambda \in \mathbb{R}$ and some smooth mapping $\alpha : \mathbb{R} \to \mathbb{R}$, where the map $\psi : \mathbb{R}^{n+1} \times \mathbb{R}^2 \to \mathbb{R}$ is *a first integral* of the vector field equations

$$\frac{\partial \psi}{\partial \tau_k} + \sum_{j=\overline{1,n}} a_j^{(k)}(x,\lambda;\tau)\frac{\partial \psi}{\partial x_j} + b^{(k)}(x,\lambda;\tau)\frac{\partial \psi}{\partial \lambda} = 0 \tag{6.22}$$

on $\mathbb{R}^{n+1} \times \mathbb{R}$ for all $(x,\lambda;\tau) \in \mathbb{R}^{n+1} \times \mathbb{R}^2$. Moreover, the value $w(x(\tau);\tau) = \lambda \in \mathbb{R}$ for all $\tau \in \mathbb{R}^2$ is constant, as follows from the condition $\alpha(\psi(x(\tau),w;\tau)) = 0$ for the $\tau \in \mathbb{R}^2$. Thus, we see that (6.22) realizing the covering relationships $J^1_{lin}(\mathbb{R}^{n+1} \times \mathbb{R}^2;\mathbb{R})|_{\mathcal{E}} \subset J^1(\mathbb{R}^{n+1} \times \mathbb{R}^2;\mathbb{R})$ for $k = \overline{1,2}$ and all $(x,\lambda;\tau) \in \mathbb{R}^{n+1} \times \mathbb{R}^2$, linearizing (6.20) and interpreting it as the corresponding Lax–Sato representation.

If *on the contrary,* (6.21) *is not compatible,* it is necessary to recover a hidden isomorphic transformation

$$J^1(\mathbb{R}^{n+1} \times \mathbb{R}^2;\mathbb{R}) \ni (x,w;\tau) \to (x,\tilde{w};\tau) \in J^1(\mathbb{R}^{n+1} \times \mathbb{R}^2;\mathbb{R}), \tag{6.23}$$

for which the resulting *a priori compatible* first order nonlinear differential relationships

$$\frac{\partial \tilde{w}}{\partial \tau_k} + \sum_{j=\overline{1,n}} \tilde{a}_j^{(k)}(x, \tilde{w}; \tau) \frac{\partial \tilde{w}}{\partial x_j} - \tilde{b}^{(k)}(x, \tilde{w}; \tau) = 0 \tag{6.24}$$

already has *a compatible adjacent system* of the corresponding flows

$$\frac{\partial x_j}{\partial \tau_k} = \tilde{a}_j^{(k)}(x, \tilde{w}; \tau) \tag{6.25}$$

on the space $\mathbb{R}^n \times \mathbb{R}$, for which any solution $\tilde{w} : \mathbb{R}^{n+2} \to \mathbb{R}$ generates a first integral $\tilde{\psi} : \mathbb{R}^{n+1} \times \mathbb{R}^2 \to \mathbb{R}$ of an adjacent compatible system

$$\frac{\partial \tilde{\psi}}{\partial \tau_k} + \sum_{j=\overline{1,n}} \tilde{a}_j^{(k)}(x, \lambda; \tau) \frac{\partial \tilde{\psi}}{\partial x_j} + \tilde{b}^{(k)}(x, \lambda; \tau) \frac{\partial \tilde{\psi}}{\partial \lambda} = 0 \tag{6.26}$$

on the space $\mathbb{R}^{n+1} \times \mathbb{R}^2 \to \mathbb{R}$ for $k = \overline{1,2}$, where $\tilde{\psi}(x, \lambda; \tau) := \tilde{\alpha}(\tilde{w}(x; \tau) - \lambda)$ for all $(x, \lambda; \tau) \in \mathbb{R}^{n+1} \times \mathbb{R}$ and a smooth map $\tilde{\alpha} : \mathbb{R} \to \mathbb{R}$. From this one easily obtains as above, a linearized covering jet-submanifold $J^1_{lin}(\mathbb{R}^{n+1} \times \mathbb{R}^2; \mathbb{R})|_{\mathcal{E}} \subset J^1(\mathbb{R}^{n+1} \times \mathbb{R}^2; \mathbb{R})$, as a compatible system of (6.26), generated by $J^1(\mathbb{R}^n \times \mathbb{R}^2; \mathbb{R})|_{\mathcal{E}} \subset J^1(\mathbb{R}^n \times \mathbb{R}^2; \mathbb{R})$ on $\mathbb{R}^n \times \mathbb{R}^2$. This determines the inverse implication (6.19) as applied to general compatible relationships (6.22), providing for the system $J^1(\mathbb{R}^n \times \mathbb{R}^2; \mathbb{R})|_{\mathcal{E}} \subset J^1(\mathbb{R}^n \times \mathbb{R}^2; \mathbb{R})$ its corresponding Lax–Sato representation.

Remark 5. The existence of the map (6.23) can be deduced as follows. Assume that the map (6.23) exists and is equivalent to

$$w(x; \tau) := \rho(x, \tilde{w}(x; \tau); \tau) \tag{6.27}$$

for some smooth $\rho : \mathbb{R}^{n+1} \times \mathbb{R}^2 \to \mathbb{R}$ and all $(x; \tau) \in \mathbb{R}^n \times \mathbb{R}^2$, where the corresponding map $\tilde{w} : \mathbb{R}^n \times \mathbb{R}^2 \to \mathbb{R}$ satisfies

$$\frac{\partial \tilde{w}}{\partial \tau_k} + \sum_{j=\overline{1,n}} a_j^{(k)}(x, \rho(x, \tilde{w}; \tau); \tau) \frac{\partial \tilde{w}}{\partial x_j} = \tilde{b}^{(k)}(x, \tilde{w}; \tau), \tag{6.28}$$

compatible for all $\tau_k \in \mathbb{R}$, $k = \overline{1,2}$, and $x \in \mathbb{R}^n$. The functions $\tilde{b}^{(k)} : \mathbb{R}^{n+1} \times \mathbb{R}^2 \to \mathbb{R}$, $k = \overline{1,2}$,

$$\tilde{b}^{(k)}(x, \tilde{w}; \tau) := \left[b^{(k)} - \sum_{j=\overline{1,n}} \left(\frac{\partial \rho}{\partial \tau_k} + a_j^{(k)} \frac{\partial \rho}{\partial x_j} \right) \right] \left(\frac{\partial \rho}{\partial x_j} \right)^{-1} \Bigg|_{w\ = \rho(x, \tilde{w}\ ; \tau)}, \tag{6.29}$$

should depend on the map (6.27) in such a way that the vector fields

$$\frac{\partial x_j}{\partial \tau_k} = a_j^{(k)}(x, \rho(x, \tilde{w}; \tau); \tau) := \tilde{a}_j^{(k)}(x, \tilde{w}; \tau) \tag{6.30}$$

are also compatible for all $j = \overline{1,n}$ and $k = \overline{1,2}$ modulo the flows (6.28). Thus, (6.28) can be equivalently represented as a compatible system of the equations

$$\frac{\partial \tilde{\psi}}{\partial \tau_k} + \sum_{j=\overline{1,n}} \tilde{a}_j^{(k)}(x,\lambda;\tau)\frac{\partial \tilde{\psi}}{\partial x_j} + \tilde{b}^{(k)}(x,\rho(x,\lambda;\tau);\tau)\frac{\partial \tilde{\psi}}{\partial \lambda} = 0 \tag{6.31}$$

on its first integral $\tilde{\psi} : \mathbb{R}^{n+1} \times \mathbb{R}^2 \to \mathbb{R}$, where $\tilde{\psi}(x,\lambda;\tau) = \alpha(\tilde{w}(x;\tau) - \lambda)$ for an arbitrarily chosen smooth mapping $\alpha : \mathbb{R} \to \mathbb{R}$, any parameter $\lambda \in \mathbb{R}$ and all $(x;\tau) \in \mathbb{R}^n \times \mathbb{R}^2$. The system (6.31) is a suitable Lax–Sato type linearization of the compatible relationships (6.17). Concerning the map (6.27) and the functions $\tilde{b}^{(k)} : \mathbb{R}^{n+1} \times \mathbb{R}^2 \to \mathbb{R}$, $k = \overline{1,2}$, which depend on it, one sees that the compatibility condition for (6.30) reduces to the a priori compatible relationships

$$\begin{aligned} &\frac{\partial a_j^{(k)}}{\partial \rho}\frac{\partial \rho}{\partial \tau_s} - \frac{\partial a_j^{(s)}}{\partial \rho}\frac{\partial \rho}{\partial \tau_k} + \left(\frac{\partial a_j^{(k)}}{\partial \rho}\tilde{b}^{(s)} - \frac{\partial a_j^{(s)}}{\partial \rho}\tilde{b}^{(k)}\right)\frac{\partial \rho}{\partial \tilde{w}} + \\ &+ \textstyle\sum_{m=\overline{1,n}} \left(\frac{\partial a_j^{(k)}}{\partial \rho} a_m^{(s)} - \frac{\partial a_j^{(s)}}{\partial \rho} a_m^{(k)}\right)\frac{\partial \rho}{\partial x_m} = 0, \end{aligned} \tag{6.32}$$

where $j = \overline{1,n}$ and $k \neq s = \overline{1,2}$, and whose solution is the desired map (6.27). As we have only two functional parameters $b^{(s)} : \mathbb{R}^{n+1} \times \mathbb{R}^2 \to \mathbb{R}$, $s = \overline{1,2}$, the system of $2n$ relationships (6.32) can be, in general, compatible only for the case $n = 1$. For all other cases $n \geq 2$ the compatibility condition for (6.32) should be checked separately.

6.2 Example: The Gibbons–Tsarev equation

As a first degenerate case of the scheme (6.19) above, we consider a compatible nonlinear first order system $J^1(\mathbb{R}^2;\mathbb{R})|_{\mathcal{E}} \subset J^1(\mathbb{R}^2;\mathbb{R})$ at $n = 0$, which was discussed in [41]:

$$\frac{\partial w}{\partial t} - \frac{1}{z_y + z_t w - w^2} = 0, \quad \frac{\partial w}{\partial y} + \frac{z_t - w}{u_y + z_t w - w^2} = 0, \tag{6.33}$$

where $(t,y;w) \in \mathbb{R}^2 \times \mathbb{R}$ and a map $u : \mathbb{R}^2 \to \mathbb{R}$ satisfies the Gibbons–Tsarev equation $\mathcal{E}[y,t;u] = 0$:

$$z_{yy} + z_t z_{ty} - z_y z_{tt} + 1 = 0, \tag{6.34}$$

derived in [32, 33]. As the system (6.33) is compatible and the adjacent system (6.25) is empty, one readily sees [58] that any solution $w : \mathbb{R}^2 \to \mathbb{R}$ to (6.33) generates a first integral $\psi : \mathbb{R} \times \mathbb{R}^2 \to \mathbb{R}$ of a system

$$\frac{\partial \psi}{\partial t} + \frac{1}{z_y + z_t\lambda - \lambda^2}\frac{\partial \psi}{\partial \lambda} = 0, \quad \frac{\partial \psi}{\partial y} - \frac{z_t - \lambda}{z_y + z_t\lambda - \lambda^2}\frac{\partial \psi}{\partial \lambda} = 0, \tag{6.35}$$

where $\psi(\lambda;y,t) := \alpha(w(t,y) - \lambda)$ for all $(\lambda;t,y) \in \mathbb{R} \times \mathbb{R}^2$ and some smooth map $\alpha : \mathbb{R} \to \mathbb{R}$. The compatible system (6.35) considered as the jet-submanifold

$J^1_{lin}(\mathbb{R}^1\times\mathbb{R}^2;\mathbb{R})|_{\mathcal{E}} \subset J^1(\mathbb{R}^1\times\mathbb{R}^2;\mathbb{R})$ solves the problem of constructing the linearizing implication (6.19).

From this starting point, we follow the above scheme complemented with Adler–Kostant–Symes theory to obtain the following system of two *a priori* compatible linear vector field equations:

$$\frac{\partial\psi}{\partial y} - \frac{\lambda+u+v}{(\lambda+u)(\lambda+v)}\frac{\partial\psi}{\partial\lambda} = 0, \qquad \frac{\partial\psi}{\partial t} - \frac{1}{(\lambda+u)(\lambda+v)}\frac{\partial\psi}{\partial\lambda} = 0 \tag{6.36}$$

for $\psi \in C^2(\mathbb{R};\mathbb{R})$ and all $(\lambda;t,y) \in \mathbb{C}\times\mathbb{R}^2$. This leads directly to the following result.

Proposition 8. *A system $J^1_{lin}(\mathbb{R}\times\mathbb{R}^2;\mathbb{R})|_{\mathcal{E}}$ of the relationships (6.36) on $J^1(\mathbb{R}\times\mathbb{R}^2;\mathbb{R})$ is compatible; i.e., it holds for any smooth solution $\psi : \mathbb{R}\times\mathbb{R}^2 \to \mathbb{R}$ at all points $(\lambda;y,t) \in \mathbb{R}\times\mathbb{R}^2$ iff $u : \mathbb{R}^2 \to \mathbb{R}$ satisfies (6.34).*

Moreover, as the flows above are Hamiltonian on $\tilde{\mathcal{G}}^*$, so are their reductions on $C^\infty(\mathbb{R}^2;\mathbb{R}^2)$.

6.3 Example: The ABC-equation

As the second example we consider a compatible system $J^1(\mathbb{R}\times\mathbb{R}^2;\mathbb{R})|_{\mathcal{E}}$ of the relationships (6.6) on $J^1(\mathbb{R}\times\mathbb{R}^2;\mathbb{R})$. It is easy to check that the adjacent system of flows

$$\frac{\partial x}{\partial t} = \frac{u_t w}{u_x k_1(k_1+k_2-1)}, \quad \frac{\partial x}{\partial y} = \frac{u_y w}{u_x k_1(w+k_1+k_2-1)}, \tag{6.37}$$

modulo the relationships (6.6), is not compatible for all $(w;t,y) \in \mathbb{R}\times\mathbb{R}^2$. Thus, we need to construct a map (6.23) such that the resulting system (6.24) has an adjacent system of vector field flows compatible for all $(\tilde{w};t,y) \in \mathbb{R}\times\mathbb{R}^2$. To do this, we first eliminate in (6.6) the strictly linear part, yielding the representation as $w(x;t,y) = (u_x(x;t,y))^\alpha \tilde{w}(x;t,y)$ for $\alpha = (k_1+k_2-1)/k_1$ and all $(x;t,y) \in \mathbb{R}\times\mathbb{R}^2$.

From this point on, we can follow the above scheme to ultimately obtain the Lax–Sato representation realization of the inverse implication (6.19)

$$\begin{aligned} \frac{\partial\psi}{\partial t} + \frac{\lambda u_x^{\alpha-1}u_t}{k_1\alpha}\frac{\partial\psi}{\partial x} - \frac{\lambda^2 u_x^\alpha(u_t u_{xx} - k_1 u_{xx}u_{xt})}{k_1^2}\frac{\partial\psi}{\partial\lambda} = 0, \\ \frac{\partial\psi}{\partial y} + \frac{k_2\lambda u_x^{\alpha-1}u_y}{k_1(\lambda u_x^\alpha+\alpha)}\frac{\partial\psi}{\partial x} - \frac{\lambda^2 k_2\alpha u_x^{\alpha-1}u_y u_{xx}}{k_1^2(\lambda u_x^\alpha+\alpha)}\frac{\partial\psi}{\partial\lambda} = 0, \end{aligned} \tag{6.38}$$

coinciding with (6.7), where $\psi(x,\lambda;t,y) := \alpha(\tilde{w}(x;t,y) - \lambda)$ for all $(x,\lambda;t,y) \in \mathbb{R}^2\times\mathbb{R}^2$ and any smooth map $\alpha : \mathbb{R} \to \mathbb{R}$. Thus, (6.38) solves the problem of constructing the inverse implication (6.19) for $J^1(\mathbb{R}\times\mathbb{R}^2;\mathbb{R})|_{\mathcal{E}}$ (6.6), thereby proving Proposition 6.4.

6.4 Example: The Manakov–Santini equation

The Manakov–Santini equations

$$u_{tx} + u_{yy} + (uu_x)_x + v_x u_{xy} - v_y u_{xx} = 0, \; v_{xt} + v_{yy} + uv_{xx} + v_x v_{xy} - v_y v_{xx} = 0, \tag{6.39}$$

where $(u,v) \in C^\infty(\mathbb{R}\times\mathbb{R}^2;\mathbb{R}^2)$, as is well known [49], are obtained as a generalization of the dispersionless reduction for the Kadomtsev–Petviashvili equation. It has the compatible nonlinear first order differential covering $J^1(\mathbb{R}\times\mathbb{R}^2;\mathbb{R})|_{\mathcal{E}} \subset J^1(\mathbb{R}\times\mathbb{R}^2;\mathbb{R})$

$$\begin{gathered}\frac{\partial w}{\partial t} + (w^2 - wv_x + u - v_y)\frac{\partial w}{\partial x} + (w - v_x)(u_x - wv_{xx}) - u_y + v_{yy} + v_x v_{xy} = 0,\\ \frac{\partial w}{\partial y} + w\frac{\partial w}{\partial x} - v_{xx}w + (u - v_y)_x = 0,\end{gathered} \tag{6.40}$$

implying for all $(x;t,y) \in \mathbb{R}\times\mathbb{R}^2$ the Manakov–Santini relationship $\mathcal{E}[x;y,t;u,v] = 0$ (6.39) as a submanifold of $J^2(\mathbb{R}\times\mathbb{R}^2;\mathbb{R}^2)$.

Upon constructing a special jet manifold isomorphism and following the above scheme, we finally obtain a jet submanifold realizing the inverse implication (6.19) as the Lax–Sato representation

$$\begin{gathered}\frac{\partial \psi}{\partial t} + (\lambda^2 + \lambda v_x + u - v_y)\frac{\partial \psi}{\partial x} + (u_y - \lambda u_x)\frac{\partial \psi}{\partial \lambda} = 0,\\ \frac{\partial \psi}{\partial y} + (v_x + \lambda)\frac{\partial \psi}{\partial x} - u_x\frac{\partial \psi}{\partial \lambda} = 0,\end{gathered} \tag{6.41}$$

thus proving Proposition 6.

7 Contact geometry linearization covering scheme

We consider two Hamilton–Jacobi type PDEs compatible for all $(x;\tau) := (x;t,y) \in \mathbb{R}\times\mathbb{R}^2$:

$$\frac{\partial z}{\partial t} + \tilde{H}^{(t)}(x,z,\partial z/\partial x;t,y) = 0, \quad \frac{\partial z}{\partial y} + \tilde{H}^{(y)}(x,z,\partial z/\partial x;t,y) = 0, \tag{7.1}$$

where $z : \mathbb{R}^3 \to \mathbb{R}$ is a so called "*action function*" and $\tilde{H}^{(t)}, \tilde{H}^{(y)} : \mathbb{R}^3\times\mathbb{R}^2 \to \mathbb{R}$ are smooth generalized Hamiltonians. The relationships 7.1 follow from the contact geometry [14, 22] interpretation of some mechanical systems, generated by vector fields. Namely, a differential one-form $\alpha^{(1)} \in \Lambda^1(\mathbb{R}^3\times\mathbb{R})$, defined by

$$\alpha^{(1)} := dz - \lambda dx, \tag{7.2}$$

is called *contact* and vector fields $X_{H^{(t)}}, X_{H^{(y)}} \in \Gamma(\,T(\mathbb{R}^3\times\mathbb{R}))$ are called *contact vector fields*, if there exist functions $\mu^{(t)}, \mu^{(y)} : \mathbb{R}^3\times\mathbb{R} \to \mathbb{R}$, such that for all $(x,z;\tau) \in \mathbb{R}^2\times\mathbb{R}^2$,

$$\begin{aligned} -i_{X_{H^{(t)}}}\alpha^{(1)} &= H^{(t)} := \tilde{H}^{(t)}|_{\partial z/\partial x=\lambda}, & -i_{X_{H^{(y)}}}\alpha^{(1)} &= H^{(y)} := \tilde{H}^{(y)}|_{\partial z/\partial x=\lambda},\\ \mathcal{L}_{H^{(t)}}\alpha^{(t)} &= \mu^{(t)}\alpha^{(t)}, & \mathcal{L}_{H^{(y)}}\alpha^{(y)} &= \mu^{(y)}\alpha^{(y)} \end{aligned} \tag{7.3}$$

hold, where $\mathcal{L}_{H^{(t)}}, \mathcal{L}_{H^{(y)}}$ are the Lie derivatives [19, 34, 22] with respect to $X_{H^{(t)}}$, $X_{H^{(y)}} \in \Gamma(\,T(\mathbb{R}^3\times\mathbb{R}))$. From (7.3) one finds [39, 74] easily that

$$\begin{aligned} X_{H^{(t)}} &= \frac{\partial H^{(t)}}{\partial \lambda}\frac{\partial}{\partial x} - (\frac{\partial H^{(t)}}{\partial x} + \lambda\frac{\partial H^{(t)}}{\partial z})\frac{\partial}{\partial \lambda} + (-H^{(t)} + \lambda\frac{\partial H^{(t)}}{\partial \lambda})\frac{\partial}{\partial z},\\ X_{H^{(y)}} &= \frac{\partial H^{(y)}}{\partial \lambda}\frac{\partial}{\partial x} - (\frac{\partial H^{(y)}}{\partial x} + \lambda\frac{\partial H^{(y)}}{\partial z})\frac{\partial}{\partial \lambda} + (-H^{(y)} + \lambda\frac{\partial H^{(y)}}{\partial \lambda})\frac{\partial}{\partial z}, \end{aligned} \tag{7.4}$$

where $H^{(t)} := \tilde{H}^{(t)}|_{\partial z/\partial x=\lambda}$, $H^{(y)} := \tilde{H}^{(y)}|_{\partial z/\partial x=\lambda}$, and the compatibility of (7.1) is equivalent to the commutativity of the vector fields:

$$[\frac{\partial}{\partial t} + X_{H^{(t)}}, \frac{\partial}{\partial y} + X_{H^{(y)}}] = 0 \tag{7.5}$$

for all $(x, z, \lambda; \tau) \in \mathbb{R}^3 \times \mathbb{R}^2$, depending parametrically on $\lambda \in \mathbb{R}$. They can be rewritten as a compatible Lax–Sato representation for the vector field equations

$$\frac{\partial \psi}{\partial t} + X_{H^{(t)}}\psi = 0, \quad \frac{\partial \psi}{\partial y} + X_{H^{(y)}}\psi = 0 \tag{7.6}$$

for smooth invariant functions $\psi \in C^2(\mathbb{R}^3 \times \mathbb{R}^2; \mathbb{R})$ and all $(x, z, \lambda; \tau) \in \mathbb{R}^3 \times \mathbb{R}^2$.

Remark 6. Note that when (7.1) do not depend on the "*action function*" $z : \mathbb{R}^3 \to \mathbb{R}$, the contact vector fields naturally reduce to

$$X_{H^{(t)}} = \frac{\partial H^{(t)}}{\partial \lambda}\frac{\partial}{\partial x} - \frac{\partial H^{(t)}}{\partial x}\frac{\partial}{\partial \lambda}, \quad X_{H^{(y)}} = \frac{\partial H^{(y)}}{\partial \lambda}\frac{\partial}{\partial x} - \frac{\partial H^{(y)}}{\partial x}\frac{\partial}{\partial \lambda}, \tag{7.7}$$

well-known [22] from symplectic geometry.

7.1 Example: The differential Toda singular manifold equation

Applications of this contact geometry linearization scheme to integrable 3*D*-dispersionless equations were recently presented by Sergyeyev [90]. We apply this scheme to a degenerate case when (7.1) is given by

$$\frac{\partial z}{\partial t} + \frac{(e^{-2u_t} - 1)}{2u_x}\frac{\partial z}{\partial x} = 0, \quad \frac{\partial z}{\partial y} - u_y u_x^{-1} e^{-2u_t}\frac{\partial z}{\partial x} = 0 \tag{7.8}$$

for smooth map $z : \mathbb{R}\times\mathbb{R}^2 \to \mathbb{R}$, whose compatibility condition [18] is the differential Toda singular manifold equation (6.12) for a smooth $u : \mathbb{R}\times\mathbb{R}^2 \to \mathbb{R}$ for all $(x; y, t) \in \mathbb{R}\times\mathbb{R}^2$, defining a relationship $J^2(\mathbb{R}^2\times\mathbb{R}^2;\mathbb{R})|_{\mathcal{E}} \subset J^2(\mathbb{R}^2\times\mathbb{R}^2;\mathbb{R})$ on $J^2(\mathbb{R}^2\times\mathbb{R}^2;\mathbb{R})$:

$$u_{xy} sh^2 u_t = u_x u_y u_{tt} \tag{7.9}$$

Even though (7.8) are linear, they contain no "spectral" parameter $\lambda \in \mathbb{R}$ subject to which one can construct the related conservation laws for (6.17) and apply the modified inverse scattering transform to construct its special exact solutions.

Yet the above contact geometry linearization scheme makes it possible to verify the following result:

Proposition 9. *The linear first order differential relationships (7.8) on* $J^2(\mathbb{R}^2 \times \mathbb{R}^2;\mathbb{R})$ *allows the following dual, quadratic in* $\lambda \in \mathbb{R}$*, Lax–Sato type linearization covering*

$$\frac{\partial \psi}{\partial t} + \frac{(e^{-2u_t} - 1)}{2u_x}\frac{\partial \psi}{\partial x} - \left[\lambda\left(\frac{e^{-2u_t} - 1}{2u_x}\right)_x + \lambda^2\left(\frac{e^{-2u_t} - 1}{2u_x}\right)_z\right]\frac{\partial \psi}{\partial \lambda} = 0,$$

$$\frac{\partial \psi}{\partial y} - \frac{u_y e^{-2u_t}}{u_x}\frac{\partial \psi}{\partial x} + \left[\lambda\left(\frac{u_y e^{-2u_t}}{u_x}\right)_x + \lambda^2\left(\frac{u_y e^{-2u_t}}{u_x}\right)_z\right]\frac{\partial \psi}{\partial \lambda} = 0 \tag{7.10}$$

for smooth invariant $\psi \in C^2(\mathbb{R}^3 \times \mathbb{R}^2; \mathbb{R})$, *all* $(x, z, \lambda; \tau) \in \mathbb{R}^3 \times \mathbb{R}^2$ *and any smooth solution* $u : \mathbb{R}^2 \times \mathbb{R}^2 \to \mathbb{R}$ *of (7.9).*

8 Integrable heavenly superflows: Their Lie-algebraic structure

We begin with the superconformal loop diffeomorphism group and its superconformal loop Lie algebra [45, 79, 36, 37, 75]. Let G denote the superconformal group of smooth loops $\{\mathbb{C} \supset \mathbb{S}^1 \to G\}$, where $G := sDiff(\mathbb{T}^{1|N})$ is the group of superconformal diffeomorphisms of an $1|N$-dimensional supertorus $\mathbb{T}^{1|N}$, $N \in \mathbb{N}$, with coordinates $(x, \vartheta) \in \mathbb{T}^{1|N} \simeq \mathbb{S}^1 \times \Lambda_1^N$ from the infinite-dimensional Grassmann algebra $\Lambda = \Lambda_0 \oplus \Lambda_1$. A superconformal vector field is defined by

$$\tilde{a} := a\frac{\partial}{\partial x} + \frac{1}{2} < Da, D >, \tag{8.1}$$

where $D := (D_{\vartheta_1}, \cdots, D_{\vartheta_N})^\top$, $D_{\vartheta_i} := \dfrac{\partial}{\partial \vartheta_i} + \vartheta_i \dfrac{\partial}{\partial x}$, $i = \overline{1, N}$, $\vartheta := (\vartheta_1, \dots, \vartheta_N)^\top \in \Lambda_1^N$, $a \in C^\infty(\mathbb{T}^{1|N}; \Lambda_0)$, is defined by the condition that the Lie superderivation $\mathcal{L}_{\tilde{a}} : \Lambda^1(\mathbb{T}^{1|N}) \to \Lambda^1(\mathbb{T}^{1|N})$ of the superdifferential 1-form $\omega^{(1)} = dx + \sum_{i=1}^{N} \vartheta_i d\vartheta_i \in \Lambda^1(\mathbb{T}^{1|N})$ is conformal, that is

$$\mathcal{L}_{\tilde{a}}\omega^{(1)} = \mu_{\tilde{a}}\omega^{(1)} \tag{8.2}$$

for some $\mu_{\tilde{a}} \in C^\infty(\mathbb{T}^{1|N}; \Lambda_0)$. From (8.2), one obtains the commutator of any two vector fields $\tilde{a}, \tilde{b} \in \tilde{\mathcal{G}}$:

$$[\tilde{a}, \tilde{b}] := \tilde{c} = c\frac{\partial}{\partial x} + \frac{1}{2} < Dc, D >, \qquad c = a\frac{\partial b}{\partial x} - b\frac{\partial a}{\partial x} + \frac{1}{2} < Da, Db >,$$

verifying that $\tilde{\mathcal{G}}$ of loop superconformal vector fields is a Lie algebra. One can identify $\tilde{\mathcal{G}}$ with a dense subspace $\tilde{\mathcal{G}}^* \simeq \Lambda^1(\mathbb{T}^{1|N})$ of the dual space via the parity

$$(\tilde{l}, \tilde{a})_s := res_{\lambda \in \mathbb{C}}\, \lambda^{-s}(l, a)_H \tag{8.3}$$

where "r" denotes the residue λ^{-1} in a Laurent series, $\tilde{l} := l dx \in \tilde{\mathcal{G}}^*$ and $s \in \mathbb{Z}$. For any superconformal vector field $\tilde{a} \in \tilde{\mathcal{G}}$ and element $\tilde{l} \in \tilde{\mathcal{G}}^*$ such as

$$\tilde{a} := a\frac{\partial}{\partial x} + \frac{1}{2} < Da, D >, \qquad \tilde{l} := l dx, \tag{8.4}$$

the bilinear form

$$(l, a)_H := \int_{\mathbb{T}^{1|N}} l(x, \vartheta) a(x, \vartheta) dx d^N\vartheta \tag{8.5}$$

is determined by means of the integration with respect to the Berezin measures

$$\int_{\mathbb{T}^{1|1}} \alpha(x)dxd\vartheta_i = 0, \quad \int_{\mathbb{T}^{1|1}} \alpha(x)dx\vartheta_i d\vartheta_i = \int_{\mathbb{S}^1} \alpha(x)dx, \tag{8.6}$$

where $\alpha \in C^\infty(\mathbb{S}^1;\mathbb{R})$, $i \in \overline{1,N}$. There are two cases for $k \in \mathbb{N}$: (i) $N = 2k-1$, $a \in C^\infty(\mathbb{T}^{1|(2k-1)};\Lambda_0)$, $l \in C^\infty(\mathbb{T}^{1|(2k-1)};\Lambda_1)$; and (ii) $N = 2k$, $a \in C^\infty(\mathbb{T}^{1|2k};\Lambda_0)$, $l \in C^\infty(\mathbb{T}^{1|2k};\Lambda_0)$.

The constructed loop Lie algebra $\mathcal{G}$ of superconformal vector fields on a supertorus $\mathbb{T}^{1|N}$ allows the natural splitting

$$\tilde{\mathcal{G}} = \tilde{\mathcal{G}}_+ \oplus \tilde{\mathcal{G}}_-, \tag{8.7}$$

where $\tilde{\mathcal{G}}_+$ and $\tilde{\mathcal{G}}_-$ are also loop Lie algebras; i.e.,$[\tilde{\mathcal{G}}_+,\tilde{\mathcal{G}}_+] \subset \tilde{\mathcal{G}}_+$, $\quad [\tilde{\mathcal{G}}_-,\tilde{\mathcal{G}}_-] \subset \tilde{\mathcal{G}}_-$.

Further, we assume that $\tilde{\mathcal{G}}_\pm$ are formed by the vector fields $\tilde{a}(\lambda)$ on $\mathbb{T}^{1|N}$, which are holomorphic in $\lambda \in \mathbb{S}^1_\pm \subset \mathbb{C}$, where $\tilde{a}(\infty) = 0$ for any $\tilde{a}(\lambda) \in \tilde{\mathcal{G}}_-$. Owing to (8.3) one can identify the following spaces: $\tilde{\mathcal{G}}_+^* \simeq \lambda^s\tilde{\mathcal{G}}_-$, $\tilde{\mathcal{G}}_-^* \simeq \lambda^s\tilde{\mathcal{G}}_+$.

The splitting (8.7) makes it possible to apply to $\tilde{\mathcal{G}}$ the Lie algebraic AKS-scheme of constructing integrable Hamiltonian flows on the dual space $\tilde{\mathcal{G}}^*$. Namely, let a $\mathcal{R}$-structure endomorphism $\mathcal{R} : \tilde{\mathcal{G}} \to \tilde{\mathcal{G}}$ be defined as $\mathcal{R} = (P_+ - P_-)/2$, where the projections $P_\pm\tilde{\mathcal{G}} := \tilde{\mathcal{G}}_\pm \subset \tilde{\mathcal{G}}$. Then the commutator $[\tilde{a},\tilde{b}]_\mathcal{R} := [\mathcal{R}\tilde{a},\tilde{b}]+[\tilde{a},\mathcal{R}\tilde{b}]$, where $\tilde{a},\tilde{b} \in \tilde{\mathcal{G}}$, defines on the linear space $\tilde{\mathcal{G}}$ a new Lie structure, satisfying the Jacobi identity, and generates the deformed Lie-Poisson bracket (see, for example, [5])

$$\{f,g\}_{s,\mathcal{R}}(\tilde{l}) := (\tilde{l},[\nabla_{r,s}f(\tilde{l}),\nabla_{l,s}g(\tilde{l})]_\mathcal{R})_s, \tag{8.8}$$

where

$$\begin{aligned} \nabla_{r,s}f(\tilde{l}) &:= \nabla_{r,s}f(l)\partial/\partial x + < D\nabla_{r,s}f(l), D > /2, \\ \nabla_{l,s}g(\tilde{l}) &:= \nabla_{l,s}g(l)\partial/\partial x + < D\nabla_{l,s}f(l), D > /2, \end{aligned} \tag{8.9}$$

f and $g \in \mathcal{D}(\tilde{\mathcal{G}}^*)$ are smooth Fréchet functionals on $\tilde{\mathcal{G}}^*$, $\nabla_{l,s}$ and $\nabla_{r,s}$ which are operators of the left and right functional gradients [95], respectively, relative to the parity (8.3).

The corresponding Casimir invariants $\mathrm{I}(\tilde{\mathcal{G}}^*)$ are generated by $h \in \mathcal{D}(\tilde{\mathcal{G}}^*)$, satisfying:

$$ad^*_{\nabla h(\tilde{l})}\tilde{l} = 0, \tag{8.10}$$

where $\nabla := \nabla_{l,1}$ and $ad^*_a l = l_x a + \dfrac{4-N}{2} l a_x + \dfrac{(-1)^{N+1}}{2} < Dl, Da >$ for any superconformal vector field $\tilde{a} \in \tilde{\mathcal{G}}$ and a fixed element $\tilde{l} \in \tilde{\mathcal{G}}^*$ in the forms (8.4).

The AKS-theorem allows us to construct the infinite hierarchy of Hamiltonian flows

$$d\tilde{l}/dt_p = -ad^*_{\nabla h^{(p)}(\tilde{l})|_+}\tilde{l}, \quad t_p \in \mathbb{R}, \quad p \in \mathbb{Z}_+, \tag{8.11}$$

or, equivalently, $d\tilde{l}/dt_p = \{\tilde{l}, h^{(p)}\}_{0,\mathcal{R}}(\tilde{l})$, where $\nabla h^{(p)}(l) = \lambda^p \nabla h(l)$, $p \in \mathbb{Z}_+$, by means of the asymptotic expansion

$$\nabla h(l) \sim \sum_{j\in\mathbb{Z}_+} \nabla h_j(x,\vartheta)\lambda^{-j} \tag{8.12}$$

for the gradient of the generating functional $h \in \mathrm{I}(\tilde{\mathcal{G}}^*)$ as $|\lambda| \to \infty$. The evolution equations (8.11) are

$$\begin{aligned} dl/dt &= -\left(\nabla h^{(p)}(l)|_+ \frac{\partial}{\partial x} + \frac{1}{2}\left\langle D\nabla h_+^{(p)}(l), D\right\rangle + \left(2 - \frac{N}{2}\right)\frac{\partial \nabla h^{(p)}(l)|_+}{\partial x}\right) \\ &= -(\tilde{A}_{\nabla h^{(p)}(l)|_+} + \tilde{B}_{\nabla\nabla h^{(p)}(l)|_+})l \end{aligned}$$

for $p \in \mathbb{Z}_+$, where $\tilde{A}_{\nabla h^{(p)}(l)_+} := (\nabla h^{(p)}(l)_+)\frac{\partial}{\partial x} + \frac{1}{2} < D\nabla h^{(p)}(l)_+, D >$, $p \in \mathbb{Z}_+$. Thus, the following result holds.

Proposition 10. *The compatibility condition for any two flows d/dt_{p_1} and d/dt_{p_2}, $p_1 \neq p_2$, stemming from the hierarchy (8.11), is equivalent to*

$$\frac{\partial}{\partial t_{p_2}}\tilde{A}_{\nabla h^{(p_1)}(l)|_+} - \frac{\partial}{\partial t_{p_1}}\tilde{A}_{\nabla h^{(p_2)}(l)|_+} = [\tilde{A}_{\nabla h^{(p_1)}(l)_+}, \tilde{A}_{\nabla h^{(p_2)}(l)|_+}] \tag{8.13}$$

or

$$\frac{\partial}{\partial t_{p_2}}\nabla h^{(p_1)}(l)|_+ - \frac{\partial}{\partial t_{p_1}}\nabla h^{(p_2)}(l)|_+ = [\nabla h^{(p_1)}(l)|_+, \nabla h(l)|_+] \tag{8.14}$$

for $p_1 \neq p_2 \in \mathbb{Z}_+$ and (8.13) is a compatibility condition for pairs of PDEs such as

$$\left(\frac{\partial}{\partial t_{p_1}} + \tilde{A}_{\nabla h_+^{(p_1)}(\tilde{l})}\right)\psi = 0, \ \left(\frac{\partial}{\partial t_{p_2}} + \tilde{A}_{\nabla h_+^{(p_2)}(\tilde{l})}\right)\psi = 0, \tag{8.15}$$

where $\psi \in C^2(\mathbb{R}^2 \times \mathbb{T}^{1|N}; \Lambda_0)$.

Proof. *Sketch*: The equivalence between the commutativity for two flows d/dt_{p_1} and d/dt_{p_2}, $p_1 \neq p_2$, and (8.13) is proven for the loop Lie algebra of vector fields on $\mathbb{T}^n$. The compatibility conditions of the PDEs (8.15) are obtained by direct calculations. ■

The procedure of reducing (8.14) on the corresponding coadjoint action orbits for different p_1 and p_2 allows us to obtain integrable superanalogs of integrable 2-dimensional systems of heavenly equations with the Lax–Sato representations in the forms (8.13).

8.1 Super-Mikhalev–Pavlov heavenly equations

Consider the Mikhalev–Pavlov heavenly equation [55, 62, 38] for $u \in C^\infty(\mathbb{R}^2 \times \mathbb{S}^1; \mathbb{R})$ and $(t, y; x) \in \mathbb{R}^2 \times \mathbb{S}^1$

$$u_{xt} + u_{yy} = u_y u_{xx} - u_x u_{xy}.$$

This is the compatibility condition for a pair of vector fields that, following the analysis above, leads, when $N \neq 4, 5$, to the following superanalog of the (Lax integrable) Mikhalev–Pavlov heavenly equation:

$$\begin{aligned} u_{xt} + u_{yy} &= u_y u_{xx} - u_x u_{yx} - \xi_x \xi_y/2 - \sum\nolimits_{i=1}^{N-1} (D_{\vartheta_i} u_x)(D_{\vartheta_i} u_y)/2, \\ \xi_{xt} + \xi_{yy} &= u_y \xi_{xx} - u_x \xi_{yx} + u_{xx} \xi_y/2 - u_{yx} \xi_x/2 + \\ &\sum\nolimits_{i=1}^{N-1} (D_{\vartheta_i} u_y)(D_{\vartheta_i} \xi_x)/2 - \sum\nolimits_{i=1}^{N-1} (D_{\vartheta_i} u_x)(D_{\vartheta_i} \xi_y)/2, \end{aligned} \tag{8.16}$$

where $t_{p_1} := y$, $t_{p_2} := t$ and $(u, \xi)^\top \in C^\infty(\mathbb{R}^2 \times \mathbb{T}^{1|(N-1)}; \Lambda_0 \times \Lambda_1)$.

Note that for $N = 4$ and $N = 5$ one cannot find the asymptotic expansions for gradients of the generating functional $h \in \mathrm{I}(\tilde{\mathcal{G}}^*)$ as $|\lambda| \to \infty$ by means of the (8.10) and so construct integrable superanalogs in the framework of the proposed Lie-algebraic approach. So we now have the following result.

Proposition 11. *When $N \neq 4, 5$, (8.16) and its corresponding Lax-Sato representations are equivalent to the commutativity of two Hamiltonian flows from a hierarchy on $\tilde{\mathcal{G}}^*$ when $p_1 = 1$ and p_2. The equations (8.16) are put into the AKS-scheme for $\tilde{\mathcal{G}}$ with the element $\tilde{l} \in \tilde{\mathcal{G}}^*$ in the forms $l := \vartheta_N ((N-5)u_x/2 + \lambda - \xi_x/2)$, if N is odd and $N \neq 5$, and $l := ((N-4)u_x/2 + \lambda) + \vartheta_N (N-4)\xi_x/2$, if N is even and $N \neq 4$.*

Proof. *Sketch of the proof.* This follows from Subsection 8.1 concerning the superanalogs of the Michael–Pavlov heavenly equations. ■

8.2 Superanalogs of the first Shabat type reduction

The expansion (8.12) for the gradient of the generating functional $h \in \mathrm{I}(\tilde{\mathcal{G}}^*)$ is suited to the above approach when a chosen fixed element $\tilde{l} \in \tilde{\mathcal{G}}^*$ is singular as $|\lambda| \to \infty$. If the element $\tilde{l}$ is singular as $|\lambda| \to 0$, the asymptotic expansion should take the form $\nabla h(l) \sim \sum_{j \in \mathbb{Z}_+} \nabla h_j(l) \lambda^j$.

Then, we take $\tilde{\mathcal{G}}_\pm$ to be the Lie subalgebras of the vector fields $\tilde{a}(\lambda)$ on $\mathbb{T}^{1|N}$, which are holomorphic in $\lambda \in \mathbb{S}^1_\pm \subset \mathbb{C}$ respectively, where $\tilde{a}(0) = 0$ for any $\tilde{a}(\lambda) \in \tilde{\mathcal{G}}_+$. Whereupon, we apply the method above, starting with Casimirs generating an infinite hierarchy of Hamiltonian flows, followed by commutativity conditions to obtain the heavenly equations

$$w_{yt} = w_t w_{yx} - w_y w_{tx} + \frac{1}{2} \sum_{i=1}^{N} (D_{\vartheta_i} w_t)(D_{\vartheta_i} w_y), \tag{8.17}$$

where $w \in C^{\infty}(\mathbb{R}^2 \times \mathbb{T}^{1|N}; \Lambda_0)$. This equation can be considered as a superanalog of the first Shabat reduction [2, 75] for every $N \in \mathbb{N}$. Its Lax–Sato representation is given by the compatibility condition for PDEs such as

$$\frac{\partial\psi}{\partial y} - \frac{1}{\lambda}\left(w_y \frac{\partial\psi}{\partial x} + \frac{1}{2}\sum_{i=1}^{N}(D_{\vartheta_i} w_y)(D_{\vartheta_i}\psi)\right) = 0,$$

$$\frac{\partial\psi}{\partial t} - \frac{1}{\lambda+1}\left(w_t \frac{\partial\psi}{\partial x} + \frac{1}{2}\sum_{i=1}^{N}(D_{\vartheta_i} w_t)(D_{\vartheta_i}\psi)\right) = 0, \tag{8.18}$$

where $\psi \in C^2(\mathbb{R}^2 \times \mathbb{T}^{1|N}; \Lambda_0)$ and $\lambda \in \mathbb{C} \setminus \{0; -1\}$.

Continuing in this manner, we arrive at the following result.

Proposition 12. *For all $N \in \mathbb{N}$ the superanalogs (8.17) of the first Shabat reduction has the Lax–Sato representations (8.18) with the "spectral" parameter $\lambda \in \mathbb{C}\setminus\{0, -1\}$, which is equivalent to the commutativity of a pair of Hamiltonian flows on $\tilde{\mathcal{G}}^*$. When $N = 1$, (8.17) are put into the ASK-scheme for the loop Lie algebra $\tilde{\mathcal{G}}$ with the element $\tilde{l} \in \tilde{\mathcal{G}}^*$.*

8.3 Superanalogs of the generalized Liouville equations

The loop Lie algebra $\widetilde{diff}(\mathbb{T}_{\mathbb{C}}^{1|N})$ superconformal vector fields on $\mathbb{T}_{\mathbb{C}}^{1|N} \simeq \mathbb{T}_{\mathbb{C}}^1 \times \Lambda_1^N$ can be used to show that the Lax–Sato integrable superanalogs of the Liouville heavenly equations can be obtained via a homeomorphism of $(z, \vartheta) \in \mathbb{T}_{\mathbb{C}}^{1|N}$.

By introducing the antiderivatives $D_{\vartheta_i} := \dfrac{\partial}{\partial\vartheta_i} + \vartheta_i \dfrac{\partial}{\partial z}$, $z \in \mathbb{T}_{\mathbb{C}}^1$, $\vartheta_i \in \Lambda_1$, $i = \overline{1, N}$, in the hyperspace $\Lambda_0 \times \Lambda_1^N$. The loop Lie algebra $\widetilde{diff}(\mathbb{T}_{\mathbb{C}}^{1|N})$ and basically following the procedures described above, one can construct the Hamiltonian flow

$$dl/dy = -l_z \nabla h^{(p_y)}(l)_+ - \frac{4-N}{2} l(\nabla h^{(p_y)}(l)_+)_z + \frac{(-1)^N}{2} < Dl, D\nabla h^{(p_y)}(l)_+ > \tag{8.19}$$

in the framework of AKS-theory. Then, a constant Casimir invariant generates the trivial flow

$$dl/dt = 0. \tag{8.20}$$

Whence, we obtain the compatibility equations

$$\frac{\partial\psi}{\partial y} + W\frac{\partial\psi}{\partial\lambda} + \frac{1}{2} < \tilde{D}W, \tilde{D}\psi >= 0, \quad \frac{\partial\psi}{\partial t} - U\frac{\partial\psi}{\partial\lambda} - \frac{1}{2} < \tilde{D}U, \tilde{D}\psi >= 0, \tag{8.21}$$

where $W := W(y, t, \tilde{\vartheta}; \lambda) = \sum_{0 \le j \le r} W_j \lambda^j$, $U := U(y, t, \tilde{\vartheta})$, $\tilde{D} := (D_{\vartheta_1}, D_{\vartheta_2}, \ldots, D_{\vartheta_N})^{\top}$ and $D_{\tilde{\vartheta}_i} := \dfrac{\partial}{\partial\tilde{\vartheta}_i} + \tilde{\vartheta}_i \dfrac{\partial}{\partial\lambda}$, $i = \overline{1, N}$.

Continuing as above, we obtain the integrable Lax–Sato equations

$$\varphi_{yt} = \exp\varphi - \frac{1}{4}\sum_{i=1}^{N}\left(\frac{\partial}{\partial\tilde{\vartheta}_i}\varphi_y\right)\left(\frac{\partial}{\partial\tilde{\vartheta}_i}\exp\varphi\right) \tag{8.22}$$

and

$$\varphi_{yt} - \varphi_{tt} == \exp\varphi - \frac{1}{4}\sum_{i=1}^{N}\left(\frac{\partial}{\partial\tilde{\vartheta}_i}(\varphi_y - \varphi_t)\right) \times \left(\frac{\partial}{\partial\tilde{\vartheta}_i}\exp\varphi\right). \tag{8.23}$$

When $\varphi := \varphi(y,t)$, one has the Lax-Sato integrable Liouville equations:

$$\varphi_{yt} = \exp\varphi, \qquad \varphi_{yt} - \varphi_{tt} = \exp\varphi. \tag{8.24}$$

Finally, the results obtained may be summarized as follows:

Proposition 13. *For all $N \in \mathbb{N}$, (8.22) and (8.23) have Lax–Sato representations (8.21), which are equivalent to the commutativity of two Hamiltonian flows (8.19) and (8.20) on $\widetilde{diff}(\mathbb{T}_{\mathbb{C}}^{1|N})^*$. When $N = 1$, (8.22) and (8.23) are put into the AKS-scheme for the loop Lie algebra $\widetilde{diff}(\mathbb{T}_{\mathbb{C}}^{1|N})$ with the element $\tilde{l} \in \widetilde{diff}(\mathbb{T}_{\mathbb{C}}^{1|N})^*$ in the form (5.28).*

9 Integrability and the Lagrange–d'Alembert principle

All evolution flows like (3.21) are Hamiltonian with respect to the second Lie–Poisson bracket (3.14) on the adjoint $\tilde{\mathcal{G}}^*$ to $\tilde{\mathcal{G}} := \widetilde{diff}(\mathbb{T}^n)$, or on the adjoint subspace $\tilde{\mathcal{G}}^*$ to $\tilde{\mathcal{G}} := \widetilde{diff}(\mathbb{T}_{\mathbb{C}}^{1+n})$. Moreover, they are poly-Hamiltonian, as the related bilinear form (3.1) is marked by integers $s \in \mathbb{Z}$. This leads [94] to an infinite hierarchy of compatible Poisson structures on the phase spaces, each isomorphic to the orbits of a chosen seed element $\tilde{l} \in \tilde{\mathcal{G}}^*$. As all these Hamiltonian flows have an infinite hierarchy of commuting nontrivial conservation laws, their complete integrability can be proved under some natural constraints. The corresponding analytical expressions for the infinite hierarchy of conservation laws can be retrieved from (3.25) for Casimir functional gradients by employing the well-known [8, 7, 59, 94] formal homotopy technique.

In "*Mecanique analytique*", v.1-2, Lagrange formulated *one of the most general, differential variational principles of classical mechanics, giving necessary and sufficient conditions for the correspondence of the motion of a system of material points, subjected to ideal constraints to the applied active forces.*

According to the d'Alembert–Lagrange principle, during a real motion of a system of $N \in \mathbb{Z}_+$ particles with masses $m_j \in \mathbb{R}_+, j = \overline{1,N}$, the sum of the elementary works performed by the active forces $F^{(j)}, j = \overline{1,N}$, and by the forces of inertia for all *the possible particle displacements* $\delta x^{(j)} \in \mathbb{E}^3, j = \overline{1,N}$, *is less than or equal to zero:*

$$\sum_{j=\overline{1,N}} < F^{(j)} - m_j\frac{d^2x^{(j)}}{dt^2}, \delta x^{(j)} > \leq 0 \tag{9.1}$$

at any time $t \in \mathbb{R}$, where $< \cdot, \cdot >$ denotes the standard scalar product in Euclidean space $\mathbb{E}^3$. *The equality in (9.1) holds for the possible reversible displacements* and the symbol $\leq$ is valid for *the possible irreversible displacements* $\delta x^{(j)} \in \mathbb{E}^3, j = \overline{1,N}$. Equation (9.1) is *the general equation of the dynamics of systems with ideal constraints*; it comprises all the equations and laws of motion, thereby reducing all dynamics to this single general formula.

Lagrange established this by generalizing the principle of virtual displacements with the aid of d'Alembert's principle. For systems subject to bilateral constraints, Lagrange used (9.1) to deduce the general properties and laws of bodies in motion, as well as the equations of motion, which he applied to solve a number of problems in dynamics *including the problems of motions of non-compressible, compressible and elastic liquids, thus combining "dynamics and hydrodynamics as branches of the same principle and as conclusions drawn from a single general formula"*.

As first demonstrated in [38], for *generalized reversible motions* of a compressible elastic liquid, located in a 1-connected open domain $\Omega_t \subset \mathbb{R}^n$ with smooth boundary $\partial\Omega_t$, $t \in \mathbb{R}$, in space $\mathbb{R}^n, n \in \mathbb{Z}_+$, (9.1) can be recast as

$$\delta W(t) := \int_{\Omega_t} < l(x(t);\lambda), \delta x(t) > d^n x(t) = 0 \tag{9.2}$$

for all $t \in \mathbb{R}$, where $l(x(t);\lambda) \in \tilde{T}^*(\mathbb{R}^n)$ is the corresponding "*reaction force*", exerted by the ambient medium on the liquid and called a *seed element*, which is here assumed to depend meromorphically on a constant complex parameter $\lambda \in \mathbb{C}$. Now supposing that the evolution of liquid points $x(t) \in \Omega_t$ is determined for any parameters $\lambda \neq \mu \in \mathbb{C}$ by

$$\frac{dx(t)}{dt} = \frac{\mu}{\mu - \lambda} \nabla h(\ l(\mu))(t; x(t)) \tag{9.3}$$

and the Cauchy data $x(t)|_{t=0} = x^{(0)} \in \Omega_0$ for any open 1-connected domain $\Omega_0 \subset \mathbb{T}^n$ with smooth boundary $\partial\Omega_0 \subset \mathbb{R}^n$ and a smooth functional $h : \tilde{T}^*(\mathbb{R}^n) \to \mathbb{R}$, the Lagrange–d'Alembert principle says: the infinitesimal virtual work (9.2) vanishes, that is $\delta W(t) = 0 = \delta W(0)$ for all $t \in \mathbb{R}$. To check this, we calculate the time derivative of (9.2):

$$\begin{aligned} \frac{d}{dt}\delta W(t) &= \frac{d}{dt}\int_{\Omega_t} \langle l(x(t);\lambda), \delta x(t)\rangle \, d^n x(t) \\ &= \int_{\Omega_0} \frac{d}{dt} \langle l(x(t);\lambda), \delta x(t)\rangle \left|\frac{\partial x(t)}{\partial x_0}\right| d^n x^{(0)} \\ &- \int_{\Omega_t} \left[\frac{d}{dt} \langle l(x(t),\lambda); \delta x(t)\rangle + \langle l(x(t),\lambda); \delta x(t)\rangle \operatorname{div} \tilde{K}(\mu)\right] d^n x(t) = 0, \end{aligned} \tag{9.4}$$

if the condition

$$\frac{d}{dt} < l(x(t);\lambda), \delta x(t) > + < l(x(t);\lambda), \delta x(t) > \operatorname{div} \tilde{K}(\mu;\lambda) = 0 \tag{9.5}$$

holds for all $t \in \mathbb{R}$, where

$$\tilde{K}(\mu;\lambda) := \frac{\mu}{\mu - \lambda} \nabla h(\ \tilde{l}(\mu)) = \frac{\mu}{\mu - \lambda} < \nabla h(\ l(\mu)), \frac{d}{dx} > \tag{9.6}$$

is a vector field on $\mathbb{R}^n$ corresponding to (9.3). Taking into account that $d/dt := \partial/\partial t + L_{\tilde{K}(\mu;\lambda)}$, where $L_{\tilde{K}(\mu;\lambda)} = i_{\tilde{K}(\mu;\lambda)}d + di_{\tilde{K}(\mu;\lambda)}$ denotes the well-known [1, 3, 7, 34] Cartan expression for the Lie derivation along (9.6), it can be represented as $\mu, \lambda \to \infty, |\lambda/\mu| < 1$ asymptotically as

$$\frac{d}{dt} \sim \sum_{j \in \mathbb{Z}_+} \mu^{-j} \frac{\partial}{\partial t_j} + \sum_{j \in \mathbb{Z}_+} \mu^{-j} L_{\tilde{K}_j(\lambda)}, \tag{9.7}$$

and (9.5) can be *recast* as an infinite hierarchy of the following evolution equations

$$\partial\, \tilde{l}(\lambda)/\partial t_j := -ad^*_{\tilde{K}_j(\lambda)_+}\, \tilde{l}(\lambda) \tag{9.8}$$

for every $j \in \mathbb{Z}_+$ on the loop space of differential 1-forms $\tilde{\Lambda}^1(\mathbb{R}^n) \simeq \tilde{\mathcal{G}}^*$, where $\tilde{l}(\lambda) :=< l(x;\lambda), dx > \in \tilde{\Lambda}^1(\mathbb{R}^n) \simeq \tilde{\mathcal{G}}^*$, $\tilde{\mathcal{G}} := \widetilde{diff}(\mathbb{R}^n)$ is the Lie algebra of the corresponding loop diffeomorphism group $\widetilde{Diff}(\mathbb{R}^n)$ [52]. It follows readily from (9.6) that

$$\tilde{K}_j(\lambda) = \nabla h^{(j)}(\,\tilde{l}) \tag{9.9}$$

for $\lambda \in \mathbb{C}$ and any $j \in \mathbb{Z}_+$, and the equations (9.8) transform into

$$\partial\, \tilde{l}(\lambda)/\partial t_j := -ad^*_{\nabla h^{(j)}(\,\tilde{l})_+}\, \tilde{l}(\lambda), \tag{9.10}$$

allowing to formulate the following Adler–Kostant–Symes type [8, 7, 9, 94, 80, 78] result.

Proposition 14. *Equations (9.10) are completely integrable commuting Hamiltonian flows on $\tilde{\mathcal{G}}^*$ for a seed element $\tilde{l}(\lambda) \in \tilde{\mathcal{G}}^*$, generated by Casimir functionals $h^{(j)} \in \mathrm{I}(\tilde{\mathcal{G}}^*)$,naturally determined by conditions $ad^*_{\nabla h^{(j)}(\,\tilde{l})}\, \tilde{l}(\lambda) = 0$, $j \in \mathbb{Z}_+$, with respect to the modified Lie-Poisson bracket on $\tilde{\mathcal{G}}^*$*

$$\{(\tilde{l}, \tilde{X}), (\tilde{l}, \tilde{Y})\} := (\tilde{l}, [\tilde{X}, \tilde{Y}]_{\mathcal{R}}),$$

defined for any $\tilde{X}, \tilde{Y} \in \tilde{\mathcal{G}}$ via the canonical $\mathcal{R}$-structure on the loop Lie algebra $\tilde{\mathcal{G}}$:

$$[\tilde{X}, \tilde{Y}]_{\mathcal{R}} := [\tilde{X}_+, \tilde{Y}_+] - [\tilde{X}_-, \tilde{Y}_-], \tag{9.11}$$

where "$\tilde{Z}$" is the positive (+)/(-)-negative part of a $\tilde{Z} \in \tilde{\mathcal{G}}$ subject to the parameter $\lambda \in \mathbb{C}$.

If, for instance, we consider the first two flows from (9.10) in the form

$$\begin{aligned} \partial\, \tilde{l}(\lambda)/\partial t_1 &:= \partial\, \tilde{l}(\lambda)/\partial y = -ad^*_{\nabla h^{(y)}(\,\tilde{l})}\, \tilde{l}(\lambda), \\ \partial\, \tilde{l}(\lambda)/\partial t_2 &:= \partial\, \tilde{l}(\lambda)/\partial t = -ad^*_{\nabla h^{(t)}(\,\tilde{l})}\, \tilde{l}(\lambda), \end{aligned} \tag{9.12}$$

where, $\nabla h^{(y)}(\tilde{l}) := \nabla h^{(1)}(\tilde{l})|_+, \nabla h^{(t)}(\tilde{l}) := \nabla h^{(t)}(\tilde{l})|_+$, which clearly commute, from their compatibility condition one obtains PDEs for the coefficients of $\tilde{l}(\lambda) \in \tilde{\mathcal{G}}^*$, which are equivalent to the Lax–Sato compatibility condition for $\nabla h^{(y)}(\tilde{l})$ and $\nabla h^{(t)}(\tilde{l}) \in \tilde{\mathcal{G}}$:

$$[\partial/\partial y + \nabla h^{(y)}(\tilde{l}),\ \partial/\partial t + \nabla h^{(t)}(\tilde{l})] = 0, \tag{9.13}$$

resulting from (9.13) - a system of nonlinear PDEs often called heavenly, analyzed in [49, 11, 60, 61, 62, 83, 84, 54, 92, 93] and recently in [13, 11, 12, 40]. These studies are closely related to the problem of constructing a hierarchy of mutually commuting vector fields analytically depending on $\lambda \in \mathbb{C}$, which was investigated and solved in general form by Pfeiffer [66, 67, 70].

Acknowledgements

The authors thank Profs. M. Błaszak, J. Cieślinski, A. Sym and A. Samoilenko for useful discussions during the Workshop "Nonlinearity and Geometry" January 2017 in Warsaw, Poland, and the International Conference in Functional Analysis dedicated to the 125th anniversary of Stefan Banach, September 2017 in Lviv, Ukraine. A.P. is greatly indebted to Profs. M. Pavlov, W.K. Schief and Ya. G. Prytula for mentioning helpful references. WE also thank Profs. V.E. Zakharov (University of Arizona) and J. Szmigelski (University of Saskatchewan) for instructive discussions during the XXXV Workshop on Geometric Methods in Physics, 26.06-2.07.2016 in Białowieża, Poland. A.P. is especially grateful to Prof. B. Kruglikov (UiT, Norway) for pointing out misprints in and suggesting references for a preliminary version of this manuscript. Finally, A.P. thanks the Department of Mathematical Sciences of New Jersey Institute of Technology (Newark NJ, USA) for the invitation to visit in the Summer of 2017, where an essential part of this review was formulated and prepared.

References

[1] R. Abraham and J. Marsden, *Foundations of Mechanics*, 2nd ed., Benjamin Cummings, NY, 1978.

[2] L. Alonso, A. Shabat, Hydrodynamic reductions and solutions of a universal hierarchy, *Theor. Math. Phys*, **104** (2004), 1073-1085.

[3] V. Arnold, *Mathematical Methods of Classical Mechanics*, Springer, NY, 1978.

[4] H. Baran, I. Krasil'shchik, O. Morozov and P. Vojcak, Integrability properties of some equations obtained by symmetry reductions, arXiv:1412.6461v1 [nlin.SI] 19 Dec 2014.

[5] F. Berezin, *Introduction to Algebra and Analysis of Anticommuting Variables*, 208 pp. Moscow State University Publisher, Moscow (1983).

[6] D. Blackmore, A.K. Prykarpatsky, Dark equations and their light integrability, *J. Nonlinear Math. Phys.*, **21**, No. 3 (2014), 407-428.

[7] D. Blackmore, A.K. Prykarpatsky and V.H. Samoylenko, *Nonlinear Dynamical Systems of Mathematical Physics*, World Scientific Publisher, NJ, USA, 2011.

[8] M. Błaszak, Classical R-matrices on Poisson algebras and related dispersionless systems, *Phys. Lett. A*, **297** (3-4) (2002) 191–195.

[9] M. Błaszak and B. Szablikowski, Classical R-matrix theory of dispersionless systems: II. (2 + 1) dimension theory, *J. Phys. A: Math. Gen.*, **35** (2002), 10345.

[10] L. Bogdanov, Interpolating differential reductions of multidimensional integrable hierarchies, TMF, 2011, Volume **167**, Number 3, 354–363.

[11] L. Bogdanov, V. Dryuma, S. Manakov, M. Dunajski, Generalization of the second heavenly equation: dressing method and the hierarchy, *J. Phys. A: Math. Theor.*, **40** (2007), 14383-14393.

[12] L. Bogdanov, B. Konopelchenko, On the heavenly equation and its reductions, *J. Phys. A, Math. Gen.*, **39** (2006), 11793-11802.

[13] L. Bogdanov, M.V. Pavlov, Linearly degenerate hierarchies of quasiclassical SDYM type, arXiv:1603.00238v2 [nlin.SI] 15 Mar 2016.

[14] A. Bruce, K. Grabowska, J. Grabowski, Remarks on contact and Jacobi geometry. *SIGMA*, **13**, 059 (2017); (arXiv:1507.05405).

[15] M.A. Buhl, Surles operateurs differentieles permutables ou non, *Bull. des Sc.Math.*, (1928), S.2, t. LII, p. 353-361.

[16] M.A. Buhl, Apercus modernes sur la theorie des groupes continue et finis, *Mem. des Sc. Math.*, fasc. XXXIII, Paris, 1928.

[17] M.A. Buhl, Apercus modernes sur la theorie des groupes continue et finis, *Mem. des Sc. Math.*, fasc. XXXIII, Paris, 1928.

[18] P. Burovskiy, E. Ferapontov, S. Tsarev, Second order quasilinear PDEs and conformal structures in projective space, *Int. J. Math.* **21** (2010), no. 6, 799–841, arXiv:0802.2626.

[19] A. Cartan, *Differential Forms.* Dover Publisher, USA, 1971.

[20] E. Coddington, N. Levinson, *Theory of Ordinary Differential Equations, International Series in Pure and Applied Mathematics*, McGraw-Hill, 1955.

[21] B. Doubrov, E. Ferapontov, On the integrability of symplectic Monge-Ampère equations, *J. Geom. Phys.*, **60** (2010), 10, 1604-1616

[22] B. Dubrovin, S. Novikov, A. Fomenko, *Modern Geometry. Methods and Applications: Part I: The Geometry of Surfaces, Transformation Groups, and Fields* (Graduate Texts in Mathematics) 2nd ed., Springer, Berlin, 1992.

[23] M. Dunajski, Anti-self-dual four-manifolds with a parallel real spinor, *Proc. Roy. Soc. A*, **458** (2002), 1205.

[24] M. Dunajski, E. Ferapontov and B. Kruglikov, On the Einstein-Weyl and conformal self-duality equations. arXiv:1406.0018v3 [nlin.SI] 29 Jun 2015.

[25] M. Dunajski, W. Kryński, Einstein–Weyl geometry, dispersionless Hirota equation and Veronese webs, arXiv:1301.0621.

[26] M. Dunajski, L. Mason, P. Tod, Einstein–Weyl geometry, the dKP equation and twistor theory, *J. Geom. Phys.*, **37** (2001), N1-2, 63-93.

[27] E. Ferapontov, B. Kruglikov, Dispersionless integrable systems in 3D and Einstein-Weyl geometry. arXiv:1208.2728v3 [math-ph] 9 May 2013.

[28] E. Ferapontov, B. Kruglikov, V. Novikov, Integrability of dispersionless Hirota type equations and the symplectic Monge-Ampère property. rXiv:1707.08070v2 [nlin.SI] 9 Jan 2018.

[29] E. Ferapontov, A. Moro, V. Sokolov, Hamiltonian systems of hydrodynamic type in 2 + 1 dimensions, *Comm. Math. Phys.*, **285** (2009), 31-65, arXiv:0710.2012.

[30] E. Ferapontov, J. Moss, Linearly degenerate PDEs and quadratic line complexes, arXiv:1204.2777v1 [math.DG] 12 Apr 2012.

[31] I. Gelfand, D. Fuchs, Cohomology of the Lie algebra of vector fields on the circle, *Funct. Anal. Appl.*, **2** (1968), 4, 342-343.

[32] J. Gibbons, S. Tsarev, Reductions of the Benney equations, *Physics Letters A*, **211**, Issue 1, Pages 19–24, 1996.

[33] J. Gibbons, S. Tsarev, Conformal maps and reductions of the Benney equations, *Phys. Lett. A*, **258** (1999) 263-270.

[34] C. Godbillon, *Geometrie Differentielle et Mecanique Analytique.* Hermann Publ., Paris, 1969.

[35] M. Gromov, *Partial differential relations, Ergebnisse der Mathematik und ihrer Grenzgebiete*, Vol. 9, Springer-Verlag, Berlin, Heidelberg, New York, 1986.

[36] O. Hentosh, The compatibly bi-Hamiltonian superconformal analogs of Lax integrable nonlinear dynamical systems, *Ukrainian Math. J.*, **58**, 7 (2006) 887–900.

[37] O. Hentosh, The Lax integrable Laberge-Mathieu hierarchy of supersymmetric nonlinear dynamical systems and its finite-dimensional Neumann type reduction, *Ukrainian Math. J.*, **61**, 7 (2009) 906–921.

[38] O.E. Hentosh, Y.A. Prykarpatsky, D. Blackmore and A.K. Prykarpatski, Lie-algebraic structure of Lax–Sato integrable heavenly equations and the Lagrange–d'Alembert principle, *Journal of Geometry and Physics*, **120** (2017), 208–227.

[39] A. A. Kirillov, Local Lie algebras, *Uspekhi Mat. Nauk,* **31** (1976), N4(190), 57–76.

[40] B. Konopelchenko, Grassmanians $Gr(N-1, N+1)$, closed differential $N-1$ forms and N-dimensional integrable systems. arXiv:1208.6129v2 [nlin.SI] 5 Mar 2013.

[41] I. Krasil'shchik, A natural geometric construction underlying a class of Lax pairs, *Lobachevskii J. Math.*, **37** (2016) no. 1, 61–66; arXiv:1401.0612.

[42] I. Krasil'shchik1, A. Sergyeyev, O.I. Morozov, Infinitely many nonlocal conservation laws for the ABC equation with $A + B + C \neq 0$, arXiv:1511.09430v1 [nlin.SI] 30 Nov 2015.

[43] I. Krichever, The τ-function of the universal Whitham hierarchy, matrix models and topological field theories, *Comm. Pure Appl. Math.* **47** (1994), 437{475, hep-th/9205110.

[44] B. Kruglikov, O. Morozov, Integrable dispersionless PDE in 4D, their symmetry pseudogroups and deformations arXiv:1410.7104v2 [math-ph] 11 Feb 2015.

[45] P. Kulish, An analogue of the Korteweg–de Vries equation for the superconformal algebra, Differential geometry, Lie groups and mechanics. Part VIII, *Zap. Nauchn. Sem. LOMI*, **155**, "Nauka", Leningrad. Otdel., Leningrad, 1986, 142–149.

[46] B. Kupershmidt, Dark equations, *J. Nonlinear Math. Phys.*, **8** (2001), 363–445.

[47] B. Kupershmidt, Mathematics of dispersive water waves, *Commun. Math. Phys.*, **99** (1985), 51–73.

[48] A. Kushner, V. Lychagin, V. Rubtsov, *Contact Geometry and Nonlinear Differential Equations*, Cambridge Univ. Press, Cambridge, 2007.

[49] S. Manakov, P. Santini, On the solutions of the second heavenly and Pavlov equations, *J. Phys. A: Math. Theor.*, **42** (2009), 404013 (11pp).

[50] S. Manakov, P. Santini, Cauchy problem on the plane for the dispersionless Kadomtsev–Petviashvili equation, *JETP Lett.*, **83** (2006), 462-466, nlin.SI/0604023.

[51] M. Manas, E. Medina, L. Martinez-Alonso, On the Whitham hierarchy: dressing scheme, string equations and additional symmetries, *J. Phys. A: Math. Gen.*, **39** (2006), 2349-2381, nlin.SI/0509017.

[52] O. Mokhov, *Symplectic and Poissonian geometry on loop spaces of smooth manifolds and integrable equations.* Inst. for Computer Studies Publ., Moscow-Izhevsk, 2004 (in Russian).

[53] O. Morozov, A two-component generalization of the integrable rd-Dym equation, *SIGMA*, **8** (2012), 051-056.

[54] O.I. Morozov, A. Sergyeyev, The four-dimensional Martinez-Alonso-Shabat equation: reductions, nonlocal symmetries, and a four-dimensional integrable generalization of the ABC equation, Preprint submitted to JGP, 2014, 11 pages.

[55] V. Mikhalev, On the Hamiltonian formalism for Korteweg–de Vries type hierarchies, *Funct. Anal. Appl.*, **26**, Issue 2 (1992), 140–142.

[56] G. Misiolek, A shallow water equation as a geodesic flow on the Bott–Virasoro group, *J. Geom. Phys.*, **24** (1998), 3, 203-208.

[57] S. Novikov (Editor), *Theory of Solitons: the Inverse Scattering Method*, Springer, 1984.

[58] A. Odesskii, V. Sokolov, Non-homogeneous systems of hydrodynamic type possessing Lax representations, *Comm. Math. Phys.*, **324** (2013), 47-62; https://arxiv.org/abs/1206.5230.

[59] P. Olver, *Applications of Lie Groups to Differential Equations*, 2nd, Springer-Verlag, New York, 1993.

[60] V. Ovsienko, Bi-Hamilton nature of the equation $u_{tx} = u_{xy}u_y - u_{yy}u_x$, arXiv:0802.1818v1 [math-ph] 13 Feb 2008.

[61] V. Ovsienko, C. Roger, Looped cotangent Virasoro algebra and nonlinear integrable systems in dimension $2+1$, *Commun. Math. Phys.*, **273** (2007), 357–378.

[62] M. Pavlov, Integrable hydrodynamic chains, *J. Math. Phys.*, **44** (2003), 4134-4156.

[63] M. Pavlov, Integrable Dispersive Chains and Energy Dependent Schrödinger Operator, arXiv:1402.3836v2 [nlin.SI] 2 Mar 2014.

[64] M. Pavlov, Classification of integrable Egorov hydrodynamic chains, *Theoret. and Math. Phys.*, **138** (2004), 45-58, nlin.SI/0603055.

[65] G. Pfeiffer, Generalisation de la methode de Jacobi pour l'integration des systems complets des equations lineaires et homogenes, Comptes Rendues de l'Academie des Sciences de l'URSS, 1930, t. 190, p. 405-409.

[66] M. G. Pfeiffer, Sur la operateurs d'un systeme complet d'equations lineaires et homogenes aux derivees partielles du premier ordre d'une fonction inconnue, Comptes Rendues de l'Academie des Sciences de l'URSS, 1930, t. 190, p. 909–911.

[67] M. G. Pfeiffer, La generalization de methode de Jacobi-Mayer, Comptes Rendues de l'Academie des Sciences de l'URSS, 1930, t. 191, p. 1107-1109.

[68] M. G. Pfeiffer, Sur la permutation des solutions s'une equation lineaire aux derivees partielles du premier ordre, *Bull. des Sc. Math.*, 1928, S.2,t.LII, p. 353-361.

[69] M. G. Pfeiffer, Quelgues additions au probleme de M. Buhl, Atti dei Congresso Internationale dei Matematici, Bologna, 1928, t.III, p. 45-46.

[70] M. G. Pfeiffer, La construction des operateurs d'une equation lineaire, homogene aux derivees partielles premier ordre, *Journal du Cycle Mathematique*, Academie des Sciences d'Ukraine, Kyiv, N1, 1931, p. 37-72 (in Ukrainian).

[71] J. Plebański, Some solutions of complex Einstein equations, *J. Math. Phys.*, **16** (1975), Issue 12, 2395-2402.

[72] C. Popovici, Sur les fonctions adjointes de M. Buhl, *Comptes Rendus*, t.145 (1907), p. 749.

[73] A. Presley, G. Segal, *Loop groups*, Oxford Math. Monographs, Oxford Univ. Press, 1986.

[74] N.K. Prykarpatska, D. Blackmore, V.H. Samoylenko, E. Wachnicki, M. Pytel-Kudela, The Cartan-Monge geometric approach to the characteristic method for nonlinear partial differential equations of the first and higher orders. *Nonlinear Oscillations*, **10**(1) (2007), 22-31.

[75] A.K. Prykarpatski A.K., O. Hentosh, Ya A.Prykarpatsky, Geometric structure of the classical Lagrange-d'Alembert principle and its application to the integrable nonlinear dynamical systems, *Mathematics*, **5**, 75, 20 pp. MDPI, Basel, Switzerland (2017).

[76] A.K. Prykarpatsky I.V. Mykytyuk, *Algebraic Integrability of Nonlinear Dynamical Systems on Manifolds: Classical and Quantum Aspects*, Kluwer, the Netherlands, 1998.

[77] Ya.A. Prykarpatsky, A.K. Prykarpatski, The integrable heavenly type equations and their Lie-algebraic structure, arXiv:1785057 [nlin.SI] 24 Jan 2017.

[78] Ya.A. Prykarpatsky, A.M. Samoilenko, Algebraic - analytic aspects of integrable nonlinear dynamical systems and their perturbations. Kyiv, Inst. Mathematics Publisher, v. **41**, 2002 (in Ukrainian).

[79] A. Radul, Lie algebras of differential operators, their central extensions and W-algebras, *Funct. Anal. Appl.*, **25**, 1 (1991) 33–49.

[80] A. Reyman, M. Semenov-Tian-Shansky, *Integrable Systems*, The Computer Research Institute Publ., Moscow-Izhvek, 2003 (in Russian).

[81] M. Sato, Y. Sato, Soliton equations as dynamical systems on infinite dimensional Grassmann manifold, in Nonlinear Partial Differential Equations in Applied Science; Proceedings of the U.S.-Japan Seminar, Tokyo, 1982, *Lect. Notes in Num. Anal.*, **5** (1982), 259–271.

[82] M. Sato, M. Noumi, Soliton equations and the universal Grassmann manifolds, Sophia Univ. Kokyuroku in Math. 18 (1984) (in Japanese).

[83] W. Schief, Self-dual Einstein spaces via a permutability theorem for the Tz itzeica equation, *Phys. Lett. A*, Volume **223**, Issues 1–2, 25 1996, 55-62.

[84] W. Schief, Self-dual Einstein spaces and a discrete Tzitzeica equation, A permutability theorem link, In *Symmetries and Integrability of Difference Equations*, P. Clarkson and F. Nijhoff, eds, London Math. Soc., Lecture Note Series 255, Cambridge Univ. Press (1999) 137-148.

[85] A. Sergyeyev and B. Szablikowski, Central extensions of cotangent universal hierarchy: (2+1)-dimensional bi-Hamiltonian systems, *Phys. Lett. A*, **372** (2008) 7016-7023.

[86] B. Szablikowski, Hierarchies of Manakov–Santini type by means of Rota-Baxter and other identities, *SIGMA* **12** (2016), 022, 14 pages.

[87] B. Szablikowski, M. Błaszak, Meromorphic Lax representations of (1+1)-dimensional multi-Hamiltonian dispersionless systems. *J. Math. Phys.*, **47**, N9, 2006, 092701.

[88] M. Semenov-Tian-Shansky, What is a classical R-matrix? *Func. Anal. Appl.*, 1983, **17**(4), 259-272.

[89] A. Sergyeyev, A simple construction of recursion operators for multidimensional dispersionless integrable systems, *J. Math. Anal. Appl.*, **454** (2017), no.2, 468–480; arXiv:1501.01955

[90] A. Sergyeyev, New integrable (3+1)-dimensional systems and contact geometry, arXiv:1401.2122v4 [math.AP] 26 Sep 2017.

[91] M. Sheftell, A. Malykh, D. Yazıcı, Recursion operators and bi-Hamiltonian structure of the general heavenly equation, arXiv:1510.03666v3 [math-ph] 25 Jun 2016.

[92] K. Takasaki, T. Takebe, SDiff(2) Toda equation – hierarchy, Tau function, and symmetries, *Letters in Math. Phys.*, **23**(3), 205–214 (1991).

[93] K. Takasaki, T. Takebe, Integrable hierarchies and dispersionless limit, *Reviews in Mathematical Physics*, **7**(05), 743–808 (1995).

[94] L. Takhtajan, L. Faddeev, *Hamiltonian Approach in Soliton Theory*, Springer, Berlin-Heidelberg, 1987.

[95] V. Vladimirov, I. Volovich, Superanalysis. I. Differential Calculus, *Theor. Math. Physics*, **59**, 1, 3–27, in Russian (1984).

[96] I. Zakharevich, Nonlinear wave equation, nonlinear Riemann problem, and the twistor transform of Veronese webs, arXiv:math-ph/0006001.

[97] V. Zakharov, A. Shabat, Integration of the nonlinear equations of mathematical physics by the method of the inverse scattering problem. II, *Funk. Anal. Prilozh.*, **13**(3) 13-22 (1979) [Funct. Anal. Appl. 13, 166-174 (1979)].

B4. Superposition formulae for nonlinear integrable equations in bilinear form

Xing-Biao Hu

LSEC, Institute of Computational Mathematics and Scientific Engineering Computing, AMSS, Chinese Academy of Sciences, P.O.Box 2719 Beijing 100190, PR China
Department of Mathematical Sciences, University of the Chinese Academy of Sciences, Beijing, PR China

Abstract

We are concerned with superposition formulae and Bianchi identities related to bilinear Bäcklund transformations for nonlinear integrable equations in bilinear form. Illustrating examples show that bilinear Bäcklund transformations and their associated superposition formulae may be utilized to generate soliton solutions, rational solutions and some other special solutions to nonlinear integrable equations under consideration.

1 Introduction

The principle of linear superposition of solutions play an essential role in the world governed by linear differential equations. Unfortunately, in the nonlinear world governed by nonlinear differential equations, this basic principle is lost. Is it possible to have nonlinear superposition of solutions for nonlinear differential equations? The answer is affirmative. For example, for the Riccati equation

$$\frac{du}{dt} = a(t) + 2b(t)u + c(t)u^2, \tag{1.1}$$

with $a(t), b(t)$ and $c(t)$ being known functions, we have the following nonlinear superposition formula for the Riccati equation (1.1)

$$\frac{u(t) - u_1(t)}{u(t) - u_2(t)} \frac{u_3(t) - u_2(t)}{u_3(t) - u_1(t)} = const., \tag{1.2}$$

where $u(t)$ is a general solution to (1.1) and $u_1(t), u_2(t)$ and $u_3(t)$ are three particular solutions to (1.1) (two of which are independent). Generally speaking, it is difficult to find nonlinear superposition formulae of nonlinear differential equations (if they exist!). However, current research in soliton theory suggests that the corresponding nonlinear superposition formula can usually be obtained for a nonlinear evolution equation (NEE) under consideration if the so-called Bäcklund transformation for the equation can be found. In this case, the equation under consideration is called integrable in the sense of having a Bäcklund transformation. In the literature, much research has been conducted on the Bäcklund transformations for integrable systems. An active area of particular interest is the Bäcklund transformation of bilinear differential equations. In this chapter, bilinear Bäcklund transformations will be discussed in detail.

Bilinear Bäcklund transformations were first introduced by Hirota [6] and developed by Hirota and Satsuma [11]. Bilinear Bäcklund transformations play an impotant role in [8, 36, 27] for

· obtaining multi-soliton solutions;

· finding new NEEs which exhibit N-soliton solutions;

· finding an inverse scattering form for the nonlinear evolution equation.

As an illustrative example, consider the KdV equation

$$u_t + 6uu_x + u_{xxx} = 0. \tag{1.3}$$

The bilinear form of the KdV equation (1.3) is given by

$$(D_x D_t + D_x^4) f \cdot f = 0, \tag{1.4}$$

with the dependent variable transformation being $u = 2(\ln f)_{xx}$. Here the bilinear D-operator is defined by

$$D_t^m D_x^n a(t,x) \cdot b(t,x) = \frac{\partial^m}{\partial s^m} \frac{\partial^n}{\partial y^n} a(t+s, x+y) b(t-s, x-y)|_{s=0,y=0},$$
$$m, n = 0, 1, 2, \cdots .$$

The following bilinear Bäcklund transformation for equation (1.4) is available in [6]:

$$(D_t + 3\lambda D_x + D_x^3) f' \cdot f = 0, \tag{1.5}$$
$$D_x^2 f' \cdot f = \lambda f' f, \tag{1.6}$$

where λ is an arbitrary constant. Substitution of the vacuum solution $f_0 = 1$ of (1.4) into the Bäcklund transformation (1.5) and (1.6) yields

$$f'_t + 3\lambda f'_x + f'_{xxx} = 0, \qquad f'_{xx} = \lambda f', \tag{1.7}$$

the particular solution to which is seen to be

$$f' = e^{\eta_1/2} + e^{-\eta_1/2} = e^{-\eta_1/2}(1 + e^{\eta_1}) \tag{1.8}$$

with $\lambda = p_1^2/4$, where

$$\eta_1 = p_1 x - p_1^3 t + \eta_1^0$$

and p_1, η_1^0 are constants. Using the fact that

$$\frac{\partial^2}{\partial x^2} \ln e^{-\eta_1/2} = 0,$$

we know that (1.8) gives a one-soliton solution of the KdV equation in the original variable:

$$u_1 = 2\frac{\partial^2}{\partial x^2}\ln(1+e^{\eta_1}) = \frac{1}{2}p_1^2 sech^2(\frac{\eta_1}{2}). \tag{1.9}$$

Furthermore, by applying the Bäcklund transformation (1.5) and (1.6) with $\lambda = p_2^2/4$ to the one-soliton solution (1.8), we can deduce the following two-soliton solution:

$$\begin{aligned} f &= (p_1-p_2)[e^{\frac{1}{2}(\eta_1+\eta_2)} + e^{-\frac{1}{2}(\eta_1+\eta_2)}] - (p_1+p_2)[e^{\frac{1}{2}(\eta_1-\eta_2)} + e^{-\frac{1}{2}(\eta_1-\eta_2)}] \\ &= (p_1-p_2)e^{-\frac{1}{2}(\eta_1+\eta_2)}[1 - \frac{p_1+p_2}{p_1-p_2}(e^{\eta_1}+e^{\eta_2}) + e^{\eta_1+\eta_2}], \end{aligned} \tag{1.10}$$

where

$$\eta_i = p_i x - p_i^3 t + \eta_i^0, \qquad i = 1, 2$$

and p_i, η_i^0 are constants. Choosing phase constants η_1^0 and η_2^0 as

$$\eta_i^0 = \ln\frac{p_2-p_1}{p_2+p_1} + \tilde{\eta}_i^0, \qquad i = 1, 2$$

with $\tilde{\eta}_1^0, \tilde{\eta}_2^0$ being constants, it can be seen that (1.10) gives a two-soliton solution of the KdV equation.

In principle, we can continue this procedure to generate an N-soliton solution from an (N-1)-soliton solution. However, as N increases, the corresponding calculations are more involved. In what follows, we shall introduce one of the most interesting results of this transformation theory, that is, the nonlinear superposition formula. Its power lies in that it may be used to generate multi-soliton solutions from single soliton solutions by purely algebraic means. Such a superposition principle can be obtained from the permutability theorem of the Bäcklund transformation.

2 Bianchi theorem of permutability and superposition formula of the KdV equation

Let us now return to the bilinear KdV equation (1.4) and see what we mean by the permutability theorem. Let f_0 be a solution of (1.4) and suppose that f_1 and f_2 are solutions of (1.4) given by the Bäcklund transformation (1.5) and (1.6) with starting solution $f = f_0$ and Bäcklund parameters $\lambda = \lambda_1$ and $\lambda = \lambda_2$, respectively. We denote $f_i = B_{\lambda_i} f_0$. A second application of the Bäcklund transformation (1.5) and (1.6) leads to two solutions $f_{12} = B_{\lambda_2} f_1 = B_{\lambda_2} B_{\lambda_1} f_0$ and $f_{21} = B_{\lambda_1} f_2 = B_{\lambda_1} B_{\lambda_2} f_0$. Generally speaking, the f_{12} and f_{21} so obtained are not unique. The permutability theorem means that the result of these two consecutive Bäcklund transformations BT_{12} and BT_{21} with matching boundary conditions for f_{12} and f_{21} is independent of the order, i.e. $f_{12} = f_{21}$. Thus a commutative Bianchi diagram may be constructed as follows:

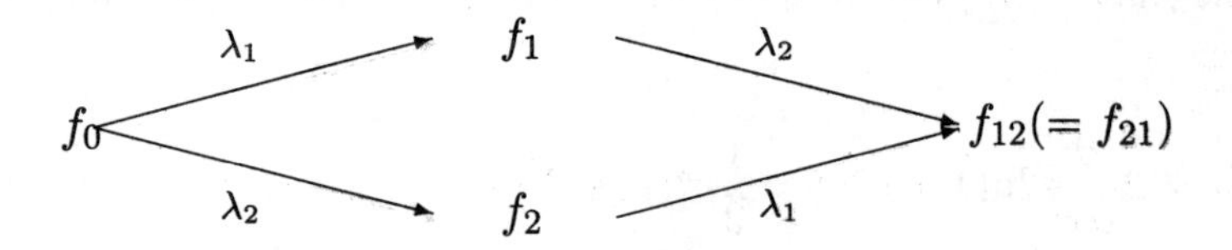

We remark that the classical permutability theorem for the sine-Gordon equation can be traced back to Bianchi's work [5].

For now we shall derive the superposition formula for the KdV equation based on the permutability theorem. From the given conditions, we have

$$(D_x^2 - \lambda_1)f_0 \cdot f_1 = 0, \qquad (D_x^2 - \lambda_2)f_0 \cdot f_2 = 0, \tag{2.1a}$$

$$(D_x^2 - \lambda_1)f_2 \cdot f_{12} = 0, \qquad (D_x^2 - \lambda_2)f_1 \cdot f_{12} = 0 \tag{2.1b}$$

from which it follows that

$$(D_x^2 f_0 \cdot f_1)f_2 f_{12} - f_0 f_1(D_x^2 f_2 \cdot f_{12}) = 0 \tag{2.2a}$$

$$(D_x^2 f_0 \cdot f_2)f_1 f_{12} - f_0 f_2(D_x^2 f_1 \cdot f_{12}) = 0. \tag{2.2b}$$

Using the bilinear operator identity

$$(D_x^2 a \cdot b)cd - abD_x^2 c \cdot d = D_x[(D_x a \cdot d) \cdot cb + ad \cdot (D_x c \cdot b)], \tag{2.3}$$

equations (2.2a) and (2.2b) become

$$D_x[(D_x f_0 \cdot f_{12}) \cdot f_2 f_1 + f_0 f_{12} \cdot (D_x f_2 \cdot f_1)] = 0 \tag{2.4}$$

and

$$D_x[(D_x f_0 \cdot f_{12}) \cdot f_1 f_2 + f_0 f_{12} \cdot (D_x f_1 \cdot f_2)] = 0, \tag{2.5}$$

respectively. The subtraction of (2.4) from (2.5) leads to

$$D_x f_0 f_{12} \cdot (D_x f_1 \cdot f_2) = 0, \tag{2.6}$$

which gives the superposition formula

$$f_0 f_{12} = cD_x f_1 \cdot f_2, \tag{2.7}$$

where c is a constant. It can be readily shown that f_{12} as given by (2.7) is indeed a solution of (1.4). Similarly, upon adding (2.4) and (2.5), we have

$$D_x(D_x f_0 \cdot f_{12}) \cdot f_2 f_1 = 0, \tag{2.8}$$

which implies that

$$f_1 f_2 = cD_x f_0 \cdot f_{12}. \tag{2.9}$$

This is another superposition formula for the KdV equation. The generalization of (2.7) and (2.9) to $(N-1)$-,N- and $(N+1)$- soliton solutions is straightforward, and the results may be expressed as

$$f_{N-1}f_{N+1} = cD_x f_N \cdot \hat{f}_N \tag{2.10}$$
$$f_N\hat{f}_N = cD_x f_{N-1} \cdot f_{N+1}. \tag{2.11}$$

Here the parameter dependencies of the four solutions $f_{N-1}, f_N, \hat{f}_N$ and f_{N+1} can be written explicitly as

$$f_{N-1} = f_{N-1}(x,t;p_1,p_2,\cdots,p_{N-1}), \tag{2.12a}$$
$$f_N = f_N(x,t;p_1,p_2,\cdots,p_{N-1},p_N), \tag{2.12b}$$
$$\hat{f}_N = \hat{f}_N(x,t;p_1,p_2,\cdots,p_{N-1},p_{N+1}), \tag{2.12c}$$
$$f_{N+1} = f_{N+1}(x,t;p_1,p_2,\cdots,p_{N-1},p_N,p_{N+1}). \tag{2.12d}$$

From these formulae, an N-soliton solution may be expressed in the form [9]

$$f_N = \sum_{\epsilon=\pm1} \frac{\prod_{j<k}^{(N)}(\epsilon_j p_j - \epsilon_k p_k)}{\prod_{j=1}^{N} p_j} \exp\left(\frac{1}{2}\sum_{j=1}^{N}\epsilon_j\eta_j\right) \tag{2.13}$$

where $\sum_{\epsilon=\pm1}$ is the summation over all possible combinations of $\epsilon_1 = \pm1, \epsilon_2 = \pm1, \cdots, \epsilon_N = \pm1$.

It should be mentioned that nonlinear superposition formulae in bilinear form were first considered by Hirota and Satsuma [9].

3 Superposition formulae for a variety of soliton equations with examples

In the previous section, we have derived superposition frormulae for the KdV equation. In what follows, we shall demonstrate superposition frormulae for a variety of soliton equations.

Example 1. The Kadomtsev-Petviashvili equation

The KP equation [26]

$$(u_t + 6uu_x + u_{xxx})_x + \alpha u_{yy} = 0 \tag{3.1}$$

is transformed into

$$(D_xD_t + D_x^4 + \alpha D_y^2)f \cdot f = 0 \tag{3.2}$$

through the variable transformation

$$u = 2(\ln f)_{xx}, \tag{3.3}$$

where the boundary condition $u \longrightarrow 0$ as $|x| \longrightarrow \infty$ is taken and $\alpha(=\pm 1)$ is a parameter depending on the dispersive property of the system. The Bäcklund transformation of the equation (3.2) is expressed as

$$(D_x^2 - bD_y)f' \cdot f = \lambda f'f, \tag{3.4}$$
$$(D_t + 3\lambda D_x + D_x^3 + 3bD_xD_y)f' \cdot f = 0, \tag{3.5}$$

where $b = \pm(\alpha/3)^{1/2}$ and λ is an arbitrary constant.

In order to get a superposition formula, we follow the same procedure as in the KdV equation. Equation (3.4) gives

$$(D_x^2 - bD_y)f_0 \cdot f_1 = \lambda_1 f_0 f_1, \tag{3.6a}$$
$$(D_x^2 - bD_y)f_0 \cdot f_2 = \lambda_2 f_0 f_1, \tag{3.6b}$$
$$(D_x^2 - bD_y)f_2 \cdot f_{12} = \lambda_1 f_2 f_{12}, \tag{3.6c}$$
$$(D_x^2 - bD_y)f_1 \cdot f_{12} = \lambda_2 f_1 f_{12}, \tag{3.6d}$$

where λ_1 and λ_2 are the corresponding parameters. Multiplying (3.6a) by $f_2 f_{12}$ and (3.6c) by $f_0 f_1$, subtracting each other, and using (2.3), we obtain

$$\begin{aligned} &D_x[(D_x f_0 \cdot f_{12}) \cdot f_1 f_2 - f_0 f_{12} \cdot (D_x f_1 \cdot f_2)] \\ &\quad -b(D_y f_0 \cdot f_1) f_2 f_{12} + b f_0 f_1 (D_y f_2 \cdot f_{12}) = 0. \end{aligned} \tag{3.7}$$

Similarly, we have from (3.6b) and (3.6d)

$$\begin{aligned} &D_x[(D_x f_0 \cdot f_{12}) \cdot f_1 f_2 - f_0 f_{12} \cdot (D_x f_2 \cdot f_1)] \\ &\quad -b(D_y f_0 \cdot f_2) f_1 f_{12} + b f_0 f_2 (D_y f_1 \cdot f_{12}) = 0. \end{aligned} \tag{3.8}$$

Subtracting (3.7) from (3.8), we get

$$D_x(f_0 f_{12}) \cdot (D_x f_1 \cdot f_2) = 0, \tag{3.9}$$

which gives a superposition formula

$$f_0 f_{12} = const. D_x f_1 \cdot f_2. \tag{3.10}$$

This is exactly the same as (2.7).

Example 2. The Boussinesq equation

The Boussinesq equation

$$u_{tt} - u_{xx} - 3(u^2)_{xx} - u_{xxxx} = 0 \tag{3.11}$$

is rewritten in the bilinear form

$$(D_t^2 - D_x^2 - D_x^4)f \cdot f = 0 \tag{3.12}$$

through the variable transformation (3.3), supposing $u \longrightarrow 0$ as $|x| \longrightarrow \infty$. It is easily seen that (3.10) (or (3.12)) is derived from the KP equation (3.1) (or (3.2)) by taking $t = x, y = t$ and $\alpha = -1$. Thus the same transformation of independent variables and parameter in (3.4) and (3.5) gives the Bäcklund transformation of (3.12)

$$(D_x^2 - i/3^{1/2}D_t)f' \cdot f = \lambda f' f, \tag{3.13}$$
$$((1+3\lambda)D_x + D_x^3 + i3^{1/2}D_xD_t)f' \cdot f = 0, \tag{3.14}$$

with $i = \sqrt{-1}$. Equation (3.13) yields the same superposition formula as (3.10).

Example 3. The modified KdV equation

We consider the modified KdV equation

$$v_t + 6v^2 v_x + v_{xxx} = 0 \tag{3.15}$$

under the boundary condition $v \longrightarrow 0$ as $|x| \longrightarrow \infty$. By transforming the dependent variable in (3.15) to

$$v = i(\ln f^*/f)_x, \tag{3.16}$$

we find that v is a solution of (3.15) if f satisfies a couple of bilinear equations,

$$(D_t + D_x^3)f^* \cdot f = 0, \tag{3.17}$$
$$D_x^2 f^* \cdot f = 0, \tag{3.18}$$

where $*$ denotes a complex conjugate. The Bäcklund transformation is given by [6]

$$D_x f^* \cdot f' = ikff'^*, \tag{3.19}$$
$$D_x f'^* \cdot f = ikf^* f', \tag{3.20}$$
$$(D_t + 3k^2 D_x + D_x^3)f' \cdot f = 0, \tag{3.21}$$
$$(D_t + 3k^2 D_x + D_x^3)f'^* \cdot f^* = 0, \tag{3.22}$$

where k is a real constant. From (3.19) and (3.20), we have the following relations:

$$\begin{aligned}(D_x^2 f' \cdot f)f'^* f^* &- f' f(D_x^2 f'^* \cdot f^*) \\ &= D_x[(D_x f' \cdot f^*) \cdot f'^* f + f' f^* \cdot (D_x f'^* \cdot f)] = 0\end{aligned} \tag{3.23}$$

$$\begin{aligned}(D_x^2 f' \cdot f)f'^* f^* &+ f' f(D_x^2 f'^* \cdot f^*) \\ &= (D_x^2 f' \cdot f'^*)f^* f + f' f'^*(D_x^2 f^* \cdot f) - 2(D_x f' \cdot f^*)(D_x f \cdot f'^*) \\ &= 2k^2 ff' f^* f'^*,\end{aligned} \tag{3.24}$$

where we have used (2.3), (3.18) and the Hirota bilinear operator identity

$$(D_x^2 a \cdot b)cd + ab(D_x^2 c \cdot d) = (D_x^2 a \cdot d)cb + ad(D_x^2 c \cdot b) - 2(D_x a \cdot c)(D_x b \cdot d).$$

Adding (3.23) and (3.24), we obtain a different expression for half of the Bäcklund transformation

$$D_x^2 f' \cdot f = k^2 f' f, \tag{3.25}$$

which is similar to (1.6). Thus we find the superposition formula of soliton solutions of the modified KdV equation is the same as that of the KdV equation, i.e.,

$$f_{N-1}f_{N+1} = const.D_x f_N \cdot \hat{f}_N, \tag{3.26}$$

or

$$f_N \hat{f}_N = const.D_x f_{N-1} \cdot f_{N+1}. \tag{3.27}$$

for the soliton solutions of (3.17) and (3.18). Here we have followed the notation of the case of the KdV equation.

Example 4. The sine-Gordon equation

We consider the sine-Gordon equation

$$\phi_{xt} = sin\phi \tag{3.28}$$

under the boundary condition $\phi \longrightarrow const.$ as $|x| \longrightarrow \infty$. By transforming the dependent variable in (3.28) to

$$\phi = 2i(\ln f^*/f), \tag{3.29}$$

we find that ϕ is a solution of (3.28) if f satisfies

$$D_x D_t f \cdot f = \frac{1}{2}(f^2 - f^{*2}) \tag{3.30}$$

and its complex conjugate counterpart. The Bäcklund transformation of (3.30) is given by [6]

$$D_x f' \cdot f = -\frac{k}{2} f'^* f^*, \tag{3.31}$$

$$D_t f' \cdot f^* = -\frac{1}{2k} f'^* f, \tag{3.32}$$

and their associated complex conjugate equation, where k is a real constant. Since the complex conjugate of equation (3.32) is similar to (3.20) in its structure, we can derive from equation (3.32) and its complex conjugate counterpart

$$D_x^2 f' \cdot f = -\frac{1}{4k^2} f' f, \tag{3.33}$$

which is similar to (1.6). Thus the following superposition formulas hold for soliton solutions of the sine-Gordon equation:

$$f_0 f_{12} = const.D_t f_1 \cdot \hat{f}_2, \tag{3.34}$$

$$D_t f_0 \hat{f}_{12} = const.f_1 \cdot f_2. \tag{3.35}$$

Next we show that the usual form of superposition formula can be recovered from equation (3.34). Indeed, multiplying (3.34) by $f_0^* f_0^*$ and using

$$D_t f_1 \cdot f_0^* = -\frac{1}{2k_1} f_1^* f_0, \tag{3.36}$$

$$D_t f_2 \cdot f_0^* = -\frac{1}{2k_2} f_2^* f_0. \tag{3.37}$$

we have

$$\begin{aligned} f_0 f_{12} f_0^* f_0^* &= const.[(D_t f_1 \cdot f_0^*) f_0^* f_2 + (D_t f_0^* \cdot f_2) f_0^* f_1] \\ &= const. f_0 f_0^* [-\frac{1}{2k_1} f_1^* f_2 + \frac{1}{2k_2} f_1 f_2^*], \end{aligned} \tag{3.38}$$

which can be reduced to

$$f_0^* f_{12} = const.[-\frac{1}{2k_1} f_1^* f_2 + \frac{1}{2k_2} f_1 f_2^*]. \tag{3.39}$$

Dividing (3.39) by its complex conjugate, we obtain

$$(f_0^*/f_0)/(f_{12}^*/f_{12}) = [k_2(f_1^*/f_1) - k_1(f_2^*/f_2)]/[k_2(f_2^*/f_2) - k_1(f_1^*/f_1)]. \tag{3.40}$$

Substitution of (3.29) into (3.40) yields

$$\exp\frac{i}{2}(\phi_{12} - \phi_0) = [k_2 - k_1 \exp\frac{i}{2}(\phi_1 - \phi_2)]/[-k_1 + k_2 \exp\frac{i}{2}(\phi_1 - \phi_2)], \tag{3.41}$$

which gives

$$\tan\frac{1}{4}(\phi_{12} - \phi_0) = \frac{k_1 + k_2}{k_1 - k_2} \tan\frac{1}{4}(\phi_1 - \phi_2). \tag{3.42}$$

Equation (3.42) is the usual form of superposition formula.

Example 5. The Benjamin-Ono equation

The Benjamin-Ono equation is given [4, 34]

$$u_t + 2uu_x + H[u_{xx}] = 0, \tag{3.43}$$

where H is the Hilbert transform operator defined by the Cauchy principal value integral

$$Hf(x) \equiv \frac{1}{\pi} P \int_{-\infty}^{\infty} \frac{f(z)}{z - x}\, dz. \tag{3.44}$$

By the dependent variable transformation

$$u(x,t) = i(\ln f'/f)_x, \tag{3.45}$$

where

$$f \propto \prod_{n=1}^{N}(x - z_n(t)), \tag{3.46}$$

$$f' \propto \prod_{n=1}^{N}(x - z_n'(t)), \tag{3.47}$$

with $N \in Z^+$ and z_n, z_n' being complex with $Im(z_n) > 0, Im(z_n') < 0, \forall n$, equation (3.43) is transformed into the bilinear form [29]

$$(iD_t - D_x^2)f' \cdot f = 0. \tag{3.48}$$

One can establish a bilinear Bäcklund transformation for equation (3.48)[29]

$$(iD_t - 2\lambda D_x - D_x^2 - \mu)f \cdot g = 0, \tag{3.49}$$
$$(iD_t - 2\lambda D_x - D_x^2 - \mu)f' \cdot g' = 0, \tag{3.50}$$
$$(D_x + i\lambda)f \cdot g' = i\nu f'g, \tag{3.51}$$

where λ, μ and ν are arbitrary constants. For the bilinear BT (3.49)-(3.51), we have the following superposition formula [29]

$$f_0 f_{12} = k[D_x + i(\lambda_2 - \lambda_1)]f_1 \cdot f_2, \tag{3.52}$$
$$f_0' f_{12}' = k[D_x + i(\lambda_2 - \lambda_1)]f_1' \cdot f_2', \tag{3.53}$$

where k is an arbitrary constant, and $f_0 \xrightarrow{\lambda_i} f_i (i = 1, 2)$ and $f_1 \xrightarrow{\lambda_2} f_{12}, f_2 \xrightarrow{\lambda_1} f_{12}$.

From the above examples, we can see that the nonlinear superposition formulae for the KdV, mKdV, Boussinesq, KP, BO, etc. possess a unified and simple structure

$$f_0 f_{12} = k(D_x + \gamma_2 - \gamma_1)f_1 \cdot f_2, \tag{3.54}$$

with γ_1 and γ_2 being Bäcklund parameters related to the solutions f_0, f_1, f_2. Other examples also possessing the above type of superposition formula include the second modified KdV equation [30], a model equation for shallow water waves [13],the KdV equation with a source and its hierarchy [14], the so-called DJKM equation (i.e. the second member of the KP hierarchy) [19], the Boussinesq hierarchy [20] and so on.

Remark 1: As mentioned above, γ_i in (3.54) are Bäcklund parameters, but why do they not appear in (2.7),(3.10) and (3.34)? The reason for this is that in order to derive a nonlinear superposition formula of the type (3.54) for the KdV, KP and sine-Gordon equations, we need to consider their two-parameter Bäcklund transformations. For example, if we consider the following two-parameter Bäcklund transformation for the KdV equation (1.4)

$$(D_t + 3\lambda D_x + D_x^3)f' \cdot f = 0, \tag{3.55}$$
$$D_x^2 f' \cdot f = \lambda f' f - \mu D_x f' \cdot f, \tag{3.56}$$

we may derive the following nonlinear superposition formula

$$f_0 f_{12} = k(D_x + \frac{1}{2}\mu_2 - \frac{1}{2}\mu_1)f_1 \cdot f_2. \tag{3.57}$$

Remark 2: We know that Hirota's method gives soliton solutions as polynomials of exponentials. But the superposition formula (2.7) gives

$$f_{12} = \frac{cD_x f_1 \cdot f_2}{f_0}. \tag{3.58}$$

Therefore if f_0 is a one soliton solution and f_1, f_2 two soliton solutions, f_0 should be a nontrivial factor of $D_x f_1 \cdot f_2$, otherwise f_{12} would not a polynomial. In Section 2, it is pointed out that this is also true for the case of multi-soliton solutions ([9]).

Besides the above type of superposition formula, there exist other types of nonlinear superposition formulae for some other soliton equations. In the following, we present another kind of nonlinear superposition formulae for the Sawada-Kotera, NNV and Ito equations, and etc.

Example 6. The Ito equation

The so-called Ito equation in bilinear form is given by [25]

$$D_t(D_t + D_x^3) f \cdot f = 0. \tag{3.59}$$

Through the dependent variable transformation $u = 2(\ln f)_{xx}$, (3.59) becomes

$$u_{tt} + u_{xxxt} + 3(2u_x u_t + uu_{xt}) + 3u_{xx} \int_{-\infty}^{x} u_t dx' = 0. \tag{3.60}$$

A Bäcklund transformation for (3.59) is given by [25, 21]

$$(D_t + 3\gamma^2 D_x - 3\gamma D_x^2 + D_x^3 - \lambda) f \cdot f' = 0, \tag{3.61}$$

$$(D_x D_t - \mu D_x - \gamma D_t + \gamma\mu) f \cdot f' = 0. \tag{3.62}$$

where λ, μ and γ are arbitrary constants. We are going to give a nonlinear superposition formula for the Ito equation. In the following discussion, we set $\gamma = \lambda = 0$ in BT (3.61) and (3.62) for the sake of convenience. In this case, (3.61) and (3.62) become

$$(D_t + D_x^3) f \cdot f' = 0, \quad (D_x D_t - \mu D_x) f \cdot f' = 0. \tag{3.63}$$

Let $f_0 \neq 0$ be a solution of the Ito equation (3.59). Let $f_i (i = 1, 2)$ be a solution of the Ito equation (3.59) and be related to f_0 under BT (3.63) with $\mu = \mu_i$, i.e., $f_0 \xrightarrow{\mu_i} f_i (i = 1, 2)$. We can establish the superposition formula [21]

$$D_x f_0 \cdot f_{12} = k D_x f_1 \cdot f_2, \tag{3.64}$$

where k is a nonzero constant. In fact, starting from

$$[(D_x D_t - \mu_1 D_x) f_0 \cdot f_1] f_2 - [(D_x D_t - \mu_2 D_x) f_0 \cdot f_2] f_1 = 0$$

we may deduce that

$$D_x[(D_t+\mu_1-\mu_2)f_1\cdot f_2+\frac{1}{k}(D_t-\mu_1-\mu_2)f_0\cdot f_{12}]\cdot f_0^2=0$$

which implies that

$$(D_t+\mu_1-\mu_2)f_1\cdot f_2+\frac{1}{k}(D_t-\mu_1-\mu_2)f_0\cdot f_{12}=c_1(t)f_0^2 \tag{3.65}$$

where $c_1(t)$ is some function of t. Next, from $[(D_t+D_x^3)f_0\cdot f_1]_x f_2-[(D_t+D_x^3)f_0\cdot f_2]_x f_1=0$, we have

$$D_x[D_tf_0\cdot f_{12}+\frac{1}{4}D_x^3f_0\cdot f_{12}+\frac{3k}{4}D_x^3f_1\cdot f_2]\cdot f_0^2=0$$

which implies that

$$D_tf_0\cdot f_{12}+\frac{1}{4}D_x^3f_0\cdot f_{12}+\frac{3k}{4}D_x^3f_1\cdot f_2=c_2(t)f_0^2 \tag{3.66}$$

where $c_2(t)$ is some function of t. In this way, we may seek particular solutions of the Ito equation according to the following steps: First, choose a given solution f_0 of (3.59). Second, from the BT (3.63) we find out f_1 and f_2 such that $f_0 \xrightarrow{\mu_i} f_i (i=1,2)$ and, further, get a particular solution $\tilde{f}_{12}$ from (3.64). Then a general solution of (3.64) is $f_{12}=\tilde{f}_{12}+c(t)f_0$ (where $c(t)$ is an arbitrary function of t). Finally we substitute f_{12} into (3.65) and (3.66). If $c(t)$ can be determined such that $c_1(t)=c_2(t)=0$, i.e. f_{12} satisfies the relations

$$(D_t+\mu_1-\mu_2)f_1\cdot f_2+\frac{1}{k}(D_t-\mu_1-\mu_2)f_0\cdot f_{12}=0, \tag{3.67}$$

$$D_tf_0\cdot f_{12}+\frac{1}{4}D_x^3f_0\cdot f_{12}+\frac{3k}{4}D_x^3f_1\cdot f_2=0, \tag{3.68}$$

then we can show that it is a new solution of the Ito equation and $f_1 \xrightarrow{\mu_2} f_{12}, f_2 \xrightarrow{\mu_1} f_{12}$. In what follows, we give one illustrative example: Choose $f_0=1$. It is easily verified that

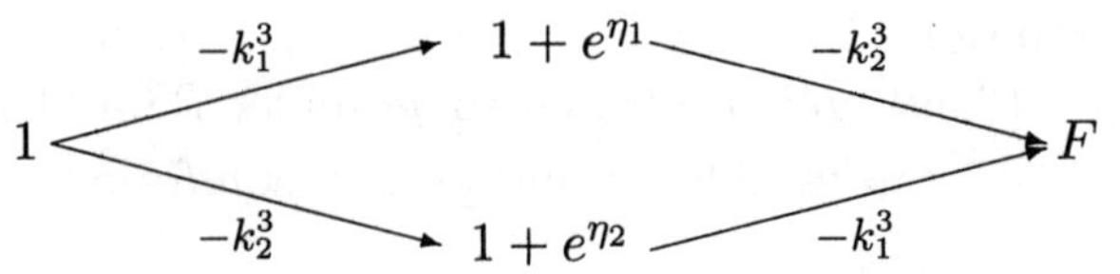

$$\begin{array}{ccccc} & \xrightarrow{-k_1^3} & 1+e^{\eta_1} & \xrightarrow{-k_2^3} & \\ 1 & & & & F \\ & \xrightarrow{-k_2^3} & 1+e^{\eta_2} & \xrightarrow{-k_1^3} & \end{array}$$

So F is a two soliton solution of the Ito equation, where

$$F\equiv\frac{k_1^3-k_2^3}{k_1^3+k_2^3}-e^{\eta_1}+e^{\eta_2}-\frac{k_1-k_2}{k_1+k_2}e^{\eta_1+\eta_2}.$$

Example 7. The Sawada-Kotera equation

The Sawada-Kotera equation in bilinear form is given by [7, 39]

$$(D_x^6 - D_x D_t) f \cdot f = 0. \tag{3.69}$$

A Bäcklund transformation for the Sawada-Kotera equation (3.69) is [38]

$$(D_x^3 - \lambda) f \cdot f' = 0, \quad (D_t + \frac{15}{2}\lambda D_x^2 + D_x^5) f \cdot f' = 0, \tag{3.70}$$

where λ is an arbitrary constant. Let $f_0 \neq 0$ be a solution of the SK equation (3.69). Let $f_i (i = 1, 2)$ be a solution of the SK equation (3.69) and be related to f_0 under BT (3.70) with $\lambda = \lambda_i$, i.e., $f_0 \xrightarrow{\lambda_i} f_i (i = 1, 2)$. We can derive the following superposition formula [15]

$$D_x f_0 \cdot f_{12} = k D_x f_1 \cdot f_2, \tag{3.71}$$

with k being a nonzero constant, together with the following two relations:

$$\frac{1}{4k} D_x^3 f_0 \cdot f_{12} + \frac{3}{4} D_x^3 f_1 \cdot f_2 - \frac{1}{k}(\lambda_1 + \lambda_2) f_0 \cdot f_{12} = c_1(t) f_0^2, \tag{3.72}$$

$$\begin{aligned} &(D_t + \frac{15}{8}(\lambda_1 + \lambda_2) D_x^2 + \frac{3}{32} D_x^5) f_0 \cdot f_{12} \\ &\quad + k(\frac{45}{8}(\lambda_2 - \lambda_1) D_x^2 + \frac{45}{32} D_x^5) f_1 \cdot f_2 = c_2(t) f_0^2, \end{aligned} \tag{3.73}$$

where $c_i(t) (i = 1, 2)$ are some functions of t. In this way, we may seek particular solutions of the Sawada-Koterra equation according to the following steps: First, choose a given solution f_0 of (3.69). Second, from the BT (3.70) we find out f_1 and f_2 such that $f_0 \xrightarrow{\lambda_i} f_i (i = 1, 2)$ and, further, get a particular solution $\tilde{f_{12}}$ from (3.71). Then a general solution of (3.71) is $f_{12} = \tilde{f_{12}} + c(t) f_0$ (where $c(t)$ is an arbitrary function of t). Finally we substitute f_{12} into (3.72) and (3.73). If $c(t)$ can be determined such that $c_1(t) = c_2(t) = 0$, i.e. f_{12} satisfies the relations

$$\frac{1}{4k} D_x^3 f_0 \cdot f_{12} + \frac{3}{4} D_x^3 f_1 \cdot f_2 - \frac{1}{k}(\lambda_1 + \lambda_2) f_0 \cdot f_{12} = 0, \tag{3.74}$$

$$\begin{aligned} &D_t + \frac{15}{8}(\lambda_1 + \lambda_2) D_x^2 + \frac{3}{32} D_x^5) f_0 \cdot f_{12} \\ &\quad + k(\frac{45}{8}(\lambda_2 - \lambda_1) D_x^2 + \frac{45}{32} D_x^5) f_1 \cdot f_2 = 0, \end{aligned} \tag{3.75}$$

then it is a new solution of the SK equation and $f_1 \xrightarrow{\lambda_2} f_{12}, f_2 \xrightarrow{\lambda_1} f_{12}$.

Example 8. A model equation for shallow water waves

Here we consider the following model equation for shallow water waves in bilinear form [12]

$$D_x(D_t - D_t D_x^2 + D_x) f \cdot f = 0. \tag{3.76}$$

We consider the following one-parameter Bäcklund transformation for (3.76) is [7, 21]

$$(D_x^3 - D_x + \lambda)f \cdot f' = 0, \quad (3D_xD_t - 1)f \cdot f' = 0, \tag{3.77}$$

where λ is an arbitrary constant. Let $f_0 \neq 0$ be a solution of the equation (3.76). Let $f_i (i = 1, 2)$ be a solution of the equation (3.76) and be related to f_0 under BT (3.77) with $\lambda = \lambda_i$, i.e., $f_0 \xrightarrow{\lambda_i} f_i (i = 1, 2)$. We can derive the following superposition formula [21]

$$D_x f_0 \cdot f_{12} = k D_x f_1 \cdot f_2, \tag{3.78}$$

with k being a nonzero constant, together with the following two relations:

$$\frac{3k}{4} D_x^3 f_1 \cdot f_2 + (\frac{1}{4} D_x^3 - D_x + \lambda_1 + \lambda_2) f_0 \cdot f_{12} = c_1(t) f_0^2, \tag{3.79}$$

$$D_t f_0 \cdot f_{12} + k D_t f_1 \cdot f_2 = c_2(t) f_0^2, \tag{3.80}$$

where $c_i(t)(i = 1, 2)$ are some functions of t. In this way, we may seek particular solutions of the model equation for the shallow water waves equation according to the following steps: First, choose a given solution f_0 of (3.76). Second, from the BT (3.77) we find out f_1 and f_2 such that $f_0 \xrightarrow{\lambda_i} f_i (i = 1, 2)$ and, further, get a particular solution $\tilde{f_{12}}$ from (3.78). Then a general solution of (3.78) is $f_{12} = \tilde{f_{12}} + c(t) f_0$ (where $c(t)$ is an arbitrary function of t). Finally we substitute f_{12} into (3.79) and (3.80). If $c(t)$ can be determined such that $c_1(t) = c_2(t) = 0$, i.e. f_{12} satisfies the relations

$$\frac{3k}{4} D_x^3 f_1 \cdot f_2 + (\frac{1}{4} D_x^3 - D_x + \lambda_1 + \lambda_2) f_0 \cdot f_{12} = 0, \tag{3.81}$$

$$D_t f_0 \cdot f_{12} + k D_t f_1 \cdot f_2 = 0, \tag{3.82}$$

then it is a new solution of the equation (3.76) and $f_1 \xrightarrow{\lambda_2} f_{12}, f_2 \xrightarrow{\lambda_1} f_{12}$.

Example 9. The Nizhnik-Novikov-Veselov equation

The Nizhnik-Novikov-Veselov equation is given by [33, 41]

$$2u_t + u_{xxx} + u_{yyy} + 3(u\partial_y^{-1} u_x)_x + 3(u\partial_x^{-1} u_y)_y = 0. \tag{3.83}$$

The NVN equation may be rewritten in the multilinear form[3, 16]

$$D_x[(D_x^3 D_y + 2D_t D_y) f \cdot f] \cdot f^2 + D_y(D_x D_y^3 f \cdot f) \cdot f^2 = 0. \tag{3.84}$$

We have the following Bäcklund transformation for the equation (3.84) [16, 17]

$$(D_x D_y - \lambda D_x) f \cdot f' = 0, \tag{3.85}$$

$$(2D_t + D_x^3 + D_y^3 + 3\lambda^2 D_x - 3\lambda D_x^2) f \cdot f' = 0, \tag{3.86}$$

where λ is an arbitrary constant. Let $f_0 \neq 0$ be a solution of the NVN equation (3.84). Suppose that $f_i (i = 1, 2)$ is a solution of the NVN equation (3.84) and is related to f_0 under BT (3.85) and (3.86) with $\lambda = \lambda_i$, i.e., $f_0 \xrightarrow{\lambda_i} f_i (i = 1, 2)$. We can derive the following superposition formula [17]

$$D_x f_0 \cdot f_{12} = k D_x f_1 \cdot f_2, \tag{3.87}$$

with k being a nonzero constant, together with the following two relations:

$$(D_x - \lambda_1 - \lambda_2) f_0 \cdot f_{12} + k(D_x + \lambda_1 - \lambda_2) f_1 \cdot f_2 = c_1(t) f_0^2, \tag{3.88}$$

$$[2D_t + \frac{1}{4}D_x^3 + \frac{1}{4}D_y^3 - \frac{3}{4}(\lambda_1 + \lambda_2)D_x^2 + \frac{3}{2}(\lambda_1^2 + \lambda_2^2)D_x] f_0 \cdot f_{12}$$
$$+ k[\frac{3}{4}D_y^3 - \frac{3}{4}D_x^3 - \frac{9}{4}(\lambda_1 - \lambda_2)D_x^2 - \frac{3}{2}(\lambda_1 + \lambda_2)^2 D_x] f_1 \cdot f_2 = c_2(t) f_0^2, \tag{3.89}$$

where $c_i(t) (i = 1, 2)$ are some functions of t. In this way, we may seek particular solutions of the NVN equation according to the following steps: First, choose a given solution f_0 of (3.84). Second, from the BT (3.85) and (3.86) we find out f_1 and f_2 such that $f_0 \xrightarrow{\lambda_i} f_i (i = 1, 2)$ and, further, get a particular solution $\tilde{f_{12}}$ from (3.87). Then a general solution of (3.87) is $f_{12} = \tilde{f_{12}} + c(t) f_0$ (where $c(t)$ is an arbitrary function of t). Finally we substitute f_{12} into (3.88) and (3.89). If $c(t)$ can be determined such that $c_1(t) = c_2(t) = 0$, i.e. f_{12} satisfies the relations

$$(D_x - \lambda_1 - \lambda_2) f_0 \cdot f_{12} + k(D_x + \lambda_1 - \lambda_2) f_1 \cdot f_2 = 0, \tag{3.90}$$

$$[2D_t + \frac{1}{4}D_x^3 + \frac{1}{4}D_y^3 - \frac{3}{4}(\lambda_1 + \lambda_2)D_x^2 + \frac{3}{2}(\lambda_1^2 + \lambda_2^2)D_x] f_0 \cdot f_{12}$$
$$+ k[\frac{3}{4}D_y^3 - \frac{3}{4}D_x^3 - \frac{9}{4}(\lambda_1 - \lambda_2)D_x^2 - \frac{3}{2}(\lambda_1 + \lambda_2)^2 D_x] f_1 \cdot f_2 = 0, \tag{3.91}$$

then f_{12} is a new solution of the NVN equation and $f_1 \xrightarrow{\lambda_2} f_{12}, f_2 \xrightarrow{\lambda_1} f_{12}$.

In examples 7-9, we have seen that these nonlinear superposition formulae have the unified form

$$D_x f_0 \cdot f_{12} = k D_x f_1 \cdot f_2. \tag{3.92}$$

Other examples exhibiting superposition formulae similar to (3.92) include a higher-order Ito equation[18], BKP [22], a coupled system derived from (5,1)-reduction of the two component BKP hierarchy [40] and so on. Finally, it is noted that a nonlinear superposition formula for the Kaup-Kupershmidt equation was given in [28], which is different from (3.54) and (3.92).

4 Superposition formulae for rational solutions

The question of rational solutions of integrable equations has been discussed widely in the literatue since the late 1970's, and different methods have been developed for the construction of rational solutions. In this section, we would like to seek rational solutions of integrable equations by using Bäcklund transformations and the corresponding nonlinear superposition formulae. As we shall see below, in most cases (especially in (1+1)dimensions) the rational solutions under consideration are linked by BTs without parameters, whereas in the literature soliton solutions are usually connected by BTs with parameters. In the example equations in the previous section, the existence of Bäcklund parameters facilitates the derivation of nonlinear superposition formulae. For rational solutions of integrable equations, because of the absence of Bäcklund parameters in most cases, it becomes more difficult to find nonlinear superposition formulae. In the following discussion, we shall give several example equations possessing superposition formulae for rational solutions.

Example 1. The KdV equation

In this example, we shall rederive the well-known rational solutions of the KdV equation (1.3) by using BT and nonlinear superposition formula. In order to obtain rational solutions of (1.3) or polynomial solutions of (1.4), it is enough to consider a special bilinear Bäcklund transformation for the bilinear KdV equation (1.4):

$$D_x^2 f \cdot f' = 0, \quad (D_t + D_x^3) f \cdot f' = 0. \tag{4.1}$$

We represent BT (4.1) symbolically by $f \longrightarrow f'$. It is evident that $f \longrightarrow f' <=> f' \longrightarrow f$. Note that in [2] Adler and Moser first discovered that polynomial solutions of (1.4) are generated by the formula

$$D_x f_{N-1} \cdot f_{N+1} = c f_N^2 \tag{4.2}$$

by considering the factorization of the Sturm-Liouville operator. In [1]Ablowitz and Satsuma recovered (1.4) by limiting the corresponding nonlinear superposition formula of the KdV soliton solutions. Summarizing the results, we have [23]

Proposition 1. Suppose $f_0 \neq 0$ and $f_1 \neq 0$ are two solutions of (1.4) being connected by (4.1). Then there exists an f_2 determined by

$$D_x f_1 \cdot f_2 = c f_0^2, \tag{4.3}$$

with c being a constant, such that f_2 is a new solution of (1.4) related to f_0 by BT (4.1), i.e. $f_0 \longrightarrow f_2$.

The above proposition easily allows us to obtain a series of polynomial solutions to KdV equation (1.4). For example, choose $f_0 = x, f_1 = 1$. By using Proposition 1, we can find that $f_2 = x^3 + 12t$ is a polynomial solution of (1.4) and $x \longrightarrow x^3 + 12t$. Next choose $f_0 = x^3 + 12t, f_1 = x$. Again by using Proposition 1, we can find that $f_2 = x^6 + 60x^3t - 720t^2$ is a polynomial solution of (1.4) and $x^3 + 12t \longrightarrow x^6 + 60x^3t - 720t^2$.

Example 2. The Boussinesq equation

Here we consider the following Boussinesq equation

$$3u_{tt} + 3(u^2)_{xx} + u_{xxxx} = 0. \tag{4.4}$$

By the dependent variable transformation $u = 2(\ln f)_{xx}$, equation (4.4) can be rewritten in the bilinear form

$$(3D_t^2 + D_x^4)f \cdot f = 0. \tag{4.5}$$

In order to obtain rational solutions of (4.4) or polynomial solutions of the bilinear Boussinesq equation (4.5), we just consider a special bilinear Bäcklund transformation for (4.5):

$$(D_t + D_x^2)f \cdot f' = 0, \quad (D_x^3 - 3D_xD_t)f \cdot f' = 0. \tag{4.6}$$

We represent BT (4.6) symbolically by $f \longrightarrow f'$. Concerning (4.6), we have the following superposition formula for polynomial solutions of (4.5)

Proposition 2. Suppose f_0, f_1 and f_{12} are three solutions of (4.5), $f_0 \longrightarrow f_1 \longrightarrow f_{12} \longrightarrow f_0$ and $f_0, f_1 \neq 0$. Then there exists an f_2 determined by

$$D_xf_1 \cdot f_2 = cf_0f_{12}, \tag{4.7}$$

with c being a constant, such that f_2 is a new solution of (4.5) and $f_0 \longrightarrow f_2 \longrightarrow f_{12}$.

For the details of the proof, see [23]. By using superposition formula (4.7), we can derive a series of polynomial solutions for (4.5).

Example 3. The KP equation

In order to obtain rational solutions of (3.6a) or polynomial solutions of the bilinear KP equation (3.6b) via superposition formulae, we first consider the following bilinear Bäcklund transformation for (3.6b) [31]:

$$(aD_y + D_x^2 + \lambda D_x)f \cdot f' = 0, \tag{4.8}$$

$$(D_t + D_x^3 - 3aD_xD_y - 3a\lambda D_y)f \cdot f' = 0, \tag{4.9}$$

where $a^2 = \frac{1}{3}\alpha$ and λ is an arbitrary constant. We represent BT (4.8) and (4.9) symbolically by $f \xrightarrow{\lambda} f'$. Concerning (3.6b), we have the following result [31]

Proposition 3. Let f_0 be a solution of (3.6b). Suppose that f_1 and f_2 are two solutions of (3.6b) such that $f_0 \xrightarrow{\lambda_i} f_i (i = 1, 2)$ and $f_j \neq 0 (j = 0, 1, 2)$. Then f_{12} as defined by

$$f_0f_{12} = c[D_x + \frac{1}{2}(\lambda_2 - \lambda_1)]f_1 \cdot f_2, \tag{4.10}$$

with c being a constant, is a new solution of (3.6b) and is related to f_1 and f_2 under the BT (4.8) and (4.9) with parameters λ_2 and λ_1 respectively.

It is noted that in [37], Satsuma and Ablowitz have obtained the so-called multi-lump solutions of the KP equation by taking limits of the corresponding soliton solutions. Here we use Proposition 3 to rederive these solutions. Setting $\theta_i = x - P_i y - \alpha P_i^2 t$, it is easily verified that $1 \longrightarrow \theta_i + \beta_i$ (where β_i is a constant). Using Proposition 3, we know that

$$\begin{aligned} f_{12} &= \frac{2}{a(P_1 - P_2)}[D_x + \frac{1}{2}(-aP_2 + aP_1)](\theta_1 + \beta_1)\cdot(\theta_2 + \beta_2) \\ &= \theta_1\theta_2 + (\beta_1 + \frac{2}{a(P_1 - P_2)})\theta_2 + (\beta_2 - \frac{2}{a(P_1 - P_2)})\theta_1 \\ &\quad + \beta_1\beta_2 + \frac{2(\beta_2 - \beta_1)}{a(P_1 - P_2)} \end{aligned}$$

is a solution of (3.6b). If $\alpha = -1, P_2 = P_1^*, \beta_1 = -2/a(P_1 - P_1^*)$ and $\beta_2 = 2/a(P_1 - P_1^*)$, then we can obtain so-called one-lump solution of (3.6b):

$$f_{12} = \theta_1\theta_1^* - \frac{12}{(P_1 - P_1^*)^2}.$$

Now we give another superposition formula to generate another series of polynomial solutions of (3.6b) [23].

Proposition 4. Suppose f_0, f_1 and f_{12} are three solutions of (3.6b), $f_0 \longrightarrow f_1 \longrightarrow f_{12} \longrightarrow f_0$ and $f_0, f_1 \neq 0$. Then there exists an f_2 determined by

$$D_x f_1 \cdot f_2 = c f_0 f_{12}, \tag{4.11}$$

with c being a constant, such that f_2 is a new solution of (3.6b) and $f_0 \longrightarrow f_2 \longrightarrow f_{12}$.

By using this result, we can derive another series of polynomial solutions of (3.6b).

Example 4. The Ito equation

In order to give a superposition formula to generate polynomial solutions of (3.59), we consider a special case of the BT (3.61), i.e.

$$D_x D_t f \cdot f' = 0, \quad (D_t + D_x^3) f \cdot f' = 0. \tag{4.12}$$

Representing (4.12) symbolically by $f \longrightarrow f'$, we have the following result:

Proposition 5. Let $f_0 \neq 0$ and $f_1 \neq 0$ be two solutions of (3.59) and let $f_0 \longrightarrow f_1$. Suppose that f_2 determined by

$$(D_x^3 - 2D_t) f_1 \cdot f_2 = c f_0^2, \tag{4.13}$$

with c being a constant, satisfies

$$D_t f_1 \cdot f_2 + \frac{1}{4} D_x^3 f_1 \cdot f_2 + \frac{3}{4} D_x^3 f_0 \cdot P = 0, \tag{4.14}$$

where P is determined by the relation

$$D_x f_0 \cdot P = D_x f_1 \cdot f_2. \tag{4.15}$$

Then f_2 is a new solution of (3.59), $f_0 \longrightarrow f_2$, and

$$D_t f_1 \cdot f_2 + D_t f_0 \cdot P = c_1(t) f_0^2. \tag{4.16}$$

$$D_t f_0 \cdot P + \frac{1}{4} D_x^3 f_0 \cdot P + \frac{3}{4} D_x^3 f_1 \cdot f_2 = c_2(t) f_0^2, \tag{4.17}$$

where $c_i(t)(i = 1, 2)$ is some suitable function of t. Furthermore, if we can choose a suitable P from (4.15) such that $c_1(t) = c_2(t) = 0$ in (4.16) and (4.17), then this P is also a new solution of (3.59) and $f_1 \longrightarrow P, f_2 \longrightarrow P$.

By using this result, we can derive a series of rational solutions of the Ito equation.

Example 5. The Ramani equation

The so-called Ramani equation is [35]

$$(D_x^6 - 5D_x^3 D_t - 5D_t^2) f \cdot f = 0. \tag{4.18}$$

A BT for equation (4.18) is given by [23]

$$(D_x^3 - D_t) f \cdot f' = 0, \quad (D_x^5 + 5D_x^2 D_t) f \cdot f' = 0. \tag{4.19}$$

We represent (4.19) symbolically by $f \longrightarrow f'$. Concerning BT (4.19), we have the following result [23]:

Proposition 6. Let $f_0 \neq 0$ and $f_1 \neq 0$ be two solutions of (4.18) and let $f_0 \longrightarrow f_1$. Suppose that f_2 determined by

$$(D_x^3 + 2D_t) f_1 \cdot f_2 = c f_0^2, \tag{4.20}$$

with c being a constant, satisfies

$$-D_t f_1 \cdot f_2 + \frac{1}{4} D_x^3 f_1 \cdot f_2 + \frac{3}{4} D_x^3 f_0 \cdot P = 0, \tag{4.21}$$

where P is determined by the relation

$$D_x f_0 \cdot P = D_x f_1 \cdot f_2. \tag{4.22}$$

Then we have

$$(D_x^3 - D_t) f_0 \cdot f_2 = 0, \tag{4.23}$$

$$\frac{1}{4} D_x^3 f_0 \cdot P + \frac{3}{4} D_x^3 f_1 \cdot f_2 - D_t f_0 \cdot P = c_1(t) f_0^2. \tag{4.24}$$

where $c_1(t)$ is some suitable function of t. Further, if f_2 and P satisfy

$$D_x^5 f_1\cdot f_2+20D_x^2D_t f_1\cdot f_2+15D_x^5 f_0\cdot P+60D_x^2D_t f_0\cdot P+60c_1(t)D_x^2 f_0\cdot f_0=0, \quad (4.25)$$

then f_2 is a solution of (4.18), $f_0 \longrightarrow f_2$, and we have

$$D_x^5 f_0\cdot P+20D_x^2D_t f_0\cdot P+15D_x^5 f_1\cdot f_2+60D_x^2D_t f_1\cdot f_2+20c_1(t)D_x^2 f_0\cdot f_0=c_2(t)f_0^2, \quad (4.26)$$

where $c_2(t)$ is some suitable function of t. Moreover, if $c_1(t)=c_2(t)=0$, then P is also a new solution of (4.18) and $f_1 \longrightarrow P, f_2 \longrightarrow P$.

By using this result, we can derive a series of rational solutions of the Ramani equation.

Besides the five examples above, we have also obtained superposition formulae for rational solutions of some other integrable equations, including the Sawada-Kotera equation, the BKP equation [23], the NVN equation [24] and so on.

5 Superposition formulae for some other particular solutions

In Sections 3 and 4, we have derived nonlinear superposition formulae for soliton solutions and rational solutions. In this section, we will show that Bäcklund transformations and the associated superposition formulae can also be utilized to generate some other particular solutions.

1) Positon-like solutions that are a mixture of exponentials and rational expression

For example, starting from superposition formula (2.12a) of the KdV equation, we have

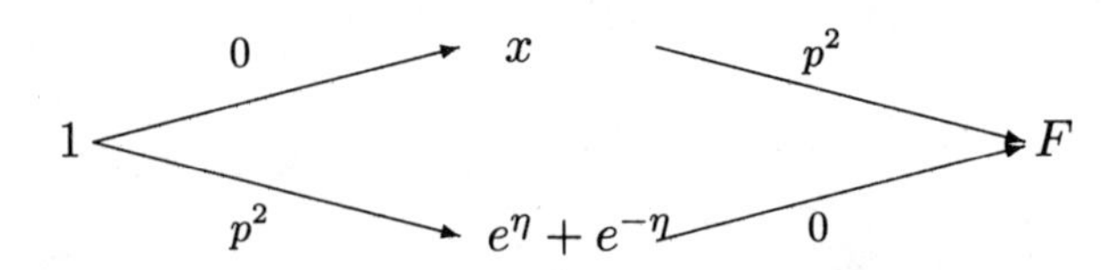

and

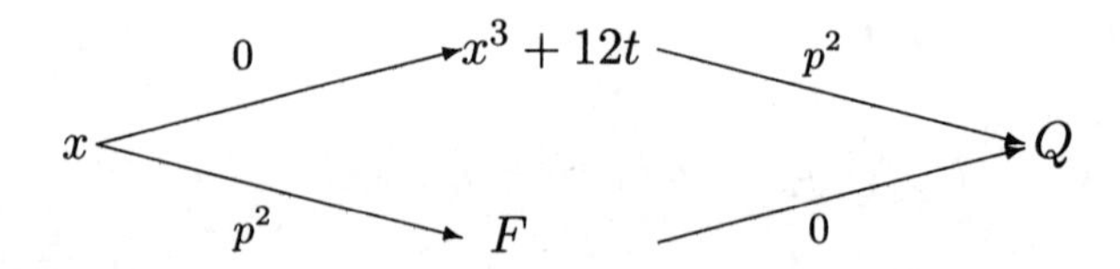

Therefore F and Q are solutions of (1.4), where $F \equiv (1-px)e^{\eta}+(1+px)e^{-\eta}$ $Q \equiv (3x-3px^2+p^2x^3+12p^2t)e^{\eta}+(3x+3px^2+p^2x^3+12p^2t)e^{-\eta}$ and $\eta = px-4p^3t+\eta^0$, with p and η^0 being constants.

2) Solutions expressed by special functions

Here we also give an example to illustrate that superposition formulae can be used to superpose solutions expressed by special functions. We consider the cylindrical KdV equation

$$u_t + 6uu_x + u_{xxx} + \frac{u}{2t} = 0. \tag{5.1}$$

By the dependent variable transformation $u = 2(\ln f)_{xx}$, equation (5.1) can be rewritten as

$$(D_xD_t + D_x^4 + \frac{1}{2t}\partial_x)f \cdot f = 0. \tag{5.2}$$

A bilinear BT is given by [10, 32]

$$(D_x^2 - \frac{x+x_1}{12t})f \cdot f' = 0, \tag{5.3}$$

$$(D_t + D_x^3 + \frac{x+x_1}{4t}D_x)f \cdot f' = 0, \tag{5.4}$$

where x_1 is an arbitrary constant parameter which physically corresponds to the initial position of the soliton. Nakamura has shown the following nonlinear superposition formula for the cKdV equation [32]

$$f_0 f_{12} = kt^{\frac{1}{2}} D_x f_1 \cdot f_2, \tag{5.5}$$

where k is a nonzero arbitrary constant. By using this result and starting from three known solutions of (5.2)

$$f_0 = 1 \qquad f_2 = t^{-\frac{1}{12}}\rho_1 Ai(z+z_i),$$
$$f_1 = (12)^{\frac{1}{3}}t^{-\frac{1}{12}}[\rho_1 Ai(z+z_i) + (x_{1'}-x_1)\frac{\pi}{\rho_1}Bi(z+z_i)],$$

where ρ_1 is an arbitrary constant, $z \equiv \frac{x}{(12t)^{\frac{1}{3}}}$, $z_i \equiv \frac{x_i}{(12t)^{\frac{1}{3}}}$ and Ai and Bi are the so-called Airy functions or two linearly independent solutions of the second ordinary differential equation

$$\frac{d^2}{dz^2}w(z) - zw(z) = 0, \tag{5.6}$$

we have

$$f_{12} = 1 + \frac{\rho_1^2}{(12t)^{\frac{1}{3}}}\int^{x} dz Ai(z+z_{1'})Ai(z+z_1). \tag{5.7}$$

Here we have chosen $k = 1/(x_{1'} - x_1)$. In the limit $x_{1'} \to x_1$, equation (5.7) reduces to

$$f_{12} = 1 + \frac{\rho_1^2}{(12t)^{\frac{1}{3}}} \int^{x} dz Ai^2(z + z_1). \tag{5.8}$$

Such solutions are called ripplons in [31]. Besides, Nakamura has given a generalized bilinear BT and its associated superposition formulae of the KP equation (3.2), consequently allowing one to produce a soliton-ripplon solution and etc. [31]

Acknowledgements

This work was supported in part by the National Natural Science Foundation of China 11331008, 11571358.

References

[1] Ablowitz M J, Satsuma J, Solitons and rational solutions of nonlinear evolution equations. *J. Math. Phys.* **19** (1978), no. 10, 2180–2186.

[2] Adler M, Moser J, On a class of polynomials connected with the Korteweg-de Vries equation. *Commun. Math. Phys.* **61** (1978), no. 1, 1–30.

[3] Athorne C, Fordy A, Integrable equations in (2+1) dimensions associated with symmetric and homogeneous spaces. *J. Math. Phys.* **28** (1987), no. 9, 2018–2024.

[4] Benjamin TB , Internal waves of permanent form in fluids of great depth, *J. Fluid Mech.* **29** (1967), 559–592.

[5] Bianchi L, Lezioni di geometria differenziale 2 (Pisa, 1902) 418

[6] Hirota R, *Prog. Theor. Phys.* **52** (1974), 1498–1512.

[7] Hirota R, Direct methods in soliton theory , in Solitons edited by R.K. Bullough and P.J. Caudrey (Springer, Berlin, 1980)

[8] Hirota R: Direct Methods in Soliton Theory (in Japanese), (Iwanami, 1992).

[9] Hirota R, Satsuma J, A simple structure of superposition formula of the Backlund transformation. *J. Phys. Soc. Japan* **45** (1978)(5), 1741–1750.

[10] Hirota R, The Backlund and inverse scattering transform of the K-dV equation with nonuniformities. *J. Phys. Soc. Japan* **46** (1979), no. 5, 1681–1682.

[11] Hirota R, Satsuma J, A variety of nonlinear network equations generated from the Backlund transformation for the Toda lattice. *Prog. Theor. Phys. Suppl.* **59** (1976), 64–100.

[12] Hirota R, Satsuma J, N-soliton solutions of model equations for shallow water waves. *J. Phys. Soc. Japan* **40** (1976), no. 2, 611–612.

[13] Hu X B, The permutability theorem for the Bäcklund transformation of the model equation for shallow water waves and the nonlinear superposition formula. (Chinese) *Sci. Exploration* **5** (1985)(4), 31–38.

[14] Hu X B, The higher-order KdV equation with a source and nonlinear superposition formula. *Chaos Solitons Fractals* **7** (1996)(2), 211–215.

[15] Hu X B, Li Y, Some results on the Caudrey-Dodd-Gibbon-Kotera-Sawada equation. *J. Phys. A* **24** (1991), no. 14, 3205–3212.

[16] Hu X B, Hirota-type equations, soliton solutions, Bäcklund transformations and conservation laws. *J. Partial Differential Equations* **3**(1990), no. 4, 87–95.

[17] Hu X B, Nonlinear superposition formula of the Novikov-Veselov equation. *J. Phys. A* **27** (1994), no. 4, 1331–1338.

[18] Hu X B, A Bäcklund transformation and nonlinear superposition formula of a higher-order Ito equation. *J. Phys. A* **26** (1993), no. 21, 5895–5903.

[19] Hu X B, Li Y, The Bäcklund transform for the DJKM equations and its nonlinear superposition formulas. (Chinese) *Acta Math. Sci.* (Chinese) **11** (1991)(2), 164–172.

[20] Hu X B, Li Y, Liu Q M, Nonlinear superposition formula of the Boussinesq hierarchy. *Acta Math. Appl. Sinica* (English Ser.) **9** (1993)1, 17–27.

[21] Hu X B, Li Y, Nonlinear superposition formulae of the Ito equation and a model equation for shallow water waves. *J. Phys. A* **24**(1991)9, 1979–1986.

[22] Hu X B, Li Y, A nonlinear superposition formula for the (1+2)-dimensional Caudrey-Dodd-Gibbon-Kotera-Sawada equation. (Chinese) *Gaoxiao Yingyong Shuxue Xuebao* **8** (1993), no. 1, Ser. A, 17–27.

[23] Hu X B, Rational solutions of integrable equations via nonlinear superposition formulae. *J. Phys. A* **30** (1997), no. 23, 8225–8240.

[24] Hu X B, Willox R, Some new exact solutions of the Novikov-Veselov equation. *J. Phys. A* **29** (1996), no. 15, 4589–4592.

[25] Ito M, An extension of nonlinear evolution equations of the K-dV (mK-dV) type to higher orders. *J. Phys. Soc. Japan* **49** (1980)(2), 771–778.

[26] Kadomtsev B B, Petviashvili V I, On the stability of solitary waves in weakly dispersive media, *Sov. Phys. Dokl.* **15** (1970), 539–541.

[27] Matsuno Y, Bilinear Transformation Method (New York: Academic), 1984.

[28] Musette M, Nonlinear superposition formulae of integrable partial differential equations by the singular manifold method, in Direct and Inverse Methods in Nonlinear Evolution Equations, Edited by R. Conte, F. Magri, M. Mussete, J. Satsuma, P. Winternitz Springer 2003.

[29] Nakamura, A, Bäcklund transform and conservation laws of the Benjamin-Ono equation, *J. Phys. Soc. Japan* **47** (1979)(4), 1335–1340.

[30] Nakamura, A, Chain of the Bäcklund transformation for the KdV equation, *J. Math. Phys.* **22** (1981)(8), 1608–1613.

[31] Nakamura A, Decay mode solution of the two-dimensional KdV equation and the generalized Backlund transformation.*J. Math. Phys.* **22** (1981), no. 11, 2456–2462.

[32] Nakamura A, Bäcklund transformation of the cylindrical KdV equation. *J. Phys. Soc. Japan* **49** (1980), no. 6, 2380–2386.

[33] Nizhnik L P, Integration of multidimensional nonlinear equations by the method of the inverse problem, *Sov. Phys. Dokl.* **25** (1980), 706–708.

[34] Ono H, Algebraic solitary waves in stratified fluids, *J. Phys. Soc. Japan* **39** (1975)(5), 1082–1091.

[35] Ramani A, Inverse scattering, ordinary differential equations of Painlev-type, and Hirota's bilinear formalism, in 4th Int. Conf. Collective Phenomena, ed. JL Lebowitz (New York, Academy of Sciences,1981), 54–67.

[36] Rogers C, Shadwick, W F, Bäcklund transformations and their applications. Mathematics in Science and Engineering, 161. Academic Press, Inc.

[37] Satsuma J, Ablowitz M J, Two-dimensional lumps in nonlinear dispersive systems. *J. Math. Phys.* **20** (1979), no. 7, 1496–1503.

[38] Satsuma J, Kaup D J, A Bäcklund transformation for a higher order Korteweg-de Vries equation. *J. Phys. Soc. Japan* **43** (1977)(2), 692–726.

[39] Sawada K, Kotera T, *Prog. Theor. Phys.* **51** (1974), 1355.

[40] Tam H W,Hu X B, Wang D L, Two integrable coupled nonlinear systems.*J. Phys. Soc. Japan* **68** (1999), no. 2, 369–379.

[41] Veselov A P, Novikov S P, *Sov. Math. Dokl.* **30** (1984), 705.

B5. Matrix solutions for equations of the AKNS system

Cornelia Schiebold [a,b]

[a] *MOD, Mid Sweden University, S-851 70 Sundsvall, Sweden*
[b] *Instytut Matematyki, Uniwersytet Jana Kochanowskiego w Kielcach, Poland*

Abstract

We will start with a detailed introduction to an operator theoretic approach for the study of soliton equations going back to Marchenko, with a certain focus on projection techniques and reduction. The specific goal will be to construct $m \times n$-matrix valued solutions for the AKNS system. The applications will mainly concern the mKdV, in comparison with existing results for the NLS. First we discuss solutions that are degenerate in the sense that particle-like waves with coinciding velocity are superposed. Then a complete asymptotic description is given for multiple pole solutions, including wave packets of weakly bound breathers. Finally the collision of vector solitons is studied.

1 Introduction

From the beginning of soliton theory, it was apparent that finding explicit expressions for solution classes can be hard, even in basic cases. Already the N-soliton, i.e. the nonlinear superposition of N solitary waves, leads to serious difficulties if one tries to find its formula by hand. This was a motivation to look for systematic approaches to the construction of solutions. Celebrated examples are the inverse scattering method, Hirota's bilinear method, the dressing method and the $\bar{\partial}$-method. Needless to say, these achievements are much more than recipes to find solution formulas and open new conceptual perspectives on integrable systems.

The present work builds on an operator theoretic approach which can be traced back to pioneering work of Marchenko [24]. The rough idea is to lift both a soliton equation and (a parameter depending family of) its solutions to the level of equations for functions with values in a (usually noncommutative) operator algebra. Working in the opposite direction, one tries then to derive solution formulas for the original equation by projection techniques. It is fair to say that our viewpoint owes much to a seminal idea of B. Carl, who paved the way for the application of advanced tools from Banach space geometry. In this spirit, the first goals in the late 1990s were to apply the new tools to problems like countable superposition of solitons [4, 14] (a topic initiated in [19]), to construct solutions with slow decay for $x \to \pm\infty$ [8, 13], and to study solution classes with a particularly transparent link between algebraic input data and qualitative properties of the solutions [35, 36], see also [15] for a survey on this period. For research of related spirit, we want to draw attention to [18, 34, 47], and references therein.

Most of this was done with classical soliton equations including lattices in mind. On the other hand, the method itself does not attribute an exceptional position to the scalar case. Obviously, its essence is to proceed via matrix or operator versions of soliton equations. A less immediate aspect is to find projection techniques that are appropriate for deriving solutions to matrix equations. Once this issue is settled (see Section 6), one arrives at a directed graph, where the vertices are operator algebras and arrows indicate projection techniques transforming solutions from one level to another. A considerable part of the present work emphasises this point of view. The underlying idea is to trade structural complexity for higher dimension. More precisely, one often succeeds in encoding the characteristic properties of a solution class in a new operator valued parameter, permitting the use of algebra and functional analysis in the further investigation.

Several aspects of our method become more conceptual when presented in the framework of the AKNS system, a family of integro-differential equations depending on a functional parameter f_0. Its extension to the noncommutative setting was discovered by Bauhardt and Pöppe [6], and then adapted by the author to a rigorous use of operator theory in [37, 39, 41]. Recall that the AKNS system first associates to every f_0 a pair of equations in two independent unknown functions R and Q, from which classical soliton equations are obtained after appropriate reduction. Our philosophy is to work as long as possible on the level before reduction, for the following reasons:

(i) The way the functions R and Q are organised in products motivates allowing them to map according to $R : F \to E$ and $Q : E \to F$. This will give us the crucial freedom needed to extract solutions to matrix equations for which the values are not square matrices, like the vector-mKdV $r_t + r_{xxx} + 3(rr^{\mathrm{T}}r_x + r_x r^{\mathrm{T}} r) = 0$, the 2-soliton of which we will study in Section 11.

(ii) When reducing the AKNS system (with f_0 given), there may be several relevant options. An example is the two noncommutative versions of mKdV considered in this article. In such cases, applying projection techniques before reduction will allow us to construct solution formulas for several noncommutative versions at one stroke.

In the applications, we will restrict ourselves to finite-dimensional techniques and focus on the relation between algebraic properties of the input data and qualitative features of the resulting solutions. To get an impression of how functional analysis can be exploited by our method, we refer the reader to [38, 41].

This article is expository in part. In what follows, we sketch its organisation and indicate where the reader may find novel elements. Section 2 gives a first introduction to our method, restricted to the two recurrent examples studied in this article, the mKdV and the NLS. In Section 3 we present the noncommutative (nc) AKNS system and derive a solution, which may be viewed as a generalisation of the 1-soliton depending on two independent operator parameters. Sections 4 – 6 are concerned with deriving solution formulas by projection techniques. We start from scalar solution formulas and progress in several steps towards the final result on matrix solutions of the AKNS system in Theorem 4. This part may also be used as a guide to the very concise treatment of the AKNS system in [39, 41]. After basic information on reductions of the AKNS system in Section 7, we collect in Section 8 technical preparations for the applications part, striving for formulas as transparent as possible in order to facilitate the comparison with the literature.

The Sections 9 – 11 contain our applications. In Section 9 we first recall how N-solitons and breathers are realised in our method. Then we continue with a discussion of solutions which are degenerate in the sense that waves with coinciding velocity are nonlinearly superposed. We make some new observations in the case of the mKdV. Section 10 is dedicated to multiple pole solutions (MPS's). After reviewing the history of the topic for the NLS, we arrive at an asymptotic description of MPS's of the mKdV, including wave packets of weakly bound breathers. This result is published here for the first time. In Section 11 we turn to solutions of matrix equations. The main result is a fairly complete asymptotic description of the 2-soliton of the vector mKdV mentioned in (i). For illustration we provide some computer plots produced with MATHEMATICA.

2 An operator approach to integrable systems

Roughly speaking, our general strategy consists of two steps: Starting from a given integrable system, the first step is to find a translation to the operator-level. Here it is crucial to find compatible translations of the system and a particular solution of the system. The aim of the second step is to use the solution on the operator-level to generate solutions to the original system. We will refer to this process (perhaps somewhat misleadingly) as projection technique, see Fig. 1.

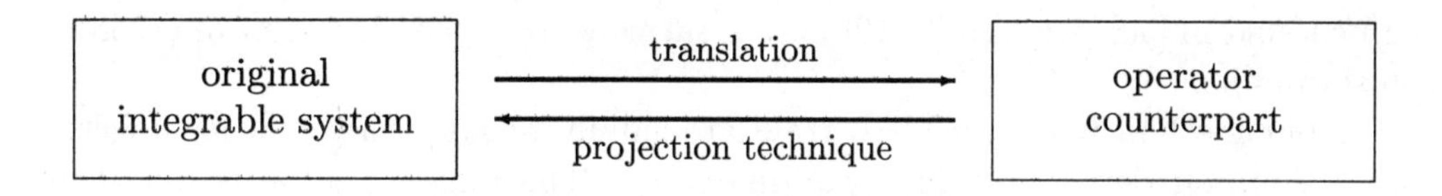

Figure 1. The general strategy.

In the present section we give an introduction to our method on the first level of generality, namely for individual soliton equations.

mKdV. Consider the modified Korteweg-de Vries equation (mKdV)

$$r_t + r_{xxx} + 6r^2 r_x = 0, \tag{2.1}$$

where one is usually interested in real-valued solutions $r : \mathbb{R}^2 \to \mathbb{R}$. Our point of departure will be the well-known 1-soliton solution of the mKdV

$$r(x,t) = \alpha \operatorname{sech}\bigl(\alpha x - \alpha^3 t + \varphi\bigr) \tag{2.2}$$

with $\alpha, \varphi \in \mathbb{R}$, $\alpha \neq 0$. We see that $r(x, 0)$ is bell-shaped with center located at $-\varphi/\alpha$. In other words, the initial-shift φ encodes where the wave is located at $t = 0$. Note that the parameter α determines both height and velocity of the soliton, the velocity being α^2 and the height α. If $\alpha < 0$, the wave is often called antisoliton.

For later use we observe that (2.2) can be rewritten as

$$r = 2\alpha \frac{\ell}{1+\ell^2} \tag{2.3}$$

with $\ell = \ell(x,t) = e^{\alpha x - \alpha^3 t + \varphi}$.

NLS. The Nonlinear Schrödinger equation (NLS),

$$\mathrm{i} r_t = r_{xx} + 2r|r|^2, \tag{2.4}$$

where one now considers complex-valued solutions $r : \mathbb{R}^2 \to \mathbb{C}$, has the 1-soliton solution

$$r(x,t) = \mathrm{Re}(\alpha)\, e^{\mathrm{i}\left(\mathrm{Im}(\alpha)x + \left(\mathrm{Im}^2(\alpha) - \mathrm{Re}^2(\alpha)\right)t + \psi\right)}\ \mathrm{sech}\Big(\mathrm{Re}(\alpha)\big(x + 2\mathrm{Im}(\alpha)t\big) + \varphi\Big) \tag{2.5}$$

with two shift parameters $\varphi, \psi \in \mathbb{R}$. Again the form of the solution is determined by α. In contrast to the mKdV, its height $|\mathrm{Re}(\alpha)|$ and its velocity $-2\mathrm{Im}(\alpha)$ can be chosen independently.

2.1 Operator counterparts of integrable systems

There is no canonic way to associate an operator counterpart to a given integrable system [43]. On the contrary, cases where a scalar equation permits several different translations are of particular interest, indicating phenomena which become invisible in the commutative setting. For nc KdV-type equations, a classification can be found in [30], see also [9, 10] for a treatment from the viewpoint of Bäcklund transformations.

For our method it is crucial to translate simultaneously the equation together with a solution depending on a paramater α. The aim is that this parameter becomes operator-valued after translation.

mKdV. For the mKdV, the translation to the operator-level described below was established in [14], see also [11] for an extension to the mKdV hierarchy. Let E be a Banach space. By $\mathcal{L}(E)$ we denote the Banach algebra of bounded linear operators from E to E. Consider the noncommutative mKdV (nc mKdV)

$$R_t + R_{xxx} + 3\{R^2, R_x\} = 0, \tag{2.6}$$

for an operator-valued function $R : \mathbb{R}^2 \to \mathcal{L}(E)$. A compatible translation of the 1-soliton solution, written as in (2.3), to the operator-level is obtained as follows: Let $A \in \mathcal{L}(E)$, and let $L : \mathbb{R}^2 \to \mathcal{L}(E)$ be an operator-valued function which solves the linear base equations $L_x = AL$, $L_t = -A^3 L$. Assume in addition that $I_E + L^2$ is invertible on an open set $\Omega \subset \mathbb{R}^2$. Then

$$R = (I_E + L^2)^{-1}(AL + LA)$$

is a solution of the nc mKdV (2.6) on Ω. The proof is a straightforward verification, where familiarity with basic operator identities is helpful.

NLS. For the NLS it is not obvious how to translate the complex conjugate to the operator level. A convenient way to postpone this question is to consider the NLS system

$$\begin{aligned} \mathrm{i}r_t &= r_{xx} - 2r^2 q, && (2.7a)\\ -\mathrm{i}q_t &= q_{xx} - 2q^2 r, && (2.7b) \end{aligned}$$

where $\bar{r}$ is replaced by an independent function q. Then (2.4) is recovered by the reduction $q = -\bar{r}$.

In the scalar case both r and q have values in $\mathbb{C}$, but on the operator-level there is a priori no reason to assume that the values of R and Q are endomorphisms on the same Banach space. More generally, we introduce the nc NLS system

$$\begin{aligned} \mathrm{i}R_t &= R_{xx} - 2RQR, && (2.8a)\\ -\mathrm{i}Q_t &= Q_{xx} - 2QRQ, && (2.8b) \end{aligned}$$

viewed as a system for operator functions $R : \mathbb{R}^2 \to \mathcal{L}(F, E)$ and $Q : \mathbb{R}^2 \to \mathcal{L}(E, F)$, where E, F are (possibly different) Banach spaces.

The following translation of the 1-soliton solution to the operator-level was given in [38]. Let E, F be a Banach spaces and $A \in \mathcal{L}(E)$, $B \in \mathcal{L}(B)$. Assume that the operator-valued functions $L : \mathbb{R}^2 \to \mathcal{L}(F, E)$, $M : \mathbb{R}^2 \to \mathcal{L}(E, F)$ solve the base equations $L_x = AL$, $L_t = -\mathrm{i}A^2 L$, $M_x = BM$, $M_t = \mathrm{i}B^2 M$ and that $I_E - LM$, $I_F - ML$ are invertible on an open set $\Omega \subset \mathbb{R}^2$. Then

$$\begin{aligned} R &= (I_E - LM)^{-1}(AL + LB),\\ Q &= (I_F - ML)^{-1}(BM + MA), \end{aligned}$$

is a solution of the nc NLS system (2.8) on Ω.

2.2 Projection techniques

The next task is to use the operator translation from the previous step to produce solutions to the original scalar equation.

To this end, a reasonable idea is to apply an appropriate functional τ to the solution R of the operator equation. To guarantee that $r = \tau(R)$ provides a solution to the original equation, τ needs to be multiplicative, at least in a certain sense.

The difficulty is that even on the smallest operator ideal $\mathcal{F}$ of finite rank operators there is no multiplicative functional. As in [4], we restrict to particular operator-solutions:

1.) We choose R one-dimensional. More precisely, we consider R of the form $a \otimes c$ with some constant nonzero functional $a \in E'$ and a vector-function $c : \mathbb{R}^2 \to E$. Recall that $a \otimes c$ operates according to $(a \otimes c)(v) = \langle v, a \rangle c$, where $\langle v, a \rangle = a(v)$ for $v \in E$.

2.) As functional τ we take the trace tr on $\mathcal{F}$.

The motivation of this choice lies in the following observation: Let E be a Banach space and $a \in E'$. Then, for all $c, \hat{c} \in E$, we have

$$\mathrm{tr}\big((a \otimes c)(a \otimes \hat{c})\big) = \mathrm{tr}\big(\langle \hat{c}, a\rangle (a \otimes c)\big) = \langle \hat{c}, a\rangle \ \mathrm{tr}\big(a \otimes c\big) = \mathrm{tr}(a \otimes c)\ \mathrm{tr}(a \otimes \hat{c}).$$

Looking at the mKdV as an example, this yields easily that $\mathrm{tr}(a \otimes c)$ solves (2.1) as soon as $a \otimes c$, where a is constant, solves (2.6).

To guarantee 1.) we can exploit the theory of elementary operators.

Theorem 1 ([3, 16]). *Let $A \in \mathcal{L}(E)$, $B \in \mathcal{L}(F)$, and let $\mathcal{A}$ be a p-Banach ideal, $0 < p \leq 1$. Then the spectrum of the operator $\Phi_{A,B} : \mathcal{A}(F, E) \to \mathcal{A}(F, E)$ defined by*

$$\Phi_{A,B}(X) = AX + XB$$

is independent of the underlying p-Banach ideal $\mathcal{A}$, and it holds

$$\mathrm{spec}(\Phi_{A,B}) = \mathrm{spec}(A) + \mathrm{spec}(B).$$

Consequently, provided that $0 \notin \mathrm{spec}(A) + \mathrm{spec}(B)$, the Sylvester equation $AX + XB = C$ with bounded right-hand side C has the unique solution $X = \Phi_{A,B}^{-1}(C)$. Furthermore, if $C \in \mathcal{A}(F, E)$, then also $\Phi_{A,B}^{-1}(C) \in \mathcal{A}(F, E)$.

mKdV. To implement the projection technique explained above for the mKdV (2.1), observe first that the base equations are satisfied by $L(x,t) = \exp(Ax - A^3 t)C$ with $C \in \mathcal{L}(E)$. Hence

$$R = \big(I_E + (e^{Ax - A^3 t} C)^2\big)^{-1} e^{Ax - A^3 t}(AC + CA),$$

and we can guarantee that R is of the right form by choosing $C = \Phi_{A,A}^{-1}(a \otimes c)$ for some $a \in E', c \in E$. This is possible if $0 \notin \mathrm{spec}(A) + \mathrm{spec}(A)$.

2.3 Solution formulas

Often our solution formulas can be improved using the theory of traces and determinants on quasi-Banach ideals [31], see also [14] for a concise summary of results relevant in the present context.

mKdV. In the particular case that E is finite-dimensional (and A a square matrix), the following argument helps us to rewrite the solution formula in a way that the computation of the matrix inverse of $I_E + L^2$ is no longer needed. Observe first that

$$\mathrm{tr}\big((I_E \pm \mathrm{i}L)^{-1} LA\big) = \mathrm{tr}\big(L(I_E \pm \mathrm{i}L)^{-1} A\big) = \mathrm{tr}\big((I_E \pm \mathrm{i}L)^{-1} AL\big),$$

where we have used successively that $(I_E \pm \mathrm{i}L)^{-1}$ and L commute, as well as the trace property $\mathrm{tr}(T_1 T_2) = \mathrm{tr}(T_2 T_1)$. Hence, making use of the base equation $L_x = AL$, we find

$$\mathrm{tr}\Big((I_E \pm \mathrm{i}L)^{-1}(AL + LA)\Big) \ = \ \mathrm{tr}\big((I_E \pm \mathrm{i}L)^{-1} AL\big) + \mathrm{tr}\big((I_E \pm \mathrm{i}L)^{-1} LA\big)$$

$$= \mp 2\mathrm{i}\,\mathrm{tr}\Big((I_E \pm \mathrm{i}L)^{-1}(\pm \mathrm{i}L)_x\Big)$$
$$= \mp 2\mathrm{i}\partial_x \log\det(I_E \pm \mathrm{i}L).$$

Finally, since $(I_E+L^2)^{-1}(AL+LA) = \frac{1}{2}\big((I_E+\mathrm{i}L)^{-1}+(I_E-\mathrm{i}L)^{-1}\big)(AL+LA)$, we can rewrite the solution formula as

$$r = \mathrm{tr}\Big((I_E+L^2)^{-1}(AL+LA)\Big) = -\mathrm{i}\frac{\partial}{\partial x}\log\frac{\det(I_E+\mathrm{i}L)}{\det(I_E-\mathrm{i}L)}.$$

Note that this argument can be made rigorous also in the general (infinite-dimensional) case. The only obstacle is that while $R = (I_E+L^2)^{-1}(AL+LA)$ is one-dimensional, and hence $\mathrm{tr}(R)$ is defined, this does not necessarily hold for $(I_E \pm \mathrm{i}L)^{-1}AL$.

Recall $L = \exp(Ax - A^3t)\Phi_{A,A}^{-1}(a\otimes c)$. Since $a \otimes c \in \mathcal{A}(E)$ for any p-Banach ideal $\mathcal{A}$, $\Phi_{A,A}^{-1}(a\otimes c) \in \mathcal{A}(E)$ by Theorem 1, and hence also $(I_E \pm \mathrm{i}L)^{-1}AL \in \mathcal{A}(E)$. Note furthermore that all traces coincide on the finite-dimensional operators. Hence, working with a p-Banach ideal $\mathcal{A}$ admitting a continuous trace τ (and corresponding determinant δ), we have

$$r = \mathrm{tr}\Big((I_E+L^2)^{-1}(AL+LA)\Big) = \tau\Big((I_E+L^2)^{-1}(AL+LA)\Big),$$

and can continue with the argument from the finite-dimensional case, replacing tr and det by τ and δ.

We sum up the above discussion in the following proposition. Note that the obtained solution formula depends on an operator-parameter $A \in \mathcal{L}(E)$.

Proposition 1. *Let $\mathcal{A}$ be a quasi-Banach operator ideal with a continuous determinant δ, let E be a Banach space, and $A \in \mathcal{L}(E)$ with $0 \notin \mathrm{spec}(A)+\mathrm{spec}(A)$. Furthermore, let $a \in E', c \in E$, and define $L(x,t) = \exp(Ax - A^3t)\,\Phi_{A,A}^{-1}(a\otimes c)$. Then*

$$r = -\mathrm{i}\frac{\partial}{\partial x}\log\frac{\delta(I_E+\mathrm{i}L)}{\delta(I_E-\mathrm{i}L)}$$

is a solution of the mKdV (2.1) *on* $\Omega = \{(x,t) \in \mathbb{R}^2 \mid \delta\big(I_E \pm \mathrm{i}L(x,t)\big) \neq 0\}$.

Remark 1. The reader may wonder how general solution formulas like the one in the above proposition are. In [8] it was shown for several scalar equations (including KdV and NLS) that solutions accessible by the inverse scattering method (with rapidly decreasing initial data) can be realised by choosing appropriate operator parameters. Moreover solutions with more general behaviour for $x \to \pm\infty$ can be constructed by using semi-group theory [13].

3 The nc AKNS system

The scalar mKdV and NLS are among the first equations for which integrability was studied [48, 50]. One of the key insights in early soliton theory was that both belong to the AKNS system [1], a family of integrable equations depending on a functional parameter. In this section we give an introduction to the nc AKNS system. This research was initiated in [6] and further developed in [37, 39].

Let E, F be Banach spaces. For given (non-zero) polynomials $f, g \in \mathbb{C}[x]$, the nc AKNS system reads

$$g(\mathcal{T}_{R,Q})\begin{pmatrix} R_t \\ Q_t \end{pmatrix} = f(\mathcal{T}_{R,Q})\begin{pmatrix} R \\ -Q \end{pmatrix} \tag{3.1}$$

where $R : \mathbb{R}^2 \to \mathcal{L}(F, E)$ and $Q : \mathbb{R}^2 \to \mathcal{L}(E, F)$ are unknown operator-valued functions, and $\mathcal{T}_{R,Q}$ denotes the operator

$$\mathcal{T}_{R,Q}\begin{pmatrix} U \\ V \end{pmatrix} = \begin{pmatrix} U_x - \Big(R\int_{-\infty}^{x}(QU + VR)d\xi + \int_{-\infty}^{x}(UQ + RV)\ d\xi R\Big) \\ -V_x + \Big(Q\int_{-\infty}^{x}(UQ + RV)d\xi + \int_{-\infty}^{x}(QU + VR)\ d\xi Q\Big) \end{pmatrix} \tag{3.2}$$

with $U : \mathbb{R}^2 \to \mathcal{L}(F, E)$, $V : \mathbb{R}^2 \to \mathcal{L}(E, F)$. Whereas $\mathcal{T}_{R,Q}$ is linear in U and V, a composite expression like $\mathcal{T}_{R,Q}(R, Q)$ is not.

Actually, R and Q need not be defined on all of $\mathbb{R}^2$. To ensure existence of the expressions in (3.1), we require that R, Q are sufficiently smooth on an open strip $\mathbb{R} \times (t_1, t_2)$ and behave sufficiently well for $x \to -\infty$. More precisely, this means for the operator-function R: For some sufficiently large n_0 (depending on the degrees of f, g), one requires that the t-dependent expressions

$$\|R(\cdot, t)\|_{\kappa,\lambda} = \sup_{x\in\mathbb{R}}(1 + |x|^{\lambda})\Big\|\frac{\partial^{\kappa}}{\partial x^{\kappa}}R(x, t)\Big\|$$

are finite for $\kappa, \lambda \leq n_0$. Moreover, the map $t \mapsto R(\cdot, t)$ has to be continuous with respect to $\|\cdot\|_{\kappa,\lambda}$, $\kappa, \lambda \leq n_0$. Finally, the t-derivative $t \mapsto R_t(\cdot, t)$ is required to have the same properties, and we make analogous assumptions on Q.

The main point is that the integrals in (3.1) can be evaluated as Bochner integrals and, by the assumption on continuity in t, both sides of (3.1) are at least continuous. However, these aspects are somewhat peripheral to our purposes, because existence will always be evident in our main results, and the solutions will even decay faster than polynomially. Therefore we do not make an effort to state optimal assumptions on smoothness and decay.

It is instructive to see how the nc NLS system (2.8) is contained in (3.1). As in the scalar case we choose $f(z) = -\mathrm{i}z^2$, $g(z) = 1$. Computing

$$\begin{aligned} \mathcal{T}_{R,Q}\begin{pmatrix} R \\ -Q \end{pmatrix} &= \begin{pmatrix} R_x - \Big(R\int_{-\infty}^{x}(QR - QR)\ d\xi + \int_{-\infty}^{x}(RQ - RQ)\ d\xi\ R\Big) \\ Q_x + \Big(Q\int_{-\infty}^{x}(RQ - RQ)\ d\xi + \int_{-\infty}^{x}(QR - QR)\ d\xi\ Q\Big) \end{pmatrix} \\ &= \begin{pmatrix} R_x \\ Q_x \end{pmatrix} \end{aligned}$$

and

$$(\mathcal{T}_{R,Q})^2\begin{pmatrix} R \\ -Q \end{pmatrix} = \mathcal{T}_{R,Q}\begin{pmatrix} R_x \\ Q_x \end{pmatrix}$$

$$= \begin{pmatrix} R_{xx} - \left(R\int_{-\infty}^{x}(QR)_x\,d\xi + \int_{-\infty}^{x}(RQ)_x\,d\xi\,R\right) \\ -Q_{xx} + \left(Q\int_{-\infty}^{x}(RQ)_x\,d\xi + \int_{-\infty}^{x}(QR)_x\,d\xi\,Q\right) \end{pmatrix}$$

$$= \begin{pmatrix} R_{xx} - 2RQR \\ -Q_{xx} + 2QRQ \end{pmatrix},$$

we find that the system associated with (3.1) becomes (2.8). Similarly, the choice $f(z) = -z^3$, $g(z) = 1$ leads to the nc mKdV system

$$R_t + R_{xxx} - 3(RQR_x + R_xQR) = 0, \tag{3.3a}$$

$$Q_t + Q_{xxx} - 3(QRQ_x + Q_xRQ) = 0, \tag{3.3b}$$

from which (2.6) is obtained via the reduction $Q = -R$. Note that this condition and the fact that RQ has to be defined forces R to be an endomorphism. In Section 7 we will study another reduction which leads to an nc mKdV for functions whose values need not be endomorphisms.

Let f_0 be the rational function f/g. The following theorem was proven in [37, 39].

Theorem 2. *Let E, F be Banach spaces and $A \in \mathcal{L}(E)$, $B \in \mathcal{L}(F)$ be constant operators such that* $\operatorname{spec}(A) \cup \operatorname{spec}(-B)$ *is contained in the domain where f_0 is holomorphic.*

Assume that the operator-valued functions $L : \mathbb{R} \times (t_1, t_2) \to \mathcal{L}(F, E)$, $M : \mathbb{R} \times (t_1, t_2) \to \mathcal{L}(E, F)$ are sufficiently smooth and behave sufficiently well for $x \to -\infty$, and solve the base equations

$$L_x = AL, \qquad L_t = f_0(A)L,$$

$$M_x = BM \qquad M_t = -f_0(-B)M.$$

Assume furthermore that $I_E - LM, I_F - ML$ are invertible on $\mathbb{R} \times (t_1, t_2)$. Then a solution of the nc AKNS system (3.1) *is given by*

$$R = (I_E - LM)^{-1}(AL + LB),$$

$$Q = (I_F - ML)^{-1}(BM + MA).$$

We should point out that the assumption on the behaviour of L and M for $x \to -\infty$ in the above theorem is included to guarantee the existence of the integrals in (3.2), and can be dropped in the treatment of important individual systems which can be written without integrals, like the nc mKdV system (3.3) and nc NLS system (2.8). Similarly the spectral condition is automatically satisfied in those examples.

4 Solution formulas for the AKNS system

In constructing a solution formula for the mKdV (2.1) from a given solution $R : \mathbb{R}^2 \to \mathcal{L}(E)$ of the nc mKdV (2.6) in Section 2, the main idea was to write R in the form $R = a \otimes c$ with $a \in E'$, $c : \mathbb{R}^2 \to E$, and apply the trace tr to R. Here the key observation is

$$R = a \otimes c \text{ solves (2.6)} \qquad \Longrightarrow \qquad r = \operatorname{tr}(R) = \langle c, a\rangle \text{ solves (2.1).}$$

Observe that taking the trace of the one-dimensional operator $R(x,t)$ means evaluation of the (constant) functional a on the vector $c(x,t)$.

Let us now consider the nc NLS system (2.8). Recall that the values of $R:\mathbb{R}^2 \to \mathcal{L}(F,E)$, $Q:\mathbb{R}^2 \to \mathcal{L}(E,F)$ need not be endomorphisms preventing us from directly taking the trace. We will overcome the difficulty by using cross-evaluation. We show

Proposition 2. *Let $R:\mathbb{R}^2 \to \mathcal{L}(F,E)$, $Q:\mathbb{R}^2 \to \mathcal{L}(E,F)$. Assume that we can write R, Q in the form $R = b \otimes c$, $Q = a \otimes d$ with constant, non-zero functionals $a \in E'$, $b \in F'$ and vector-functions $c:\mathbb{R}^2 \to E$, $d:\mathbb{R}^2 \to F$. Then*

$$\left\{ \begin{array}{c} R = b\otimes c \\ Q = a \otimes d \end{array} \right\} \text{ solves (2.8)} \quad \Longrightarrow \quad \left\{ \begin{array}{c} r = \langle c, a\rangle \\ q = \langle d, b\rangle \end{array} \right\} \text{ solves (2.7).}$$

Proof. Let us start from (2.8a). Since b is constant, we have $R_t = b \otimes c_t$, $R_{xx} = b \otimes c_{xx}$, and by successive evaluation we get

$$\begin{aligned} (RQR)(v) &= RQ\big((b\otimes c)(v)\big) = \langle v,b\rangle (RQ)(c) = \langle v,b\rangle\, R\big((a\otimes d)(c)\big) \\ &= \langle v,b\rangle\langle c,a\rangle R(d) = \langle v,b\rangle\langle c,a\rangle\big((b\otimes c)(d)\big) = \langle v,b\rangle\langle c,a\rangle\langle d,b\rangle c \end{aligned}$$

for all $v \in F$. This shows that $RQR = rq(b\otimes c)$, and (2.8a) becomes $ib\otimes c_t = b \otimes c_{xx} - 2rq(b\otimes c)$. In other words, the operator

$$T = b \otimes \big(- \mathrm{i} c_t + c_{xx} - 2rq\, c\big)$$

vanishes identically, and hence also the functional aT on F. This means that

$$0 = (aT)(v) = \langle Tv, a\rangle = \langle v,b\rangle\langle -\mathrm{i}c_t + c_{xx} - 2rq\, c, a\rangle$$

for all $v \in F$. Since $b \neq 0$, we have $\langle v_0, b\rangle \neq 0$ for some $v_0 \in F$, and we can conclude that $0 = \langle -\mathrm{i}c_t + c_{xx} - 2rq\, c, a\rangle = -\mathrm{i}\langle c,a\rangle_t + \langle c,a\rangle_{xx} - 2rq\,\langle c,a\rangle = -\mathrm{i}r_t + r_{xx} - 2r^2q$, i.e. that (2.7a) is satisfied. ■

Next we indicate how to exploit the idea of Proposition 2 for the AKNS system.

Starting with the solutions R, Q derived in Theorem 2, we first observe that $L = \hat{L}C$, $M = \hat{M}D$, where $\hat{L}(x,t) = \exp(Ax + f_0(A)t)$, $\hat{M}(x,t) = \exp(Bx - f_0(-B)t)$, satisfy the base equations. Assume $0 \notin \operatorname{spec}(A) + \operatorname{spec}(B)$, and define

$$C = \Phi_{A,B}^{-1}(b\otimes c), \quad D = \Phi_{B,A}^{-1}(a \otimes d),$$

for (non-zero) $a \in E'$, $b \in F'$, $c \in E$, $d \in F$. Then

$$R = (I_E - LM)^{-1}\hat{L}(AC + CB) = (I_E - LM)^{-1}\hat{L}(b\otimes c) = b \otimes \big((I_E - LM)^{-1}\hat{L}c\big),$$

and, analogously, $Q = a \otimes \big((I_F - ML)^{-1}\hat{M}d\big)$. Now we can apply cross-evaluation like in the proof of Proposition 2, and obtain the following solution formula for the AKNS system

$$r = \langle (I_E - LM)^{-1}\hat{L}c, a\rangle,$$

$$q = \langle (I_F - ML)^{-1}\hat{M}d, b\rangle.$$

In order to avoid the computation of inverses, we rewrite r and q in terms of a determinant δ on some quasi-Banach ideal $\mathcal{A}$. Exemplarily, we give the argument for r. Let τ be the trace corresponding to δ according to the Trace-determinant theorem [31]. Then

$$\begin{aligned}
r &= \langle (I_E - LM)^{-1}\hat{L}c, a\rangle \\
&= \mathrm{tr}\Big(a \otimes \big((I_E - LM)^{-1}\hat{L}c\big)\Big) \\
&= \mathrm{tr}\Big((I_E - LM)^{-1}\hat{L}(a \otimes c)\Big) \\
&= \tau\Big((I_E - LM)^{-1}\hat{L}(a \otimes c)\Big) \\
&= 1 - \delta\Big(I_E - (I_E - LM)^{-1}\hat{L}(a \otimes c)\Big) \\
&= 1 - \delta\Big((I_E - LM)^{-1}\big(I_E - LM - \hat{L}(a \otimes c)\big)\Big) \\
&= 1 - \frac{\delta\big(I_E - LM - \hat{L}(a \otimes c)\big)}{\delta(I_E - LM)} \\
&= 1 - \delta\begin{pmatrix} I_E - \hat{L}(a \otimes c) & L \\ M & I_F \end{pmatrix} \Big/ \delta\begin{pmatrix} I_E & L \\ M & I_F \end{pmatrix}.
\end{aligned}$$

In the argument above we have used the fact that, when restricted to the finite-dimensional operators, τ coincides with the usual trace tr, and that $\delta(I_E + a \otimes c) = 1 + \tau(a \otimes c) = 1 + \langle c, a\rangle$. For the transition to generalized determinants of double size we refer to [37, Section 2.4.1].

The following theorem [37, 41] summarizes the discussion of this section.

Theorem 3. *Let $\mathcal{A}$ be a quasi-Banach ideal admitting a continuous determinant δ, and E, F Banach spaces. Let $A \in \mathcal{L}(E)$, $B \in \mathcal{L}(F)$ such that* $\mathrm{spec}(A) \cup \mathrm{spec}(-B)$ *is contained in the domain where f_0 is holomorphic, and* $0 \notin \mathrm{spec}(A) + \mathrm{spec}(B)$. *Assume furthermore that* $\exp(Ax)$, $\exp(Bx)$ *behave sufficiently well for $x \to -\infty$.*

For non-zero $a \in E'$, $b \in F'$, and $c \in E$, $d \in F$, define

$$\begin{aligned}
L &= \hat{L}\ \Phi_{A,B}^{-1}(b \otimes c), & \hat{L}(x,t) &= \exp\big(Ax + f_0(A)t\big), \\
M &= \hat{M}\ \Phi_{B,A}^{-1}(a \otimes d), & \hat{M}(x,t) &= \exp\big(Bx - f_0(-B)t\big).
\end{aligned}$$

Then a solution of the scalar AKNS system, on strips $\mathbb{R} \times (t_1, t_2)$ on which the denominator does not vanish, is given by

$$\begin{aligned}
r &= 1 - \delta\begin{pmatrix} I_E - \hat{L}(a \otimes c) & L \\ M & I_F \end{pmatrix} \Big/ \delta\begin{pmatrix} I_E & L \\ M & I_F \end{pmatrix}, \\
q &= 1 - \delta\begin{pmatrix} I_E & L \\ M & I_F - \hat{M}(b \otimes d) \end{pmatrix} \Big/ \delta\begin{pmatrix} I_E & L \\ M & I_F \end{pmatrix}.
\end{aligned}$$

5 Projection techniques revisited

The last two decades saw a strongly increasing interest in integrable matrix and vector equations. The reasons may be diverse, certainly ideas from physics but also the fact that crucial methods from soliton theory, like Lax pairs or operator approaches, make a transition to matrix equations appear natural. For more information and references we refer to the monographs [2, 21].

So far we were concerned with projection techniques to *scalar* equations and got as far as to the AKNS system. Now we want to start the construction of solutions to *matrix* equations, and go back to the most accessible case, the nc mKdV (2.6), considered as an equation for a function R which takes its values in the $\mathsf{d}\times\mathsf{d}$-matrices $\mathcal{M}_{\mathsf{d},\mathsf{d}}(\mathbb{C})$ over $\mathbb{C}$. We pursue an analogous strategy as for the scalar case.

The key step to is to find a (continuous, linear) map σ, with values in the $\mathsf{d}\times\mathsf{d}$-matrices, which is multiplicative at least on a certain class of operators. To this end, we introduce the Banach algebra

$$\mathcal{S}_{a_1,\ldots,a_\mathsf{d}}(E) = \Big\{ \sum_{j=1}^{\mathsf{d}} a_j \otimes c_j \;\Big|\; c_1,\ldots,c_\mathsf{d} \in E \Big\},$$

for fixed, linearly independent functionals $a_1,\ldots,a_\mathsf{d} \in E'$. Obviously, $\mathcal{S}_{a_1,\ldots,a_\mathsf{d}}(E)$ is a left ideal in $\mathcal{L}(E)$.

Note that the representation $\sum_{j=1}^{\mathsf{d}} a_j \otimes c_j$ in terms of the given functionals a_j is unique. Hence the map $\sigma : \mathcal{S}_{a_1,\ldots,a_\mathsf{d}}(E) \to \mathcal{M}_{\mathsf{d},\mathsf{d}}(\mathbb{C})$ given by

$$\sigma\Big(\sum_{j=1}^{\mathsf{d}} a_j \otimes c_j\Big) = \Big(\langle c_k, a_j\rangle\Big)_{j,k=1}^{\mathsf{d}}$$

is well-defined.

Lemma 1. *Let E be a Banach space and $a_1,\ldots,a_\mathsf{d} \in E'$ linearly independent. Then, for all $c_1,\ldots,c_\mathsf{d}$, $\hat{c}_1,\ldots,\hat{c}_\mathsf{d} \in E$, we have*

$$\sigma\Big(\Big(\sum_{j=1}^{\mathsf{d}} a_j \otimes c_j\Big)\Big(\sum_{j=1}^{\mathsf{d}} a_j \otimes \hat{c}_j\Big)\Big) = \sigma\Big(\sum_{j=1}^{\mathsf{d}} a_j \otimes c_j\Big)\, \sigma\Big(\sum_{j=1}^{\mathsf{d}} a_j \otimes \hat{c}_j\Big)$$

Proof.

$$\begin{aligned}
\sigma\Big(\Big(\sum_{j=1}^{\mathsf{d}} a_j \otimes c_j\Big)\Big(\sum_{j=1}^{\mathsf{d}} a_j \otimes \hat{c}_j\Big)\Big) &= \sigma\Big(\sum_{j,k=1}^{\mathsf{d}} (a_k \otimes c_k)(a_j \otimes \hat{c}_j)\Big) \\
= \sigma\Big(\sum_{j,k=1}^{\mathsf{d}} \langle \hat{c}_j, a_k\rangle a_j \otimes c_k\Big) &= \sigma\Big(\sum_{j=1}^{\mathsf{d}} a_j \otimes \Big(\sum_{k=1}^{\mathsf{d}} \langle \hat{c}_j, a_k\rangle c_k\Big)\Big) \\
= \Big(\langle \sum_{k=1}^{\mathsf{d}} \langle \hat{c}_j, a_k\rangle c_k, a_i\rangle\Big)_{i,j=1}^{\mathsf{d}} &= \Big(\sum_{k=1}^{\mathsf{d}} \langle \hat{c}_j, a_k\rangle \langle c_k, a_i\rangle\Big)_{i,j=1}^{\mathsf{d}}
\end{aligned}$$

$$= \Big(\langle c_k, a_i\rangle\Big)_{i,k=1}^{\mathsf{d}} \Big(\langle \hat{c}_j, a_k\rangle\Big)_{k,j=1}^{\mathsf{d}} = \sigma\Big(\sum_{j=1}^{\mathsf{d}} a_j \otimes c_j\Big)\, \sigma\Big(\sum_{j=1}^{\mathsf{d}} a_j \otimes \hat{c}_j\Big)$$

■

This motivates us to use the map σ for projection instead of the trace, and to assume that the solution we start from is d-dimensional instead of one-dimensional. Generalising what we have done in the scalar case, we arrive at the following solution formula.

Proposition 3. *Let E be a Banach space and $A \in \mathcal{L}(E)$ with $0 \notin \operatorname{spec}(A)+\operatorname{spec}(A)$. Furthermore, let $a_1, \dots, a_{\mathsf{d}} \in E'$ be linearly independent and $c_1, \dots, c_{\mathsf{d}} \in E$. Define*

$$L = \hat{L}\,\Phi_{A,A}^{-1}\Big(\sum_{j=1}^{\mathsf{d}} a_j \otimes c_j\Big), \qquad \hat{L}(x,t) = \exp(Ax - A^3 t).$$

Then, on $\Omega = \{(x,t) \in \mathbb{R}^2 \mid I_E + L^2(x,t) \text{ is invertible}\}$, a solution of the nc mKdV (2.6) with values in the space of $\mathsf{d} \times \mathsf{d}$-matrices over $\mathbb{C}$ is given by

$$R = \Big(\langle (I_E + L^2)^{-1}\hat{L}c_k, a_j\rangle\Big)_{j,k=1}^{\mathsf{d}}.$$

Similarly as the solution r in the scalar case, the entries of the matrix R in the above proposition can be rewritten in terms of a continuous determinant δ on a quasi-Banach ideal. Analogously as in the argument before Theorem 3 we get

$$\begin{aligned} \langle (I_E + L^2)^{-1}\hat{L}c_k, a_j\rangle &= 1 - \frac{\delta(I_E + L^2 - \hat{L}(a_j \otimes c_k))}{\delta(I_E + L^2)} \\ &= 1 - \delta\begin{pmatrix} I_E - \hat{L}(a_j \otimes c_k) & \mathrm{i}L \\ \mathrm{i}L & I_E \end{pmatrix} \Big/ \delta\begin{pmatrix} I_E & \mathrm{i}L \\ \mathrm{i}L & I_E \end{pmatrix}. \end{aligned}$$

Note also that since the operator $I_E + L^2$ is invertible if and only if $0 \neq \delta(I_E + L^2) = \delta(I_E + \mathrm{i}L)\delta(I_E - \mathrm{i}L)$, the set Ω in the above proposition is equal to

$$\Omega = \{(x,t) \in \mathbb{R}^2 \mid \delta\big(I_E \pm \mathrm{i}L(x,t)\big) \neq 0\}.$$

6 Matrix- and vector-AKNS systems

We have seen in Section 4 how the nc AKNS system naturally leads us to consider systems for functions with values in $\mathcal{L}(F,E)$ and $\mathcal{L}(E,F)$. Hence we expect that combining this with the projection techniques from Sections 4 and 5 allows us to construct solutions values in matrices of arbitrary size. For the sake of concreteness, we start with a detailed treatment of the nc NLS system (2.8).

Proposition 4. *Let $R : \mathbb{R}^2 \to \mathcal{L}(F,E)$, $Q : \mathbb{R}^2 \to \mathcal{L}(E,F)$. Assume that we can write R, Q in the form $R = \sum_{j=1}^{\mathsf{d}_1} b_j \otimes c_j$, $Q = \sum_{j=1}^{\mathsf{d}_2} a_j \otimes d_j$ with constant, linearly* *independent functionals* $a_{j_2} \in E'$ *and* $b_{j_1} \in F'$, *and vector-functions* $c_{j_1} : \mathbb{R}^2 \to E$ *and* $d_{j_2} : \mathbb{R}^2 \to F$ $(j_1 = 1, \ldots, \mathsf{d}_1,\ j_2 = 1, \ldots, \mathsf{d}_2)$. *Then*

$$\left\{ \begin{array}{l} R = \sum\limits_{j=1}^{\mathsf{d}_1} b_j \otimes c_j \\ Q = \sum\limits_{j=1}^{\mathsf{d}_2} a_j \otimes d_j \end{array} \right\} \text{ solves (2.8)} \quad \Longrightarrow \quad \left\{ \begin{array}{l} r = \Big(\langle c_k, a_j \rangle \Big)_{\substack{j=1,\ldots,\mathsf{d}_2 \\ k=1,\ldots,\mathsf{d}_1}} \\ q = \Big(\langle d_k, b_j \rangle \Big)_{\substack{j=1,\ldots,\mathsf{d}_1 \\ k=1,\ldots,\mathsf{d}_2}} \end{array} \right\} \text{ solves (2.8).}$$

Note that in Proposition 4 the nc NLS system (2.8) is viewed as a system for functions R, Q with values in $\mathcal{L}(F, E)$, $\mathcal{L}(E, F)$ on the one hand, and as a system for functions r, q with values in $\mathcal{M}_{\mathsf{d}_2,\mathsf{d}_1}(\mathbb{C})$, $\mathcal{M}_{\mathsf{d}_1,\mathsf{d}_2}(\mathbb{C})$ on the other.

Proof. Let us assume that (2.8a) is satisfied for R, Q. Since the functionals b_j, $j = 1, \ldots, \mathsf{d}_1$, are constant, we have $R_t = \sum_{j=1}^{\mathsf{d}_1} b_j \otimes c_{j,t}$, $R_{xx} = \sum_{j=1}^{\mathsf{d}_1} b_j \otimes c_{j,xx}$, and, by successive evaluation,

$$\begin{aligned} (RQR)(x) &= RQ\Big(\big(\sum_{k=1}^{\mathsf{d}_1} b_k \otimes c_k \big)(x) \Big) = R \sum_{k=1}^{\mathsf{d}_1} \langle x, b_k \rangle \big(Q(c_k) \big) \\ &= R \sum_{k=1}^{\mathsf{d}_1} \langle x, b_k \rangle \sum_{k_2=1}^{\mathsf{d}_2} \big(a_{k_2} \otimes d_{k_2} \big)(c_k) = \sum_{k=1}^{\mathsf{d}_1} \langle x, b_k \rangle \sum_{k_2=1}^{\mathsf{d}_2} \langle c_k, a_{k_2} \rangle \, \big(R(d_{k_2}) \big) \\ &= \sum_{k=1}^{\mathsf{d}_1} \langle x, b_k \rangle \sum_{k_2=1}^{\mathsf{d}_2} \langle c_k, a_{k_2} \rangle \sum_{k_1=1}^{\mathsf{d}_1} \big(b_{k_1} \otimes c_{k_1} \big)(d_{k_2}) \\ &= \sum_{k=1}^{\mathsf{d}_1} \langle x, b_k \rangle \sum_{k_2=1}^{\mathsf{d}_2} \langle c_k, a_{k_2} \rangle \sum_{k_1=1}^{\mathsf{d}_1} \langle d_{k_2}, b_{k_1} \rangle \, c_{k_1} \\ &= \Big(\sum_{k=1}^{\mathsf{d}_1} b_k \otimes \Big(\sum_{k_1=1}^{\mathsf{d}_1} \sum_{k_2=1}^{\mathsf{d}_2} \langle d_{k_2}, b_{k_1} \rangle \langle c_k, a_{k_2} \rangle \, c_{k_1} \Big) \Big)(x) \end{aligned}$$

for all $x \in F$. Observe next that $\sum_{k_2=1}^{\mathsf{d}_2} \langle d_{k_2}, b_{k_1} \rangle \langle c_k, a_{k_2} \rangle = (qr)_{k_1 k}$, the (k_1, k)-th entry of the $\mathsf{d}_1 \times \mathsf{d}_1$-matrix qr. Hence

$$RQR = \sum_{k=1}^{\mathsf{d}_1} b_k \otimes \Big(\sum_{k_1=1}^{\mathsf{d}_1} (qr)_{k_1 k} \, c_{k_1} \Big).$$

This shows that if (2.8a) is satisfied, the operator

$$T = \sum_{l=1}^{\mathsf{d}_1} b_l \otimes \Big(-\mathrm{i} c_{l,t} + c_{l,xx} - 2 \sum_{k_1=1}^{\mathsf{d}_1} (qr)_{k_1 l} \, c_{k_1} \Big)$$

vanishes identically. Since $b_1, \ldots, b_{\mathsf{d}_1}$ are linearly independent functionals on F, there are vectors $x_1, \ldots, x_{\mathsf{d}_1} \in F$ such that $b_l(x_k) = \langle x_k, b_l \rangle = \delta_{lk}$ (the latter denoting the Krondecker delta), and we can conclude that

$$0 = \langle T(x_k), a_j \rangle = \sum_{l=1}^{\mathsf{d}_1} \langle x_k, b_l \rangle \big\langle -\mathrm{i} c_{l,t} + c_{l,xx} - 2 \sum_{k_1=1}^{\mathsf{d}_1} (qr)_{k_1 l} \, c_{k_1}, a_j \big\rangle$$

$$
\begin{aligned}
&= \Big\langle -\mathrm{i}c_{k,t} + c_{k,xx} - 2\sum_{k_1=1}^{\mathsf{d}_1} (qr)_{k_1 k}\, c_{k_1}, a_j \Big\rangle \\
&= -\mathrm{i}\langle c_k, a_j\rangle_t + \langle c_k, a_j\rangle_{xx} - 2\sum_{k_1=1}^{\mathsf{d}_1} (qr)_{k_1 k}\, \langle c_{k_1}, a_j\rangle \\
&= \big(-\mathrm{i}r_t + r_{xx} - 2rqr\big)_{jk}
\end{aligned}
$$

for all $k = 1,\dots,\mathsf{d}_1$, $j = 1,\dots,\mathsf{d}_2$, i.e. (2.8a) is satisfied for r, q. ∎

Combining Theorem 2 with an extension of the proof of Proposition 4 one can prove the following generalisation of Theorem 3 to matrix- and vector-AKNS systems [41].

Theorem 4. *Let E, F be Banach spaces, and $A \in \mathcal{L}(E)$, $B \in \mathcal{L}(F)$ such that* $\mathrm{spec}(A) \cup \mathrm{spec}(-B)$ *is contained in the domain where f_0 is holomorphic, and* $0 \notin \mathrm{spec}(A) + \mathrm{spec}(B)$. *Assume in addition that* $\exp(Ax)$, $\exp(Bx)$ *behave sufficiently well for $x \to -\infty$.*

Let $a_{j_2} \in E'$, $b_{j_1} \in F'$ be linearly independent, and $c_{j_1} \in E$, $d_{j_2} \in F$ for $j_1 = 1,\dots,\mathsf{d}_1$, $j_2 = 1,\dots,\mathsf{d}_2$. Define

$$
L = \hat{L}\, \Phi_{A,B}^{-1}\Big(\sum_{j=1}^{\mathsf{d}_1} b_j \otimes c_j\Big), \qquad \hat{L}(x,t) = \exp\big(Ax + f_0(A)t\big),
$$

$$
M = \hat{M}\, \Phi_{B,A}^{-1}\Big(\sum_{j=1}^{\mathsf{d}_2} a_j \otimes d_j\Big), \qquad \hat{M}(x,t) = \exp\big(Bx - f_0(-B)t\big).
$$

Then, on strips $\mathbb{R} \times (t_1, t_2)$ on which the inverses exist, a solution r (with values in the space of $\mathsf{d}_2 \times \mathsf{d}_1$-matrices), q (with values in the space of $\mathsf{d}_1 \times \mathsf{d}_2$-matrices) of the matrix-AKNS system (3.1) *is given by*

$$
\begin{aligned}
r &= \Big(\langle (I_E - LM)^{-1}\hat{L}c_k, a_j\rangle\Big)_{\substack{j=1,\dots,\mathsf{d}_2\\ k=1,\dots,\mathsf{d}_1}}, \\
q &= \Big(\langle (I_F - ML)^{-1}\hat{M}d_k, b_j\rangle\Big)_{\substack{j=1,\dots,\mathsf{d}_1\\ k=1,\dots,\mathsf{d}_2}}.
\end{aligned}
$$

Remark 2. Note that assuming directly the existence of solutions of

$$
AX + XB = \sum_{j=1}^{\mathsf{d}_1} b_j \otimes c_j, \qquad BX + XA = \sum_{j=1}^{\mathsf{d}_2} a_j \otimes d_j,
$$

the assumption $0 \notin \mathrm{spec}(A) + \mathrm{spec}(B)$ can be dropped. We merely keep it for the simplicity of the exposition.

To avoid the computation of inverses, we may rewrite r and q in terms of a continuous determinant δ on a quasi-Banach ideal $\mathcal{A}$. Using a similar argument as in the end of Section 5, we obtain

$$
r = \left(1 - \delta\begin{pmatrix} I_E - \hat{L}(a_j \otimes c_k) & L \\ M & I_F \end{pmatrix} \Big/ \delta\begin{pmatrix} I_E & L \\ M & I_F \end{pmatrix}\right)_{\substack{j=1,\dots,\mathsf{d}_2\\ k=1,\dots,\mathsf{d}_1}}
$$

$$q = \left(1 - \delta\begin{pmatrix} I_E & L \\ M & I_F - \hat{M}(b_j \otimes d_k)\end{pmatrix} \Big/ \delta\begin{pmatrix} I_E & L \\ M & I_F\end{pmatrix}\right)_{\substack{j=1,\dots,d_1 \\ k=1,\dots,d_2}}$$

Note that invertibility of $I_E - LM$ (and $I_F - ML$) is equivalent to $\delta\begin{pmatrix} I_E & L \\ M & I_F\end{pmatrix} \neq 0$.

Remark 3. Since solutions of the nc mKdV (2.6) are obtained for $f_0(z) = -z^3$ after the reduction $R = -Q$, Proposition 3 is essentially a corollary of Theorem 4. However Theorem 4 gives much more on matrix mKdV's, see in Section 7.

Remark 4. For a unifying point of view on projection techniques (treating the AKNS system as a single matrix equation), we refer to [39, 41].

7 Reduction

For the scalar AKNS system in general even soliton-like solutions may explode in finite time (see [1], p.278). Equations with globally regular solitons can be obtained by appropriate reductions. In the sequel we will focus on the following two reductions for the nc AKNS system (3.1) with values in the $\mathsf{d}_2 \times \mathsf{d}_1$-matrices.

$\mathbb{C}$-reduction: Assume that, for $z \in \mathbb{C}$, $\overline{f(z)} = -\epsilon f(-\overline{z})$ and $\overline{g(z)} = \epsilon g(-\overline{z})$ with $\epsilon = \pm 1$. Then the $\mathbb{C}$-reduced AKNS system reads

$$g(\mathcal{T}_{r,q})\begin{pmatrix} r_t \\ q_t \end{pmatrix} = f(\mathcal{T}_{r,q})\begin{pmatrix} r \\ -q \end{pmatrix}, \tag{7.1a}$$

$$q = -r^{\mathrm{H}}, \tag{7.1b}$$

where r^{H} denotes the conjugate transpose of r.

$\mathbb{R}$-reduction: Assume that, for $z \in \mathbb{C}$, $f(z) = -\epsilon f(-z)$ and $g(z) = \epsilon g(-z)$ with $\epsilon = \pm 1$. Then the $\mathbb{R}$-reduced AKNS system reads

$$g(\mathcal{T}_{r,q})\begin{pmatrix} r_t \\ q_t \end{pmatrix} = f(\mathcal{T}_{r,q})\begin{pmatrix} r \\ -q \end{pmatrix}, \tag{7.2a}$$

$$q = -r^{\mathrm{T}}. \tag{7.2b}$$

Since one is mainly interested in real solutions in this context, we assume in addition that $f_0 = f/g$ is real in the sense that $f_0(x) \in \mathbb{R}$ for all $x \in \mathbb{R}$ where f_0 is defined.

Note that we consider both (7.1), (7.2) as a single equation for the matrix-function r (with values in the $\mathsf{d}_2 \times \mathsf{d}_1$-matrices). Here an explanatory remark is in order: For the $\mathbb{C}$-reduction one evaluates powers of $\mathcal{T}_{r,-r^{\mathrm{H}}}$ on $(r_t, -r_t^{\mathrm{H}})^{\mathrm{T}}$ respectively $(r, r^{\mathrm{H}})^{\mathrm{T}}$. Observing that

$$\mathcal{T}_{r,-r^{\mathrm{H}}}\begin{pmatrix} u \\ \pm u^{\mathrm{H}} \end{pmatrix} = \begin{pmatrix} v \\ \mp v^{\mathrm{H}} \end{pmatrix}$$

and taking into account what the reduction assumption implies for the coefficients of f, g, it is straightforward to verify that the two equations in (7.1a) are equivalent. Note also that for the meromorphic function $f_0 = f/g$, which plays a central role in the solution formulas in Theorem 4, the reduction conditions on f, g imply $\overline{f_0(z)} = -f_0(-\overline{z})$ for all $z \in \mathbb{C}$ where f_0 is holomorphic. Analogous observations apply to the $\mathbb{R}$-reduction.

To illustrate the $\mathbb{C}$-reduction, consider the nc NLS system (2.8). We have $f(z) = -\mathrm{i}z^2$, $g(z) = 1$, which clearly satisfy the assumptions on f, g with $\epsilon = 1$, and (7.1) becomes the matrix NLS

$$\mathrm{i}r_t = r_{xx} + 2rr^{\mathrm{H}}r. \tag{7.3}$$

In the scalar case this is the *focusing* NLS (2.4), for which the 1-soliton (2.5) is regular. In contrast, the similar looking reduction $q = r^{\mathrm{H}}$ leads to the *defocusing* NLS $\mathrm{i}r_t = r_{xx} - 2r|r|^2$ in the scalar case, where the 1-soliton

$$r(x,t) = -\mathrm{Re}(\alpha)\ e^{\mathrm{i}\left(\mathrm{Im}(\alpha)x + \left(\mathrm{Im}^2(\alpha) - \mathrm{Re}^2(\alpha)\right)t + \psi\right)}\ \mathrm{csch}\Big(\mathrm{Re}(\alpha)\big(x + 2\mathrm{Im}(\alpha)t\big) + \varphi\Big),$$

csch denoting the hyperbolic cosecant, is singular.

As an illustration for the $\mathbb{R}$-reduction, we look at the nc mKdV system (3.3). Here $f(z) = -z^3$, $g(z) = 1$, satisfying the assumptions on f, g with $\epsilon = 1$, and (7.2) becomes the matrix mKdV

$$r_t + r_{xxx} + 3(rr^{\mathrm{T}}r_x + r_x r^{\mathrm{T}} r) = 0. \tag{7.4}$$

In the vector case ($\mathsf{d}_1 = 1$, $\mathsf{d}_2 = \mathsf{d}$) we study soliton collisions for (7.4) in Section 11. Note also that (7.4) is already the second variant of a matrix mKdV we meet after (2.6). In [10] the reader finds much more on nc KdV-type equations.

8 The finite-dimensional case

Here we collect technical preparations for the applications in Sections 9 – 11. From now on we restrict to the finite-dimensional case $E = \mathbb{C}^n$, $F = \mathbb{C}^m$. For a reader, who wants to get to the applications as quickly as possible, we summarise the main results of the present section:

- In the finite-dimensional setting we can explicitly restate the solution formulas from Theorem 4 in terms of elementary linear algebra, see Subsection 8.1.

- The solution formula in Proposition 5 depends on two matrices $A \in \mathcal{M}_{n,n}(\mathbb{C})$, $B \in \mathcal{M}_{m,m}(\mathbb{C})$ of possibly different size. In Subsection 8.2 we explain why only their Jordan normal form is relevant.

- In Subsections 8.3 – 8.4 we turn to the $\mathbb{C}$- and $\mathbb{R}$-reduced AKNS system and explain by which choices the reductions can be achieved.

8.1 Reformulation of the solution formulas in terms of elementary linear algebra

Let us briefly recall the transition from the general to the finite-dimensional case. With respect to the standard basis, a linear operator $S : \mathbb{C}^n \to \mathbb{C}^n$ is uniquely expressed by a matrix $[S]$ satisfying $S(x) = [S] \cdot x$. Any functional β on $\mathbb{C}^n$ is of the form $\beta : x \mapsto b^{\mathrm{T}} x$ for some $b \in \mathbb{C}^n$, where b^{T} denotes the transpose of b, and we can identify β with b. Since $(\beta \otimes c)(x) = \beta(x)c = (b^{\mathrm{T}}x)c = (cb^{\mathrm{T}})x$, we have $[\beta \otimes c] = cb^{\mathrm{T}}$. Similarly, d-dimensional operators can be written as

$$\Big[\sum_{k=1}^{\mathsf{d}} \beta_k \otimes c_k\Big] = \sum_{k=1}^{\mathsf{d}} c_k b_k^{\mathrm{T}} = \begin{pmatrix} c_1 \dots c_{\mathsf{d}} \end{pmatrix} \begin{pmatrix} b_1^{\mathrm{T}} \\ \vdots \\ b_{\mathsf{d}}^{\mathrm{T}} \end{pmatrix}.$$

We need another elementary observation.

Lemma 2. *Let R be a $n \times n$-matrix and let c_{j_1}, $j_1 = 1, \dots, \mathsf{d}_1$, and a_{j_2}, $j_2 = 1, \dots \mathsf{d}_2$, be n-vectors over $\mathbb{C}$. Then*

$$\Big(\langle Rc_{j_1}, a_{j_2}\rangle\Big)_{\substack{j_2=1,\dots,\mathsf{d}_2 \\ j_1=1,\dots \mathsf{d}_1}} = \begin{pmatrix} a_1^{\mathrm{T}} \\ \vdots \\ a_{\mathsf{d}_2}^{\mathrm{T}} \end{pmatrix} R \begin{pmatrix} c_1 \dots c_{\mathsf{d}_1} \end{pmatrix}.$$

Thus we can reformulate the solution formulas in Theorem 4.

Proposition 5. *Let $A \in \mathcal{M}_{n,n}(\mathbb{C})$, $B \in \mathcal{M}_{m,m}(\mathbb{C})$ such that* $\operatorname{spec}(A) \cup \operatorname{spec}(-B)$ *is contained in the domain where f_0 is holomorphic, and* $0 \notin \operatorname{spec}(A) + \operatorname{spec}(B)$. *Assume in addition that* $\exp(Ax)$, $\exp(Bx)$ *behave sufficiently well for $x \to -\infty$.*

Let $a_{j_2} \in \mathbb{C}^n$, $b_{j_1} \in \mathbb{C}^m$ be linearly independent, and $c_{j_1} \in \mathbb{C}^n$, $d_{j_2} \in \mathbb{C}^m$ for $j_1 = 1, \dots, \mathsf{d}_1$, $j_2 = 1, \dots, \mathsf{d}_2$.

Set $L = \hat{L}\, C$, $M = \hat{M}\, D$, where $\hat{L}(x,t) = \exp\big(Ax + f_0(A)t\big)$, $\hat{M}(x,t) = \exp\big(Bx - f_0(-B)t\big)$, and where C, D denote the (unique) solutions of the respective Sylvester equations

$$AX + XB = \begin{pmatrix} c_1 \dots c_{\mathsf{d}_1} \end{pmatrix} \begin{pmatrix} b_1^{\mathrm{T}} \\ \vdots \\ b_{\mathsf{d}_1}^{\mathrm{T}} \end{pmatrix}, \qquad BX + XA = \begin{pmatrix} d_1 \dots d_{\mathsf{d}_2} \end{pmatrix} \begin{pmatrix} a_1^{\mathrm{T}} \\ \vdots \\ a_{\mathsf{d}_2}^{\mathrm{T}} \end{pmatrix}.$$

Then, on strips $\mathbb{R} \times (t_1, t_2)$ on which the inverses exist, a solution r (with values in the space of $\mathsf{d}_2 \times \mathsf{d}_1$-matrices), q (with values in the space of $\mathsf{d}_1 \times \mathsf{d}_2$-matrices) of the matrix-AKNS system (3.1) *is given by*

$$r = \begin{pmatrix} a_1^{\mathrm{T}} \\ \vdots \\ a_{\mathsf{d}_2}^{\mathrm{T}} \end{pmatrix} (I_n - LM)^{-1} \hat{L} \begin{pmatrix} c_1 \dots c_{\mathsf{d}_1} \end{pmatrix},$$

$$q = \begin{pmatrix} b_1^{\mathrm{T}} \\ \vdots \\ b_{\mathsf{d}_1}^{\mathrm{T}} \end{pmatrix} (I_m - ML)^{-1} \hat{M} \begin{pmatrix} d_1 \dots d_{\mathsf{d}_2} \end{pmatrix}.$$

In the end of Section 6 the solution r, q in Theorem 4 was expressed in terms of a continuous determinant δ on a quasi-Banach ideal $\mathcal{A}$. This reformulation translates to the finite-dimensional setting as follows.

Corollary 1. For the solution r, q in Proposition 5 it holds

$$r = \left(1 - \det\begin{pmatrix} I_n - \hat{L}c_{j_1}a_{j_2}^{\mathrm{T}} & \hat{L}C \\ \hat{M}D & I_m \end{pmatrix} \Big/ \det\begin{pmatrix} I_n & \hat{L}C \\ \hat{M}D & I_m \end{pmatrix}\right)_{\substack{j_2=1,\dots,\mathrm{d}_2 \\ j_1=1,\dots,\mathrm{d}_1}},$$

$$q = \left(1 - \det\begin{pmatrix} I_n & \hat{L}C \\ \hat{M}D & I_m - \hat{M}d_{j_2}b_{j_1}^{\mathrm{T}} \end{pmatrix} \Big/ \det\begin{pmatrix} I_n & \hat{L}C \\ \hat{M}D & I_m \end{pmatrix}\right)_{\substack{j_1=1,\dots,\mathrm{d}_1 \\ j_2=1,\dots,\mathrm{d}_2}}.$$

Remark 5. For results like in Proposition 5 (or Corollary 1), several simplifications are possible, an aspect often neglected in the literature. A first simplification is the transition to input data A, B in Jordan form in Proposition 6. Further information can be found in Remark 10.

8.2 Transition to Jordan data

The following proposition explains why the two matrices $A \in \mathcal{M}_{n,n}(\mathbb{C})$, $B \in \mathcal{M}_{m,m}(\mathbb{C})$ in Proposition 5 can be assumed to be in Jordan form.

Proposition 6. *Let $A \in \mathcal{M}_{n,n}(\mathbb{C})$, $B \in \mathcal{M}_{m,m}(\mathbb{C})$, and let U, V be matrices transforming A, B into Jordan form J_A, J_B, i.e. $A = U^{-1}J_AU$, $B = V^{-1}J_BV$. Then the solution in Proposition 5 is not altered if we replace simultaneously A, B by J_A, J_B, and a_{j_2}, b_{j_1}, c_{j_1}, d_{j_2} by $(U^{-1})^{\mathrm{T}}a_{j_2}$, $(V^{-1})^{\mathrm{T}}b_{j_1}$, Uc_{j_1}, Vd_{j_2} for $j_1 = 1,\dots,\mathrm{d}_1$, $j_2 = 1,\dots,\mathrm{d}_2$.*

Proof. We use the representation of r and q given in Corollary 1. We only treat the numerator in r, the arguments for the other determinants being similar.

Note first that $\hat{L} = U^{-1}\hat{L}_0U$, $\hat{M} = V^{-1}\hat{M}_0V$, where $\hat{L}_0 = \exp\big(J_Ax - f_0(A)t\big)$ and $\hat{M}_0 = \exp\big(J_Bx - f_0(-J_B)t\big)$. Hence

$$\det\begin{pmatrix} I_n - \hat{L}c_{j_1}a_{j_2}^{\mathrm{T}} & \hat{L}C \\ \hat{M}D & I_m \end{pmatrix} = \det\left(I_{n+m} + \begin{pmatrix} \hat{L} & 0 \\ 0 & \hat{M} \end{pmatrix}\begin{pmatrix} -c_{j_1}a_{j_2}^{\mathrm{T}} & C \\ D & 0 \end{pmatrix}\right)$$

$$= \det\left(I_{n+m} + \begin{pmatrix} U^{-1} & 0 \\ 0 & V^{-1} \end{pmatrix}\begin{pmatrix} \hat{L}_0 & 0 \\ 0 & \hat{M}_0 \end{pmatrix}\begin{pmatrix} U & 0 \\ 0 & V \end{pmatrix}\begin{pmatrix} -c_{j_1}a_{j_2}^{\mathrm{T}} & C \\ D & 0 \end{pmatrix}\right)$$

$$= \det\left(I_{n+m} + \begin{pmatrix} \hat{L}_0 & 0 \\ 0 & \hat{M}_0 \end{pmatrix}\begin{pmatrix} U & 0 \\ 0 & V \end{pmatrix}\begin{pmatrix} -c_{j_1}a_{j_2}^{\mathrm{T}} & C \\ D & 0 \end{pmatrix}\begin{pmatrix} U^{-1} & 0 \\ 0 & V^{-1} \end{pmatrix}\right)$$

$$= \det\left(I_{n+m} + \begin{pmatrix} \hat{L}_0 & 0 \\ 0 & \hat{M}_0 \end{pmatrix}\begin{pmatrix} -Uc_{j_1}a_{j_2}^{\mathrm{T}}U^{-1} & UCV^{-1} \\ VDU^{-1} & 0 \end{pmatrix}\right).$$

Obviously, $Uc_{j_1}a_{j_2}^{\mathrm{T}}U^{-1} = (Uc_{j_1})\big((U^{-1})^{\mathrm{T}}a_{j_2}\big)^{\mathrm{T}}$ and, to focus on one of the involved Sylvester equations, since

$$J_A(UCV^{-1}) + (UCV^{-1})J_B = (UA)CV^{-1} + UC(BV^{-1})$$

$$= \quad U(AC+CB)V^{-1} \;=\; U\Big(\sum_{j=1}^{\mathsf{d}_1} c_j b_j^{\mathrm{T}}\Big)V^{-1} \;=\; \sum_{j=1}^{\mathsf{d}_1}(Uc_j)\big((V^{-1})^{\mathrm{T}}b_j\big)^{\mathrm{T}},$$

we have that UCV^{-1} is the (unique) solutions of

$$J_A X + X J_B \;=\; \Big(Uc_1 \dots Uc_{\mathsf{d}_1}\Big)\begin{pmatrix} \big((V^{-1})^{\mathrm{T}}b_1\big)^{\mathrm{T}} \\ \vdots \\ \big((V^{-1})^{\mathrm{T}}b_{\mathsf{d}_1}\big)^{\mathrm{T}} \end{pmatrix}.$$

This completes the proof. ∎

8.3 $\mathbb{C}$-reduction

Recall that, for $a \in \mathbb{C}^n$ ($\mathcal{A} \in \mathcal{M}_{n,\mathsf{d}}(\mathbb{C})$), we denote by $\overline{a}$ ($\overline{\mathcal{A}}$) the vector (matrix) with complex conjugate entries, and by a^{H} ($\mathcal{A}^{\mathrm{H}}$) the conjugated transpose, or Hermitian, of a ($\mathcal{A}$).

For the solution r, q in Proposition 5, the reduction $q = -r^{\mathrm{H}}$ can be achieved by choosing

$$B = A^{\mathrm{H}}, \quad \text{and} \quad b_{j_1}^{\mathrm{T}} = c_{j_1}^{\mathrm{H}}, \quad d_{j_2}^{\mathrm{T}} = -a_{j_2}^{\mathrm{H}}$$

for $j_1 = 1, \dots, \mathsf{d}_1$, $j_2 = 1, \dots, \mathsf{d}_2$. Note that this in particular implies $m = n$.

The assumption on f_0 for the $\mathbb{C}$-reduction then implies $-f_0(-A^{\mathrm{H}}) = f_0(A)^{\mathrm{H}}$, and thus $\hat{M}(x,t) = \hat{L}(x,t)^{\mathrm{H}}$. Furthermore, C and D satisfy

$$AC + CA^{\mathrm{H}} = \mathcal{C}\mathcal{C}^{\mathrm{H}}, \qquad A^{\mathrm{H}}D + DA = -\mathcal{A}^{\mathrm{H}}\mathcal{A},$$

where we have introduced the $n \times \mathsf{d}_1$-matrix $\mathcal{C} = \big(c_1 \dots c_{\mathsf{d}_1}\big)$ and the $\mathsf{d}_2 \times n$-matrix $\mathcal{A}$ such that $\mathcal{A}^{\mathrm{T}} = \big(a_1 \dots a_{\mathsf{d}_2}\big)$. Observe that $C^{\mathrm{H}} = C$, $D^{\mathrm{H}} = D$ by the unique solvability of the Sylvester equation. This shows

$$\begin{aligned}
q \;&=\; -\mathcal{C}^{\mathrm{H}}(I_n - \hat{L}^{\mathrm{H}}D\hat{L}C)^{-1}\hat{L}^{\mathrm{H}}\,\mathcal{A}^{\mathrm{H}} \;=\; -\Big(\mathcal{A}\hat{L}(I_n - C^{\mathrm{H}}\hat{L}^{\mathrm{H}}D^{\mathrm{H}}\hat{L})^{-1}\mathcal{C}\Big)^{\mathrm{H}} \\
&=\; -\Big(\mathcal{A}\Big((I_n - C\hat{L}^{\mathrm{H}}D\hat{L})\hat{L}^{-1}\Big)^{-1}\mathcal{C}\Big)^{\mathrm{H}} \;=\; -\Big(\mathcal{A}\Big(\hat{L}^{-1}(I_n - \hat{L}C\hat{L}^{\mathrm{H}}D)\Big)^{-1}\mathcal{C}\Big)^{\mathrm{H}} \\
&=\; -\Big(\mathcal{A}(I_n - \hat{L}C\hat{L}^{\mathrm{H}}D)^{-1}\hat{L}\,\mathcal{C}\Big)^{\mathrm{H}} \;=\; -r^{\mathrm{H}}.
\end{aligned}$$

We sum up.

Proposition 7. *Let $A \in \mathcal{M}_{n,n}(\mathbb{C})$ such that* $\mathrm{spec}(A) \cup \mathrm{spec}(-A^{\mathrm{H}})$ *is contained in the domain where f_0 is holomorphic,* $0 \notin \mathrm{spec}(A) + \mathrm{spec}(A^{\mathrm{H}})$, *and* $\exp(Ax)$ *behaves sufficiently well for $x \to -\infty$.*

Furthermore let $a_{j_2} \in \mathbb{C}^n$, $j_2 = 1, \dots, \mathsf{d}_2$, be linearly independent and $c_{j_1} \in \mathbb{C}^n$, $j_1 = 1, \dots, \mathsf{d}_1$. Set $\mathcal{A}^{\mathrm{T}} = \big(a_1 \dots a_{\mathsf{d}_2}\big)$ and $\mathcal{C} = \big(c_1 \dots c_{\mathsf{d}_1}\big)$.

Then, on strips $\mathbb{R} \times (t_1, t_2)$ on which the inverse exists, a solution r (with values in the space of $\mathsf{d}_2 \times \mathsf{d}_1$-matrices) of the $\mathbb{C}$-reduced matrix-AKNS system (7.1) is given by

$$r = \mathcal{A}\Big(I_n - \hat{L}C\hat{L}^{\mathrm{H}}D\Big)^{-1}\hat{L}\,\mathcal{C}, \tag{8.1}$$

where $\hat{L}(x,t) = \exp\big(Ax + f_0(A)t\big)$, *and* C, D *are determined by* $AC + CA^{\mathrm{H}} = \mathcal{C}\mathcal{C}^{\mathrm{H}}$ *and* $A^{\mathrm{H}}D + DA = -\mathcal{A}^{\mathrm{H}}\mathcal{A}$.

If we focus on the matrix NLS (7.3), fewer assumptions are needed. On the one hand f_0 is holomorphic on the whole of $\mathbb{C}$; on the other hand the assumption that $\exp(Ax)$ behaves sufficiently well can be dropped, since the actual system (2.8) does not involve integrals. Expressing the resulting solution formula in terms of determinants, we get

Corollary 2. Let $A \in \mathcal{M}_{n,n}(\mathbb{C})$ with $0 \notin \operatorname{spec}(A) + \operatorname{spec}(A^{\mathrm{H}})$, and let $a_{j_2} \in \mathbb{C}^n$, $j_2 = 1, \ldots, \mathsf{d}_2$, be linearly independent, and $c_{j_1} \in \mathbb{C}^n$, $j_1 = 1, \ldots, \mathsf{d}_1$. In addition, let C, D be determined by $AC + CA^{\mathrm{H}} = \mathcal{C}\mathcal{C}^{\mathrm{H}}$ and $A^{\mathrm{H}}D + DA = -\mathcal{A}^{\mathrm{H}}\mathcal{A}$, where $\mathcal{A}^{\mathrm{T}} = (a_1 \ldots a_{\mathsf{d}_2})$ and $\mathcal{C} = (c_1 \ldots c_{\mathsf{d}_1})$, and set $\hat{L}(x,t) = \exp\big(Ax - \mathrm{i}A^2 t\big)$. Then

$$r = \left(1 - \frac{\det\begin{pmatrix} I_n - \hat{L}c_{j_1}a_{j_2}^{\mathrm{T}} & \hat{L}C \\ \hat{L}^{\mathrm{H}}D & I_n \end{pmatrix}}{\det\begin{pmatrix} I_n & \hat{L}C \\ \hat{L}^{\mathrm{H}}D & I_n \end{pmatrix}}\right)_{\substack{j_2=1,\ldots,\mathsf{d}_2 \\ j_1=1,\ldots,\mathsf{d}_1}}$$

is a solution of the matrix NLS (7.3) with values in the space of $\mathsf{d}_2 \times \mathsf{d}_1$-matrices on the open subset of $\mathbb{R}^2$ on which the determinant in the denominator is not zero.

Remark 6. In the case of the matrix NLS (7.3), where fewer assumptions are needed, the solution formula (8.1) was also derived in [17].

8.4 $\mathbb{R}$-reduction

Analogously as in Section 8.3, the $\mathbb{R}$-reduction $q = -r^{\mathrm{T}}$ can be achieved choosing

$$B = A^{\mathrm{T}} \quad \text{and} \quad b_{j_1} = c_{j_1}, \quad d_{j_2} = -a_{j_2}.$$

The main remaining question concerns reality of the solution r. Of course this can be guaranteed choosing A and c_{j_1}, a_{j_2} with real entries, but in view of applications we want to include A with pairs of complex conjugate eigenvalues (see Section 9).

Proposition 8. *Let* $A \in \mathcal{M}_{n,n}(\mathbb{C})$ *such that* $\operatorname{spec}(A) \cup \operatorname{spec}(-A)$ *is contained in the domain where* f_0 *is holomorphic,* $0 \notin \operatorname{spec}(A) + \operatorname{spec}(A)$, *and* $\exp(Ax)$ *behaves sufficiently well for* $x \to -\infty$.

Furthermore let $a_{j_2} \in \mathbb{C}^n$, $j_2 = 1, \ldots, \mathsf{d}_2$, *be linearly independent and* $c_{j_1} \in \mathbb{C}^n$, $j_1 = 1, \ldots, \mathsf{d}_1$. *Set* $\mathcal{A}^{\mathrm{T}} = (a_1 \ldots a_{\mathsf{d}_2})$ *and* $\mathcal{C} = (c_1 \ldots c_{\mathsf{d}_1})$, *and assume that there is an invertible* $\Pi \in \mathcal{M}_{n,n}(\mathbb{C})$ *such that*

$$\overline{A} = \Pi A \Pi^{-1}, \qquad \overline{\mathcal{A}} = \mathcal{A}\Pi^{-1}, \quad \overline{\mathcal{C}} = \Pi \mathcal{C}. \tag{8.2}$$

Then, on strips $\mathbb{R} \times (t_1, t_2)$ *on which the inverse exists, a real solution* r *(with values in the space of* $\mathsf{d}_2 \times \mathsf{d}_1$*-matrices) of the* $\mathbb{R}$*-reduced matrix-AKNS system* (7.2) *is given by*

$$r = \mathcal{A}\Big(I_n - \hat{L}C\hat{L}^{\mathrm{T}}D\Big)^{-1}\hat{L}\mathcal{C},$$

where $\hat{L}(x,t) = \exp\big(Ax + f_0(A)t\big)$, *and* C, D *are determined by* $AC + CA^{\mathrm{T}} = \mathcal{C}\mathcal{C}^{\mathrm{T}}$ *and* $A^{\mathrm{T}}D + DA = -\mathcal{A}^{\mathrm{T}}\mathcal{A}$.

Proof. We need only to verify the reality of r. Note that $\overline{A} = \Pi A \Pi^{-1}$ implies $\overline{\hat{L}} = \Pi \hat{L} \Pi^{-1}$ and $\overline{\hat{L}}^{\mathrm{T}} = (\Pi^{\mathrm{T}})^{-1}\hat{L}^{\mathrm{T}}\Pi^{\mathrm{T}}$. Hence

$$\begin{aligned}\overline{r} &= \overline{\mathcal{A}}\Big(I_n - \overline{\hat{L}}\overline{C}\overline{\hat{L}}^{\mathrm{T}}\overline{D}\Big)^{-1}\overline{\hat{L}}\overline{\mathcal{C}} \\ &= (\mathcal{A}\Pi^{-1})\Big(I_n - \Pi\hat{L}\Pi^{-1}\overline{C}(\Pi^{\mathrm{T}})^{-1}\hat{L}^{\mathrm{T}}\Pi^{\mathrm{T}}\overline{D}\Big)^{-1}(\Pi\mathcal{C}) \\ &= \mathcal{A}\Big(I_n - \hat{L}(\Pi^{-1}\overline{C}(\Pi^{\mathrm{T}})^{-1})\hat{L}^{\mathrm{T}}(\Pi^{\mathrm{T}}\,\overline{D}\Pi)\Big)^{-1}\mathcal{C}.\end{aligned}$$

To conclude the proof, we show that $\Pi^{-1}\overline{C}(\Pi^{\mathrm{T}})^{-1} = C$ and $\Pi^{\mathrm{T}}\,\overline{D}\,\Pi = D$. Since the arguments are comparable, we only show the first equation.

Taking the complex conjugate of $A^{\mathrm{T}}D + DA = -\mathcal{A}^{\mathrm{T}}\mathcal{A}$, and using (8.2), we get $(\Pi A\Pi^{-1})^{\mathrm{T}}\overline{D} + \overline{D}\,\Pi A\Pi^{-1} = -(\mathcal{A}\Pi^{-1})^{\mathrm{T}}(\mathcal{A}\Pi^{-1})$, which, after multiplication with Π^{T} from the left and Π from the right, simplifies to $A^{\mathrm{T}}(\Pi^{\mathrm{T}}\overline{D}\Pi) + (\Pi^{\mathrm{T}}\overline{D}\Pi)A = -\mathcal{A}^{\mathrm{T}}\mathcal{A}$. This shows that both $\Pi^{\mathrm{T}}\overline{D}\Pi$ and D solve the Sylvester equation $A^{\mathrm{T}}X + XA = -\mathcal{A}^{\mathrm{T}}\mathcal{A}$, and by uniqueness of the solution we infer $\Pi^{\mathrm{T}}\overline{D}\Pi = D$. ■

In view of later applications, we state

Corollary 3. Let $A \in \mathcal{M}_{n,n}(\mathbb{C})$ with $0 \notin \operatorname{spec}(A) + \operatorname{spec}(A)$, and let $a_{j_2} \in \mathbb{C}^n$, $j_2 = 1, \dots, \mathsf{d}_2$ be linearly independent and $c_{j_1} \in \mathbb{C}^n$, $j_1 = 1, \dots, \mathsf{d}_1$. In addition, let C, D be determined by $AC + CA^{\mathrm{T}} = \mathcal{C}\mathcal{C}^{\mathrm{T}}$ and $A^{\mathrm{T}}D + DA = -\mathcal{A}^{\mathrm{T}}\mathcal{A}$, where $\mathcal{A}^{\mathrm{T}} = (a_1 \dots a_{\mathsf{d}_2})$ and $\mathcal{C} = (c_1 \dots c_{\mathsf{d}_1})$, and let the reality condition (8.2) be met. Set $\hat{L}(x,t) = \exp\big(Ax - A^3t\big)$. Then

$$r = \left(1 - \frac{\det\begin{pmatrix} I_n - \hat{L}c_{j_1}a_{j_2}^{\mathrm{T}} & \hat{L}C \\ \hat{L}^{\mathrm{T}}D & I_n \end{pmatrix}}{\det\begin{pmatrix} I_n & \hat{L}C \\ \hat{L}^{\mathrm{T}}D & I_n \end{pmatrix}}\right)_{\substack{j_2=1,\dots,\mathsf{d}_2 \\ j_1=1,\dots,\mathsf{d}_1}}$$

is a real solution of the matrix mKdV (7.4) with values in the space of $\mathsf{d}_2 \times \mathsf{d}_1$-matrices on the open subset of $\mathbb{R}^2$ on which the determinant in the denominator is not zero.

Remark 7. In [37], the reductions $q = -\overline{r}$ and $q = -r$ for $\mathsf{d}_1 = \mathsf{d}_2 =: \mathsf{d}$ are used. Note that for $\mathsf{d} = 1$, this gives sufficiently transparent solution formulas for an asymptotic analysis of the multiple-pole solutions of the NLS (2.4) in [42].

9 Solitons, strongly bound solitons (breathers), degeneracies

The first part of the applications is about scalar equations. We will focus on the prototypical examples of the $\mathbb{C}$- and $\mathbb{R}$-reduction, the NLS (2.4) and the mKdV (2.1).

For the NLS, the first observation is that diagonal matrices A (with N different complex eigenvalues) give rise to N-solitons. Since their height and velocity can be chosen independently, different solitons can move with the same velocity, appearing as a new solution of particle character [5, Example 7.2]. Note however that such constellations in general are not stable under collision with other solitons (see [42, Example 9.1] for a study of the aforementioned example in [5] from this point of view, and a subsequent asymptotic analysis).

In the sequel we report on the situation for the mKdV. Here also the reality condition (8.2) has to be taken into account: real eigenvalues lead to solitons/antisolitons, and pairs of complex conjugated eigenvalues to breathers (classically interpreted as bound states of a soliton and an antisoliton [28]). We also briefly discuss degeneracies.

9.1 Finite superpositions of solitons and breathers

For the mKdV (2.1) we can work directly with Proposition 1 (In fact one can show that Corollary 3 gives the same solution formula in the case $\mathsf{d}_1 = \mathsf{d}_2 = 1$).

Let $A = \operatorname{diag}\{\alpha_1, \dots, \alpha_N\}$ with pairwise different $\alpha_j \in \mathbb{C}$. Note that the spectral condition $0 \notin \operatorname{spec}(A) + \operatorname{spec}(A)$ is satisfied if we assume in addition that $\alpha_i + \alpha_j \neq 0$ for all i, j. Furthermore, one can show that the assumption that all α_j have positive real parts does not cause any loss of generality, see also Remark 10.

In order to evaluate the solution formula in Proposition 1, we observe that $\exp(Ax - A^3t) = \operatorname{diag}\{\ell_1, \dots, \ell_N\}$, where $\ell_j = \ell_j(x,t) = \exp(\alpha_j x - \alpha_j^3 t)$, and that it is straightforward to verify that

$$\Phi_{A,A}^{-1}(ca^{\mathrm{T}}) = \left(\frac{a_k c_j}{\alpha_j + \alpha_k}\right)_{j,k=1}^{N}.$$

Hence

$$\begin{aligned}
p_{\pm} \;:=\; \det(I_N \pm \mathrm{i}L) &= \det\left(\delta_{jk} \pm \mathrm{i}\frac{a_k c_j}{\alpha_j + \alpha_k}\ell_j\right)_{j,k=1}^{N} \\
&= 1 + \sum_{k=1}^{N} \sum_{\substack{j_1,\dots,j_k=1 \\ j_1<\dots<j_k}}^{N} (\pm\mathrm{i})^k \det\left(\left(\frac{a_{j_{\kappa'}} c_{j_\kappa}}{\alpha_{j_\kappa} + \alpha_{j_{\kappa'}}}\ell_{j_\kappa}\right)_{\kappa,\kappa'=1}^{k}\right) \\
&= 1 + \sum_{k=1}^{N} \sum_{\substack{j_1,\dots,j_k=1 \\ j_1<\dots<j_k}}^{N} (\pm\mathrm{i})^k \prod_{\kappa=1}^{k} a_{j_\kappa} c_{j_\kappa} \ell_{j_\kappa} \,\det\left(\left(\frac{1}{\alpha_{j_\kappa} + \alpha_{j_{\kappa'}}}\right)_{\kappa,\kappa'=1}^{k}\right) \qquad (9.1) \\
&= 1 + \sum_{k=1}^{N} \sum_{\substack{j_1,\dots,j_k=1 \\ j_1<\dots<j_k}}^{N} (\pm\mathrm{i})^k \prod_{\kappa=1}^{k} f_{j_\kappa} \prod_{\substack{\kappa,\kappa'=1 \\ \kappa<\kappa'}}^{k} \left(\frac{\alpha_{j_\kappa} - \alpha_{j_{\kappa'}}}{\alpha_{j_\kappa} + \alpha_{j_{\kappa'}}}\right)^2,
\end{aligned}$$

where $f_j = \big(a_j c_j/(2\alpha_j)\big)\exp(\alpha_j x - \alpha_j^3 t)$. In the computation above, the explicit evaluation of the determinant in (9.1) is an exercise from [32] (see also [40] for a generalization relevant for the study of MPS's).

Obviously the solution constructed above is not necessarily real-valued. However, an argument analogous to that in the proof of Proposition 8, compare also [12, Proposition 15], can be used to verify reality under the additional assumption that for each j, either α_j and $\exp(\varphi_j)$ are real, or there is a unique index $\bar{\jmath}$ with $\alpha_{\bar{\jmath}} = \overline{\alpha}_j$ and $a_{\bar{\jmath}} c_{\bar{\jmath}} = \overline{a}_j \overline{c}_j$.

The resulting solution is a superposition of solitons and breathers. In particular, the 1-soliton is obtained for $N = 1$ and α_1, $a_1 c_1 \in \mathbb{R}$. Writing $\alpha_1 = \alpha$ and $a_1 c_1/(2\alpha) = \epsilon \exp(\varphi)$ with $\epsilon = \pm 1$, $\varphi \in \mathbb{R}$,

$$r(x,t) = -\mathrm{i}\frac{\partial}{\partial x} \log \frac{p(x,t)}{\overline{p(x,t)}} \text{ with } p(x,t) = 1 + \mathrm{i}\epsilon\, e^{\Gamma(x,t)} \text{ and } \Gamma(x,t) = \alpha x - \alpha^3 t + \varphi, \tag{9.2}$$

and we have

$$r(x,t) = \epsilon \alpha \operatorname{sech}\big(\alpha x - \alpha^3 t + \varphi\big).$$

As for the KdV equation, solitons are bell-shaped, moving with constant velocity $\alpha^2 > 0$. With increasing velocity, their amplitude increases and their width decreases. In contrast to the KdV however, not only solitons (corresponding to $\epsilon = 1$) but also antisolitons (corresponding to $\epsilon = -1$) are regular.

Next we consider $N = 2$ and $\alpha_2 = \overline{\alpha}_1$, $a_2 c_2 = \overline{a}_1 \overline{c}_1$. Let us first write $\alpha_1 = \alpha = |\alpha| \exp(\mathrm{i}\beta)$ with $0 < \beta \leq \pi/2$ (which can be done without loss of generality, compare Remark 10) and set $\gamma = \mathrm{Im}(\alpha)/\mathrm{Re}(\alpha) > 0$. Note that $(\alpha_1 - \alpha_2)/(\alpha_1 + \alpha_2) = \mathrm{i}\gamma$. Next we write $\gamma\, a_1 c_1/(2\alpha) = \exp(\varphi)$ with $\varphi \in \mathbb{C}$. Then

$$r(x,t) = -\mathrm{i}\frac{\partial}{\partial x} \log \frac{p(x,t)}{\overline{p(x,t)}} \text{ with } p(x,t) = 1 + \mathrm{i}\gamma^{-1}\big(e^{\Gamma(x,t)} + e^{\overline{\Gamma(x,t)}}\big) + e^{\left(\Gamma(x,t) + \overline{\Gamma(x,t)}\right)}, \tag{9.3}$$

where $\Gamma(x,t) = \alpha x - \alpha^3 t + \varphi$. Introducing the notation $e^{\Gamma} = R \exp(i\Phi)$, where $R = e^{\mathrm{Re}(\Gamma)}$, $\Phi = \mathrm{Im}(\Gamma)$, and using $\gamma = \sin(\beta)/\cos(\beta)$, we get

$$r(x,t) = \gamma^{-1} |\alpha| \, \frac{R^{-1} \cos(\Phi + \beta) - R \cos(\Phi - \beta)}{\frac{1}{4}\big(R + R^{-1}\big)^2 + \big(\gamma^{-1} \cos(\Phi)\big)^2},$$

the breather. Its height is $2\mathrm{Re}(\alpha)$, and its velocity $\mathrm{Re}(\alpha)^2 - 3\mathrm{Im}(\alpha)^2$. Hence, in contrast to solitons, breathers can also be stationary or move to the left.

In fact, eigenvalues $\alpha = d + \mathrm{i}\sqrt{(d^2 - v)/3}$ give rise to breathers with the same velocity v. Observe that this means that $d > \sqrt{v}$ for $v > 0$ and $d > 0$ for $v \leq 0$. It may be instructive to reflect about the physical meaning of the relation between height, velocity and frequency of breathers.

Example. In Figure 2, the plot on the left shows the solution of the mKdV equation obtained from Proposition 1 with $N = 3$, $n_1 = n_2 = n_3 = 1$,

$$\alpha_1 = \tfrac{1}{3}\Big(1 + \mathrm{i}\sqrt{\tfrac{10}{3}}\Big), \quad \alpha_2 = \overline{\alpha}_1, \quad \alpha_3 = 1$$

and $a_1c_1 = \frac{2}{3} - \mathrm{i}\sqrt{\frac{2}{15}}$, $a_2c_2 = \overline{a_1c_1}$, $a_3c_3 = -2$. In the resulting solution a breather of height $2/3$, which moves with velocity -1, and an antisoliton of height 1, which moves with velocity 1, collide. It is depicted for $-20 \leq x \leq 20$, $-10 \leq t \leq 10$.

In Figure 2, the plot on the right shows the solution of the NLS equation obtained from Corollary 2 with $N = 2$, $n_1 = n_2 = 1$,

$$\alpha_1 = \tfrac{1}{2} + \mathrm{i}\tfrac{2}{5}, \quad \alpha_2 = \overline{\alpha}_1$$

and $a_1c_1 = 1$, $a_2c_2 = \overline{a_1c_1}$. In the resulting solution two solitons with velocities $4/5$ resp. $-4/5$ collide. Its modulus is depicted for $-30 \leq x \leq 30$, $-20 \leq t \leq 20$.

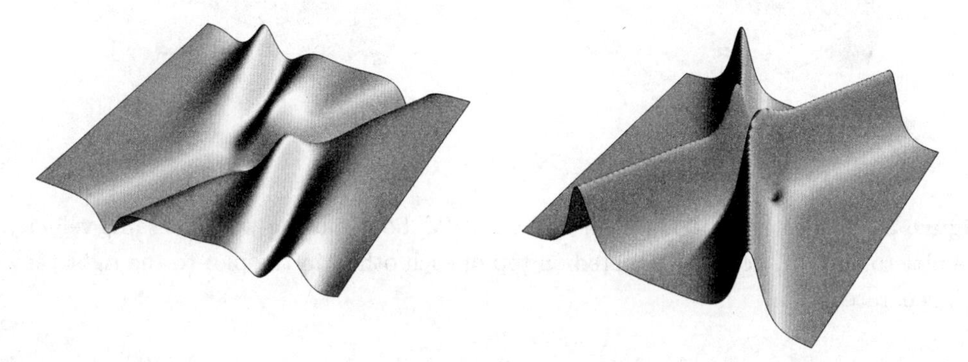

Figure 2. The plot on the left shows the collision of a breather and an antisoliton of the mKdV, which move with opposite velocities. Note that, whereas solitons of the mKdV always move to the right, breathers may move to the left. The plot to the right shows the collision of two solitons of the NLS moving with opposite velocities (two complex conjugated eigenvalues), the modulus of the solution being plotted.

9.2 Degeneracies

For the KdV, solitons are the only smooth solutions with particle character (stable shape, rapid decrease for $x \to \pm\infty$). If several of them are superposed, they must have different velocities. For the mKdV, solitons/antisolitons have analogous properties, but breathers are further solutions with particle character. In addition one can superpose solitons and breathers with coinciding velocity.

Example. Both plots in Figure 3 show the solution of the mKdV equation obtained from Proposition 1 with $N = 3$, $n_1 = n_2 = n_3 = 1$,

$$\alpha_1 = 1 + \mathrm{i}\frac{1}{\sqrt{6}}, \quad \alpha_2 = \overline{\alpha}_1, \quad \alpha_3 = \frac{1}{\sqrt{2}}.$$

For the plot on the left, we have

$$a_1c_1 = \frac{4\sqrt{2}}{3}\left(1 + \mathrm{i}\frac{1}{\sqrt{6}}\right), \quad a_2c_2 = \overline{a_1c_1}, \quad a_3c_3 = \sqrt{2}\frac{(1 + \frac{1}{\sqrt{2}})^2 + \frac{1}{6}}{(1 - \frac{1}{\sqrt{2}})^2 + \frac{1}{6}},$$

and for the plot on the right the modification $a_3c_3 \mapsto e^{-5}a_3c_3$ was made. In both cases, the resulting solution is a constellation of a breather of height 2 and a soliton of height $1/\sqrt{2}$, each moving with velocity $1/2$. They are depicted for $-20 \leq x \leq 20$, $-20 \leq t \leq 20$.

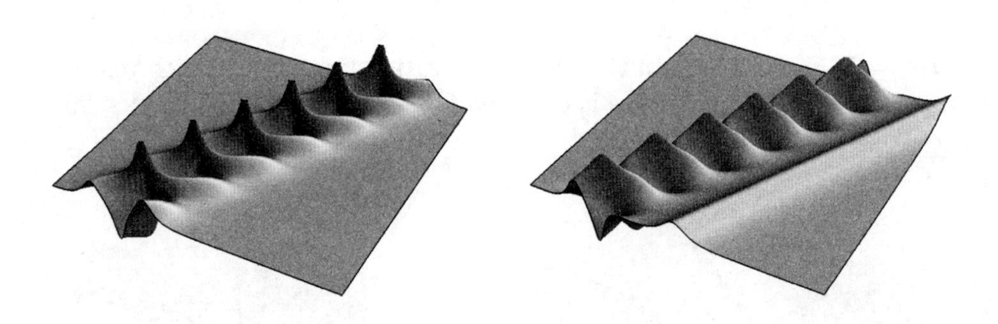

Figure 3. A breather and a soliton of the mKdV, both moving with the same velocity. In the plot to the left, both are situated on top of each other, in the plot to the right they are well separated.

Recall that, for a fixed velocity $v > 0$, there is a 1-parameter family of breathers characterised (up to translations in time and space) by their height $h > 2\sqrt{v}$. Hence in a solution combining a soliton and a breather of the same velocity, the breather is at least twice as high as the soliton, explaining why the breather dominates the picture.

Note furthermore that one can also superpose several breathers of the same (possibly negative) velocity.

Remark 8. As said before, constellations as in the example above are not stable. If one superposes them with another soliton, the relative position of the soliton and the breather with the same velocity can change as a result of the collision. It is clear that the asymptotic description of such solutions becomes subtler. First general results are obtained in [22, 42] for the NLS.

10 Multiple pole solutions

In the framework of the inverse scattering method, N-solitons of the NLS are obtained in the reflectionless case, if the transmission coefficient has N simple poles in the upper half plane. It is natural to ask what happens if there are multiple poles. A first answer was given by Zakharov and Shabat who discovered the solutions corresponding to a double-pole. In [50] they construct a wave packet of two solitons by 'coalescence' of (two simple) poles. The solitons in the packet are weakly bound, in the sense that they approach and drift apart logarithmically. A systematic study was initiated by Olmedilla. In [29] he studies solutions which consist of a single wave packet with L solitons. He gives an asymptotic analysis for $L = 2$ or 3, refers to computer evidence for $L \leq 9$, and conjectures a formula for L arbitrary.

In our method, MPS's can be realised by matrices A with non-trivial Jordan structure, the size of the Jordan block corresponding to the number of solitary waves grouped in the respective wave packet. In particular, superposition of several packets is obtained naturally. In [42], the author was able to give a complete and rigorous asymptotic analysis, confirming Olmedilla's conjecture in the case of one wave packet.

In this section we examine the asymptotic description of MPS's for the mKdV. Compared with the NLS case reviewed above the novelty is that also breathers can be included in the picture. For the proof, which is beyond the scope of the present article, we refer to [37, Theorem 5.2.2].

Remark 9. a) MPS's for various equations appear sporadically in the literature, see e.g. [44, 49], mostly in cases of low complexity. A renewed interest in their asymptotic description was aroused by Matveev, see [25, 26] and references therein, who studied positions, a class of slowly decreasing solutions with certain common features with MPS's (compare e.g. [33]).

b) For complete asymptotic descriptions of MPS's for other soliton equations we refer to [15] (for the KdV), [35] (for the sine-Gordon equation), [37] (for a simultaneous result for the reduced equations of the AKNS system) and [36] (for the Toda lattice). We mention that the result on the sine-Gordon equation also takes breathers into account.

c) To the best of the author's knowledge, there are no comparable studies of MPS's for 2-dimensional equations. In the forthcoming [27], first examples are studied for the 2-dimensional Toda lattice, including an asymptotic analysis in the spirit of [7].

10.1 Asymptotic results for MPS's

In this subsection we present the asymptotic behaviour of the solutions of the mKdV (2.1) provided by Corollary 3 in the case $\mathsf{d}_1 = \mathsf{d}_2 = 1$. We make two technical assumptions:

Assumption 1. The matrix $A \in \mathcal{M}_{n,n}(\mathbb{C})$ is in Jordan form with N Jordan blocks A_j of sizes $n_j \times n_j$ corresponding to eigenvalues α_j.

We adapt our notation to the given Jordan structure of A: For $v \in \mathbb{C}^n$, $T \in \mathcal{M}_{n,n}(\mathbb{C})$, we write $v = (v_j)_{j=1}^N$ with $v_j = (v_j^{(\mu)})_{\mu=1}^{n_j} \in \mathbb{C}^{n_j}$, $T = (T_{jk})_{j,k=1}^N$ with $T_{jk} = (t_{jk}^{(\mu\nu)})_{\substack{\mu=1,\ldots,n_j \\ \nu=1,\ldots,m_k}} \in \mathcal{M}_{n_j,n_k}(\mathbb{C})$.

Note that $n = \sum_{j=1}^N n_j$.

Assumption 2. The vectors $a, c \in \mathbb{C}^n$ satisfy $a_j^{(1)} c_j^{(n_j)} \neq 0$.

In fact, Assumptions 1 and 2 can be made without loss of generality. For Assumption 1, this follows from Proposition 6, and we refer to [42, Lemma 4.3] for

Assumption 2, where a proof is given in the case of the NLS. The argument can be easily transferred to the case of the mKdV.

The next assumption guarantees that reality condition (8.2) in Corollary 3 is met.

Assumption 3. For all j, either α_j is real, a_j, $c_j \in \mathbb{R}^{n_j}$ or there is a unique index $\bar{\jmath} \neq j$ with $n_{\bar{\jmath}} = n_j$ such that $\alpha_{\bar{\jmath}} = \overline{\alpha}_j$ and $a_{\bar{\jmath}} = \overline{a}_j$, $c_{\bar{\jmath}} = \overline{c}_j$.

Finally we need to explain what we mean by saying that two functions coincide asymptotically: Two functions $f = f(x,t)$, $g = g(x,t)$ have the same asymptotic behaviour for $t \to \infty$ ($t \to -\infty$), if for every $\epsilon > 0$ there is t_ϵ such that for $t > t_\epsilon$ ($t < t_\epsilon$) we have $|u(x,t) - v(x,t)| < \epsilon$ uniformly in x. In this case we write $u(x,t) \approx v(x,t)$ for $t \approx \infty$ ($t \approx -\infty$).

Now we can formulate the main result.

Theorem 5. *Let Assumptions 1, 2 and 3 be met, and suppose that*

a) the $\mathrm{Re}(\alpha_j)$ *are positive,*

b) the $v_j = \mathrm{Re}(\alpha_j^3)/\mathrm{Re}(\alpha_j)$ *are pairwise different for* $j \in \{k \mid \mathrm{Im}(\alpha_k) \geq 0\}$.

To these data we associate, for $\mathrm{Im}(\alpha_j) = 0$, *the (soliton-like) functions* $r^{\pm}_{jj'}$ *and, for* $\mathrm{Im}(\alpha_j) > 0$, *the (breather-like) functions* $R^{\pm}_{jj'}$, $j' = 0, \ldots, n_j - 1$, *where*

$$r^{\pm}_{jj'} = -\mathrm{i}\frac{\partial}{\partial x}\log\frac{p^{\pm}_{jj'}}{\overline{p^{\pm}_{jj'}}} \quad \textit{with } p^{\pm}_{jj'} = 1 + \mathrm{i}(-1)^{n_j - 1 + j'}\epsilon_j e^{\Gamma^{\pm}_{jj'}(x,t)},$$

$$R^{\pm}_{jj'} = -\mathrm{i}\frac{\partial}{\partial x}\log\frac{P^{\pm}_{jj'}}{\overline{P^{\pm}_{jj'}}}$$

$$\textit{with } P^{\pm}_{jj'}(x,t) = 1 + \mathrm{i}\gamma_j^{-1}\left(e^{\Gamma^{\pm}_{jj'}(x,t)} + e^{\overline{\Gamma^{\pm}_{jj'}(x,t)}}\right) + e^{\left(\Gamma^{\pm}_{jj'}(x,t) + \overline{\Gamma^{\pm}_{jj'}(x,t)}\right)},$$

where

$$\Gamma^{\pm}_{jj'}(x,t) = \alpha_j x - \alpha_j^3 t \mp J' \log|t| + \varphi_j + \varphi_j^{\pm} + \varphi^{\pm}_{jj'},$$

and where we have set $J' = -(n_j - 1) + 2j'$. *Moreover, the quantities* φ_j, $\varphi_j^{\pm}$, $\varphi^{\pm}_{jj'}$ *are determined as follows:*

For $\mathrm{Im}(\alpha_j) = 0$, $a_j^{(1)} c_j^{(n_j)}/(2\alpha_j)^{n_j} = \epsilon_j \exp(\varphi_j)$ *with* $\epsilon_j = \pm 1$, $\varphi_j \in \mathbb{R}$, *and for* $\mathrm{Im}(\alpha_j) > 0$, $\gamma_j^{n_j} a_j^{(1)} c_j^{(n_j)}/(2\alpha_j)^{n_j} = \exp(\varphi_j)$ *with* $\varphi_j \in \mathbb{C}$. *Furthermore*

$$\exp(\varphi_j^{\pm}) = \prod_{k:\, v_k \lessgtr v_j} \left[\frac{\alpha_j - \alpha_k}{\alpha_j + \alpha_k}\right]^{2n_k},$$

$$\exp\left(\pm\varphi^{\pm}_{jj'}\right) = d_j^{-J'}\frac{j'!}{(j' - J')!},$$

where $d_j = 4\alpha_j^3$ *for* $\mathrm{Im}(\alpha_j) = 0$ *and* $d_j = 2\alpha_j\gamma_j^{-1}(v_j - 3\alpha_j^2)$ *for* $\mathrm{Im}(\alpha_j) > 0$.

Then the asymptotic behaviour of the solution $q(x,t)$ associated to A, a, c by Corollary 3 is

$$r(x,t) \approx \sum_{\substack{j=1 \\ \mathrm{Im}(\alpha_j)=0}}^{N} \sum_{j'=0}^{n_j-1} r^{\pm}_{jj'}(x,t) + \sum_{\substack{j=1 \\ \mathrm{Im}(\alpha_j)>0}}^{N} \sum_{j'=0}^{n_j-1} R^{\pm}_{jj'}(x,t) \quad \textit{for } t \approx \pm\infty.$$

The notation soliton-like and breather-like function in the theorem is motivated by (9.2), (9.3). For a comment on assumptions a), b) in the above theorem see Remark 10.

10.2 Geometric interpretation and examples

Here we discuss the geometric content of Theorem 5 and give some concrete examples.

(1) One Jordan block with real eigenvalue. For a single Jordan block A_1 of dimension n_1 with eigenvalue $\alpha_1 > 0$, the resulting solution is a wave packet consisting of n_1 soliton and antisolitons (with shapes determined by α_1).

The first observation is that this wave packet (more precisely, its geometric center) moves with constant velocity $v_1 = \alpha_1^2 > 0$, and the involved solitons and antisolitons drift apart logarithmically. In fact, for large negative times, we can imagine each soliton/antisoliton to be located on one side of the center, approaching it logarithmically. At some time it changes over to the other side of the center, moving away from the center again logarithmically. In the process the involved solitons and antisolitons collide elastically, but those internal collisions do not affect the path of the geometric center of the wave packet.

According to this interpretation – which is confimed by our computer plots – the solitons/antisolitons appear exactly in reversed order in the asymptotic forms for $t \approx -\infty$ and $t \approx +\infty$. Moreover, in the asymptotic form solitons and antisolitons always alternate.

(2) Two Jordan blocks with complex conjugated eigenvalues. If A consists of two Jordan blocks A_1, $\overline{A_1}$ of the same dimension n_1 and with eigenvalues α_1, $\overline{\alpha_1}$, where $\mathrm{Re}(\alpha_1) > 0$, the solution forms a wave packet containing n_1 breathers. The geometric center of the wave packet now moves with (not necessarily positive) velocity $v_1 = \mathrm{Re}(\alpha_1)^2 - 3\,\mathrm{Im}(\alpha_1)^2$, but otherwise the interpretation in **(1)** remains valid. Essentially, this even holds for the statement that in the asymptotic form solitons and antisolitons always alternate, since breathers can be viewed as a bound state of a soliton and an antisoliton.

(3) The general case. Finally, if A consists of N_1 real and N_2 pairs of complex conjugated Jordan blocks, the resulting solution is the superposition of N_1 wave packets as described in **(1)** and N_2 as described in **(2)**. The wave packets collide elastically, the only effect detectable by asymptotic analysis being that their geometric centers suffer a position shift. Note that the formula for the position shifts neatly generalises the well-known formula in the soliton case.

Remark 10. Assumption b) in Theorem 5 ensures that the wave packets have different velocities, and hence can be neatly distinguished by the asymptotic analysis.

In contrast, assumption a) in Theorem 5 can be made without loss of generality (see [42, Proposition B.2] for an argument in the related case of the NLS). Furthermore we can assume that the eigenvalues of A have geometric multiplicity 1, i.e. each eigenvalue appears in exactly one Jordan block ([42, Lemma B.1]).

Remark 11. Since MPS's were first described in the framework of the ISM, the reader may be interested in the link between the two approaches. For the NLS (2.4), the details can be found in [42, Section 5].

Example. Both plots in Figure 4 show solutions of the mKdV. For the plot on the left, the parameters are $N = 3$, $n_1 = 3$,

$$\alpha_1 = 1$$

and $a_1^{(1)} = 1$, $a_1^{(2)} = a_1^{(3)} = 0$, $c_1^{(1)} = 200$, $c_1^{(2)} = 10$, $c_1^{(3)} = 8$. The resulting solution is a wave packet in which two solitons and an antisoliton are weakly bound. As a whole the wave packet moves with velocity 1. It is plotted for $-15 \leq x \leq 15$, $-15 \leq t \leq 25$, in the coordinates $(x+t, t)$.

For the plot on the right the parameters are $N = 4$, $n_1 = n_2 = 2$,

$$\alpha_1 = 1 + \mathrm{i}\frac{1}{\sqrt{3}}, \quad \alpha_2 = \overline{\alpha_1}$$

and $a_1^{(1)} = a_2^{(1)} = 1$, $a_1^{(2)} = a_2^{(2)} = 0$, $c_1^{(1)} = 4(1 - 5\sqrt{3}\mathrm{i}) = \overline{c}_2^{(1)}$, $c_1^{(2)} = -8(1 + \mathrm{i}\sqrt{3}) = \overline{c}_2^{(2)}$. Here the solution is a wave packet consisting of two breathers, which as a whole is stationary. It is plotted for $-15 \leq x \leq 15$, $-20 \leq t \leq 20$.

Example. Both plots in Figure 5 show solutions of the mKdV. For the plot on the left, the parameters are $N = 4$, $n_1 = n_2 = 1$, $n_3 = 2$,

$$\alpha_1 = \tfrac{1}{\sqrt{2}}(1 + \mathrm{i}), \ \alpha_2 = \overline{\alpha}_1, \quad \alpha_3 = \tfrac{1}{2},$$

and $a_1^{(1)} = a_2^{(1)} = a_3^{(1)} = 1$, $a_3^{(2)} = 0$, $c_1^{(1)} = c_2^{(1)} = 2\sqrt{2}$, $c_3^{(1)} = 10$, $c_3^{(2)} = 1$. In the resulting solution a breather collides with a wave packet in which a soliton and an antisoliton are weakly bound. It is plotted for $-25 \leq x \leq 35$, $-35 \leq t \leq 65$.

For the plot on the right, the parameters are $N = 4$, $n_1 = n_2 = 2$,

$$\alpha_1 = 0.5, \quad \alpha_2 = 1,$$

and $a_1^{(1)} = a_2^{(1)} = 1$, $a_1^{(2)} = a_2^{(2)} = 0$, $c_1^{(1)} = c_2^{(1)} = 0$, $c_1^{(2)} = 1$, $c_2^{(2)} = 2$. In the solution two wave packets collide, each consisting of a soliton and an antisoliton which are weakly bound. The two wave packets collide elastically. The solution is plotted for $-30 \leq x \leq 30$, $-25 \leq t \leq 35$.

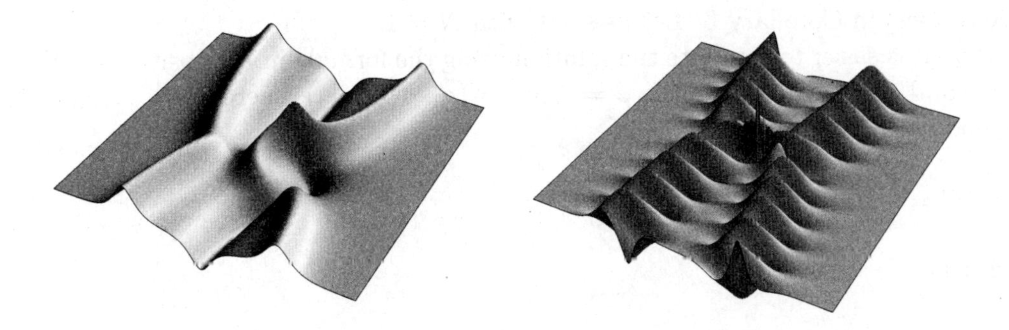

Figure 4. The plot on the left shows a 3-pole of the mKdV, which consists of two solitons and an antisoliton, and the plot to the right a 2-pole consisting of two breathers. Note that the latter is stationary, whereas the former has velocity 1 and is plotted in the coordinates $(x+t,t)$.

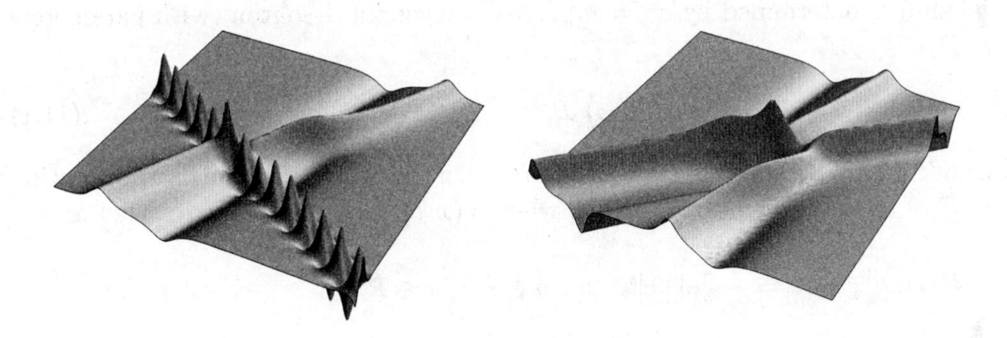

Figure 5. The plot on the left shows the collision of a breather of the mKdV with a 2-pole, which consists of a soliton and an antisoliton; the plot on the right shows the collision of two 2-poles, each consisting of a soliton and an antisoliton.

11 Solitons of matrix- and vector-equations

The final section contains our applications to matrix integrable systems. From the early discovery of the Manakov system [23], a vector version of the NLS, matrix systems attracted considerable interest, see [2, 45] and references therein. After first information on matrix and vector solitons, we focus on the interaction behaviour of two solitons. We intentionally do this for the vector version of the mKdV which is not treated in [46].

11.1 On matrix solitons

Recall that Propositions 7 and 8 also give access to the $\mathbb{C}$- and $\mathbb{R}$-reduced $\mathsf{d}_2 \times \mathsf{d}_1$-matrix AKNS systems (7.1) and (7.2). In the sequel we focus on vector-equations ($\mathsf{d}_1 = 1$, $\mathsf{d}_2 = \mathsf{d}$), specifically on the vector-mKdV (7.4).

A first aim is to derive N-solitons. For this a natural ansatz, motivated by our work on scalar equations, is to use $N \times N$-diagonal matrices A (with different eigenvalues) in Corollary 3. Let us start with $N = 1$.

Here it is easier to compute the solution using the formula in Proposition 8. With $n = 1$ and $A = \alpha > 0$, as well as $\mathcal{C} = c$, $\mathcal{A}^{\mathrm{T}} = \begin{pmatrix} a_1 & \dots & a_{\mathsf{d}} \end{pmatrix}$ where $c, a_1, \dots, a_{\mathsf{d}} \in \mathbb{R}$, we find

$$C = \frac{c^2}{2\alpha}, \quad D = -\frac{\|a\|^2}{2\alpha},$$

and hence

$$r = \begin{pmatrix} a_1 \\ \vdots \\ a_{\mathsf{d}} \end{pmatrix} \Big(1 + \frac{c^2\|a\|^2}{(2\alpha)^2}\ell^2\Big)^{-1} \ell c,$$

where $\ell(x,t) = \exp(\alpha x - \alpha^3 t)$. Without loss of generality we can take $c = 1$. Introducing the unit vector $P = \frac{1}{\|a\|}a$ (sometimes called polarisation vector) and the initial shift φ determined by $e^{\varphi} = \|a\|/(2\alpha)$, we get the 1-soliton (with parameters α, φ, P)

$$r(x,t) = \alpha \operatorname{sech}\big(\alpha x - \alpha^3 t + \varphi\big)\, P. \tag{11.1}$$

Remark 12. Note that an analogous argument as above also yields the solution $R = rP$ of the $\mathsf{d}_2 \times \mathsf{d}_1$-matrix mKdV, where $r(x,t) = \alpha \operatorname{sech}\big(\alpha x - \alpha^3 t + \varphi\big)$ and

$$P = ac^{\mathrm{T}}, \quad e^{\varphi} = \frac{1}{2\alpha}\|a\|\|c\| \quad \text{for } a \in \mathbb{R}^{\mathsf{d}_2},\ c \in \mathbb{R}^{\mathsf{d}_1}.$$

On the other hand, any function R of the form rP, where r is a solution of the scalar mKdV (2.1) and P a constant $\mathsf{d}_2 \times \mathsf{d}_1$-matrix satisfying $PP^{\mathrm{T}}P = P$ obviously solves the $\mathsf{d}_2 \times \mathsf{d}_1$-matrix mKdV (7.4). This shows that there is no need to assume that P is one-dimensional.

A way to obtain solutions with more general polarisation matrices from Proposition 8 is to set $A = \alpha I_{\mathsf{d}}$, where $\mathsf{d} = \min(\mathsf{d}_1, \mathsf{d}_2)$, leading to[1]

$$\begin{aligned} R &= \mathcal{A}\Big(I_{\mathsf{d}} + \frac{\ell^2}{(2\alpha)^2}(\mathcal{C}\mathcal{C}^{\mathrm{T}}\mathcal{A}^{\mathrm{T}})\mathcal{A}\Big)^{-1}\ell\,\mathcal{C} = \Big(I_{\mathsf{d}} + \frac{\ell^2}{(2\alpha)^2}\mathcal{A}(\mathcal{C}\mathcal{C}^{\mathrm{T}}\mathcal{A}^{\mathrm{T}})\Big)^{-1}\ell\,\mathcal{A}\mathcal{C} \\ &= \Big(I_{\mathsf{d}} + \frac{\ell^2}{(2\alpha)^2}\mathcal{B}\mathcal{B}^{\mathrm{T}}\Big)^{-1}\ell\,\mathcal{B} \end{aligned}$$

for the $\mathsf{d}_2 \times \mathsf{d}_1$-matrix $\mathcal{B} = \mathcal{A}\mathcal{C}$. Note also that these solutions need not be the product of a function and a constant matrix.

Remark 13. In [12, Section VIII] it was shown that the N-solitons of the $\mathsf{d} \times \mathsf{d}$-matrix KdV $r_t = r_{xxx} + 3\{r, r_x\}$, which are constructed in [20] by means of the inverse scattering method, can be realised by block diagonal matrices $A = \operatorname{diag}\{\alpha_1 I_{\mathsf{d}}, \dots, \alpha_N I_{\mathsf{d}}\}$.

[1]Recall the general identity $S(I_E + TS)^{-1} = (I_F + ST)^{-1}S$ for $S \in \mathcal{L}(E,F)$, $T \in \mathcal{L}(F,E)$, where E, F are Banach spaces.

11.2 Asymptotics for 2-solitons of the vector mKdV ($d_1 = 1$, $d_2 = d$).

We conclude this section by an outline of the asymptotic analysis of the solution r generated by setting $n = 2$ and $A = \operatorname{diag}\{\alpha_1, \alpha_2\}$, as well as $c = (c^{(i)})_{i=1}^2$, $a_j = (a_j^{(i)})_{i=1}^2$, $j = 1, \ldots, d$, in Corollary 3.

It is not difficult to see that one can choose $c = \begin{pmatrix} 1 & 1 \end{pmatrix}^T$ without loss of generality. In addition, we may assume $\alpha_2 > \alpha_1 > 0$.

Let us introduce the functions $\ell_i(x,t) = \exp(\alpha_i x - \alpha_i^3 t)$ and the vectors $v_i = \begin{pmatrix} a_1^{(i)} & \ldots & a_d^{(i)} \end{pmatrix}^T$ for $i = 1, 2$. Then

$$r = \left(1 - q_j/q\right)_{j=1}^{d}$$

with

$$q_j = \det\left(I_4 + \begin{pmatrix} -LC_j & LC \\ LD & 0 \end{pmatrix}\right), \quad q = \det\left(I_4 + \begin{pmatrix} 0 & LC \\ LD & 0 \end{pmatrix}\right),$$

where $L = \operatorname{diag}\{\ell_1, \ell_2\}$ and

$$C = \left(\frac{1}{\alpha_i + \alpha_j}\right)_{i,j=1}^{2}, \quad D = -\left(\frac{v_i^T v_j}{\alpha_i + \alpha_j}\right)_{i,j=1}^{2}, \quad C_j = \begin{pmatrix} 1 \\ 1 \end{pmatrix} \begin{pmatrix} a_j^{(1)} & a_j^{(2)} \end{pmatrix}.$$

Behaviour for $t \to -\infty$. To move with the second soliton, we consider curves $x - \alpha_2^2 t = c$. On such a curve, $\ell_1(x,t) \to 0$ for $t \to -\infty$. The arguments for q_j, $j = 1, \ldots d$, and q being parallel, we look at q_j and find

$$q_j \to \det \begin{pmatrix} 1 & 0 & 0 & 0 \\ -a_j^{(1)} \ell_2 & 1 - a_j^{(2)} \ell_2 & \frac{1}{\alpha_1+\alpha_2} \ell_2 & \frac{1}{2\alpha_2} \ell_2 \\ 0 & 0 & 1 & 0 \\ -\frac{v_2^T v_1}{\alpha_1+\alpha_2} \ell_2 & -\frac{v_2^T v_2}{2\alpha_2} \ell_2 & 0 & 1 \end{pmatrix} = \det \begin{pmatrix} 1 - a_j^{(2)} \ell_2 & \frac{1}{2\alpha_2} \ell_2 \\ -\frac{v_2^T v_2}{2\alpha_2} \ell_2 & 1 \end{pmatrix}$$

Using this argument, we find the soliton (11.1) with the parameters α_2, φ_2^- given by $e^{\varphi_2^-} = \|v_2\|/(2\alpha_2)$ and $P_2^- = \frac{1}{\|v_2\|} v_2$.

Moving with the first soliton, on curves $x - \alpha_1^2 t = c$, we have $1/\ell_2(x,t) \to 0$ for $t \to -\infty$. In this case we divide in the determinants q_j and q the second and the fourth row by ℓ_2. Note that this does not affect the quotient q_j/q. Again we focus on the more complicated determinant q_j, for which we get $q_j \to \tilde{q}_j$, where

$$\tilde{q}_j = \det \begin{pmatrix} 1 - a_j^{(1)} \ell_1 & -a_j^{(2)} \ell_1 & \frac{1}{2\alpha_1} \ell_1 & \frac{1}{\alpha_1+\alpha_2} \ell_1 \\ -a_j^{(1)} & -a_j^{(2)} & \frac{1}{\alpha_2+\alpha_1} & \frac{1}{2\alpha_2} \\ -\frac{v_1^T v_1}{2\alpha_1} \ell_1 & -\frac{v_1^T v_2}{\alpha_1+\alpha_2} \ell_1 & 1 & 0 \\ -\frac{v_2^T v_1}{\alpha_2+\alpha_1} & -\frac{v_2^T v_2}{2\alpha_2} & 0 & 0 \end{pmatrix}.$$

To identify the solution, we

(1) multiply the second row by $-\ell_1(2\alpha_2)/(\alpha_1+\alpha_2)$ and add it to the first row,
(2) multiply the second column by $-((2\alpha_2)/(\alpha_1+\alpha_2))((v_2^{\mathrm{T}}v_1)/(v_2^{\mathrm{T}}v_2)$ and add it to the first column.

After expansion, we get

$$\tilde{q}_j = \gamma \det(I_2 + T) \quad \text{with} \quad T = \begin{pmatrix} -(v_1^-)^{(j)} A\ell_1 & \frac{1}{2\alpha_1} A^2 \ell_1 \\ -\frac{v_1^{\mathrm{T}} v_1}{2\alpha_2} \Delta\, \ell_1 & 0 \end{pmatrix}, \quad \gamma = \frac{v_2^{\mathrm{T}} v_2}{(2\alpha_2)^2},$$

with

$$A = \frac{\alpha_2 - \alpha_1}{\alpha_2 + \alpha_1} > 0, \qquad \Delta = 1 - \frac{4\alpha_1\alpha_2}{(\alpha_1+\alpha_2)^2} \frac{(v_1^{\mathrm{T}} v_2)^2}{\|v_1\|^2 \|v_2\|^2},$$

and $(v_1^-)^{(j)}$ the j-th component of the vector

$$v_1^- = -(v_1 - \frac{2\alpha_2}{\alpha_1+\alpha_2} \frac{v_1^{\mathrm{T}} v_2}{v_2^{\mathrm{T}} v_2} v_2).$$

Observe that

$$\Delta\, v_1^{\mathrm{T}} v_1 = (v_1^-)^{\mathrm{T}} v_1^-.$$

Now, factorising $T = T_1 T_2$, where $T_1 = \operatorname{diag}\{A, 1\}$, $T_2 = T_1^{-1} T$, and using the identity $\det(I + T_1 T_2) = \det(I + T_2 T_1)$, we get

$$\tilde{q}_j = \gamma \det \begin{pmatrix} 1 - (v_1^-)^{(j)} (A\ell_1) & \frac{1}{2\alpha_1}(A\ell_1) \\ -\frac{(v_1^-)^{\mathrm{T}} v_1^-}{2\alpha_1}(A\ell_1) & 1 \end{pmatrix}.$$

The factor γ appears also in front of the other determinants and hence cancels out in the quotients. As a result, we find the soliton (11.1) with the parameters α_1, φ_1^- given by $e^{\varphi_1^-} = A\|v_1^-\|/(2\alpha_1)$ and $P_1^- = \frac{1}{\|v_1^-\|} v_1^-$.

On curves $x + \beta t = c$ with $\beta \notin \{\alpha_1, \alpha_2\}$, one checks that there is no asymptotic contribution.

Behaviour for $t \to \infty$. By analogous arguments as for $t \to -\infty$ we obtain

1. the soliton (11.1) with the parameters α_1, φ_1^+ given by $e^{\varphi_1^+} = \|v_1\|/(2\alpha_1)$ and $P_1^+ = \frac{1}{\|v_1\|} v_1$ on curves $x - \alpha_1^2 t = c$,

2. the soliton (11.1) with the parameters α_2, φ_2^+ given by $e^{\varphi_2^+} = A\|v_2^+\|/(2\alpha_2)$ and $P_2^+ = \frac{1}{\|v_2^+\|} v_2^+$ on curves $x - \alpha_2^2 t = c$, where

$$v_2^+ = v_2 - \frac{2\alpha_1}{\alpha_1+\alpha_2} \frac{v_1^{\mathrm{T}} v_2}{v_1^{\mathrm{T}} v_1} v_1.$$

3. no contribution on curves $x + \beta t = c$ with $\beta \notin \{\alpha_1, \alpha_2\}$.

Example. The plots in Figure 6 show the two components of a 2-soliton solution of the 2×1-vector mKdV (7.4). The parameters used in Corollary 3 are $\mathsf{d}_1 = 1$, $\mathsf{d}_2 = 2$ (specifying the dimensions of the solution), $n = 2$ and $\alpha_1 = 1$, $\alpha_2 = 2$. Furthermore, $c = a_1 = (1,1)^{\mathrm{T}}$ and $a_2 = (0,1)^{\mathrm{T}}$. The solution is plotted for $-20 \leq x \leq 20$, $-10 \leq t \leq 10$ with a plot range from $-\frac{1}{2}$ to 2 in both plots. The example nicely confirms the asymptotic as the table below shows.

		$t \to -\infty$	$t \to \infty$
first soliton:	$\alpha_1 = 1$	$P_1^- = \frac{1}{\sqrt{5}}\begin{pmatrix}-1\\2\end{pmatrix}$	$P_1^+ = \begin{pmatrix}1\\0\end{pmatrix}$
second soliton:	$\alpha_2 = 2$	$P_2^- = \frac{1}{\sqrt{2}}\begin{pmatrix}1\\1\end{pmatrix}$	$P_2^+ = \frac{1}{\sqrt{10}}\begin{pmatrix}1\\3\end{pmatrix}$

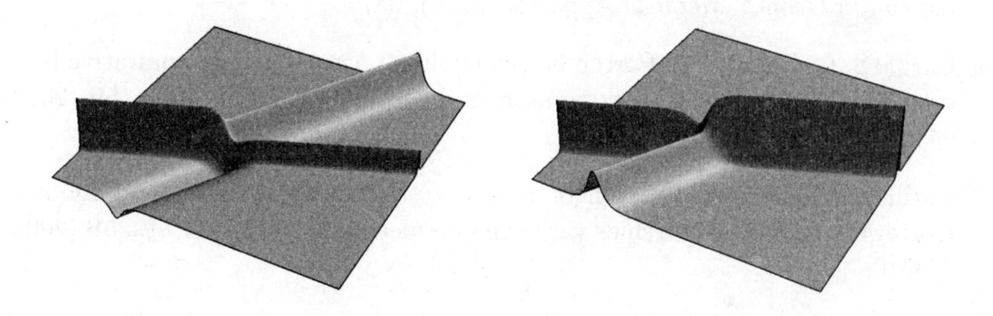

Figure 6. A 2-soliton of the vector-mKdV for $\mathsf{d}_1 = 1$, $\mathsf{d}_2 = 2$. The plot to the left shows the first component, the plot to the right the second component of the solution.

Acknowledgement

This work was partially supported by a research grant from the Jan Kochanowski University in Kielce.

References

[1] Ablowitz M J, Kaup D J, Newell A C, Segur H, The inverse scattering transformation - Fourier analysis for nonlinear problems, *Stud. Appl. Math.* **53** (1974), 249–315.

[2] Ablowitz M J, Prinari B, Trubatch A D, *Discrete and Continuous Nonlinear Schrödinger Systems*, Cambridge University Press, Cambridge 2004.

[3] Aden H, Elementary operators and solutions of the Korteweg-de Vries equation, Thesis, Jena 1996.

[4] Aden H, Carl B, On realizations of solutions of the KdV equation by determinants on operator ideals, *J. Math. Phys.* **37** (1996), 1833–1857.

[5] Aktosun T, Demontis F, van der Mee C, Exact solutions to the focusing nonlinear Schrödinger equation *Inverse Problems* **23** (2007), 2171–2195.

[6] Bauhardt W, Pöppe, The Zakharov-Shabat inverse spectral problem for operators, *J. Math. Phys.* **34** (1993), 3073–3086.

[7] Biondini G, Wang D, On the soliton solutions of the two-dimensional Toda lattice, *J. Phys. A* **43** (2010), 434007.

[8] Blohm, H, Solution of nonlinear equations by trace methods, *Nonlinearity* **13** (2000), 1925–1964.

[9] Carillo S, Lo Schiavo M, Schiebold C, Bäcklund transformations and nonabelian nonlinear evolution equations: A novel Bäcklund chart, *Symmetry Integrability Geom.: Methods Appl.* **12** (2016), 087.

[10] Carillo S, Lo Schiavo M, Porten E, Schiebold C, A novel noncommutative KdV-type equation, its recursion operator, and solitons, *J. Math. Phys.* **59** (2018), 043501.

[11] Carillo S, Schiebold C, Noncommutative Korteweg-de Vries and modified Korteweg-de Vries hierarchies via recursion methods, *J. Math. Phys.* **50** (2009), 073510.

[12] Carillo S, Schiebold C, Matrix Korteweg-de Vries and modified Korteweg-de Vries hierarchies: Noncommutative soliton solutions, *J. Math. Phys.* **52** (2011), 053507.

[13] Carl B, Huang S-Z, On realizations of solutions of the KdV equations by the C_0-semigroup method, *Amer. J. Math.* **122** (2000), 403–438.

[14] Carl B, Schiebold C, Nonlinear equations in soliton physics and operator ideals, *Nonlinearity* **12** (1999), 333–364.

[15] Carl B, Schiebold C, Ein direkter Ansatz zur Untersuchung von Solitonengleichungen, *Jahresber. Deutsch. Math.-Verein.* **102** (2000), 102–148.

[16] Dash A T, Schechter M, Tensor products and joint spectra, *Israel J. Math.* **8** (1970), 191–193.

[17] Demontis F, van der Mee C, Explicit solutions of the cubic matrix nonlinear Schrödinger equation, *Inverse Problems* **24** (2008), 025020.

[18] Dimakis A, Müller-Hoissen F, Solutions of matrix NLS systems and their discretisations: a unified treatment, *Inverse Problems* **26** (2010), 095007.

[19] Gesztesy F, Karwowski W, Zhao Z, Limits of soliton solutions, *Duke Math. J.* **68** (1992), 101–150.

[20] Goncharenko V M, On multisoliton solutions of the matrix KdV equation, *Theoret. Math. Phys.* **126** (2001), 81–91.

[21] Kupershmidt B A, *KP or mKP. Noncommutative Mathematics of Lagrangian, Hamiltonian, and Integrable Systems*, Mathematical Surveys and Monographs **78**, American Mathematical Society, Providence RI, 2000.

[22] Li S, Biondini, G and Schiebold C, On the degenerate soliton solutions of the focusing nonlinear Schrödinger equation, *J. Math. Phys.* **58** (2017), 033507.

[23] Manakov S V, On the theory of two-dimensional stationary self-focusing of electromagnetic waves, *Sov. Phys. JETP* **38** (1974), 248–253.

[24] Marchenko V A, *Nonlinear Equations and Operator Algebras*, Reidel, Dordrecht, 1988.

[25] Matveev V B, Asymptotics of the multipositon-soliton function of the Korteweg-de Vries equation and the supertransparency, *J. Math. Phys.* **35** (1994), 2955–2970.

[26] Matveev V B, Positons: slowly decreasing analogues of solitons, *Theor. Math. Phys.* **131** (2002), 483–497.

[27] Nilson T, Schiebold C, Solution formulas for the two-dimensional Toda lattice and particle-like solutions with unexpected asymptotic behaviour. Preprint 2018.

[28] Novikov S, Manakov S V, Pitaevskii L P, Zakharov V E, *Theory of Solitons. The Inverse Scattering Method*, Consultants Bureau, New York, 1984.

[29] Olmedilla E, Multiple pole solutions of the non-linear Schrödinger equation, *Physica D* **25** (1986), 330–346.

[30] Olver P J, Sokolov V V, Integrable evolution equations on nonassociative algebras, *Commun. Math. Phys.* **193** (1998), 245–268.

[31] Pietsch A, *Eigenvalues and s-Numbers*, Cambridge University Press, Cambridge 1987.

[32] Pólya G, and Szegö G, *Aufgaben und Lehrsätze zur Analysis II*, Springer, New York, 1976.

[33] Rasinariu C, Sukhatme U, Khare A, Negaton and positon solutions of the KdV and mKdV hierarchy, *J. Phys. A* **29** (1996), 1803–1823.

[34] Sakhnovich A L, Sakhnovich L A, Roitberg I Ya, *Inverse Problems and Nonlinear Evolution Equations: Solutions, Darboux Matrices and Weyl-Titchmarsh Functions*, De Gruyter Studies in Mathematics 47, De Gruyter, Berlin 2013.

[35] Schiebold C, Solitons of the sine-Gordon equation coming in clusters, *Revista Mat. Complut.* **15** (2002), 265–325.

[36] Schiebold C, On negatons of the Toda lattice, *J. Nonlin. Math. Phys.* **10** (2003),181–193.

[37] Schiebold C, *Integrable Systems and Operator Equations*, Habilitation thesis, Jena 2005. `http://apachepersonal.miun.se/~corsch/Habilitation.PDF`

[38] Schiebold C, A non-abelian Nonlinear Schrödinger equation and countable superposition of solitons, *J. Gen. Lie Theory Appl.* **2** (2008), 245–250.

[39] Schiebold C, Noncommutative AKNS systems and multisoliton solutions to the matrix sine-Gordon equation, *Discrete and Continuous Dynamical Systems* **2009** (2009), 678–690.

[40] Schiebold C, Cauchy-type determinants and integrable systems, *Linear Algebra Appl.* **433** (2010), 447–475.

[41] Schiebold C, The noncommutative AKNS systems: Projection to matrix systems, countable superposition, and soliton-like solutions, *J. Phys. A* **43** (2010), 434030.

[42] Schiebold C, Asymptotics for the multiple poles solutions of the Nonlinear Schrödinger equation, *Nonlinearity* **30** (2017), 2930–2981.

[43] Svinolupov S I, Sokolov V V, Vector-matrix generalizations of classical integrable systems, *Theor. Math. Physics* **100** (1994), 959–962.

[44] Tanaka S, Non-linear Schrödinger equation and modified Korteweg-de Vries equation; construction of solutions in terms of scattering data, *Publ. RIMS Kyoto Univ.* **10** (1975), 329–357.

[45] Tsuchida T, *N-soliton collision in the Manakov model*, Prog. Theor. Phys. **111** (2004), 151–182.

[46] Tsuchida T, Wadati M, *The coupled modified Korteweg-de Vries equations*, J. Phys. Soc. Japan **67** (1998), 1175–1187.

[47] van der Mee C, *Nonlinear Evolution Models of Integrable Type*, SIMAI e-Lecture Notes Vol. 11, 2013.

[48] Wadati M, The modified Korteweg-de Vries equation, *J. Phys. Soc. Japan* **34** (1973), 1289–1296.

[49] Wadati M, Ohkuma K, Multiple pole solutions of the modified Korteweg-de Vries equation, *J. Phys. Soc. Japan* **51**(1981), 2029–2035.

[50] Zakharov V E, Shabat A, Exact theory of two-dimensional self-focusing and one-dimensional self-modulation of waves in nonlinear media, *Sov. Phys. JETP* **34** (1972), 62–69.

B6. Algebraic traveling waves for the generalized KdV-Burgers equation and the Kuramoto-Sivashinsky equation

Claudia Valls

Departamento de Matemática, Instituto Superior Técnico
Universidade de Lisboa, 1049-001 Lisboa, Portugal

Abstract

In this paper, using a new method provided in [9] we characterize all traveling wave solutions of the Generalized Korteweg-de-Vries-Burgers equation as well as for the Kuramoto-Sivashinsky equation. In particular we recover the traveling wave solutions for the well-known Korteweg-de-Vries-Burgers equation.

1 Introduction and statement of the main results

Looking for traveling wave solutions to nonlinear evolution equations has long been the major problem for mathematicians and physicists. These solutions may well describe various phenomena in physics and other fields and thus they may give more insight into the physical aspects of the problems. Many methods for obtaining traveling wave solutions have been established [6, 7, 23, 24, 32, 33] with more or less success. When the degree of the nonlinearity is high most of the methods fail or can only lead to a kind of special solution and the solution procedures become very complex and do not lead to an efficient way to compute them. In this paper we will first study the algebraic traveling wave solutions to the Kuramoto-Sivashinsky equation of the form

$$u_t + u_{xxxx} + u_{xx} + uu_x = 0. \tag{1.1}$$

This equation has attracted a great deal of interest as a model for complex spatiotemporal dynamics in spatially extended systems, and as a paradigm for finite-dimensional dynamics in a partial differential equation. The Kuramoto-Sivashinsky equation describes the dynamics near long-wave-length primary instabilities in the presence of appropriate (translational, parity and Galilean) symmetries [28] and it has been independently derived in the context of several extended physical systems driven far from equilibrium by intrinsic instabilities, including instabilities of dissipative trapped ion modes in plasmas. It has been independently derived in the context of several extended physical systems driven far from equilibrium by intrinsic instabilities, including instabilities of dissipative trapped ion modes in plasmas [5, 21], instabilities in laminar frame fronts [30], phase dynamics in reaction-diffusion systems [20], and fluctuations in fluid films on inclines [29].

The term u_{xx} is responsible for an instability at large scales; the dissipative term u_{xxxx} provides damping at small scales; and the non-linear term uu_x (which has the same form as that in the Burgers or one-dimensional Navier Stokes equations) stabilizes by transferring energy between large and small scales.

The existence of steady solutions was shown in [26, 31] where the authors using numerical studies and analytical solutions obtained partial results concerning the existence of bounded solutions for equation (1.1) under periodic boundary conditions. In this paper we focus on proving the existence of traveling wave solutions for equation (1.1). More precisely, we will look for what we will call algebraic traveling wave solutions (see below for a precise definition).

Second, we will focus on obtaining algebraic traveling wave solutions to the generalized Korteweg-de-Vries-Burgers equation (GKdVB) of the form

$$au_{xxx} + bu_{xx} + du^n u_x + u_t = 0 \tag{1.2}$$

where $n = 1, 2$ and a, b, d are real constants with $abd \neq 0$. When $n = 1$ is the well-known Korteweg-de-Vries-Burgers equation (KdVB) that has been intensively investigated. When $n = 2$ we will call it the generalized Korteweg-de-Vries-Burgers equation (GKdVB). These equations are widely used in fields such as solid-states physics, plasma physics, fluid physics and quantum field theory (see, for instance [13, 38] and the references therein). They mainly appear when seeking the asymptotic behavior of complicated systems governing physical processes in solid and fluid mechanics.

Special attention is paid to the KdVB, often considered as a combination of the Burgers equation and KdV equation since in the limit $a \to 0$, the equation reduces to the Burgers equation (named after its use by Burgers [3] for studying turbulence in 1939), and taking the limit as $b \to 0$ we get the KdV equation (first suggested by Korteweg de Vries [19] who used it as a nonlinear model to study the change of forms of long waves advancing in a rectangular channel).

The KdVB equation is the simplest form of the wave equation in which the nonlinear term uu_x, the dispersion u_{xxx} and the dissipation u_{xx} all occur. It arises in many physical contexts such as the undular bores in shallow water [1, 17], the flow of liquids containing gas bubbles [34], the propagation of waves in an elastic tube filled with a viscous fluid [16], weakly nonlinear plasma waves with certain dissipative effects [10, 12], the cascading down process of turbulence [8] and atmospheric dynamics [18].

It is nonintegrable in the sense that its spectral problem is nonexistent. The existence of traveling wave solutions for the (KdVB) was obtained for the first time in [36] and after that many other papers computing the traveling wave of the KdVB appeared (see for instance [11, 14, 15, 25, 32, 35, 37]), but most of them did not obtain all the possible traveling wave solutions. However, despite the attention paid to the (KdVB), nothing is known of the existence of traveling wave solutions for the (GKdVB). This is due to the presence of high nonlinear terms. In this paper we will fill in this gap. We will use a method that will supply the already known traveling wave solution for the (KdVG) and will allows us to prove that there are no algebraic traveling wave solutions for the KdVG (i.e., equation (1.2) with $n = 2$).

As explained above, there are various approaches for constructing traveling wave solutions, but these methods become more and more useless as the degree of the nonlinear terms increase. However, in [9] the authors gave a technique to prove the existence of traveling wave solutions for general n-th order partial differential equations by showing that traveling wave solutions exist if and only if the associated n-dimensional first order ordinary differential equation has some invariant algebraic curve.

More precisely, consider the n-th order partial differential equations of the form

$$\frac{\partial^n u}{\partial x^n} = F\left(u, \frac{\partial u}{\partial x}, \frac{\partial u}{\partial t}, \frac{\partial^2 u}{\partial x^2}, \frac{\partial^2 u}{\partial x \partial t}, \frac{\partial^2 u}{\partial t^2}, \ldots, \frac{\partial^{n-1} u}{\partial x^{n-1}}, \frac{\partial^{n-1} u}{\partial x^{n-2}\partial t}, \ldots, \frac{\partial^{n-1} u}{\partial x \partial t^{n-2}}, \frac{\partial^{n-1} u}{\partial t^{n-1}}\right), \tag{1.3}$$

where x and t are real variables and F is a smooth map. The traveling wave solutions of system (1.3) are particular solutions of the form $u = u(x,t) = U(x - ct)$ where $U(s)$ satisfies the boundary conditions

$$\lim_{s\to-\infty} U(s) = A \quad \text{and} \quad \lim_{s\to\infty} U(s) = B \tag{1.4}$$

where A and B are solutions, not necessarily different, of $F(u, 0, \ldots, 0) = 0$. Plugging $u(x,t) = U(x - ct)$ into (1.2) we get that $U(s)$ has to be a solution, defined for all $s \in \mathbb{R}$ of the 3rd order ordinary differential equation

$$\begin{aligned} U^{(n)} = F(&U, U', -cU', U'' - cU'', c^2U''', \ldots, \\ &U^{(n-1)}, -cU^{(n-1)}, \ldots, (-c)^{n-2}U^{(n-1)}, (-c)^{n-1}U^{(n-1)}), \end{aligned} \tag{1.5}$$

where $U(s)$ and the derivatives are taken with respect to s. The parameter c is called the *speed* of the traveling wave solution.

We say that $u(x,y) = U(x - ct)$ is an *algebraic traveling wave solution* if $U(s)$ is a nonconstant function that satisfies (1.4) and (1.5) and there exists a polynomial p such that $p(U(s), U'(s), U''(s), \ldots, U^{(n-1)}(s)) = 0$.

We recall that for irreducible polynomials, we have the following algebraic characterization of invariant algebraic surfaces: Given an irreducible polynomial of degree m, $g(x_1, \ldots, x_n)$, we have that $g(x_1, \ldots, x_n) = 0$ is an algebraic invariant surface for the system

$$\begin{aligned} x_1' &= P_1(x_1, \ldots, x_n), \\ x_2' &= P_2(x_1, \ldots, x_n), \\ \vdots' &= \vdots \\ x_n' &= P_n(x_1, \ldots, x_n), \end{aligned}$$

for $P_1, \ldots, P_n \in \mathbb{C}[x_1, \ldots, x_n]$, if there exists a polynomial $K = K(x_1, \ldots, x_n)$ of degree at most $m - 1$, called the *cofactor* of g such that

$$P_1(x_1, \ldots, x_n)\frac{\partial g}{\partial x_1} + \ldots + P_n(x_1, \ldots, x_n)\frac{\partial g}{\partial x_n} = K(x_1, \ldots, x_n)g. \tag{1.6}$$

The main result that we will use is the following theorem; see [9] for its proof.

Theorem 1. *The partial differential equation* (1.3) *has an algebraic traveling wave solution with respect to c if and only if the first order differential system*

$$\begin{cases} y_1' &= y_2, \\ y_2' &= y_3, \\ \vdots &= \vdots \\ y_{n-1}' &= y_n, \\ y_n' &= G_c(y_1, y_2, \dots, y_n), \end{cases} \tag{1.7}$$

where

$$G_c(y_1, y_2, \dots, y_n)$$
$$= F(y_1, y_2, -cy_2, y_3, -cy_3, c^2y_3, \dots, y_n, -cy_n, \dots, (-c)^{n-2}y_n, (-c)^{n-1}y_n)$$

has an invariant algebraic curve containing the critical points $(A, 0, \dots, 0)$ *and* $(B, 0, \dots, 0)$ *and no other critical point between them.*

The first main result of this paper is, with the techniques in [9] to obtain all algebraic traveling wave solutions of the Kuramoto-Sivashinsky equation.

Theorem 2. *System* (1.1) *has no algebraic traveling wave solutions.*

The proof of Theorem 2 is given in Section 2. In Section 2 we have also included some preliminary results that will be used to prove the second main result in the paper which, with the techniques in [9], obtains all algebraic traveling wave solutions of the (KdVB) and (GKdVB), i.e., all explicit traveling wave solutions of equation (1.2) when $n = 1$ and when $n = 2$.

Theorem 3. *The following holds for system* (1.2)*:*

(i) If $n = 1$ *(KdVB), it has the algebraic traveling wave solution*

$$u(x,t) = -\frac{12b^2}{25da}\left(\frac{1}{1+\kappa_1 e^{b(x-vt)/(5a)}}\right)^2 + \frac{6b^2}{25da} + \frac{v}{d},$$

where

$$v^2 = \frac{36b^4 - 1250da^3\kappa_2}{625a^2},$$

being κ_1, κ_2 *arbitrary constants with* $\kappa_1 > 0$.

(ii) If $n = 2$ *(GKdVB), it has no algebraic traveling wave solutions.*

The proof of Theorem 3 is given in Section 3 when $n = 1$ and in Section 4 when $n = 2$.

2 Proof of Theorem 2 and some preliminary results

In this section we first focus on obtaining the proof of Theorem 2.

Proof of Theorem 2. We first note that if there exists a solution of the form $u(x,t) = U(x-ct)$ of equation (1.1), then substituting in (1.1) and performing one integration yield

$$U''' + U' + \frac{1}{2}U^2 - cU = \theta,$$

where θ is the integration constant. Therefore, we will look for the invariant algebraic surfaces of the system

$$\begin{aligned} x' &= y, \\ y' &= z, \\ z' &= \theta - y + cx - \frac{1}{2}x^2, \end{aligned} \tag{2.1}$$

where $x(s) = U(s)$ and $\theta \in \mathbb{R}$. Now we consider the change of variables

$$X = x - c, \quad Y = y, \quad Z = z,$$

in equation (2.1) we get

$$\begin{aligned} X' &= Y, \\ Y' &= Z, \\ Z' &= \bar{\theta} - Y - \frac{1}{2}X^2, \end{aligned} \tag{2.2}$$

where $\bar{\theta} = \theta + c^2/2$. If $\bar{\theta} < 0$ then system (2.2) has no critical points. In view of Theorem 1, system (1.1) cannot have algebraic traveling wave solutions; hence we can assume that $\bar{\theta} \geq 0$. We write it as $\bar{\theta} = \psi^2$. In this way, system (2.2) is the well-known Michelson system. It was proven in [22] that system (2.2) has no invariant algebraic surfaces and so, in view of Theorem 1 system (1.1) cannot have algebraic traveling wave solutions. This concludes the proof of Theorem 2. ■

In the rest of the section we introduce some notations and results that will be used during the proof of Theorem 3.

The first one is based on the previous works of Seidenberg [28] and was stated and proved in [4]. In the next theorem we included only the results from [4] that will be used in the paper.

Theorem 4. *Let $g(x,y) = 0$ be an invariant algebraic curve of a planar system with corresponding cofactor $K = K(x,y)$. Assume that $p = (x_0, y_0)$ is one of the critical points of the system. If $g(x_0, y_0) \neq 0$, then $K(x_0, y_0) = 0$. Moreover, assume that λ and μ are the eigenvalues of such critical point. If either $\mu \neq 0$ and λ and μ are rationally independent or $\lambda\mu < 0$, or $\mu = 0$, then $K(x_0, y_0) \in \{\lambda, \mu, \lambda + \mu\}$.*

A polynomial $g(x,y)$ is said to be a *weight homogeneous polynomial* if there exist $s=(s_1,s_2)\in\mathbb{N}^2$ and $m\in\mathbb{N}$ such that for all $\alpha\in\mathbb{R}\setminus\{0\}$, we have

$$g(\alpha^{s_1}x,\alpha^{s_2}y)=\alpha^m g(x,y),$$

where $\mathbb{R}$ denotes the set of real numbers and $\mathbb{N}$ is the set of positive integers. We shall refer to $s=(s_1,s_2)$ as the weight of g, to m as the weight degree and to $x=(x_1,x_2)\mapsto(\alpha^{s_1}x_1,\alpha^{s_2}x_2)$ as the weight change of variables.

We first note that if there exists a solution of the form $u(x,t)=U(x-ct)$ then substituting in (1.2) and performing one integration yield

$$U''=-\beta U'-\gamma U^{n+1}+\delta U+\theta,$$

where $\beta=b/a$, $\gamma=d/(a(n+1))$, $\delta=c/a$ and θ is the integration constant. Therefore, we will look for the invariant algebraic curves of the system

$$\begin{aligned} x'&=y,\\ y'&=-\beta y-\gamma x^{n+1}+\delta x+\theta, \end{aligned} \tag{2.3}$$

where $x(s)=U(s)$ and $\beta,\gamma,\delta,\theta\in\mathbb{R}$ with $\beta\gamma\delta\neq 0$.

When $n=1$, the solution of $\gamma x^2-\delta x-\theta=0$, that is,

$$x_{1,2}=\frac{\delta}{2\gamma}\pm\frac{\sqrt{\delta^2+4\gamma\theta}}{2\gamma}$$

must be real; otherwise there would not be algebraic traveling wave solutions. Hence $\delta^2+4\gamma\theta\geq 0$. Set $x=\bar{x}+x_1$, and $y=\bar{y}$, Then we can rewrite system (2.3) with $n=1$ in the variables $(\bar{x},y)$ as

$$\begin{aligned} \bar{x}'&=\bar{y},\\ \bar{y}'&=-\beta\bar{y}-\gamma(\bar{x}+x_1)^2+\delta(\bar{x}+x_1)+\theta\\ &=-\beta\bar{y}-\gamma\bar{x}^2-2\gamma x_1\bar{x}-\gamma x_1^2+\delta\bar{x}+\delta x_1+\theta\\ &=-\beta\bar{y}-\gamma\bar{x}^2+\bar{\delta}\bar{x}, \end{aligned} \tag{2.4}$$

where $\bar{\delta}=\delta-2\gamma x_1=\sqrt{\delta^2+4\gamma\theta}$.

When $n=2$, the solution of $\gamma x^3-\delta x-\theta=0$ has at least one real solution, that we denote by x_1. Set $x=\bar{x}+x_1$, and $y=\bar{y}$. Then we rewrite system (2.3) with $n=2$ in the variables $(\bar{x},\bar{y})$ as

$$\begin{aligned} \bar{x}'&=\bar{y},\\ \bar{y}'&=-\beta\bar{y}-\gamma(\bar{x}+x_1)^3+\delta(\bar{x}+x_1)-\theta\\ &=-\beta\bar{y}-\gamma\bar{x}^3-3\gamma x_1\bar{x}^2-3\gamma x_1^2\bar{x}-\gamma x_1^3+\delta\bar{x}+\delta x_1-\theta\\ &=-\beta\bar{y}-\gamma\bar{x}^3-\bar{\gamma}\bar{x}^2+\bar{\delta}\bar{x}, \end{aligned} \tag{2.5}$$

where $\bar{\gamma}=3\gamma x_1$ and $\bar{\delta}=\delta-3\gamma x_1^2$.

3 Proof of Theorem 3 with $n = 1$

In this section we consider system (2.3) with $n = 1$. By the results in Section 2 this is equivalent to work with system (2.4).

Theorem 5. *System* (2.4) *has an invariant algebraic curve* $g(\bar{x}, \bar{y}) = 0$ *if and only if*

$$\beta = \pm\frac{5\sqrt{\delta}}{\sqrt{6}}.$$

More precisely, if $\beta = 5\sqrt{\delta}/\sqrt{6}$ *then*

$$g(\bar{x}, \bar{y}) = \frac{\bar{y}^2}{2} - \frac{\sqrt{2}}{\sqrt{3}}\frac{\sqrt{\delta}}{\gamma}(\bar{\delta} - \gamma\bar{x})\bar{y} + \frac{\bar{x}}{3\gamma}(\delta - \gamma\bar{x})^2,$$

and if $\beta = -5\sqrt{\delta}/\sqrt{6}$ *then*

$$g(\bar{x}, \bar{y}) = \frac{\bar{y}^2}{2} + \frac{\sqrt{2}}{\sqrt{3}}\frac{\sqrt{\delta}}{\gamma}(\bar{\delta} - \gamma\bar{x})\bar{y} + \frac{\bar{x}}{3\gamma}(\delta - \gamma\bar{x})^2.$$

Proof. Let $g = g(\bar{x}, \bar{y}) = 0$ be an invariant algebraic curve of system (2.4). First we show that the cofactor $K(\bar{x}, \bar{y})$ must be constant. Indeed, since system (2.4) has degree two, the cofactor has degree at most one, i.e., $K(\bar{x}, \bar{y}) = k_0 + k_1\bar{x} + k_2\bar{y}$ with $k_0, k_1, k_2 \in \mathbb{C}$. Then equation (1.6) writes as

$$-\bar{y}\frac{\partial g}{\partial \bar{x}} - (\beta\bar{y} + \gamma\bar{x}^2 - \bar{\delta}\bar{x})\frac{\partial g}{\partial \bar{y}} = (k_0 + k_1\bar{x} + k_2\bar{y})g. \tag{3.1}$$

Now we write g as sum of its homogeneous parts in the form

$$g = \sum_{j=0}^{n} g_j(\bar{x}, \bar{y}), \quad \text{where } g_j \text{ are homogenous polynomials of degree } j.$$

Assuming that the higher order term in expression (3.1) vanishes, we get the partial differential equation

$$-\gamma\bar{x}^2\frac{\partial g_n}{\partial \bar{y}} = (k_1\bar{x} + k_2\bar{y})g_n.$$

Solving it we get

$$g_n = \bar{g}_n(\bar{x})e^{-\frac{k_1}{\gamma\bar{x}}\bar{y} - \frac{k_2}{2\gamma\bar{x}^2}\bar{y}^2},$$

where $\bar{g}_n$ is a smooth function in x. Since g_n has to be a polynomial we get that $k_2 = k_1 = 0$ and so $K(\bar{x}, \bar{y}) = k_0 \in \mathbb{R}$.

Now we introduce the weight-change of variables of the form

$$\bar{x} = \mu^{-2}X, \quad \bar{y} = \mu^{-3}Y, \quad t = \mu\tau.$$

In this form, system (2.4) becomes

$$\begin{aligned} X' &= Y, \\ Y' &= -\gamma X^2 - \mu\beta X + \bar{\delta}\mu^2 X, \end{aligned} \tag{3.2}$$

where the prime denotes the derivative in τ. Now we set

$$G(X,Y) = \mu^N g(\mu^{-2}X, \mu^{-3}Y) \quad \text{and} \quad K = \mu k_0,$$

where N is the highest weight degree in the weight homogeneous components of g in x, y with weight $(2,3)$.

We note that $G = 0$ is an invariant algebraic curve of system (3.2) with cofactor μk_0. Indeed

$$\frac{dG}{d\tau} = \mu^{N+1}\frac{dg}{d\tau} = \mu^{N+1}k_0 g = \mu k_0 G.$$

Assume that $G = \sum_{i=0}^{\ell} G_i$ where G_i is a weight homogeneous polynomial in X, Y with weight degree $\ell - i$ for $i = 0, \ldots, \ell$ and $\ell \geq N$. Obviously

$$g = G|_{\mu=1}.$$

From the definition of invariant algebraic curve we have

$$Y\sum_{i=0}^{\ell}\mu^i\frac{\partial G_i}{\partial X} - (\gamma X^2 + \mu\beta Y - \delta\mu^2 X)\sum_{i=0}^{\ell}\frac{\partial G_i}{\partial Y} = \mu k_0\sum_{i=0}^{\ell}\mu^i G_i.$$

Equating terms with μ^0 and μ^1 we get

$$L[G_0] = 0, \quad L[G_1] = \beta Y\frac{\partial G_0}{\partial X} + k_0 G_0, \tag{3.3}$$

where

$$L = Y\frac{\partial}{\partial X} - \gamma X^2\frac{\partial}{\partial Y}.$$

The characteristic equations associated with the first linear partial differential equation of system (3.3) are

$$\frac{dX}{dY} = -\gamma\frac{Y}{X^2}.$$

This system has the general solution $Y^2/2 + \gamma X^3/3 = \kappa$, where κ is a constant.

According with the method of characteristics (see [2]) we make the change of variables

$$u = Y^2/2 + \gamma X^3/3, \quad v = X. \tag{3.4}$$

Its inverse transformation is

$$Y = \pm\sqrt{2u - 2\gamma v^3/3}, \quad X = v. \tag{3.5}$$

In the following for simplicity we only consider the case $Y = +\sqrt{2u - 2\gamma v^3/3}$. Under the changes (3.4) and (3.5), the first equation in (3.3) becomes the following ordinary differential equation (for fixed u):

$$\sqrt{2u - 2\gamma v^3/3}\frac{d\tilde{G}_0}{dv} = 0,$$

where $\tilde{G}_0$ is G_0 written in the variables u, v. In what follows we always write $\tilde{F}$ to denote a function $F = F(X, Y)$ written in the (u, v) variables, that is, $\tilde{F} = \tilde{F}(u, v)$. The above equation has the general solution

$$\tilde{G}_0 = \tilde{F}_0(u),$$

where $\tilde{F}_0$ is an arbitrary smooth function in the variable u. In order that G_0 be a weight homogeneous polynomial with weight degree m, since X and Y have weight degrees 2 and 3, respectively, we get that G_0 should be of weight degree $\ell = 6m$ for some convenient $m \in \mathbb{N}$. So, G_0 has the form

$$G_0 = a_\ell\Big(\frac{Y^2}{2} + \gamma\frac{X^3}{3}\Big)^\ell, \quad a_\ell \in \mathbb{C} \setminus \{0\}.$$

Substituting G_0 into the second equation in (3.3) and doing some computations we obtain

$$\begin{aligned}
L[G_1] &= \beta a_\ell \ell Y^2\Big(\frac{Y^2}{2} + \gamma\frac{X^3}{3}\Big)^{\ell-1} + k_0 a_\ell\Big(\frac{Y^2}{2} + \gamma\frac{X^3}{3}\Big)^\ell \\
&= \beta a_\ell \ell\Big(2\Big(\frac{Y^2}{2} + \gamma\frac{X^3}{3}\Big) - \frac{2}{3}\gamma X^3\Big)\Big(\frac{Y^2}{2} + \gamma\frac{X^3}{3}\Big)^{\ell-1} + k_0 a_\ell\Big(\frac{Y^2}{2} + \gamma\frac{X^3}{3}\Big)^\ell \\
&= a_\ell(2\beta\ell + k_0)\Big(\frac{Y^2}{2} + \gamma\frac{X^3}{3}\Big)^\ell - \frac{2}{3}\beta a_\ell \ell\gamma X^3\Big(\frac{Y^2}{2} + \gamma\frac{X^3}{3}\Big)^{\ell-1}.
\end{aligned}$$

Using the transformations in (3.4) and (3.5) and working in a similar way to solve $\tilde{G}_0$ we get the following ordinary differential equation

$$\sqrt{2u - 2\gamma v^3/3}\frac{d\tilde{G}_1}{dv} = a_\ell(2\beta\ell + k_0)u^\ell - \frac{2}{3}\beta a_\ell \ell\gamma v^3 u^{\ell-1}.$$

Integrating this equation with respect to v we get

$$\tilde{G}_1 = \tilde{F}_1(u) + \frac{2\beta\ell u^{\ell-1}}{5\sqrt{3}}v\sqrt{2u - \gamma v^3/3} + \frac{1}{5\sqrt{2}}u^{\ell-1/2}v(6\beta\ell + 5k_0)_2F_1\Big(\frac{1}{2}, \frac{1}{3}, \frac{4}{3}, \frac{\gamma v^3}{6u}\Big),$$

where $\tilde{F}_1(u)$ is a smooth function in the variable u and

$$_2F_1(a, b, c, y) = \sum_{k=0}^{\infty}\frac{a(a+1)\cdots(a+k-1)}{b(b+1)\cdots(b+k-1)c(c+1)\cdots(c+k-1)}\frac{y^k}{k!} \tag{3.6}$$

is the hypergeometric function that is well defined if b, c are not negative integers. In particular, it is a polynomial if and only if a is a negative integer. Note that in our case $a=1/2$, $b=1/3$ and $c=4/3$. Consequently 2_{F_1} is never a polynomial. Since $G_1(X,Y)=\tilde{F}_1(u)=F_1(Y^2/2+\gamma X^3/3)$ in order that $\tilde{G}_1$ is a weight homogeneous polynomial of weight degree $6m-1$ we must have $\tilde{F}_1=0$ and

$$6\beta\ell+5k_0=0, \quad \text{that is} \quad k_0=-\frac{6\beta\ell}{5}.$$

Now we apply Theorem 4. We recall that since $k_0\neq 0$ is a constant, in view of Theorem 4, g must vanish in the critical points of system (2.4), which are $(0,0)$ and $(\bar{\delta}/\gamma,0)$. Moreover, the critical point $(0,0)$ has the eigenvalues

$$\lambda^+=-\frac{\beta}{2}+\frac{\sqrt{\beta^2+4\bar{\delta}}}{2} \quad \text{and} \quad \lambda^-=-\frac{\beta}{2}-\frac{\sqrt{\beta^2+4\bar{\delta}}}{2}$$

while the critical point $(\bar{\delta}/\gamma,0)$ has the eigenvalues

$$\mu^+=-\frac{\beta}{2}+\frac{\sqrt{\beta^2-4\bar{\delta}}}{2} \quad \text{and} \quad \mu^-=-\frac{\beta}{2}-\frac{\sqrt{\beta^2-4\bar{\delta}}}{2}.$$

Note that since $\bar{\delta}>0$, $(0,0)$ is a saddle. In view of Theorem 4 we must have that $k_0\in\{\lambda^+,\lambda^-,\lambda^++\lambda^-\}=\{\lambda^+,\lambda^-,-\beta\}$. Note that if $k_0=-\beta$ then

$$-\frac{6\beta\ell}{5}=-\beta \quad \text{that is} \quad \beta\frac{5-6\ell}{5}=0,$$

which is not possible. So, either $k_0=\lambda^+$, or $k_0=\lambda^-$.

Moreover, if $\beta^2-4\bar{\delta}<0$ then μ^+ and μ^- would be rationally independent and in view of Theorem 4, then $k_0\in\{\mu^+,\mu^-,\mu^++\mu^-\}=\{\mu^+,\mu^-,-\beta\}$. Since $k_0\neq-\beta$, then k_0 would be either μ^- or μ^+. However, this would imply that either $\lambda^+=\mu^+$, or $\lambda^+=\mu^-$, or $\lambda^-=\mu^+$, or $\lambda^-=\mu^-$ and all cases are not possible. Hence, $\beta^2>4\bar{\delta}$. Note that $k_0\in\{\lambda^+,\lambda^-\}$ can be written as

$$-\beta\frac{12\ell-5}{5}=\pm\sqrt{\beta^2+4\bar{\delta}}.$$

Note that $\bar{\delta}=\sqrt{\delta^2+4\gamma\theta}>0$. Hence,

$$\beta=\pm\frac{5\sqrt{\bar{\delta}}}{\sqrt{6\ell}\sqrt{6\ell-5}}.$$

Assuming that

$$0<\beta^2-4\bar{\delta}=-\frac{\bar{\delta}}{6\ell(6\ell-5)}(144\ell^2-120\ell-25),$$

and taking into account that ℓ is an integer, we must have $\ell=1$. Therefore,

$$k_0=-\frac{6\beta}{5}, \quad \beta=\pm\frac{5\sqrt{\bar{\delta}}}{\sqrt{6}}, \quad G_0=a_1\left(\frac{Y^2}{2}+\frac{\gamma X^3}{3}\right).$$

This implies that the highest order terms in the invariant algebraic curve $g(\bar{x}, \bar{y})$ in the variable $\bar{x}$ is $\gamma\bar{x}^3/3$ and in the variable $\bar{y}$ is $\bar{y}^2/2$. Hence

$$g(\bar{x}, \bar{y}) = \frac{\gamma}{3}\bar{x}^3 + \frac{1}{2}\bar{y}^2 + g_1(\bar{x}, \bar{y})$$

where

$$g_1(\bar{x}, \bar{y}) = a_1\bar{x}^2 + a_2\bar{x}\bar{y} + a_3\bar{x} + a_4\bar{y} + a_5, \quad a_i \in \mathbb{C}, \ i = 1, \ldots, 5.$$

Assuming that g satisfies the definition of invariant algebraic curve with $k_0 = -6\beta/5$ and $\beta = 5\sqrt{\bar{\delta}}/\sqrt{6}$ we get that the unique invariant algebraic curve $g(\bar{x}, \bar{y}) = 0$ is of the form

$$g(\bar{x}, \bar{y}) = \frac{\bar{y}^2}{2} - \frac{\sqrt{2}}{\sqrt{3}}\frac{\sqrt{\bar{\delta}}}{\gamma}(\bar{\delta} - \gamma\bar{x})\bar{y} + \frac{\bar{x}}{3\gamma}(\bar{\delta} - \gamma\bar{x})^2, \tag{3.7}$$

and supposing that y satisfies the definition of invariant algebraic curve with $k_0 = -6\beta/5$ and $\beta = -5\sqrt{\bar{\delta}}/\sqrt{6}$ we get that the unique invariant algebraic curve $g(\bar{x}, \bar{y}) = 0$ is of the form

$$g(\bar{x}, \bar{y}) = \frac{\bar{y}^2}{2} + \frac{\sqrt{2}}{\sqrt{3}}\frac{\sqrt{\bar{\delta}}}{\gamma}(\bar{\delta} - \gamma\bar{x})\bar{y} + \frac{\bar{x}}{3\gamma}(\bar{\delta} - \gamma\bar{x})^2. \tag{3.8}$$

This concludes the proof of the theorem. ■

Proof of Theorem 3. Consider first the case $\beta = \frac{5\sqrt{\bar{\delta}}}{\sqrt{6}}$. It follows from Theorem 5 that the invariant algebraic curve is given in (3.7). The branch of $g(\bar{x}, \bar{y}) = 0$ that contains the origin is

$$\bar{y} = \frac{\sqrt{2}}{\sqrt{3}\gamma}(\bar{\delta} - \gamma\bar{x})\left(\sqrt{\bar{\delta}} - \sqrt{\bar{\delta} - \gamma\bar{x}}\right).$$

Since $\bar{x}' = \bar{y}$ we obtain

$$\bar{x}' = \frac{\sqrt{2}}{\sqrt{3}\gamma}(\bar{\delta} - \gamma\bar{x})\left(\sqrt{\bar{\delta}} - \sqrt{\bar{\delta} - \gamma\bar{x}}\right) = \frac{\sqrt{2}\bar{\delta}^{3/2}}{\sqrt{3}\gamma}\left(1 - \frac{\gamma}{\bar{\delta}}\bar{x}\right)\left(1 - \sqrt{1 - \frac{\gamma}{\bar{\delta}}\bar{x}}\right).$$

Set $U(s) = x(s) = \bar{x}(s) + x_1$ and take $W(s) - \sqrt{1 - \frac{\gamma}{\delta}(U(s) - x_1)}$. Then

$$W'(s) = -\frac{\gamma}{\bar{\delta}}\frac{U'(s)}{2\sqrt{1 - \frac{\gamma}{\delta}(U(s) - x_1)}} = -\frac{\sqrt{\bar{\delta}}}{\sqrt{6}}W(s)(1 - W(s)).$$

Its non-constant solutions that are defined for all $s \in \mathbb{R}$ are

$$W(s) = \frac{1}{1 + \kappa e^{\sqrt{\bar{\delta}}s/\sqrt{6}}}, \quad \kappa > 0.$$

Hence,

$$U(s) = x_1 + \frac{\bar{\delta}}{\gamma}\Big(1 - \Big(\frac{1}{1+\kappa e^{\sqrt{\bar{\delta}}s/\sqrt{6}}}\Big)^2\Big), \qquad \kappa > 0.$$

This together with the definition of $x_1, \bar{\delta}, \delta, \gamma$ and β, yields the traveling wave solution in the statement of the theorem.

If we take the branch of $g(\bar{x}, \bar{y}) = 0$ that does not contain the origin then

$$\bar{y} = \frac{\sqrt{2}}{\sqrt{3}\gamma}(\bar{\delta} - \gamma\bar{x})\big(\sqrt{\bar{\delta}} + \sqrt{\bar{\delta} - \gamma\bar{x}}\big).$$

Proceeding exactly as above we get that

$$W(s) = \frac{1}{1 - \kappa e^{\sqrt{\bar{\delta}}s/\sqrt{6}}}, \qquad \kappa > 0,$$

which is not a global solution. So, in this case there are no traveling wave solutions.

Now take $\beta = -\frac{5\sqrt{\delta}}{\sqrt{6}}$. It follows from Theorem 5 that the invariant algebraic curve is given in (3.8). The branch of $g(\bar{x}, \bar{y}) = 0$ that contains the origin is

$$\bar{y} = -\frac{\sqrt{2}}{\sqrt{3}\gamma}(\bar{\delta} - \gamma\bar{x})\big(\sqrt{\bar{\delta}} - \sqrt{\bar{\delta} - \gamma\bar{x}}\big).$$

Since $\bar{x}' = \bar{y}$ we obtain

$$\bar{x}' = -\frac{\sqrt{2}}{\sqrt{3}\gamma}(\bar{\delta} - \gamma\bar{x})\big(\sqrt{\bar{\delta}} - \sqrt{\bar{\delta} - \gamma\bar{x}}\big) = -\frac{\sqrt{2}\bar{\delta}^{3/2}}{\sqrt{3}\gamma}\Big(1 - \frac{\gamma}{\bar{\delta}}\bar{x}\Big)\Big(1 - \sqrt{1 - \frac{\gamma}{\bar{\delta}}\bar{x}}\Big).$$

Set $U(s) = x(s) = \bar{x}(s) + x_1$ and take $W(s) = \sqrt{1 - \frac{\gamma}{\bar{\delta}}(U(s) - x_1)}$. Then

$$W'(s) = \frac{\gamma}{\bar{\delta}}\frac{U'(s)}{2\sqrt{1 - \frac{\gamma}{\bar{\delta}}(U(s) - x_1)}} = \frac{\sqrt{\bar{\delta}}}{\sqrt{6}}W(s)(1 - W(s)).$$

Its non-constant solutions that are defined for all $s \in \mathbb{R}$ are

$$W(s) = \frac{1}{1 + \kappa e^{-\sqrt{\bar{\delta}}s/\sqrt{6}}}, \qquad \kappa > 0.$$

Hence,

$$U(s) = x_1 + \frac{\bar{\delta}}{\gamma}\Big(1 - \Big(\frac{1}{1+\kappa e^{-\sqrt{\bar{\delta}}s/\sqrt{6}}}\Big)^2\Big), \qquad \kappa > 0.$$

This together with the definition of $x_1, \bar{\delta}, \delta, \gamma$ and β, yields the traveling wave solution in the statement of the theorem.

If we take the branch of $g(\bar{x}, \bar{y}) = 0$ that does not contain the origin then

$$\bar{y} = -\frac{\sqrt{2}}{\sqrt{3}\gamma}(\bar{\delta} - \gamma\bar{x})\big(\sqrt{\bar{\delta}} + \sqrt{\bar{\delta} - \gamma\bar{x}}\big).$$

Proceeding exactly as above we get that

$$W(s) = \frac{1}{1 - \kappa e^{-\sqrt{\bar{\delta}}s/\sqrt{6}}}, \qquad \kappa > 0,$$

which is not a global solution. So, in this case there are no traveling wave solutions and the proof of the theorem is complete. ■

4 Proof of Theorem 3 with $n - 2$

In this section we consider system (2.3) with $n = 2$. By the results in Section 2 this is equivalent to work with system (2.5).

The proof of Theorem 3 with $n = 2$ follows directly from the following theorem which states that system (2.5) has no invariant algebraic curves.

Theorem 6. *System* (2.5) *has no invariant algebraic curves.*

Proof. Let $g = g(\bar{x}, \bar{y}) = 0$ be an invariant algebraic curve of system (2.5) with cofactor K. We write both g and K in their power series in the variable y as

$$K(\bar{x}, \bar{y}) = \sum_{j=0}^{2} K_j(\bar{x})\bar{y}^j, \quad g = \sum_{j=0}^{\ell} g_j(\bar{x})\bar{y}^\ell,$$

for some integer ℓ and where K_j is a polynomial in the variable x of degree j. Without loss of generality, since $g \neq 0$ we can assume that $g_\ell = g_\ell(\bar{x}) \neq 0$. Moreover, note that if system (2.5) has an invariant algebraic curve then

$$\bar{y}\frac{\partial g}{\partial \bar{x}} - (\beta\bar{y} + \gamma\bar{x}^3 + \bar{\gamma}\bar{x}^2 - \bar{\delta}\bar{x})\frac{\partial g}{\partial \bar{y}} = Kg. \tag{4.1}$$

We compute the coefficient of $\bar{y}^{2+\ell}$ in (4.1) and we get

$$g_\ell K_2 = 0 \quad \text{that is} \quad K_2 = 0$$

because $g_\ell \neq 0$. So, $K(\bar{x}) = K_0(\bar{x}) + K_1(\bar{x})\bar{y}$. Computing the coefficient of $\bar{y}^{\ell+1}$ in (4.1) we get

$$g_\ell'(\bar{x}) = K_1 g_\ell$$

which yields $g_\ell = \kappa e^{\int K_1(\bar{x})\, d\bar{x}}$, for $\kappa \in \mathbb{C} \setminus \{0\}$. Since g_ℓ must be a polynomial then $K_1 = 0$. This implies that $K(\bar{x}) = K_0(\bar{x})$, that we write as

$$K(\bar{x}) = K_0(\bar{x}) = \sum_{j=0}^{2} k_j\bar{x}^j, \quad k_j \in \mathbb{R}.$$

Now equation (4.1) writes as

$$\bar{y}\frac{\partial g}{\partial \bar{x}} - (\beta\bar{y} + \gamma\bar{x}^3 + \bar{\gamma}\bar{x}^2 - \bar{\delta}\bar{x})\frac{\partial g}{\partial \bar{y}} = \sum_{j=0}^{2} k_j\bar{x}^j g. \tag{4.2}$$

Now we introduce the weight-change of variables of the form

$$\bar{x} = \mu^{-2}X, \quad \bar{y} = \mu^{-4}Y, \quad t = \mu^2\tau.$$

In this form, system (2.5) becomes

$$\begin{aligned} X' &= Y, \\ Y' &= -\gamma X^3 - \mu^2\beta Y - \mu^2\bar{\gamma}X^2 + \bar{\delta}\mu^4 X \end{aligned} \tag{4.3}$$

where the prime denotes the derivative in τ. Now we set

$$G(X,Y) = \mu^N g(\mu^{-2}X, \mu^{-4}Y)$$

and

$$\bar{K} = \mu^2 K = \mu^2 k_0 + k_1 X + \mu^{-2}X^2, \quad k_0, k_1, k_2 \in \mathbb{C},$$

where N is the highest weight degree in the weight homogeneous components of g in $\bar{x}$, $\bar{y}$ with weight $(2,4)$.

We note that $G = 0$ is an invariant algebraic curve of system (4.3) with cofactor $\mu^2 K$. Indeed,

$$\frac{dG}{d\tau} = \mu^N \frac{dg}{d\tau} = \mu^N \mu^2 K g = \mu^n \bar{K} G.$$

Assume that $G = \sum_{i=0}^{\ell} G_i$ where G_i is a weight homogeneous polynomial in the variables X and Y with weight degree $\ell - i$ for $i = 0, \ldots, \ell$ and $\ell \geq N$. From the definition of invariant algebraic curve we have

$$\begin{aligned} &Y\sum_{i=0}^{\ell}\mu^i\frac{\partial G_i}{\partial X} - (\gamma X^3 + \mu^2\beta Y + \mu^2\bar{\gamma}X^2 - \bar{\delta}\mu^4 X)\sum_{i=0}^{\ell}\mu^i\frac{\partial G_i}{\partial Y} \\ &= (\mu^2 k_0 + k_1 X + \mu^{-2}k_2 X^2)\sum_{i=0}^{\ell}\mu^i G_i. \end{aligned} \tag{4.4}$$

Computing the terms with μ^{-2} we get that $k_2 = 0$. Now the terms of μ^0 in (4.4) becomes

$$L[G_0] = k_1 G_0, \quad L = Y\frac{\partial}{\partial X} - \gamma X^3 \frac{\partial}{\partial Y}. \tag{4.5}$$

The characteristic equations associated with the first linear partial differential equation of system (4.4) are

$$\frac{dX}{dY} = -\gamma \frac{Y}{X^3}.$$

This system has the general solution $u = Y^2/2 + \gamma X^4/4 = \kappa$, where κ is a constant. According to the method of characteristics we make the change of variables

$$u = \frac{Y^2}{2} + \frac{\gamma}{4}X^4, \quad v = X. \tag{4.6}$$

Its inverse transformation is

$$Y = \pm\sqrt{2u - 2\gamma v^4/2}, \quad X = v. \tag{4.7}$$

In the following for simplicity we only consider the case $Y = +\sqrt{2u - \gamma v^4/2}$. Under the changes (4.6) and (4.7), equation (4.5) becomes the following ordinary differential equation (for fixed u)

$$\sqrt{2u - \gamma v^4/2}\frac{d\tilde{G}_0}{dv} = k_1 \tilde{G}_0,$$

where $\tilde{G}_0$ is G_0 written in the variables u, v. The above equation has the general solution

$$\tilde{G}_0 = u^\ell \tilde{F}_0(u) \exp\left(\frac{k_1}{\sqrt{2u}} {}_2F_1\left(\frac{1}{2}, \frac{1}{4}, \frac{5}{4}, \frac{\gamma v^4}{4u}\right)\right),$$

where $\tilde{F}_0$ is an arbitrary smooth function in the variable u and ${}_2F_1$ is the hypergeometric function introduced in (3.6). Note that in this case ${}_2F_1$ is never a polynomial. Since $G_0(X, Y) = F_0(Y^2/2+\gamma X^4/4)$ in order that G_0 is a weight homogeneous polynomial of weight degree ℓ, since X and Y have weight degrees 2 and 4, respectively, we get that G_0 should be of weight degree $N = 8\ell$ and that $k_1 = 0$. Hence,

$$G_0 = a_\ell\left(\frac{Y^2}{2} + \gamma\frac{X^4}{4}\right)^\ell, \quad a_\ell \in \mathbb{R} \setminus \{0\}.$$

Computing the terms with μ in (4.4) using G_0, we get

$$L[G_1] = 0.$$

Using the transformations in (4.6) and (4.7) and working in a similar way to solve G_0 we get the following ordinary differential equation

$$\sqrt{2u - \gamma v^4/2}\frac{d\tilde{G}_1}{dv} = 0,$$

that is $\tilde{G}_1 = \tilde{G}_1(u)$. Since $\tilde{G}_1$ is a weight homogeneous polynomial of weight degree $N - 1 = 8\ell - 1$ and u has an even weight degree, we must have $\tilde{G}_1 = 0$ and so $G_1 = 0$.

Computing the terms with μ^2 in (4.4) using the expression of G_0 and the fact that $G_1 = 0$ we get

$$\begin{aligned}
L[G_2] &= \beta a_\ell \ell\left(\frac{Y^2}{2} + \gamma\frac{X^4}{4}\right)^{\ell-1} + \bar{\gamma} a_\ell \ell X^2 Y\left(\frac{Y^2}{2} + \gamma\frac{X^4}{4}\right)^{\ell-1} + k_0 a_\ell\left(\frac{Y^2}{2} + \gamma\frac{X^4}{4}\right)^\ell \\
&= \beta a_\ell \ell\left(2\left(\frac{Y^2}{2} + \gamma\frac{X^4}{4}\right) - \frac{2}{3}\gamma X^4\right)\left(\frac{Y^2}{2} + \gamma\frac{X^4}{4}\right)^{\ell-1} \\
&\quad + \bar{\gamma} a_\ell \ell X^2 Y\left(\frac{Y^2}{2} + \gamma\frac{X^4}{4}\right)^{\ell-1} + k_0 a_\ell\left(\frac{Y^2}{2} + \gamma\frac{X^4}{4}\right)^\ell \\
&= a_\ell(2\beta\ell + k_0)\left(\frac{Y^2}{2} + \gamma\frac{X^4}{4}\right)^\ell - \frac{1}{2}\beta a_\ell \ell \gamma X^4\left(\frac{Y^2}{2} + \gamma\frac{X^4}{4}\right)^{\ell-1} \\
&\quad + \bar{\gamma} a_\ell \ell X^2 Y\left(\frac{Y^2}{2} + \gamma\frac{X^4}{4}\right)^{\ell-1}.
\end{aligned}$$

Using the transformations in (4.6) and (4.7) and working in a similar way to solve $\tilde{G}_0$ we get the following ordinary differential equation

$$\sqrt{2u-\gamma v^4/2}\frac{d\tilde{G}_2}{dv} = a_\ell(2\beta\ell+k_0)u^\ell - \frac{1}{2}\beta a_\ell \ell\gamma v^4 u^{\ell-1} + \bar{\gamma} a_\ell \ell v^2\sqrt{2u-\gamma v^4/2}u^{\ell-1}.$$

Integrating this equation with respect to v we get

$$\begin{aligned}\tilde{G}_2 = \tilde{F}_2(u) &+ \frac{\beta\ell u^{\ell-1}}{6}v\sqrt{2u-\gamma v^4/2} + \frac{\bar{\gamma}a_\ell \ell}{3}v^3u^{\ell-1}\\ &+ \frac{1}{3\sqrt{2}}u^{\ell-1/2}v(4\beta\ell+3k_0)_2F_1\left(\frac{1}{2},\frac{1}{4},\frac{5}{4},\frac{\gamma v^4}{8u}\right),\end{aligned}$$

where $\tilde{F}_2$ is a smooth function in the variable u and ${}_2F_1$ is the hypergeometric function introduced in (3.6). Hence, ${}_2F_1$ is never a polynomial. Since G_2 should be a polynomial in the variable X we must have that

$$4\beta\ell + 3k_0 = 0, \quad \text{that is} \quad k_0 = -\frac{4\beta\ell}{3}.$$

Now we apply Theorem 4. We recall that since $k_0 \neq 0$ is a constant, in view of Theorem 4, g must vanish in the critical points of system (2.5), which are $(0,0)$, $(\psi_+,0)$ and $(\psi_-,0)$ where

$$\psi_\pm = \frac{-\bar{\gamma}\pm\sqrt{\bar{\gamma}^2+4\bar{\delta}\gamma}}{2\gamma}.$$

Moreover, the critical point $(0,0)$ has the eigenvalues

$$\lambda^+ = -\frac{\beta}{2}+\frac{\sqrt{\beta^2+4\delta}}{2} \quad \text{and} \quad \lambda^- = -\frac{\beta}{2}-\frac{\sqrt{\beta^2+4\delta}}{2},$$

the critical point $(\psi_+,0)$ has the eigenvalues

$$\mu^+ = -\frac{\beta}{2}+\frac{\sqrt{\beta^2+4T_+}}{2} \quad \text{and} \quad \mu^- = -\frac{\beta}{2}-\frac{\sqrt{\beta^2+4T_+}}{2},$$

being

$$T_+ = \frac{(\bar{\gamma}-\sqrt{\bar{\gamma}^2+4\bar{\delta}\gamma})\sqrt{\bar{\gamma}^2+4\bar{\delta}\gamma}}{2\gamma}$$

and the critical point $(\psi_-,0)$ has the eigenvalues

$$\nu^+ = -\frac{\beta}{2}+\frac{\sqrt{\beta^2+4T_-}}{2} \quad \text{and} \quad \nu^- = -\frac{\beta}{2}-\frac{\sqrt{\beta^2+4T_-}}{2},$$

being

$$T_- = \frac{(-\bar{\gamma}-\sqrt{\bar{\gamma}^2+4\bar{\delta}\gamma})\sqrt{\bar{\gamma}^2+4\bar{\delta}\gamma}}{2\gamma}.$$

We consider different cases.

Case 1: $\bar{\delta}\gamma > 0$ *and* $\gamma < 0$. In this case both $(\psi_+, 0)$ and $(\psi_-, 0)$ are saddles. In view of Theorem 4 we must have that $k_0 \in \{\mu^+, \mu^-, \mu^+ + \mu^-\} = \{\mu^+, \mu^-, -\beta\}$ and $k_0 \in \{\nu^+, \nu^-, \nu^+ + \nu^-\} = \{\nu^+, \nu^-, -\beta\}$. Note that if $k_0 = -\beta$ then

$$-\frac{4\beta\ell}{3} = -\beta \quad \text{that is} \quad \beta\frac{3-4\ell}{3} = 0,$$

which is not possible because $\beta \neq 0$ and ℓ is an integer with $\ell \geq 1$. So, $k_0 \in \{\mu^+, \mu^-\}$ and $k_0 \in \{\nu^+, \nu^-\}$. The only possibility is that $\bar{\gamma} = 0$. In this case

$$-\frac{4\beta\ell}{3} = -\frac{\beta}{2} \pm \frac{\sqrt{\beta^2 - 8\bar{\delta}}}{2}$$

which yields

$$\beta = \pm\frac{3\sqrt{-\bar{\delta}}}{\sqrt{14}}.$$

Moreover, the eigenvalues on $(0,0)$ are λ^+ and λ^-. If $\beta^2 + 4\bar{\delta} < 0$ then λ^+ and λ^- would be rationally independent. In view of Theorem 4 then $k_0 \in \{\lambda^+, \lambda^-, \lambda^+ + \lambda^-\} = \{\lambda^+, \lambda^-, -\beta\}$. But this would imply that

$$\sqrt{-\bar{\delta}}(i\sqrt{47} \pm (8\ell + 3)) = 0,$$

which is not possible. Hence, $\beta^2 + 4\bar{\delta} > 0$. However

$$\beta^2 + 4\bar{\delta} = \frac{47\bar{\delta}}{14} < 0$$

and so this case is not possible.

Case 2: $\bar{\delta}\gamma > 0$ *and* $\gamma > 0$. In this case $(0,0)$ is a saddle. In view of Theorem 4 we must have that $k_0 \in \{\lambda^+, \lambda^-, \lambda^+ + \lambda^-\} = \{\lambda^+, \lambda^-, -\beta\}$. As in Case 1, we cannot have $k_0 = -\beta$. So, assuming that $k_0 \in \{\lambda^+, \lambda^-\}$ we conclude that

$$\beta = \pm\frac{3\sqrt{\bar{\delta}}}{2\sqrt{\ell(3+4\ell)}}.$$

Moreover if $\beta^2 + 4T_+ < 0$ we would have that μ^+ a nd μ^- are rationally independent and so $k_0 \in \{\mu^+, \mu^-, -\beta\}$. However, $\mu^+ = \lambda^+$ (respectively $\mu^- = \lambda^-$) if and only if

$$\bar{\gamma} = \frac{3i\sqrt{\bar{\delta}\gamma}}{\sqrt{2}},$$

which is not possible. So $\beta^2 + 4T_+ > 0$. Equivalently, if $\beta^2 + 4T_- < 0$ we would have that ν^+ and ν^- are rationally independent and so $k_0 \in \{\nu^+, \nu^-, -\beta\}$. However, $\nu^+ = \lambda^+$ (respectively $\nu^- = \lambda^-$) if and only if

$$\bar{\gamma} = \frac{3i\sqrt{\bar{\delta}\gamma}}{\sqrt{2}},$$

which is not possible. So, $\beta^2 + 4T_- > 0$. This implies that

$$\frac{9\bar{\delta}}{2\ell(3+4\ell)} > \frac{2}{\gamma}\sqrt{\bar{\gamma}^2 + 4\bar{\delta}\gamma}(\bar{\gamma} + \sqrt{\bar{\gamma}^2 + 4\bar{\delta}\gamma})$$

and

$$\frac{9\bar{\delta}}{2\ell(3+4\ell)} > \frac{2}{\gamma}\sqrt{\bar{\gamma}^2 + 4\bar{\delta}\gamma}(-\bar{\gamma} + \sqrt{\bar{\gamma}^2 + 4\bar{\delta}\gamma})$$

or, in short,

$$\frac{9\bar{\delta}}{2\ell(3+4\ell)} > \frac{2}{\gamma}\sqrt{\bar{\gamma}^2 + 4\bar{\delta}\gamma}(|\bar{\gamma}| + \sqrt{\bar{\gamma}^2 + 4\bar{\delta}\gamma}) = 8\bar{\delta} + \frac{2}{\gamma}(|\bar{\gamma}|\sqrt{\bar{\gamma}^2 + 4\bar{\delta}\gamma} + \bar{\gamma}^2),$$

being $|\bar{\gamma}|$ the absolute value of $\bar{\gamma}$. Note that in particular this implies that

$$-\frac{\bar{\delta}(64\ell^2 + 48\ell - 9)}{2\ell(3+4\ell)} > \frac{2}{\gamma}(|\bar{\gamma}|\sqrt{\bar{\gamma}^2 + 4\bar{\delta}\gamma} + \bar{\gamma}^2) > 0,$$

which is not possible because $\bar{\delta} > 0$ and $\ell \geq 1$. So, this case is not possible.

Case 3*: $\bar{\delta}\gamma < 0$ and $\gamma < 0$.* In this case $(0,0)$ is a saddle. In view of Theorem 4 we must have $k_0 \in \{\lambda^+, \lambda^-, \lambda^+ + \lambda^-\} = \{\lambda^+, \lambda^-, -\beta\}$. As in Case 1, we cannot have $k_0 = -\beta$. So, assuming that $k_0 \in \{\lambda^+, \lambda^-\}$ we conclude that

$$\beta = \pm\frac{3\sqrt{\bar{\delta}}}{2\sqrt{\ell(3+4\ell)}}.$$

Now we assume that $\bar{\gamma} \leq 0$ (otherwise we will do the argument with T_- instead of T_+). Since T_+ is a saddle we must have $k_0 \in \{\mu^+, \mu^-, \mu^+ + \mu^-\} = \{\mu^+, \mu^-, -\beta\}$. Proceeding as in Case 2, we cannot have $k_0 = -\beta$ and equating it to either μ^+ or μ^- we obtain that

$$\bar{\gamma} = \frac{3i\sqrt{\delta\gamma}}{\sqrt{2}} = -\frac{3\sqrt{|\delta\gamma|}}{\sqrt{2}}.$$

Now proceeding as in Case 1 we have that $\mu^+ = \nu^+$ (respectively $\mu^- = \nu^-$) if and only if $\bar{\gamma} = 0$, which in this case is not possible because then $\bar{\delta} = \delta$ and $\delta\gamma \neq 0$. So, $\beta^2 + 4T_- > 0$; otherwise we would have that ν^+ and ν^- would be rationally independent and so $k_0 \in \{\nu^+, \nu^-, -\beta\}$ which we already showed is not possible. However, using that $\mu^+ = \lambda^+$ and $\mu^- = \lambda^-$ (that is, $T_+ = \bar{\delta}$) we get that

$$\bar{\gamma}\sqrt{\bar{\gamma}^2 + 4\bar{\delta}\gamma} = 2\gamma\bar{\delta} + \bar{\gamma}^2 + 4\bar{\delta}\gamma$$

and so

$$\begin{aligned}\beta^2 + 4T_- &= \frac{9\bar{\delta}}{4(\ell(3+4\ell))} - \frac{4}{2\gamma}(2\bar{\gamma}^2 + 10\bar{\delta}\gamma) = \frac{9\bar{\delta}}{4(\ell(3+4\ell))} + \frac{2}{\gamma}|\bar{\delta}\gamma| \\ &= \frac{9\bar{\delta}}{4(\ell(3+4\ell))} - 2\bar{\delta} = \frac{\bar{\delta}}{4(\ell(3+4\ell))}(9 - 24\ell - 32\ell^2) < 0,\end{aligned}$$

because $\ell \geq 1$. In short, this case is not possible.

Case 4: $\bar{\delta}\gamma < 0$ and $\gamma > 0$. We consider the case $\bar{\gamma} \geq 0$ because the case $\bar{\gamma} < 0$ is the same working with T_- instead of T_+. Since $\bar{\gamma} \geq 0$ we have that $(\psi_+, 0)$ is a saddle. In view of Theorem 4 we must have that $k_0 \in \{\lambda^+, \lambda^-, \lambda^+ + \lambda^-\} = \{\lambda^+, \lambda^-, -\beta\}$. As in Case 1, we cannot have $k_0 = -\beta$. So, assuming that $k_0 \in \{\lambda^+, \lambda^-\}$ we conclude that

$$\beta = \pm \frac{3\sqrt{T_+}}{2\sqrt{\ell(3+4\ell)}}.$$

Now proceeding as in Case 1, it follows from Theorem 4 that we have either $\mu^+ = \nu^+$ (respectively $\mu^- = \nu^-$) in the case in which $\beta^2 + 4T_- < 0$ (because they will be rationally independent), or $\beta^2 + 4T_- > 0$. In the first case, proceeding as in Case 1 we must have $\bar{\gamma} = 0$. So assume $\bar{\gamma} > 0$. Then

$$\begin{aligned} \beta^2 + 4T_- &= \frac{1}{4\ell(3+4\ell)}(9T_+ + 16\ell(3+4\ell)T_-) \\ &= \frac{1}{8\gamma\ell(3+4\ell)}(\bar{\gamma}\sqrt{\bar{\gamma}^2 + 4\bar{\delta}\gamma(9 - 16\ell(3+4\ell))} \\ &\quad - (\bar{\gamma}^2 + 4\bar{\delta}\gamma)((9 + 16\ell(3+4\ell))) < 0, \end{aligned}$$

which is not possible. In short, we must have $\bar{\gamma} = 0$. Then

$$\beta = \pm \frac{3\sqrt{-4\bar{\delta}}}{\sqrt{2}\sqrt{\ell(3+4\ell)}}.$$

Note that

$$\beta^2 + 4\bar{\delta} = \frac{9}{2\ell(3+4\ell)}|\bar{\delta}| - 4|\bar{\delta}| = \frac{|\bar{\delta}|}{2\ell(3+4\ell)}(9 - 8\ell(3+4\ell)) < 0.$$

So, again proceeding as in Case 1 we must have $k_0 \in \{\lambda^+, \lambda^-\}$. Assuming it, we conclude that $\bar{\delta} = 0$ which is not possible because $\bar{\delta} = \delta \neq 0$ whenever $\bar{\gamma} = 0$. This concludes the proof of the theorem. ■

5 Final comments

Due to the importance of the traveling wave solutions in the study of nonlinear evolution equations it is desirable to obtain new methods or to generalize the existing ones to cover as many partial differential equations as possible, in particular the ones with the presence of high nonlinear terms or the ones that are of high order. This is due to the fact that the existing methods become more and more useless or hard to implement as either the order of the nonlinearity increases or the order of the partial differential equations also increases. Not only this, but the method used in this paper detects the so-called algebraic traveling wave solutions but fails to detect any other type of traveling wave solution that may also exist for a concrete partial differential equation.

Acknowledgements

Partially supported by FCT/Portugal through the project UID/MAT/04459/2013

References

[1] Benney D J, Long waves in liquid films, *J. Math. Phys. A*, **45** (1966) 422–443.

[2] Bleecker D, Csordas G, *Basic Partial Differential Equations*, Van Nostrand Reinhold New York, 1992.

[3] Burgers J M, Mathematical examples illustrating relations occurring in the theory of turbulence fluid motion, *Trans. Roy. Neth. Acad. Sci. Amsterdam*, **17** (1939) 1–53.

[4] Chavarriga J, Giacomini H, Grau M, Necessary conditions for the existence of invariant algebraic curves for planar polynomial systems, *Bull. Sci. Math.*, **129** (2005) 99–126.

[5] Cohen B, Krommes J, Tang W, Rosenbluth M, Nonlinear saturation of the dissipative trapped-ion mode by mode coupling, *Nucl. Fus.*, **16** (1976) 971–992.

[6] Drazin P G, Johnson R S, *Solitons: An Introduction*, Cambridge University Press, 1989.

[7] Fisher R A, The wave of advance of advantageous genes, *Ann. Eugenics*, **7** (1937) 355–369.

[8] Gao G, A theory of interaction between dissipation and dispersion of turbulence, **Sci. Sinica Ser. A.**, **28** (1985) 616–627.

[9] Gasull A, Giacomini H, Explicit travelling waves and invariant algebraic curves, *Nonlinearity*, **28** (2015) 1597–1606.

[10] Grad H, Hu P.N. Unified shock profile in a plasma, *Phys. Fluids*, **10** (1967) 2596–2602.

[11] Halford W D, Vlieg-Hulstman M, The Korteweg-de-Vries equation and Painlevé property, *J. Phys. A*, **25** (1992) 2375–2379.

[12] Hu P N, Collisional theory of shock and nonlinear waves in a plasma, *Phys. Fluids*, **15** (1972) 854–864.

[13] Jeffrey A, Kakutani T, Weak nonlinear dispersive waves: A discussion centered around the Korteweg-de Vries equation, *SIAM Rev.*, **14** (1972) 582–643.

[14] Jeffrey A, Exact solutions to the Korteweg-de-Vries-Burgers equation, *Wave Motion*, **11** (1989) 559–564.

[15] Jeffrey A, Mohamad M N, Exact solutions to the KdV-Burgers equation, *Wave Motion*, **14** (1991) 369–375.

[16] Johnson R S, A nonlinear equation incorporating damping and dispersion, *J. Fluid Mech.*, **42** (1970) 49–60.

[17] Johnson R S, Shallow water waves on a viscous fluid-the undular bore, *Phys. Fluids*, **15** (1972) 1693–1699.

[18] Johnson R S, *A Modern Introduction to the Mathematical Theory of Water Waves*, Cambridge University Press, 1997.

[19] Korteweg D J, De Vries G, On the change of form of long waves advancing in a rectangular channel, and on a new type of long stationary waves, *Phil. Mag.*, **39** (1895) 422–443.

[20] Kuramoto Y, Tsuzuki T, Persistent propagation of concentration waves in dissipative media far from thermal equilibrium, *Progr. Theoret. Phys.*, **55** (1976) 356–369.

[21] LaQuey R, Mahajan S, Rutherford P, Tang W, Nonlinear saturation of the trappedion mode, *Phys. Rev. Lett.*, **34** (1975) 391–394.

[22] Llibre J, Valls C, The Michelson system is neither global analytic, nor Darboux integrable, *Physica D*, 239 (2010) 414–419.

[23] Lu B Q, Xiu B Z, Pang A L, Jiang X F, Exact traveling wave solution of one class of nonlinear diffusion equations, *Phys. Lett. A*, **175** (1993) 113–115.

[24] Lu B Q, Pan Z L, Qu B Z, Jiang X F, Solitary wave solutions for some systems of coupled nonlinear equations, *Phys. Lett. A*, **180** (1993) 61–64.

[25] McIntosh I, Simple phase averaging and traveling wave solutions of the modified Burgers- Korteweg-de-Vries equation, *Phys. Lett. A*, **143** (1990) 57–61.

[26] Michelson D, Steady solutions of the Kuramoto-Sivashinsky equation, *Physica D*, **19** (1986) 89–111.

[27] Misbah C, Valance A, Secondary instabilities in the stabilized Kuramoto-Sivashinsky equation, *Phys. Rev. E*, **49** (1994) 166–183.

[28] Seidenberg A, Reduction of singularities of the differential equation $A\,dy = B\,dx$, *Amer. J. Math.*, **90** (1968) 248–269.

[29] Sivashinsky G, Michelson D, On irregular wavy ow of a liquid film down a vertical plane, *Progr. Theoret. Phys.*, **63** (1980) 2112–2114.

[30] Sivashinsky G, Nonlinear analysis of hydrodynamic instability in laminar ames I. Derivation of basic equations, *Acta Astron.*, **4** (1977) 1177-1206.

[31] Troy W, The existence of steady solutions of the Kuramoto-Sivashinsky equation, *J. Differential Equations*, **82** (1989) 269–313.

[32] Wang M L, Exact solutions for a compound KdV-Burgers equation, *Phys. Lett. A*, **213** (1996) 279–287.

[33] Whitham G B, *Linear and Nonlinear Waves*, Springer-Verlag, New York, 1974.

[34] Wijngaarden L V, On the motion of gas bubbles in a perfect fluid, *Ann. Rev. Fluid Mech.*, **4** (1972) 369–373.

[35] Xie X Y, Tang J S, New solitary wave solutions to the Kdv-Burgers equation, *Int. J. Theor. Phys.*, **44** (2005) 293–301.

[36] Xiong S L, An analytic solution of Burgers-KdV equation, *Chin. Sci. Bull.*, **34** (1989) 1158–1162.

[37] Zhang W G, Exact solutions of the Burgers-combined KdV mixed type equation, *Acta Math. Sci.*, **16** (1996) 241–248.

[38] Zhang J F, New exact solitary wave solutions of the KS equation, *Int. J. Theor. Phys.*, **38** (1999) 1829–1834.

C1. Nonlocal invariance of the multipotentialisations of the Kupershmidt equation and its higher-order hierarchies

Marianna Euler and Norbert Euler

Division of Mathematics, Department of Engineering Sciences and Mathematics, Luleå University of Technology, SE-971 87 Luleå, Sweden

Abstract

The term multipotentialisation of evolution equations in $1+1$ dimensions refers to the process of potentialising a given evolution equation, followed by at least one further potentialisation of the resulting potential equation. For certain equations this process can be applied several times to result in a finite chain of potential equations, where each equation in the chain is a potential equation of the previous equation. By a potentialisation of an equation with dependent variable u to an equation with dependent variable v, we mean a differential substitution $v_x = \Phi^t$, where Φ^t is a conserved current of the equation in u. The process of multipotentialisation may lead to interesting nonlocal transformations between the equations. Remarkably, this can, in some cases, result in nonlocal invariance transformations for the equations, which then serve as iteration formulas by which solutions can be generated for all the equations in the chain. In the current paper we give a comprehensive introduction to this subject and report new nonlocal invariance transformations that result from the multipotentialisation of the Kupershmidt equation and its higher-order hierarchies. The recursion operators that define the hierarchies are given explicitly.

1 Introduction: symmetry-integrable equations and multipotentialisations

Assume that the following is an nth-order symmetry-integrable evolution equation

$$E := u_t - F(x, u, u_x, u_{xx}, \dots, u_{nx}) = 0. \tag{1.1}$$

Throughout this paper subscripts t and x denote partial derivatives. That is, for the dependent variable $u = u(x,t)$, we have

$$u_t = \frac{\partial u}{\partial t}, \quad u_x = \frac{\partial u}{\partial x}, \quad u_{xx} = \frac{\partial^2 u}{\partial x^2}, \quad u_{xxx} = \frac{\partial^3 u}{\partial x^3}, \quad u_{qx} = \frac{\partial^q u}{\partial x^q} \quad \text{for} \quad q \geq 4.$$

For the dependent variable $u_1 = u_1(x,t)$ we use the notation $u_{1,x}$, $u_{1,xx}$, $u_{1,xxx}$, $u_{1,qx}$ for $q \geq 4$. By a symmetry-integrable evolution equation of the form (1.1), we mean an equation that admits a recursion operator R_u that generates an infinite set of commuting Lie-Bäcklund symmetries

$$Z_j = \eta_j \frac{\partial}{\partial u}, \qquad j = 1, 2, \dots \tag{1.2}$$

with

$$\eta_j = R_u^j\, u_t \quad \text{or} \quad \eta_k = R_u^k\, u_x \qquad j = 1, 2, \ldots, \qquad k = 1, 2, \ldots . \tag{1.3}$$

Here η_j satisfies the symmetry condition

$$L_E[u]\eta_j \Big|_{E=0} = 0 \tag{1.4}$$

for every j for (1.1), where $L_E[u]$ is the linear operator

$$L_E[u] = \frac{\partial E}{\partial u} + \frac{\partial E}{\partial u_t} D_t + \frac{\partial E}{\partial u_x} D_x + \frac{\partial E}{\partial u_{xx}} D_x^2 + \cdots + \frac{\partial E}{\partial u_{nx}} D_x^n . \tag{1.5}$$

Here and below, D_x denotes the total x-derivative and D_t the total t-derivative. For an extensive list of symmetry-integrable evolution equations we refer to [17]. We use integro-differential recursion operators of the form

$$R_u = \sum_{j=0}^{m} G_j D_x^j + \sum_{i=1}^{3} I_i(u_x, u_t) D_x^{-1} \circ \Lambda_i, \tag{1.6}$$

where G_j are functions of u and x-derivatives of u, D_x^{-1} is the integral operator and Λ_i are integrating factors of the equation. The condition for a recursion operator R_u of (1.1) is

$$[L_E[u],\, R_u] = D_t R_u, \tag{1.7}$$

where $[\;,\;]$ is the standard Lie bracket $[A, B] := AB - BA$. A hierarchy of symmetry-integrable equations then follow for (1.1), namely the hierarchy of evolution equations

$$u_{t_j} = R_u^j\, F, \qquad u_{\tau_j} = R_u^j u_x \qquad j = 1, 2, \ldots \tag{1.8}$$

where every equation in the hierarchy admits the infinite set of commuting symmetries (1.2) generated by R_u^j. Assume that (1.1) admits a conservation law with conserved current Φ^t and flux Φ^x. That is

$$\left(D_t \Phi^t + D_x \Phi^x\right)\Big|_{u_t = F} = 0. \tag{1.9}$$

Then Φ^t must satisfy the relation (see, for example, [11] and [1])

$$\Lambda = \hat{E}[u]\, \Phi^t, \tag{1.10}$$

where Λ is the corresponding integrating factor (or multiplier) of (1.1) and $\hat{E}[u]$ is the Euler operator

$$\hat{E}[u] = \frac{\partial}{\partial u} - D_x \circ \frac{\partial}{\partial u_x} - D_t \circ \frac{\partial}{\partial u_t} + D_x^2 \circ \frac{\partial}{\partial u_{xx}} - D_x^3 \circ \frac{\partial}{\partial u_{xxx}} + \cdots . \tag{1.11}$$

The necessary and sufficient condition on Λ is

$$\hat{E}[u]\,(\Lambda E) = 0, \tag{1.12}$$

or, equivalently, the conditions

$$L_E^*\Lambda\Big|_{E=0} = 0 \quad \text{and} \quad L_\Lambda E = L_\Lambda^* E, \tag{1.13}$$

where L_E^* is the adjoint operator of L_E, namely

$$L_E^* = \frac{\partial E}{\partial u} - D_x \circ \frac{\partial E}{\partial u_x} - D_t \circ \frac{\partial E}{\partial u_t} + D_x^2 \circ \frac{\partial E}{\partial u_{xx}} - D_x^3 \circ \frac{\partial E}{\partial u_{xxx}} + \cdots$$

$$+(-1)^k D_x^k \circ \frac{\partial E}{\partial u_{kx}}. \tag{1.14}$$

Note that the first condition in (1.13) is the necessary and sufficient condition for an adjoint symmetry Λ for (1.1) (see [1] for details on adjoint symmetries).

Following the description above, we can calculate a conservation law of (1.1) by first calculating Λ by conditions (1.13) and then Φ^t by condition (1.10). The flux then follows from

$$\Phi^x = -D_x^{-1}\left(D_t\Phi^t\Big|_{E=0}\right). \tag{1.15}$$

Using integration by parts, (1.15) results in the following useful formula for Φ^x:

Proposition 1. *Assume that $u_t = F(x, u, u_x, \ldots, u_{nx})$ admits an integrating factor Λ and a conserved current $\Phi^t = \Phi^t(x, y, u_x, \ldots, u_{mx})$, such that $\Lambda = \hat{E}[u]\Phi^t$. Then the corresponding flux Φ^x is given by*

$$\Phi^x = -D_x^{-1}(\Lambda F) + \sum_{k=1}^{m}\sum_{j=0}^{m-k}(-1)^k \left(D_x^j F\right) D_x^{k-1}\left(\frac{\partial \Phi^t}{\partial u_{(j+k)x}}\right). \tag{1.16}$$

As an example for Proposition 1, consider $\Phi^t = \Phi^t(x, u, u_x, u_{xx}, u_{xxx})$. Then (1.15) takes the form

$$\Phi^x = -D_x^{-1}(\Lambda F) - \frac{\partial \Phi^t}{\partial u_x} F - \frac{\partial \Phi^t}{\partial u_{xx}} D_x F - \frac{\partial \Phi^t}{\partial u_{xxx}} D_x^2 F \tag{1.17}$$

$$+F D_x\left(\frac{\partial \Phi^t}{\partial u_{xx}}\right) - F D_x^2\left(\frac{\partial \Phi^t}{\partial u_{xxx}}\right) + (D_x F)\, D_x\left(\frac{\partial \Phi^t}{\partial u_{xxx}}\right). \tag{1.18}$$

Example: Consider the Schwarzian Korteweg-de Vries equation (SKdV)

$$u_t = u_{xxx} - \frac{3}{2}\frac{u_{xx}^2}{u_x} \equiv u_x\{u, x\}, \tag{1.19}$$

where $\{u, x\}$ it the Schwarzian derivative defined by

$$\{u,x\} := \frac{u_{xxx}}{u_x} - \frac{3}{2}\left(\frac{u_{xx}}{u_x}\right)^2 \equiv \left(\frac{u_{xx}}{u_x}\right)_x - \frac{1}{2}\left(\frac{u_{xx}}{u_x}\right)^2.$$

(See [24] where (1.19) was introduced as well as [23] and [3] for more details on its applications in view of the Painlevé Test). Using condition (1.13) we calculate all integrating factors for (1.19) up to order 4 and obtain

$$\Lambda_1 = -\frac{2u_{xx}}{u_x^3}, \quad \Lambda_2 = -\frac{2u^2u_{xx}}{u_x^3} + \frac{4u}{u_x}, \quad \Lambda_3 = \alpha\left(\frac{u_{4x}}{u_x^2} - \frac{4u_{xx}u_{xxx}}{u_x^3} + \frac{3u_{xx}^3}{u_x^4}\right)$$

$$\Lambda_4 = \alpha\left(\frac{uu_{xx}}{u_x^3} - \frac{1}{u_x}\right),$$

where α is an arbitrary non-zero constant. Then, using condition (1.10) and (1.16), the corresponding conserved current and flux for each integrating factor take the following form:

$$\Phi_1^t = \frac{1}{u_x}, \qquad \Phi_2^t = \frac{u^2}{u_x}, \qquad \Phi_3^t = \frac{\alpha}{2}\left(\frac{u_{xx}}{u_x}\right)^2, \qquad \Phi_4^t = -\frac{\alpha}{2}\frac{u}{u_x}$$

$$\Phi_1^x = \frac{u_{xxx}}{u_x^2} - \frac{1}{2} + \frac{u_{xx}^2}{u_x^3}, \qquad \Phi_2^x = \frac{u^2u_{xxx}}{u_x^2} - \frac{1}{2}\frac{u^2u_{xx}^2}{u_x^3} - 4\frac{uu_{xx}}{u_x} + 4u_x$$

$$\Phi_3^x = \alpha\left(-\frac{u_{xx}u_{4x}}{u_x^2} + \frac{1}{2}\frac{u_{xxx}^2}{u_x^2} + \frac{2u_{xx}^2u_{xxx}}{u_x^3} - \frac{9}{8}\frac{u_{xx}^4}{u_x^4}\right)$$

$$\Phi_4^x = \alpha\left(-\frac{1}{2}\frac{uu_{xxx}}{u_x^2} + \frac{1}{4}\frac{uu_{xx}^2}{u_x^3} + \frac{u_{xx}}{u_x}\right).$$

By the recursion operator Ansatz (1.6) with $m = 2$, condition (1.7) results in the following recursion operator for (1.19) ([21], [4]):

$$R_u = D_x^2 - \left(\frac{2u_{xx}}{u_x}\right)D_x + \frac{u_{xxx}}{u_x} - \left(\frac{u_{xx}}{u_x}\right)^2 - u_xD_x^{-1}\circ\Lambda_3,$$

where Λ_3 is the integrating factor given above with $\alpha = 1$. Now

$$\eta_1 = R_u\,u_x = u_{xxx} - \frac{3}{2}\frac{u_{xx}^2}{u_x}$$

$$\eta_2 = R_u^2\,u_x = u_{5x} - \frac{5u_{xx}u_{4x}}{u_x} - \frac{5}{2}\frac{u_{xxx}^2}{u_x} + \frac{25}{2}\frac{u_{xx}^2u_{xxx}}{u_x^2} - \frac{45}{8}\frac{u_{xx}^4}{u_x^3},$$

which are two commuting symmetries $Z_1 = \eta_1\frac{\partial}{\partial u}$ and $Z_2 = \eta_2\frac{\partial}{\partial u}$ of (1.19), i.e. $[Z_1,\ Z_2] = 0$. This provides the first two members of the symmetry-integrable

Schwarzian Korteweg-de Vries hierarchy (in terms of the "time" variables t_1 and t_2), namely

$$u_{t_1} = u_{xxx} - \frac{3}{2}\frac{u_{xx}^2}{u_x}, \quad u_{t_2} = u_{5x} - \frac{5u_{xx}u_{4x}}{u_x} - \frac{5}{2}\frac{u_{xxx}^2}{u_x} + \frac{25}{2}\frac{u_{xx}^2 u_{xxx}}{u_x^2} - \frac{45}{8}\frac{u_{xx}^4}{u_x^3}.$$

In the current paper we are interested in the potentialisation of equations. To potentialise equation (1.1) we need to introduce a new variable $v(x,t)$ such that (1.1) maps to a new equation,

$$v_t = G(x, v_x, v_{xx}, \ldots, v_{nx}), \tag{1.21}$$

called the potential equation of (1.1). This may be achieved by using an appropriate conserved current Φ_1^t and flux Φ_1^x of (1.1), whereby we let

$$v_x = \Phi_1^t(x, t, u, u_x, \ldots, u_{qx}), \qquad v_t = -\Phi_1^x(x, t, u, u_x, \ldots, u_{qx}).$$

This is illustrated in Diagram 1.

Diagram 1: Potentialisation of $u_t = F$

$$\boxed{u_t = F(x, u, u_x, \ldots, u_{nx})}$$

$$\Big\downarrow{\scriptstyle v_x = \Phi_1^t}$$

$$\boxed{v_t = G(x, v_x, v_{xx}, \ldots, v_{nx})}$$

There is a general relation between the order of the equation (1.1) and the functional dependence of Φ^t on u_{qx}. In particular, to achieve a potentialisation of an nth-order equation (1.1) we need

$$\frac{\partial \Phi^t}{\partial u_{qx}} \neq 0 \quad \text{with} \quad q \leq n-1, \tag{1.22}$$

where the corresponding integrating factor Λ satisfies

$$\frac{\partial \Lambda}{\partial u_{rx}} \neq 0 \quad \text{with} \quad r \leq n+1 \;\; \text{(even)} \tag{1.23}$$

and $r = 2q$. The Euler operator $\hat{E}$ is then of order q.

We now discuss the multipotentialisation of evolution equations. Equation (1.1) can be multipotentialised if (1.21) further admits a potentialisation

$$w_t = H(x, w_x, \ldots, w_{nx}), \tag{1.24}$$

with

$$w_x = \Phi_2^t(x, v, v_x, \ldots, v_{qx}), \qquad w_t = -\Phi_2^x(x, v, v_x, \ldots, v_{qx}).$$

Here Φ_2^t is a conserved current and Φ_2^x a flux for (1.21). This multipotentialisation then results in a chain of three equations connected by their potentialisations as illustrated in Diagram 2.

A given equation (1.1) may admit more than one potentialisation, in which case the equation would be related to different potential equations. Also an equation may potentialise to the same equation by different potentialisations which can lead to interesting invariance transformations. Examples of multipotentialisations and their applications are given in [9], [5], [10], [8] and [7].

Diagram 2: A multipotentialisation of $u_t = F$

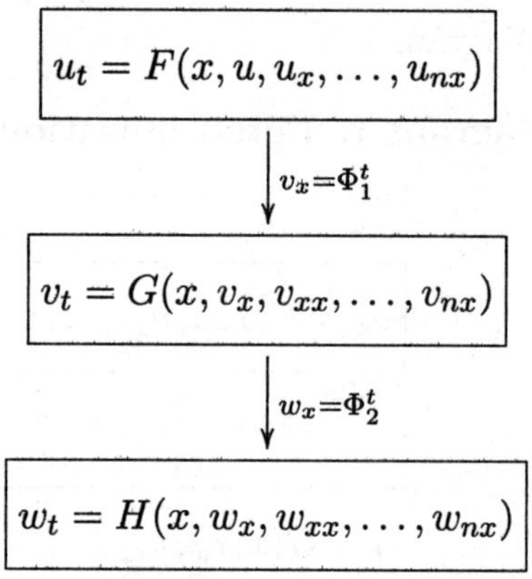

$$u_t = F(x, u, u_x, \ldots, u_{nx})$$

$$\Big\downarrow\, v_x = \Phi_1^t$$

$$v_t = G(x, v_x, v_{xx}, \ldots, v_{nx})$$

$$\Big\downarrow\, w_x = \Phi_2^t$$

$$w_t = H(x, w_x, w_{xx}, \ldots, w_{nx})$$

Example: We again consider the Schwarzian Korteweg-de Vries equation (1.19), namely

$$u_t = u_{xxx} - \frac{3}{2}\frac{u_{xx}^2}{u_x}.$$

Using the integrating factor $\Lambda_1 = -\dfrac{2u_{xx}}{u_x^3}$ we can potentialise (1.19) with

$$w_{1,x} = \frac{1}{u_x}$$

to the same equation (1.19), albeit in terms of the variable w_1. Also, the integrating factor $\Lambda_2 = -\dfrac{2u^2u_{xx}}{u_x^3} + \dfrac{4u}{u_x}$ leads to

$$w_{2,x} = \frac{u^2}{u_x}$$

which again satisfies (1.19), now in terms of w_2. Hence we have a multipotentialisation of (1.19) to itself. This is illustrated in Diagram 3. The potentialisations give the relation $w_{2,x} = u^2 w_{1,x}$ which leads to a nonlocal invariance transformation $u \mapsto \{w_1, w_2\}$ for (1.19), namely

$$w_2 = \int \left(\frac{u^2}{u_x}\right) dx + f_2(t) \tag{1.25a}$$

$$w_1 = \int \left(\frac{1}{u_x}\right) dx + f_1(t), \tag{1.25b}$$

where f_1 and f_2 have to be determined such that w_1 and w_2 satisfy (1.19) for any solution u of (1.19) for which $u_x \neq 0$.

Diagram 3: Multipotentialisation of SKdV to itself.

$$\boxed{w_{1,t} = w_{1,xxx} - \frac{3}{2}\frac{w_{1,xx}^2}{w_{1,x}}}$$

$$\Lambda_1 = -\frac{2u_{xx}}{u_x^3} \quad \uparrow \quad w_{1,x} = \frac{1}{u_x}$$

$$\boxed{u_t = u_{xxx} - \frac{3}{2}\frac{u_{xx}^2}{u_x}}$$

$$\Lambda_2 = -\frac{2u^2 u_{xx}}{u_x^3} + \frac{4u}{u_x} \quad \downarrow \quad w_{2,x} = \frac{u^2}{u_x}$$

$$\boxed{w_{2,t} = w_{2,xxx} - \frac{3}{2}\frac{w_{2,xx}^2}{w_{2,x}}}$$

Applying the integrating factor

$$\Lambda_3 = \alpha\left(\frac{u_{4x}}{u_x^2} - \frac{4u_{xx}u_{xxx}}{u_x^3} + \frac{3u_{xx}^3}{u_x^4}\right)$$

we can use the corresponding conserved current Φ_3^t (given in the previous example) to define the potential variable w_3 as follows:

$$w_{3,x} = \frac{\alpha}{2}\frac{u_{xx}^2}{u_x^2}.$$

This leads to

$$w_{3,t} = w_{3,xxx} - \frac{3}{4}\frac{w_{3,xx}^2}{w_{3,x}} - \frac{3}{2\alpha}w_{3,x}^2.$$

Also

$$\Lambda_4 = \alpha\left(\frac{u u_{xx}}{u_x^3} - \frac{1}{u_x}\right)$$

leads to a new potential variable w_4, where

$$w_{4,x} = -\frac{\alpha}{2}\frac{u}{u_x},$$

which results in the equation

$$w_{4,t} = w_{4,xxx} - \frac{3}{2}\frac{w_{4,xx}^2}{w_{4,x}} + \frac{3\alpha^2}{8}\frac{1}{w_{4,x}}.$$

These potentialisations are illustrated in Diagram 4.

Diagram 4: Potentialisations of SKdV to two equations.

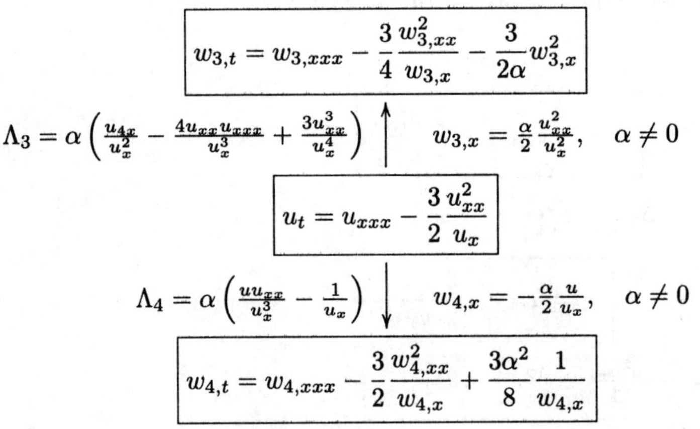

$$\boxed{w_{3,t} = w_{3,xxx} - \frac{3}{4}\frac{w_{3,xx}^2}{w_{3,x}} - \frac{3}{2\alpha}w_{3,x}^2}$$

$$\Lambda_3 = \alpha\left(\frac{u_{4x}}{u_x^2} - \frac{4u_{xx}u_{xxx}}{u_x^3} + \frac{3u_{xx}^3}{u_x^4}\right) \quad \uparrow \quad w_{3,x} = \frac{\alpha}{2}\frac{u_{xx}^2}{u_x^2}, \quad \alpha \neq 0$$

$$\boxed{u_t = u_{xxx} - \frac{3}{2}\frac{u_{xx}^2}{u_x}}$$

$$\Lambda_4 = \alpha\left(\frac{uu_{xx}}{u_x^3} - \frac{1}{u_x}\right) \quad \downarrow \quad w_{4,x} = -\frac{\alpha}{2}\frac{u}{u_x}, \quad \alpha \neq 0$$

$$\boxed{w_{4,t} = w_{4,xxx} - \frac{3}{2}\frac{w_{4,xx}^2}{w_{4,x}} + \frac{3\alpha^2}{8}\frac{1}{w_{4,x}}}$$

To demonstrate the use of (1.25a) – (1.25b) to iterate solutions for the Schwarzian KdV equation (1.19), we use the seed solution

$$w = e^{x-t/2} + a_1$$

where a_1 is an arbitrary constant. By (1.25b) we have

$$w_1 = e^{-x+t/2} + f_1(t),$$

which satisfies (1.19) iff $f_1(t) = a_2$, where a_2 is an arbitrary constant. By (1.25a) we then have

$$w_2 = -a_1^2 e^{-x+t/2} + e^{x-t/2} + 2a_1 x + f_2(t),$$

which satisfies (1.19) iff $f_2(t) = 3a_1 t + a_3$, where a_3 is an arbitrary constant. Thus the given seed solution w leads to two new solutions

$$w_1 = e^{-x+t/2} + a_2$$

$$w_2 = -a_1^2 e^{-x+t/2} + e^{x-t/2} + (3t + 2x)a_1 + a_3.$$

We can now use the above w_2 as a new seed solution to generate two more solutions of (1.19) by again applying (1.25a) – (1.25b), etc.

We remark that SKdV (1.19) plays a central role in the derivation of iterating solution formulas for the 3rd-order Krichever-Novikov equation. This has been reported in [9].

For some cases an appropriate multipotentialisation can lead to a nonlocal linearisation of the equation. A remarkable example is the linearisation of the Calogero-Degasperis-Ibragimov-Shabat (CDIS) equation (see [10] and [8])

$$u_t = u_{xxx} + 3u^2 u_{xx} + 9uu_x^2 + 3u^4 u_x. \tag{1.26}$$

Using the integrating factor $\Lambda = 1$, we introduce

$$v_x = u^2, \tag{1.27}$$

by which (1.26) can be potentialised to

$$v_t = v_{xxx} - \frac{3}{4}\frac{v_{xx}^2}{v_x} + 3v_x v_{xx} + v_x^3. \tag{1.28}$$

Equation (1.28) admits the integrating factor $\Lambda = e^v$, by which we can achieve a potentialisation of (1.28) with

$$w_x = v_x^{1/2} e^v \tag{1.29}$$

to the linear equation

$$w_t = w_{xxx}. \tag{1.30}$$

Combining now (1.27) and (1.29), we obtain

$$w_x = ue^{\int u^2 dx} \tag{1.31}$$

or, equivalently,

$$u_x = \left(\frac{w_{xx}}{w_x}\right) u - u^3, \tag{1.32}$$

which is a Bernoulli equation that can easily be solved in general to give the linearising transformation

$$u = w_x \left(2\int w_x^2\, dx + f(t)\right)^{-1/2} \tag{1.33}$$

of (1.26). Here $f(t)$ must be obtained for every solution w of (1.30), such that u in (1.32) satisfies (1.26). This multipotentialisation is illustrated in Diagram 5.

Diagram 5: Linearisation of the CDIS equation

$$\boxed{u_t = u_{xxx} + 3u^2 u_{xx} + 9uu_x^2 + 3u^4 u_x}$$

$$v_x = u^2 \quad \downarrow \quad \Lambda = 1$$

$$\boxed{v_t = v_{xxx} - \frac{3}{4}\frac{v_{xx}^2}{v_x} + 3v_x v_{xx} + v_x^3}$$

$$w_x = v_x^{1/2} e^v \quad \downarrow \quad \Lambda = e^v$$

$$\boxed{w_t = w_{xxx}}$$

It is interesting to point out that (1.26) does not admit a recursion operator of the form (1.6) with local symmetries I and local integrating factors Λ (see [21]). In [18] we found that (1.26) admits a recursion operator that generates local symmetries, and therefore a local hierarchy of equations, all of which are linearisable under the same transformation (1.32), namely the operator

$$\begin{aligned}
R_u &= D_x^2 + 2u^2 D_x + 10uu_x + u^4 \\
&\quad +2\left(u_{xx} + 2u^2 u_x + 2ue^{-2\int u^2\,dx}\int\left(e^{2\int u^2\,dx}u_x^2\right)dx\right)D_x^{-1}\circ u \\
&\quad -2ue^{-2\int u^2\,dx}D_x^{-1}\circ\left[\left(u_{xx}+2u^2u_x\right)e^{2\int u^2\,dx} + 2u\int\left(e^{2\int u^2\,dx}u_x^2\right)dx\right]
\end{aligned}$$

The second member of the higher-order hierarchy is then

$$\begin{aligned}
u_t &= u_{5x} + 5u^2 u_{4x} + 40uu_x u_{xxx} + 25uu_{xx}^2 + 50u_x^2 u_{xx} + 10u^4 u_{xxx} \\
&\quad +120u^3 u_x u_{xx} + 140u^2 u_x^3 + 10u^6 u_{xx} + 70u^5 u_x^2 + 5u^8 u_x.
\end{aligned}$$

In the current paper we study a chain of symmetry-integrable hierarchies of evolution equations, namely the chain of the multipotentialisation of the Kupershmidt hierarchy. This chain of hierarchies is obtained from the Kupershmidt equation,

$$K_t = K_{5x} - 5(K_x K_{xxx} + K_{xx}^2) - 5(K^2 K_{xxx} + 4KK_x K_{xx} + K_x^3 + K^4 K_x),$$

which is a well-known 5th-order symmetry-integrable evolution equation ([12], [17]). The paper is organized as follows: in Section 1 we perform a multipotentialisation of the Kupershmidt equation which results in a chain of five potential equations. We also make use of a Miura transformations to map the Kupershmidt equation.

In Section 3 we establish the nonlocal invariance properties for each equation in the chain that was obtained in Section 2. In Section 4 we discuss the hierarchies of higher-order equations associated with each equation in the chain. We explicitly give the 7th-order equations of the hierarchies. In Section 5 we make some concluding remarks and point out some open problems. The recursion operators that define the hierarchies of the Kupershmidt chain are given in Appendix A, and in Appendix B we give an example of a symmetry-integrable equation that cannot be potentialised.

2 The multipotentialisation of the Kupershmidt equation

In this section we derive a chain of potential Kupershmidt equations by a multipotentialisation of the Kupershmidt equation (see equation (2.1)). We also make use of a result by Fordy and Gibbons [12] and use Miura transformations to obtain the Sawada-Kotera equation and an equation that we name the k-equation, which is similar to the Sawada-Kotera equation but with different coefficients.

The Kupershmidt equation in the dependent variable K is of the form

$$\boxed{K_t = K_{5x} - 5K_xK_{xxx} - 5K_{xx}^2 - 5K^2K_{xxx} - 20KK_xK_{xx} - 5K_x^3 + 5K^4K_x.} \tag{2.1}$$

Equation (2.1) is symmetry-integrable and admits a 6th-order recursion operator R_K ([20], [5]) given in Appendix A.

By now multipotentialising (2.1) we obtain a chain of symmetry-integrable evolution equations of order five, all of which admit recursion operators of order six. Those are listed in Appendix A.

Using the integrating factor $\Lambda = 1$ we introduce the potentialisation

$$U_x = K. \tag{2.2}$$

This leads to the **1st Potential Kupershmidt equation** in the variable U (see Diagram 6), namely

$$\boxed{U_t = U_{5x} - 5U_{xx}U_{xxx} - 5U_x^2U_{xxx} - 5U_xU_{xx}^2 + U_x^5.} \tag{2.3}$$

We now continue with the 1st Potential Kupershmidt equation (2.3). With the integrating factor $\Lambda = -e^{2U}$ we introduce

$$u_x = -\frac{1}{2}e^{2U}, \tag{2.4}$$

which leads to the **2nd Potential Kupershmidt equation** in the variable u (see Diagram 6), namely

$$\boxed{u_t = u_{5x} - 5\frac{u_{xx}u_{4x}}{u_x} - \frac{15}{4}\frac{u_{xxx}^2}{u_x} + \frac{65}{4}\frac{u_{xx}^2u_{xxx}}{u_x^2} - \frac{135}{16}\frac{u_{xx}^4}{u_x^3}.} \tag{2.5}$$

We remark that the 2nd potential Kupershmidt equation plays a important role in the nonlocal invariance of the Kaup-Kupershmidt equation (see [19] and [7]). Equation (2.5) potentialises in the **3rd Potential Kupershmidt equation** in the variable v (see Diagram 6)

$$v_t = v_{5x} - 5\frac{v_{xx}v_{4x}}{v_x} + 5\frac{v_{xx}^2 v_{xxx}}{v_x^2} \tag{2.6}$$

where

$$v_x = u_x^{-1/2} \quad \text{or} \quad v_x = uu_x^{-1/2}. \tag{2.7}$$

These are obtained by the use of the integrating factors

$$\Lambda = \frac{3}{4}u_x^{-5/3}u_{xx} \quad \text{or} \quad \Lambda = \frac{3}{2}u_x^{-1/2} - \frac{3}{4}uu_x^{-5/2}u_{xx},$$

respectively. Moreover, (2.6) potentialises back into (2.5) by

$$u_x = v_x^{-2} \quad \text{or} \quad u_x = v^4 v_x^{-2}, \tag{2.8}$$

which correspond to the integrating factors

$$\Lambda = -6v_x^{-4}v_{xx} \quad \text{or} \quad \Lambda = 12v^3 v_x^{-2} - 6v^4 v_x^{-4}v_{xx},$$

respectively. This fact will be exploited to iterate solutions for these equations (see Propositions 2 and Proposition 3 below).

Continuing with (2.6) we find that this equation potentialises to the **4th Potential Kupershmidt equation** in the variable w (see Diagram 6)

$$\begin{aligned} w_t = w_{5x} &- 5\frac{w_{xx}w_{4x}}{w_x} - \frac{15}{4}\frac{w_{xxx}^2}{w_x} + \frac{65}{4}\frac{w_{xx}^2 w_{xxx}}{w_x^2} - \frac{135}{16}\frac{w_{xx}^4}{w_x^3} \\ &+ \frac{5\beta}{6}\left(\frac{w_{xxx}}{w_x} - \frac{7}{4}\frac{w_{xx}^2}{w_x^2}\right) - \frac{5\beta^2}{36}\frac{1}{w_x}, \end{aligned} \tag{2.9}$$

where

$$w_x = -\frac{\beta}{6}v^2 v_x^{-2} \quad \text{with} \quad \Lambda = -\beta v v_x^{-2} + \frac{\beta}{2}v^2 v_x^{-4}v_{xx}. \tag{2.10}$$

Diagram 6: A chain of potential Kupershmidt equations

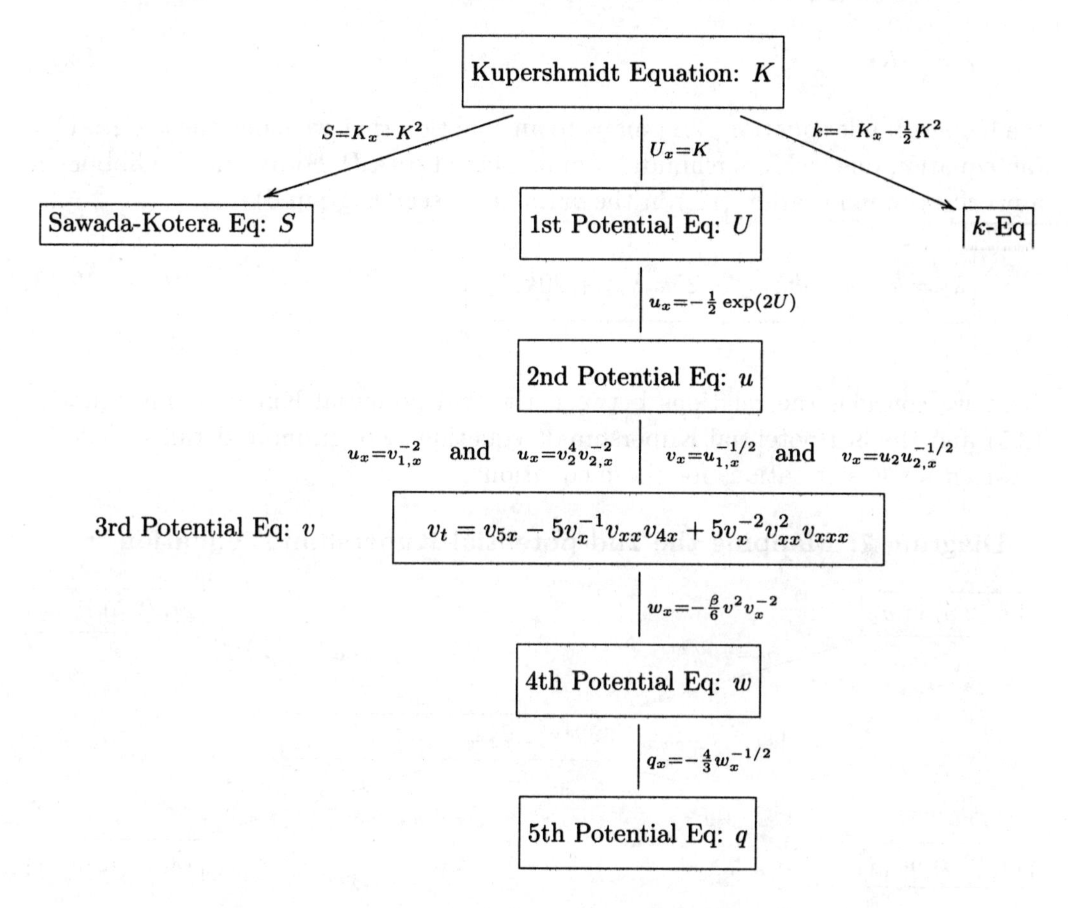

Finally we obtain the **5th Potential Kupershmidt equation** in the variable q (see Diagram 6)

$$\boxed{q_t = q_{5x} - 5\frac{q_{xx}q_{4x}}{q_x} + 5\frac{q_{xx}^2 q_{xxx}}{q_x^2} + \frac{15\beta}{32}q_x^2 q_{xxx} - \frac{15\beta}{32}q_x q_{xx}^2 + \left(\frac{3\beta}{32}\right)^2 q_x^5} \tag{2.11}$$

by the potentialisation

$$q_x = -\frac{4}{3}w_x^{-1/2} \quad \text{with} \quad \Lambda = w_x^{-5/2}w_{xx}. \tag{2.12}$$

Applying the Miura transformation

$$S = K_x - K^2 \tag{2.13}$$

we map the Kupershmidt equation (2.1) to the **Sawada-Kotera equation** [12] in the variable S (see Diagram 6)

$$\boxed{S_t = S_{5x} + 5SS_{3x} + 5S_xS_{xx} + 5S^2S_x.} \tag{2.14}$$

Also, by applying the Miura transformation

$$k = -K_x - \frac{1}{2}K^2 \tag{2.15}$$

the Kupershmidt equation (2.1) maps to an equation that we name the k-**equation** (an equation due to Kupershmidt, communicated to A.P. Fordy and J. Gibbons in a private communication [12]) in the variable k (see Diagram 6)

$$\boxed{k_t = k_{5x} + 10kk_{xxx} + 25k_xk_{xx} + 20k^2k_x.} \tag{2.16}$$

Next we consider the relations between the 2nd potential Kupershmidt equation (2.5) and the 3rd potential Kupershmidt equation (2.6) in more detail and derive invariance transformations for those equations.

Diagram 7: Mapping the 2nd potential Kupershmidt equation

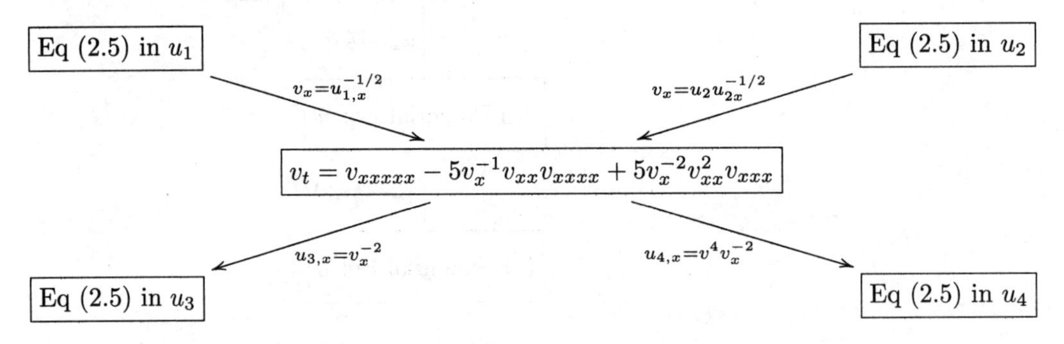

Diagram 7 depicts the potentialisation of the 2nd potential Kupershmidt equation (2.5) to the 3rd potential Kupershmidt equation (2.6), and back again to the 2nd potential Kupershmidt equation; in both cases this is achieved by two different potentialisations as indicated in Diagram 7. This leads to

Proposition 2. *[6] The 2nd potential Kupershmidt equation (2.5), i.e.*

$$u_t = u_{5x} - 5\frac{u_{xx}u_{4x}}{u_x} - \frac{15}{4}\frac{u_{xxx}^2}{u_x} + \frac{65}{4}\frac{u_{xx}^2u_{xxx}}{u_x^2} - \frac{135}{16}\frac{u_{xx}^4}{u_x^3},$$

is invariant under the transformation $u(x,t) \mapsto \bar{u}(x,t)$, *where*

$$\bar{u} = u + c_1 \quad or \tag{2.17a}$$

$$\bar{u} = -\frac{1}{u} + c_2 \quad or \tag{2.17b}$$

$$\bar{u} = \int u_x \left[\int u_x^{-1/2}\, dx + f_1(t)\right]^4 dx + f_2(t). \tag{2.17c}$$

Here c_1 and c_2 are arbitrary constants, whereas f_1 and f_2 have to be determined such that $\bar{u}$ of (2.17c) satisfies (2.5).

Proof: We refer to Diagram 7. Using relations between u_1 and u_3, we obtain (2.17a) with $\bar{u} = u_3$ and $u = u_1$. Using relations between u_2 and u_3, we obtain (2.17b) by eliminating v_x with $\bar{u} = u_3$ and $u = u_2$. Similarly, by the given relations between u_1 and u_4 we eliminate v by differentiation to obtain

$$\bar{u}_{xx} = \left(\frac{u_{xx}}{u_x}\right)\bar{u}_x + 4\left(u_x^{-1/4}\right)\bar{u}_x^{3/4}. \tag{2.18}$$

where $\bar{u} = u_4$ and $u = u_1$. Equation (2.17c) is a first-order Bernoulli equation of the form

$$Y_x = \left(\frac{u_{xx}}{u_x}\right)Y + 4\left(u_x^{-1/4}\right)Y^{3/4},$$

where $Y(x,t) = \bar{u}_x(x,t)$, which can easily be linearised by the substitution $Z(x,t) = Y^{1/4}(x,t)$. The general solution is (2.17c). Note that the condition that results when we make use of the relations between u_2 and u_4 is the same as the condition which results if we insert (2.17b) in (2.17c). □

Example: We start with the seed solution

$$u_1 = -\frac{1}{x}$$

of (2.5). Applying now (2.17c) we obtain

$$\bar{u}_1 = \frac{1}{112}x^7 + \frac{1}{10}f_1(t)x^5 + \frac{1}{2}f_1^2(t)x^3 + 2f_1^3(t)x - f_1^4(t)\left(\frac{1}{x}\right) + f_2(t)$$

and insering this $\bar{u}_1$ into (2.5) we obtain

$$f_1(t) = c_{01}, \quad f_2(t) = 72c_{01}\,t + c_{02},$$

where c_{01} and c_{02} are arbitrary constants. We could then continue by applying, for example, (2.17b), i.e.

$$\bar{u}_2 = -\frac{1}{\bar{u}_1} + c_{03}$$

and use the solution $\bar{u}_2$ in (2.17c) to obtain yet another solution $\bar{u}_3$, etc.

Diagram 8: Mapping the 3rd potential Kupershmidt equation

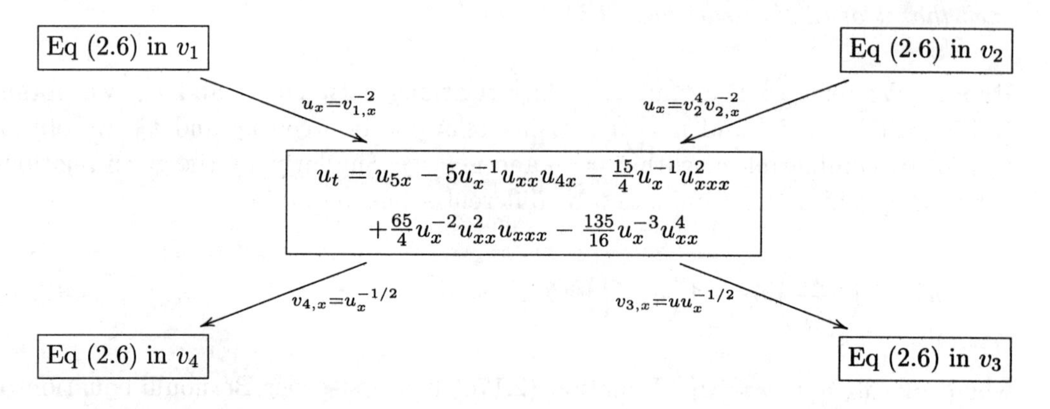

Diagram 8 depicts the potentialisation of the 3rd potential Kupershmidt equation (2.6) to the 2nd potential Kupershmidt equation (2.5), and back again to the 3rd potential Kupershmidt equation; in both cases this is achieved by two different potentialisations as indicated in Diagram 8. This leads to

Proposition 3. *[6] The 3rd potential Kupershmidt equation (2.6), i.e.*

$$v_t = v_{5x} - 5\frac{v_{xx}v_{4x}}{v_x} + 5\frac{v_{xx}^2 v_{xxx}}{v_x^2},$$

is invariant under the transformation $v(x,t) \mapsto \bar{v}(x,t)$, where

$$\bar{v} = v + c_1, \quad or \tag{2.19a}$$

$$\bar{v} = -\frac{1}{v} + c_2, \quad or \tag{2.19b}$$

$$\bar{v} = v\int\left(\frac{1}{v_x^2}\right)\,dx - \int\left(\frac{v}{v_x^2}\right)\,dx + f_1(t)v + f_2(t). \tag{2.19c}$$

Here c_1 and c_2 are arbitrary constants, whereas f_1 and f_2 must be determined such that $\bar{v}$ of (2.19c) satisfies (2.6).

Proof: We refer to Diagram 8. Using relations between v_1 and v_4, we obtain (2.19a). Using the relations between v_1 and v_2, we obtain (2.19b) by eliminating u_x with $\bar{v} = v_2$ and $v = v_1$. Similarly, by the given relations between v_1 and v_3 we eliminate u by differentiation to obtain

$$\bar{v}_{xx} = \left(\frac{v_{xx}}{v_x}\right)\bar{v}_x + \frac{1}{v_x}, \tag{2.20}$$

where $\bar{v} = v_3$ and $v = v_1$. The general solution of the linear equation (2.20) is (2.19c). Note that no additional condition results when we make use of the relations between v_2 and v_3. □

Example: We start with the seed solution

$$v_1 = \frac{1}{x^2}$$

of (2.6). Applying now (2.19c) we obtain

$$\bar{v}_1 = \frac{1}{70x^2}\left(- x^7 + 70x^2 f_2(t) + 70 f_1(t)\right)$$

and insering this $\bar{v}_1$ into (2.6) we obtain

$$f_1(t) = c_{01}, \quad f_2(t) = -36\,t + c_{02},$$

where c_{01} and c_{02} are arbitrary constants. We could then continue by applying, for example, (2.19b), i.e.

$$\bar{v}_2 = -\frac{1}{\bar{v}_1} + c_{03}$$

and use the solution $\bar{v}_2$ in (2.19c) to obtain yet another solution $\bar{v}_3$, etc.

3 Invariance of the Kupershmidt equation and its chain of potentialisations

Using Proposition 2 and Proposition 3 we obtain nonlocal invariance relations for all the equations depicted in Diagram 6.

Proposition 4.

a) *The Kupershmidt equation (2.1) is invariant under $K(x,t) \mapsto \bar{K}(x,t)$, in which*

$$\bar{K} = K + 2\,[\ln(v)]_x \quad \text{and} \quad K = -\,[\ln(v_x)]_x\,, \tag{3.1}$$

where v is a solution of (2.6) such that $v_x \neq 0$, which can be iterated by the relations in Proposition 3.

b) *The 1st potential Kupershmidt equation (2.3) is invariant under $U(x,t) \mapsto \bar{U}(x,t)$, in which*

$$\bar{U} = U + 2\ln(v) \quad \text{and} \quad U = \ln\sqrt{2} - \ln(v_x), \tag{3.2}$$

where v is any solution of (2.6) such that $v_x \neq 0$, which can be iterated by the relations in Proposition 3.

c) *The 2nd potential Kupershmidt equation (2.5) is invariant under* $u(x,t) \mapsto \bar{u}(x,t)$, *in which*

$$\bar{u} = v^4 u - 4\int uv^3 v_x\,dx + f_2(t) \quad and \quad u = \int \left(\frac{1}{v_x^2}\right) dx + f_1(t), \tag{3.3}$$

where f_1 *and* f_2 *must be determined such that* u *and* $\bar{u}$ *satisfy (2.5). Here* v *is any solution of (2.6) such that* $v_x \neq 0$, *which can be iterated by the relations in Proposition 3.*

d) *The 3rd potential Kupershmidt equation (2.6) is invariant under* $v(x,t) \mapsto \bar{v}(x,t)$, *in which*

$$\bar{v} = uv - \int \left(\frac{v}{v_x^2}\right) dx + f_2(t) \quad and \quad v = \int u_x^{-1/2}\,dx + f_1(t), \tag{3.4}$$

where f_1 *must be determined such that* v *satisfies (2.6) and* f_2 *such that* $\bar{v}$ *satisfies (2.6). Here* u *is any solution of (2.5) such that* $u_x \neq 0$, *which can be iterated by the relations in Proposition 2.*

e) *The 4th potential Kupershmidt equation (2.9) is invariant under* $w(x,t) \mapsto \bar{w}(x,t)$, *in which*

$$\bar{w} = \left(\frac{\bar{v}}{uv}\right)^2 w - \int w\left[\left(\frac{\bar{v}}{uv}\right)^2\right]_x dx + f_4(t) \tag{3.5a}$$

$$w = -\frac{\beta}{6}uv^2 + \frac{\beta}{3}\int uvv_x\,dx + f_3(t) \tag{3.5b}$$

$$u = \int \left(\frac{1}{v_x^2}\right) dx + f_1(t) \tag{3.5c}$$

$$and \quad \bar{v} = \int uu_x^{-1/2}\,dx + f_2(t) \tag{3.5d}$$

where f_1 *must be determined such that* u *satisfies (2.5),* f_2 *such that* $\bar{v}$ *satisfies (2.6),* f_3 *such that* w *satisfies (2.9), and* f_4 *such that* $\bar{w}$ *satisfies (2.9). Here* v *is any solution of (2.6) such that* $v_x \neq 0$, *which can be iterated by the relations in Proposition 3.*

f) *The 5th potential Kupershmidt equation (2.11) is invariant under* $q(x,t) \mapsto \bar{q}(x,t)$, *in which*

$$\bar{q} = q\left(\frac{uv}{\bar{v}}\right) - \int q\left(\frac{uv}{\bar{v}}\right)_x dx + f_3(t), \quad q = -\frac{4}{3}\left(\frac{-6}{\beta}\right)^{1/2}\ln(v) \tag{3.6a}$$

$$u = \int \left(\frac{1}{v_x^2}\right) dx + f_1(t) \quad and \quad \bar{v} = \int uu_x^{-1/2}\,dx + f_2(t) \tag{3.6b}$$

where f_1 *must be determined such that* u *satisfies (2.5),* f_2 *such that* $\bar{v}$ *satisfies (2.6), and* f_3 *such that* q *satisfies (2.11). Here* $\beta < 0$ *and* v *is any solution of (2.6) such that* $v_x \neq 0$, *which can be iterated by the relations in Proposition 3.*

g) [7] The Sawada-Kotera equation (2.14) is invariant under $S(x,t) \mapsto \bar{S}(x,t)$*, in which*

$$\bar{S} = S + 6\,[\ln(v)]_{xx} \quad \textit{and} \quad S = -\frac{v_{xxx}}{v_x}, \tag{3.7}$$

where v *is any solution of (2.6) such that* $v_x \neq 0$*, which can be iterated by the relations in Proposition 3.*

h) The k-equation (2.16) is invariant under $k(x,t) \mapsto \bar{k}(x,t)$*, in which*

$$\bar{k} = k + \frac{3}{2}\,[\ln(u)]_{xx} \tag{3.8a}$$

$$\textit{and} \quad k = \frac{3}{2}\,[\ln(u)]_{xx} - \frac{1}{2}\,[\ln(u_x)]_{xx} - \frac{1}{8}\,[\ln(u_x)]_x^2, \tag{3.8b}$$

where u *is any solution of (2.5) such that* $u_x \neq 0$*, which can be iterated by the relations in Proposition 2.*

Proof: We start with the statement (3.2) for the 1st potential Kupershmidt equation (2.3). Following the potentialisations as depicted in Diagram 6, we consider

$$u_{1,x} = \frac{1}{2}e^{2U_1}, \qquad u_{2,x} = \frac{1}{2}e^{2U_2},$$

with

$$u_{1,x} = v_x^{-2}, \qquad u_{2,x} = v^4 v_x^{-2}.$$

This leads to

$$U_2 = \ln\left(\sqrt{2}v^2 v_x^{-1}\right), \qquad U_1 = \ln\left(\sqrt{2}v_x^{-1}\right),$$

so that

$$U_2 = U_1 + \ln(v^2), \qquad U_1 = \ln\sqrt{2} - \ln(v_x).$$

With $U_2 \equiv \bar{U}$ and $U_1 \equiv U$ we obtain (3.2). Since $U_x = K$, we differentiate (3.2) to obtain (3.1). Also

$$u_{2,x} = u_{1,x}v^4, \qquad u_{1,x} = v_x^{-2}$$

which becomes (3.3) after integration. Integrating the relations

$$v_{1,x} = u_x^{-1/2}, \qquad v_{3,x} = u u_x^{-1/2}$$

we obtain

$$v_1 = \int u_x^{-1/2}\,dx + f_1(t)$$

$$v_3 = \int u v_{1,x}\, dx + f_2(t) = \int u v_1\, dx - \int v v_x^{-2}\, dx + f_2(t).$$

With $v_3 \equiv \bar{v}$ and $v_1 \equiv v$ we obtain (3.4). For the 4th potential Kupershmidt equation (2.9) we consider

$$w_{1,x} = -\frac{\beta}{6} v_1^2 v_{1,x}^{-2}, \qquad w_{2,x} = -\frac{\beta}{6} v_3^2 v_{3,x}^{-2}$$

$$v_{1,x} = u_x^{-1/2}, \qquad v_{3,x} = u u_x^{-1/2},$$

which leads to

$$w_{2,x} = w_{1,x} v_1^{-2} v_3^2 u^{-2}, \qquad w_{1,x} = -\frac{\beta}{6} v_1^2 u_x.$$

Integrating the above expressions for $w_{2,x}$ and $w_{1,x}$ leads to (3.5a) – (3.5d), where $w_2 \equiv \bar{w}$, $w_1 \equiv w$, $v_3 \equiv \bar{v}$ and $v_1 \equiv v$. For the 5th potential Kupershmidt equation (2.11) we consider

$$q_{1,x} = -\frac{4}{3} w_{1,x}^{-1/2}, \qquad q_{2,x} = -\frac{4}{3} w_{2,x}^{-1/2}$$

$$w_{1,x} = -\frac{\beta}{6} v_1^2 v_{1,x}^{-2}, \qquad w_{2,x} = -\frac{\beta}{6} v_3^2 v_{3,x}^{-2}$$

$$v_{1,x} = u_x^{-1/2}, \qquad v_{3,x} = u u_x^{-1/2}.$$

Combining the above relations, we obtain

$$q_{1,x} = A v_1^{-1} u_x^{-1/2}, \qquad q_{2,x} = A v_3^{-1} u u_x^{-1/2}, \qquad A = -\frac{4}{3}\left(-\frac{\beta}{6}\right)^{-1/2},$$

so that

$$q_{2,x} = u v_1 v_3^{-1} q_{1,x} \quad \text{where} \quad q_{1,x} = A v_1^{-1} v_{1,x}.$$

Integrating the above expressions for $q_{2,x}$ and $q_{1,x}$, we obtain (3.6a) – (3.6b), where $q_2 \equiv \bar{q}$, $q_1 \equiv q$, $v_3 \equiv \bar{v}$ and $v_1 \equiv v$. For the Sawada-Kotera equation (2.14) we consider

$$S_1 = K_{1,x} - K_1^2, \qquad S_2 = K_{2,x} - K_2^2$$

with

$$K_2 = K_1 + 6\,[\ln(v)]_x\,, \qquad K_1 = -\,[\ln(v_x)]_x\,.$$

This leads to

$$S_2 = S_1 + 6\,[\ln(v)]_{xx}\,, \qquad S_1 = -\frac{v_{xxx}}{v_x},$$

which is (3.7) where $S_2 \equiv \bar{S}$ and $S_1 \equiv S$. Finally, for the k-equation (2.16) we have

$$k = -K_x - \frac{1}{2}K^2, \qquad K = -\left[\ln(v_x)\right]_x.$$

This leads to

$$k = \frac{v_{xxx}}{v_x} - \frac{3}{2}\left(\frac{v_{xx}}{v_x}\right)^2 \equiv \{v,\, x\},$$

where $\{v,\, x\}$ is the Schwarzian derivative. We now consider

$$k_1 = \{v_1,\, x\}, \qquad k_2 = \{v_2,\, x\}$$
$$v_{1,x} = u u_x^{-1/2}, \qquad v_{2,x} = u_x^{-1/2},$$

which leads to (3.8a) – (3.8b), where $k_2 \equiv \bar{k}$ and $k_1 \equiv k$. □

Example: We apply Proposition 4 e) for the 4th potential Kupershmidt equation (2.9). We consider the seed solution

$$v = \frac{\sqrt{5}x^4}{20} + \frac{36\sqrt{5}t}{x}$$

for the 3rd potential Kupershmidt equation (2.6). By requiring that

$$u = \int v_x^{-2}\, dx + f_1(t)$$

satisfies (2.5) we obtain $f_1(t) = c_1$, where c_1 is an arbitrary constant. Then

$$u = -\frac{1}{x^5 - 180t} + c_1.$$

We then calculate $\bar{v}$ by the formula

$$\bar{v} = \int u u_x^{-1/2}\, dx + f_2(t)$$

and, by requiring that $\bar{v}$ satisfies (2.6) we obtain $f_2(t) = c_2$, where c_2 is an arbitrary constant. Then

$$\bar{v} = \frac{-\sqrt{5}(c_1 x^5 + 720 c_1 t + 4)}{20x} + c_2.$$

A solution for (2.9) is now obtained by the formula

$$w = -\frac{\beta}{6}v^2 u + \frac{\beta}{6}\int u v v_x\, dx + f_3(t),$$

from which it follows that $f_3(t) = c_3$, where c_3 is an arbitrary constant. This gives

$$w = \frac{\beta x^8 - 288 c_3 x^5 - 2880\beta t x^3 + 51840 c_3 t}{-288x^5 + 51840t}.$$

One more solution for (2.9) then follows by applying the formula

$$\bar{w} = \left(\frac{\bar{v}}{uv}\right)^2 w - \int w\left(\frac{\bar{v}^2}{u^2v^2}\right)_x dx + f_4(t).$$

Again $f_4(t) = c_4$, where c_4 is an arbitrary constant. This solution for (2.9) then takes the following form:

$$\bar{w} = \Big(\beta c_1^2x^8 - 288c_1^2c_5x^5 + 24\sqrt{5}\beta c_1c_2x^4 - 16\beta c_1x^3 - 2880\beta c_1^2x^3t$$
$$+ 51840c_1^2c_5t - 48\beta c_2^2\Big)\Big(- 288c_1^2x^5 + 51840c_1^2t + 288c_1\Big)^{-1},$$

where we have set $c_3 + c_4 = c_5$.

4 The hierarchies

All equations in Diagram 6 admit six-order integro-differential recursion operators of the form

$$R_u = \sum_{j=0}^{6} G_jD_x^j + \sum_{i=1}^{3} I_i(u_x, u_t)D_x^{-1} \circ \Lambda_i,$$

where G_j are functions of u and x-derivatives of u. Λ_i are integrating factors of the equations.

Acting a recursion operator R_u of a symmetry-integrable evolution equation $u_t = F(u, u_x, \ldots, u_{5x})$ on its t-translation symmetry $u_t\dfrac{\partial}{\partial u}$ or its x-translation symmetry $u_x\dfrac{\partial}{\partial u}$, results in a hierarchy of higher-order commuting symmetries and these flows define a higher-order hierarchy of symmetry-integrable evolution equations. The equations of Diagram 6 result in two symmetry-integrable hierarchies for each equation. For example, for the Kupershmidt equation (2.1) we obtain the following two hierarchies of equations

$$K_{t_j} = R_K^j\, K_t \quad \text{is a hierarchy of order } 6j+5, \text{ and}$$
$$K_{\tau_j} = R_K^j\, K_x \quad \text{is a hierarchy of order } 6j+1,$$

where $j = 1, 2, \ldots$ and

$$K_t = K_{5x} - 5K_xK_{xxx} - 5K_{xx}^2 - 5K^2K_{xxx} - 20KK_xK_{xx} - 5K_x^3 + 5K^4K_x.$$

The recursion operator R_K for (2.1) is given in Appendix A. This is the case for all the equations in Diagram 6, i.e. the Kupershmidt equation, its 5 potential equations, the Sawada-Kotera equation, and the k-equation. The recursion operators of all these equations are listed explicitly in Appendix A.

The same potential variables U, u, v, w, and q that were introduced in Section 2 for the Kupershmidt equation (2.1), illustrated in Diagram 6, can now be introduced for all the corresponding hierarchies. This is illustrated in Diagram 10. In Diagram 9 we illustrate the Sawada-Kotera hierarchy and the k-equation hierarchy related to the Kupershmidt hierarchy by Miura transformations. The flows corresponding to the different "times", t_j and τ_j, all commute for all equations in Diagram 9 and Diagram 10. That is

$$[Z_t,\ Z_{t_j}] = 0, \quad [Z_t,\ Z_{\tau_j}] = 0, \quad [Z_{t_i},\ Z_{t_j}] = 0$$

$$[Z_{\tau_i},\ Z_{\tau_j}] = 0, \quad [Z_{t_i},\ Z_{\tau_j}] = 0,$$

where, for the equation $u_t = F(u, u_x, u_{xx}, \ldots, u_{5x})$, we have

$$Z_t = u_t \frac{\partial}{\partial u} \equiv F \frac{\partial}{\partial u}, \quad Z_{t_j} = \left(R_u^j\, u_t\right) \frac{\partial}{\partial u}, \quad Z_{\tau_j} = \left(R_u^j\, u_x\right) \frac{\partial}{\partial u}.$$

The same invariance transformations as given in Proposition 2, Proposition 3 and Proposition 4 now also hold for the respective higher-order hierarchies of equations illustrated in Diagram 9 and Diagram 10.

Diagram 9: The Sawada-Kotera and k-equation hierarchies

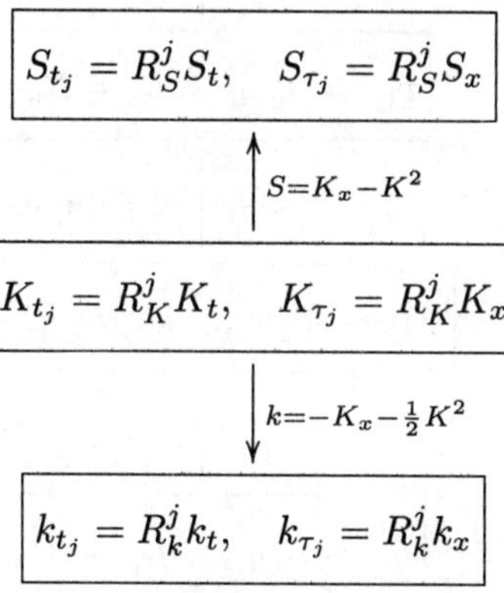

$$S_{t_j} = R_S^j S_t, \quad S_{\tau_j} = R_S^j S_x$$

$\uparrow$ $S = K_x - K^2$

$$K_{t_j} = R_K^j K_t, \quad K_{\tau_j} = R_K^j K_x$$

$\downarrow$ $k = -K_x - \frac{1}{2}K^2$

$$k_{t_j} = R_k^j k_t, \quad k_{\tau_j} = R_k^j k_x$$

Below we explicitly list the evolution equations of order 7 which belongs to the hierarchies show in Diagram 9 and Diagram 10, i.e. the 7th-order equation

$$u_{\tau_1} = R_u^1\, u_x,$$

corresponding to all the 5th-order equations $u_t = F$ in Diagram 6.

The **7th-order Kupershmidt equation,** $K_{\tau_1} = R_K\, K_x$, where R_K is the recursion operator of the Kupershmidt equation (2.1) given in Appendix A, has the following form:

$$K_{\tau_1} = K_{7x} - 7K_xK_{5x} - 7K^2K_{5x} - 21K_{xx}K_{4x} - 42KK_xK_{4x} - 14K_{xxx}^2$$

$$-70KK_{xx}K_{xxx} - 56K_x^2K_{3x} + 14K^2K_xK_{xxx} + 14K^4K_{xxx} - 77K_xK_{xx}^2$$

$$+14K^2K_{xx}^2+56KK_x^2K_{xx}+112K^3K_xK_{xx}+\frac{28}{3}K_x^4+84K^2K_x^3$$

$$-\frac{28}{3}K^6K_x. \tag{4.2}$$

Diagram 10: A chain of potential Kupershmidt hierarchies

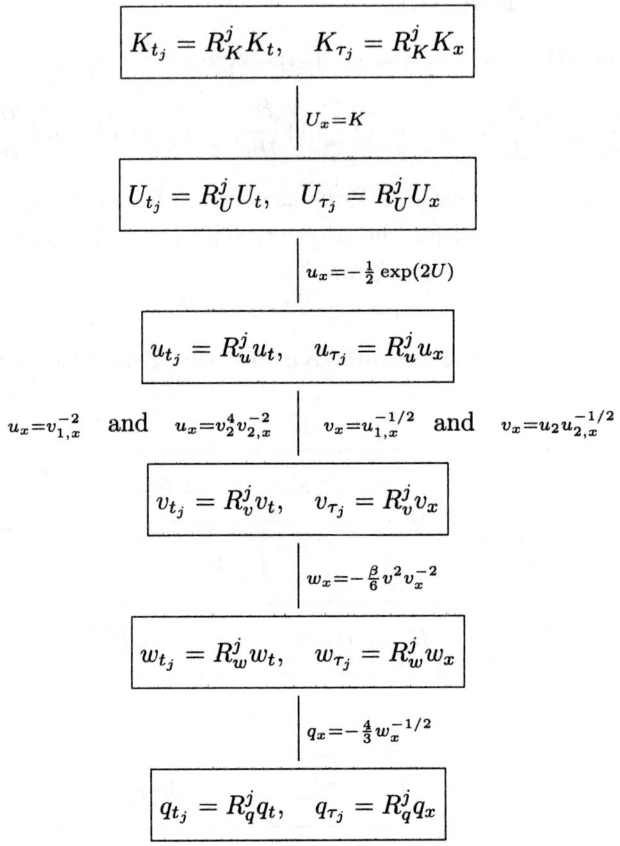

$$K_{t_j}=R_K^jK_t,\quad K_{\tau_j}=R_K^jK_x$$

$\Big\downarrow$ $U_x=K$

$$U_{t_j}=R_U^jU_t,\quad U_{\tau_j}=R_U^jU_x$$

$\Big\downarrow$ $u_x=-\frac{1}{2}\exp(2U)$

$$u_{t_j}=R_u^ju_t,\quad u_{\tau_j}=R_u^ju_x$$

$u_x=v_{1,x}^{-2}$ and $u_x=v_2^4v_{2,x}^{-2}$ $\Big\updownarrow$ $v_x=u_{1,x}^{-1/2}$ and $v_x=u_2u_{2,x}^{-1/2}$

$$v_{t_j}=R_v^jv_t,\quad v_{\tau_j}=R_v^jv_x$$

$\Big\downarrow$ $w_x=-\frac{\beta}{6}v^2v_x^{-2}$

$$w_{t_j}=R_w^jw_t,\quad w_{\tau_j}=R_w^jw_x$$

$\Big\downarrow$ $q_x=-\frac{4}{3}w_x^{-1/2}$

$$q_{t_j}=R_q^jq_t,\quad q_{\tau_j}=R_q^jq_x$$

The **7th-order 1st Potential Kupershmidt equation**, $U_{\tau_1}=R_U\,U_x$, where R_U is the recursion operator of the 1st Potential Kupershmidt equation (2.3) given in Appendix A, has the following form:

$$U_{\tau_1}=U_{7x}-7U_{xx}U_{5x}-7U_x^2U_{5x}-14U_{xxx}U_{4x}-28U_xU_{xx}U_{4x}-21U_xU_{xxx}^2$$

$$-28U_{xx}^2U_{xxx}+14U_x^2U_{xx}U_{xxx}+14U_x^4U_{xxx}+\frac{28}{3}U_xU_{xx}^3+28U_x^3U_{xx}^2$$

$$-\frac{4}{3}U_x^7. \tag{4.3}$$

The **7th-order 2nd Potential Kupershmidt equation**, $u_{T_1} = R_u\, u_x$, where R_u is the recursion operator of the 2nd Potential Kupershmidt equation (2.5) given in Appendix A, has the following form:

$$\begin{aligned} u_{T_1} &= u_{7x} - 7u_x^{-1}u_{xx}u_{6x} - \frac{35}{2}u_x^{-1}u_{xxx}u_{5x} + \frac{147}{4}u_x^{-2}u_{xx}^2u_{5x} - \frac{49}{4}u_x^{-1}u_{4x}^2 \\ &+ \frac{301}{2}u_x^{-2}u_{xx}u_{xxx}u_{4x} - \frac{609}{4}u_x^{-3}u_{xx}^3u_{4x} + \frac{217}{6}u_x^{-2}u_{xxx}^3 \\ &- \frac{1365}{4}u_x^{-3}u_{xx}^2u_{xxx}^2 + \frac{3675}{8}u_x^{-4}u_{xx}^4u_{xxx} - \frac{2457}{16}u_x^{-5}u_{xx}^6. \end{aligned} \tag{4.4}$$

The **7th-order 3rd Potential Kupershmidt equation**, $v_{T_1} = R_v\, v_x$, where R_v is the recursion operator of the 3rd Potential Kupershmidt equation (2.6) given in Appendix A, has the following form:

$$\begin{aligned} v_{T_1} &= v_{7x} - 7v_x^{-1}v_{xx}v_{6x} - 7v_x^{-1}v_{xxx}v_{5x} + 21v_x^{-2}v_{xx}^2v_{5x} - 7v_x^{-1}v_{4x}^2 \\ &+ 56v_x^{-2}v_{xx}v_{xxx}v_{4x} - 42v_x^{-3}v_{xx}^3v_{4x} + \frac{14}{3}v_x^{-2}v_{xxx}^3 - 63v_x^{-3}v_{xx}^2v_{xxx}^2 \\ &+ 42v_x^{-4}v_{xx}^4v_{xxx}. \end{aligned} \tag{4.5}$$

The **7th-order 4th Potential Kupershmidt equation**, $w_{T_1} = R_w\, w_x$, where R_w is the recursion operator of the 4th Potential Kupershmidt equation (2.6) given in Appendix A, has the following form:

$$\begin{aligned} w_{T_1} &= w_{7x} - 7w_x^{-1}w_{xx}w_{6x} - \frac{35}{2}w_x^{-1}w_{xxx}w_{5x} + \frac{147}{4}w_x^{-2}w_{xx}^2w_{5x} + \frac{7\beta}{6}w_x^{-1}w_{5x} \\ &- \frac{49}{4}w_x^{-1}w_{4x}^2 + \frac{301}{2}w_x^{-2}w_{xx}w_{xxx}w_{4x} - \frac{609}{4}w_x^{-3}w_{xx}^3w_{4x} - \frac{91\beta}{12}w_x^{-2}w_{xx}w_{4x} \\ &+ \frac{217}{6}w_x^{-2}w_{xxx}^3 - \frac{1365}{4}w_x^{-3}w_{xx}^2w_{xxx}^2 - \frac{119\beta}{24}w_x^{-2}w_{xxx}^2 + \frac{3675}{8}w_x^{-4}w_{xx}^4w_{xxx} \\ &+ 28\beta w_x^{-3}w_{xx}^2w_{xxx} + \frac{7\beta^2}{18}w_x^{-2}w_{xxx} - \frac{2457}{16}w_x^{-5}w_{xx}^6 - \frac{539\beta}{32}w_x^{-4}w_{xx}^4 \\ &- \frac{7\beta^2}{8}w_x^{-3}w_{xx}^2 - \frac{7\beta^3}{324}w_x^{-2}. \end{aligned} \tag{4.6}$$

The **7th-order 5th Potential Kupershmidt equation**, $q_{T_1} = R_q\, q_x$, where R_q is the recursion operator of the 5th Potential Kupershmidt equation (2.11) given in Appendix A, has the following form:

$$q_{T_1} = q_{7x} - 7q_x^{-1}q_{xx}q_{6x} + 21q_x^{-2}q_{xx}^2q_{5x} - 7q_x^{-1}q_{xxx}q_{5x} + \frac{21\beta}{32}q_x^2q_{5x} - 7q_x^{-1}q_{4x}^2$$

$$+56q_x^{-2}q_{xx}q_{xxx}q_{4x} - 42q_x^{-3}q_{xx}^3q_{4x} - \frac{21\beta}{16}q_xq_{xx}q_{4x} + \frac{14}{3}q_x^{-2}q_{xxx}^3$$

$$-63q_x^{-3}q_{xx}^2q_{xxx}^2 + \frac{21\beta}{32}q_xq_{xxx}^2 + 42q_x^{-4}q_{xx}^4q_{xxx} + \frac{63\beta^2}{512}q_x^4q_{xxx}$$

$$+\frac{9\beta^3}{8192}q_x^7. \tag{4.7}$$

The **7th-order Sawada-Kotera equation,** $S_{\tau_1} = R_S\, S_x$, where R_S is the recursion operator of the Sawada-Kotera equation (2.14) given in Appendix A, has the following form:

$$S_{\tau_1} = S_{7x} + 7SS_{5x} + 14S_xS_{4x} + 14S^2S_{xxx} + 21S_{xx}S_{xxx} + 42SS_xS_{xx}$$

$$+7S_x^3 + \frac{28}{3}S^3S_x. \tag{4.8}$$

Note that (4.8) appears in the literature [15] and has been named the **Sawada-Kotera-Ito equation** (see [14], equation (3.10) with condition (3.11)).

The **7th-order k-equation,** $k_{\tau_1} = R_k\, k_x$, where R_k is the recursion operator of the k-equation (2.16) given in Appendix A, has the following form:

$$k_{\tau_1} = k_{7x} + 14kk_{5x} + 49k_xk_{4x} + 56k^2k_{xxx} + 84k_{xx}k_{xxx} + 252kk_xk_{xx}$$

$$+70k_x^3 + \frac{244}{3}k^3k_x. \tag{4.9}$$

Note that (4.9) appears in the literature, namely in [14] (see equation (3.10) with condition (3.12)).

5 Concluding remarks

In this paper we review some of our earlier results on the multipotentialisation of symmetry-integrable equations, but we also report several new results on the multipotentialisation of the Kupershmidt equation. In fact, Proposition 4 is new, except for case g) which was reported earlier in [7] (where the result was derived in a different manner). As far as we know, the recursion operators listed in Appendix A have not been reported before, except for the recursion operators of the Kupershmidt equation (2.1) [20] [5], the 1st potential Kupershmidt equation (2.3) [5], and the Sawada-Kotera equation (2.14) [13]. The 7th-order equations that result as the second member of the Kupershmidt chain of multipotentialisations are listed explicitly. As far as we know, only two of the 7th-order equations that are listed in Section 4 appeared earlier in the literature, namely the 7th-order equation of the Sawada-Kotera hierarchy (also known as the Sawada-Kotera-Ito equation) and the 7th-order equation of the k-equation hierarchy.

It is clear that this multipotentialisation procedure for symmetry-integrable equations is useful, as it can easily lead to interesting nonlocal invariance relations for all the equations in the chain that can be applied to generate solutions. This was demonstrated here for the Kupershmidt chain of multipotentialisations. We should however point out that not every symmetry-integrable equation in $1+1$ dimensions can be potentialised. In Appendix B we exemplify the statement of a non-potentialisable equation for a special Krichever-Novikov equation. It may therefore be of interest to study the procedure of multipotentialisation for the whole class of symmetry-integrable equations of dimension $1+1$ and classify the equations accordingly.

The connections between the multipotentialisation process and nonlocal symmetries also needs to be studied in more detail. In [7] we have shown this connection for the Kaup-Kupershmidt equation and for the Sawada-Kotera equation (see also [19] regarding nonlocal symmetries of the Kaup-Kupershmidt equation). Furthermore, it may be of interest to investigate this multipotentialisation procedure for non-evolutionary equations, systems of equations, and of course for higher-dimensional equations.

Appendix A: A list of recursion operators

In this Appendix we list the recursion operators of the equations in Diagram 6. All equations in Diagram 6 admit six-order integro-differential recursion operators of the form

$$R_u = \sum_{j=0}^{6} G_j D_x^j + \sum_{i=1}^{3} I_i(u_x, u_t) D_x^{-1} \circ \Lambda_i, \tag{A.1}$$

where G_j are functions of u and x-derivatives of u for the equation $u_t = F$.

The Kupershmidt equation (2.1) in the dependent variable K admits the recursion operator R_K of the form (A.1) ([20], [5]) with

$$G_0 = -K_{5x} - 12KK_{4x} - 23K_xK_{xxx} + 3K^2K_{xxx} - 15K_{xx}^2 + 38KK_xK_{xx}$$
$$+38K^3K_{xx} + 6K_x^3 + 74K^2K_x^2 - 4K^6$$

$$G_1 = -6K_{4x} - 30KK_{xxx} - 63K_xK_{xx} + 9K^2K_{xx} + 18KK_x^2 + 54K^3K_x$$

$$G_2 = -14K_{xxx} - 40KK_{xx} - 31K_x^2 + 6K^2K_x + 9K^4$$

$$G_3 = -15K_{xx} - 30KK_x$$

$$G_4 = -6K_x - 6K^2, \qquad G_5 = 0, \qquad G_6 = 1$$

$$\Lambda_1 = -2K_{4x} + 10K_xK_{xx} + 10K^2K_{xx} + 10KK_x^2 - 2K^5$$

$$\Lambda_2 = -2K, \quad \Lambda_3 = 0, \quad I_1 = K_x, \quad I_2 = K_t.$$

The **1st Potential Kupershmidt equation** (2.3) in the variable U admits the recursion operator R_U of the form (A.1) [5] with

$$G_0 = -4U_xU_{5x} + 20U_xU_{xx}U_{xxx} + 20U_x^3U_{xxx} + 20U_x^2U_{xx}^2 - 4U_x^6$$

$$G_1 = -U_{5x} - 8U_xU_{4x} - 15U_{xx}U_{xxx} + 3U_x^2U_{xxx} + 6U_xU_{xx}^2 + 18U_x^3U_{xx}$$

$$G_2 = -5U_{4x} - 22U_xU_{xxx} - 13U_{xx}^2 + 9U_x^4 + 6U_x^2U_{xx}$$

$$G_3 = -9U_{xxx} - 18U_xU_{xx}, \qquad G_4 = -6U_{xx} - 6U_x^2, \qquad G_5 = 0, \qquad G_6 = 1$$

$$\Lambda_1 = U_{6x} - 5U_{xx}U_{4x} - 5U_x^2U_{4x} - 5U_{xxx}^2 - 20U_xU_{xx}U_{xxx}$$

$$-5U_{xx}^3 + 5U_x^4U_{xx}$$

$$\Lambda_2 = U_{xx}, \quad \Lambda_3 = 0, \quad I_1 = 2U_x, \quad I_2 = 2U_t.$$

The **2nd Potential Kupershmidt equation** (2.5) admits the recursion operator R_u of the form (A.1) with

$$G_0 = \frac{1}{2}u_x^{-1}u_{7x} - 3u_x^{-2}u_{xx}u_{6x} + \frac{1}{8}\left(121u_x^{-3}u_{xx}^2u_{5x} - 66u_x^{-2}u_{xxx}u_{5x}\right)$$

$$-\frac{25}{4}u_x^{-2}u_{4x}^2 - \frac{125}{2}u_x^{-4}u_{xx}^3u_{4x} + \frac{135}{2}u_x^{-3}u_{xx}u_{xxx}u_{4x} + \frac{65}{4}u_x^{-3}u_{xxx}^3$$

$$-\frac{575}{4}u_x^{-4}u_{xx}^2u_{xxx}^2 + \frac{2935}{16}u_x^{-5}u_{xx}^4u_{xxx} - \frac{945}{16}u_x^{-6}u_{xx}^6$$

$$G_1 = -\frac{3}{2}u_x^{-1}u_{6x} + 13u_x^{-2}u_{xx}u_{5x} + \frac{1}{8}\left(246u_x^{-2}u_{xxx}u_{4x} - 649u_x^{-3}u_{xx}^2u_{4x}\right)$$

$$-\frac{257}{2}u_x^{-3}u_{xx}u_{xxx}^2 + \frac{1313}{4}u_x^{-4}u_{xx}^3u_{xxx} - 162u_x^{-5}u_{xx}^5$$

$$G_2 = -5u_x^{-1}u_{5x} + \frac{95}{2}u_x^{-2}u_{xx}u_{4x} + \frac{137}{4}u_x^{-2}u_{xxx}^2 - \frac{895}{4}u_x^{-3}u_{xx}^2u_{xxx}$$

$$+\frac{2457}{16}u_x^{-4}u_{xx}^4$$

$$G_3 = -\frac{25}{2}u_x^{-1}u_{4x} + 78u_x^{-2}u_{xx}u_{xxx} - \frac{159}{2}u_x^{-3}u_{xx}^3$$

$$G_4 = -12u_x^{-1}u_{xxx} + \frac{51}{2}u_x^{-2}u_{xx}^2, \qquad G_5 = -6u_x^{-1}u_{xx}, \quad G_6 = 1$$

$$\Lambda_1 = u_x^{-2}u_{8x} - 8u_x^{-3}u_{xx}u_{7x} - \frac{47}{2}u_x^{-3}u_{xxx}u_{6x} + \frac{197}{4}u_x^{-4}u_{xx}^2u_{6x} - \frac{85}{2}u_x^{-3}u_{4x}u_{5x}$$

$$+255u_x^{-4}u_{xx}u_{xxx}u_{5x} - 255u_x^{-5}u_{xx}^3u_{5x} + \frac{355}{2}u_x^{-4}u_{xx}u_{4x}^2 + \frac{995}{4}u_x^{-4}u_{xxx}^2u_{4x}$$

$$-1565u_x^{-5}u_{xx}^2u_{xxx}u_{4x} + \frac{16665}{16}u_x^{-6}u_{xx}^4u_{4x} - 760u_x^{-5}u_{xx}u_{xxx}^3$$

$$+\frac{12415}{4}u_x^{-6}u_{xx}^3u_{xxx}^2 - \frac{12435}{4}u_x^{-7}u_{xx}^5u_{xxx} + \frac{14175}{16}u_x^{-8}u_{xx}^7$$

$$\Lambda_2 = u_x^{-2}u_{4x} - 4u_x^{-3}u_{xx}u_{xxx} + 3u_x^{-4}u_{xx}^3, \quad \Lambda_3 = 0, \quad I_1 = -\frac{1}{2}u_x, \quad I_2 = -\frac{1}{2}u_t$$

The **3rd Potential Kupershmidt equation** (2.6) in the variable v admits the recursion operator R_v of the form (A.1) with

$$G_0 = 2v_x^{-1}v_{7x} - 12v_x^{-2}v_{xx}v_{6x} + 38v_x^{-3}v_{xx}^2v_{5x} - 18v_x^{-2}v_{xxx}v_{5x} - 10v_x^{-2}v_{4x}^2$$

$$-70v_x^{-4}v_{xx}^3v_{4x} + 90v_x^{-3}v_{xx}v_{xxx}v_{4x} + 20v_x^{-3}v_{xxx}^3 - 110v_x^{-4}v_{xx}^2v_{xxx}^2$$

$$+70v_x^{-5}v_{xx}^4v_{xxx}$$

$$G_1 = -3v_x^{-1}v_{6x} + 16v_x^{-2}v_{xx}v_{5x} + 21v_x^{-2}v_{xxx}v_{4x} - 41v_x^{-3}v_{xx}^2v_{4x}$$

$$-46v_x^{-3}v_{xx}v_{xxx}^2 + 53v_x^{-4}v_{xx}^3v_{xxx}$$

$$G_2 = -2v_x^{-1}v_{5x} + 19v_x^{-2}v_{xx}v_{4x} + 2v_x^{-2}v_{xxx}^2 - 31v_x^{-3}v_{xx}^2v_{xxx}$$

$$G_3 = 24v_x^{-2}v_{xx}v_{xxx} - 12v_x^{-3}v_{xx}^3 - 8v_x^{-1}v_{4x}$$

$$G_4 = -3v_x^{-1}v_{xxx} + 12v_x^{-2}v_{xx}^2, \quad G_5 = -6v_x^{-1}v_{xx}, \quad G_6 = 1$$

$$\Lambda_1 = v_x^{-2}v_{8x} - 8v_x^{-3}v_{xx}v_{7x} - 16v_x^{-3}v_{xxx}v_{6x} + 38v_x^{-4}v_{xx}^2v_{6x} - 20v_x^{-3}v_{4x}v_{5x}$$

$$+120v_x^{-4}v_{xx}v_{xxx}v_{5x} - 120v_x^{-5}v_{xx}^3v_{5x} + 65v_x^{-4}v_{xx}v_{4x}^2 + 80v_x^{-4}v_{xxx}^2v_{4x}$$

$$-440v_x^{-5}v_{xx}^2v_{xxx}v_{4x} + 240v_x^{-6}v_{xx}^4v_{4x} - 160v_x^{-5}v_{xx}v_{xxx}^3 + 460v_x^{-6}v_{xx}^3v_{xxx}^2$$

$$-240v_x^{-7}v_{xx}^5v_{xxx}$$

$$\Lambda_2 = v_x^{-2}v_{4x} - 4v_x^{-3}v_{xx}v_{xxx} + 3v_x^{-4}v_{xx}^3, \quad \Lambda_3 = 0, \quad I_1 = -2v_x, \quad I_2 = -2v_t.$$

The **4th Potential Kupershmidt equation** (2.9) in the variable w admits the recursion operator R_w of the form (A.1) with

$$G_0 = \frac{1}{2}w_x^{-1}w_{7x} - 3w_x^{-2}w_{xx}w_{6x} - \frac{33}{4}w_x^{-2}w_{xxx}w_{5x} + \frac{121}{8}w_x^{-3}w_{xx}^2w_{5x} + \frac{\beta}{3}w_x^{-2}w_{5x}$$

$$+\frac{135}{2}w_x^{-3}w_{xx}w_{xxx}w_{4x}-\frac{125}{2}w_x^{-4}w_{xx}^3w_{4x}-\frac{13\beta}{6}w_x^{-3}w_{xx}w_{4x}-\frac{25}{4}w_x^{-2}w_{4x}^2$$

$$+\frac{65}{4}w_x^{-3}w_{xxx}^3-\frac{575}{4}w_x^{-4}w_{xx}^2w_{xxx}^2-\frac{29\beta}{24}w_x^{-3}w_{xxx}^2+\frac{2935}{16}w_x^{-5}w_{xx}^4w_{xxx}$$

$$+\frac{115\beta}{16}w_x^{-4}w_{xx}^2w_{xxx}-\frac{5\beta^2}{72}w_x^{-3}w_{xxx}-\frac{945}{16}w_x^{-6}w_{xx}^6-\frac{25\beta}{6}w_x^{-5}w_{xx}^4$$

$$+\frac{5\beta^2}{144}w_x^{-4}w_{xx}^2+\frac{\beta^3}{54}w_x^{-3}$$

$$G_1=-\frac{3}{2}w_x^{-1}w_{6x}+13w_x^{-2}w_{xx}w_{5x}+\frac{123}{4}w_x^{-2}w_{xxx}w_{4x}-\frac{649}{8}w_x^{-3}w_{xx}^2w_{4x}$$

$$-\frac{9\beta}{4}w_x^{-2}w_{4x}-\frac{257}{2}w_x^{-3}w_{xx}w_{xxx}^2+\frac{1313}{4}w_x^{-4}w_{xx}^3w_{xxx}+\frac{44\beta}{3}w_x^{-3}w_{xx}w_{xxx}$$

$$-162w_x^{-5}w_{xx}^5-\frac{49\beta}{3}w_x^{-4}w_{xx}^3-\frac{2\beta^2}{3}w_x^{-3}w_{xx}$$

$$G_2=-5w_x^{-1}w_{5x}+\frac{95}{2}w_x^{-2}w_{xx}w_{4x}+\frac{137}{4}w_x^{-2}w_{xxx}^2-\frac{895}{4}w_x^{-3}w_{xx}^2w_{xxx}$$

$$-\frac{49\beta}{12}w_x^{-2}w_{xxx}+\frac{2457}{16}w_x^{-4}w_{xx}^4+\frac{143\beta}{12}w_x^{-3}w_{xx}^2+\frac{\beta^2}{4}w_x^{-2}$$

$$G_3=-\frac{25}{2}w_x^{-1}w_{4x}+78w_x^{-2}w_{xx}w_{xxx}-\frac{159}{2}w_x^{-3}w_{xx}^3-5\beta w_x^{-2}w_{xx}$$

$$G_4=-12w_x^{-1}w_{xxx}+\frac{51}{2}w_x^{-2}w_{xx}^2+\beta w_x^{-1},\quad G_5=-6w_x^{-1}w_{xx},\quad G_6=1$$

$$\Lambda_1=w_x^{-2}w_{8x}-8w_x^{-3}w_{xx}w_{7x}-\frac{47}{2}w_x^{-3}w_{xxx}w_{6x}+\frac{197}{4}w_x^{-4}w_{xx}^2w_{6x}+\beta w_x^{-3}w_{6x}$$

$$-\frac{85}{2}w_x^{-3}w_{4x}w_{5x}+255w_x^{-4}w_{xx}w_{xxx}w_{5x}-255w_x^{-5}w_{xx}^3w_{5x}-9\beta w_x^{-4}w_{xx}w_{5x}$$

$$+\frac{355}{2}w_x^{-4}w_{xx}w_{4x}^2+\frac{995}{4}w_x^{-4}w_{xxx}^2w_{4x}-1565w_x^{-5}w_{xx}^2w_{xxx}w_{4x}$$

$$-15\beta w_x^{-4}w_{xxx}w_{4x}+\frac{553\beta}{12}w_x^{-5}w_{xx}^2w_{4x}+\frac{16665}{16}w_x^{-6}w_{xx}^4w_{4x}+\frac{5\beta^2}{18}w_x^{-4}w_{4x}$$

$$-760w_x^{-5}w_{xx}w_{xxx}^3+\frac{12415}{4}w_x^{-6}w_{xx}^3w_{xxx}^2+\frac{373\beta}{6}w_x^{-5}w_{xx}w_{xxx}^2$$

$$-\frac{12435}{4}w_x^{-7}w_{xx}^5w_{xxx}-\frac{965\beta}{6}w_x^{-6}w_{xx}^3w_{xxx}-\frac{20\beta^2}{9}w_x^{-5}w_{xx}w_{xxx}$$

$$+\frac{14175}{16}w_x^{-8}w_{xx}^7+\frac{605\beta}{8}w_x^{-7}w_{xx}^5+\frac{25\beta^2}{9}w_x^{-6}w_{xx}^3+\frac{\beta^3}{27}w_x^{-5}w_{xx}$$

$$\Lambda_2=w_x^{-2}w_{4x}-4w_x^{-3}w_{xx}w_{xxx}+3w_x^{-4}w_{xx}^3+\frac{2\beta}{3}w_x^{-3}w_{xx},\quad \Lambda_3=0$$

$$I_1=-\frac{1}{2}w_x,\quad I_2=-\frac{1}{2}w_t.$$

The **5th Potential Kupershmidt equation** (2.11) in the variable q admits the recursion operator R_q of the form (A.1) with

$$G_0=2q_x^{-1}q_{7x}-12q_x^{-2}q_{xx}q_{6x}-18q_x^{-2}q_{xxx}q_{5x}+38q_x^{-3}q_{xx}^2q_{5x}+\frac{21\beta}{16}q_xq_{5x}$$

$$-10q_x^{-2}q_{4x}^2+90q_x^{-3}q_{xx}q_{xxx}q_{4x}-70q_x^{-4}q_{xx}^3q_{4x}-\frac{15\beta}{16}q_{xx}q_{4x}+20q_x^{-3}q_{xxx}^3$$

$$-110w_x^{-4}q_{xx}^2q_{xxx}^2+\frac{15\beta}{16}q_{xxx}^2+70q_x^{-5}q_{xx}^4q_{xxx}-\frac{45\beta}{16}q_x^{-1}q_{xx}^2q_{xxx}$$

$$+\frac{9\beta^2}{32}q_x^3q_{xxx}+\frac{15\beta}{8}q_x^{-2}q_{xx}^4+\frac{9\beta^2}{128}q_x^2q_{xx}^2+\frac{27\beta^3}{8192}q_x^6$$

$$G_1=-3q_x^{-1}q_{6x}+16q_x^{-2}q_{xx}q_{5x}+21q_x^{-2}q_{xxx}q_{4x}-41q_x^{-3}q_{xx}^2q_{4x}-\frac{27\beta}{32}q_xq_{4x}$$

$$-46q_x^{-3}q_{xx}q_{xxx}^2+53q_x^{-4}q_{xx}^3q_{xxx}-\frac{15\beta}{32}q_{xx}q_{xxx}+\frac{15\beta}{16}q_x^{-1}q_{xx}^3$$

$$-\frac{27\beta^2}{512}q_x^3q_{xx}$$

$$G_2=-2q_x^{-1}q_{5x}+19q_z^{-2}q_{xx}q_{4x}+2q_x^{-2}q_{xxx}^2-31q_x^{-3}q_{xx}^2q_{xxx}+\frac{39\beta}{32}q_xq_{xxx}$$

$$-\frac{15\beta}{32}q_{xx}^2+\frac{81\beta^2}{1024}q_x^4$$

$$G_3=-8q_x^{-1}q_{4x}+24q_x^{-2}q_{xx}q_{xxx}-12q_x^{-3}q_{xx}^3-\frac{9\beta}{8}q_xq_{xx}$$

$$G_4=-3q_x^{-1}q_{xxx}+12q_x^{-2}q_{xx}^2+\frac{9\beta}{16}q_x^2,\quad G_5=-6q_x^{-1}q_{xx},\quad G_6=1$$

$$\Lambda_1=q_x^{-2}q_{8x}-8q_x^{-3}q_{xx}q_{7x}-16q_x^{-3}q_{xxx}q_{6x}+38q_x^{-4}q_{xx}^2q_{6x}+\frac{9\beta}{16}q_{6x}$$

$$-20q_x^{-3}q_{4x}q_{5x}+120q_x^{-4}q_{xx}q_{xxx}q_{5x}-120q_x^{-5}q_{xx}^2q_{5x}+65q_x^{-4}q_{xx}q_{4x}^2$$

$$+80q_x^{-4}q_{xxx}^2q_{4x}-440q_x^{-5}q_{xx}^2q_{xxx}q_{4x}+240q_x^{-6}q_{xx}^4q_{4x}-\frac{15\beta}{16}q_x^{-2}q_{xx}^2q_{4x}$$

$$+\frac{45\beta^2}{512}q_x^2 q_{4x} - 160q_x^{-5}q_{xx}q_{xxx}^3 + 460q_x^{-6}q_{xx}^3 q_{xxx}^2 - \frac{15\beta}{8}q_x^{-2}q_{xx}q_{xxx}^2$$

$$-240q_x^{-7}q_{xx}^5 q_{xxx} + \frac{15\beta}{4}q_x^{-3}q_{xx}^3 q_{xxx} + \frac{45\beta^2}{128}q_x q_{xx} q_{xxx} - \frac{45\beta}{32}q_x^{-4}q_{xx}^5$$

$$+\frac{45\beta^2}{512}q_{xx}^3 + \frac{135\beta^3}{32768}q_x^4 q_{xx}$$

$$\Lambda_2 = q_x^{-2}q_{4x} - 4q_x^{-3}q_{xx}q_{xxx} + 3q_x^{-4}q_{xx}^3 + \frac{3\beta}{32}q_{xx}$$

$$\Lambda_3 = 0, \quad I_1 = -2q_x, \quad I_2 = -2q_t.$$

The **Sawada-Kotera equation** (2.14) in the variable S admits the recursion operator R_S of the form (A.1) ([22], [2], [13]) with

$$G_0 = 5S_{4x} + 16SS_{xx} + 6S_x^2 + 4S^3, \; G_1 = 10S_{3x} + 21SS_x, \; G_2 = 11S_{xx} + 9S^2$$

$$G_3 = 9S_x, \quad G_4 = 6S, \quad G_5 = 0, \quad G_6 = 1$$

$$\Lambda_1 = 2S_{xx} + S^2, \quad \Lambda_2 = 1, \quad \Lambda_3 = 0, \quad I_1 = S_x, \quad I_2 = S_t.$$

The **k-equation** (2.16) in the variable k admits the recursion operator R_k of the form (A.1) with

$$G_0 = 13k_{4x} + 82kk_{xx} + 69k_x^2 + 32k^3, \quad G_1 = 35k_{xxx} + 120kk_x$$

$$G_2 = 36k^2 + 49k_{xx}, \quad G_3 = 36k_x, \quad G_4 = 12k, \quad G_5 = 0, \quad G_6 = 1$$

$$\Lambda_1 = k_{xx} + 4k^2, \quad \Lambda_2 = 1, \quad \Lambda_3 = 0, \quad I_1 = k_x, \quad I_2 = k_t.$$

Appendix B: An equation that does not potentialise

Not all symmetry-integrable equations can be potentialised. We demonstrate this claim for the following special Krichever-Novikov equation:

$$u_t = u_{xxx} - \frac{3}{2}\frac{u_{xx}^2}{u_x} + \frac{u^3}{u_x}. \tag{B.1}$$

Equation (B.1) is a special case of the symmetry-integrable Krichever-Novikov equation [16] (see also [17])

$$u_t = u_{xxx} - \frac{3}{2}\frac{u_{xx}^2}{u_x} + \frac{P(u)}{u_x}, \qquad P^{(5)}(u) = 0. \tag{B.2}$$

In [9] we report some multipotentialisations of (B.2), namely for the case

$$P(u) = k_2(u^2 + k_1 u + k_0)^2,$$

where k_0, k_1 and k_2 are arbitrary constants with $k_2 \neq 0$. Equation (B.1) was not included in our study, as (B.1) does not admit a potentialisation, which we'll now establish explicitly.

Calculating all integrating factors for (B.1) up to order four, we find that (B.1) admits no zero-order and no second-order integrating factors. In fact, the lowest integrating factor that (B.1) admits is a fourth-order integrating factor, namely the following:

$$\Lambda = \frac{u_{4x}}{u_x^2} - \frac{4u_{xx}u_{3x}}{u_x^3} + \frac{3u_{xx}^3}{u_x^4} - \frac{2u^3u_{xx}}{u_x^4} + \frac{3u^2}{u_x^2}.$$

This gives the following conserved current and flux (with $\Phi^x = \Phi^x(x, u, u_x, \ldots, u_{qx})$, $q = 4$ as the lowest order):

$$\Phi^t = -\frac{1}{2}\frac{u_{xx}^2}{u_x^2} + \frac{1}{3}\frac{u^3}{u_x^2}$$

$$\Phi^x = \frac{u_{4x}u_{xx}}{u_x^2} + \frac{1}{2}\frac{u_{xxx}^2}{u_x^2} + \frac{2u_{xx}^2u_{xxx}}{u_x^3} + \frac{2}{3}\frac{u^3u_{xxx}}{u_x^3}$$
$$-\frac{9}{8}\frac{u_{xx}^4}{u_x^4} + \frac{1}{2}\frac{u^3u_{xx}^2}{u_x^4} - \frac{3u^2u_{xx}}{u_x^2} + \frac{1}{6}\frac{u^6}{u_x^4}.$$

For a potential variable v we need to consider

$$v_x = \Phi^t$$

and replace u and the x-derivatives of u in

$$v_t = -\Phi^x$$

to obain a local equation of the form

$$v_t = G(x, v_x, v_{xx}, v_{xxx}).$$

It is now easy to show that this is not possible for the above given Φ^t and Φ^x. The same is true for higher-order integrating factors and currents and fluxes that depend on higher-order derivatives. We therefore conclude that (B.1) cannot be potentialised.

Acknowledgement

It is our pleasure to thank Enrique Reyes for useful comments on a draft version of this paper that led to the current improved version.

References

[1] Anco C S and Bluman G W, Direct construction method for conservation laws of partial differential equations Part II: General treatment, *Euro. Jnl. of Applied Mathematics* **13**, 567–585, 2002.

[2] Caudrey P J, Dodd R K and Gibbon J D, A new hierarchy of Korteweg-de Vries equation, *Proc. Roy. Soc. London Ser. A* **351**, 407–422, 1976.

[3] Conte R (Ed), *The Painlevé Property One Century Later*, Springer, New York, 1999.

[4] Euler M and Euler N, Second-order recursion operators of third-order evolution equations with fourth-order integrating factors, J. Nonlinear Math. Phys. **14**, 321–323, 2007.

[5] Euler M and Euler N, A class of semilinear fifth-order evolution equations: Recursion operators and multipotentialisations, *J. Nonlinear Math. Phys.* **18** Suppl. 1, 61–75, 2011.

[6] Euler M and Euler N, Invariance of the Kaup-Kupershmidt equation and triangular auto-Bäcklund transformations, *J. Nonlinear Math. Phys* **19**, 1220001-1-7, 2012.

[7] Euler M, Euler N and Reyes E G, Multipotentialisation and nonlocal symmetries: Kupershmidt, Kaup-Kupershmidt and Sawada-Kotera equations, *J. Nonlinear Math. Phys.* **24**, 303–314, 2017.

[8] Euler N and Euler M, On nonlocal symmetries, nonlocal conservation laws and nonlocal transformations of evolution equations: Two linearisable hierarchies, *J. Nonlinear Math. Phys.* **16**, 489–504, 2009.

[9] Euler N and Euler M, Multipotentialisation and iterating-solution formulae: The Krichever-Novikov equation, *J. Nonlinear Math. Phys.* **16 Suppl. 1**, 93–106, 2009.

[10] Euler N and Euler M, The converse problem for the multipotentialisation of evolution equations and systems *J. Nonlinear Math. Phys.* **18 Suppl. 1**, 77–105, 2011.

[11] Fokas A S and Fuchssteiner B, On the structure of symplectic operators and hereditary symmetries, *Lett. Nuovo Cimento* **28**, 299–303, 1980.

[12] Fordy A P and Gibbons J, Some remarkable nonlinear transformations, *Physics Letters* **75A** (5), 325, 1980.

[13] Fuchssteiner B and Oevel W, The bi-Hamiltonian structure of some nonlinear fifth- and seventh-order differential equations and recursion formulas for their symmetries and conserved covariants, *J. Math. Phys.* **23** (3), 358–363, 1982.

[14] Göktaş Ü and Hereman W, Symbolic computation of conserved densities for systems of nonlinear evolution equations, *J. Symbolic Computation* **24**, 591–622, 1997.

[15] Ito M, An extension of nonlinear evolution equations of the K-dV (mmK-dV) type to higher orders, *J. Phys. Soc. Jpn.* **49** 771–778, 1980.

[16] Krichever I M and Novikov S P, Holomorphic bundles over algebraic curves, and nonlinear equations, *Russ. Math. Surv.* **35**, 53–80, 1980.

[17] Mikhailov A V, Shabat A B and Sokolov V V, The symmetry approach to classification of integrable equations, in *What is Integrability?*, Zhakarov E V (Ed), Springer, Berlin, 115–184, 1991

[18] Petersson N, Euler N, and Euler M, Recursion Operators for a Class of Integrable Third-Order Evolution Equations, *Stud. Appl. Math.* **112**, 201–225, 2004.

[19] Reyes E G, Nonlocal symmetries and the Kaup-Kupershmidt equation, *J Math. Phys.* **46**, 073507, 19 pp, 2005.

[20] Sanders J A and Wang J P, On the integrability of homogeneous scalar evolution equations, *J. Diff. Eqs.* **147**, 410–434, 1998.

[21] Sanders J A and Wang J P, Integrable systems and their recursion operators, *Nonlinear Analysis* **47**, 5213–5240, 2001.

[22] Sawada K and Kotera T, A method of finding N-soliton solutions of the KdV and KdV-like equation, *Progr. Theoret. Phys.* **51**, 1355–1367, 1974.

[23] Steeb W-H and Euler N, *Nonlinear Evolution Equations and Painlevé Test*, World Scientific, Singapore, 1988.

[24] Weiss J, The Painlevé property for partial differential equations. II: Bäcklund transformation, Lax pairs, and the Schwarzian derivative, *J. Math. Phys.* **24**, 1405–1413, 1983.

C2. Geometry of Normal Forms for Dynamical Systems

Giuseppe Gaeta

Dipartimento di Matematica, Università degli Studi di Milano, via Saldini 50, 20133 Milano (Italy); e-mail: giuseppe.gaeta@unimi.it

Abstract

We discuss several aspects of the geometry of vector fields in (Poincaré-Dulac) *normal form*. Our discussion relies substantially on *Michel theory* and aims at a constructive approach to simplify the analysis of normal forms via a splitting based on the action of certain groups. The case, common in physics, of systems enjoying an *a priori* symmetry is also discussed in some detail.

1 Introduction

Most applications of mathematics in natural sciences go through differential equations. These are generically *nonlinear*, and nonlinear differential equations as a rule cannot be solved. Thus the only way to get some analytical information about their behavior is through *perturbation theory* – in particular for systems which are in some sense close to integrable (e.g. linear) ones, or for solutions which are in some sense close to exactly known ones.

Henri Poincaré (1854-1912) set at the basis of perturbation theory his method of *normal forms* [90, 91] (for the life and work of Poincaré, see [102]). Here we will be concerned in particular with normal forms for finite dimensional dynamical systems , but we stress that Poincaré's approach also extends to evolution PDEs, see e.g. [27, 33, 85].

We will be specially interested in some *geometric* feature of the normal form approach; it is maybe worth stressing that the motivation and goal for this has not to be traced to a desire of mathematical abstraction, or to a preference for the geometric (rather than analytic) approach, but rather to concrete computational tasks. I hope I will convince the reader of the advantages of having (also) a geometrical view of this topic.

The *plan of the paper* is as follows. In Section 2 we give a short account of (the basics of) the normal forms construction (which can be skipped by the reader having some basic knowledge of this); in Section 3 we discuss the symmetry properties of systems in normal form. In Section 4 we mention some feature of the Michel theory[1] [80] of symmetric vector fields and potential; these concerns how these objects can be retraced to the *orbit space*, which – when can be properly defined, see the discussion there – is in general a *stratified manifold* [57]. We can then combine the two, which

[1]The name of Michel is associated to this theory in particular for physics applications; the mathematically oriented reader will associate to it the names of Hilbert, Schwarz, Procesi, Bredon, Bierstone, Thom among others. See also Sections 4 and 11, as well as [38].

we do in Section 5, discussing how the peculiar features of systems in normal form allow, through the use of Michel theory and more generally of invariants theory [86] (and the separation of vector fields in parts along and transversal to the group action [19, 20, 72, 74]) obtaining a very effective splitting.

In many relevant cases in physics, the systems under study have some symmetry property (e.g. under space rotations, Lorentz boost, etc.); in this case the physical symmetry and the one built in the normal form construction can combine in different ways, and the procedure discussed so far can be further enhanced; this is discussed in Section 6 for general symmetries, and in Section 7 in the case where the physical symmetries act *linearly*. Under certain circumstances, the symmetry properties enforce a *finite* normal form, as discussed in Section 8; and it may even happen that – again due to symmetry properties – the normal form (or even any symmetric vector field) has a *gradient* structure, see Section 9, which in turn may lead to *spontaneous linearization*, i.e. to dynamics being asymptotically linear, see Section 10.

Needless to say, many topics remain outside this treatment; some of these are briefly mentioned in the final Section 11, where we also summarize and discuss our findings.

The paper is completed by several Appendices. In Appendix A we recall the basic features of the normal forms construction; Appendices B and C are devoted to applications of the unfolding procedure described in Section 5, respectively one to some illustrative examples and the other to the case of Hopf and Hamiltonian Hopf bifurcations.

Finally, albeit our discussion will mostly be conducted at the formal level, leaving the issue of converge of the involved series (which in practical applications are of course always truncated to some finite order) to a case-by.case discussion[2], it makes of course a lot of sense to try having information about this beforehand, i.e. in general terms. This is the subject of Appendix D.

I would like to stress two points concerning matters *not* discussed here:

(A) Another relevant case in physics is of course that where the system under study is *Hamiltonian*. The theory is of course well developed in this case, and actually it has the advantage of dealing more economically with a scalar object (the Hamiltonian) rather than with a vector one (the dynamical vector field). We have chosen not to deal specifically with this case, for two reasons: (a) Hamiltonian vector fields are a special type of vector fields, i.e. our general discussion will also cover the Hamiltonian case; (b) adding a specific discussion of the Hamiltonian case would have made the article even longer, while this is already beyond the limits assigned by the Editor, whom we thank both for the invitation to contribute to this volume and for the patience in this respect.

(B) Similar considerations apply to study of the dynamics near a *relative equilibrium*

[2]The theory is constructive, and all transformations are explicitly determined; so one can determine explicitly also the radius of convergence of the resulting (infinite or truncated) series of transformations; see below.

(e.g. a periodic orbit or an invariant torus) rather than a simple one; most of the approach and results described below are extended to this more general setting, but this would cause the article to grow far too much.

The symbol $\odot$ will mark the end of a remark, while $\triangle$ the end of a proof. Summation over repeated indices will be routinely assumed, except in certain formulas where sums are explicitly indicated.

2 Normal forms

Let us consider a dynamical system

$$\dot{x} = f(x) \tag{1}$$

in a smooth n-dimensional manifold M. We assume that there is some equilibrium point $x_0 \in M$, thus $f(x_0) = 0$, and as we are specially interested in the behavior near x_0, we will consider a local chart with origin x_0 (so from now on $x_0 = 0$), coordinates x^i and Euclidean metric. We can thus write (1) in components as

$$\dot{x}^i = f^i(x) \; ; \tag{2}$$

moreover, as we wish to study the situation nearby the origin, we expand $f(x)$ in a Taylor series, and write

$$\dot{x}^i = \sum_{k=0}^{\infty} f_k^i(x) \; , \tag{3}$$

where the f_k are homogeneous of degree $k+1$ in the x (the reason for this "notational shift" will be apparent in the following),

$$f_k(ax) = a^{k+1} \, f_k(x) \; .$$

The term $f_0(x)$ is linear, and will have a special role in the following; we will also write

$$f_0(x) = Ax \; . \tag{4}$$

As well known, the constant matrix A can always be decomposed into a semisimple and nilpotent part (Jordan normal form), and the two commute with each other:

$$A = A_s + A_n \; ; \; [A_s, A_n] = 0 \; .$$

In the following, we will always assume (i) and (ii) below, and be mostly interested in the case where (iii) also holds:

(i) $A_s \neq 0$;
(ii) The local coordinates have been chosen so that A_s is diagonal,

$$A_s = \mathtt{diag}(\lambda_1, ..., \lambda_n) \; .$$

(iii) $A_n = 0$.

Remark 1. As well known, the eigenvalues of A coincide with those of A_s, hence (for the choice of coordinates mentioned above) with the entries on the diagonal of A_s. It is also well known that if all the λ_i are distinct, then necessarily $A_n = 0$; on the other hand, one can have $A_n = 0$ even with multiple eigenvalues. For normal forms with $A_n \neq 0$, see e.g. [58, 59, 106]. ⊙

Remark 2. Note that in many (but not all) cases of physical interest, including the (special, but relevant) case of Hamiltonian systems near a non-degenerate elliptic equilibrium point, $A_n = 0$; moreover in that case the eigenvalues λ_i come in pairs of complex conjugate ones. For the case of an elliptic equilibrium, we have purely imaginary eigenvalues, $\lambda_m = \pm i\,\omega_k$. ⊙

Remark 3. We stress that here we are considering a *given* A. For system depending on an external parameter – as e.g. those met in studying *phase transitions* – it is more appropriate to consider families of matrices $A(\mu)$ depending on such parameters. In cases of interest for physics, as indeed in phase transitions, these go through $A = 0$ (or at least $A_s = 0$) and hence our hypothesis, in particular (i) above, are necessarily violated. See in this respect, and in a concrete physical application, the discussion in [47]. ⊙

It is obvious that the linearized system

$$\dot{x} \;=\; A\,x \tag{5}$$

can be solved as $x(t) = \exp[At]x_0$. One would expect that – at least until $|x|$ remains small – the solutions to the full system (3) are approximated by those to (5). How good is this approximation will of course depend on the nonlinear terms; in particular we expect that if the first nonlinear terms $f_1, f_2, \ldots$ are actually vanishing, the approximation will be better.

If we were able – without altering the linear term – to find coordinates which would make the nonlinear terms vanishing up to some finite but arbitrary order N, the solution $x(t) = \exp[At]x_0$ to the linear equation (in the new coordinates) would approximate the full solution (in the new coordinates) with arbitrary precision.

Poincaré showed that – subject to a relevant *non-resonance condition* on the spectrum of A_s, see below – not only do such changes exist[3], but they can be determined algorithmically [90, 91].

The work by Poincaré was then extended by his pupil Henri Dulac (1870-1955) who studied what happens when the non-resonance condition is violated [31]; he showed that albeit in this case nonlinear terms can in general not be eliminated, they can be "simplified" (in a sense to be explained below), or – as one now says – *normalized.* One can indeed reduce the system (up to some finite but arbitrary

[3]Here by "exist" we mean they exist *formally*. More precisely, they are described by a series, which is in general only formal; criteria for the convergence of the series (at least in some small neighborhood of the origin) have of course been widely studied, see e.g. [26] and Appendix D below.

order) to one which contains only *resonant* terms, in the sense to be discussed in a moment.

Remark 4. In the case of Hamiltonian vector fields one can work directly on the Hamiltonian (one scalar function) rather than on the associated vector field (with n components, i.e. n scalar functions). This situation was studied by George David Birkhoff[4] (1884-1944) for the non-resonant case [11], and by Fred Gustavson for the resonant one [65]. ⊙

We will present the normal forms construction in an Appendix, for the reader not already familiar with it (standard references for it are [5, 34]. Here we will just characterize the vector fields which are obtained as a result of the normalization procedure.

Definition 1. *A vector with components $x^\mu \mathbf{e}_i$ is of order m if $\mu_1 + ... + \mu_n = m$, and it is* resonant *(with A_s) if*

$$\mu \cdot \lambda = \sum \mu_k \lambda_k = \lambda_i . \tag{6}$$

A vector field $X_f = f^i \partial_i$ is resonant (with A_s) if its components are resonant.

Obviously the set of resonant vectors of a given order is a linear space of finite (possibly zero) dimension, and the same holds – except for the finite dimension, in general – if we consider resonant vectors of any order. Our discussion in Appendix A can be summarized as

Proposition 2. *Let the vector field $X_f = f^i \partial_i$ admitting a zero in the origin have linear part $X_0 = (Ax)^i \partial_i$, with $A = A_s$. Then X_f is in normal form if and only if it is resonant with A_s.*

3 Normal forms and symmetry

The discussion of the previous section allows to characterize (vector fields in) normal forms in terms of their *symmetry properties*: in fact, the vector fields X_k associated to all the nonlinear terms F_k do commute with X_0, the one associated to the linear part[5] of the system:

$$X_k := F_k^\alpha(x) \frac{\partial}{\partial x^\alpha} ; \ [X_0, X_k] = 0 . \tag{7}$$

This condition also provides a characterization of *resonant vector fields.*[6]

[4]Not to be mistaken with his son Garrett Birkhoff (1911-1996); he is also associated with Poincaré through the so called (Capelli)-Poincaré-Birkhoff-Witt theorem.

[5]We recall once again we are assuming, for the sake of simplicity, that this linear part is semisimple, $A = A_s$.

[6]At first sight this is an invariant, i.e.coordinate-independent, characterization. However, note that it depends on what is the linear part of the dynamics, and this is dependent on the choice of coordinates, albeit will not change under well-behaved coordinate changes.

It is immediate to observe that (7), together with the Jacobi identity, implies that:

Lemma 1. *Vector fields which are resonant with a given linear one X_0, span a Lie algebra.*

It may be useful to consider the transposition of the Lie bracket (i.e. the commutator) in terms of components of the vector fields. This is the Lie-Poisson bracket between vector functions $f, g : \mathbf{R}^n \to \mathbf{R}^n$, defined as

$$\{f,g\}^i := (f^j \partial_j), g^i - (g^j \partial_j) f^i ; \tag{8}$$

equivalently,

$$\{f,g\} := (f \cdot \nabla) g - (g \cdot \nabla) f . \tag{9}$$

It is immediate to check that if $X = f^i \partial_i$, $Y = g^i \partial_i$, then

$$Z = [X,Y] = h^i \partial_i ; \quad h = \{f,g\} . \tag{10}$$

We will thus consider the set $V \equiv V^{(A)}$ of (polynomial) equivariant vector functions, i.e. of functions $f : \mathbf{R}^n \to \mathbf{R}^n$ such that

$$\{Ax, f\} = 0 . \tag{11}$$

In particular, we will consider the (linear space) of equivariant functions homogeneous of degree $k+1$, denoted as V_k.

It will also be natural to consider the ring $I \equiv I^{(A)}$ of (polynomial) scalar functions invariant under the linear part of X, i.e. functions $\beta : \mathbf{R}^n \to \mathbf{R}$ such that $X_0(\beta) = 0$. In particular, we will consider the (linear space) of invariant functions homogeneous of degree k, denoted as I_k.

It is rather obvious that $V^{(A)}$ has the structure of a *Lie module* over $I^{(A)}$. It is also obvious that for any $\beta \in I_m$, $f \in V_k$, we have $\beta f \in \mathcal{V}_{k+m}$. Further details are provided e.g. in Chapter III of [25]. A full characterization of normal forms in terms of symmetry is provided by the following Lemma; see [103] for its proof:

Lemma 2. *If $A = (Df)(0)$, then the NF $\widehat{f}$ for f can be written in the form*

$$\widehat{f}(x) = \sum_{j=0}^{s} \mu_j(x) \, M_j \, x , \tag{12}$$

where M_j are a basis for the linear space of real n-dimensional matrices commuting with A, and $\mu_j(x)$ are scalar (polynomial or possibly rational) functions for the linear flow $\dot{x} = Ax$.

Remark 5. With the notation used in this Lemma, it is natural to choose one of the M_j, say M_0, to coincide with A; note that $s \geq 0$, and that if $s = 0$ we have $\widehat{f}(x) = [1 + \alpha(x)]Ax$ with α an invariant function; we are thus in the framework of what is known as "condition α", see Appendix D. $\odot$

Summarizing, vector fields in normal form are characterized by their symmetry property under the vector field X_0. Recalling that normal forms are by definition also polynomial, the task of describing the most general normal form – and its dynamics – for a given linear part is then reduced to the task of studying the most general polynomial vector field commuting with (hence covariant w.r.t.) a given linear one.

4 Michel theory

We are thus led to consider, in full generality, polynomial (or, for that matter, C^∞) vector fields which commute with a given linear one X_0; equivalently, which are symmetric (that is, equivariant) under the action of a linear Lie group G_0, generated by X_0.

It should be noted that in many physical situations the system (even before the reduction to normal form) will be required to have some symmetry properties on the basis of the physics it describes. The more common ones are of course symmetries under translations, rotations and inversions (Euclidean group) or in the relativistic context under the Lorentz or the full Poincaré group.

Moreover, it may happen that the original system (1) has some special symmetry beyond (or instead of) those mentioned above; in full generality we assume that the original system has a Lie symmetry described by a group G with Lie algebra $\mathcal{G}$. In this case it is well known that the whole normalization procedure can be performed preserving such symmetries; see e.g. [25] and references given there. The normal form will then correspond to vector fields which are symmetric under G *and* also under G_0[7]

Thus we consider vector fields in $\mathbf{R}^n$ which are G-equivariant for G a general Lie group acting in $\mathbf{R}^n$.

We will consider the *orbit space* $\Omega = M/G$. Its elements are the G-orbits ω in M, i.e. the sets

$$\omega_x \ = \ \{y \in M \ : \ y = gx \text{ for some } g \in G\} \ = \ Gx \ . \tag{13}$$

Note that here (and below) we think the G-action (i.e. the representation T through which G acts) in M to be given, and identify gx with $T_g x$, etc.

The distance between two orbits ω_x and ω_y is defined as

$$\delta(\omega_x, \omega_y) \ = \ \min \left[d(\xi, \eta), \ \xi \in \omega_x, \ \eta \in \omega_y \right]$$

with d the standard distance in M.

Assumption. We will from now on assume that G acts *regularly* in M.

Remark 6. This is automatically satisfied if G is a *compact* Lie group, but typically G_0 (see notation above) is not compact, at least for generic dynamical systems

[7] Note that it may happen that $G_0 \subset G$. E.g., consider the case where we have a dynamical system in $\mathbf{R}^2$ required to be rotationally invariant and whose linear part is just a rotation.

(we have a compact G_0, actually $G_0 = \mathbf{T}^\ell$, for the special but relevant case of a Hamiltonian system near an elliptic fixed point). ⊙

Remark 7. In many respects, the requirement of a compact group G can be replaced by a weaker one, i.e. that $D(H) := N(H)/H$ is compact for maximal isotropy subgroups $H \subseteq G$. Here we denote by $N(H) = N(H,G)$ the normalizer of H in G, that is the greater subgroup of G in which H is a normal subgroup; obviously $H \subseteq N(H)$. Then $D(H) := N(H)/H$, the quotient being well defined since H is actually by definition a normal subgroup in $N(H)$. ⊙

With this hypothesis, $\delta(\omega_x, \omega_y) = 0$ if and only if x, y belong to the same orbit, i.e. if and only if $y = gx$ for some $g \in G$. (A counterexample when the assumption is not satisfied is provided, as usual, by the irrational flow on the torus.)

The orbit space $\Omega = M/G$ is then a *stratified manifold* [57] in the sense of algebraic geometry, i.e. the union of smooth manifolds with manifolds of smaller dimension lying at the border of those of greater dimension.[8]

There is also a different (in principles) stratification of Ω, based on symmetry properties; we will refer to this as its *isotropy stratification.*

Given the G-action on M, we can associate with any $x \in M$ its isotropy subgroup

$$G_x \ = \ \{g \in G \ : \ gx = x\} \ \subseteq \ G \ . \tag{14}$$

It is quite clear that points on the same G-orbit have isotropy subgroups which are conjugated in G. In fact, if $y = gx$ with $g \in G$, then $G_y \ = \ g\, G_x\, g^{-1}$.

If we define an *isotropy type* as the set of points in M which have isotropy subgroups which are G-conjugated, it is then clear that points on ω_x all belongs to the same isotropy type $[G_x]$, and we can assign to $\omega = \omega_x$ an isotropy class (i.e. an isotropy type).

There is a natural inclusion relation among (isotropy) subgroups of G, and this also naturally extends to a relation among isotropy types: we say that $[G_y] \subseteq [G_x]$ if there are subgroups $G_1 \in [G_y]$ and $G_2 \in [G_x]$ such that $G_1 \subseteq G_2$.

Then we can stratify Ω (and actually also M) on the basis of the isotropy properties of orbits $\omega \in \Omega$; there will be a generic stratum with lower isotropy G_0 (usually – and surely if the G-action in M is effective – just $G_0 = \{e\}$), and then higher and higher strata with isotropy types corresponding to larger and larger isotropy subgroups of G.[9]

It was realized by L.Michel [79, 80, 81] that the geometric stratification of Ω (this is also called its *Whitney stratification*) is coherent with its *isotropy stratification.*[10]

[8]Note that this does *not* imply that Ω itself is a manifold. A familiar example of a stratified manifold which is not a manifold is provided by a cube. The interior of the cube is a three-dimensional manifold M^3, the (interior of the) faces are two-dimensional manifolds $M^2 \subset \partial M^3$, the (interior of the) edges are one-dimensional manifolds $M^1 \subset \partial M^2$, and the vertices are zero-dimensional manifolds $M^0 \in \partial M^1$.

[9]We stress that while the subgroups of G are defined independently of the way G acts in M, the lattice of isotropy subgroups depends on the G-action. For example, if G acts via the trivial representation, all points have isotropy G.

[10]E.g., if we consider R^3 and on it $G = Z_2 \times Z_2 \times Z_2$ acting as $R_x \times R_y \times R_z$, where R_α is the

In the case of an equivariant dynamics, i.e. where a G-covariant vector field X is defined in M, this has a very relevant consequence. That is, the vector field is everywhere tangent to strata in M, and hence: (i) it can be projected to a vector field X_ω in Ω; (ii) strata in M and in Ω are invariant under the flow defined by X and respectively X_ω.[11]

Thus, we conclude that symmetric dynamics can – under rather mild conditions on the geometry (topology) of the relevant group action – be projected to the orbit space. Needless to say, this is in general of smaller (sometimes much smaller) dimension and hence hopefully more easily studied. We will see in the following that actually *if* we are able to solve this "simpler" – but nevertheless in general *nonlinear* – dynamics, the dynamics of systems in normal form can be reconstructed by solving *linear* (albeit non-autonomous) equations.

Remark 8. A very readable introduction to Michel theory and its (original) physical applications is provided by Abud and Sartori [2] (see also [92]); for a more comprehensive discussion, see [81]. For an extension to *gauge theories*, see [50] (the geometry of gauge orbit space is discussed e.g. in [1, 73]). Dynamical aspects are discussed in [37, 39, 38] and in [19, 20, 72, 74]. For the original issues leading physicists to consider these problems, see [17, 79, 82]. The work of R. Palais on the "symmetric criticality principle" [87, 88, 89] could be seen as an attempt to extend this theory to the infinite dimensional case. ⊙

Remark 9. As for the mathematical aspects of (or counterpart to) Michel theory, this would lead to a long discussion, and we will just refer to [36, 38]. ⊙

5 Unfolding of normal forms

We can now go back to dynamical systems in (Poincaré-Dulac) normal form. The idea we want to pursue is to *increase* the dimension of the system by embedding it into a larger system carrying the same information. The goal is to have a larger system with simpler properties (this idea was successfully carried on in the famous paper by Kazhdan, Konstant and Sternberg [70] on the Calogero integrable system [18]); in this context we refer to the larger system as an *unfolding* of the original one.

In order to do this, we should look more carefully into resonances. These can be of two types, i.e. those corresponding to *invariance relations* on the one hand, and *sporadic resonances* on the other. We now define these concepts.

reflection in the reflection in the coordinate α, the orbit space Ω is made of a octant in R^3, say the first one. Points with three non-zero coordinates have isotropy type $[e]$, points on one of the faces, say the one with the α coordinate equal to zero, have isotropy type $[R_\alpha]$; points on one of the edges, say the one with both α and β coordinate equal to zero, have isotropy type $[R_\alpha \times R_\beta]$; and the vertex in the origin has full isotropy $[G]$.

[11] A finer analysis in this respect is contained in [20, 72, 74]; see also [19, 21] for a comprehensive discussion.

5.1 Resonances, invariance relations, sporadic resonances

Recalling that we denote by λ_i the eigenvalues of A_s (see section 2), it is clear that if there are non-negative integers σ_i such that

$$\sum_{i=1}^{n} \sigma_i \, \lambda_i \; = \; 0 \; , \tag{15}$$

say with $s_1 + ...\sigma_n = |\sigma| \neq 0$, these σ_i can always be added (term by term) to any resonance vector μ_i (of order $|\mu|$) to produce new resonant vectors (of order $|\mu| + k|\sigma|$, with any $k \in \mathbf{N}$).

We say that (15) identifies an *invariance relation.* Having invariance relations is the only way to have infinitely many resonances (and hence infinitely many terms in a normal form) in a finite dimensional system [103].

Any nontrivial resonance (6) such that there is no σ with $\sigma_i \leq \mu_i$ (for all $i = 1, ..., n$) providing an invariance relation, is said to be a *sporadic resonance.* Sporadic resonances are always in finite number (possibly zero) in a finite dimensional system [103].

5.2 Invariance relations and invariant functions

It is clear that invariance relations are associated with invariant functions under the action of G_0, or equivalently of (its generator, i.e.) the linear vector field $X_0 = (A_s x)\nabla$; and conversely, any polynomial scalar function which is invariant under G_0 is associated with an invariance relation. In fact, if

$$I(x) = x_1^{\sigma_1} ... x_n^{\sigma_n} \; ,$$

we immediately have

$$X_0(I) \; = \; \lambda_i x_i \, \frac{\partial}{\partial x_i} I(x) \; = \; \left(\sum_{i=1}^{n} \sigma_i \, \lambda_i \right) \, I(x) \; = \; 0 \; .$$

We assume there are $r \geq 0$ independent invariance relations (here "independent" means that $I_1(x), ..., I_r(x)$ are functionally independent).

If we look at the full dynamics of $I(x)$, it follows from $[X, X_0] = 0$ that $I(x)$ remains always G_0-invariant; hence when we write

$$\frac{dI(x)}{dt} \; = \; \frac{\partial I(x)}{\partial x_i} \, \frac{dx_i}{dt} \; = \; \frac{\partial I(x)}{\partial x_i} \, f^i(x) \; := \; Z(x) \tag{16}$$

the function $Z(x)$ must be itself invariant under G_0; for what we have said above, this means $Z(x)$ can be written as

$$Z(x) \; = \; \Phi[I_1(x), ..., I_r(x)] \; , \tag{17}$$

i.e. that the set of generators for the ring of G_0-invariant functions evolves in time according to

$$\frac{dI_a(x)}{dt} \; = \; \Phi_a[I_1(x), ..., I_r(x)] \quad (a = 1, ..., r) \; . \tag{18}$$

Thus if we introduce auxiliary variables φ_a ($\alpha = 1, ..., r$) and let them evolve according to

$$\frac{d\varphi_a}{dt} = \Phi_a(\varphi_1, ..., \varphi_r) , \tag{19}$$

the relation $\varphi_a = I_a(x_1, ..., x_m)$ – if satisfied at $t = 0$ – will be preserved by the flow.

5.3 Sporadic resonances and auxiliary variables

We will now consider auxiliary variables w_i associated with sporadic resonances [43]; if (6) identifies a sporadic resonance, we define

$$R_i = x_1^{\mu_1} ... x_n^{\mu_n} .$$

In this case we immediately have

$$X_0(R_i) = \left(\sum_k \lambda_k \mu_k\right) R_i = \lambda_i R_i .$$

Thus the functions $R_i[x(t)]$ are covariant, in the sense they evolve as the x_i involved in the (sporadic) resonance relation. Note in particular this means they evolve *linearly* – despite being *nonlinear* functions of the x. Moreover, again assuming R is associated with (6) and considering the linear dynamic, by construction

$$X_0[x_i - R_i(x)] = \lambda_i x_i - \sum_k \frac{\partial R_i}{\partial x_k} \lambda_k x_k = \lambda_i x_i - \left(\sum_k \lambda_k \mu_k\right) R_i = \lambda_i (x_i - R_i) .$$

In particular, the manifold identified by

$$x_i = R_i(x)$$

is by construction invariant.

Remark 10. Note that we can have two different sporadic resonances (with μ and $\mu' \neq \mu$) involve the same distinguished variable x_i only if they are actually related by an invariance relation, as follows immediately from noting that $\mu' - \mu = 0$. Thus we can have at most one independent sporadic resonance for degree of freedom, and the notation w_i is convenient to identify the distinguished variable involved in this. $\odot$

We will let these auxiliary variables w_i – which we see as independent variables – evolve according to

$$\frac{dw_i}{dt} = \frac{\partial R_i}{\partial x_k} \frac{dx^k}{dt} = \frac{\partial R_i}{\partial x_k} f^k(x) . \tag{20}$$

5.4 Unfolding of normal form

Summarizing, we have three types of variables: the n natural coordinates x^i, the $m \leq n$ auxiliary variables w^i associated with sporadic resonances, and the r auxiliary variables ϕ^a associated with invariance relations. Correspondingly, we have a dynamics in a $(n+m+r)$-dimensional space, whose general form is

$$\begin{aligned} \dot{x}^i &= f^i(x,w,\phi) \\ \dot{w}^i &= g^i(x,w,\phi) \\ \dot{\phi}^a &= h^a(x,w,\phi) \,. \end{aligned} \tag{21}$$

In order for this to represent our original dynamics (1), the functions f, g, h should be suitably assigned. Obviously f should reproduce the F appearing in (1), and g, h should be compatible with the identification of w, ϕ given by (18) and (20) respectively. Moreover, precisely these identifications introduce some ambiguity in the writing of f, g, h in terms of the enlarged set of variables. This ambiguity can be used to write the equations in a convenient form – which is precisely the reason to introduce the auxiliary variables.

In fact, we have the following result, which is a restatement of those given in our previous work [54].

Lemma 3. *The function h can be written as an analytic function of the ϕ variables alone; the functions f and g can be written as analytic functions, linear in the x and the w variables.*

Proof. Let us start by considering f; the function $F^s(x)$ appearing in (1) is made of resonant terms only, and these can always be written in terms of invariance relations and sporadic resonances as

$$F^s(x) \;=\; \alpha_s[\phi_1(x), \ldots \phi_r(x)]\, x_s \;+\; \beta_s[\phi_1(x), \ldots \phi_r(x)]\, w_s \;;$$

thus it suffices to define (no sum on i)

$$f^i(x,w,\phi) \;=\; \alpha_i(\phi) x_i \;+\; \beta_i(\phi) w_i \;. \tag{22}$$

Let us now consider the g^i. In this case we would have

$$\dot{w}^i \;=\; \frac{\partial w^i}{\partial x^j}\, \dot{x}^j \;=\; \frac{\partial w^i}{\partial x^j}\, f^j(x,w,\phi) \;;$$

but all terms on the r.h.s. are resonant with x^j, hence can be written in terms of resonant monomials and x^j, or w^j itself.

Finally, we have seen above that the time evolution of ϕ^a is written in terms of invariant functions only, hence of the ϕ themselves. $\triangle$

This is a remarkable result in that it allows to identify the main obstacle to the analysis of systems in normal form and tells how to proceed in this task. We will write its relevant consequences in the form of a Corollary.

Corollary. *The equations (21) can be written as*

$$\begin{aligned} \dot{x}^i &= F^i{}_j(\phi)\, x^j \\ \dot{w}^i &= G^i{}_j(\phi)\, w^j \\ \dot{\phi}^a &= h^a(\phi) \,. \end{aligned} \tag{23}$$

If the (in general, nonlinear) last set of equations is solved, providing $\phi = \phi(t)$, then the first two sets reduce to

$$\begin{aligned} \dot{x}^i &= \widehat{F}^i{}_j(t)\, x^j \\ \dot{w}^i &= \widehat{G}^i{}_j(t)\, w^j \end{aligned} \tag{24}$$

where of course $\widehat{F}(t) = F[\phi(t)]$, $\widehat{G}(t) = G[\phi(t)]$, and we only have to solve linear *(in general, non autonomous) equations.*

Some examples of application of our construction are given in Appendix B. Application to the analysis of the Hopf and the Hamiltonian Hopf bifurcations is given in Appendix C.

Remark 11 Further developments of this approach are discussed in [48, 55, 63], see also [61, 62, 96, 97]; the reader is referred to the original papers for detail. ⊙

6 Normal forms in the presence of symmetry

In the case where the original system (1) has some symmetry – possibly, but not necessarily, dictated by the physics it describes (e.g. covariance under rotation or the Lorentz group) – it is well known that the whole Poincaré-Dulac normalization procedure can be carried out remaining within the class of covariant objects: the Poincaré transformations at each step will have covariant generating functions, and the normalized vector fields will be covariant. Moreover, the vector fields in normal form will have the extra symmetry defined by X_A.[12]

In this case, we do *not* have to study the most general resonant vector field, but the most general *resonant and covariant* one. Needless to say, this is in general much less general than requiring just resonance, i.e. in general a covariant normal form will be simpler (in the sense of admitting fewer terms) than a generic one.

The discussion of Section 5 and the construction described there still apply, except that now the role of G_0 is played in general by a larger group G (in practice, this is most often a group with a *linear* action; but this is not necessarily the case). In particular, the invariant functions ϕ^a will now be the functions which are invariant under *both* the linear dynamical group G_0 *and* the group G of "physical" symmetries (in other words, only G-symmetric invariance relations will have a role). Similarly, only sporadic resonances which respect the G symmetry will correspond to resonant terms present in the normal form and hence to relevant auxiliary variables w^i.

[12] We stress this can induce a larger reduction, see example 2 and Remark B.1 in Appendix B.

Remark 12. From this point of view, we should stress that G_0 acts in general as a group of *non-linear* transformations, but its generator is associated to a linear function of the (adapted) variables. That is, if $A = \text{diag}(\lambda_1, ...\lambda_n)$ and s is the group parameter, then

$$g = \exp[sA] : (x_1, .., x_n) \to \left(e^{s\lambda_1} x_1, ..., e^{s\lambda_n} x_n \right) .$$

Having a generator which depends linearly on the x simplifies in many ways the situation to be studied. ⊙

Thus the extension to the symmetric case is essentially trivial from the theoretical point of view; but it can lead to relevant simplifications in practice. This is possibly better illustrated by considering directly a concrete example, related to one of those considered above.

7 Normal forms and classical Lie groups

In many physical applications, one meets systems with a symmetry described by simple compact Lie groups, and in particular by the classical groups.

The relevant point here is that if the "physical" symmetry G acts regularly (which is definitely the case for a linear representation of a compact Lie group), we can forget about the non-compact nature of the G_0 action, and perform reduction to orbit space *only* under the G-action.

This amounts to consider general (polynomial) vector fields which are covariant under the G-action, and these can be studied in general terms for the simple Lie groups.

The basic classification result here is the (general version of) Schur Lemma and a simple consequence of this, which we quote from Kirillov [71].

Lemma 4 (Schur Lemma). *Let the dimension of the irreducible group representation T in a linear space over the field* **K** *be at most countable; denote by $C(T)$ the centralizer of T, by $c(T)$ the intertwining number $c(T) = \dim_{\mathbf{K}}[C(T)]$. Then if* $\mathbf{K} = \mathbf{C}$, $C(T) \simeq \mathbf{C}$, $c(T) = 1$; *if* $\mathbf{K} = \mathbf{R}$, $C(T)$ *is isomorphic to either* **R** *or* **C** *or* **H** *and correspondingly* $c(T) = 1, 2, 4$.

In the case $\mathbf{K} = \mathbf{R}$, the representation T is said to be of real, complex or quaternionic type according to the form of $C(T)$, see above.

Lemma 5. *Let $T_{\mathbf{C}}$ be the complexification of the real irreducible representation T. If T is of real type then $T_{\mathbf{C}}$ is irreducible; if T is of complex type then $T_{\mathbf{C}}$ is the sum of two inequivalent irreducible representations; if T is of quaternionic type then $T_{\mathbf{C}}$ is the sum of two equivalent irreducible representations.*

We can thus classify symmetric normal forms in R^n according to the type of the irreducible representation describing the symmetry [46]. Note that if we have an irreducible *orthogonal* representation, this is necessarily transitive on the unit sphere S^{n-1} of the carrier space R^n, hence the only invariant is r; as we want a *polynomial* invariant we should consider $\rho = r^2 = x_1^2 + ... + x_n^2$. In this case polynomial covariant vector fields are of the form

$$\dot{x} = \sum_{j=0}^{s} \mu_j(\rho) \, K_j \, x \, , \tag{25}$$

where the K_j are a basis for the set $C(T)$ of n-dimensional real matrices commuting with T; we can and will always choose $K_0 = I$.

7.1 Real type

In this case $C(T) \simeq R$, hence is given by multiples of the identity, and (25) is just

$$\dot{x} = \mu_0(\rho) \, x \; ; \tag{26}$$

obviously this evolves towards spheres with radius ρ^* corresponding to the zeros of $\mu_0(\rho)$; more precisely towards those with $\mu_0(\rho^*) = 0$, $\mu_0'(\rho^*) < 0$.

In more detail, ρ evolves according to

$$\dot{\rho} = 2 \, \rho \, \mu_0(\rho) \; ;$$

note this is a separable equation and can hence be solved computing a rational (as μ_0 is a polynomial) integral,

$$\int \frac{1}{\rho \, \mu_0(\rho)} \, d\rho = 2 \, (t - t_0) \, .$$

The linear part is a multiple of the identity: no resonances are present, and the normal form is just linear. Note that the Poincaré criterion applies (see Appendix D), thus there is a convergent normalizing transformation.

7.2 Complex type

In this case $C(T) \simeq \mathbf{C}$; in other words we have two independent matrices commuting with T, one of them is of course the identity I, while the other will be denoted as J. Note that T is irreducible over R but as a complex representation it will be given by $T = T_0 \oplus \widehat{T}_0$, by Schur lemma. This implies $n = 2m$.

Now, using coordinates adapted to this decomposition, (25) reads just

$$\dot{x} = \mu_0(\rho) \, I x \, + \, \mu_1(\rho) \, J x \; ; \tag{27}$$

in these coordinates, J is written in block form as (the standard symplectic matrix)

$$J = \begin{pmatrix} 0 & -I \\ I & 0 \end{pmatrix} .$$

Now the linear part reads

$$A = c_0 \, I \, + \, c_1 \, J \; ;$$

different cases are possible depending on the vanishing of the constants c_0 and c_1. We exclude the fully degenerate case $c_0 = 0 = c_1$, where we have $A = 0$.

1. If $c_1 = 0$, $c_0 \neq 0$, we are in the same situation as in the real case: the linear part is a multiple of the identity and no resonance is present; the normal form is linear (with a convergent normalizing transformation).

2. If $c_0 \neq 0$, $c_1 \neq 0$, the eigenvalues of A are equal to $\lambda_\pm = c_0 \pm i c_1$ (each of these with multiplicity m). Again no resonances are present; hence the normal form is linear, and again the spectrum belongs to a Poincaré domain and hence there is a convergent normalizing transformation.

3. If $c_0 = 0$, $c_1 \neq 0$, the eigenvalues of A are $\lambda_\pm = \pm i c_1$ (each with multiplicity m). In this case there is an invariance relation $\lambda_+ + l a_- = 0$, hence an infinite number of resonances. Moreover the spectrum does *not* belong to a Poincaré domain; hence we are not guaranteed there exists a convergent normalizing transformation.

7.3 Quaternionic type

The only fundamental representation of a simple Lie group realizing this case occurs for $G = SU(2)$, i.e. the quaternion group itself; we thus discuss directly this case in concrete terms.[13]

The basis matrices of the Lie algebra $su(2)$ can be taken to be

$$H_1 = \begin{pmatrix} 0 & 0 & 1 & 0 \\ 0 & 0 & 0 & 1 \\ -1 & 0 & 0 & 0 \\ 0 & -1 & 0 & 0 \end{pmatrix}, \ H_2 = \begin{pmatrix} 0 & 0 & 0 & -1 \\ 0 & 0 & 1 & 0 \\ 0 & -1 & 0 & 0 \\ 1 & 0 & 0 & 0 \end{pmatrix}, \ H_3 = \begin{pmatrix} 0 & -1 & 0 & 0 \\ 1 & 0 & 0 & 0 \\ 0 & 0 & 0 & 1 \\ 0 & 0 & -1 & 0 \end{pmatrix}.$$

With these, $C(T)$ is spanned by the identity $I = K_0$ and by the three matrices

$$K_1 = \begin{pmatrix} 0 & 1 & 0 & 0 \\ -1 & 0 & 0 & 0 \\ 0 & 0 & 0 & 1 \\ 0 & 0 & -1 & 0 \end{pmatrix}, \ K_2 = \begin{pmatrix} 0 & 0 & 0 & 1 \\ 0 & 0 & 1 & 0 \\ 0 & -1 & 0 & 0 \\ -1 & 0 & 0 & 0 \end{pmatrix}, \ K_3 = \begin{pmatrix} 0 & 0 & 1 & 0 \\ 0 & 0 & 0 & -1 \\ -1 & 0 & 0 & 0 \\ 0 & 1 & 0 & 0 \end{pmatrix};$$

these do of course span another, not equivalent, $su(2)$ representation (see again the Schur Lemma).

The general form of (25) is thus

$$\dot{x} = \sum_{j=0}^{3} \mu_j(\rho)\, K_j\, x\ . \tag{28}$$

The linear part of this is $A = \sum c_j K_j$, and we write

$$\omega = \sqrt{c_1^2 + c_2^2 + c_3^2}\ .$$

The eigenvalues of A are $\lambda_\pm = c_0 \pm i\omega$, each with multiplicity two. Several subcases are possible, as in the complex case (we again exclude the fully degenerate case $c_0 = 0 = \omega$, where we have $A = 0$).

[13] There are higher representations of other simple Lie groups of this type, see e.g. [24], but these appear to be of little physical interest.

1. If $\omega = 0$, $c_0 \neq 0$, the linear part is a multiple of the identity; the normal form is linear with a convergent normalizing transformation.

2. If $c_0 \neq 0$, $\omega \neq 0$, no resonances are present, the normal form is linear, and there is a convergent normalizing transformation.

3. If $c_0 = 0$, $\omega \neq 0$, then $\lambda_\pm = \pm i\omega$. There is an invariance relation, hence an infinite number of resonances, and we are not guaranteed there exists a convergent normalizing transformation.

Remark 13. In all the three (R, C, H) cases, one can discuss in rather general terms *further normalization* [44] (more precisely *Lie renormalized forms* [45]); we refer to [46] for this. $SU(2)$-related dynamics are also studied in [51]. $\odot$

8 Finite normal forms

As discussed above, for systems enjoying an external, "physical" symmetry G, the normal form corresponds to polynomial vectors which are symmetric under *both* G and the symmetry G_0 identified by the linear part of the system itself (it may happen that $G_0 \subseteq G$). This condition can, in some cases, be quite restrictive, and in particular it can happen that there is only a finite dimensional linear space of vectors satisfying it. In this case, we have a *finite normal form*; i.e. the most general normal form will have a finite number of terms.

We note that this can be enforced already by the G_0 symmetry alone, as for systems whose linear part satisfies the Poincaré condition (see Appendix D), but here we discuss – following [53] – cases where it is the interplay of G_0 and G to produce this effect.

We denote by M the algebra of ($n \times n$, real) matrices in $G_0 \oplus G$; with this is associated an algebra $\mathcal{M}$ of linear vector fields: with any $B \in M$ is associated the vector field $X_B = (Bx)\nabla$. We are thus interested in the centralizer $C(\mathcal{M})$ of this algebra in the set of vector fields. Consider a scalar polynomial function $\Phi : R^n \to R$. This is a (polynomial) *relative invariant* of $\mathcal{M}$ if for all $B \in M$ it results

$$X_B(\Phi) \;=\; \mu(B)\,\Phi\,; \tag{29}$$

conversely, the set of functions for which this holds with a given function $\mu : M \to R$ is denoted as $I_\mu(\mathcal{M})$. Obviously $I_0(\mathcal{M})$ corresponds to (polynomial) usual, or *absolute*, invariants.

It is easy to prove [53] that:

Proposition 3. *If $I_0(\mathcal{M})$ is not the full algebra of polynomials in R^n, then $C(\mathcal{M})$ has infinite dimension. If $C(\mathcal{M})$ has infinite dimension, then some $I_\mu(\mathcal{M})$ has infinite dimension.*

Proposition 4. *If $C(\mathcal{M})$ is infinite-dimensional, then $\mathcal{M}$ admits nontrivial rational invariants. If moreover either $[\mathcal{M}, \mathcal{M}] = \mathcal{M}$ or $\mathcal{M}$ is solvable, then $I_0(\mathcal{M})$ is nontrivial.*

Proposition 5. *Let $\mathcal{M}$ be such that $I_0(\mathcal{M}) \neq R$ and $\mathcal{L}$ be the Lie algebra of a compact linear Lie group such that $[\mathcal{L}, \mathcal{M}] \subseteq \mathcal{M}$. Then $I_0(\mathcal{M} + \mathcal{L})$ is nontrivial.*

These results can be used to characterize situations in which $C(\mathcal{M})$ fails to be infinite dimensional; see [53] for applications.

9 Gradient property

We say that a group representation has a *gradient property* [52] if all the (polynomial) vector functions which are covariant w.r.t. it can be expressed as gradients – ordinary or generalized, i.e. symplectic w.r.t. some symplectic structure – of invariant (polynomial) scalar functions.[14]

This means that although the system has not by itself a variational nature, it can nevertheless be analyzed with the tools of variational analysis, with an obvious advantage.

It may happen that a full system is not variational and its symmetry does not have a gradient property, but the reduced equations corresponding to its normal form near a critical point (or a bifurcation equation describing the change of stability of this) have the gradient property.

There is a simple way to ascertain if a group representation has this property. In fact, as is well known, the number of polynomial invariants s_k and covariants v_k of any degree k can be computed in terms of the (power expansion of the) *Molien function.*

If $T = T(g)$ is a representation of the Lie group G, and $(T^{\otimes n})_s$ its symmetrized n-fold tensor product, the number of invariants is given by the coefficient c_n^0 in the Molien series

$$\sum_n c_n^0 \, z^n \;=\; \frac{1}{|G|} \int_G \det \left[\frac{1}{(1 - zT(g))} \right] \, d\nu(g) \; , \tag{30}$$

where $d\nu(g)$ is the Haar measure on G; the function $1/(1 - zT(g))$ is called the *Molien function.*

Similarly, the number of covariants is given by the coefficient c_n^1 in the series

$$c_n^1 \, z^n \;=\; \frac{1}{|G|} \int_G \det \left[\frac{1}{(1 - zT(g))} \right] \, \overline{\chi^1(g)} \, d\nu(g) \; , \tag{31}$$

where $\chi(g)$ is the character of g in the T representation.

Remark 14. More generally,

$$|G|^{-1} \int_G \det \left[\frac{1}{(1 - zT^\mu(g))} \right] \, \overline{\chi^\sigma(g)} \, d\nu(g)$$

measures the multiplicity of the representation T^σ in $[(T^\mu)^{\otimes n}]_s$; in this respect, see e.g. [93, 94, 95]. ⊙

[14]We also speak of gradient property *at order* N if this holds for all polynomial covariant vector function of degree up to N.

We denote by s the number of *linear* covariants for T; it is then clear that the number γ_n of covariants of order n which can be obtained as generalized gradients of invariant functions – i.e. as $\psi = K \nabla \Phi$ with K a matrix and Φ an invariant of degree $n+1$ – is just $\gamma_n = s \cdot c_{n+1}^0$; in general we have $\gamma_n \le c_n^1$, and the gradient property (at order N) is equivalent to having

$$c_n^1 = s \cdot c_{n+1}^0 \tag{32}$$

at all orders (for all orders $n \le N$) [52].

It turns out that gradient property holds at all orders for the defining representation of all the $SO(n)$ and $SU(n)$ groups (and more generally whenever you have a transitive representation). The results is surely not extendible beyond the defining representation, as already for $SO(3)$ it is known not to hold at order $N = 4$ for other representations; on the other hand, it holds up to order $N = 3$ for all representations [41].

10 Spontaneous linearization

We speak of *spontaneous linearization* when the dynamics of a system evolves towards an asymptotic regime governed by linear (autonomous or non-autonomous) equations, and this for whatever initial conditions or at least for whatever initial conditions in a certain range (possibly, all those not leading to unbounded solutions).

It happens that this kind of behavior can be guaranteed on the basis of symmetry considerations alone, i.e. can be present for *all* dynamical systems with certain symmetry properties.

In particular, consider the case where the group representation is transitive on the unit sphere S^{n-1} of the carrier space R^n. In this case we have only one polynomial invariant, which is just $\rho = x_1^2 + ... + x_n^2$. According to our discussion in Section 5, the evolution of ρ is hence governed by a function of ρ itself alone,

$$\dot{\rho} = h(\rho) . \tag{33}$$

As we have only one variable, either the solutions $\rho(t)$ diverge or they reach some fixed point

$$\rho_k^* = \lim_{t \to \infty} \rho(t) ;$$

note that there can be different limit points for different initial conditions $\rho(0)$.

Recalling now Lemma 2, in this case the most general system in normal form will be

$$\dot{x} = \sum_{j=0}^{s} \mu_j(\rho) \, M_j x . \tag{34}$$

It is obvious that if we look at the asymptotic behavior for $t \to \infty$, for all initial data such that the solution does not diverge (and we recall that the normal form is relevant to the actual dynamics only in a neighborhood of the origin), this is given by an equation of the form

$$\dot{x} = \sum_{j=0}^{s} \mu_j(\rho_k^*) \, M_j x = \sum_{j=0}^{s} \mu_{jk}^* \, M_j x \, , \tag{35}$$

which is indeed linear.

11 Discussion and conclusions

After introducing the basic ideas – going back to Poincaré – in normal forms theory, we have considered several geometric aspects of vector fields in normal form. In particular, we have considered how these have built-in symmetry properties, the relevant group G_0 being associated to the linear part of the vector field itself, and how they can be reduced to the G_0-orbit space. This in turn means that we can introduce new auxiliary variables associated with the basic invariants for the G_0 action, and their evolution will depend only on the invariants themselves.

The same approach can be used in connection with resonant terms of the vector field – which in view of the normal form construction do represent *all* the nonlinear terms – and in this way we are led to introduce two sets of auxiliary variables, one associated with sporadic resonances and one with invariance relations.

The relevant point is that the evolution of the enlarged set of variables (x, w, ϕ) is governed by the equations (24), which are *linear* for the x and w. That is, *if* we are able to solve – or at least to determine the asymptotic form of solutions – the autonomous system of the equation governing the ϕ evolution (this the system in orbit space), then we are left with a set of – generally non autonomous – *linear* equations for the x and w. A number of examples, showing this approach can be implemented in practice, have been presented in Section 5, which is the heart of this work.

We have then considered the case where the system has an "external", in particular a physical, symmetry G. In this case the considerations presented above can be extended to consider $G_0 \times G$, and a reduction of normal forms follows. We have considered in particular the case where G corresponds to one of the *classical groups.* We have also remarked that special situations – enforced by symmetry alone – can be present; in particular we have briefly considered the case where normal forms are necessarily *finite*, that in which they enjoy the *gradient property*, and that where we get *spontaneous linearization.*

Having discussed the behavior of vector fields in normal form, we have noted that the correspondence between the original system and that in normal form is in general an actual – and not just formal – one, only in a small neighborhood of the origin, if any. The properties of convergence of the series defining the normalizing transformation can be checked at hand in any concrete application and for any *finite order* (the full series being in general not convergent, and at best *asymptotic*); but nevertheless one would like to have some general notions of, and results about, convergence available with no need to actually perform the detailed computation (that is, available *before* embarking on a generally complex concrete computation). We have discussed this, providing some general results – and focusing on those based on the symmetry properties of the vector field – in Appendix D, thus setting all the previous discussion on a firmer theoretical basis.

We would now like to mention some topics which, due to limited space (and time), have not been included in the present discussion but which would be of interest in this context. Here we can only give brief hints at them, with a few references.

1. First of all, as stressed in the introduction, we have not considered the specific features of Hamiltonian vector fields. In this case one can deal directly with the Hamiltonian (a single scalar function) rather than with the vector field ($2n$ coefficients).

2. We have also not considered ways to make more efficient the normalization steps. In particular, in the "standard" procedure (the one described here) one has to invert certain operators, which is a serious problem in concrete computations and can present convergence problems. Both problems can be circumvented by considering *Lie series*, i.e. transformations corresponding to the finite-time (e.g. $t = 1$) flow of a vector field. Besides the computational advantages of this approach, in the present context it has to be noted that it introduces a further geometrization of the whole procedure. The approach via Lie series is discussed e.g. in [12, 25]; see also [35] for the Hamiltonian case.

3. As briefly mentioned in Appendix A (see footnote 16), one can to some extent control the effect of normalization on higher order terms, and attempt to obtain a "further normalization". In the most complete (and hence generally only formal) outcome, this will make all terms commute among themselves. The theory is connected with both geometrical and Lie algebraic aspects, and introduces further constants of motion (existing in more and more restricted neighborhood of the origin). For further detail see e.g. [25] and references therein.

4. We have worked with fixed vector fields (1); but in many cases of physical interest one is interested in vector fields depending on (one or more) external parameters. One often wishes to study situations in which as the parameters are varied the considered fixed point undergoes a *bifurcation* [29, 30, 42, 56, 64, 67, 68, 101]. In this case the basic assumption of the normalization approach – i.e. non-degeneration of the linear part – fails, and one has to consider a generalization of the theory. We cannot consider this here, but we mention that in this case too the presence of symmetries introduces several interesting (and helpful for practical purposes) aspects [93, 94, 95].

5. We would also like to mention that the normal forms approach was initially devised (and can be used in the non-resonant case) to obtain a *perturbative linearization* of the system around a given fixed point. It is natural to wonder – in particular in cases where this can be reached – if one could operate to obtain the same result outside perturbation theory, i.e. to investigate

non-perturbative linearization. For a normalization-related approach to this problem, see e.g. [9, 23, 49].

6. The normal forms approach was of course created in the context of classical mechanics. It is natural to wonder if this approach does also extend – and is effective – also in the quantum mechanics realm. The answer is positive, and in a way surprisingly so. Note that in this context one could consider "quantum normal forms" [3, 32, 66, 77], or quantize classical normal forms. In this context, one would expect correspondence with experimental data to be limited to a small neighborhood of the origin; in the case of an atom or a molecule this means the fundamental and maybe some of the first excited levels. It is very remarkable that instead the normalization approach provides very good quantitative results up to near the ionization or dissociation level [16, 69, 98].

7. Finally, as we have briefly mentioned above, the normal form approach has been extended to evolution PDEs [27, 33, 85], including Hamiltonian PDEs [28, 60, 75, 76]. As far as I know, the geometric approach sketched here has never been considered in this context.

APPENDICES

A The normal forms construction

The construction of normal forms is a classical topic, but nevertheless we will show the basic computation leading to establishing them [5, 34] (see also [4, 6]), as this is simple, compact, and at the basis of our discussion. On the other hand, as already stressed in the introduction, we will not deal specifically with the Hamiltonian case, limiting to discuss things at the level of vector fields. Moreover, we will just consider explicit computations in the case where (iii) in Section 2 above is satisfied.

A.1 Normalization

The key idea is to consider *near-identity* changes of coordinates, and to proceed sequentially normalizing terms of order two, three, etc. in the natural order.

Thus, let us suppose the system has been normalized up to f_{m-1}, and let us see how (and in which sense) the term f_m can be simplified. We will consider a change of coordinates of the form

$$x^i \ = \ y^i \ + \ h^i_m(y) \ , \tag{36}$$

with h_m homogeneous of degree $m+1$ in the x.

All we have to do is to insert this change of coordinates into (3), keeping track of what happens at order $m+1$; we will happily lose track of the effects at higher order.[15]

[15]Obviously we can lose track of these for the sake of theoretical considerations, but in actual applications we will need to keep carefully track of them! See also footnote 16 in this respect.

As for the l.h.s. of (3), we just have

$$\dot{x}^i \;=\; \dot{y}^i \;+\; \left(\frac{\partial h^i_m}{\partial y^j}\right)\,\dot{y}^j \;:=\; B^i_{\ j}\,\dot{y}^j\ ,$$

where we have of course defined $B = (I + \partial h_m/\partial y)$.

The computations referring to the r.h.s. of (3) are also elementary:

$$f^i_k(x) \;=\; f^i_k(y + h_m(y)) \;=\; f^i_k(y) \;+\; \left(\frac{\partial f^i_k}{\partial y^j}\right)\,h^j_m(y) \;+\; \texttt{h.o.t.}\ .$$

Note that $\partial f_k/\partial y$ is of order k, hence the term we have written explicitly is of order $m+k+1$; similarly the higher order terms would start with a term $(\partial^2 f_k/\partial y^p \partial y^q) h^p_m h^q_m$ of order $[(k+1-2)+(m+1)+(m+1)] = (2m+k+1)$. Recall also that $m \geq 1$.

Thus, up to h.o.t., eq. (3) reads in the new coordinates as

$$\dot{y}^i \;=\; (B^{-1})^i_j\,\left[\sum_k f^j_k(y) \;+\; \sum_k \left(\frac{\partial f^j_k}{\partial y^\ell}\right)\,h^\ell_m(y)\right]\ . \tag{37}$$

To have this in explicit form, we only have to note that

$$B^{-1} \;=\; I \;-\; (\partial h_m/\partial y) \;+\; \texttt{h.o.t.}\ ,$$

where now the h.o.t. are of order $2m$ (if $B = I + \beta$, then $B^{-1} = I - \beta + \beta^2/2 - ...$). Note also that the term $(\partial f^j_k/\partial y^\ell) h^\ell_m$ is of order $m+k+1$; so only the term with $k=0$ is relevant (as we work at order $m+1$).

Thus, in the end, keeping only terms up to order $m+1$, we get

$$\dot{y}^i \;=\; A^i_{\ j}\,y^j \;+\; \sum_{k=1}^m f^i_k(y) \;+\; A^i_{\ j} h^j_m(y) \;-\; \left(\frac{\partial h^i_m}{\partial y^j}\right)\,A^j_{\ \ell}\,y^\ell \;+\; \texttt{h.o.t.}\ . \tag{38}$$

In other words, all terms f_k with $k < m$ are unchanged, the term f_m changes according to

$$f_m \;\to\; \widetilde{f}_m \;=\; f_m \;+\; A^i_{\ j} h^j_m(y) \;-\; \left(\frac{\partial h^i_m}{\partial y^j}\right)\,A^j_{\ \ell}\,y^\ell\ , \tag{39}$$

and higher order terms change in a way we are not explicitly describing here.[16]

A.2 The homological operator

We define the linear operator

$$\mathcal{L} \;:=\; \left(A^j_{\ \ell}\,y^\ell\right)\,\frac{\partial}{\partial y^j} \;-\; A \;=\; Ax\cdot\nabla \;-\; A\ ; \tag{40}$$

with this, (39) reads

$$f_m \;\to\; \widetilde{f}_m \;=\; f_m \;-\; \mathcal{L}(h_m)\ . \tag{41}$$

This operator $\mathcal{L}$ is also known as the *homological operator* associated to A.

It is quite obvious that:

[16] Keeping control on these can of course be quite interesting, and leads to a further reduction of the normal form (one also speaks of "further normalization"); see e.g. [7, 8, 40] and the discussion in [44, 45].

1. denoting by V_m the set of vector functions homogeneous of order $m+1$, $\mathcal{L} : V_m \to V_m$; thus we can consider the restriction $\mathcal{L}_m$ of $\mathcal{L}$ to V_m, and work at each order with the finite dimensional linear operators (that is, matrices) $\mathcal{L}_m$;

2. terms δh_m in $\mathrm{Ker}(\mathcal{L}_m)$ have no effect whatsoever, at least at order m;

3. by a suitable choice of h_m we can eliminate all terms f_m in $\mathrm{Ran}(\mathcal{L}_m)$;

4. hence we can reduce to have (up to a finite but arbitrary order N) *only* nonlinear terms f_m in a space complementary to $\mathrm{Ran}(\mathcal{L}_m)$, in which case we say that the system is in *normal form* (to order N);

5. the "suitable choice" mentioned above is given by $h_m = \mathcal{L}_m^*(\pi_m f_m)$, where π_m is the projector on $\mathrm{Ran}(\mathcal{L}_m)$, and $\mathcal{L}_m^*$ is the pseudo-inverse to $\mathcal{L}_m$; the h_m thus determined is not unique, being defined up to an element in $\mathrm{Ker}(\mathcal{L}_m)$.

Remark A.1. It is maybe also worth stressing that the transformation at order m will in general produce new terms of all orders $\ell > m$. Thus even if we start with only one nonlinear term, the normalization procedure will produce (even just at first step) nonlinear terms of all higher orders. Some of these can be eliminated at later stages, but there can be also resonant terms (see below) which cannot be eliminated by normalization at their order. ⊙

In the case where $A_n = 0$, i.e. $A = A_s$, the operator $\mathcal{L}$ is specially simple. To see this, it is convenient to write h_m as

$$h_m(y) \;=\; \sum_{i=1}^{n} \sum_{|J|=m+1} c^i_{j_1 \dots j_n} y_1^{j_1} \dots y_n^{j_n} \; \mathbf{e}_i \;:=\; \sum_{i=1}^{n} \sum_{|J|=m+1} c^i_J Y^J \, e_i \;, \tag{42}$$

where $\mathbf{e}_i$ is the i-th basis vector in $\mathbf{R}^n$, $|J| = \sum_{i=1}^n j_i$, the j_i are non-negative integers, and the sum over $J = (j_1, ..., j_n)$ is on all the J satisfying $|J| = m+1$. Then, with a compact but intuitive notation,

$$\frac{\partial h^i_m}{\partial y^\ell} \;=\; \sum_{i=1}^{n} \sum_{|J|=m+1} j_\ell \, c^i_J Y^{J-e_\ell} \; \mathbf{e}_i \;.$$

On the other hand, in this case

$$Ay \;=\; \sum_{i=1}^{n} \lambda_i y^i \, e_i \;.$$

Therefore, in the end

$$\mathcal{L}(Y^J \mathbf{e}_i) \;=\; \left(\sum_\ell \lambda_\ell j_\ell \;-\; \lambda_i \right) \; \mathbf{e}_i \;. \tag{43}$$

It is hence clear that $\mathrm{Ker}(\mathcal{L}_k)$ is made of the vectors whose i-th component has a monomial Y^J with $\lambda_\ell j_\ell = \lambda_i$, and conversely that by choosing suitably the coefficients c^i_J we can generate all terms except those in the kernel.

A.3 Scalar product; adjoint homological operator

It would be convenient to have a notion of orthogonality in the spaces of vector functions. The natural scalar product in V_k [5] is defined as follows. We take a basis $\xi_{\mu,i} = Y^\mu \mathbf{e}_i$ (with μ a multi-index, $|\mu| = k$) in each of the spaces V_k, and define

$$(\xi_{\mu,i}, \xi_{\nu,\ell}) = \delta_{\mu,\nu} \, \delta_{i\ell} \; . \tag{44}$$

It would be even better if we could choose a scalar product such that the range and the kernel of $\mathcal{L}_k$ are orthogonal. This is possible by choosing the Bargman scalar product [10, 34]. Denoting

$$\langle y^\mu, y^\nu \text{Rangle} = \partial_\mu y^\nu := \frac{\partial^k}{\partial y_1^{\mu_1} ... \partial y_n^{\mu_n}} \, y_1^{\nu_1} ... y_n^{\nu_n} \; , \tag{45}$$

and $\mu! = (\mu_1!) ... (\mu_n!)$, we define the scalar product as

$$(\xi_{\mu,i}, \xi_{\nu,\ell}) := \delta_{\mu,\nu} \, \delta_{i,\ell} \, \mu! \; . \tag{46}$$

The real advantage of this scalar product is embodied in the following

Proposition 1. *If $\mathcal{L}$ is the homological operator associated with the matrix A, then its adjoint $\mathcal{L}^+$ w.r.t. the scalar product (46) is the homological operator associated to the matrix A^+.*

Now there is a natural choice for the space complementary to $\text{Ran}(\mathcal{L}_k)$, namely $[\text{Ran}(\mathcal{L}_k)]^c = [\text{Ran}(\mathcal{L}_k)]^+$; moreover – with the choice (46) for the scalar product, we have that

$$[\text{Ran}(\mathcal{L}_k)]^+ = \text{Ker} \left[\mathcal{L}^+ \right] \; . \tag{47}$$

B Examples of unfolding

In this Appendix we briefly illustrate the construction of Section 5 by some examples (the last one will actually concern the discussion of Section 6). Here it will always be meant that we consider generic perturbations to a given linear part, and systems in normal form with respect to (the semisimple part of) this linear part. We will of course use coordinates adapted to the linear part (that is, eigencoordinates for the matrix A), and as we work in small dimension we denote these as x, y, z rather than x_1, x_2, x_3, for ease of notation.

Example 1.

For $k > 1$ a positive integer and

$$A = \begin{pmatrix} 1 & 0 \\ 0 & k \end{pmatrix} \; ,$$

the only resonant vector is $v = (0, x^k)$ (this is associated to a sporadic resonance), thus the most general system in normal form reads

$$\begin{aligned} \dot{x} &= x \\ \dot{y} &= k y + c x^k \; , \end{aligned}$$

with c an arbitrary real constant. Introducing $w \simeq x^k$, the system unfolding reads

$$\begin{aligned} \dot{x} &= x \\ \dot{y} &= k y + c w \\ \dot{w} &= k w \; . \end{aligned}$$

Note that the manifold M identified by $\psi := w - x^k = 0$ is obviously invariant under this flow; in fact,

$$\frac{d\psi}{dt} = \dot{w} - k\,x^{k-1}\,\dot{x} = k\,w - k\,x^k = k\,\psi\,.$$

The solution to this three-dimensional system is immediately obtained,

$$x(t) = x_0\,e^t\,,\quad y(t) = y_0\,e^{kt} + (ckw_0)\,t\,e^{kt}\,,\quad w(t) = w_0\,e^{kt}\,;$$

restricting this to the manifold M and projecting to the two-dimensional space spanned by x and y, we get

$$x(t) = x_0\,e^t\,,\quad y(t) = [y_0 + (c_1 k x_0)t]\,e^{kt}\,.$$

Example 2.

Let us consider (the perturbation of) a system of two oscillators with irrational frequencies, i.e.

$$A = \begin{pmatrix} 0 & -1 & 0 & 0 \\ 1 & 0 & 0 & 0 \\ 0 & 0 & 0 & -\omega \\ 0 & 0 & \omega & 0 \end{pmatrix}$$

with ω real and irrational. Now we have eigenvalues $(\lambda_1, \lambda_2, \lambda_3, \lambda_4) = (-i, +i, -i\omega, +i\omega)$. There are no sporadic resonances apart from the trivial ones of order one, and two invariance relations, given obviously by $\lambda_1+\lambda_2 = 0$, $\lambda_3+\lambda_4 = 0$; correspondingly we have two invariant functions, $\rho_1 = x_1^2 + x_2^2$, $\rho_2 = x_3^2 + x_4^2$. Thus the general normal form for perturbations of this system is written, in vector notation, as

$$\dot{\xi} = M\,\xi\,,$$

where $\xi = (x_1, x_2, x_3, x_4)$ and M is a 4×4 matrix of the form

$$M = \begin{pmatrix} \alpha & -\beta & 0 & 0 \\ \beta & \alpha & 0 & 0 \\ 0 & 0 & \gamma & -\eta \\ 0 & 0 & \eta & \gamma \end{pmatrix}$$

with $\alpha, \beta, \gamma, \eta$ polynomial functions of ρ_1, ρ_2.

The evolution of these is given by

$$\begin{aligned} \dot{\rho}_1 &= 2\,\alpha(\rho_1, \rho_2)\,\rho_1\,, \\ \dot{\rho}_2 &= 2\,\gamma(\rho_1, \rho_2)\,\rho_2\,; \end{aligned} \tag{48}$$

if we are able to solve this system, then we write $\alpha(t) := \alpha[\rho_1(t), \rho_2(t)]$ and the like, and we are reduced to studying a linear (time-dependent) four-dimensional system, $\dot{\xi} = M(t)\xi$ (which actually decouples into two two-dimensional ones).

Note that even if we are not able to solve the above system (48), its nature could (depending on the functions σ and γ) allow us, via the Poincaré-Bendixson theorem, to be sure it will go asymptotically either to a constant or to a periodic motion. Correspondingly, the four-dimensional linear time-dependent system to be solved would become (asymptotically) a time-independent or time-periodic one.

Remark B.1. The linear part of this system generates the irrational flow on the torus $\mathbf{T}^2$, so the (compact!) closure of this group is the full $\mathbf{T}^2$; correspondingly we have invariants associated with the full $\mathbf{T}^2$ action. ⊙

Example 3.

Consider now two oscillators in 1:1 resonance, i.e.

$$A = \begin{pmatrix} 0 & -1 & 0 & 0 \\ 1 & 0 & 0 & 0 \\ 0 & 0 & 0 & -1 \\ 0 & 0 & 1 & 0 \end{pmatrix} .$$

The eigenvalues are $(\lambda_1, \lambda_2, \lambda_3, \lambda_4) = (-i, +i, -i, +i)$. There are no sporadic resonances apart from the trivial ones of order one, but four invariance relations, given obviously by $\lambda_1 + \lambda_2 = 0$, $\lambda_3 + \lambda_4 = 0$, $\lambda_1 + \lambda_4 = 0$, $\lambda_2 + \lambda_3 = 0$; note that these are linearly dependent, but none of this can be written polynomially in terms of the others. Correspondingly we have four (algebraically but not functionally independent) invariant functions, which can be chosen in a non-unique way; e.g. we choose $\rho_1 = x_1^2 + x_2^2$, $\rho_2 = x_3^2 + x_4^2$, $\rho_3 = x_1 x_3 + x_2 x_4$, $\rho_4 = x_1 x_4 - x_2 x_3$. The centralizer of A is now an eight-dimensional algebra; in terms of the two-dimensional matrices

$$I = \begin{pmatrix} 1 & 0 \\ 0 & 1 \end{pmatrix} , \quad J = \begin{pmatrix} 0 & -1 \\ 1 & 0 \end{pmatrix} ,$$

its generators can be written in block notation as

$$B_1 = \begin{pmatrix} I & 0 \\ 0 & 0 \end{pmatrix} , \quad B_2 = \begin{pmatrix} 0 & 0 \\ 0 & I \end{pmatrix} , \quad B_3 = \begin{pmatrix} 0 & I \\ I & 0 \end{pmatrix} , \quad B_4 = \begin{pmatrix} 0 & J \\ -J & 0 \end{pmatrix} ;$$

$$S_1 = \begin{pmatrix} J & 0 \\ 0 & 0 \end{pmatrix} , \quad S_2 = \begin{pmatrix} 0 & 0 \\ 0 & J \end{pmatrix} , \quad S_3 = \begin{pmatrix} 0 & -I \\ I & 0 \end{pmatrix} , \quad S_4 = \begin{pmatrix} 0 & J \\ J & 0 \end{pmatrix} .$$

Note that $B_i^+ = B_i$, $S_i^+ = -S_i$; actually our choice of the ρ_i corresponds to $\rho_i = (\xi, B_i \xi)$, where $(.,.)$ is the standard scalar product in R^4.

If now we consider (with the same notation as in the previous example) perturbations of the linear system $\dot{\xi} = A\xi$, the normal form for these will be written as $\dot{\xi} = M(\rho)\xi$, where

$$M = \begin{pmatrix} \alpha & -\beta & \gamma & -\eta \\ \beta & \alpha & \eta & -\gamma \\ \mu & -\nu & \sigma & -\tau \\ \nu & \mu & \tau & \sigma \end{pmatrix} = \begin{pmatrix} \alpha I + \beta J & \gamma I + \eta J \\ \mu I + \nu J & \sigma I + \tau J \end{pmatrix} ,$$

and the $\alpha, ..., \tau$ are polynomial functions of $\rho_1, ..., \rho_4$.

Note that while the linear system is invariant under the exchange of the two oscillators, or under a simultaneous rotation in the $(x_1 x_3)$ and the $(x_2 x_4)$ planes, the general normal form does not enjoy any property of this type.

The evolution of the invariant functions is compactly written as

$$\dot{\rho}_a = (\xi , (B_a M + M^+ B_a) \xi) ; \tag{49}$$

as M is in general a nonlinear function of the ρ, this is a system of four nonlinear equations and in general we are not able to solve them, nor we can use the Poincaré-Bendixson theorem as in the previous Example in order to get information about its asymptotic behavior. If we are able by some means to determine some (stable) stationary or periodic solution to (49), we can then determine the corresponding solutions to the linear system $\dot{\xi} = M[\rho(t)]\xi$.

Example 4.

Our last example concerns the situation discussed in Section 6. We consider a system which is the perturbation of two identical oscillators, coupled in such a way to have an exchange symmetry (that is, the system is covariant under the simultaneous exchanges $x_1 \leftrightarrow x_3$, $x_2 \leftrightarrow x_4$). Then the linear part is again

$$A = \begin{pmatrix} 0 & -1 & 0 & 0 \\ 1 & 0 & 0 & 0 \\ 0 & 0 & 0 & -1 \\ 0 & 0 & 1 & 0 \end{pmatrix}$$

as in Example 3, and the same considerations as in Example 3 above apply, *but* only nonlinear resonant terms respecting the exchange symmetry are allowed.

Note that under this exchange, ρ_3 is invariant, while $\rho_1 \leftrightarrow \rho_2$ (thus we should consider as invariant $\rho_1 + \rho_2$), and $\rho_4 \to -\rho_4$. Thus we have only two invariants, say $\phi_1 = \rho_1 + \rho_2$ and $\phi_2 = \rho_3$.

Correspondingly, we should consider the simultaneous centralizer of A and of the exchange matrix

$$E = \begin{pmatrix} 0 & 0 & 1 & 0 \\ 0 & 0 & 0 & 1 \\ 1 & 0 & 0 & 0 \\ 0 & 1 & 0 & 0 \end{pmatrix} = \begin{pmatrix} 0 & I \\ I & 0 \end{pmatrix} .$$

This is a four-dimensional algebra, spanned by the matrices

$$B_1 + B_2 \ , \ B_3 \ ; \ S_1 + S_2 \ , \ S_4 \ .$$

Correspondingly, the most general normal form $\dot{\xi} = M(\rho)\xi$ is now written in terms of matrices

$$M = \begin{pmatrix} \alpha & -\beta & \gamma & -\eta \\ \beta & \alpha & \eta & -\gamma \\ \gamma & -\eta & \alpha & -\beta \\ \eta & \gamma & \beta & \alpha \end{pmatrix} = \begin{pmatrix} \alpha I + \beta J & \gamma I + \eta J \\ \gamma I + \eta J & \alpha I + \beta J \end{pmatrix} ,$$

and the $\alpha, ..., \eta$ are polynomial functions of ϕ_1, ρ_2.

This should be compared with the situations described in Example 3.

C Hopf and Hamiltonian Hopf bifurcations

The approach described in Section 5 can also be used to study systems at a Hopf or Hamiltonian Hopf bifurcation. Note that in these cases, as opposed to what happens e.g. in a pitchfork bifurcation, the linear part of the system at the bifurcation does not vanish; this allows us to apply the normal form approach (and hence also our method) also at the bifurcation.

C.1 Hopf bifurcation

In the case of Hopf bifurcation [29, 30, 42, 56, 64, 67, 68, 101], the linear part of the system at the bifurcation is described by

$$A = \begin{pmatrix} 0 & -\omega_0 \\ \omega_0 & 0 \end{pmatrix} , \tag{50}$$

where $\omega_0 \neq 0$ is a real parameter representing the frequency (of the bifurcating periodic solutions) at the bifurcation.

The eigenvalues are obviously $\lambda_\pm = \pm i\omega_0$. So we have no sporadic resonances, and one invariance relation, $\lambda_+ + \lambda_- = 0$; the associated invariant is $\rho = x^2 + y^2 \geq 0$. The most general system in normal form is hence

$$\begin{aligned} \dot{x} &= \alpha(\rho,\mu)\, x \ - \ \beta(\rho,\mu)\, y \\ \dot{y} &= \beta(\rho,\mu)\, x \ + \ \alpha(\rho,\mu)\, y \ , \end{aligned}$$

where we have allowed the system to also depend on an external parameter (driving the bifurcation) μ. In this case we get immediately

$$\dot{\rho} \ = \ 2\, \rho\, \alpha(\rho,\mu)\ .$$

In the standard setting for Hopf bifurcation,

$$\alpha(\rho,\mu) \ = \ \mu - c\,\rho\ ; \ \ \beta(\rho,\mu) \ = \ \omega_0 \ + \ b(\rho,\mu)$$

where $b(0,0) = 0$ (here the bifurcation takes place in the origin at $\mu = 0$).

In our approach, writing $\rho = \phi$ to recover our general notation, we pass to study a three-dimensional system

$$\begin{aligned} \dot{x} &= \mu\, x \ - \ \omega_0\, y \ - \ c(\phi)\, x \ - \ b(\phi,\mu)\, y \\ \dot{y} &= \omega_0\, x \ + \ \mu\, y \ + \ b(\phi,\mu)\, x \ - \ c(\phi)\, y \\ \dot{\phi} &= 2\, \alpha(\phi,\mu)\, \phi\ . \end{aligned}$$

The invariant $\phi(t)$ can grow indefinitely; but if this is not the case, it will approach one of the zeros of the function $\alpha(\phi,\mu)$, call it ϕ_0. Then we get asymptotically the linear system

$$\begin{aligned} \dot{x} &= - \ [\omega_0 \ + \ b(\phi_0,\mu)]\ y \\ \dot{y} &= [\omega_0 \ + \ b(\phi_0,\mu)]\ x. \end{aligned}$$

The standard analysis of Hopf bifurcation is thus recovered.

C.2 Hamiltonian Hopf bifurcation

In the case of Hamiltonian Hopf bifurcation [99, 100] the linear part of the system at the bifurcation is

$$A \ = \ \begin{pmatrix} \mu & -\omega & 0 & 0 \\ \omega & \mu & 0 & 0 \\ 0 & 0 & -\mu & -\omega \\ 0 & 0 & \omega & -\mu \end{pmatrix} \ = \ \begin{pmatrix} \mu I + \omega J & 0 \\ 0 & -\mu I + \omega J \end{pmatrix}\ , \tag{51}$$

where $\mu \neq 0$ and $\omega \neq 0$ are real parameters (note that $\mu = 0$ corresponds to a pair of oscillators in 1:1 resonance; in applications, μ is the external control parameter and when it goes through zero we have the bifurcation).

We thus have eigenvalues $\lambda_{\pm\pm} = \pm\mu \pm i\omega$; for generic μ there are no sporadic resonances, and there are two invariance relations,

$$\lambda_{++} \ + \ \lambda_{--} \ = \ 0\ ; \ \ \lambda_{+-} \ + \ \lambda_{-+} \ = \ 0\ ;$$

the associated invariants are $\phi_1 = x_1 x_3 + x_2 x_4$ and $\phi_2 = x_1 x_4 - x_2 x_3$.

It is convenient to introduce the two-dimensional vectors $\eta_1 = (x_1, x_2)$, $\eta_2 = (x_3, x_4)$, $\phi = (\phi 1, \phi_2)$. The matrices in the centralizer of A are written in terms of real constants α_k, β_k as block-diagonal ones, of the form $M = \mathtt{diag}(\alpha_1 I + \beta_1 J, \alpha_2 I + \beta_2 J)$. Correspondingly, systems in normal form will be given by

$$\dot{\eta} = \begin{pmatrix} \alpha_1(\phi)\, I \, + \, \beta_1(\phi)\, J & 0 \\ 0 & \alpha_2(\phi)\, I \, + \, \beta_2(\phi)\, J \end{pmatrix} \eta \, .$$

The functions α_k, β_k (which can also depend on μ, ω, albeit we omitted to write explicitly this dependence) can be written as

$$\alpha_k \; = \; (-1)^{k+1} \mu \, + \, a(\phi) \; , \quad \beta_k \; = \; \omega \, + \, b_k(\phi) \; ; \quad a_k(0) \; = \; 0 \; = \; b_k(0) \; .$$

Note that the system is Hamiltonian, i.e. preserves the symplectic form $dx^1 \wedge dx^2 + dx^3 \wedge dx^4$, if and only if $\alpha_2 = -\alpha_1$ and $\beta_2 = \beta_1$ (these relations are always satisfied at the linear level).

The system is hence described, proceeding with our unfolding procedure, as

$$\begin{aligned} \dot{\eta}_1 &= [(\mu + a_1(\phi))\, I \, + \, (\omega - b_1(\phi))\, J]\; \eta_1 \, , \\ \dot{\eta}_2 &= [(-\mu + a_2(\phi))\, I \, + \, (\omega - b_2(\phi))\, J]\; \eta_2 \, ; \\ \dot{\phi} &= [(a_1(\phi) + a_2(\phi))\, I \, + \, (b_2(\phi) - b_1(\phi))\, J]\; \phi \, . \end{aligned}$$

Note that if α_k, β_k are such that the system is Hamiltonian, ϕ is constant and we are always reduced to a linear system on each level set of $\phi = (\phi_1, \phi_2)$; if the system is not Hamiltonian we deal however with a two dimensional system and under the standard conditions for a bifurcation to take place – i.e. if $|\phi|$ cannot grow indefinitely – the Poincaré-Bendixson theorem applies.

At the bifurcation point $\mu = 0$ we are in the framework of Example B.4 above.

D Symmetry and convergence for normal forms

We have seen that vector fields in normal forms can be characterized in terms of a symmetry property. However, symmetry is independent of the coordinate description. This suffices to conclude, as first emphasized by J. Moser [83, 84], that a vector field which lacks symmetry cannot be conjugated to its normal form (that is, in this case the series describing the normalization procedure are only formal ones).

It is somewhat surprising that – modulo certain, not always trivial, conditions – the converse is also true. That is, a suitable symmetry of the vector field suffices, together with other assumptions, to guarantee the convergence of the normalizing transformation. This theory has been developed by Bruno, Markhashov, Walcher and Cicogna [13, 14, 15, 22, 78, 104, 105]; see also works by Ito, Russmann and Vey in the Hamiltonian case [26], which we will not touch upon.

The matter is discussed in some length by Cicogna and Walcher [26]; here we will just report the main results, using freely the notation established above – in particular we give for understood that we are dealing with transformation of a system (1) into normal form near an equilibrium, and that $A = (Df)(0)$ is the matrix describing the linearization of the system at the given equilibrium.

First of all, we recall a classical result.

Poincaré criterion. *Let $\lambda_1, ..., \lambda_n$ be the eigenvalues of the matrix A; if the convex hull of these does not contain zero, then the normalizing transformation is convergent and the normal form is analytic.*

Remark D.1. Note that with this assumption on the spectrum of A (which are also stated saying that A belongs to a *Poincaré domain*), we have no invariance relations, and only a finite number (possibly zero) of sporadic resonances. ⊙

We will then introduce two rather general conditions[17], first considered by Bruno [13, 14] (here summation over the dummy index i is implied, for ease of writing).

Condition α. *The vector field $f(x)$ with linear part Ax is said to satisfy* condition α *if*

$$f(x) \;=\; [1+\alpha(x)]\; A\,x \tag{52}$$

for some scalar-valued power series $\alpha(x)$.

Condition ω. *Let $\Lambda = \{\lambda_1, ..., \lambda_n\}$ be the spectrum of $A = (Df)(0)$. Denote by ω_k (with $k > 0$ any integer) the minimum of $|q^i\lambda_i \;-\; \lambda_j|$ for all $j = 1, .., n$ and all n-tuples of non negative integers q^i such that $q^i\lambda_i \neq \lambda_j$ and $|q| = q^1 + ... + q^n$ satisfies $1 < |q| < 2^k$. Then if*

$$\sum_{k=1}^{\infty} 2^{-k} \, \log\left(\omega_k^{-1}\right) \;<\; \infty \tag{53}$$

we say that condition ω *is satisfied.*

Remark D.2. Condition ω is a mild one, guaranteeing there is no accumulation of small denominators. We will always assume this is satisfied in the forthcoming discussion. ⊙

Proposition D.1. [13]. *If A satisfies condition α, and f can be taken by means of a sequence of Poincaré transformations to a normal form which satisfies condition α, then there is a convergent normalizing transformation for f.*

We will now consider the symmetric case. In this case one can aim at normalizing both the dynamical vector field and the symmetry one; moreover, favorable properties of the latter one can extend their benefit also on the former one.

Proposition D.2. [26] *Let the dynamical system (1), with $f(x) = Ax + F(x)$, admit an analytic symmetry vector field $Y = g^i(x)\partial_i$, with $g(x) = Bx + G(x)$. Assume that either (i) B belongs to a Poincaré domain, or (ii) there is a coordinate transformation taking g into a normal form satisfying condition α. Then there is a convergent normalizing transformation which takes g into normal form and moreover maps f into $\widetilde{f} = Ax + \widetilde{F}(x)$, where $\widetilde{F}$ is resonant with B (not necessarily with A).*

We now recalling the statement – and the notation – of Lemma 2 above, see Section 3. As mentioned there, if $s = 0$ we are in the case where assumption α is surely satisfied, and hence there is a convergent normalizing transformation. The following theorem considers the more general setting.

Proposition D.3. [22] *Let the normal form for f be written in the form*

$$\widehat{f}(x) \;=\; [1\,+\,\alpha(x)]\,A\,x \;+\; \sum_{j=1}^{s} \mu_j(x)\,M_j\,x \;:=\; [1\,+\,\alpha(x)]\,A\,x \;+\; \widehat{F}(x)\;,$$

with μ_j and M_j as in Lemma 2, $s \neq 0$, and $\widehat{F} \neq 0$. Assume that:

[17] Their names, taken together, have a slight taste of blasphemy; unfortunately these names are by now well established in the literature.

1. *f admits an analytic symmetry $g(x) = Bx + G(x)$ with $B = kA$ for some (possibly zero) constant k, and G fully nonlinear;*

2. *if $S = \sum_{j=1}^{s} \nu_j(x) M_j$ (with $X_0(\nu) = 0$) nd $(DS)(0) = 0$, the equation $\{\widehat{F}, S\} = 0$ has only trivial solutions $S = c\widehat{F}(x)$.*

Then f can be taken to normal form by means of a convergent normalizing transformation.

A simple yet relevant and clarifying application of these result is provided by the problem of the *isochronous center*, see the discussion in [26].

References

[1] M. C. Abbati, R. Cirelli, and A. Manià, The orbit space of the action of gauge transformation group on connections, *J. Geom. Phys.* **6** (1989), 537-558.

[2] M. Abud and G. Sartori, The geometry of spontaneous symmetry breaking, *Ann. Phys.* **150** (1983), 307-372.

[3] M.K. Ali, The quantum normal form and its equivalents, *J. Math. Phys.* **26** (1985), 2565-2572.

[4] V.I. Arnold, *Equations differentielles ordinaires*, M.I.R., Moscow, 1974, 2nd ed. 1990; V.I. Arnold: *Ordinary Differential Equations*, Springer 1992.

[5] V. Arnold, *Chapitres supplementaires de la théorie des equations differentielles ordinaires*, M.I.R., Moscow, 1980; V.I. Arnold, *Geometrical methods in the theory of ordinary differential equations*, Springer 1983.

[6] V.I. Arnold and Yu.S. Il'yashenko, *Ordinary differential equations*; in: Encyclopaedia of Mathematical Sciences vol. 1 – Dynamical Systems I, (D.V. Anosov and V.I. Arnold eds.), pp. 1-148, Springer 1988.

[7] A. Baider, Unique normal forms for vector fields and hamiltonians, *J. Diff. Eqs.* **78** (1989), 33-52.

[8] A. Baider and R.C. Churchill, Uniqueness and non-uniqueness of normal forms for vector fields, *Proc. R. Soc. Edinburgh A* **108** (1988), 27-33.

[9] D. Bambusi, G. Cicogna, G. Gaeta, and G. Marmo, Normal forms, symmetry, and linearization of dynamical systems, *J. Phys. A*, **31** (1998) 5065-5082.

[10] V. Bargmann, On a Hilbert space of analytic functions and an associated integral transform, *Comm. Pure Appl. Math.* **14** (1961), 187-214 & **20** (1967), 1-101.

[11] G.D. Birkhoff, *Dynamical systems*, A.M.S. 1927.

[12] H.W. Broer, *Bifurcations of singularities in volume preserving vector fields*, Ph.D. Thesis, Groningen 1979; Formal normal form theorems for vector fields and some consequences for bifurcations in the volume preserving case, in *Dynamical systems and turbulence* (LNM 898), D.A. Rand and L.S. Young eds., Springer 1981.

[13] A.D. Bruno, Analytical form of differential equations, *Trans. Moscow Math. Soc.* **25** (1971), 131-288.

[14] A.D. Bruno, *Local Methods in Nonlinear Differential Equations*, Springer 1989.

[15] A.D. Bruno and S. Walcher, Symmetries and converegence of normalizing transformations, *J. Math. Anal. Appl.* **183** (1994), 571-576.

[16] S. Buyukdagli and M. Joyeux, On the application of canonical perturbation theory up to the dissociation threshold, *Chem. Phys. Lett.* **412** (2005), 200-205.

[17] N. Cabibbo and L. Maiani, Weak interactions and the breaking of hadronic symmetry, in *Evolution of Particle Physics (E. Amaldi Festschrift)*, M. Conversi ed., pp. 50-80, Academic Press 1970.

[18] F. Calogero, Solution of the one-dimensional N-body problem with quadratic and/or inversely quadratic potentials, *J. Math. Phys.* **12** (1971), 419-436; Exactly solvable one-dimensional many-body problems, *Lett. Nuovo Cim.* **13** (1975), 411-416.

[19] P. Chossat, The reduction of equivariant dynamics to the orbit space for compact group actions, *Acta Appl. Math.* **70** (2002), 71-94.

[20] P. Chossat and M. Koenig, Characterization of bifurcations for vector fields which are equivariant under the action of a compact Lie group, *C. R. Acad. Sci. Paris* **318** (1994), 31-36.

[21] P. Chossat and R. Lauterbach, *Methods in Equivariant Bifurcations and Dynamical Systems with Applications*, World Scientific, 1999.

[22] G. Cicogna, Symmetries of dynamical systems and convergent normal forms, *J. Phys. A* **28** (1995), L179-L182; On the convergence of the normalizing transformation in the presence of symmetries, *J. Math. Anal. Appl.* **199** (1996), 243-255; Convergent normal forms of symmetric dynamical systems, *J. Phys. A* **30** (1997), 6021-6028.

[23] G. Cicogna, Multiple-periodic bifurcation and resonance in dynamical systems, *Nuovo Cimento B* **113** (1998), 1425-1430.

[24] G. Cicogna and G. Gaeta, Bifurcation, symmetries and maximal isotropy subgroups; *Nuovo Cimento B* **102** (1988), 451-470.

[25] G. Cicogna and G. Gaeta, *Symmetry and perturbation theory in nonlinear dynamics*, Springer (LNP m57) 1999.

[26] G. Cicogna and S. Walcher, Convergence of normal form transformation: the role of symmetries, *Acta Appl. Math.* **70** (2002), 95-111.

[27] P. Collet and J.P. Eckmann, Space-time behaviour in problems of hydrodynamic type: a case study, *Nonlinearity* **5** (1992), 1265-1302.

[28] W. Craig, Birkhoff normal forms for water waves, *Cont. Math.* **200** (1996), 57-74.

[29] J.D. Crawford, Introduction to bifurcation theory, *Rev. Mod. Phys.* **63** (1991), 991-1037.

[30] J.D. Crawford and E. Knobloch, 'Symmetry and symmetry-breaking bifurcations in fluid dynamics, *Ann. Rev. Fluid Mech.* **23** (1991), 341-387.

[31] H. Dulac, *Points singuliers des équations différentielles*, Gauthier-Villars 1903 (available on the web via *http://www.numdam.org*).

[32] B. Eckhardt, Birkhoff-Gustavson normal form in classical and quantum mechanics, *J. Phys. A* **19** (1986), 2961-2972.

[33] J.P. Eckmann, H. Epstein and C.E. Wayne, Normal forms for parabolic partial differential equations, *Ann. I.H.P. (Phys. Thèo.)* **58** (1993), 287-308.

[34] C. Elphick, E. Tirapegui, M.E. Brachet, P. Coullet P., and G. Iooss, A simple global characterization for normal forms of singular vector fields, *Physica D* **29** (1987), 95-127; addendum *Physica D* **32** (1988), 488.

[35] F. Fassò, Lie series method for vector fields and Hamiltonian perturbation theory, *ZAMP* **41** (1990), 843-864.

[36] M.J. Field, Equivariant dynamical systems, *Trans. A.M.S.* **259** (1980), 185205.

[37] M. Field, Symmetry breaking for compact Lie groups, *Mem. A.M.S.* **574** (1996), 1-170.

[38] M. Field, *Dynamics and Symmetry*, Imperial College Press 2007.

[39] M.J. Field and R.W. Richardson, Symmetry breaking in equivariant bifurcation problems, *Bull. Am. Math. Soc.* **22** (1990), 79-84; Symmetry breaking and branching patterns in equivariant bifurcation theory, *Arch. Rat. Mech. Anal.* **118** (1992), 297-348 & **120** (1992), 147-190.

[40] E. Forest and D. Murray, Freedom in minimal normal forms, *Physica D* **74** (1994), 181-196.

[41] G. Gaeta, Gradient property of low order covariants and truncated bifurcation equations for SO(N) symmetries, *Phys. Lett. A* **113** (1985), 114-116.

[42] G. Gaeta, Bifurcation and symmetry breaking, *Phys. Rep.* **189** (1990), 1-87.

[43] G. Gaeta, A splitting lemma for equivariant dynamics, *Lett. Math. Phys.* **33** (1995), 313-320.

[44] G. Gaeta, Reduction of Poincaré normal forms, *Lett. Math. Phys.* **42** (1997), 103-114; Poincaré renormalized forms, *Ann. Inst. H. Poincaré (Phys. Theo.)* **70** (1999), 461-514; Poincaré normal and renormalized forms, *Acta Appl. Math.* **70** (2002), 113-131.

[45] G. Gaeta, Algorithmic reduction of Poincaré-Dulac normal forms and Lie algebraic structure, *Lett. Math. Phys.* **57** (2001), 41-60.

[46] G. Gaeta, Poincaré normal forms and simple compact Lie groups, *Int. J. Mod. Phys. A* **17** (2002), 3571-3587.

[47] G. Gaeta, Lie-Poincaré transformations and a reduction criterion in Landau theory, *Ann. Phys.* **312** (2004), 511-540; Poincaré-like approach to Landau theory, *J. Math. Phys.* **56** (2015), 083504 & **56** (2015), 083505; On the isotropic-biaxial phase transition in nematic liquid crystals, *EuroPhys. Lett.* **112** (2015), 46002.

[48] G. Gaeta, F.D. Grosshans, J. Scheurle and S. Walcher, Reduction and reconstruction for symmetric ordinary differential wquations, *J. Diff. Eqs.* **244** (2008), 1810-1839.

[49] G.Gaeta and G.Marmo, Nonperturbative linearization of dynamical systems, *J. Phys. A* **29** (1996), 5035-5048.

[50] G. Gaeta and P. Morando, Michel theory of symmetry breaking and gauge theories, *Ann. Phys.* **260** (1997), 149-170.

[51] G. Gaeta and M.A. Rodriguez, *Lectures on HyperHamiltonian Dynamics*, Springer 2017.

[52] G. Gaeta and P. Rossi, Gradient property of bifurcation equation for systems with rotational symmetry, *J. Math. Phys.* **25** (1984), 1671-1673; Gradient property of standard representation for classical orthogonal groups, *Nuovo Cimento B* **93** (1986), 66-72; G. Cicogna, G. Gaeta and P. Rossi, Remarks on bifurcation with symmetry, gradient property, and reducible representations, *J. Math. Phys.* **27** (1986), 447-450.

[53] G. Gaeta and S. Walcher, Lie algebras with finite-dimensional polynomial centralizer, *J. Math. Anal. Appl.* **269** (2002), 578-587.

[54] G. Gaeta and S. Walcher, Dimension increase and splitting for Poincaré-Dulac normal forms, *J. Nonlin. Math. Phys.* **12** (2005), S1.327-S1.342.

[55] G. Gaeta and S. Walcher: Embedding and splitting ordinary differential equations in normal form, *J. Diff. Eqs.* **224** (2006), 98-119.

[56] P. Glendinning, *Stability, Instability and Chaos*, Cambridge UP 1994.

[57] M Goresky and R. MacPherson, *Stratified Morse Theory*, Springer 1988.

[58] T. Gramchev and S. Walcher, Normal forms of maps: formal and algebraic aspects, *Acta Appl. Math.* **87** (2005), 123-146.

[59] T. Gramchev and M. Yoshino, Rapidly convergent iteration methods for simultaneous normal forms of commuting maps, *Math. Z.* **231** (1999), 745-770.

[60] B. Grébert, Birkhoff normal forms and Hamiltonian PDEs, *SMF Séminaires & Congrès* **15** (2007), 1-46.

[61] F.D. Grosshans, Localization and Invariant Theory, *Adv. Math.* **21** (1976), 50-60.

[62] F.D. Grosshans, J. Scheurle and S. Walcher, Invariant sets forced by symmetry, *J. Geom. Mech.* **4** (2012), 281-296.

[63] F.D. Grosshans and S. Walcher, Modules of higher order invariants, *Proc. A.M.S.* **143** (2015), 531-542.

[64] J. Guckenheimer and P. Holmes, *Nonlinear Oscillations, Dynamical Systems, and Bifurcation of Vector Fields*, Springer 1983.

[65] F.G. Gustavson, On constructing formal integrals of a Hamiltonian system near an equilibrium point, *Astronomical J.* **71** (1966), 670-686.

[66] G.A. Hagedorn, Classification and normal forms for avoided crossings of quantum-mechanical energy levels, *J. Phys. A* **31** (1998), 369-383.

[67] G. Iooss and M. Adelmeyer, *Topics in Bifurcation Theory and Applications*, World Scientific 1992.

[68] G. Iooss and D.D. Joseph, *Elementary Stability and Bifurcation Theory*, Springer 1990.

[69] M. Joyeux and D. Sugny, Canonical perturbation theory for highly excited dynamics, *Canad. J. Phys.* **80** (2002), 1459-1480; M. Joyeux, S.C. Farantos and R. Schinke, Highly excited motion in molecules: Saddle-node bifurcations and their fingerprints in vibrational spectra, *J. Phys. Chem. A* **106** (2002), 5407-5421; J. Robert and M. Joyeux, Canonical perturbation theory versus Born-Oppenheimer-type separation of motions: The vibrational dynamics of C_3, *J. Chem. Phys.* **119** (2003), 8761-8762.

[70] D. Kazhdan, B. Kostant and S. Sternberg, Hamiltonian group action and dynamical systems of Calogero type, *Comm. Pure Appl. Math.* **31** (1978), 481-508.

[71] A.A. Kirillov, *Elements of the Theory of Representations*, Springer 1984.

[72] M. Koenig, Linearization of vector fields on the orbit space of the action of a compact Lie group, *Math. Proc. Cambridge Philos. Soc.* **121** (1996), 401-424.

[73] W. Kondracki and P. Sadowski, Geometric structure on the orbit space of gauge connections, *J. Geom. Phys.* **3** (1986), 421-434; W. Kondracki and J. S. Rogulski, On the stratification of orbit space for the action of automorphisms on connections, *Diss. Math.* **250** (1986), 1-62.

[74] M. Krupa, Bifurcations of relative equilibria, *SIAM J. Math. Anal.* **21** (1990), 1453-1486.

[75] S.B. Kuksin, *Analysis of Hamiltonian PDEs*, Clarendon Press 2000.

[76] S.B. Kuksin, *Nearly Integrable Infinite-Dimensional Hamiltonian Systems*, Springer 2006.

[77] R.G. Littlejohn, The semiclassical evolution of wave packets, *Phys. Rep.* **138** (1986), 193-291; Hamiltonian perturbation theory in noncanonical coordinates, *J. Math. Phys.* **23** (1982), 742-747.

[78] L.M. Markhashov, On the reduction of an analyrtic system of differential equations to the normal form by an analytic transformation, *J. Appl. Math. Mech.* **38** (1974), 788-790.

[79] L. Michel, Points critiques de fonctions invariantes sur une G-variété, *C. R. Acad. Sci. Paris A* **272** (1971), 433-436.

[80] L. Michel, Symmetry defects and broken symmetry. Configurations. Hidden symmetry, *Rev. Mod. Phys.* **52** (1980), 617.

[81] L. Michel ed., Physics, Invariants, Topology, *Phys. Rep.* **341** (2001), 1-395.

[82] L. Michel and L. Radicati, Properties of the breaking of hadronic internal symmetry, *Ann. Phys. (N.Y.)* **66** (1971), 758-783; The geometry of the octet, *Ann. I.H.P.* **18** (1973), 185-214.

[83] J.K. Moser: New aspects in the theory of stability of Hamiltonian systems, *Comm. Pure Appl. Math.* **11** (1958), 81-114.

[84] J.K. Moser, Lectures on Hamiltonian systems, *Mem. Amer. Math. Soc.* **81** (1968), 1-60.

[85] N. V. Nikolenko, The method of Poincaré normal forms in problems of integrability of equations of evolution type, *Russ. Math. Surv.* **41** (1986), 109-152.

[86] P.J. Olver, *Classical Invariant Theory*, Cambridge University Press 1999.

[87] R. S. Palais, The principle of symmetric criticality, *Comm. Math. Phys.* **69** (1979), 19-30.

[88] R. S. Palais, Applications of the symmetric criticality principle in mathematical physics and differential geometry, in *Proceedings, 1981 Shanghai Symposium on Differential Geometry and Differential Equations*, Gu Chaohao ed., Science Press Beijing 1984.

[89] R.S. Palais and C.L. Terng, *Critical Point Theory and Submanifold Geometry* (LNM 1353), Springer 2006.

[90] H. Poincaré, Sur le problème des trois corps et les équations de la dynamique, *Acta Math.* **13** (1890), 1-270.

[91] H. Poincaré, *Les Méthodes Nouvelles de la Mécanique Céleste* (3 voll.), Gauthier-Villars 1892, 1893, 1899.

[92] G. Sartori, Geometric invariant theory in a model-independent analysis of spontaneous symmetry and supersymmetry breaking, *Acta Appl. Math.* **70** (2002), 183-207; G. Sartori and V. Talamini, Universality in orbit spaces of compact linear groups, *Comm. Math. Phys.* **139** (1991), 559-588.

[93] D.H. Sattinger, *Group Theoretic Methods in Bifurcation Theory* (LNM 762), Springer 1979.

[94] D.H.Sattinger, *Branching in the Presence of Symmetry*, SIAM 1983.

[95] D.H. Sattinger and O.L. Weaver, *Lie Groups and Algebras with Applications to Physics, Geometry, and Mechanics*, Springer 1986.

[96] J. Scheurle and S. Walcher, Minima of invariant functions: the inverse problem, *Acta Appl. Math.* **137** (2015), 233-252.

[97] R. Schroeders and S. Walcher, Orbit space reduction and localizations, *Indag. Math.* **27** (2016), 12651278.

[98] D. Sugny and M. Joyeux, On the application of canonical perturbation theory to floppy molecules, *J. Chem. Phys.* **112** (2000), 31-39; D. Sugny, M. Joyeux and E.L. Siber III, Investigation of the vibrational dynamics of the HCN/CNH isomers through high order canonical perturbation theory, *J. Chem. Phys.* **113** (2000), 7165-7177.

[99] J.C. Van der Meer, *The Hamiltonian Hopf Bifurcation* (LNM 1160), Springer 1985.

[100] J.C. Van der Meer, Hamiltonian Hopf bifurcation with symmetry, *Nonlinearity* **3** (1990) 1041-1056.

[101] F. Verhulst, *Nonlinear Differential Equations and Dynamical Systems*, Springer 2006.

[102] F. Verhulst, *Henri Poincaré. Impatient gGnius*, Springer 2012.

[103] S. Walcher, On differential equations in normal form, *Math. Ann.* **291** (1991), 293-314.

[104] S. Walcher, On transformation into normal form, *J. Math. Anal. Appl.* **180** (1993), 617-632.

[105] S. Walcher, On convergent normal form transformations in presence of symmetries, *J. Math. Anal. Appl.* **244** (2000), 17-26; On the Poincaré problem, *J. Diff. Eqs.* **166** (2000), 51-78.

[106] M. Yoshino, Simultaneous normal forms of commuting maps and vector fields, in *Symmetry and Perturbation Theory – SPT98*, A. Degasperis and G. Gaeta eds., World Scientific 1999.

C3. Computing symmetries and recursion operators of evolutionary super-systems using the SsTools environment

Arthemy V Kiselev [a], *Andrey O Krutov* [b] *and Thomas Wolf* [c]

[a] *Johann Bernoulli Institute for Mathematics and Computer Science, University of Groningen, P.O.Box 407, 9700 AK Groningen, The Netherlands.*

[b] *Independent University of Moscow, Bolshoj Vlasyevskij Pereulok 11, 119002, Moscow, Russia.*

[c] *Department of Mathematics and Statistics, Brock University, 500 Glenridge Avenue, St.Catharines, Ontario, Canada L2S 3A1*

Abstract

At a *very* informal but practically convenient level, we discuss the step-by-step computation of nonlocal recursions for symmetry algebras of nonlinear coupled boson-fermion $N = 1$ supersymmetric systems by using the SsTools environment.

The principle of symmetry plays an important role in modern mathematical physics. The differential equations that constitute integrable models practically always admit symmetry transformations. The presence of symmetry transformation in a system yields two types of explicit solutions: those which are invariant under a transformation (sub)group and the solutions obtained by propagating a known solution by the same group. The recursion operator is a (pseudo)differential operator which maps symmetries of a given system into symmetries of the same system. The recursion operators allow obtaining new symmetries for a given seed symmetry.

It is common for important equations of mathematical physics not to have local recursion operators other than the identity id: $\varphi \mapsto \varphi$. Instead, they often admit nonlocal recursions which involve integrations such as taking the inverse of the total derivative D_x with respect to the independent variable x. To describe such nonlocal structures we use the approach of nonlocalities. By nonlocalities we mean an extension of the initial system by new fields such that the initial fields are differential consequences of the new ones. In the case of recursion operators such fields often arise from conservation laws. We refer to a recursion operator for the Korteweg–de Vries equations as a motivating example of a nonlocal recursion operator; see Example 7 on page 400.

The supersymmetric integrable systems, i.e. systems involving commuting (bosonic, or even) and anticommuting (fermionic, or odd) independent variables and/or unknown functions, have found remarkable applications in modern mathematical physics (for example supergravity models, perturbed conformal field theory [12]; we refer to [1, 4] for a general overview). When dealing with supersymmetric models of theoretical physics, it is often hard to predict whether a certain mathematical approximation will be truly integrable or not. Therefore we apply the symbolic computation to exhibit necessary integrability features. In what follows we restrict ourselves to the case of $N = 1$ (where N refers to the number of odd anticommuting independent variables θ_i). Nevertheless, the techniques and computer programs described below could be easily applied to the case of arbitrary N. Usually, the N is not bigger than 8; see for example [11] and [3]. It is an interesting open problem to establish criteria that set a limit on N in "N-extended" supersymmetric equations of mathematical physics.

The latest version of SsTools can be found at [16]; see also [8]. We refer to [1, 4] and [2, 5, 6, 9, 13] for reviews of the geometry and supergeometry of partial differential equations. We refer to [10] for an overview of other software that could be used for similar computational tasks.

1 Notation and definitions

We fix notation first. Let x be the independent variable, $u = (u^1, \dots, u^m)$ denote the unknown functions irrespective of their (anti-)commutation properties, and $u_k = \{u_k^j | u_k^j = \partial^k u^j / \partial x^k\}$ denote the partial derivatives of u^j of order $k \in \mathbb{N}$. We extend the independent variable x by the pair (x, θ), where θ is the Grassmann variable such that $\theta^2 = 0$. The super-derivative is defined as

$$\mathcal{D} = \partial_\theta + \theta \partial_x.$$

Its square power is the spatial derivative, $\mathcal{D}^2 = \partial_x$.

Fields $u(x,t)$ now become $N = 1$ superfields $u(x,t;\theta)$. Provided that $\theta^2 = 0$, they have a very simple Taylor expansion in θ:

$$u(x,t;\theta) = u^0(x,t) + \theta u^1(x,t),$$

here u^0 has the same parity as u and u^1 has the opposite parity. The bosonic fields (those commuting with everything) are denoted by $b(x,t;\theta)$, and the fermionic fields, which anti-commute between themselves and with θ, will be denoted by $f(x,t;\theta)$. Further, we write $u_{k/2} = \mathcal{D}^k u$ for the kth order super-derivative of u. Note that the super-derivatives $\mathcal{D}^{2k+1} b$ are fermionic and $\mathcal{D}^{2k+1} f$ are bosonic for any $k \in \mathbb{N}$.

Computer input will be shown in text font, for example, `f(i)` for f^i, `b(j)` for b^j, `df(b(j),x)` for the derivative of b^j with respect to x, `d(1,f(i))` for the super-derivative $\mathcal{D}_{\theta^1} f^i$ written as $\mathcal{D} f^i$, because we will have only one θ and one $\mathcal{D}$.

Let $k, r < \infty$ be fixed integers. Suppose that $F^\alpha(x, t, u, \mathcal{D}(u), \dots, u_k, u_t)$ is a smooth function for any integer $\alpha \le r$. In what follows, we consider systems of differential equations,

$$\{F^\alpha(x, u, \mathcal{D}(u), u_1, \mathcal{D}(u_1), \dots, u_k, u_t) = 0\}, \quad \alpha = 1, \dots, r, \tag{1.1}$$

of order k and, especially, the autonomous translation-invariant evolutionary systems,

$$\mathcal{E} = \{F^\alpha = u_t^\alpha - \Phi^\alpha(u, \dots, u_k) = 0\}, \quad \alpha = 1, \dots, r,$$

which are resolved w.r.t. the time derivatives, and the systems obtained by extending the evolutionary systems with some further differential relations upon u's and their (super-)derivatives.

The weight technique

Physically meaningful equations often have several symmetries, among them one or more scaling symmetries. Suppose that to each superfield u^j and to the derivative ∂_t one can assign a real number (the weight), which is denoted by $[A]$ for any object A. By definition, $[\partial_x] \equiv 1$. The weight of a product of two objects is the sum of the weights of the factors, whence $[\mathcal{D}] = \frac{1}{2}$. The weight of any nonzero constant equals 0, but the weight of a zero-valued constant can be arbitrary.

From now on, we consider differential equations (1.1) with differential-polynomials F^α that admit the introduction of weights for all variables and derivations such that the weights of all monomials in each equation $F^\alpha = 0$ coincide. These equations are scaling-invariant, or homogeneous.

Example 1. Consider the Burgers equation

$$b_t = b_{xx} - 2bb_x, \quad b = b(x,t). \tag{1.2}$$

The weights are uniquely defined,

$$[b] = 1, \quad [\partial_t] = 2 \Leftrightarrow [t] = -2, \quad [\partial_x] \equiv 1.$$

Indeed, equation (1.2) is homogeneous w.r.t. these weights,

$$[b_t] = 1 + 2 = [b_{xx}] = 1 + 1 + 1 = [2bb_x] = 0 + 1 + 1 + 1 = 3$$

and, clearly, this is the only way to choose the weights.

A system of differential equations could be homogeneous w.r.t. to different weight systems. For a given system of differential equations these weight systems can be found by using the `FindSSWeight` function from SsTools:

`FindSSWeights(N,nf,nb,exli,zerowei,verbose)`

where

`N` ... the number of superfields θ_i;
`nf` ... number of fermion fields `f(1), f(2), ..., f(nf)`;
`nb` ... number of boson fields `b(1), b(2), ..., b(nb)`;
`exli` ... list of equations or expressions;
`zerowei` ... list of constants or other kernels that should have zero weight;
`verbose` ... (=t(true)/`nil`(false)) whether detailed comments shall be made.

The program returns a list of homogeneities, each homogeneity being a list `fh, bh, hi` of

`fh` ... a list of the weights of `f(1),b(2), ..., f(nf)`;

bh ... a list of the weights of b(1),b(2), ..., b(nb);
hi ... a list of the weights of equations/expressions in the input.

Weights are scaled such that weight of $[\partial_x]$ is 2; i.e. the weight of any $\mathcal{D}$ is 1. So the computer weights will be twice the "usual" weight. Input expressions can be in field form or coordinate form.

Example 2. Consider the nonlinear Schrödinger equation

$$b_t^1 = b_{xx}^1 + 2(b^1)^2 b^2, \qquad b^1 = b^1(x,t),$$
$$b_t^2 = -b_{xx}^2 - 2(b^2)^2 b^1, \qquad b^2 = b^2(x,t).$$

We compute all possible weight systems of this system of equations:

```
FindSSWeights(0,0,2,{df(b(1),t) =   df(b(1),x,2) + 2*b(1)**2*b(2),
                     df(b(2),t) = - df(b(2),x,2) - 2*b(2)**2*b(1) },
             {},t)$
```

The output contains

```
This system has the following homogeneities:
W[t] = -4
W[b(2)] =  - arbcomplex(1) + 4
W[b(1)] = arbcomplex(1)
W[x] = -2
```

which gives us the following family of weight systems for (1.3)

$$[\partial_t] = 2, \quad [b^2] = -c + 4, \quad [b^1] = c, \quad [\partial_x] = 1,$$

where c is an arbitrary constant.

2 Symmetries

Definition 1. An lth order symmetry of an evolutionary system $\mathcal{E} = \{u_t = \Phi\}$ is another autonomous evolutionary system $\mathcal{E}' = \{u_s = \varphi(x, u, \mathcal{D}(u), \ldots, u_l)\}$ upon $u(s, t, x; \theta)$ such that a solution of the Cauchy problem for $\mathcal{E}'$ propagates solutions of $\mathcal{E}$ to solutions of $\mathcal{E}$. A necessary and sufficient condition for a vector φ to be a symmetry of $\mathcal{E}$ is that solutions of the system satisfy

$$D_t D_s u = \pm D_s D_t u \tag{2.1}$$

where the minus sign applies if both t and s are fermionic. Because (2.1) is to be satisfied by solutions u of the system $\mathcal{E}$, i.e. u_t is replaced by Φ giving the so-called linearization $\operatorname{Lin}\mathcal{E}$ of the system $\mathcal{E}$:

$$\operatorname{Lin}_\Phi(\varphi) := D_t\varphi = \pm D_s\Phi.$$

This is a linear system for φ. If s is a bosonic variable then the symmetry is a system $\mathcal{E}' = \{f_s = F,\ b_s = B\}$ and if s is fermionic then the symmetry is $\mathcal{E}' = \{f_s = B,\ b_s = F\}$.

The understanding of linearized systems $\operatorname{Lin}\mathcal{E}$ from a computational viewpoint is as follows; we consider the differential polynomial case since this is what SsTools can be applied to. Let us first consider the case of bosonic s.

Given a system $\mathcal{E}$ of super-equations, formally assign the new 'linearized' fields $F^i =$ `f(nf+i)` and $B^j =$ `b(nb+j)` to $f^i =$ `f(i)` and $b^j =$ `b(j)`, respectively, with $i = 1, \ldots,$ `nf` and $j = 1, \ldots,$ `nb`. Pass through all equations, and whenever a power of a derivative of a variable f^i or b^j is met, differentiate (in the usual sense) this power with respect to its base, multiply the result from the right by the same order derivative of F^i or B^j, respectively, and insert the product in the position where the power of the derivative was met. Now proceed by the Leibniz rule. The final result, when all equations in the system $\mathcal{E}$ are processed, is the linearized system $\operatorname{Lin}\mathcal{E}$.

If s is fermionic then proceed in the same way, except we get extra factors of -1:

- An overall factor -1 appears in (2.1) if t is fermionic due to anticommuting D_t and D_s.
- When differentiating a factor in a product then a factor -1 appears for each fermionic factor to the left of the differentiated factor.
- When changing the order of D_s and $\mathcal{D}$ in differentiating then a factor of -1 appears as well.

The second difference between bosonic and fermionic s is the number of new 'linearized' fields F^i, B^j that are introduced in the linearized equation. For bosonic s these are $F =$ `f(nf+1)...f(nf+nf)`, $B =$ `b(nb+1)...b(nb+nb)` whereas for fermionic s these are $F =$ `f(nf+1)...f(nf+nb)`, $B =$ `b(nb+1)...b(nb+nf)`.

To summarize, the linearization is obtained by a complete differentiation $D_s\Phi$ applying the Leibniz rule and chain rule (in place) and substituting $u_s = \varphi$.

Example 3. The linearized counterpart of $b_t = b(\mathcal{D}f)^2$ is $B_t = B(\mathcal{D}f)^2 + 2b\mathcal{D}f\mathcal{D}F$ for bosonic symmetry parameter s and $F_t = F(\mathcal{D}f)^2 - 2b\mathcal{D}f\mathcal{D}B$ for fermionic $\bar{s}$. Likewise, for $b_t = bff_x(Df)^2$ and parity-odd $\bar{s}$, the linearization is $F_t = Fff_x(Df)^2 + bBf_x(Df)^2 - bfB_x(Df)^2 - 2bff_xDfDB$.

The scaling weights of the new fields are always set by $[F] = [f]$, $[B] = [b]$ for bosonic s and $[F] = [b]$, $[B] = [f]$ for fermionic $\bar{s}$.

The linearization $\operatorname{Lin}\mathcal{E}$ for a system of evolution equations $\mathcal{E}$ is obtained using the procedure `linearize`:

```
linearize(pdes, nf, nb, tpar, spar);
```

where

`pdes` ... list of equations $b^i_t = \Phi^i_b$, $f^j_t = \Phi^j_f$;
`nf` ... number of the fermion fields `f(1), f(2), ..., f(nf)`;
`nb` ... number of the boson fields `b(1), b(2), ..., b(nb)`;
`tpar` ... (=`t`(true)/`nil`(false)) whether t is parity changing or not
`spar` ... (=`t`(true)/`nil`(false)) whether s is parity changing or not

Example 4. The linearizations of the system with parity reversing time $\bar{t}$,

$$f_{\bar{t}} = b^2 + \mathcal{D}f, \quad b_{\bar{t}} = fb + \mathcal{D}b, \tag{2.2}$$

are obtained as follows,

```
depend {b(1),f(1)},x,t;
linearize({df(f(1),t)=b(1)**2+d(1,f(1)),
           df(b(1),t)=f(1)*b(1)+d(1,b(1))},1,1,t,nil);
```

for bosonic s, and the same call with `t` as last parameter instead of `nil` for fermionic $\bar{s}$.

The result is the new system involving twice as many variables as the original equation. The linearization correspondence between the fields is

$$f \mapsto F \quad (\mathtt{f(1)} \mapsto \mathtt{f(2)}), \qquad b \mapsto B \quad (\mathtt{b(1)} \mapsto \mathtt{b(2)})$$

for bosonic s and

$$f \mapsto B \quad (\mathtt{f(1)} \mapsto \mathtt{b(2)}), \qquad b \mapsto F \quad (\mathtt{b(1)} \mapsto \mathtt{f(2)})$$

for fermionic $\bar{s}$.

The procedure to compute the linearization is the same for normal times t and for parity reversing times $\bar{t}$ except of a factor (-1) of the rhs's B, F if both $\bar{t}, \bar{s}$ are fermionic (because of the anticommutativity of $D_{\bar{t}}, D_{\bar{s}}$ in that case).

The linearized system incorporates

- the initial system:
  ```
  df(f(1),t) = b(1)**2 + d(1,f(1)),
  df(b(1),t) = f(1)*b(1) + d(1,b(1)),
  ```
- and its linearizations:
  ```
  df(f(2),t) = 2*b(2)*b(1) + d(1,f(2)),
  df(b(2),t) = d(1,b(2)) + f(2)*b(1) + f(1)*b(2)
  ```
  for s; respectively,
  ```
  df(b(2),t) = d(1,b(2)) - 2*f(2)*b(1),
  df(f(2),t) = - b(2)*b(1) + d(1,f(2)) - f(2)*f(1)
  ```
  for $\bar{s}$.

One does not need to compute the linearizations in order to obtain a symmetry of a differential equation. However, the explicit computation of the linearizations will be required for finding the recursions, which are "symmetries of symmetries."

For computing symmetries of any system $\mathcal{E}$, use the procedure `ssym` with the call `ssym(N,tw,sw,afwlist,abwlist,eqnlist,fl,inelist,flags);`

`N` ... the number of superfields θ_i;

`tw` ... 2×the weight of ∂_t;

`sw` ... 2×the weight of ∂_s;

`afwlist` ... list of 2×weights of the fermion fields `f(1),f(2),...,f(nf)`;
`abwlist` ... list of 2×weights of the boson fields `b(1),b(2),...,b(nb)`;
`eqnlist` ... list of extra conditions on the undetermined coefficients;
`fl` ... extra unknowns in `eqnlist` to be determined;
`inelist` ... a list, each element of it is a non-zero expression or a list with at least one of its elements being non-zero;
`flags` ... list of flags:
`init`: only initialization of global data,
`zerocoeff`: all coefficients = 0 which do not appear in inelist,
`tpar`: if the time variable t changes parity,
`spar`: if the symmetry variable s changes parity,
`lin`: if symmetries of a linearization are to be computed,
`filter`: if a symmetry should satisfy homogeneity weights defined in `hom_wei`

Note that the computer representation of the weights is twice the standard notation; thus we avoid half-integer values for convenience. For more details on other flags run the command `sshelp()`.

3 Recursions

Definition 2. A recursion operator for the symmetry algebra of an evolutionary system

$$\mathcal{E} = \{f_t^i = \Phi_f^i,\ b_t^i = \Phi_b^j\},\quad i = 1,\dots,n_f,\quad j = 1,\dots,n_b$$

is the vector expression

$$\mathcal{R} = (\mathcal{R}_f^1(\varphi),\dots,\mathcal{R}_f^{n_f}(\varphi),\mathcal{R}_b^1(\varphi),\dots,\mathcal{R}_b^{n_b}(\varphi))^T,$$

which is linear w.r.t. the new fields φ, which for bosonic s are $\varphi = (F^1,\dots,F^{n_f},B^1,\dots,B^{n_b})^T$ and for fermionic $\bar{s}$ are $\varphi = (B^1,\dots,B^{n_f},F^1,\dots,F^{n_b})^T$, and their derivatives and which is a (right-hand side of a) symmetry of $\mathcal{E}$ whenever

$$\varphi = (F^1,\dots,F^{n_f},B^1,\dots,B^{n_b})^T \quad \text{for } s, \text{ or}$$

$$\varphi = (B^1,\dots,B^{n_f},F^1,\dots,F^{n_b})^T \quad \text{for } \bar{s}$$

is a symmetry of $\mathcal{E}$. In other words, $\mathcal{R}$: $\operatorname{sym}\mathcal{E} \to \operatorname{sym}\mathcal{E}$ is a linear operator that generates a symmetry of $\mathcal{E}$ when applied to a symmetry $\varphi \in \operatorname{sym}\mathcal{E}$.

The weight $[\mathcal{R}]$ of recursion $\mathcal{R}$ is the difference $[\mathcal{R}(\varphi)] - [\varphi]$ of the weights of the (time derivatives $\partial_{s'}$, ∂_s of u, i.e. of the) resulting and the initial symmetries, here $\varphi \in \operatorname{sym}\mathcal{E}$.

Different recursions can have the same weight. Experiments show that in this case operators may have different properties, e.g., a majority of them is nilpotent, several zero-order operators act through multiplication by a differential-functional expression and do not increase the differential orders of the flows, and only few recursions construct higher-order symmetries and reveal the integrability.

If $[\mathcal{R}]$ is the weight of a recursion $\mathcal{R}$, then, clearly, at least one recursion is found with weight $2 \times [\mathcal{R}]$. Indeed, this is $\mathcal{R}^2$. At the same time, other recursions may appear with weight $\mathbb{N} \times [\mathcal{R}]$. If $\mathcal{R}$ is nonlocal (see below), then its powers are also nonlocal, but it remains a very delicate matter to predict the form of their nonlocalities, and strong theoretical assertions can be formulated for some particular integrable system.

Remark 1. The derivation of the weight of a recursion is constructive in the following sense. To attempt finding a recursion for an evolutionary system, it is beneficial to know already many symmetries φ_{s_i} of different weights $[s_i]$. Then one tries first the weights $[\mathcal{R}] := [s_i] - [s_j]$ for various i, j. However, the recursions obtained this way can be nilpotent, i.e. $\mathcal{R}^q(\varphi) \equiv 0$ for some q and any φ. More promising are weights for which there exist i, j, k with $[\mathcal{R}] = [s_i] - [s_j] = [s_j] - [s_k]$ and s_i, s_j, s_k being elements of the infinite hierarchies of symmetries with low weights. Still, the actual weights of unknown recursions can turn out to be larger than the weight differences of the lowest order symmetries.

Finally, non-trivial recursions may only appear in nonlocal settings. We discuss this in Section 4.

The crucial point is that $\mathcal{R}$ is a symmetry of the linearized system $\operatorname{Lin}\mathcal{E}$. The original system $\mathcal{E}$ is only used for substitutions. Hence we use `ssym` for finding recursions of the linearizations, which are previously calculated by `linearize`.

Example 5. Let us construct a recursion for equation (1.2). We obtain the linearization using the procedure `linearize`,

```
linearize({df(b(1),t)=df(b(1),x,2)-2*b(1)*df(b(1),x)},0,1,nil,nil);
```

The new system depends on the two fields $b =$ `b(1)` and $B =$ `b(2)`.

```
df(b(2),t) = -2*b(2)*df(b(1),x)-2*b(1)*df(b(2),x)+df(b(2),x,2),
df(b(1),t) = df(b(1),x,2)-2*b(1)*df(b(1),x).
```

The recursion of weight 0 is obtained as follows (thus `sw` $= 2[\mathcal{R}] = 0$).

```
ssym(1, 4, 0, {}, {2, 2},
  {df(b(2),t)= -2*b(2)*df(b(1),x)-2*b(1)*df(b(2),x)+df(b(2),x,2),
   df(b(1),t)=> df(b(1),x,2)-2*b(1)*df(b(1),x)}, {}, {}, {lin});
```

By writing `df(b(1),t)=> ...` we require that the original system is only used for substitutions.

The output contains

```
df(b(2),s)=b(2)
1 solution was found.
```

This is the identity transformation

$$\varphi \mapsto \mathcal{R} = \mathrm{id}\,(\varphi) \equiv \varphi;$$

it maps symmetries to themselves. Clearly, the identity is a recursion for any system!

Example 6 (A recursion for the Korteweg–de Vries equation)**.** The KdV equation upon the bosonic field $u(x,t)$ is

$$u_t = -u_{xxx} + uu_x. \tag{3.1}$$

Equation (3.1) is homogeneous w.r.t. the weights

$$[u] = 2, \quad [t] = -3, \quad [x] = -1.$$

The linearization of (3.1) is constructed using the procedure `linearize`,

```
linearize({df(b(1),t)= -df(b(1),x,3) + b(1)*df(b(1),x)},0,1,nil,nil)
```

Thus we obtain the new system that depends on the fields $u =$ `b(1)`, $U =$ `b(2)`:

```
df(b(1),t) = - df(b(1),x,3) + b(1)*df(b(1),x)
df(b(2),t) = b(2)*df(b(1),x) + b(1)*df(b(2),x) - df(b(2),x,3).
```

As a first guess we are looking for the recursion operator of weight 0. The recursion of weight 0 (hence `sw` $= -2[\mathcal{R}] = 0$) is constructed as follows.

```
ssym(1, 6, 0, {}, {4, 4},
     {df(b(2),t) =  b(2)*df(b(1),x)+b(1)*df(b(2),x)-df(b(2),x,3),
      df(b(1),t) => -df(b(1),x,3)+b(1)*df(b(1),x)},{},{},{lin});
```

The output contains

```
df(b(2),s)=b(2)
1 solution was found.
```

Again, this operator $\mathcal{R}$ of weight 0 is the identity,

$$\varphi \mapsto \mathcal{R}(\varphi) = (\mathrm{id})\,(\varphi).$$

The well-known explanation for this result is that, as a rule, one needs to introduce nonlocalities first and only then obtains nontrivial recursions in the nonlocal setting.

4 Nonlocalities

The nonlocal variables for $N = 1$ super-systems are constructed by trivializing [7, 9] conservation laws

$$\partial_t(\text{density}) \doteq \mathcal{D}(\text{super-flux});$$

that is, in each case the above equality holds by virtue ($\doteq$) of the system at hand and all possible differential consequences from it. The standard procedure [9] suggests that every conserved current determines the new nonlocal variable, say v, whose derivatives are set to

$$v_t = \text{super-flux}, \quad \mathcal{D}v = \text{density} \tag{4.1a}$$

if the time t preserves the parities and

$$v_{\bar{t}} = -\text{super-flux}, \quad \mathcal{D}v = \text{density} \tag{4.1b}$$

if the time $\bar{t}$ is parity-reversing. Note that in the classical case the nonlocality v can be specified through

$$v_t = \text{flux}, \quad v_x = \text{density} \tag{4.1c}$$

for the conservation law $\partial_t(\text{density}) \doteq \partial_x(\text{flux})$. Each nonlocality thus makes the conserved current trivial because the cross derivatives of v coincide in this case, $[D_t, \mathcal{D}](v) = 0$, where $[\,,\,]$ stands for the commutator if t is parity-preserving and for the *anti*commutator whenever the time $\bar{t}$ is parity-reversing. The new variables can be bosonic or fermionic; the parities are immediately clear from the formulae for their derivatives.

Hence, starting with an equation $\mathcal{E}$, one calculates several conserved currents for it and *trivializes* them by introducing a *layer* of nonlocalities whose derivatives are still local differential functions. This way the number of fields is increased and the system is extended by new substitution rules. Moreover, it may acquire new conserved currents that depend on the nonlocalities and thus specify the second layer of nonlocal variables with nonlocal derivatives. At each step the number of variables will increase by 2 compared with the previous layer (a new nonlocal variable plus the corresponding linearized field). Clearly, the procedure is self-reproducing.

So, one keeps computing conserved currents and adding the layers of nonlinearities until an extended system $\tilde{\mathcal{E}}$ is achieved such that its linearization $\operatorname{Lin}\tilde{\mathcal{E}}$ has a symmetry $\mathcal{R}$; this symmetry of $\operatorname{Lin}\tilde{\mathcal{E}}$ is a recursion for the extended system $\tilde{\mathcal{E}}$.

The calculation of conservation laws for evolutionary super-systems with homogeneous polynomial right-hand sides is performed by using the procedure `ssconl`:

```
ssconl(N,tw,mincw,maxcw,afwlist,abwlist,pdes);
```

where

`N` ... the number of superfields θ^i;

`tw` ... 2×the weight $[\partial_t]$;
`mincw` ... minimal weight of the conservation law;
`maxcw` ... maximal weight of the conservation law;
`afwlist` ... list of weights of the fermionic fields `f(1)`,...,`f(nf)`;
`abwlist` ... list of weights of the bosonic fields `b(1)`,...,`b(nb)`;
`pdes` ... list of the equations for which a conservation law must be found.

For positive weights of bosonic variables, the ansatz is fully determined through the weight `mincw, ..., maxcw` of the conservation law. If a boson weight is non-positive then the global variable `max_deg` must have a positive integer value which is the highest degree of such a variable or any of its derivatives in any ansatz. The conservation law condition leads to an algebraic system for the undetermined coefficients, which is further solved automatically by CRACK.

Having obtained a conserved current, one defines the new bosonic or fermionic dependent variable (the nonlocality) using the standard rules (4.1).

We illustrate the general scheme of fixing the derivatives of a nonlocal variable by several examples. Further information on the SsTOOLS environment is contained in [8] and the `sshelp()` function in SsTOOLS. The algebraic structures that describe the geometry of recursion operators for super-PDE are described in detail in [9]. Some more examples and their applications are also found in [7].

Example 7 (A nonlocal recursion for the KdV equation)**.** Consider the Korteweg–de Vries equation (3.1) again,

$$\partial_t(u) \doteq \partial_x\Big(-u_{xx} + \frac{1}{2}u^2\Big).$$

We declare that the conserved density u is the spatial derivative $w_x = u$ of a new nonlinear variable w and the flux is its derivative w.r.t. the time, $w_t = -u_{xx} + \frac{1}{2}u^2$. Then $w_{xt} = w_{tx}$ by virtue of (3.1). Thus we introduce the bosonic nonlocality w by trivializing the conserved current. Let us remember that

$$w_x = u,$$
$$w_t = -w_{xxx} + \frac{1}{2}w_x^2$$

and the weight of w is $[w] = 1$ because $[w] + [\partial_x] = [u] = 2$.

Next, we compute the linearization of equation (3.1) and of the relations that specify the new variable,

```
linearize({df(b(1),t)= -df(b(1),x,3) + b(1)*df(b(1),x),
           df(b(2),x)= b(1),
           df(b(2),t)= -df(b(2),x,3) + df(b(2),x)**2}, 0, 2);
```

The linearization correspondence between the fields is

$$u \mapsto U \quad (\texttt{b(1)} \mapsto \texttt{b(3)}), \qquad w \mapsto W \quad (\texttt{b(2)} \mapsto \texttt{b(4)}).$$

The linearized system is

```
df(b(3),t) = b(3)*df(b(1),x) + b(1)*df(b(3),x) - df(b(3),x,3),
df(b(4),x) = b(3),
df(b(4),t) = -df(b(4),x,3) + df(b(2),x)*df(b(4),x)
```

In this nonlocal setting, we obtain the nonlocal recursion of weight sw $= 2[\mathcal{R}] = 4$ as follows,

```
ssym(1, 6, 4, {}, {4, 2, 4, 2},{
    df(b(3),t) - b(3)*df(b(1),x) + b(1)*df(b(3),x) - df(b(3),x,3),
    df(b(1),t) => b(1)*df(b(1),x) - df(b(1),x,3),
    df(b(2),x) => b(1),
    df(b(2),t) => -df(b(2),x,3) + 1/2 * df(b(2),x)**2,
    df(b(4),x) => b(3),
    df(b(4),t) => -df(b(4),x,3) + df(b(2),x)*df(b(4),x)
}, {}, {}, {lin});
```

We recall here that only the linearized system should be written as equations, and all other relations, including the nonlocalities, should be written as substitutions.

This yields the solution

```
df(b(3),s) = -3*df(b(3),x,2) + 2*b(1)*b(3) + df(b(1),x)*b(4),
```

which is the well-known nonlocal recursion operator for KdV,

$$\varphi \mapsto \mathcal{R} = \left(-3D_x^2 + 2u + u_x \cdot D_x^{-1}\right)(\varphi). \tag{4.2}$$

This recursion generates the hierarchy of *local* symmetries starting from the translation $\varphi_0 = u_x$. The powers $\mathcal{R}^2$, $\mathcal{R}^3$, ... of the recursion operator are also nonlocal.

Example 8. Consider the Burgers equation (1.2) and introduce the bosonic nonlocality w of weight $[w] = [b] - [\partial_x] = 0$ by trivializing the conserved current $\partial_t(b) \doteq \partial_x\left(b_x - b^2\right)$. We therefore, set

$$w_x = b, \quad w_t = w_{xx} - w_x^2. \tag{4.3}$$

The linearization of the extended system is obtained through

```
linearize({df(b(1),t)=df(b(1),x,2) - 2*b(1)*df(b(1),x),
           df(b(2),x)=b(1),
           df(b(2),t)=df(b(2),x,2) - df(b(2),x)**2}, 0, 2);
```

The correspondence between the bosonic fields is

$$b \mapsto B \quad (\texttt{b(1)} \mapsto \texttt{b(3)}), \qquad w \mapsto W \quad (\texttt{b(2)} \mapsto \texttt{b(4)}).$$

The entire linearized system is (1.2) and (4.3) together with the relations

```
df(b(3),t)= -2*b(3)*df(b(1),x)-2*b(1)*df(b(3),x)+df(b(3),x,2);
df(b(4),x)= b(3);
df(b(4),t)= df(b(4),x,2) - 2*df(b(4),x)*df(b(2),x).
```

The difference between weights of the first-order and the second-order symmetries is 1. Hence, the recursion operator could have weight 1; see Remark 1. The nonlocal recursion of weight 1, sw $= 2[\mathcal{R}] = 2$, is obtained by

```
max_deg:=1;
ssym(1, 4, 2, {}, {2, 0, 2, 0},{
    df(b(3),t) = -2*b(3)*df(b(1),x)-2*b(1)*df(b(3),x)+df(b(3),x,2),
    df(b(4),x) => b(3),
    df(b(4),t) => df(b(4),x,2)-2*df(b(4),x)*df(b(2),x),
    df(b(1),t) => -2*b(1)*df(b(1),x)+df(b(1),x,2),
    df(b(2),x) => b(1),
    df(b(2),t) => df(b(2),x,2)-df(b(2),x)**2}, {},{},{lin});
```

We finally get the recursion

```
df(b(3),s) = -df(b(3),x) + b(1)*b(3) + df(b(1),x)*b(4),
```

which is nonlocal,

$$\varphi \mapsto \mathcal{R} = \left(-D_x + b + b_x D_x^{-1}\right)(\varphi).$$

Example 9. Consider the superfield representation [7] of the Burgers equation, see (1.2),

$$f_t = \mathcal{D}b, \quad b_t = \mathcal{D}f + b^2;$$

its weights are $|f| = |b| = \frac{1}{2}$, $|\partial_x| = 1$, and $|\partial_t| = \frac{1}{2}$.

We introduce the nonlocal bosonic field $w(x,t;\theta)$ of weight $[w] = [f] - [\mathcal{D}] = 1 - 1 = 0$ such that

$$\mathcal{D}w = -f, \quad w_t = -b.$$

We get the linearized system by

```
linearize({df(f(1),t)= d(1, b(1)),
           df(b(1),t)= df(f(1),x) + b(1)**2,
           d(1,b(2)) = -f(1),
           df(b(2),t)= -b(1)}, 1, 2);
```

For the linearization correspondence between the fields is $f \to F$, $b \to B$, $w \to W$, and we have

$$F_t = \mathcal{D}B, \quad B_t = 2bB + \mathcal{D}F, \quad \mathcal{D}W = -F, \quad W_t = -B,$$

that is,

```
df(f(2),t)= d(1, b(3)),
df(b(3),t)= 2*b(1)*b(3)+d(1,f(2)),
d(1,b(4)) = -f(2),
df(b(4),t)= -b(3).
```

In this setting, we obtain the nonlocal recursion of weight $\frac{1}{2}$, sw $= 2[\mathcal{R}] = 1$: the input is

```
max_deg:=1;
ssym(1, 1, 1, {1, 1}, {1, 0, 1, 0},
     {df(f(2),t) = d(1, b(3)),
      df(b(3),t) = 2*b(3)*b(1) + d(1, f(2)),
      d(1, b(4)) => - f(2),
      df(b(4),t) => - b(3),
      df(f(1),t) => d(1, b(1)),
      df(b(1),t) => d(1, f(1)) + b(1)**2,
      d(1, b(2)) => -f(1),
      df(b(2),t) => -b(1)}, {}, {}, {lin});
```

The recursion is

```
df(f(2),s)=d(1,b(3)) + d(1,b(1))*b(4)  - f(2)*b(1),
df(b(3),s)=b(4)*b(1)**2 + b(3)*b(1) + d(1,f(2)) + d(1,f(1))*b(4),
```

in other words,

$$\begin{pmatrix} F \\ B \end{pmatrix} \mapsto \mathcal{R}\begin{pmatrix} F \\ B \end{pmatrix} = \begin{pmatrix} \mathcal{D}B - \mathcal{D}b\mathcal{D}^{-1}F - Fb \\ -b^2\mathcal{D}^{-1}F + Bb + \mathcal{D}F - \mathcal{D}f\mathcal{D}^{-1}F \end{pmatrix}.$$

Example 10. Consider the fifth order evolution superequation found by Tian and Liu (Case F in [14], see also [6, 15]):

$$f_t = f_{5x} + 10(f_{xx}\mathcal{D}f)_x + 5(f_x\mathcal{D}f_x)_x + 15f_x(\mathcal{D}f)^2 + 15f(\mathcal{D}f_x)(\mathcal{D}f). \tag{4.4}$$

In what follows, we are considering this equation in components. Substituting the $\xi + \theta u$ for f in (4.4), for example, using SsTools, we obtain

$$u_t = u_{5x} + 10uu_{xxx} + 20u_xu_{xx} + 30u^2u_x - 5\xi_{xxx}\xi_x + 15u\xi_{xx}\xi + 15u_x\xi_x\xi, \tag{4.5a}$$

$$\xi_t = \xi_{5x} + 10u\xi_{xxx} + 15u_x\xi_{xx} + 5u_{xx}\xi_x + 15u^2\xi_x + 15uu_x\xi, \tag{4.5b}$$

where u is a bosonic field and ξ is a fermionic field.

Observe that the bosonic limit ($\xi := 0$) is the fifth order symmetry of Korteweg–de Vries equation (3.1). However, a direct computation shows that the equation (4.5) has local symmetries of the orders $1 + 6k$ and $5 + 6k$, where $k \in \mathbb{N}$, and does not have any local symmetries of order $3 + 6k$, where $k \in \mathbb{N}$. Therefore, the recursion operator for (4.5) should be at least of order 6. Let us also assume that the bosonic limit of the recursion operator for (4.5) is the third power $\mathcal{R}^3$ of the recursion operator (4.2) for the Korteweg–de Vries equation.

It is easy to check that for the construction of the 3rd power of the recursion operator (4.2) we should "trivialise" the following conserved densities of the Korteweg–de Vries equation: u, $u^2 + u_{xx}$ and $u_{4x} + 6uu_{xx} + 5u_x^2 + 2u^3$.

Let S and Q satisfy the linearized equation for (4.5). The correspondence between fields is the following $u \mapsto S$, $\xi \mapsto Q$. The linearized system of nonlocalities for the generalisation of those conservation laws for the supersymmetric equation (4.5) is the following:

1) the layer of nonlocalities corresponding to the generalisation of the conserved density u of the Korteweg–de Vries equation

$$\begin{aligned} W_{1;x} &= S, \\ W_{1;t} &= -15Qu\xi_x + S(30u^2 + 10u_{xx} - 15\xi\xi_x) - 5Q_{xx}\xi_x + Q_x(5\xi_{xx} + 15\xi u) + S_{4x} \\ &\quad + 10S_{xx}u + 10S_x u_x, \end{aligned}$$

2) the layer of nonlocalities corresponding to the generalisation of the conserved density $u^2 + u_{xx}$ of the Korteweg–de Vries equation

$$\begin{aligned} W_{2;x} &= 2Su + 2Q_x\xi, \\ Q_{2;x} &= Qu + S\xi, \\ W_{2;t} &= Q(-30u^2\xi_x + 30\xi uu_x) + S(60u^3 + 40uu_{xx} + 2u_{4x} - 30\xi_{xx}\xi_x - 120\xi u\xi_x \\ &\quad - 20\xi\xi_{xxx}) + 2Q_{5x}\xi - 2Q_{4x}\xi_x + Q_{xxx}(2\xi_{xx} + 20\xi u) + Q_{xx}(-30u\xi_x - 2\xi_{xxx} \\ &\quad + 30\xi u_x) + Q_x(30u\xi_{xx} + 2\xi_{4x} + 60\xi u^2 + 10\xi u_{xx}) + 2S_{4x}u - 2S_{xxx}u_x \\ &\quad + S_{xx}(20u^2 + 2u_{xx} - 10\xi\xi_x) + S_x(-2u_{xxx} - 30\xi\xi_{xx}), \\ Q_{2;t} &= Q(15u^3 + 10uu_{xx} + u_{4x} + 5u_x^2 - 5\xi_{xx}\xi_x) + S(20u\xi_{xx} + \xi_{4x} - 5\xi_x u_x + 45\xi u^2 \\ &\quad + 10\xi u_{xx}) + Q_{4x}u - Q_{xxx}u_x + Q_{xx}(10u^2 + u_{xx} + 5\xi\xi_x) + Q_x(-5uu_x - u_{xxx} \\ &\quad - 5\xi\xi_{xx}) + S_{4x}\xi - S_{xxx}\xi_x + S_{xx}(\xi_{xx} + 10\xi u) + S_x(-5u\xi_x - \xi_{xxx} + 10\xi u_x), \end{aligned}$$

3) the layer of nonlocalites corresponding to the generalisation of the conserved densities $u_{4x} + 6uu_{xx} + 5u_x^2 + 2u^3$ of the Korteweg–de Vries equation

$$\begin{aligned} W_{3;x} &= -9Q_2\xi u + 3Qu\xi_x + S(6u^2 + 3\xi\xi_x) - Q_{xxx}\xi - 3Q_x\xi u + 2S_{xx}u, \\ Q_{3;x} &= Q_2(6u^2 - 6\xi\xi_x) - 6W_2\xi u + 14Su\xi_x + 2Q_{xxx}u + 7Q_xu^2 - 2S_{xxx}\xi, \\ W_{3;t} &= Q_2(-90u^2\xi_{xx} - 9u\xi_{4x} + 45u\xi_xu_x + 9\xi_{xxx}u_x - 9\xi_{xx}u_{xx} + 9\xi_xu_{xxx} - 135\xi u^3 \\ &\quad - 90\xi uu_{xx} - 9\xi u_{4x} - 45\xi u_x^2 + 45\xi\xi_{xx}\xi_x) + Q(45u^3\xi_x + 9u^2\xi_{xxx} + 3u\xi_{5x} \\ &\quad - 42u\xi_{xx}u_x + 21u\xi_xu_{xx} - 3\xi_{4x}u_x + 3\xi_{xxx}u_{xx} - 3\xi_{xx}u_{xxx} + 3\xi_xu_{4x} + 42\xi_xu_x^2 \\ &\quad - 15\xi uu_{xxx} - 45\xi u_{xx}u_x) + S(180u^4 + 300u^2u_{xx} + 32uu_{4x} + 138u\xi_{xx}\xi_x \\ &\quad - 32u_{xxx}u_x + 16u_{xx}^2 + 16\xi_{4x}\xi_x - 16\xi_{xxx}\xi_{xx} + 135\xi u^2\xi_x + 78\xi u\xi_{xxx} + 13\xi\xi_{5x} \\ &\quad + 93\xi\xi_{xx}u_x + 111\xi\xi_xu_{xx}) + Q_{6x}\xi_x + Q_{5x}(-\xi_{xx} - 13\xi u) + Q_{4x}(6u\xi_x + \xi_{xxx} \\ &\quad - 32\xi u_x) + Q_{xxx}(-26u\xi_{xx} - \xi_{4x} + 26\xi_xu_x - 24\xi u^2 - 48\xi u_{xx}) + Q_{xx}(69u^2\xi_x \\ &\quad + 16u\xi_{xxx} - 15\xi_{xx}u_x + 17\xi_xu_{xx} - 48\xi uu_x - 22\xi u_{xxx}) + Q_x(-69u^2\xi_{xx} \\ &\quad + 4u\xi_{4x} + 45u\xi_xu_x - \xi_{xxx}u_x - 2\xi_{xx}u_{xx} + 5\xi_xu_{xxx} - 45\xi u^3 - 66\xi uu_{xx} \\ &\quad - 8\xi u_{4x} - 87\xi u_x^2) + 2S_{6x}u - 2S_{5x}u_x + S_{4x}(26u^2 + 2u_{xx} + 8\xi\xi_x) \\ &\quad + S_{xxx}(28uu_x - 2u_{xxx} + 22\xi\xi_{xx}) + S_{xx}(120u^3 + 112uu_{xx} + 2u_{4x} - 28u_x^2 \\ &\quad + 22\xi_{xx}\xi_x + 81\xi u\xi_x + 48\xi\xi_{xxx}) + S_x(180u^2u_x + 48uu_{xxx} - 16u_{xx}u_x \\ &\quad + 16\xi_{xxx}\xi_x + 3\xi u\xi_{xx} + 32\xi\xi_{4x} + 174\xi\xi_xu_x) + \xi Q_{7x}, \end{aligned}$$

$$
\begin{aligned}
Q_{3;t} = {} & Q_2(90u^4 + 120u^2u_{xx} + 12uu_{4x} - 180u\xi_{xx}\xi_x - 12u_{xxx}u_x + 6u_{xx}^2 - 12\xi_{4x}\xi_x \\
& + 12\xi_{xxx}\xi_{xx} - 270\xi u^2\xi_x - 60\xi u\xi_{xxx} - 6\xi\xi_{5x} - 90\xi\xi_{xx}u_x - 30\xi\xi_x u_{xx}) \\
& + Q(45u^3u_x + 24u^2u_{xxx} + 84uu_{xx}u_x - 12u\xi_{xxx}\xi_x - 28u_x^3 - 84\xi_{xx}\xi_x u_x \\
& + 60\xi u^2\xi_{xx} - 24\xi u\xi_{4x} - 30\xi u\xi_x u_x - 96\xi\xi_{xxx}u_x - 84\xi\xi_{xx}u_{xx} - 36\xi\xi_x u_{xxx}) \\
& + W_2(-60u^2\xi_{xx} - 6u\xi_{4x} + 30u\xi_x u_x + 6\xi_{xxx}u_x - 6\xi_{xx}u_{xx} + 6\xi_x u_{xxx} - 90\xi u^3 \\
& - 60\xi uu_{xx} - 6\xi u_{4x} - 30\xi u_x^2 + 30\xi\xi_{xx}\xi_x) + S(660u^3\xi_x + 288u^2\xi_{xxx} + 34u\xi_{5x} \\
& - 28u\xi_{xx}u_x + 590u\xi_x u_{xx} - 34\xi_{4x}u_x + 34\xi_{xxx}u_{xx} - 34\xi_{xx}u_{xxx} + 34\xi_x u_{4x} \\
& + 28\xi_x u_x^2 + 135\xi u^2u_x - 102\xi uu_{xxx} - 20\xi u_{5x} - 366\xi u_{xx}u_x - 72\xi\xi_{xxx}\xi_x) \\
& - 2Q_{6x}u_x + Q_{5x}(27u^2 + 2u_{xx} - 10\xi\xi_x) + Q_{4x}(36uu_x - 2u_{xxx} - 20\xi\xi_{xx}) \\
& + Q_{xxx}(106u^3 + 124uu_{xx} + 2u_{4x} - 36u_x^2 - 20\xi_{xx}\xi_x + 72\xi u\xi_x) \\
& + Q_{xx}(121u^2u_x + 36uu_{xxx} - 12u_{xx}u_x + 10\xi_{xxx}\xi_x - 30\xi u\xi_{xx} + 20\xi\xi_{4x} \\
& + 144\xi\xi_x u_x) + Q_x(165u^4 + 295u^2u_{xx} + 24uu_{4x} + 28uu_x^2 - 30u\xi_{xx}\xi_x \\
& - 24u_{xxx}u_x + 12u_{xx}^2 + 10\xi_{4x}\xi_x - 10\xi_{xxx}\xi_{xx} - 42\xi u\xi_{xxx} + 10\xi\xi_{5x} \\
& - 114\xi\xi_{xx}u_x + 30\xi\xi_x u_{xx}) + 2S_{6x}\xi_x + S_{5x}(-2\xi_{xx} - 20\xi u) + S_{4x}(44u\xi_x \\
& + 2\xi_{xxx} - 80\xi u_x) + S_{xxx}(16u\xi_{xx} - 2\xi_{4x} + 36\xi_x u_x - 36\xi u^2 - 140\xi u_{xx}) \\
& + S_{xx}(325u^2\xi_x + 104u\xi_{xxx} - 82\xi_{xx}u_x + 94\xi_x u_{xx} - 306\xi uu_x - 140\xi u_{xxx}) \\
& + S_x(91u^2\xi_{xx} + 56u\xi_{4x} + 386u\xi_x u_x - 22\xi_{xxx}u_x - 12\xi_{xx}u_{xx} + 46\xi_x u_{xxx} \\
& + 45\xi u^3 - 246\xi uu_{xx} - 80\xi u_{4x} - 414\xi u_x^2 - 174\xi\xi_{xx}\xi_x) + 2uQ_{7x} - 2\xi S_{7x}.
\end{aligned}
$$

Here W_1, W_2, W_3 are bosonic fields and Q_2, Q_3 are fermionic fields. Let us note that the generalisation of the conservation law of (3.1) with the density $u_{4x} + 6uu_{xx} + 5u_x^2 + 2u^3$ is no longer a local conservation law for (4.5).

The weights of fields are as follows

$$
\begin{aligned}
& |x| = -1,\ |t| = -5,\ |u| = |S| = 2,\ |\xi| = |Q| = \tfrac{3}{2}, \\
& |W_1| = 1,\ |W_2| = 3,\ |W_3| = 5,\ |Q_2| = \tfrac{5}{2},\ |Q_3| = \tfrac{11}{2}.
\end{aligned}
$$

Using the technique described above we obtain the following recursion operator (cf. [14])

$$
\mathcal{R}\left(\begin{pmatrix} S \\ Q \end{pmatrix}\right) = \begin{pmatrix} S' \\ Q' \end{pmatrix},
$$

where

$$
\begin{aligned}
S' = {} & 3Q_3\xi_x + Q_2(-42u\xi_{xx} - 6\xi_{4x} - 42\xi_x u_x - 18\xi u^2) + Q(-51u^2\xi_x - 12u\xi_{xxx} \\
& + 2\xi_{5x} - 26\xi_{xx}u_x - 22\xi_x u_{xx} + 18\xi uu_x - 2\xi u_{xxx}) + 4W_3u_x + W_2(24uu_x \\
& + 4u_{xxx} + 6\xi\xi_{xx}) + W_1(120u^2u_x + 40uu_{xxx} + 4u_{5x} + 80u_{xx}u_x - 20\xi_{xxx}\xi_x \\
& - 60\xi u\xi_{xx} - 60\xi\xi_x u_x) + S(128u^3 + 192uu_{xx} + 24u_{4x} + 144u_x^2 - 16\xi_{xx}\xi_x \\
& - 180\xi u\xi_x - 30\xi\xi_{xxx}) - 12Q_{4x}\xi_x + Q_{xxx}(-6\xi_{xx} + 36\xi u) + Q_{xx}(-12u\xi_x \\
& + 8\xi_{xxx} + 58\xi u_x) + Q_x(-12u\xi_{xx} + 8\xi_{4x} - 22\xi_x u_x + 72\xi u^2 + 30\xi u_{xx}) + 2S_{6x} \\
& + 24S_{4x}u + 60S_{xxx}u_x + S_{xx}(96u^2 + 80u_{xx} - 30\xi\xi_x) \\
& + S_x(280uu_x + 60u_{xxx} - 54\xi\xi_{xx}),
\end{aligned}
$$

$$\begin{aligned}Q' = {} & 3Q_3u + Q_2(42uu_x + 6u_{xxx}) + Q(35u^3 + 48uu_{xx} + 2u_{4x} + 40u_x^2 - 6\xi_{xx}\xi_x \\ & + 4\xi\xi_{xxx}) + 4W_3\xi_x + W_2(-12u\xi_x - 2\xi_{xxx} - 6\xi u_x) + W_1(60u^2\xi_x + 40u\xi_{xxx} \\ & + 4\xi_{5x} + 60\xi_{xx}u_x + 20\xi_x u_{xx} + 60\xi uu_x) + S(114u\xi_{xx} + 22\xi_{4x} + 104\xi_x u_x \\ & + 72\xi u^2 + 30\xi u_{xx}) + 2Q_{6x} + 24Q_{4x}u + 48Q_{xxx}u_x + Q_{xx}(54u^2 + 38u_{xx} + 4\xi\xi_x) \\ & + Q_x(126uu_x + 14u_{xxx}) + 12S_{xxx}\xi_x + S_{xx}(42\xi_{xx} + 42\xi u) \\ & + S_x(100u\xi_x + 46\xi_{xxx} + 54\xi u_x).\end{aligned}$$

In [6] this system of nonlocalities was used to construct a zero-curvature representation of (4.5) to prove its integrability.

Acknowledgements

This work was supported in part by an NSERC grant to T. Wolf who is thanked by A. V. K. for warm hospitality. This research was done in part while A. O. K. was visiting at New York University Abu Dhabi; the hospitality and warm atmosphere of this institution are gratefully acknowledged.

References

[1] Berezin F A, *Introduction to superanalysis. Mathematical Physics and Applied Mathematics*, D. Reidel Publishing Co., Dordrecht–Boston, MA, 1987.

[2] Bocharov A V, Chetverikov V N, Duzhin S V, Khor'kova N G, Krasil'shchik I S, Samokhin A V, Torkhov Y N, Verbovetsky A M, Vinogradov A M, *Symmetries and conservation laws for differential equations of mathematical physics, Translations of Mathematical Monographs*, volume 182. American Mathematical Society, Providence, RI, 1999, xiv+333 pp.

[3] Delduc F, Gallot L, Ivanov E, New super KdV system with the $N = 4$ SCA as the Hamiltonian structure. *Phys. Lett. B*, **396**:1-4(1997), 122–132.

[4] Deligne P, Etingof P, Freed D S, Jeffrey L C, Kazhdan D, Morgan J W, Morrison D R, Witten E, *Quantum Fields and Strings: a Course for Mathematicians*, volume 1-2. AMS, Providence, RI; Institute for Advanced Study (IAS), Princeton, NJ, 1999, Vol. 1: xxii+723 pp.; Vol. 2: pp. i–xxiv and 727–1501 pp.

[5] Kiselev A V, The twelve lectures in the (non)commutative geometry of differential equations. Preprint IHÉS/M/12/13 (Bures-sur-Yvette, France), 2012, `http://preprints.ihes.fr/2012/M/M-12-13.pdf`.

[6] Kiselev A V, Krutov A O, On the (non)removability of spectral parameters in $\mathbb{Z}_2$-graded zero-curvature representations and its applications. *Acta Appl. Math.*, 2018, 39 pp., `arXiv:math.DG/1301.7143 [math.DG]`.

[7] Kiselev A V, Wolf T, Supersymmetric representations and integrable fermionic extensions of the Burgers and Boussinesq equations. *SIGMA Symmetry Integrability Geom. Methods Appl.*, **2** (2006), Paper 030, 19 pp., `arXiv:math-ph/0511071`.

[8] Kiselev A V, Wolf T, Classification of integrable super-systems using the SsTools environment. *Comput. Phys. Comm.*, **177**:3(2007), 315–328, `arXiv:nlin/0609065 [nlin.SI]`.

[9] Krasil'shchik I S, Kersten P H M, *Symmetries and recursion operators for classical and supersymmetric differential equations*, *Mathematics and its Applications*, volume 507. Kluwer Academic Publishers, Dordrecht, 2000, xvi+384 pp.

[10] Krasil'shchik J, Verbovetsky A, Vitolo R, *Symbolic Computation of Integrability Structures for Partial Differential Equations*, *Texts & Monographs in Symbolic Computation*, Springer, Cham, 2017, xv+263 pp.

[11] Krivonos S, Pashnev A, Popowicz Z, Lax pairs for $N = 2, 3$ supersymmetric KdV equations and their extensions. *Modern Phys. Lett. A*, **13**:18(1998), 1435–1443.

[12] Kupershmidt B A, Mathieu P, Quantum Korteveg-de Vries like equations and perturbed conformal field theories. *Phys. Lett. B*, **227**:2(1989), 245–250.

[13] Olver P J, *Applications of Lie groups to differential equations*, *Graduate Texts in Mathematics*, volume 107. Springer-Verlag, New York, second edition, 1993, xxviii+513 pp.

[14] Tian K, Liu Q P, Supersymmetric fifth order evolution equations. *AIP Conf. Proc.*, **1212**(2010), pp. 81–88. Nonlinear and modern mathematical physics. (July 15–21, 2009; Beijing, China).

[15] Tian K, Wang J P, Symbolic representation and classification of $N = 1$ supersymmetric evolutionary equations. *Stud. Appl. Math.*, **138**:4(2017), pp. 467–498. `arXiv:1607.03947 [nlin.SI]`.

[16] Wolf T, Schrüfer E, SsTools: Computations with supersymmetric algebraic and differential expressions. `http://lie.math.brocku.ca/crack/susy/sstools.red`.

C4. Symmetries of Itô stochastic differential equations and their applications

Roman Kozlov

Department of Business and Management Science, Norwegian School of Economics, Helleveien 30, 5045, Bergen, Norway
e-mail: Roman.Kozlov@nhh.no

Abstract

Lie point symmetries of Itô stochastic differential equations (SDEs) are considered. They correspond to Lie group transformations of the independent variable (time) and dependent variables, which preserve the differential form of the SDEs and properties of Brownian motion. In the considered framework transformations of Brownian motion are generated by random time change. There are provided some properties of the SDEs symmetries: the symmetries form a Lie algebra, there is a relation between symmetries and first integrals, and for some classes of SDEs there are results on maximal dimensionality of the admitted symmetry algebras. The symmetries can be used to construct Lie symmetry group classifications and to integrate SDEs with the help of quadratures. The relation to symmetries of the associated Fokker-Planck (FP) equation is also mentioned.

1 Introduction

Lie group theory of differential equations is well understood [19, 33, 34]. It studies transformations taking solutions of differential equations into other solutions of the same equations, while leaving the set of all solutions invariant. Now this theory is a very general and useful tool for finding analytical solutions of large classes of differential equations.

Recently there appeared applications of Lie group theory to stochastic differential equations (SDEs). First, restricted cases of point transformations were considered: transformations which do not change time $\bar{t} = t$ [1, 31, 32] and fiber-preserving transformations $\bar{t} = \bar{t}(t)$ [12]. Then, the theory for general point transformations was developed [10, 11, 18, 36, 37, 39]. In the latter case the transformation of the Brownian motion needs to be more deeply specified. One can find detailed description of symmetry development in the review article [16].

The purpose of this chapter is to provide a short presentation of symmetries of SDEs, their properties and applications. In order to keep the text short only Itô SDE will be considered. For the same reason computation of symmetries will be omitted.

The topic of the chapter stands on the intersection of two fields: Lie point symmetries and stochastic differential equations. We assume that the reader is familiar with Lie group symmetries [19, 33, 34] and mention important theoretical results concerning SDEs [2, 9, 17, 20, 40].

The chapter is organized as follows: Section 2 gives a simple illustrating example of a symmetry admitted by a scalar SDE and how this symmetry can be used for integration by quadratures. Section 3 gives background information on SDEs and describes determining equations for their Lie point symmetries. We consider symmetries which preserve the differential form of the SDEs and properties of Brownian motion. Then, Section 4 provides some properties of SDEs symmetries: There are provided some properties of the SDEs symmetries: the symmetries form a Lie algebra, there is a relation between symmetries and first integrals, and for some classes of SDEs there are results on maximal dimensionality of the admitted symmetry algebras. Section 5 presents applications of SDEs symmetries. They can be used to construct Lie symmetry group classifications and to integrate SDEs with the help of quadratures. The relation to symmetries of the associated Fokker-Planck (FP) equation is also mentioned.

2 Illustrating example

Let us consider a scalar SDE

$$dx = \beta x dt + \sigma x dW(t), \qquad \sigma \neq 0. \tag{2.1}$$

This equation is known as *geometric Brownian motion* [40]. It is important as a model equation for stochastic prices in economics.

SDE (2.1) admits a symmetry

$$X = x\frac{\partial}{\partial x}. \tag{2.2}$$

One can easily see that the corresponding Lie group transformation

$$\bar{t} = e^{\varepsilon X}(t) = t, \qquad \bar{x} = e^{\varepsilon X}(x) = e^{\varepsilon}x, \tag{2.3}$$

which is a scaling of the dependent variable x, does not change the form of the SDE (2.1). We consider two applications of the symmetry (2.2).

1. Integration by quadratures.

   Let us show how one can integrate SDE (2.1) with initial value $x(t_0) = x_0 > 0$ using the symmetry (2.2). Variable change

   $$y = \ln x \tag{2.4}$$

   performs stratification of the symmetry and brings it to the translation form

   $$\bar{X} = \frac{\partial}{\partial y}. \tag{2.5}$$

   It also transforms the SDE into the form (see Itô formula in point 3.1)

   $$dy = \left(\beta - \frac{1}{2}\sigma^2\right) dt + \sigma dW(t), \qquad y(t_0) = y_0 = \ln x_0, \tag{2.6}$$

which can be easily integrated as

$$y(t) = y_0 + \left(\beta - \frac{1}{2}\sigma^2\right)(t - t_0) + \sigma(W(t) - W(t_0)). \tag{2.7}$$

Finally, we find the solution of the original equation (2.1) as

$$x(t) = x_0 \mathrm{exp}\left(\left(\beta - \frac{1}{2}\sigma^2\right)(t - t_0) + \sigma(W(t) - W(t_0))\right). \tag{2.8}$$

2. Symmetries of the associated Fokker-Planck equation.

   The Fokker-Planck equation [35]

$$u_t = -(\beta x u)_x + \frac{1}{2}(\sigma^2 x^2 u)_{xx}, \tag{2.9}$$

   which corresponds to SDE (2.1), also admits symmetry (2.2).

3 Itô SDEs and Lie point symmetries

Let us consider a system of Itô stochastic differential equations

$$dx_i = f_i(t, \mathbf{x})dt + g_{i\alpha}(t, \mathbf{x})dW_\alpha(t), \qquad i = 1, ..., n, \qquad \alpha = 1, ..., m \tag{3.1}$$

where $f_i(t, \mathbf{x})$ is a drift vector, $g_{i\alpha}(t, \mathbf{x})$ is a diffusion matrix and $W_\alpha(t)$ is a vector Wiener process (vector Brownian motion) [40]. We assume summation for repeated indexes and use notation $\mathbf{x} = (x_1, ..., x_n)$. Note that $W_\alpha(t)$, $\alpha = 1, ..., m$ are independent one-dimensional Brownian motions.

3.1 Itô formula

Itô formula describes the transformation of the dependent variables in stochastic calculus. Given SDEs (3.1), we perform variable change for the dependent variables $\mathbf{y} = \mathbf{y}(t, \mathbf{x})$ with the help of Itô formula (see, for example, [40])

$$dy_i = \frac{\partial y_i}{\partial t}dt + \frac{\partial y_i}{\partial x_j}dx_j + \frac{1}{2}\frac{\partial^2 y_i}{\partial x_j \partial x_k}dx_j dx_k, \qquad i = 1, ..., n,$$

where $dx_j dx_k$ are computed according to the substitution rules

$$dt \cdot dt = 0, \tag{3.2a}$$

$$dt \cdot dW_\alpha = dW_\alpha \cdot dt = 0, \tag{3.2b}$$

$$dW_\alpha \cdot dW_\beta = \delta_{\alpha\beta}dt. \tag{3.2c}$$

This results in the formula for taking differentials of functions in stochastic calculus

$$dF(t, \mathbf{x}) = D_0(F)dt + D_\alpha(F)dW_\alpha(t), \tag{3.3}$$

where

$$D_0 = \frac{\partial}{\partial t} + f_j\frac{\partial}{\partial x_j} + \frac{1}{2}g_{j\alpha}g_{k\alpha}\frac{\partial^2}{\partial x_j \partial x_k}, \qquad D_\alpha = g_{j\alpha}\frac{\partial}{\partial x_j}.$$

3.2 First integrals

Some systems of stochastic differential equations possess first integrals.

Definition 1. A quantity $I(t, \mathbf{x})$ is a first integral of a system of SDEs (3.1) if it remains constant on the solutions of the SDEs.

The Itô differential (3.3) of the first integral

$$dI(t, \mathbf{x}) = D_0(I)dt + D_\alpha(I)dW_\alpha(t) = 0$$

leads to partial differential equations

$$D_0(I) = 0, \tag{3.4a}$$
$$D_\alpha(I) = 0, \tag{3.4b}$$

to which a conserved quantity should satisfy.

Let us note that SDEs with diffusion matrices of full rank have no nontrivial first integrals (only constants satisfy Eqs. (3.4)). In particular, this results holds for the scalar SDEs.

3.3 Determining equations

Let us consider a system of Itô SDEs (3.1). The derivation of the determining equations for the admitted symmetries follows [37].

We are interested in infinitesimal group transformations (near identity changes of variables)

$$\bar{t} = \bar{t}(t, \mathbf{x}, \varepsilon) = t + \varepsilon\tau(t, \mathbf{x}) + O(\varepsilon^2), \tag{3.5a}$$
$$\bar{x}_i = \bar{x}_i(t, \mathbf{x}, \varepsilon) = x_i + \varepsilon\xi_i(t, \mathbf{x}) + O(\varepsilon^2), \tag{3.5b}$$

which leave Eqs. (3.1) and the framework of Itô calculus, namely the substitution rules (3.2), invariant. Here ε is a group parameter. Such transformations can be represented by generating operators of the form

$$X = \tau(t, \mathbf{x})\frac{\partial}{\partial t} + \xi_i(t, \mathbf{x})\frac{\partial}{\partial x_i}. \tag{3.6}$$

We will also consider transformations restricted to the space of the dependent variables ($\bar{t} = t$), which are generated by operators of the form

$$Y = \xi_i(t, \mathbf{x})\frac{\partial}{\partial x_i}. \tag{3.7}$$

It should be noted that transformations (3.5) do not change the rank of the diffusion matrix. In particular, if the original system of SDEs has a diffusion matrix of full rank, then the diffusion matrix of the transformed system also has full rank.

The transformations (3.5) provide us with the transformations of the differentials

$$d\bar{t} = D_0(\bar{t})dt + D_\alpha(\bar{t})dW_\alpha = dt + \varepsilon[D_0(\tau)dt + D_\alpha(\tau)dW_\alpha] + O(\varepsilon^2), \qquad (3.8a)$$
$$d\bar{x}_i = D_0(\bar{x}_i)dt + D_\alpha(\bar{x}_i)dW_\alpha = dx_i + \varepsilon[D_0(\xi_i)dt + D_\alpha(\xi_i)dW_\alpha] + O(\varepsilon^2). \qquad (3.8b)$$

We emphasize that the differentials are taken according to stochastic calculus rules.

Transformation of Brownian motions (actually differentials of Brownian motions)

$$d\bar{W}_\alpha = dW_\alpha + \varepsilon X(dW_\alpha) + O(\varepsilon^2) \qquad (3.9)$$

is induced by the random time change and remains to be found.

First we consider invariance of the substitution rules (3.2) for transformations governed by (3.8a) and (3.9). Here we are interested in infinitesimal conditions (terms of the first order in the parameter ε). The first relation (3.2a) holds as an identity. The second substitution rule (3.2b) gives

$$D_\alpha(\tau) = 0. \qquad (3.10)$$

Finally, using Eqs. (3.2c) and (3.10)

$$(dW_\alpha + \varepsilon X(dW_\alpha) + O(\varepsilon^2))(dW_\beta + \varepsilon X(dW_\beta) + O(\varepsilon^2)) = \delta_{\alpha\beta}(dt + \varepsilon D_0(\tau)dt + O(\varepsilon^2))$$

for $\alpha = \beta$, we obtain

$$X(dW_\alpha) = \frac{1}{2}D_0(\tau)dW_\alpha. \qquad (3.11)$$

For $\alpha \neq \beta$ we do not get any additional conditions.

Requiring invariance of the SDEs (3.1), we get equations

$$\begin{aligned} &dx_i + \varepsilon[D_0(\xi_i)dt + D_\alpha(\xi_i)dW_\alpha] + O(\varepsilon^2) \\ &= (f_i + \varepsilon X(f_i) + O(\varepsilon^2))(dt + \varepsilon[D_0(\tau)dt + D_\alpha(\tau)dW_\alpha] + O(\varepsilon^2)) \\ &\qquad + (g_{i\alpha} + \varepsilon X(g_{i\alpha}) + O(\varepsilon^2))(dW_\alpha + \varepsilon X(dW_\alpha) + O(\varepsilon^2)), \end{aligned} \qquad (3.12)$$

which should hold on the solutions of the SDEs. We substitute (3.10) and (3.11) and split the equations (3.12) into infinitesimal conditions for independent differentials dt and dW_α as

$$D_0(\xi_i) = X(f_i) + f_i D_0(\tau), \qquad (3.13a)$$

$$D_\alpha(\xi_i) = X(g_{i\alpha}) + \frac{1}{2}g_{i\alpha}D_0(\tau). \qquad (3.13b)$$

Finally, we present the determining equations as

$$D_0(\xi_i) - X(f_i) - f_i D_0(\tau) = 0, \qquad (3.14a)$$

$$D_\alpha(\xi_i) - X(g_{i\alpha}) - \frac{1}{2}g_{i\alpha}D_0(\tau) = 0, \qquad (3.14b)$$

$$D_\alpha(\tau) = 0. \qquad (3.14c)$$

It is interesting to note that the determining equations are deterministic even though they describe symmetries of the stochastic differential equations. In the general case, when functions $f_i(t, \mathbf{x})$ and $g_{i\alpha}(t, \mathbf{x})$ are arbitrary, the determining equations (3.14) have no non-trivial solutions; i.e. there are no symmetries.

Here we will consider only infinitesimal transformations. Discussion of the corresponding finite transformations can be found in [10]. Recently symmetries (3.6) defined by the determining equations (3.14) were called *deterministic* symmetries to distinguish them from more general *random* symmetries, whose coefficients also depend on the Brownian motion [13, 16].

4 Properties of symmetries of Itô SDEs

In this section there are presented several properties of Lie point symmetries admitted by SDEs.

4.1 Lie algebra of symmetries of Itô SDEs

First we prove that symmetries of Itô SDEs form a Lie algebra [25].

Lemma 1. *Operator X is a symmetry of the SDEs (3.1) if and only if*

$$[D_0, X] = D_0(\tau)D_0 \qquad \text{and} \qquad [D_\alpha, X] = \frac{1}{2}D_0(\tau)D_\alpha. \tag{4.1}$$

The lemma is proved by direct computation. Let us mention that we can rewrite commutator relations (4.1) as

$$D_0X = (X + D_0(\tau))\,D_0, \qquad D_\alpha X = \left(X + \frac{1}{2}D_0(\tau)\right) D_\alpha. \tag{4.2}$$

Theorem 1. *Symmetries of Itô SDEs, given by determining equation (3.14), form a Lie algebra.*

Proof. Using relations (4.2), it is straightforward to check that for two symmetries of the SDEs

$$X_1 = \tau^1(t, \mathbf{x})\frac{\partial}{\partial t} + \xi_i^1(t, \mathbf{x})\frac{\partial}{\partial x_i} \qquad \text{and} \qquad X_2 = \tau^2(t, \mathbf{x})\frac{\partial}{\partial t} + \xi_i^2(t, \mathbf{x})\frac{\partial}{\partial x_i}$$

their commutator

$$X = [X_1, X_2] = \left(X_1(\tau^2) - X_2(\tau^1)\right)\frac{\partial}{\partial t} + \left(X_1(\xi_i^2) - X_2(\xi_i^1)\right)\frac{\partial}{\partial x_i}$$

is also a symmetry.

The other conditions of a Lie algebra [33], namely bi-linearity, skew-symmetry and Jacobi identity of the commutator, are evident. □

4.2 Symmetries and first integrals

Symmetries transform first integrals into another first integrals.

Theorem 2. *If $I(t, \mathbf{x})$ is a first integral and X is a symmetry of SDEs (3.1), then $X(I)$ is also a first integral.*

Proof. The result follows from the relations (4.2). □

The first versions of this property were proven for symmetries with $\tau = 0$ in [1, 31]. Another property concerns relation between first integrals and linearly connected symmetries, i.e. symmetry operators whose coefficients are proportional with a nonconstant coefficient of proportionality [23].

Theorem 3. *A system of SDEs (3.1) with a non-zero diffusion matrix admits two linearly connected symmetries*

$$X_1 = \tau(t, \mathbf{x})\frac{\partial}{\partial t} + \xi_i(t, \mathbf{x})\frac{\partial}{\partial x_i}$$

and

$$X_2 = I(t, \mathbf{x})\tau(t, \mathbf{x})\frac{\partial}{\partial t} + I(t, \mathbf{x})\xi_i(t, \mathbf{x})\frac{\partial}{\partial x_i}$$

if and only if function $I(t, \mathbf{x})$ is a first integral of the system.

The proof is based on the comparison of the determining equations (3.14) for symmetries X_1 and X_2. According to Theorem 3, if a system of SDEs (3.1) has at least one symmetry and one first integral, it admits an infinite-dimensional symmetry group.

Corollary 1. If a system of SDEs (3.1) has no first integrals, then it does not admit linearly connected symmetry operators.

Corollary 1 facilitates construction of Lie group classifications. It allows discarding some realizations of Lie algebras by vector fields, which cannot be admitted as symmetries.

4.3 Dimensions of Lie symmetry groups

For several classes of SDEs there were obtained results concerning maximal dimensionality of the admitted Lie symmetry group. These SDEs must have no first integrals. Otherwise they have no or infinitely many symmetries.

At this point we assume that all functions $f_i(t, \mathbf{x})$ and $g_{i\alpha}(t, \mathbf{x})$, describing the SDEs, as well as coefficients $\tau(t, \mathbf{x})$ and $\xi_i(t, \mathbf{x})$ of the symmetry operators are analytic.

4.3.1 Scalar SDEs

First, let us examine scalar SDEs

$$dx = f(t,x)dt + g(t,x)dW(t), \qquad g(t,x)\neq 0, \tag{4.3}$$

which have no first integrals.

The determining equations (3.14) for the admitted symmetries

$$X = \tau(t,x)\frac{\partial}{\partial t} + \xi(t,x)\frac{\partial}{\partial x} \tag{4.4}$$

take the form

$$\xi_t + f\xi_x + \frac{1}{2}g^2\xi_{xx} - \tau f_t - \xi f_x - f\left(\tau_t + f\tau_x + \frac{1}{2}g^2\tau_{xx}\right) = 0, \tag{4.5a}$$

$$g\xi_x - \tau g_t - \xi g_x - \frac{g}{2}\left(\tau_t + f\tau_x + \frac{1}{2}g^2\tau_{xx}\right) = 0, \tag{4.5b}$$

$$g\tau_x = 0. \tag{4.5c}$$

The last determining equation (4.5c) can be solved as

$$\tau = \tau(t). \tag{4.6}$$

Therefore, the symmetries admitted by Eq. (4.3) are fiber–preserving symmetries

$$X = \tau(t)\frac{\partial}{\partial t} + \xi(t,x)\frac{\partial}{\partial x} \tag{4.7}$$

that substantially simplify further consideration. In particular, we are restricted to equivalence transformations

$$\bar{t} = \bar{t}(t), \qquad \bar{x} = \bar{x}(t,x), \qquad \bar{t}_t \neq 0, \qquad \bar{x}_x \neq 0, \tag{4.8}$$

where change of time is not random.

Example. SDE with constant drift and diffusion coefficients

$$dx = \mu dt + \sigma dW(t), \qquad \sigma \neq 0 \tag{4.9}$$

admits a symmetry group given by the operators

$$X_1 = \frac{\partial}{\partial t}, \qquad X_2 = 2t\frac{\partial}{\partial t} + (x+\mu t)\frac{\partial}{\partial x}, \qquad X_3 = \frac{\partial}{\partial x}.$$

Example. The equation of Brownian motion (a particular case of SDE (4.9))

$$dx = dW(t) \tag{4.10}$$

admits three symmetries

$$X_1 = \frac{\partial}{\partial t}, \qquad X_2 = 2t\frac{\partial}{\partial t} + x\frac{\partial}{\partial x}, \qquad X_3 = \frac{\partial}{\partial x}. \tag{4.11}$$

There exists a bound on the number of symmetries which can be admitted by the scalar SDEs [21].

Theorem 4. *The dimension of the Lie algebra of symmetries admitted by equation (4.3) can be equal to 0, 1, 2 or 3.*

Proof. Let us take into account the result (4.6) and write down a simplified version of the determining equations (4.5) as

$$\xi_t + f\xi_x + \frac{1}{2}g^2\xi_{xx} - \tau f_t - \xi f_x - f\tau_t = 0, \tag{4.12a}$$

$$g\left(\xi_x - \frac{1}{2}\tau_t\right) = \tau g_t + \xi g_x, \tag{4.12b}$$

where $\tau(t)$ and $\xi(t,x)$.

The equation (4.12b) can be resolved as

$$\xi_x - \frac{1}{2}\tau_t = \varphi, \qquad \varphi \in \text{span}(\tau, \xi). \tag{4.13}$$

By $\text{span}(\tau, \xi)$ we mean functions which are linear in τ and ξ with coefficients depending on some functions of t and x.

From (4.13) we obtain

$$\xi_{xx} = \chi, \qquad \chi \in \text{span}(\tau, \tau_t, \xi). \tag{4.14}$$

Substitution of (4.13) and (4.14) into equations (4.12a) provides us with

$$\xi_t = \psi, \qquad \psi \in \text{span}(\tau, \tau_t, \xi). \tag{4.15}$$

Finally, from (4.13) and (4.15) we conclude that all derivatives of τ and ξ are linear homogeneous functions of τ, ξ and τ_t. The total number of unconstrained derivatives is at most 3. Thus, the space of the solutions is at most three-dimensional. A detailed justification of this reasoning can be found is Section 48 of [6].

We refer to Eq. (4.9) to show that there are scalar SDEs which admit three symmetries. □

Similarly (see [21]), one can prove the following result for symmetries

$$Y = \xi(t,x)\frac{\partial}{\partial x}, \tag{4.16}$$

acting only on the dependent variable.

Theorem 5. *The dimension of the Lie algebra of symmetries (4.16) admitted by equation (4.3) can be equal to 0 or 1.*

We will later see from the Lie group classification of the scalar SDEs that the scalar SDEs admitting maximal possible symmetry groups for symmetries (4.7) or (4.16) can be transformed into the Brownian motion equation (4.10).

In the next points we will consider two generalizations of scalar first-order SDEs: systems of first-order SDEs and scalar SDEs of higher order.

4.3.2 Systems of SDEs with diffusion matrices of full rank

Now we consider systems of Itô stochastic differential equations (3.1) with diffusion matrices of full rank, i.e. with diffusion matrices satisfying

$$\operatorname{rank}\{g_{i\alpha}(t,\mathbf{x})\} = n. \tag{4.17}$$

In particular, this implies $m \geq n$. The system of SDEs (3.1),(4.17) has no first integrals. Note that all scalar SDEs (4.3), considered in the previous point, belong to this class.

Under the condition (4.17) the last set of determining equations (3.14c) can be solved as $\tau = \tau(t)$. Therefore, the symmetries admitted by system (3.1),(4.17) are fiber-preserving symmetries

$$X = \tau(t)\frac{\partial}{\partial t} + \xi_i(t,\mathbf{x})\frac{\partial}{\partial x_i} \tag{4.18}$$

that simplify their investigation. Particularly, we are restricted to equivalence transformations

$$\bar{t} = \bar{t}(t), \qquad \bar{\mathbf{x}} = \bar{\mathbf{x}}(t,\mathbf{x}), \qquad \bar{t}_t \neq 0, \qquad \det\left(\frac{\partial \bar{\mathbf{x}}}{\partial \mathbf{x}}\right) \neq 0, \tag{4.19}$$

where change of time is not random.

It should be noted that transformations (4.19) do not change the rank of the diffusion matrix. In particular, if the original system of SDEs has a diffusion matrix of full rank, then the diffusion matrix of the transformed system also has full rank.

The bounds on the maximal dimension of the admitted symmetry groups were established in [23].

Theorem 6. *The maximal dimension of a group of symmetries (3.6) (actually (4.18)) admitted by system of SDEs (3.1),(4.17) is $n+2$.*

Theorem 7. *Let us consider group transformations generated by the operators of the form (3.7). The maximal dimension of such symmetry group admitted by system of SDEs (3.1),(4.17) is n.*

Example. Let us examine the systems of SDEs (we assume full rank for the diffusion matrix)

$$dx_i = C_i dt + C_{i\alpha} dW_\alpha(t), \qquad i = 1,\dots,n, \qquad \alpha = 1,\dots,m \tag{4.20}$$

with constant drift and diffusion coefficients. The symmetry algebra of SDEs (4.20) is given by the operators

$$X_1 = \frac{\partial}{\partial t}, \qquad X_2 = 2t\frac{\partial}{\partial t} + (x_i + C_i t)\frac{\partial}{\partial x_i}, \qquad X_{2+i} = \frac{\partial}{\partial x_i}, \qquad i = 1,\dots,n. \tag{4.21}$$

This symmetry algebra has dimension $n+2$. If we consider symmetries acting only on the dependent variables, then we get only n symmetries. For both types of symmetries (3.6) (actually (4.18)) and (3.7) the SDEs (4.20) achieve maximal possible dimensions for a system of n SDEs with diffusion matrix of full rank.

By change of variables $\bar{x}_i = x_i - C_i t$ we can always remove the drift terms. We obtained the system

$$d\bar{x}_i = C_{i\alpha} dW_\alpha(t), \qquad i = 1, ..., n, \qquad \alpha = 1, ..., m. \tag{4.22}$$

For $m = n$ this system can be split into n separate equation of Brownian motion

$$d\bar{x}_i = dW_i(t), \qquad i = 1, ..., n$$

by an appropriate linear transformation of the dependent variables.

4.3.3 Higher order scalar SDEs

Another generalization of scalar SDEs (4.3) leads to the stochastic differential equations of higher order

$$dx^{(n-1)} = f(t, x, \dot{x}, ..., x^{(n-1)})dt + g(t, x, \dot{x}, ..., x^{(n-1)})dW(t),$$
$$g(t, x, \dot{x}, ..., x^{(n-1)}) \not\equiv 0. \tag{4.23}$$

We assume $n \geq 2$ because first order scalar SDEs were already considered. The equation (4.23) can be rewritten as the system of the first-order SDEs

$$d\begin{pmatrix} x \\ \dot{x} \\ \vdots \\ x^{(n-2)} \\ x^{(n-1)} \end{pmatrix} = \begin{pmatrix} \dot{x} \\ \ddot{x} \\ \vdots \\ x^{(n-1)} \\ f \end{pmatrix} dt + \begin{pmatrix} 0 \\ 0 \\ \vdots \\ 0 \\ g \end{pmatrix} dW(t). \tag{4.24}$$

The SDE (4.23) has no first integrals because first integrals $I = I(t, x, \dot{x}, ..., x^{(n-1)})$ should satisfy the equations

$$\frac{\partial I}{\partial t} + \dot{x}\frac{\partial I}{\partial x} + \ddot{x}\frac{\partial I}{\partial \dot{x}} + ... + f\frac{\partial I}{\partial x^{(n-1)}} = 0, \qquad g\frac{\partial I}{\partial x^{(n-1)}} = 0,$$

which have only constant solutions.

We will be interested in infinitesimal group transformations generating operators of the form

$$X = \tau(t, x)\frac{\partial}{\partial t} + \xi(t, x)\frac{\partial}{\partial x}. \tag{4.25}$$

The determining equations for symmetries (4.25) admitted by the SDE (4.23) were obtained in [39]. It is convenient to present them using the prolonged vector fields given recursively

$$X^{[k]} = X^{[k-1]} + \xi^{[k]}\frac{\partial}{\partial x^{(k)}}, \qquad X^{[0]} = X$$

with coefficients computed according to standard prolongation formulas for derivatives

$$\xi^{[k]} = D(\xi^{[k-1]}) - x^{(k)} D(\tau), \qquad \xi^{[0]} = \xi.$$

Here D is the total differentiation operator

$$D = \frac{\partial}{\partial t} + \dot{x}\frac{\partial}{\partial x} + \ddot{x}\frac{\partial}{\partial \dot{x}} + \ldots$$

The determining equations for symmetries (4.25) admitted by SDE (4.23) can be given as

$$X^{[n]}(x^{(n)} - f)\Big|_{x^{(n)}=f} = \begin{cases} g^2 \tau_x, & n = 2, \\ 0, & n \geq 3, \end{cases} \tag{4.26a}$$

$$X^{[n-1]}(g) + g\left[\left(n + \frac{1}{2}\right) D(\tau) - \tau_t - \xi_x\right] = 0. \tag{4.26b}$$

Example. The SDEs

$$dx^{(n-1)} = \sigma dW(t), \qquad \sigma \neq 0 \tag{4.27}$$

admits symmetries

$$X_1 = \frac{\partial}{\partial t}, \qquad X_2 = t\frac{\partial}{\partial t} + \left(n - \frac{1}{2}\right) x \frac{\partial}{\partial x}, \qquad X_{2+i} = t^{i-1}\frac{\partial}{\partial x}, \quad i = 1, \ldots, n. \tag{4.28}$$

It should be mentioned that SDE (4.27) can be scaled to get $\sigma = 1$.

Second-order SDEs were consideted in [39], SDEs of order $n \geq 3$ in [24]. Combining the results given in these papers, one obtains the following statement.

Theorem 8. *Stochastic differential equation (4.23) of any order $n \geq 2$ admits at most $(n + 2)$-dimensional group of Lie point symmetries. The SDEs of order n admitting $(n+2)$-dimensional symmetry groups are reducible to SDE (4.27), which admits symmetry group (4.28).*

Similarly, one can establish sharp upper bounds on the dimensions of the admitted symmetry groups acting in the space of the dependent variable [24].

Theorem 9. *Let us consider group transformations generated by the operators of the form*

$$Y = \xi(t, x)\frac{\partial}{\partial x}. \tag{4.29}$$

The maximal dimension of such symmetry group admitted by SDEs (4.23) of any order $n \geq 2$ is n.

For symmetries (4.29) there is no uniqueness (up to equivalence) of the symmetry groups of maximal dimension for $n \geq 2$. We refer to the SDE (4.27) as a SDE which has maximal dimensionalities for symmetries (4.25) and (4.29).

5 Symmetry applications

5.1 Lie group classifications of SDEs

A standard application of Lie symmetries is Lie group classifications of equations. Let us consider the scalar SDE (4.3). We recall that the scalar SDE (4.3) admits only fiber-preserving symmetries (4.7) and that the equivalence transformations are given in (4.8).

It is always possible to simplify the SDE (4.3) with the help of the variable change

$$y = \int \frac{dx}{g(t,x)}, \tag{5.1}$$

which brings this SDE into the form

$$dy = h(t,y)dt + dW(t). \tag{5.2}$$

Therefore, Lie group classification of the SDE (4.3) can be given by representative equations of the form (5.2). It should be noted that any diffusion process can be reduced to a process governed by equation (5.2) with the help of a random replacement of time [8].

One can obtain Lie group classification using different approaches. In [21] a direct method was used. First, the SDE was simplified under an assumption that there exists one symmetry admitted by the equation. Then, all particular cases leading to existence of additional symmetries were identified.

Alternatively, the Lie group classification of the scalar SDE is obtained with the help of Lie algebra realizations by vector fields [22]. First, it was proved that the admitted symmetry group can be at most three-dimensional (Theorem 4). Then, there were considered realizations of Lie algebras by fiber-preserving vector fields (4.7). All SDEs which admit such symmetry algebras were found. Here it is useful to exploit Theorem 3, which excludes realizations with linearly connected operators. The Lie group classification of the scalar SDE (4.3) is presented in Table 1.

The structure of the Lie algebras present in the group classification allows stating the following results [21].

Theorem 10. *If SDE (4.3) admits a three-dimensional symmetry group, it can be transformed into the Brownian motion equation (4.10) by change of variables (4.8).*

Theorem 11. *The stochastic differential equation (4.3) admits a three-dimensional symmetry group if and only if it admits a symmetry of the form (4.16). A SDE admits such symmetry if and only if functions $f(t,x)$ and $g(t,x)$ satisfy the condition*

$$\left(\frac{g_t}{g} - g\left(\frac{f}{g}\right)_x + \frac{1}{2} g g_{xx} \right)_x = 0. \tag{5.3}$$

Table 1. Lie group classification of a scalar stochastic differential equation.

Group dimension	Basis operators	Equation
0	No symmetries	$dx = f(t,x)dt + dW(t)$
1	$X_1 = \frac{\partial}{\partial t}$	$dx = f(x)dt + dW(t)$
2	$X_1 = \frac{\partial}{\partial t}, \quad X_2 = 2t\frac{\partial}{\partial t} + x\frac{\partial}{\partial x}$	$dx = \frac{\alpha}{x}dt + dW(t), \quad \alpha \neq 0$
3	$X_1 = \frac{\partial}{\partial t}, \quad X_2 = 2t\frac{\partial}{\partial t} + x\frac{\partial}{\partial x}, \quad X_3 = \frac{\partial}{\partial x}$	$dx = dW(t)$

Proof. Taking into account how symmetries of the representative equations, given in Table 1, are transformed by changes of variables (4.8), we see that equivalence class of the Brownian motion equation can be characterized as SDE admitting symmetry (4.16). Condition (5.3) is obtained from the determining equations. □

It can be useful to have a criterion which does not assume any knowledge concerning the admitted symmetry group. Such criterion for invariance under two- and three-dimensional symmetry groups was established in [21].

It is stated in Theorem 11 that if the stochastic equation (4.3) admits a symmetry (4.16), then it admits a three-dimensional symmetry group. This observation can be used to establish a criterion when the admitted symmetry group is two-dimensional. Such groups consist of two symmetry operators X_1 and X_2 which satisfy the condition $\tau(t) \not\equiv 0$.

Theorem 12. *The stochastic differential equation (5.2) admits a two-dimensional symmetry group if and only if*

$$h(t,y) = \frac{C}{y + H(t)} + F(t)(y + H(t)) - H'(t), \qquad C \neq 0 \tag{5.4}$$

and a three-dimensional symmetry group if and only if

$$h(t,y) = F(t)y + G(t). \tag{5.5}$$

Here $F(t)$, $G(t)$ and $H(t)$ are arbitrary functions and C is a constant.

Let us reformulate the condition (5.5), which is given for the simplified SDE (5.2), for the original scalar SDE. The original SDE (4.3) can be transformed into an equation of the simplified form (5.2) by the change of variable (5.1). We obtain the equation

$$dy = \left(-\int \frac{g_t}{g^2} dx + \frac{f}{g} - \frac{g_x}{2} \right) dt + dW(t).$$

Applying condition (5.5) to the transformed equation, we get

$$-\int \frac{g_t}{g^2} dx + \frac{f}{g} - \frac{g_x}{2} = F(t) \int \frac{dx}{g(t,x)} + G(t).$$

Excluding the arbitrary functions by differentiation with respect to x, we conclude the following result.

Corollary 2. The stochastic differential equation (4.3) admits a three-dimensional symmetry group if and only if its drift and diffusion coefficient satisfy the equation (5.3).

This condition was already obtained in Theorem 11 by other reasoning.

It is possible to provide changes of variables which transform stochastic differential equation (5.2) for drift coefficients (5.4) and (5.5) to the canonical forms of SDEs admitting two- and three-dimensional symmetry groups, presented in Table 1. These transformations are

$$\bar{t}(t) = \int \left[e^{-2\int F(t)dt} \right] dt, \qquad \bar{y}(t,y) = (y + H(t))e^{-\int F(t)dt} \tag{5.6}$$

and

$$\bar{t}(t) = \int \left[e^{-2\int F(t)dt} \right] dt, \qquad \bar{y}(t,y) = ye^{-\int F(t)dt} - \int \left[e^{-\int F(t)dt} \right] G(t)dt, \tag{5.7}$$

respectively.

To conclude this point we mention that Lie group classifications were obtained for the systems of two SDEs with diffusion matrices of full rank [23]; the second-order scalar SDE [39], which is equivalent to the system of two SDEs with a deficient diffusion matrix and no first integrals; the third-order scalar SDE [24]; and the ODE-SDE system [26], which is equivalent to the system of two SDEs with a first integral.

5.2 Integration of scalar SDEs

In the general case we can use only symmetries acting in the space of the dependent variables for integration by quadratures. This can be illustrated already for scalar SDEs.

Example. Let us consider a scalar SDE (4.3), which can admit fiber-preserving symmetries (4.7).

If the equation admits a symmetry with $\tau(t) \equiv 0$, i.e. symmetry

$$Y = \xi(t,x)\frac{\partial}{\partial x}, \tag{5.8}$$

it can be transformed into the form

$$dx = f(t)dt + g(t)dW(t), \tag{5.9}$$

which can be solved as

$$x(t) = x(t_0) + \int_{t_0}^{t} f(s)ds + \int_{t_0}^{t} g(s)dW(s).$$

The transformation of Eq. (4.3) into Eq. (5.9) is performed by the variable change

$$\bar{t} = t, \qquad \bar{x} = \int \frac{dx}{\xi(t,x)},$$

which also provides stratification of the symmetry (5.8), bringing it to the translation of x.

However, if the admitted symmetry is

$$X = \tau(t)\frac{\partial}{\partial t} + \xi(t,x)\frac{\partial}{\partial x}, \qquad \tau(t) \not\equiv 0, \tag{5.10}$$

then the equation can be brought to the form

$$dx = f(x)dt + g(x)dW(t), \tag{5.11}$$

which is not integrable in the general case.

Transformation of SDE (4.3) into SDE (5.11) is performed with the help of variable change

$$\bar{t} = \int \frac{dt}{\tau(t)}, \qquad \bar{x}(t,x) \quad \text{solves} \quad X(\bar{x}(t,x)) = \tau(t)\frac{\partial \bar{x}}{\partial t} + \xi(t,x)\frac{\partial \bar{x}}{\partial x} = 0,$$

which brings the symmetry (5.10) into the translation of t.

The example leads to the following statement.

Theorem 13. *The stochastic differential equation (4.3) is integrable by quadratures if it admits a symmetry (4.16).*

Comparing the discussion concerning integrability by quadratures and the results of the Lie group classification, given in Table 1, we can reformulate the sufficient condition of integrability by quadratures as follows.

Corollary 3. The stochastic differential equation (4.3) is integrable by quadratures if it admits a three-dimensional symmetry group.

Remark 1. A wider family of stochastic differential equations can be integrated if we consider realizations of SDEs by decoupled systems [17, 27]. Since this integration method goes beyond transformations (4.8), it is not considered here. It is worth mentioning that besides quadratures there are other means to present closed–form solutions of stochastic differential equations [29].

In the rest of this point we will consider a practical example.

Example. Interest rate models

Let us consider the generalized one-factor interest rate model

$$dx = (\alpha + \beta x)dt + \sigma x^{\gamma} dW(t), \qquad \sigma \neq 0, \tag{5.12}$$

where the parameters α, β, γ and σ are constants. It is reasonable to restrict ourselves to the case $x(t) > 0$. The description of particular models of this form can be found in [3, 28]. Such models are widely used in financial mathematics. An empirical comparison of eight short–term interest rate models of the form (5.12) was performed in [4].

The model equation (5.12) can be transformed into Eq. (5.2) by the variable change

$$y = \int \frac{dx}{\sigma x^{\gamma}} = \begin{cases} \dfrac{1}{\sigma}\dfrac{x^{1-\gamma}}{1-\gamma}, & \gamma \neq 1, \\ \dfrac{1}{\sigma}\ln x, & \gamma = 1. \end{cases}$$

For these subcases the transformed equation takes the forms

$$dy = \left(\alpha\sigma^{\frac{1}{\gamma-1}}((1-\gamma)y)^{\frac{\gamma}{\gamma-1}} + \beta(1-\gamma)y + \frac{\gamma}{2(\gamma-1)}\frac{1}{y}\right)dt + dW(t), \qquad \gamma \neq 1$$

and

$$dy = \left(\frac{\alpha}{\sigma}e^{-\sigma y} + \left(\frac{\beta}{\sigma} - \frac{\sigma}{2}\right)\right)dt + dW(t), \qquad \gamma = 1.$$

Using Theorem 12, we conclude the dimensionality of the admitted symmetry group without the actual computation of the symmetry operators. The results are the following:

1. Three-dimensional symmetry group

   (a) $\gamma = 0$,

   (b) $\gamma = \frac{1}{2}$ and $\alpha = \frac{\sigma^2}{4}$,

   (c) $\gamma = 1$ and $\alpha = 0$.

2. Two-dimensional symmetry group

   (a) $\gamma \neq \{0, \frac{1}{2}, 1\}$ and $\alpha = 0$,

   (b) $\gamma = \frac{1}{2}$ and $\alpha \neq \frac{\sigma^2}{4}$.

3. One-dimensional symmetry group

   All other cases of coefficients α, β and γ. (It is easy to see that for any values of the coefficients equation (5.12) admits time translations.)

Among the cases which are singled out one can see some popular models of interest rates: Vasicek model [38]: $\gamma = 0$; its particular case – Merton's model [30]: $\gamma = \beta = 0$; Cox–Ingersoll–Ross model [5]: $\gamma = \frac{1}{2}$; and Dothan's model [7]: $\gamma = 1$ and $\alpha = 0$.

Let us consider the equations admitting three-dimensional symmetry groups in greater details. For $\gamma = 0$, the equation (5.12) takes the form

$$dx = (\alpha + \beta x)dt + \sigma dW(t). \tag{5.13}$$

It admits the following symmetry groups depending on the value of parameter β:

1. $\beta \neq 0$

$$X_1 = \frac{\partial}{\partial t}, \qquad X_2 = e^{2\beta t}\frac{\partial}{\partial t} + (\alpha + \beta x)e^{2\beta t}\frac{\partial}{\partial x}, \qquad X_3 = e^{\beta t}\frac{\partial}{\partial x};$$

2. $\beta = 0$

$$X_1 = \frac{\partial}{\partial t}, \qquad X_2 = 2t\frac{\partial}{\partial t} + (x + \alpha t)\frac{\partial}{\partial x}, \qquad X_3 = \frac{\partial}{\partial x}.$$

By the change of variable

$$y = e^{-\beta t}x$$

equation (5.13) is transformed into the equation

$$dy = \alpha e^{-\beta t}dt + \sigma e^{-\beta t}dW(t), \tag{5.14}$$

which can be easily integrated. For the original equation we get the solution

$$x(t) = e^{\beta(t-t_0)}x(t_0) + \frac{\alpha}{\beta}\left(e^{\beta(t-t_0)} - 1\right) + \sigma\int_{t_0}^{t} e^{\beta(t-s)}dW(s), \qquad \beta \neq 0$$

and

$$x(t) = x(t_0) + \alpha(t - t_0) + \sigma(W(t) - W(t_0)), \qquad \beta = 0.$$

For the case $\gamma = \frac{1}{2}$ and $\alpha = \frac{\sigma^2}{4}$ we get the equation

$$dx = \left(\frac{\sigma^2}{4} + \beta x\right)dt + \sigma\sqrt{x}dW(t), \tag{5.15}$$

which admits the following symmetries:

1. $\beta \neq 0$

$$X_1 = \frac{\partial}{\partial t}, \qquad X_2 = e^{\beta t}\frac{\partial}{\partial t} + \beta x e^{\beta t}\frac{\partial}{\partial x}, \qquad X_3 = \sqrt{x}e^{\beta t/2}\frac{\partial}{\partial x};$$

2. $\beta = 0$

$$X_1 = \frac{\partial}{\partial t}, \qquad X_2 = t\frac{\partial}{\partial t} + x\frac{\partial}{\partial x}, \qquad X_3 = \sqrt{x}\frac{\partial}{\partial x}.$$

Change of variable

$$y = e^{-\beta t/2}\sqrt{x}$$

brings equation (5.15) into the form

$$dy = \frac{\sigma}{2}e^{-\beta t/2}dW(t). \tag{5.16}$$

One can easily integrate it and find the solution of the original problem as

$$x(t) = e^{\beta(t-t_0)}\left(\sqrt{x(t_0)} + \frac{\sigma}{2}\int_{t_0}^{t} e^{\beta(t_0-s)/2}dW(s)\right)^2.$$

In the last case with $\gamma = 1$ and $\alpha = 0$ we obtain the equation

$$dx = \beta x dt + \sigma x dW(t), \tag{5.17}$$

which is well–known as the equation of geometric Brownian motion. It admits symmetries

$$X_1 = \frac{\partial}{\partial t}, \qquad X_2 = 2t\frac{\partial}{\partial t} + \left(\ln x + \left(\beta - \frac{\sigma^2}{2}\right)t\right)x\frac{\partial}{\partial x}, \qquad X_3 = x\frac{\partial}{\partial x}.$$

This SDE was solved in Section 2.

It is interesting to note that there are the same cases with two- and three-dimensional symmetry groups if we consider equation

$$dx = (\alpha(t) + \beta(t)x)dt + \sigma(t)x^\gamma dW(t), \qquad \sigma(t) \neq 0 \tag{5.18}$$

instead of (5.12). Here $\alpha \neq \frac{\sigma^2}{4}$ should be understood as $\alpha(t) = \rho\sigma^2(t)$, where $\rho \neq \frac{1}{4}$ is a constant. The cases with a one-dimensional symmetry group cannot be characterized with the help of Theorem 12. Generally, there is one admitted symmetry for other values of functions $\alpha(t)$, $\beta(t)$ and $\sigma(t)$ which satisfy additional conditions. These conditions can be obtained from the determining equations. There are cases when equation (5.18) has no symmetries.

5.3 Integrability by quadratures for systems of SDEs

It is known (see, for example, [33]) that knowledge of an r-parameter solvable group of symmetries allows reducing the order of a system of first order ODEs by r. An analogous result is valid for the system of SDEs. However, in the general case we can use only symmetries acting in the space of the dependent variables as it was illustrated for the scalar SDEs in the previous point.

The classical results concerning systems of first order ODEs take the following form [23].

Theorem 14. *Suppose system (3.1) admits a symmetry of the form (3.7); then there exists a change of variables*

$$\bar{t} = t, \qquad \bar{\mathbf{x}} = \bar{\mathbf{x}}(t, \mathbf{x}), \qquad det\left(\frac{\partial \bar{\mathbf{x}}}{\partial \mathbf{x}}\right) \neq 0 \tag{5.19}$$

which transform the system into the form

$$d\bar{x}_i = \bar{f}_i(\bar{t}, \bar{x}_1, ..., \bar{x}_{n-1})d\bar{t} + \bar{g}_{i\alpha}(\bar{t}, \bar{x}_1, ..., \bar{x}_{n-1})d\bar{W}_\alpha(\bar{t}). \tag{5.20}$$

Thus the system gets reduced to a system of $n-1$ SDEs for $\bar{x}_1$, ..., $\bar{x}_{n-1}$. The solution of the last equation can be given by quadratures

$$\bar{x}_n(\bar{t}) = \bar{x}_n(\bar{t}_0) + \int_{\bar{t}_0}^{\bar{t}} \bar{f}_i(s, \bar{x}_1(s), ..., \bar{x}_{n-1}(s))ds$$

$$+ \int_{\bar{t}_0}^{\bar{t}} \bar{g}_{i\alpha}(s, \bar{x}_1(s), ..., \bar{x}_{n-1}(s))d\bar{W}_\alpha(s). \tag{5.21}$$

Theorem 15. *Suppose system (3.1) admits an r-parameter solvable group of symmetries (3.7), acting regularly with r-dimensional orbits. Then the solution can be obtained by quadratures from the solution of a reduced system of order $n-r$. If the system (3.1) admits an n-parameter solvable group, its general solution can be found by quadratures.*

Let us consider a system of two stochastic differential equations

$$\begin{aligned} dx_1 &= f_1(t, x_1, x_2)dt + g_{1\alpha}(t, x_1, x_2)dW_\alpha(t), \\ dx_2 &= f_2(t, x_1, x_2)dt + g_{2\alpha}(t, x_1, x_2)dW_\alpha(t), \end{aligned} \qquad \alpha = 1, ..., m \tag{5.22}$$

with diffusion matrix of full rank

$$\text{rank}\begin{pmatrix} g_{11} & g_{12} & \cdots & g_{1m} \\ g_{21} & g_{22} & \cdots & g_{2m} \end{pmatrix} = 2 \tag{5.23}$$

in detail. Condition (5.23) requires $m \geq 2$.

The following discussion relies on the Lie group classification of SDEs (5.22),(5.23) given in [23]. In the space of two variables (x_1, x_2) two-dimensional Lie algebras have five non-equivalent realizations by vector fields of the form

$$Y = \xi_1(t, x_1, x_2)\frac{\partial}{\partial x_1} + \xi_2(t, x_1, x_2)\frac{\partial}{\partial x_2}. \tag{5.24}$$

Three realizations, which are given by linearly connected operators, cannot be symmetries of system (5.22),(5.23). Thus, we obtain the system

$$\begin{aligned} dx_1 &= f_1(t)dt + g_{1\alpha}(t)dW_\alpha(t), \\ dx_2 &= f_2(t)dt + g_{2\alpha}(t)dW_\alpha(t), \end{aligned} \tag{5.25}$$

which is invariant with respect to operators

$$Y_1 = \frac{\partial}{\partial x_1}, \qquad Y_2 = \frac{\partial}{\partial x_2}, \tag{5.26}$$

and the system

$$\begin{aligned} dx_1 &= f_1(t)x_2dt + g_{1\alpha}(t)x_2dW_\alpha(t), \\ dx_2 &= f_2(t)x_2dt + g_{2\alpha}(t)x_2dW_\alpha(t), \end{aligned} \tag{5.27}$$

which admits symmetries

$$Y_1 = \frac{\partial}{\partial x_1}, \qquad Y_2 = x_1\frac{\partial}{\partial x_1} + x_2\frac{\partial}{\partial x_2}. \tag{5.28}$$

It is easy to see that these systems are integrable by quadratures. The equations of system (5.25) can be integrated independently. In system (5.27) we can integrate the second equation as

$$x_2(t) = x_2(t_0)\exp\left(\int\limits_{t_0}^{t}\left(f_2(s) - \frac{1}{2}g_{2\alpha}^2(s)\right)ds + \int\limits_{t_0}^{t} g_{2\alpha}(s)dW(s)\right)$$

and use this solution to integrate the first equation.

Example. Let us consider the system

$$\begin{aligned} dx_1 &= (a_1 + b_1x_1 + c_1x_2)dt + C_{11}dW_1(t) + C_{12}dW_2(t), \\ dx_2 &= (a_2 + b_2x_1 + c_2x_2)dt + C_{21}dW_1(t) + C_{22}dW_2(t), \end{aligned} \tag{5.29}$$

satisfying full rank condition (5.23). We will look for symmetries of the form (5.24). Resolving determining equation (3.14b)-(3.14c), we obtain

$$\xi_1 = \xi_1(t), \qquad \xi_2 = \xi_2(t).$$

Substitution into the last set (3.14a) gives

$$\xi_1'(t) = b_1\xi_1(t) + c_1\xi_2(t),$$
$$\xi_2'(t) = b_2\xi_1(t) + c_2\xi_2(t). \tag{5.30}$$

The solution of the system is two-dimensional. By Theorem 15 system (5.30) is integrable by quadratures. The system can be transformed to the form (5.25) because the symmetry group is Abelian. The solution of system (5.30) can always be given as

$$\begin{pmatrix} \xi_1(t) \\ \xi_2(t) \end{pmatrix} = \alpha \begin{pmatrix} A_1(t) \\ B_1(t) \end{pmatrix} + \beta \begin{pmatrix} A_2(t) \\ B_2(t) \end{pmatrix}$$

We will not provide detailed expressions for the functions $A_1(t)$, $B_1(t)$, $A_2(t)$ and $B_2(t)$ because it would require considering three different cases for roots of the characteristic polynomial. The admitted symmetries

$$Y_1 = A_1(t)\frac{\partial}{\partial x_1} + B_1(t)\frac{\partial}{\partial x_2}, \qquad Y_2 = A_2(t)\frac{\partial}{\partial x_1} + B_2(t)\frac{\partial}{\partial x_2}$$

can be transformed to the form (5.26) by change of variables

$$\bar{x}_1 = \frac{B_2(t)x_1 - A_2(t)x_2}{\Delta}, \qquad \bar{x}_2 = \frac{-B_1(t)x_1 + A_1(t)x_2}{\Delta},$$

where

$$\Delta = A_1(t)B_2(t) - A_2(t)B_1(t).$$

The transformation brings system (5.29) to the form

$$d\bar{x}_1 = f_1(t)dt + g_{11}(t)dW_1(t) + g_{12}(t)dW_2(t),$$
$$d\bar{x}_2 = f_2(t)dt + g_{21}(t)dW_1(t) + g_{22}(t)dW_2(t)$$

with

$$f_1(t) = \frac{B_2(t)a_1 - A_2(t)a_2}{\Delta}, \qquad f_2(t) = \frac{-B_1(t)a_1 + A_1(t)a_2}{\Delta},$$
$$g_{11}(t) = \frac{B_2(t)C_{11} - A_2(t)C_{21}}{\Delta}, \qquad g_{21}(t) = \frac{-B_1(t)C_{11} + A_1(t)C_{21}}{\Delta},$$
$$g_{12}(t) = \frac{B_2(t)C_{12} - A_2(t)C_{22}}{\Delta}, \qquad g_{22}(t) = \frac{-B_1(t)C_{12} + A_1(t)C_{22}}{\Delta}.$$

The subsequent integration is straightforward.

Although there are more powerful methods for SDEs integration based on decoupling, symmetry methods suggest changes of variables which lead to simplification of systems of SDEs. In many cases it can be sufficient for integration.

The integration of SDEs with the help of symmetries acting in the space of the dependent variables was extended to more general, namely random, symmetries in [14, 15].

5.4 Symmetries of SDE versus symmetries of the associated Fokker–Planck equation

The relation of the symmetries of SDEs to the symmetries of the associated Fokker–Planck equations was studied in [12, 25, 26, 37]. Here we will follow [25]. In this point we use notations

$$F_{,t} = \frac{\partial F}{\partial t}, \qquad F_{,u} = \frac{\partial F}{\partial u} \qquad \text{and} \qquad F_{,i} = \frac{\partial F}{\partial x_i}$$

(the last notation uses indexes different from t and u).

We investigate the relation between the Lie point symmetries of SDEs and Lie point symmetries admitted by the associated Fokker–Planck equation [35]

$$u_t + A_{ij}u_{x_ix_j} + B_k u_{x_k} + Cu = 0, \tag{5.31}$$

where

$$A_{ij} = -\frac{1}{2}g_{i\alpha}g_{j\alpha}, \qquad B_k = f_k + 2A_{kl,l}, \qquad C = f_{i,i} + A_{kl,kl}.$$

This PDE describes the evolution of the probability measure. In what follows we will assume that A_{ij} are not all zero.

Let us find Lie point symmetries

$$X = \tau(t,\mathbf{x},u)\frac{\partial}{\partial t} + \xi_i(t,\mathbf{x},u)\frac{\partial}{\partial x_i} + \eta(t,\mathbf{x},u)\frac{\partial}{\partial u} \tag{5.32}$$

which are admitted by the FP equation.

The infinitesimal invariance criterion [19, 33, 34] states that the application of the second prolongation of the operator X to the second-order PDE (5.31) should be zero on the solutions of this PDE:

$$\mathbf{pr}^{(2)}X(u_t + A_{ij}u_{x_ix_j} + B_k u_{x_k} + Cu)\Big|_{(5.31)} = 0. \tag{5.33}$$

Let us briefly describe the derivation of the determining equations for symmetries of the FP equation. Eq. (5.33) splits for different spatial derivatives of u. We obtain

$$\tau_{,u} = 0$$

for products of third derivatives with first derivatives and

$$A_{ij}\tau_{,i} = 0 \tag{5.34}$$

for third derivatives. Further, we get

$$\xi_{i,u} = 0$$

for products of second derivatives with first derivatives and

$$\Delta_{ij} = \tau A_{ij,t} + \xi_k A_{ij,k} + A_{ij}\left(\tau_{,t} + B_p\tau_{,p} + A_{pq}\tau_{,pq}\right) - A_{ik}\xi_{j,k} - A_{kj}\xi_{i,k} = 0 \tag{5.35}$$

for second derivatives.

Products of first derivatives provides us with

$$\eta_{,uu} = 0.$$

Substituting

$$\eta = \varphi(t, \mathbf{x})u + \psi(t, \mathbf{x})$$

into the rest of Eq. (5.33), we split into

$$\Delta_i = \xi_{i,t} + B_p\xi_{i,p} + A_{pq}\xi_{i,pq} - \tau B_{i,t} - \xi_p B_{i,p} - 2A_{ij}\varphi_{,j} \\ - B_i\left(\tau_{,t} + B_p\tau_{,p} + A_{pq}\tau_{,pq}\right) = 0, \quad (5.36)$$

$$\Delta = \varphi_{,t} + B_k\varphi_{,k} + A_{ij}\varphi_{,ij} + \tau C_{,t} + \xi_k C_{,k} + C\left(\tau_{,t} + B_p\tau_{,p} + A_{pq}\tau_{,pq}\right) = 0 \quad (5.37)$$

and

$$\psi_t + A_{ij}\psi_{,ij} + B_k\psi_{,k} + C\psi = 0.$$

Using the differential consequences of (5.34), we can simplify Eqs. (5.36) and (5.37) as

$$\tilde{\Delta}_i = (\Delta_i + 2\Delta_{ij,j})|_{A_{ij}\tau_{,i}=0} = -2A_{ij}Q_{,j} + (D_0(\xi_i) - X(f_i) - f_i D_0(\tau)) = 0 \quad (5.38)$$

and

$$\tilde{\Delta} = (\Delta + \Delta_{i,i} + \Delta_{ij,ij})|_{A_{ij}\tau_{,i}=0} = D_0(Q) = 0, \quad (5.39)$$

where

$$Q = \varphi + \xi_{i,i} + \tau_{,t} - D_0(\tau). \quad (5.40)$$

Let us stress that in contrast to [37] we simplified the determining equation for symmetries of the FP equation without use of the determining equations for symmetries of the SDEs. We summarize the obtained results.

Theorem 16. *Lie point symmetries of FP equation (5.31) are given by*

1. *vector fields of the form*

$$X = \tau(t, \mathbf{x})\frac{\partial}{\partial t} + \xi_i(t, \mathbf{x})\frac{\partial}{\partial x_i} + \varphi(t, \mathbf{x})u\frac{\partial}{\partial u} \quad (5.41)$$

with coefficients satisfying Eqs. (5.34),(5.35),(5.40), where function $Q(t, \mathbf{x})$ is a solution of Eqs. (5.38),(5.39),

and

2. *trivial symmetries*

$$X_* = \psi(t, \mathbf{x}) \frac{\partial}{\partial u}, \tag{5.42}$$

where the coefficient is an arbitrary solution of the FP equation, corresponding to the linear superposition principle.

Remark 2. Let us note that the determining equations always have a solution

$$\tau = 0, \qquad \xi_i = 0, \qquad Q = \text{const},$$

which provides us with symmetry

$$X_0 = u \frac{\partial}{\partial u}, \tag{5.43}$$

corresponding to linearity of the FP equation.

Now we are in a position to relate symmetries of the SDEs and symmetries of the associated FP equation. Let us note that with the help of (5.34) we can rewrite Eq. (5.35) as

$$\begin{aligned}\tilde{\Delta}_{ij} = \Delta_{ij}|_{A_{ij}\tau_{,i}=0} = \frac{1}{2} g_{i\alpha} \left(D_\alpha(\xi_j) - X(g_{j\alpha}) - \frac{1}{2} g_{j\alpha} D_0(\tau) \right) \\ + \frac{1}{2} g_{j\alpha} \left(D_\alpha(\xi_i) - X(g_{i\alpha}) - \frac{1}{2} g_{i\alpha} D_0(\tau) \right) = 0. \end{aligned} \tag{5.44}$$

Theorem 17. *Let operator X of the form (3.6) be a symmetry of the SDEs (3.1); then the associated FP equation admits symmetry*

$$\bar{X} = X + (D_0(\tau) - \tau_{,t} - \xi_{i,i}) u \frac{\partial}{\partial u} \tag{5.45}$$

Proof. From the determining equations (3.14a)-(3.14c) it follows that Eqs. (5.34) and Eqs. (5.44), which are equivalent to Eqs. (5.35) if Eqs. (5.34) holds, are satisfied. Choosing $Q \equiv 0$, which is always a solution of Eqs. (5.38) (if Eqs. (3.14a) hold) and (5.39), we get $\bar{X}$ as a symmetry of the FP equation. □

It is possible to relate some symmetries of the FP equation to first integrals of the SDEs.

Theorem 18. *Let SDEs (3.1) possess a first integral $I(t, \mathbf{x})$; then the associated FP equation admits symmetry*

$$Y = I(t, \mathbf{x}) u \frac{\partial}{\partial u}. \tag{5.46}$$

Proof. It follows from Eqs. (3.4) that the determining equations for symmetries of the FP equation are satisfied. □

Let us state the opposite results.

Theorem 19. *If FP equation (5.31), which corresponds to SDEs (3.1), admits a symmetry $\bar{X}$ of the form (5.45) with coefficients satisfying equations (3.14b) and (3.14c), then the truncated symmetry X is admitted by the SDEs.*

Theorem 20. *If FP equation (5.31), which corresponds to SDEs (3.1), admits a symmetry of the form (5.46) and function $I(t,\mathbf{x})$ satisfies the equations (3.4b), then $I(t,\mathbf{x})$ is a first integral of the SDEs.*

The additional conditions of Theorems 19 and 20, specifying the particular SDEs, are not surprising: the same FP equation can correspond to different SDEs, which have the same drift coefficients f_i and $A_{ij} = \frac{1}{2} g_{i\alpha} g_{j\alpha}$.

Finally, we summarize the results of this section by presenting four types of Lie point symmetries of Fokker–Planck equations. They are

1. symmetries (5.42) and (5.43) corresponding to linearity of FP equation
2. symmetries (5.45) which are related to symmetries of SDEs
3. symmetries (5.46) which are related to first integrals of SDEs
4. the other symmetries, which are not related to SDEs

This relation suggests the following approach to use symmetries of SDEs to classify FP equations. Starting from Lie group classification of SDEs, we identify corresponding FP equations. Then we can find some symmetries of these FP equations using symmetries and first integrals of SDEs. Generally, there can be symmetries which are not related to SDEs. They cannot be found by the presented approach, but can be obtained by direct computation. We stress that we cannot obtain Lie group classifications of FP equations by using these steps.

This procedure was applied to the scalar SDE and the systems of two SDEs which correspond to $(1+1)$- and $(1+2)$-dimensional FP equations, respectively. We refer to [26] for the obtained results and their discussion.

References

[1] Albeverio S, Fei S M, A remark on symmetry of stochastic dynamical systems and their conserved quantities, *J. Phys. A* **28** (1995), 6363–6371.

[2] Arnold L, *Stochastic Differential Equations: Theory and Applications*, Wiley–Interscience New York–London–Sydney, 1974.

[3] Cairns A J G, *Interest Rate Models: An Introduction*, Princeton University Press Princeton, N.J., 2004.

[4] Chan K C, Karolyi G A, Longstaff F A, Sanders A B, An empirical comparison of alternative models of the short-term interest rate, *Journal of Finance* **47** (1992), 1209–1227.

[5] Cox J C, Ingersoll J E, Ross S A, A theory of the term structure of interest rates, *Econometrica* **53** (1985), 385–407.

[6] Dickson L E, Differential equations from the group standpoint, *Ann. Math.* **25** (1924), 287–378.

[7] Dothan L, On the term structure of interest rates, *Journal of Financial Economics* **6** (1978), 59–69.

[8] Dynkin E B, *Markov Processes* Vols. I, II; Academic Press Inc. Publishers New York, Springer–Verlag Berlin–Göttingen–Heidelberg, 1965.

[9] Evans L C, *An Introduction to Stochastic Differential Equations*, A.M.S., 2013.

[10] Fredericks E, Mahomed F M, Symmetries of first-order stochastic ordinary differential equations revisited, *Math. Methods Appl. Sci.* **30**, (2007), 2013–2025.

[11] Fredericks E, Mahomed F M, A formal approach for handling Lie point symmetries of scalar first-order Itô stochastic ordinary differential equations, *J. Nonlinear Math. Phys.* **15**, suppl. 1 (2008), 44–59.

[12] Gaeta G, Quintero N R, Lie–point symmetries and stochastic differential equations, *J. Phys. A* **32** (1999), 8485–8505.

[13] Gaeta G, Spadaro F, Random Lie-point symmetries of stochastic differential equations, *J. Math. Phys.* **58** (2017), 053503.

[14] Gaeta G, Lunini C, On Lie-point symmetries for Itô stochastic differential equations, *J. Nonlinear Math. Phys.* **24-S1** (2017), 90–102.

[15] Gaeta G, Lunini C, Symmetry and integrability for stochastic differential equations, *J. Nonlinear Math. Phys.* **25** (2018), 262–289.

[16] Gaeta G, Symmetry of stochastic non-variational differential equations, *Physics Reports* **686** 2017, 1–62.

[17] Gard T C, *Introduction to Stochastic Differential Equations*, Marcel Dekker, Inc. New York, 1988.

[18] Grigoriev Y N, Ibragimov N H, Meleshko S V, Kovalev V F, *Symmetries of Integro-Differential Equations with Applications in Mechanics and Plasma Physics*, Springer, 2010.

[19] Ibragimov N H, *Transformation Groups Applied to Mathematical Physics*, D. Reidel Publishing Co. Dordrecht, 1985.

[20] Ikeda N, Watanabe S, *Stochastic Differential Equations and Diffusion Processes*, North Holland, 1981.

[21] Kozlov R, Group classification of a scalar stochastic differential equation, *J. Phys. A: Math. Theor.* **43** (2010), 055202.

[22] Kozlov R, On Lie group classification of a scalar stochastic differential equation, *J. Nonlinear Math. Phys.* **18** suppl. 1 (2011), 177–187.

[23] Kozlov R, Symmetries of systems of stochastic differential equations with diffusion matrices of full rank, *J. Phys. A* **43** (2010), 245201.

[24] Kozlov R, On maximal Lie point symmetry groups admitted by scalar stochastic differential equations, *J. Phys. A* **44** (2011), 205202.

[25] Kozlov R, On symmetries of stochastic differential equations *Commun. Nonlinear Sci. Numer. Simul.* **17** (2012), 4947–4951.

[26] Kozlov R, On symmetries of the Fokker–Planck equation, *J. Engrg. Math.* **82** (2013), 39–57.

[27] Krener A J, Lobry C, The complexity of stochastic differential equations, *Stochastics* **4** (1980/81), 193–203.

[28] Kwok Y K, *Mathematical Models of Financial Derivatives*, Springer Singapore, 1998.

[29] Lanconelli A, Proske F, On explicit strong solution of Itô–SDE's and the Donsker delta function of a diffusion, *Infin. Dimens. Anal. Quantum Probab. Relat. Top.* **7** (2004), 437–447.

[30] Merton R C, Theory of rational option pricing, *Bell Journal of Economics* **4** (1973), 141–183.

[31] Misawa T, Conserved quantities and symmetry for stochastic dynamical systems, *Phys. Lett. A* **195** (1994), 185–189.

[32] Misawa T, New conserved quantities derived from symmetry for stochastic dynamical systems, *J. Phys. A* **27** (1994), L777–L782.

[33] Olver P J, *Applications of Lie Groups to Differential Equations*, Springer–Verlag New York, 1993.

[34] Ovsiannikov L V, *Group Analysis of Differential Equations*, Academic Press, Inc. New York–London, 1982.

[35] Risken H, *The Fokker-Planck Equation: Methods of Solution and Applications*, Springer-Verlag Berlin Heidelberg, 1996.

[36] Srihirun B, Meleshko S, Schulz E, On the definition of an admitted Lie group for stochastic differential equations, *Commun. Nonlinear Sci. Numer. Simul.* **12** (2007), 1379–1389.

[37] Ünal G, Symmetries of Itô and Stratonovich dynamical systems and their conserved quantities, *Nonlinear Dynam.* **32** (2003), 417–426.

[38] Vasicek O, An equilibrium characterization of the term structure, *J. Financ. Econ.* **5** (1977), 177–188.

[39] Wafo Soh C, Mahomed F M, Integration of stochastic ordinary differential equations from a symmetry standpoint, *J. Phys. A* **34** (2001), 177–194.

[40] Øksendal B, *Stochastic Differential Equations*, Springer–Verlag Berlin, 2003.

C5. Statistical symmetries of turbulence

Martin Oberlack [a], *Marta Wacławczyk* [b] *and Vladimir Grebenev* [c]

[a] *Chair of Fluid Dynamics, Department of Mechanical Engineering, Otto-Berndt-Strasse 2, 64287 Darmstadt, Germany*

[b] *Institute of Geophysics, Faculty of Physics, University of Warsaw, Pasteura 7, 02-093 Warsaw, Poland*

[c] *Institute of Computational Technologies, Russian Academy of Sciences, Lavrentjev ave. 6, 630090 Novosibirsk, Russia*

Abstract

In the last decade it has been shown by the present authors and co-workers, that all of the complete statistical approaches of turbulence based on Navier-Stokes equations i.e. the infinite set of multi-point moment equations, the infinite hierarchy of multi-point probability-density function equations and the Hopf functional equation admit more symmetries compared to the original Navier-Stokes equations. Hence, these equations admit symmetries which go beyond the classical Galilean group of Navier-Stokes equations, and were named statistical symmetries. For the generic three-dimensional case these symmetries mirror important properties such as intermittency and non-gaussianity. The above findings are important consequences for our understanding of the statistics of turbulence such as intermittency and non-gaussianity.

1 Foreword

Many high-dimensional physical processes are subject to strong stochastic fluctuations and some of these processes are defined precisely by this behavior. These include first and foremost the motion of molecules in gases and liquids on microscopic length scales, where the statistics of the system are significantly responsible for many mechanical or thermodynamic effects such as viscosity or thermal conductivity. On macroscopic length scales it is mainly turbulence, the seemingly stochastic motion of fluids at large Reynolds numbers, whose high-dimensional behaviour can be traced back to the properties of the Navier-Stokes equations, and which are simultaneously responsible for many effects of turbulent flows. These are e.g. the mostly undesired strong increase of the frictional drag on overflowed surfaces in comparison to laminar, i.e. smooth undisturbed flow, to turbulent flow or the ability of turbulence to generate a strong mixing, which is central e.g. for the function of IC engines.

It was O. Reynolds' idea to describe turbulence for the first time as a statistical process. Here he has divided velocity and pressure into an average and a fluctuation part. Since the Navier-Stokes equations form a non-linear and, through the Poisson equation of pressure, a non-local system of partial differential equations (PDE), an infinite sequence of equations results from a statistical description. If the series is broken off at one point, an unclosed system is created. This is one of the central difficulties in the treatment of turbulence and Lie symmetries offer here the possibility to obtain interesting properties and solutions for the infinite or truncated system.

To date, essentially three all-encompassing descriptions of turbulence have been developed: the infinite hierarchy of the multi-point correlation (MPC) equations approach (Friedmann-Keller hierarchy) [10], the infinite hierarchy of the multi-point probability density function (PDF) equations (Lundgren-Monin-Novikov - LMN - equations) [14, 18, 19] and, finally, the Hopf functional approach [8]. All three approaches will be addressed in the following chapters, but the focus will be on the MPC equations.

The main body of results presented are based on the habilitation thesis of the first author [20], several dissertation theses [15], [26], as well as a variety of journal publications: [21], [25], [16], [2], [31], [32], [23] and [30].

This chapter is structured as follows: in the next section the basic concepts for statistical description are laid out; i.e. PDF, moments and characteristic function are defined, each using the multi-point concept, whereby the Burgers equation is to serve as an example. Furthermore, the concept of statistical symmetries is introduced for the first time. In Section 3 the above concepts are then applied to the Navier-Stokes equations and their significance for turbulence is discussed. In particular, the statistical symmetries of intermittency and the non-Gaussian PDF are derived. The concluding section will look at the current challenges and discuss future open problems.

2 Stochastic behavior and symmetries of differential equations - an introduction

The stochastic and chaotic behaviour of dynamic processes as they occur in natural, technical or economic processes has initiated in recent decades to the development of mathematical descriptions for these phenomena. For spatio-temporal processes classical ordinary and partial differential equations were very often used as a base model, which, extended accordingly, then describe the stochastic processes. There are three global process variants: (i) The stochastic process is modeled in the differential equations (DE) by introducing stochastic elements, e.g. by adding additional forcing terms. The resulting DE are named stochastic DE and are abbreviated accordingly, SDE; (ii) the stochastic properties of the system are introduced into the system by the initial or boundary conditions and, finally, (iii) the system itself has the property that stochastic behavior is induced. This behavior is archetypal for turbulence and will be the focus of this work.

In the case (i) SDE are currently an intensively researched scientific field in which question on symmetries of SDE and the related Fokker-Planck (FP) equation are intensively explored e.g. in [28, 12, 13] or [6]. As an example, in [12, 13] four types of symmetry transformations of FP equations are identified, namely (a) symmetries corresponding to the linearity of the FP equation, (b) symmetries related to symmetries emerging from the orginal SDEs, (c) symmetries related to first integrals of SDEs, i.e. such functions which remain constant on the solutions of SDE and (d) other symmetries, which have no direct counterpart in the original SDEs. The presence of symmetries of the latter type may follow from the fact that the same FP equation can correspond to different SDEs.

SDEs can be viewed as a generalization of the dynamical systems theory to models with noise. This is an important generalization because real systems cannot be completely isolated from their environments and for this reason always experience external stochastic influence. For various examples of symmetries admitted by SDE, see e.g. [7] for a review.

Similarly, in case (ii) the stochastic properties of the system are initiated, where the stochastic properties are not part of the dynamic system but are implied e.g. by initial conditions. A typical example is the statistics of the Burgers equation, which e.g. is considered in [22, 31]. The rationale therein is that the statistical behavior is induced by the stochastic variation of the initial conditions.

The focus of the present work will be limited to cases (ii) and (iii) where no stochastic element will be present as part of the differential equation. To this end, the next section focuses on the Burgers equation as an educational example in terms of its statistical properties. The Burgers equation itself has no stochastic term and does not induce stochastic behavior, even at the limit of vanishing viscosity ν. Rather, shock-like structures are created in this limiting case. Looking at the statistics of the Burgers equation, it is rather implied that the initial conditions are stochastically varied. If statistics of the processes are subsequently formed, statistical equations are to be treated analogously, e.g. as the statistics of the Navier-Stokes equation. In particular, the focus of the two central concepts such as moments/correlations and multi-point statistics is treated analogously.

2.1 Statistics and statistical description of the Burgers equation

In the present sub-section we will employ the Burgers equation

$$M(U(x,t)) = \frac{\partial U}{\partial t} + U\frac{\partial U}{\partial x} - \nu\frac{\partial^2 U}{\partial x^2} = 0 \tag{2.1}$$

as a simple pedagogical example of the theory and methods that we will discuss and used in detail in Section 3 on the statistics of Navier-Stokes equations.

In order to describe stochastic processes in a first elementary step, a statistically averaged quantity is to be defined by an ensemble operator

$$\overline{Z}(\boldsymbol{x},t) = \mathcal{K}\left[Z(\boldsymbol{x},t)\right] = \lim_{N\to\infty}\left(\frac{1}{N}\sum_{n=1}^{N} Z_n(\boldsymbol{x},t)\right) , \tag{2.2}$$

where each Z_n is a single realization of the process, while a more precise definition based on Probability Density Function (PDF) will be given subsequently beginning with equation (2.8). If the operator (2.2) is applied the Burgers equation (2.1) we obtain the following unclosed equation

$$\bar{M}(u(x,t)) = \frac{\partial \overline{U}}{\partial t} + \frac{1}{2}\frac{\partial \overline{U^2}}{\partial x} - \nu \frac{\partial^2 \overline{U}}{\partial x^2} = 0 \; . \tag{2.3}$$

The closure problem due to $\overline{U^2}$ apparently appeared due to the non-linear term in equation (2.1), which cannot be recast in terms of $\overline{U}$.

A way to rectify this problem has already been suggested by O. Reynolds namely multiplying the basic equation, here equation (2.1), by U^n and averaging thereafter. For $n = 1$ this results in the equation

$$\overline{U(x,t)M(U(x,t))} = \frac{1}{2}\frac{\partial \overline{U^2}}{\partial t} + \frac{1}{3}\frac{\partial \overline{U^3}}{\partial x} - \nu \left(\frac{\partial^2 \overline{U^2}}{\partial x^2} - \overline{\frac{\partial U}{\partial x}\frac{\partial U}{\partial x}} \right) = 0. \tag{2.4}$$

Apparently this procedure seems to work out for the unsteady and the non-linear terms of the Burgers equation (2.1) and leads to an infinite series of PDEs for the moments $\overline{U^n}$, where the equation for $\overline{U^n}$ contains a term of the form $\overline{U^{n+1}}$. This, in brief, is the *moment or correlation approach* also very common in turbulence, which due to the presence of non-linear terms apparently leads to an infinite series of PDEs.

In fact, however, the third term in the Burgers equation (2.1) does not fit into this scheme. Already in the equation for the second moment (2.4) it becomes visible that a new unclosed term of the form $\overline{\frac{\partial U}{\partial x}\frac{\partial U}{\partial x}}$ is created, which is typically interpreted as dissipation. This additional closure problem is primarily caused by differential operators of order two or more, which indicate to the elliptic or parabolic character of a PDE, and also occurs in analog form for the Navier-Stokes equation.

To eliminate this difficulty we will introduce the concept of *multi-point correlation*. The original Burgers equation (2.4) is multiplied by U, but U is not taken at the location x, but rather at a new variable location y. For this purpose we define the two-point correlation functions

$$\begin{aligned} H_2(x,y,t) &= \overline{U(x,t)U(y,t)}, \\ H_{3a}(x,y,t) &= \overline{U(x,t)U(x,t)U(y,t)}, \text{ and} \\ H_{3b}(x,y,t) &= \overline{U(x,t)U(y,t)U(y,t)} \; . \end{aligned} \tag{2.5}$$

Based on this, and in analogy to equation (2.4), the equation for the two-point correlation H is generated by linking U and M at different locations x and y

$$\begin{aligned} &\overline{U(x,t)M(U(y,t)) + M(U(x,t))U(y,t)} = \\ &\qquad \frac{\partial H_2}{\partial t} + \frac{1}{2}\frac{\partial H_{3a}}{\partial x} + \frac{1}{2}\frac{\partial H_{3b}}{\partial y} - \nu \left(\frac{\partial^2 H_2}{\partial x^2} + \frac{\partial^2 H_2}{\partial y^2} \right) = 0 \; . \end{aligned} \tag{2.6}$$

It is apparent, that the knowledge of the two-point correlation $H_2(x,y,t)$ may immediately reduced to the second moment $\overline{U^2}$ by taking the limit $x = y$, i.e.

$$\overline{U^2}(x,t) = H_2(x,x,t) \; , \tag{2.7}$$

which therefore is also called a one-point moment.

This chain of two-point correlation can easily be extended to multi-point correlations. For this, the three-point triple correlation $H_3 = \overline{U(x,t)U(y,t)U(z,t)}$ must be defined in the next step, from which the two-point triple correlations defined in (2.5) can apparently be derived directly. In this way, an infinite chain of multi-point correlation equations can be generated. With this description form, derivatives larger than one can also be treated without generating further unclosed terms. This, however, is obvious at the expense that each additional moment adds another spatial dimension.

Up to this point only the multi-point moments description has been introduced, but without defining the moments directly via the PDF. In order to do so, let us define multi-point PDFs in the usual way: the 1-point PDF $f_1(x, v; t)$ is defined such that $f_1(x, v; t)\,\mathrm{d}v$ expresses the probability of measuring a velocity v in an infinitesimal interval $\mathrm{d}v$ around v at position x. The 1-point PDF can be written as:

$$f_1(x; v, t) = \overline{\delta(U(x,t) - v)} \ . \tag{2.8}$$

where $\overline{\delta(U(x,t) - v)}$ is the Dirac delta function. Analogously to (2.8), the 2-point PDF, which denotes the joint probability of measuring two given velocities v and w at two defined points x and y in space at the same time t, can be expressed as:

$$f_2(v, w; x, y, t) = \overline{\delta(U(x,t) - v)\delta(U(y,t) - w)} \ . \tag{2.9}$$

This can be continued so forth to the definition of the n point PDF. Based on this, and following the procedure given in [14], we may employ the Burgers equation (2.1) to derive the one-point PDF equation for the Burgers equation

$$\frac{\partial f_1}{\partial t} + v\frac{\partial f_1}{\partial x} = -\frac{\partial}{\partial v}\left[\lim_{|y-x|\to 0} \nu \frac{\partial^2}{\partial y \partial y} \int w f_2 \,\mathrm{d}w\right] \tag{2.10}$$

which apparently contains the 2-point PDF (2.9). Employing (2.9) in a similar fashion we may derive the equation for the 2-point PDF

$$\begin{aligned} &\frac{\partial f_2}{\partial t} + v\frac{\partial f_2}{\partial x} + w\frac{\partial f_2}{\partial y} = \\ &-\frac{\partial}{\partial v}\left[\lim_{|z-x|\to 0} \nu \frac{\partial^2}{\partial z \partial z} \int w' f_3 \,\mathrm{d}w'\right] - \frac{\partial}{\partial w}\left[\lim_{|z-y|\to 0} \nu \frac{\partial^2}{\partial z \partial z} \int w' f_3 \,\mathrm{d}w'\right] \end{aligned} \tag{2.11}$$

which contains the 3-point PDF $f_3 = f_3(v, w, w'; x, y, z, t)$.

Further, all multi-point PDF need to obey normalization constraints

$$\int f_1 \,\mathrm{d}v = 1 \ , \ \int f_2 \,\mathrm{d}w = f_1, \tag{2.12}$$

which essentially corresponds to the fact that the sum of the probabilities of all events adds up to one.

As indicated above, the concept of statistical moments can now be defined more precisely. Assuming the PDFs (2.8) or (2.9) are known, then every possible moment can be defined, e.g. from the one-point PDF we find

$$\overline{U^m(x,t)} = \int v^m f_1(x;v,t)\,\mathrm{d}v \ , \tag{2.13}$$

while the related two-point moments are defined as

$$\overline{U^m(x,t)U^n(y,t)} = \int v^m w^n f_2(v,w;x,y,t)\,\mathrm{d}v\,\mathrm{d}w \ . \tag{2.14}$$

A final probabilistic concepts that connects both the one- and multi-point moments (2.5) and the related PDFs in (2.8) and (2.9) is the characteristic function Φ. It is essentially the Fourier transform of the PDF, which in one- and two-point form reads

$$\begin{aligned}\Phi_1(\hat{v}) &= \int \mathrm{e}^{\mathrm{i}v\hat{v}} f_1(v;x,t)\,\mathrm{d}v \ \text{ and}\\ \Phi_2(\hat{v},\hat{w}) &= \int \mathrm{e}^{\mathrm{i}(v\hat{v}+w\hat{w})} f_2(v,w;x,y,t)\,\mathrm{d}v\,\mathrm{d}w \ .\end{aligned} \tag{2.15}$$

The connection of the moments, PDF and characteristic function becomes apparent by taking the first derivative of Φ_1 with respect to $\hat{v}$ to get

$$\frac{\partial\Phi_1(\hat{v},t)}{\partial\hat{v}} = \int iv\mathrm{e}^{\mathrm{i}v\hat{v}} f_1(v;x,t)\,\mathrm{d}v \ . \tag{2.16}$$

Employing $\hat{v}=0$ and using (2.13) we obtain

$$\left.\frac{\partial\Phi_1(\hat{v},t)}{\partial\hat{v}}\right|_{\hat{v}=0} = i\overline{U}(x,t) \ . \tag{2.17}$$

This may be readily extended so that together with (2.14) we obtain

$$\left.\frac{\partial^{m+n}\Phi_2(\hat{v},\hat{w},t)}{\partial\hat{v}^m\partial\hat{w}^n}\right|_{\hat{v}=\hat{w}=0} = i^{m+n}\overline{U^m(x,t)U^n(y,t)} \ . \tag{2.18}$$

Since we do not need the multi-point equation for characteristic function for the Burgers equation, we will omit it for the sake of brevity.

2.2 Symmetries and statistical symmetries of the Burgers equation

In order to better understand the concepts of statistical symmetries for the Navier-Stokes equations, we will first take up all classical symmetries of the Burgers equation again in this section. In the second step we show that all these symmetries of the Burger equation remain in the context of the statistical description, i.e. for the moment description as well as for the PDF and the characteristic function.

Below we have listed all classical point-symmetries of the Burgers equation (2.1). It should be noted here that the Cole-Hopf transformation, which transforms the Burgers equation into the heat conduction equation, is not listed because it only becomes visible if the Burgers equation (2.1) is rewritten in its potential form. Hence, the list of symmetries of equation (2.1) is limited to the following cases:

$$X_1 = \frac{\partial}{\partial t}, \tag{2.19}$$

$$X_2 = \frac{\partial}{\partial x}, \tag{2.20}$$

$$X_3 = \frac{\partial}{\partial U}, \tag{2.21}$$

$$X_4 = x\frac{\partial}{\partial x} + 2t\frac{\partial}{\partial t} - U\frac{\partial}{\partial U}, \tag{2.22}$$

$$X_5 = t\frac{\partial}{\partial x} + \frac{\partial}{\partial U}, \tag{2.23}$$

$$X_6 = tx\frac{\partial}{\partial x} + t^2\frac{\partial}{\partial t} + (x - tU)\frac{\partial}{\partial U}, \tag{2.24}$$

It can be easily shown, that the symmetries (2.19-2.24) of the Burgers equation (2.1) are transferred to all statistical descriptions; i.e. that statistical equations (2.3), (2.6) and (2.10)-(2.11) are all invariant under the above groups when symmetries are reformulated accordingly. To show this at least exemplarily we focus here on the scale symmetry (2.22) and write this for our purpose in global form

$$T_4: \ t^* = \mathrm{e}^{2k}t\,, \ x^* = \mathrm{e}^{k}x\,, \ U^* = \mathrm{e}^{-k}U \ . \tag{2.25}$$

For the statistical single- and multi-point moments defined in (2.5), the scaling symmetry can be easily converted and one obtains

$$\bar{T}_4: \ t^* = \mathrm{e}^{2k}t\,, \ x^* = \mathrm{e}^{k}x\,, \ \overline{U^n}^* = \mathrm{e}^{-kn}\overline{U^n}\,, \ H_n^* = \mathrm{e}^{-kn}H_n \ . \tag{2.26}$$

For the single-point and multi-point PDF descriptions in (2.10) and (2.11) the scaling symmetry looks very similar, whereby the PDFs needs to be rescaled due to the normalization condition (2.12)

$$\begin{aligned}\tilde{T}_4: \ & t^* = \mathrm{e}^{2k}t\,, \ x^* = \mathrm{e}^{k}x\,, \ y^* = \mathrm{e}^{k}y\,, \ z^* = \mathrm{e}^{k}z\,, \\ & v^* = \mathrm{e}^{-k}v\,, \ w^* = \mathrm{e}^{-k}w\,, \ w'^* = \mathrm{e}^{-k}w' \\ & f_1^* = \mathrm{e}^{k}f_1\,, \ f_2^* = \mathrm{e}^{2k}f_2,\,, \ f_3^* = \mathrm{e}^{3k}f_3 \ .\end{aligned} \tag{2.27}$$

From the definition of the characteristic function in (2.15) and employing the scaling symmetry for the PDF in (2.27) we deduce that the characteristic function Φ stays unaltered but the wave numbers rescale

$$\begin{aligned}\hat{T}_4: \ & t^* = \mathrm{e}^{2k}t\,, \ x^* = \mathrm{e}^{k}x\,, \ y^* = \mathrm{e}^{k}y\,, \ z^* = \mathrm{e}^{k}z\,, \\ & \hat{v}^* = \mathrm{e}^{k}\hat{v}\,, \ \hat{w}^* = \mathrm{e}^{k}\hat{w}\,, \ \Phi_1^* = \Phi_1\,, \ \Phi_2^* = \Phi_2 \ .\end{aligned} \tag{2.28}$$

Completely analogous to X_4, for each of the symmetries (2.19)-(2.24) a corresponding symmetry can be derived in each of the three statistical forms, i.e. moments, PDF and characteristic function.

Especially interesting for the next section on the statistics of the Navier-Stokes equations is that the statistical equations (2.3), (2.6) and (2.10/2.11) have other statistical symmetries that cannot be traced back to one of the six symmetries (2.19)-(2.24) of the original Burgers equation.

At this point we want to consider two symmetries in particular, which also play a central role in the following chapter on turbulence: (i) statistical scaling symmetry and (ii) translation symmetry in the correlation functions, both of which can also be converted into the PDF and the characteristic function form.

Independent of the order of the moment n the statistical scaling symmetry scales all statistical single-point and multi-point moments equally

$$\bar{T}_{s1} : t^* = t\,, \;\; x^* = x\,, \;\; y^* = y\,, \;\; z^* = z\,, \;\; \overline{U^n}^* = \mathrm{e}^{k_{s1}}\overline{U^n}\,, \;\; H_n^* = \mathrm{e}^{k_{s1}} H_n\,. \tag{2.29}$$

which can directly verified in the moment equations (2.3) and (2.6).

The second central statistical symmetry results from the translation of the individual moments, since all terms in the moment equations, as shown in (2.3) and (2.6), are placed under a differential, i.e. form divergence expressions. This results in the symmetries

$$\bar{T}_{s1} : t^* = t\,, x^* = x\,, y^* = y\,, z^* = z\,, \overline{U^n}^* = \overline{U^n} + C_n\,, H_n^* = H_n + C_n\,, \tag{2.30}$$

where, strictly speaking, a separate translation symmetry exists for each moment and thus a n-parametric symmetry group is defined.

The forms of description of statistics developed within this chapter for the statistics of the Burgers equation are extended in the following chapter to the significantly more complicated Navier-Stokes equation, on the basis of which a better understanding of turbulence will be developed.

3 Statistics of the Navier-Stokes equations and its symmetries

3.1 Classical symmetries of Euler and Navier-Stokes equations

3.1.1 Navier-Stokes equations

The starting point of the analysis to follow is the three dimensional Navier-Stokes equations for an incompressible fluid assuming Newtonian fluid behavior under constant density and viscosity conditions. In Cartesian tensor notation we have the momentum and the continuity equation

$$\mathcal{M}_i(\boldsymbol{x}) = \frac{\partial U_i}{\partial t} + U_k \frac{\partial U_i}{\partial x_k} + \frac{\partial P}{\partial x_i} - \nu \frac{\partial^2 U_i}{\partial x_k \partial x_k} = 0\;, \quad \frac{\partial U_k}{\partial x_k} = 0 \tag{3.1}$$

where $t \in \mathbb{R}^+$, $\boldsymbol{x} \in \mathbb{R}^3$, $\boldsymbol{U} = \boldsymbol{U}(\boldsymbol{x}, t)$ and $P = P(\boldsymbol{x}, t)$ represent time, position vector, instantaneous velocity vector and pressure respectively, while pressure has been normalized by a constant density. As we consider turbulence in 3D, indices will vary accordingly $i, k = 1, 2, 3$.

3.1.2 Symmetries of the Euler and Navier-Stokes equations

The Euler equations, i.e. equation (3.1) with $\nu = 0$, admit a ten-parameter symmetry group,

$$
\begin{aligned}
T_1 &: t^* = t + k_1, \ \boldsymbol{x}^* = \boldsymbol{x}, \ \boldsymbol{U}^* = \boldsymbol{U}, \ P^* = P, \\
T_2 &: t^* = t, \ \boldsymbol{x}^* = \mathrm{e}^{k_2}\boldsymbol{x}, \ \boldsymbol{U}^* = \mathrm{e}^{k_2}\boldsymbol{U}, \ P^* = \mathrm{e}^{2k_2}P, \\
T_3 &: t^* = \mathrm{e}^{k_3}t, \ \boldsymbol{x}^* = \boldsymbol{x}, \ \boldsymbol{U}^* = \mathrm{e}^{-k_3}\boldsymbol{U}, \ P^* = \mathrm{e}^{-2k_3}P, \\
T_4 - T_6 &: t^* = t, \ \boldsymbol{x}^* = \mathbf{a} \cdot \boldsymbol{x}, \ \boldsymbol{U}^* = \mathbf{a} \cdot \boldsymbol{U}, \ P^* = P, \\
T_7 - T_9 &: t^* = t, \ \boldsymbol{x}^* = \boldsymbol{x} + \boldsymbol{f}(t), \ \boldsymbol{U}^* = \boldsymbol{U} + \frac{\mathrm{d}\boldsymbol{f}}{\mathrm{d}t}, \ P^* = P - \boldsymbol{x} \cdot \frac{\mathrm{d}^2\boldsymbol{f}}{\mathrm{d}t^2}, \\
T_{10} &: t^* = t, \ \boldsymbol{x}^* = \boldsymbol{x}, \ \boldsymbol{U}^* = \boldsymbol{U}, \ P^* = P + f_4(t) \ ,
\end{aligned} \tag{3.2}
$$

where k_1-k_3 are independent group-parameters, $\mathbf{a}$ denotes a constant rotation matrix with the properties $\mathbf{a} \cdot \mathbf{a}^\mathsf{T} = \mathbf{a}^\mathsf{T} \cdot \mathbf{a} = \mathbf{I}$ and $|\mathbf{a}| = 1$. Moreover $\boldsymbol{f}(t) = (f_1(t), f_2(t), f_3(t))^\mathsf{T}$ with twice differentiable functions f_1-f_3 and $f_4(t)$ may have arbitrary time dependence.

From (3.2) we may derive the corresponding infinitesimal generators

$$
\begin{aligned}
\mathrm{X}_1 &= \frac{\partial}{\partial t} \ , \\
\mathrm{X}_2 &= x_i\frac{\partial}{\partial x_i} + U_j\frac{\partial}{\partial U_j} + 2P\frac{\partial}{\partial P} + 2\nu\frac{\partial}{\partial \nu} \ , \\
\mathrm{X}_3 &= t\frac{\partial}{\partial t} - U_i\frac{\partial}{\partial U_i} - 2P\frac{\partial}{\partial P} - \nu\frac{\partial}{\partial \nu} \ , \\
\mathrm{X}_4 &= -x_2\frac{\partial}{\partial x_1} + x_1\frac{\partial}{\partial x_2} - U_2\frac{\partial}{\partial U_1} + U_1\frac{\partial}{\partial U_2} \ , \\
\mathrm{X}_5 &= -x_3\frac{\partial}{\partial x_2} + x_2\frac{\partial}{\partial x_3} - U_3\frac{\partial}{\partial U_2} + U_2\frac{\partial}{\partial U_3} \ , \\
\mathrm{X}_6 &= -x_3\frac{\partial}{\partial x_1} + x_1\frac{\partial}{\partial x_3} - U_3\frac{\partial}{\partial U_1} + U_1\frac{\partial}{\partial U_3} \ , \\
\mathrm{X}_7 &= f_1(t)\frac{\partial}{\partial x_1} + \frac{\mathrm{d}f_1(t)}{\mathrm{d}t}\frac{\partial}{\partial U_1} - x_1\frac{\mathrm{d}^2 f_1(t)}{\mathrm{d}t^2}\frac{\partial}{\partial P} \ , \\
\mathrm{X}_8 &= f_2(t)\frac{\partial}{\partial x_2} + \frac{\mathrm{d}f_2(t)}{\mathrm{d}t}\frac{\partial}{\partial U_2} - x_2\frac{\mathrm{d}^2 f_3(t)}{\mathrm{d}t^2}\frac{\partial}{\partial P} \ , \\
\mathrm{X}_9 &= f_3(t)\frac{\partial}{\partial x_3} + \frac{\mathrm{d}f_3(t)}{\mathrm{d}t}\frac{\partial}{\partial U_3} - x_3\frac{\mathrm{d}^2 f_3(t)}{\mathrm{d}t^2}\frac{\partial}{\partial P} \ , \\
\mathrm{X}_{10} &= f_4(t)\frac{\partial}{\partial P} \ .
\end{aligned} \tag{3.3}
$$

Each of the symmetries has a distinct physical meaning. T_1 means time translation i.e. any physical experiment is independent of the actual starting point. T_4-T_6 designate rotation invariance which refers to the possibility of letting an experiment undergo a fixed rotation without changing physics. Note, that this does *not* mean moving into a rotating system since this does significantly change physics and hence is not a symmetry. The symmetries T_7-T_9 comprise translational invariance in space for constant f_1-f_3 as well as the classical Galilei group if f_1-f_3 are linear in time. These are key properties of classical mechanics referring to the fact that physics is independent of the location or if moved at a constant speed. In its rather general form T_7-T_9 and T_{10} are direct consequences of an incompressible fluid and do not have a counterpart in the case of compressible fluids.

Invoking a formal transfer from Euler to the Navier-Stokes equations symmetry properties change and a recombination of the two scaling symmetries T_2 and T_3 in (3.2) is observed

$$T_{NS}: \; t^* = \mathrm{e}^{2k_{NS}} t, \;\; \boldsymbol{x}^* = \mathrm{e}^{k_{NS}} \boldsymbol{x}, \;\; \boldsymbol{U}^* = \mathrm{e}^{-k_{NS}} \boldsymbol{U}, \;\; P^* = \mathrm{e}^{-2k_{NS}} P, \tag{3.4}$$

with $k_3 = 2k_{NS}$ and $k_2 = k_{NS}$, while the remaining groups stay unaltered. The corresponding generator reads

$$\mathrm{X}_{NS} = 2t\frac{\partial}{\partial t} + x_i\frac{\partial}{\partial x_i} - U_j\frac{\partial}{\partial U_j} - 2P\frac{\partial}{\partial P} \; . \tag{3.5}$$

It should be noted that additional symmetries exist for dimensional restricted cases such as plane or axisymmetric flows (see e.g. in [1, 4]).

3.2 Introduction to statistical descriptions of turbulence

3.2.1 Statistical averaging

In the following we define the classical Reynolds averaging. The quantity Z represents an arbitrary statistical variable, i.e. $\boldsymbol{U}$ and P, which in the following we also denote as *instantaneous value.* According to the classic definition by O. Reynolds, all instantaneous quantities are decomposed into their mean and their fluctuation value

$$Z = \overline{Z} + z \; . \tag{3.6}$$

Here, the overbar denotes a statistically averaged quantity whereas the lower-case z denotes the fluctuation value of Z. The statistically averaging denoted by an overbar has already been defined in (2.2). More generally, any statistical quantity such as the mean value or higher order moments may also be defined more precisely by employing the concept of a probability density function (PDF) which will be detailed in Subsection 3.2.4.

3.2.2 Reynolds averaged transport equations

After $\boldsymbol{U}$ and P are decomposed according to the Reynolds decomposition, i.e. $\boldsymbol{U} = \overline{\boldsymbol{U}} + \boldsymbol{u}$ and $P = \overline{P} + p$, we gain an averaged versions of the momentum and continuity equations

$$\frac{\partial \overline{U}_i}{\partial t} + \overline{U}_k \frac{\partial \overline{U}_i}{\partial x_k} = -\frac{\partial \overline{P}}{\partial x_i} + \nu \frac{\partial^2 \overline{U}_i}{\partial x_k \partial x_k} - \frac{\partial \overline{u_i u_k}}{\partial x_k} , \quad \frac{\partial \overline{U}_k}{\partial x_k} = 0 . \tag{3.7}$$

At this point we observe the well-known closure problem of turbulence since, compared to the original set of equations, the unknown Reynolds stress tensor $\overline{u_i u_k}$ appeared. However, rather different from the classical approach we will not proceed with deriving the Reynolds stress tensor transport equation, which contains additional four unclosed tensors. Instead, the multi-point correlation approach as already defined in Section 2.1 is put forward the reason being twofold.

First, if the infinite set of correlation equations is considered, the closure problem is somewhat bypassed. Second, the multi-point correlation delivers additional information on the turbulence statistics such as length scale information which may not be gained from the Reynolds stress tensor, which is a single-point quantity.

For this we need the equations for the fluctuating quantities $\boldsymbol{u}$ and p, which are derived by taking the differences between the averaged and the non-averaged equations, i.e. (3.1) and (3.7) and employing (3.6). The resulting fluctuation equations read

$$\mathcal{N}_i(\boldsymbol{x}) = \frac{\bar{\mathrm{D}} u_i}{\bar{\mathrm{D}} t} + u_k \frac{\partial \overline{U}_i}{\partial x_k} - \frac{\partial \overline{u_i u_k}}{\partial x_k} + \frac{\partial u_i u_k}{\partial x_k} + \frac{\partial p}{\partial x_i} - \nu \frac{\partial^2 u_i}{\partial x_k \partial x_k} = 0 , \quad \frac{\partial u_k}{\partial x_k} = 0 , \tag{3.8}$$

with $\bar{\mathrm{D}}/\bar{\mathrm{D}}t = \partial/\partial t + \overline{U}_k \partial/\partial x_k$.

3.2.3 Multi-point correlation equations

It is surmised that the idea of two- and multi-point equations in turbulence was first established in [10]. At the time it was assumed that all correlation equations of orders higher than two may be negligible. Theoretical considerations and measurements showed that all higher order correlations have to be accounted for. Consequently, all multi-point correlation equations have been considered in the symmetry analysis that follows below.

Two different sets of multi-point correlation (MPC) equations will be derived below. The first is based on the instantaneous values of $\boldsymbol{U}$ and P while the second follows the classical notation based on the fluctuating quantities $\boldsymbol{u}$ and p.

MPC equations: instantaneous approach In order to write the MPC equations in a very compact form, we introduce the following notation. The multi-point velocity correlation tensor of order $n+1$ is defined as follows:

$$H_{i_{\{n+1\}}} = H_{i_{(0)} i_{(1)} \cdots i_{(n)}} = \overline{U_{i_{(0)}}(\boldsymbol{x}_{(0)}, t) \cdot \ldots \cdot U_{i_{(n)}}(\boldsymbol{x}_{(n)}, t)} , \tag{3.9}$$

where the index i of the farthermost left quantity refers to its tensor character, while its superscript in curly brackets denotes the tensor order. The second term exemplifies this since a list of $n+1$ tensor indices is given, where the index in parenthesis is a counter for the tensor order. It is important to mention that the index counter starts with 0 which is an advantage when introducing a new coordinate system based on the Euclidean distance of two or more space points with (3.9). The mean velocity is given by the first order tensor as $H_{i_{\{1\}}} = H_{i_{(0)}} = \overline{U}_i$.

In some cases the list of indices is interrupted by one or more other indices, which is pointed out by attaching the replaced value in square brackets to the index

$$
\begin{aligned}
H_{i_{\{n+1\}}[i_{(l)} \mapsto k_{(l)}]} = &\overline{U_{i_{(0)}}(\boldsymbol{x}_{(0)},t)\cdot \ldots \cdot U_{i_{(l-1)}}(\boldsymbol{x}_{(l-1)},t)*} \\
&\overline{*U_{k_{(l)}}(\boldsymbol{x}_{(l)},t)U_{i_{(l+1)}}(\boldsymbol{x}_{(l+1)},t)\cdot \ldots \cdot U_{i_{(n)}}(\boldsymbol{x}_{(n)},t)}\ . \qquad (3.10)
\end{aligned}
$$

where the $*$ under the overbar denotes that the averaging includes both lines. This notation is also continued below in the equations (3.12) and (3.13). This is further extended by

$$
H_{i_{\{n+2\}}[i_{(n+1)} \mapsto k_{(l)}]}[\boldsymbol{x}_{(n+1)} \mapsto \boldsymbol{x}_{(l)}] = \overline{U_{i_{(0)}}(\boldsymbol{x}_{(0)},t)\cdot \ldots \cdot U_{i_{(n)}}(\boldsymbol{x}_{(n)},t)U_{k_{(l)}}(\boldsymbol{x}_{(l)},t)}\ , \qquad (3.11)
$$

where not only that index $i_{(n+1)}$ is replaced by $k_{(l)}$, but also the independent variable $\boldsymbol{x}_{(n+1)}$ is replaced by $\boldsymbol{x}_{(l)}$. If indices are missing e.g. between $i_{(l-1)}$ and $i_{(l+1)}$ we define

$$
\begin{aligned}
H_{i_{\{n\}}[i_{(l)} \mapsto \emptyset]} = &\overline{U_{i_{(0)}}(\boldsymbol{x}_{(0)},t)\cdot \ldots \cdot U_{i_{(l-1)}}(\boldsymbol{x}_{(l-1)},t)*} \\
&\overline{*U_{i_{(l+1)}}(\boldsymbol{x}_{(l+1)},t)\cdot \ldots \cdot U_{i_{(n)}}(\boldsymbol{x}_{(n)},t)}\ . \qquad (3.12)
\end{aligned}
$$

Finally, if pressure is involved we write

$$
\begin{aligned}
I_{i_{\{n\}}[l]} = &\overline{U_{i_{(0)}}(\boldsymbol{x}_{(0)},t)\cdot \ldots \cdot U_{i_{(l-1)}}(\boldsymbol{x}_{(l-1)},t)*} \\
&\overline{*P(\boldsymbol{x}_{(l)},t)U_{i_{(l+1)}}(\boldsymbol{x}_{(l+1)},t)\cdot \ldots \cdot U_{i_{(n)}}(\boldsymbol{x}_{(n)},t)}\ , \qquad (3.13)
\end{aligned}
$$

which is, considering all the above definitions, sufficient to derive the MPC equations from the equations of instantaneous velocity and pressure i.e. equations (3.1).

Applying the Reynolds averaging operator (2.2) according to the sum below

$$
\begin{aligned}
&\mathcal{S}_{i_{\{n+1\}}}(\boldsymbol{x}_{(0)},\ldots,\boldsymbol{x}_{(n)},t) \\
&= \overline{\mathcal{M}_{i_{(0)}}(\boldsymbol{x}_{(0)},t)U_{i_{(1)}}(\boldsymbol{x}_{(1)},t)\cdot \ldots \cdot U_{i_{(n)}}(\boldsymbol{x}_{(n)},t)} \\
&+ \overline{U_{i_{(0)}}(\boldsymbol{x}_{(0)},t)\mathcal{M}_{i_{(1)}}(\boldsymbol{x}_{(1)},t)U_{i_{(2)}}(\boldsymbol{x}_{(2)},t)\cdot \ldots \cdot U_{i_{(n)}}(\boldsymbol{x}_{(n)},t)} \\
&+ \ldots \\
&+ \overline{U_{i_{(0)}}(\boldsymbol{x}_{(0)},t)\cdot \ldots \cdot U_{i_{(n-2)}}(\boldsymbol{x}_{(n-2)},t)\mathcal{M}_{i_{(n-1)}}(\boldsymbol{x}_{(n-1)},t)U_{i_{(n)}}(\boldsymbol{x}_{(n)},t)} \\
&+ \overline{U_{i_{(0)}}(\boldsymbol{x}_{(0)},t)\cdot \ldots \cdot U_{i_{(n-1)}}(\boldsymbol{x}_{(n-1)},t)\mathcal{M}_{i_{(n)}}(\boldsymbol{x}_{(n)},t)}\ , \qquad (3.14)
\end{aligned}
$$

we obtain the $\mathcal{S}$-equation which writes

$$
\mathcal{S}_{i_{\{n+1\}}} = \frac{\partial H_{i_{\{n+1\}}}}{\partial t} + \sum_{l=0}^{n} \left[\frac{\partial H_{i_{\{n+2\}}[i_{(n+1)} \mapsto k_{(l)}]}[\boldsymbol{x}_{(n+1)} \mapsto \boldsymbol{x}_{(l)}]}{\partial x_{k_{(l)}}} \right.
$$

$$+\frac{\partial I_{i_{\{n\}}[l]}}{\partial x_{i_{(l)}}}-\nu\frac{\partial^2 H_{i_{\{n+1\}}}}{\partial x_{k_{(l)}}\partial x_{k_{(l)}}}\Bigg]=0 \text{ for } n=1,\ldots,\infty\ . \qquad (3.15)$$

In general, equation (3.15) implies the full multi-point statistical information of the Navier-Stokes equations at the expense of dealing with an infinite dimensional chain of differential equations starting with order 1 i.e. $n = 0$. It is a remarkable fact that (3.15) is a linear equation which considerably simplifies the finding of Lie symmetries to be pointed out below.

From the second equation of (3.1) a continuity equation for $H_{i_{\{n+1\}}}$ and $I_{i_{\{n\}}[l]}$ can be derived. This leads to

$$\frac{\partial H_{i_{\{n+1\}}[i_{(l)}\mapsto k_{(l)}]}}{\partial x_{k_{(l)}}}=0 \quad \text{for} \quad l=0,\ldots,n\ ,$$

$$\frac{\partial I_{i_{\{n\}}[k][i_{(l)}\mapsto m_{(l)}]}}{\partial x_{m_{(l)}}}=0 \quad \text{for} \quad k,l=0,\ldots,n \text{ and } k\neq l\ . \qquad (3.16)$$

At this point we adopt the classic notation of distance vectors. Accordingly the usual position vector $\boldsymbol{x}$ is employed and the remaining independent spatial variables are expressed as the difference of two position vectors $\boldsymbol{x}_{(l)}$ and $\boldsymbol{x}_{(0)}$. The coordinate transformations are

$$\boldsymbol{x}=\boldsymbol{x}_{(0)}\ ,\quad \boldsymbol{r}_{(l)}=\boldsymbol{x}_{(l)}-\boldsymbol{x}_{(0)} \quad \text{with} \quad l=1,\ldots,n \quad \text{and}$$

$$\frac{\partial}{\partial x_{k_{(0)}}}=\frac{\partial}{\partial x_k}-\sum_{l=1}^{n}\frac{\partial}{\partial r_{k_{(l)}}}\ ,\quad \frac{\partial}{\partial x_{k_{(l)}}}=\frac{\partial}{\partial r_{k_{(l)}}} \quad \text{for} \quad l\geq 1\ . \qquad (3.17)$$

For consistency the first index $i_{(0)}$ is replaced by i. Thus, the indices of the tensor $\mathbf{H}_{\{n+1\}}$ are $H_{ii_{(1)}i_{(2)}\cdots i_{(n)}}$. Using the rules of transformation (3.17) the $\mathcal{S}$-equation leads to

$$\begin{aligned}
&\mathcal{S}_{i_{\{n+1\}}}\\
&=\frac{\partial H_{i_{\{n+1\}}}}{\partial t}+\frac{\partial H_{i_{\{n+2\}}[i_{(n+1)}\mapsto k]}[\boldsymbol{x}_{(n+1)}\mapsto \boldsymbol{x}]}{\partial x_k}\\
&-\sum_{l=1}^{n}\left[\frac{\partial H_{i_{\{n+2\}}[i_{(n+1)}\mapsto k_{(l)}]}[\boldsymbol{x}_{(n+1)}\mapsto \boldsymbol{x}]}{\partial r_{k_{(l)}}}-\frac{\partial H_{i_{\{n+2\}}[i_{(n+1)}\mapsto k_{(l)}]}[\boldsymbol{x}_{(n+1)}\mapsto \boldsymbol{r}_{(l)}]}{\partial r_{k_{(l)}}}\right]\\
&+\frac{\partial I_{i_{\{n\}}[0]}}{\partial x_i}+\sum_{l=1}^{n}\left(-\left.\frac{\partial I_{i_{\{n\}}[0]}}{\partial r_{m_{(l)}}}\right|_{[m_{(l)}\mapsto i]}+\frac{\partial I_{i_{\{n\}}[l]}}{\partial r_{i_{(l)}}}\right)\\
&-\nu\left[\frac{\partial^2 H_{i_{\{n+1\}}}}{\partial x_k\partial x_k}+\sum_{l=1}^{n}\left(-2\frac{\partial^2 H_{i_{\{n+1\}}}}{\partial x_k\partial r_{k_{(l)}}}\right.\right.\\
&\left.\left.+\sum_{m=1}^{n}\frac{\partial^2 H_{i_{\{n+1\}}}}{\partial r_{k_{(m)}}\partial r_{k_{(l)}}}+\frac{\partial^2 H_{i_{\{n+1\}}}}{\partial r_{k_{(l)}}\partial r_{k_{(l)}}}\right)\right]=0 \text{ for } n=1,\ldots,\infty\ ,
\end{aligned} \qquad (3.18)$$

and the two continuity equations become

$$\frac{\partial H_{i_{\{n+1\}}[i_{(0)}\mapsto k]}}{\partial x_k} - \sum_{j=1}^{n} \frac{\partial H_{i_{\{n+1\}}[i_{(0)}\mapsto k_{(j)}]}}{\partial r_{k_{(j)}}} = 0 \ ,$$

$$\frac{\partial H_{i_{\{n+1\}}[i_{(l)}\mapsto k_{(l)}]}}{\partial r_{k_{(l)}}} = 0 \quad \text{for} \quad l = 1, \ldots, n \tag{3.19}$$

and

$$\frac{\partial I_{i_{\{n\}}[k][i\mapsto m]}}{\partial x_m} - \sum_{j=1}^{n} \frac{\partial I_{i_{\{n\}}[k][i\mapsto m_{(j)}]}}{\partial r_{m_{(j)}}} = 0 \ \text{for} \ k = 1, \ldots, n$$

$$\frac{\partial I_{i_{\{n\}}[k][i_{(l)}\mapsto m_{(l)}]}}{\partial r_{m_{(l)}}} = 0 \ \text{for} \ k = 0, \ldots, n, l = 1, \ldots, n, k \neq l \ . \tag{3.20}$$

MPC equations: fluctuation approach In the present subsection we adopt the classical approach; i.e. all correlation functions are based on the fluctuating quantities $\boldsymbol{u}$ and p as originally introduced by Reynolds and not on the full instantaneous quantities $\boldsymbol{U}$ and P as in the previous subsection. Hence, similar to (3.9) we have the multi-point correlation for the fluctuation velocity

$$R_{i_{\{n+1\}}} = R_{i_{(0)}i_{(1)}\ldots i_{(n)}} = \overline{u_{i_{(0)}}(\boldsymbol{x}_{(0)}) \cdot \ldots \cdot u_{i_{(n)}}(\boldsymbol{x}_{(n)})} \ . \tag{3.21}$$

All other correlations defined in Subsection 3.2.3 are defined accordingly; i.e. equivalent to the definitions (3.10)-(3.13) we respectively define $R_{i_{\{n+1\}}[i_{(l)}\mapsto k_{(l)}]}$, $R_{i_{\{n\}}[i_{(l)}\mapsto \emptyset]}$, $R_{i_{\{n+2\}}[i_{(n+1)}\mapsto k_{(l)}]}[\boldsymbol{x}_{(n+1)} \mapsto \boldsymbol{x}_{(l)}]$ and $P_{i_{\{n\}}[l]}$.

Finally, we define the correlation equation in analogy to (3.14) where $\mathcal{M}_i$ is replaced by the equation for the fluctuations (3.8) denoted by $\mathcal{N}_i$ and U_i and P are substituted by u_i and p. The resulting equation is denoted by $\mathcal{T}_{i_{\{n+1\}}}$

$$\begin{aligned} \mathcal{T}_{i_{\{n+1\}}} = & \frac{\partial R_{i_{\{n+1\}}}}{\partial t} + \sum_{l=0}^{n} \Bigg[\overline{U}_{k_{(l)}}(\boldsymbol{x}_{(l)}) \frac{\partial R_{i_{\{n+1\}}}}{\partial x_{k_{(l)}}} + R_{i_{\{n+1\}}[i_{(l)}\mapsto k_{(l)}]} \frac{\partial \overline{U}_{i_{(l)}}(\boldsymbol{x}_{(l)})}{\partial x_{k_{(l)}}} \\ & + \frac{\partial P_{i_{\{n\}}[l]}}{\partial x_{i_{(l)}}} - \nu \frac{\partial^2 R_{i_{\{n+1\}}}}{\partial x_{k_{(l)}} \partial x_{k_{(l)}}} - R_{i_{\{n\}}[i_{(l)}\mapsto \emptyset]} \frac{\partial \overline{u_{i_{(l)}} u_{k_{(l)}}}(\boldsymbol{x}_{(l)})}{\partial x_{k_{(l)}}} \\ & + \frac{\partial R_{i_{\{n+2\}}[i_{(n+1)}\mapsto k_{(l)}]}[\boldsymbol{x}_{(n+1)} \mapsto \boldsymbol{x}_{(l)}]}{\partial x_{k_{(l)}}} \Bigg] = 0, \ \text{for} \ n = 1, \ldots, \infty \ . \end{aligned} \tag{3.22}$$

From the second of the equations (3.8) a continuity equation for $R_{i_{\{n+1\}}}$ and $P_{i_{\{n\}}[l]}$ can be derived which have identical form to (3.16) for $H_{i_{\{n+1\}}}$ and $I_{i_{\{n\}}}$ and complete the system.

The first tensor equation of the infinite chain (3.22) propagates $R_{i_{\{2\}}}$ which has a close link to the usual Reynolds stress tensor, i.e.

$$\overline{u_{i_{(0)}} u_{i_{(1)}}}(\boldsymbol{x}_{(l)}) = \lim_{x_{(k)}\to x_{(l)}} R_{i_{\{2\}}} = \lim_{x_{(k)}\to x_{(l)}} R_{i_{(0)}i_{(1)}} \quad \text{for} \quad k \neq l \ , \tag{3.23}$$

which is the key unclosed quantity in the Reynolds stress transport equation (3.7). Here $x_{(k)}$ and $x_{(l)}$ can be arbitrary vectors out of $x_{(0)}, \ldots, x_{(n)}$.

Similar to (3.15), also equation (3.22) implies all statistical information of the Navier-Stokes equations. However, apart from the latter simple relation to the Reynolds stress tensor in (3.23), (3.22) possesses the key disadvantage of being a *non-linear* infinite dimensional system of differential equations, which has a strong *cross-coupling* between tensors as the first and the second moments are present in all higher order moment equations. This makes the extraction of Lie symmetries from this equation rather cumbersome and essentially impossible. In contrast, in equation (3.15) each moment of the order n is only coupled to the tensor of the order $n+1$, which simplifies the analysis considerably.

Of course, there is a unique relation between the instantaneous (**H**, **I**) and the fluctuation approach (**R**, **P**) though the actual crossover is complex in particular with increasing tensor order because they may only be given in recursive form. Since needed later we give the first relations

$$H_{i_{(0)}} = \overline{U}_{i_{(0)}} \ , \tag{3.24}$$

$$H_{i_{(0)}i_{(1)}} = \overline{U}_{i_{(0)}}\overline{U}_{i_{(1)}} + R_{i_{(0)}i_{(1)}} \ , \tag{3.25}$$

$$\begin{aligned} H_{i_{(0)}i_{(1)}i_{(2)}} &= \overline{U}_{i_{(0)}}\overline{U}_{i_{(1)}}\overline{U}_{i_{(2)}} \\ &\quad + R_{i_{(0)}i_{(1)}}\overline{U}_{i_{(2)}} + R_{i_{(0)}i_{(2)}}\overline{U}_{i_{(1)}} + R_{i_{(1)}i_{(2)}}\overline{U}_{i_{(0)}} + R_{i_{(0)}i_{(1)}i_{(2)}}, \ \ldots \end{aligned} \tag{3.26}$$

where the indices also refer to the spatial points as indicated.

3.2.4 Lundgren-Monin-Novikov hierarchy for PDF functions

As already stated in the introductory Section 2, in the LMN approach one assumes that for the velocity field there exists a probability density function (PDF), which describes the joint probabilities of measuring contemporarily sets of velocities at multiple points in space. When calculating mean values thereof, these PDFs play the role of the weighting measure.

For this, let us define multi-point PDFs in the usual way: the 1-point PDF $f_1(\boldsymbol{x}_{(0)}, \boldsymbol{v}_{(0)}; t)$ is such that $f_1(\boldsymbol{x}_{(0)}, \boldsymbol{v}_{(0)}; t)\,\mathrm{d}\boldsymbol{v}_{(0)}$ expresses the probability to measure a velocity in an infinitesimal interval $\mathrm{d}\boldsymbol{v}_{(0)}$ around $\boldsymbol{v}_{(0)}$ at position $\boldsymbol{x}_{(0)}$ (or, equivalently, the fraction of systems in the ensemble such that the given condition is satisfied). The 1-point PDF can be written as:

$$f_1(\boldsymbol{x}_{(0)}; \boldsymbol{v}_{(0)}, t) = \overline{\delta(\boldsymbol{U}(\boldsymbol{x}_{(0)}, t) - \boldsymbol{v}_{(0)})} \ . \tag{3.27}$$

where $\delta(\boldsymbol{U}(\boldsymbol{x}_{(0)}, t) - \boldsymbol{v}_{(0)})$ is the Dirac delta function and can be understood as a velocity distribution function for a single element of the ensemble, whereas f_1 is the average distribution function. Analogously to (3.27), and following the extension from (2.8) to the 2-point PDF in (2.9), which denotes the joint probability to measure two given velocities $\boldsymbol{v}_{(0)}$ and $\boldsymbol{v}_{(1)}$ at two defined points $\boldsymbol{x}_{(0)}$ and $\boldsymbol{x}_{(1)}$ in space at the same time t, this concept can be continued so forth to the definition of the n point PDF

$$f_{n+1}(\boldsymbol{v}_{(0)},\ldots,\boldsymbol{v}_{(n)};\boldsymbol{x}_{(0)},\ldots,\boldsymbol{x}_{(n)},t) = \overline{\delta(\boldsymbol{U}(\boldsymbol{x}_{(0)},t)-\boldsymbol{v}_{(0)})\cdot\cdots\cdot\delta(\boldsymbol{U}(\boldsymbol{x}_{(n)},t)-\boldsymbol{v}_{(n)})}\ . \quad (3.28)$$

Sometimes the following abbreviation will be used in this paper, cf. [14]:

$$f_{n+1}\equiv f_{n+1}(0,\ldots,n)\equiv f_{n+1}\left(\boldsymbol{v}_{(0)},\ldots,\boldsymbol{v}_{(n)};\boldsymbol{x}_{(0)},\ldots,\boldsymbol{x}_{(n)},t\right)\ .$$

In the LMN language it is the moments associated with the PDFs that correspond to the multi-point moments (3.9) of the MPC approach, i.e. to the components of the tensor $\mathbf{H}$:

$$H_{i_{\{n+1\}}}\equiv\overline{U_{i_{(0)}}(\boldsymbol{x}_{(0)},t)\ldots U_{i_{(n)}}(\boldsymbol{x}_{(n)},t)}=\int f_{n+1}\,v_{i_{(0)}}\cdots v_{i_{(n)}}\,\mathrm{d}\boldsymbol{v}_{(0)}\ldots\mathrm{d}\boldsymbol{v}_{(n)}. \quad (3.29)$$

Properties of the PDFs In order for the previously defined PDFs to be well defined from a physical point of view, they are required to satisfy four conditions as detailed in [14]:

I. The reduction or normalization property imposed by the concept of probability:

$$\int f_1(\boldsymbol{v}_{(0)};\boldsymbol{x}_{(0)},t)\,\mathrm{d}\boldsymbol{v}_{(0)}=1;$$

$$\int f_2\left(\boldsymbol{v}_{(0)},\boldsymbol{v}_{(1)};\boldsymbol{x}_{(0)},\boldsymbol{x}_{(1)},t\right)\,\mathrm{d}\boldsymbol{v}_{(1)}=f_1\left(\boldsymbol{v}_{(0)};\boldsymbol{x}_{(0)},t\right);\ \ldots \quad (3.30)$$

II. An infinite number of "continuity" conditions dictated by the incompressibility of the fluid:

$$\frac{\partial}{\partial x_{k_{(i)}}}\int v_{k_{(i)}}\,f_{n+1}\,\mathrm{d}\boldsymbol{v}_{(i)}=0,\quad\forall i\in\{0,\ldots,n\}; \quad (3.31)$$

III. The "coincidence" property, required by the condition for the velocity field to be well defined:

$$\lim_{\left|\boldsymbol{x}_{(0)}-\boldsymbol{x}_{(1)}\right|\to 0} f_2(0,1)=f_1(0)\,\delta\left(\boldsymbol{v}_{(1)}-\boldsymbol{v}_{(0)}\right);$$

$$\lim_{\left|\boldsymbol{x}_{(0)}-\boldsymbol{x}_{(2)}\right|\to 0} f_3(0,1,2)=f_2(0,1)\,\delta\left(\boldsymbol{v}_{(2)}-\boldsymbol{v}_{(0)}\right);\ \ldots \quad (3.32)$$

IV. The "separation" property (here shown only for the 2-point PDF) which expresses the fact that the velocities of two fluid elements tends to become independent if the two points are set far apart from each other:

$$\lim_{\left|\boldsymbol{x}_{(0)}-\boldsymbol{x}_{(1)}\right|\to\infty} f_2(0,1)=f_1(0)f_1(1). \quad (3.33)$$

The LMN hierarchy Analogously to the MPCE hierarchy in Section 3.2.3 an infinite chain of equations for the multipoint PDF's can be derived based on the Navier-Stokes equations. In the paper [14] the continuity equation was first used in order to eliminate the pressure term from the Navier-Stokes equations (3.1) and present it in the following form

$$\frac{\partial U_{i_{(0)}}\left(\boldsymbol{x}_{(0)},t\right)}{\partial t}+U_{k_{(0)}}\left(\boldsymbol{x}_{(0)},t\right)\frac{\partial U_{i_{(0)}}\left(\boldsymbol{x}_{(0)},t\right)}{\partial x_{k_{(0)}}}=$$
$$-\frac{1}{4\pi}\frac{\partial}{\partial x_{i_{(0)}}}\int\frac{1}{\left|\boldsymbol{x}_{(1)}-\boldsymbol{x}_{(0)}\right|}\frac{\partial}{\partial x_{k_{(1)}}}\left[U_{l_{(1)}}(\boldsymbol{x}_{(1)},t)\frac{\partial U_{k_{(1)}}\left(\boldsymbol{x}_{(1)},t\right)}{\partial x_{l_{(1)}}}\right]\mathrm{d}\boldsymbol{x}_{(1)}$$
$$+\nu\frac{\partial^2}{\partial x_{k_{(0)}}\partial x_{k_{(0)}}}U_{i_{(0)}}(\boldsymbol{x}_{(0)},t). \tag{3.34}$$

In order to derive the first equation in the hierarchy, the time derivative of formula (3.27) should be calculated, leading to

$$\frac{\partial f_1(0)}{\partial t}=\overline{\frac{\partial}{\partial t}\delta(\boldsymbol{U}(\boldsymbol{x}_{(0)},t)-\boldsymbol{v}_{(0)})}=-\overline{\frac{\partial U_{i_{(0)}}}{\partial t}\frac{\partial}{\partial v_{i_{(0)}}}\delta(\boldsymbol{U}(\boldsymbol{x}_{(0)},t)-\boldsymbol{v}_{(0)})}\ . \tag{3.35}$$

Next, the Navier-Stokes equations (3.34) were substituted for the time derivative $\partial\boldsymbol{U}/\partial t$ in equation (3.35) leading, after additional transformations have been implemented, to the following first equation of the LMN hierarchy (for details see [14])

$$\frac{\partial f_1(0)}{\partial t}+v_{i_{(0)}}\frac{\partial f_1(0)}{\partial x_{i_{(0)}}}=$$
$$\frac{1}{4\pi}\frac{\partial}{\partial v_{i_{(0)}}}\iint\left(\frac{\partial}{\partial x_{i_{(0)}}}\frac{1}{\left|\boldsymbol{x}_{(0)}-\boldsymbol{x}_{(1)}\right|}\right)\left(v_{j_{(1)}}\frac{\partial}{\partial x_{j_{(1)}}}\right)^2 f_2(0,1)\,\mathrm{d}\boldsymbol{v}_{(1)}\,\mathrm{d}\boldsymbol{x}_{(1)}$$
$$-\frac{\partial}{\partial v_{i_{(0)}}}\left[\lim_{\left|\boldsymbol{x}_{(1)}-\boldsymbol{x}_{(0)}\right|\to 0}\nu\frac{\partial^2}{\partial x_{j_{(1)}}\partial x_{j_{(1)}}}\int v_{i_{(1)}}f_2(0,1)\,\mathrm{d}\boldsymbol{v}_{(1)}\right]. \tag{3.36}$$

In a similar way, equations for the $n+1$ PDF function may be derived, starting from the time derivative of equation (3.28), substituting (3.34) and leading to

$$\frac{\partial f_{n+1}}{\partial t}+\sum_{k=0}^{n}v_{i_{(k)}}\frac{\partial f_{n+1}}{\partial x_{i_{(k)}}}=$$
$$-\frac{1}{4\pi}\sum_{k=0}^{n}\frac{\partial}{\partial v_{i_{(k)}}}\iint\left(\frac{\partial}{\partial x_{i_{(k)}}}\frac{1}{\left|\boldsymbol{x}_{(k)}-\boldsymbol{x}_{(n+1)}\right|}\right)*$$
$$*\left(v_{j_{(n+1)}}\frac{\partial}{\partial x_{j_{(n+1)}}}\right)^2 f_{n+2}\,\mathrm{d}\boldsymbol{v}_{(n+1)}\,\mathrm{d}\boldsymbol{x}_{(n+1)}$$
$$-\sum_{k=0}^{n}\frac{\partial}{\partial v_{i_{(k)}}}\left[\lim_{\left|\boldsymbol{x}_{(n+1)}-\boldsymbol{x}_{(k)}\right|\to 0}\nu\frac{\partial^2}{\partial x_{j_{(n+1)}}\partial x_{j_{(n+1)}}}\int v_{i_{(n+1)}}f_{n+2}\,\mathrm{d}\boldsymbol{v}_{(n+1)}\right]. \tag{3.37}$$

Hence, the LMN hierarchy constitutes an infinite chain of equations where, on the $n+1$ level, the unknown $n+2$-point PDF is needed. As it was pointed out in [5], the chain can be formally truncated at the $n+1$ level by replacing the terms with f_{n+2} by conditional averages. For the one-point PDF equations and in the case of homogeneous, isotropic turbulence, these conditionally averaged quantities where estimated based on the DNS data in [33, 34] in order to study the deviations of the PDF from Gaussianity.

3.2.5 Hopf functional approach

E. Hopf in [8] introduced another very general approach to the description of turbulence. He considered the case where the number of points in the PDF goes to infinity, so that the PDF becomes a probability density functional $F([\boldsymbol{v}(\boldsymbol{x})], t)$, where, instead of the vector of sample space variables $\boldsymbol{v}_0, \ldots, \boldsymbol{v}_n$ at points $\boldsymbol{x}_0, \ldots, \boldsymbol{x}_n$, one deals with a continuous set of sample space variables $[\boldsymbol{v}(\boldsymbol{x})]$. Further, instead of dealing with the probability density functional itself it was more convenient to consider its functional Fourier transform, called the characteristic functional

$$\Phi([\boldsymbol{y}(\boldsymbol{x})], t) = \int \mathrm{e}^{i(\boldsymbol{y},\boldsymbol{v})} F([\boldsymbol{v}(\boldsymbol{x})], t) D\boldsymbol{v}(\boldsymbol{x}) = \overline{\mathrm{e}^{i(\boldsymbol{y},\boldsymbol{v})}} \ , \tag{3.38}$$

where the integration is with respect to the probability measure $F([\boldsymbol{v}(\boldsymbol{x})], t) D\boldsymbol{v}(\boldsymbol{x})$ and $(\boldsymbol{y}, \boldsymbol{v}) = \int_G y_\alpha v_\alpha \mathrm{d}\mathbf{x}$ is a scalar product of two vector fields. We recall here that in the probability theory the $n+1$-point characteristic function is defined as the following $n+1$-dimensional inverse Fourier transform of the $n+1$-point PDF f_{n+1}

$$\Phi_{n+1} = \int \mathrm{e}^{\mathrm{i}\boldsymbol{v}_{(0)}\cdot\boldsymbol{y}_{(0)}} \ldots \mathrm{e}^{\mathrm{i}\boldsymbol{v}_{(n)}\cdot\boldsymbol{y}_{(n)}} * \\ * f_{n+1}(\boldsymbol{v}_{(0)}, \ldots, \boldsymbol{v}_{(n)}; \boldsymbol{x}_{(0)}, \ldots, \boldsymbol{x}_{(n)}, t)\, \mathrm{d}\boldsymbol{v}_{(0)} \ldots \mathrm{d}\boldsymbol{v}_{(n)}. \tag{3.39}$$

Hence, Φ may be treated as a functional analogue of the characteristic function for $n \to \infty$. The functional embodies the statistical properties of the fluid flow in a more concise form than the infinite set of functions Φ_n.

Solutions of Φ are admitted only, if at any time t the following conditions are fulfilled $\Phi^*([y(x)], t) = \Phi(-[y(x)], t)$, $\Phi(0, t) = 1$ and $|\Phi([y(x)], t)| \leq 1$. These conditions follow from the properties of probability density functional, which is strictly positive $F([v(x)], t) \geq 0$ and its integral over the entire sample space equals 1.

An introduction to the functional approach and definitions of functional derivatives have been presented in [22, 31], where also the extension of Lie group analysis towards such equations was detailed. Here we note that with the definition (3.38) moments of the velocity can be calculated as the functional derivatives of the characteristic functional at the origin $\boldsymbol{y} = 0$. The first functional derivative of Φ is given by

$$\frac{\delta\Phi([\boldsymbol{y}(\boldsymbol{x})], t)}{\delta y_{i_{(0)}}(\boldsymbol{x}_{(0)})} = \int_\Omega i v_{i_{(0)}}(\boldsymbol{x}_{(0)}) \mathrm{e}^{i(\boldsymbol{y},\boldsymbol{v})} F([\boldsymbol{v}(\boldsymbol{x})], t) \mathrm{D}\boldsymbol{v}(\boldsymbol{x}) = \overline{i U_{i_{(0)}}(\boldsymbol{x}_{(0)}, t) \mathrm{e}^{i(\boldsymbol{y},\boldsymbol{u})}} \ ,$$

(3.40)

and, hence, at $\boldsymbol{y}=0$ we find

$$\left.\frac{\delta\Phi([\boldsymbol{y}(\boldsymbol{x})],t)}{\delta y_{i_{(0)}}(\boldsymbol{x}_{(0)})}\right|_{\boldsymbol{y}=0} = i\overline{U_{i_{(0)}}(\boldsymbol{x}_{(0)},t)}. \tag{3.41}$$

The $n+1$ order derivative of Φ give

$$\left.\frac{\delta^{n+1}\Phi([\boldsymbol{y}(\boldsymbol{x})],t)}{\delta y_{i_{(0)}}(\boldsymbol{x}_{(0)})\delta y_{i_{(1)}}(\boldsymbol{x}_1)\cdots\delta y_{i_{(n)}}(\boldsymbol{x}_{(n)})}\right|_{\boldsymbol{y}=0}$$
$$= i^{n+1}\overline{U_{i_{(0)}}(\boldsymbol{x}_{(0)},t)U_{i_{(1)}}(\boldsymbol{x}_{(1)},t)\cdots U_{i_{(n)}}(\boldsymbol{x}_{(n)},t)}. \tag{3.42}$$

Based on the Navier-Stokes equations, E. Hopf derived an evolution equation for the characteristic functional (3.38). It is only one equation (not a hierarchy) and all turbulence statistics can formally be calculated from the solution of the Hopf equation. We do not present here the derivation of this equation, and an interested reader is referred to the works [8] and [17]. The Hopf equation for velocity in the physical space reads

$$\frac{\partial\Phi}{\partial t} = \int_R y_k(\boldsymbol{x})\left[i\frac{\partial}{\partial x_l}\frac{\delta^2\Phi}{\delta y_l(\boldsymbol{x})\delta y_k(\boldsymbol{x})} + \nu\nabla_x^2\frac{\delta\Phi}{\delta y_k(\boldsymbol{x})} - \frac{\partial\Pi}{\partial x_k}\right]\mathrm{d}\boldsymbol{x} \tag{3.43}$$

where, in order to eliminate pressure functional Π from the equation, vector field $\widetilde{\boldsymbol{y}}$ such that $\boldsymbol{y}(\boldsymbol{x}) = \widetilde{\boldsymbol{y}}(\boldsymbol{x}) + \nabla\phi$ was introduced in [8]. The scalar ϕ is chosen such that $\widetilde{\boldsymbol{y}} = \boldsymbol{0}$ at the boundary B and the continuity equation is satisfied $\nabla\cdot\widetilde{\boldsymbol{y}} = 0$. With this, the Hopf functional equation reads

$$\frac{\partial\Phi}{\partial t} = \int_R \widetilde{y}_k(\boldsymbol{x})\left[i\frac{\partial}{\partial x_l}\frac{\delta^2\Phi}{\delta y_l(\boldsymbol{x})\delta y_k(\boldsymbol{x})} + \nu\nabla_x^2\frac{\delta\Phi}{\delta y_k(\boldsymbol{x})}\right]\mathrm{d}\boldsymbol{x}. \tag{3.44}$$

The first RHS term describes convection of velocity and the second appears due to the presence of the viscosity.

The compactness of the form is an advantage of the Hopf approach; however making this equation of practical use for applications or finding meaningful solutions is challenging. Recently, a link between the functional approach and simulation of turbulence in terms of the many-particle dynamics has been established in [9]. Solutions of the Hopf equation for the statistically stationary case allowing calculating higher-order and multipoint statistics from the known first-order moments have been proposed in [27]. General forms of solutions to equation (3.44) were discussed in [29] and in the paper of [8]. Such a solution was presented as a regular Taylor expansion

$$\Phi = 1 + C_1 + C_2 + \cdots, \tag{3.45}$$

where C_{n+1} is a polynomial functional of degree $n+1$ in $\boldsymbol{y}(\boldsymbol{x})$ of the form

$$C_{n+1} = \int\cdots\int K_{i_{(0)}\ldots i_{(n)}}(\boldsymbol{x}_{(0)},\ldots,\boldsymbol{x}_{(n)},t)*$$
$$* y_{i_{(0)}}(\boldsymbol{x}_{(0)})\cdots y_{i_{(n)}}(\boldsymbol{x}_{(n)})\,\mathrm{d}\boldsymbol{x}_{(0)}\ldots\mathrm{d}\boldsymbol{x}_{(n)} \tag{3.46}$$

and the kernel function K is defined by

$$K_{i_{(0)}\dots i_{(n)}}(\boldsymbol{x}_{(0)},\dots,\boldsymbol{x}_{(n)},t)=\frac{1}{(n+1)!}\frac{\delta^{n+1}\Phi}{\delta y_{i_{(0)}}(\boldsymbol{x}_{(0)})\dots y_{i_{(n)}}(\boldsymbol{x}_{(n)})}\bigg|_{\boldsymbol{y}=0}. \tag{3.47}$$

This, according to the relation (3.42) is related to the $n+1$-point velocity correlation function. Hence, the subsequent terms in the proposed form of solution (3.45) represent the contributions of different velocity statistics, up to infinite order.

3.3 Symmetries of statistical equations of turbulence

3.3.1 Symmetries of MPC equations

Symmetries of the MPC equations implied by Euler and Navier-Stokes symmetries Adopting the classical Reynolds notation first, where the instantaneous quantities are split into mean and fluctuating values, we may directly derive from (3.2)

$$\begin{aligned}
\bar{T}_1:\ & t^*=t+k_1,\ \boldsymbol{x}^*=\boldsymbol{x},\ \boldsymbol{r}^*_{(l)}=\boldsymbol{r}_{(l)},\ \overline{\boldsymbol{U}}^*=\overline{\boldsymbol{U}},\\
& \overline{P}^*=\overline{P},\ \mathbf{R}^*_{\{n\}}=\mathbf{R}_{\{n\}},\ \mathbf{P}^*_{\{n\}}=\mathbf{P}_{\{n\}},\\
\bar{T}_2:\ & t^*=t,\ \boldsymbol{x}^*=\mathrm{e}^{k_2}\boldsymbol{x},\ \boldsymbol{r}^*_{(l)}=\mathrm{e}^{k_2}\boldsymbol{r}_{(l)},\ \overline{\boldsymbol{U}}^*=\mathrm{e}^{k_2}\overline{\boldsymbol{U}},\\
& \overline{P}^*=\mathrm{e}^{2k_2}\overline{P},\ \mathbf{R}^*_{\{n\}}=\mathrm{e}^{nk_2}\mathbf{R}_{\{n\}},\mathbf{P}^*_{\{n\}}=\mathrm{e}^{(n+2)k_2}\mathbf{P}_{\{n\}},\\
\bar{T}_3:\ & t^*=\mathrm{e}^{k_3}t,\ \boldsymbol{x}^*=\boldsymbol{x},\ \boldsymbol{r}^*_{(l)}=\boldsymbol{r}_{(l)},\ \overline{\boldsymbol{U}}^*=\mathrm{e}^{-k_3}\overline{\boldsymbol{U}},\\
& \overline{P}^*=\mathrm{e}^{-2k_3}\overline{P},\ \mathbf{R}^*_{\{n\}}=\mathrm{e}^{-nk_3}\mathbf{R}_{\{n\}},\ \mathbf{P}^*_{\{n\}}=\mathrm{e}^{-(n+2)k_3}\mathbf{P}_{\{n\}},\\
\bar{T}_4\text{–}\bar{T}_6:\ & t^*=t,\ \boldsymbol{x}^*=\mathbf{a}\cdot\boldsymbol{x},\ \boldsymbol{r}^*_{(l)}=\boldsymbol{r}_{(l)},\ \overline{\boldsymbol{U}}^*=\mathbf{a}\cdot\overline{\boldsymbol{U}},\ \overline{P}^*=\overline{P},\\
& \mathbf{R}^*_{\{n\}}=\mathbf{A}_{\{n\}}\otimes\mathbf{R}_{\{n\}},\ \mathbf{P}^*_{\{n\}}=\mathbf{A}_{\{n\}}\otimes\mathbf{P}_{\{n\}},\\
\bar{T}_7\text{–}\bar{T}_9:\ & t^*=t,\ \boldsymbol{x}^*=\boldsymbol{x}+\boldsymbol{f}(t),\ \boldsymbol{r}^*_{(l)}=\boldsymbol{r}_{(l)},\ \overline{\boldsymbol{U}}^*=\overline{\boldsymbol{U}}+\frac{\mathrm{d}\boldsymbol{f}}{\mathrm{d}t},\\
& \overline{P}^*=\overline{P}-\boldsymbol{x}\cdot\frac{\mathrm{d}^2\boldsymbol{f}}{\mathrm{d}t^2},\ \mathbf{R}^*_{\{n\}}=\mathbf{R}_{\{n\}},\ \mathbf{P}^*_{\{n\}}=\mathbf{P}_{\{n\}},\\
\bar{T}_{10}:\ & t^*=t\ ,\ \boldsymbol{x}^*=\boldsymbol{x},\ \boldsymbol{r}^*_{(l)}=\boldsymbol{r}_{(l)},\ \overline{\boldsymbol{U}}^*=\overline{\boldsymbol{U}},\\
& \overline{P}^*=\overline{P}+f_4(t),\ \mathbf{R}^*_{\{n\}}=\mathbf{R}_{\{n\}},\ \mathbf{P}^*_{\{n\}}=\mathbf{P}_{\{n\}},
\end{aligned} \tag{3.48}$$

where all function and parameter definitions are adopted from 3.1.2 and $\mathbf{A}$ is a concatenation of rotation matrices as $A_{i_{(0)}j_{(0)}i_{(1)}j_{(1)}\dots i_{(n)}j_{(n)}}=a_{i_{(0)}j_{(0)}}a_{i_{(1)}j_{(1)}}\dots a_{i_{(n)}j_{(n)}}$.

The latter symmetries may also be transformed into the $\mathbf{H}$-notation according to equation (3.15).

Statistical symmetries of the MPC equations The concept of an extended set of symmetries for the MPC equation in the form (3.15) or (3.22) may e.g. be taken from [20] and [11]. Its importance was not observed therein - rather it was stated that they may be mathematical artifacts of the averaging process and probably physically irrelevant. The set of new symmetries was first presented and its key importance for turbulence recognised in [21] and later extended in [25].

The actual finding of symmetries of the non-rotating MPC is rather difficult since an infinite system of equations has to be analyzed. For this task, however, it is considerably easier to investigate the linear **H**-**I**-system (3.15)-(3.16) rather than the non-linear **R**-**P**-system (3.22). However, since the latter formulation is more common, the symmetries will finally be re-written in **R**-**P** notation.

This new set of symmetries for the **H**-**I**-system (3.15)-(3.16) can be separated into three distinct sets of symmetries

$$\bar{T}_1' : t^* = t\,,\ \boldsymbol{x}^* = \boldsymbol{x}\,,\ \boldsymbol{r}_{(l)}^* = \boldsymbol{r}_{(l)} + \boldsymbol{k}_{(l)}\,,\ \mathbf{H}_{\{n\}}^* = \mathbf{H}_{\{n\}}\,,\ \mathbf{I}_{\{n\}}^* = \mathbf{I}_{\{n\}}\,, \tag{3.49}$$

$$\bar{T}_{2_{\{n\}}}' : t^* = t, \boldsymbol{x}^* = \boldsymbol{x}, \boldsymbol{r}_{(l)}^* = \boldsymbol{r}_{(l)}, \mathbf{H}_{\{n\}}^* = \mathbf{H}_{\{n\}} + \mathbf{C}_{\{n\}}, \mathbf{I}_{\{n\}}^* = \mathbf{I}_{\{n\}} + \mathbf{D}_{\{n\}}, \tag{3.50}$$

$$\bar{T}_s' : t^* = t\,,\ \boldsymbol{x}^* = \boldsymbol{x}\,,\ \boldsymbol{r}_{(l)}^* = \boldsymbol{r}_{(l)}\,,\ \mathbf{H}_{\{n\}}^* = \mathrm{e}^{k_s}\mathbf{H}_{\{n\}}\,,\ \mathbf{I}_{\{n\}}^* = \mathrm{e}^{k_s}\mathbf{I}_{\{n\}}\,, \tag{3.51}$$

all of which are of purely statistical nature. In particular the symmetries (3.50) and (3.51) have their counterpart (2.30) and (2.29) admitted by the moment equations for the Burgers equations. Subsequently, we will call them statistical symmetries. In the translation of the relative coordinates (3.49) $\boldsymbol{k}_{(l)}$ represents the related set of group parameters. Note that this group is not related to the classical Galilean group in usual $\boldsymbol{x}$-space (here $T_7 - T_9$ in equation (3.2) with $\boldsymbol{f} = const.$), which is translational in the velocity.

The second set of statistical symmetries (3.50) was in fact already partially identified in [20], however, mistakenly taken for the Galilean group. In the above general form it was identified in [21], where $\mathbf{C}_{\{n\}}$ and $\mathbf{D}_{\{n\}}$ refer to group parameters and further extended in [25], so that $\mathbf{C}_{\{n\}}$ is a function of time $\mathbf{C}_{\{n\}}(t)$ and then a temporal derivative of $\mathbf{C}_{\{n\}}$ appears also for the transformation of $\mathbf{I}_{\{n\}}^*$.

The third statistical group (3.51) that has been identified denotes simple scaling of all MPC tensors with the same factor e^{k_s}.

Furthermore there exists at least one more symmetry, which consists of a combination of multi-point velocity and of pressure-velocity correlations (see e.g. in [25]). Its concrete form is omitted at this point because it is not needed for the further considerations.

It should finally be added that due to the linearity of the MPC equation (3.15) another generic symmetry is admitted. This is in fact featured by all linear differential equations (see e.g. in [3]). It merely reflects the super-position principle of linear differential equations though usually cannot directly be adopted for the practical derivation of group invariant solutions.

We may transform (3.49)-(3.51) into classical **R**-**P**-notation to obtain

$$\begin{aligned} \bar{T}_1' : t^* = t,\ \boldsymbol{x}^* = \boldsymbol{x},\ \boldsymbol{r}_{(l)}^* = \boldsymbol{r}_{(l)} + \boldsymbol{k}_{(l)},\ \overline{\boldsymbol{U}}^* = \overline{\boldsymbol{U}}, \\ \overline{P}^* = \overline{P},\ \mathbf{R}_{\{n\}}^* = \mathbf{R}_{\{n\}},\ \mathbf{P}_{\{n\}}^* = \mathbf{P}_{\{n\}}, \end{aligned} \tag{3.52}$$

$$\bar{T}_{2_{\{1\}}}' : t^* = t,\ \boldsymbol{x}^* = \boldsymbol{x},\ \boldsymbol{r}_{(l)}^* = \boldsymbol{r}_{(l)},\ \overline{U}_{i_{(0)}}^* = \overline{U}_{i_{(0)}} + C_{i_{(0)}},$$

$$R^*_{i_{(0)}i_{(1)}} = R_{i_{(0)}i_{(1)}} + \overline{U}_{i_{(0)}}\overline{U}_{i_{(1)}} - \left(\overline{U}_{i_{(0)}} + C_{i_{(0)}}\right)\left(\overline{U}_{i_{(1)}} + C_{i_{(0)}}\right), \cdots \tag{3.53}$$

$$\bar{T}'_{2_{\{2\}}} : t^* = t,\ \ \boldsymbol{x}^* = \boldsymbol{x},\ \ \boldsymbol{r}^*_{(l)} = \boldsymbol{r}_{(l)},\ \overline{U}^*_{i_{(0)}} = \overline{U}_{i_{(0)}},$$
$$R^*_{i_{(0)}i_{(1)}} = R_{i_{(0)}i_{(1)}} + C_{i_{(0)}i_{(1)}},\ \ \cdots \tag{3.54}$$

$$\bar{T}'_s : t^* = t,\ \ \boldsymbol{x}^* = \boldsymbol{x},\ \ \boldsymbol{r}^*_{(l)} = \boldsymbol{r}_{(l)},\ \overline{U}^*_{i_{(0)}} = \mathrm{e}^{k_s}\overline{U}_{i_{(0)}},$$
$$R^*_{i_{(0)}i_{(1)}} = \mathrm{e}^{k_s}\left[R_{i_{(0)}i_{(1)}} + \left(1 - \mathrm{e}^{k_s}\right)\overline{U}_{i_{(0)}}\overline{U}_{i_{(1)}}\right],\ \ \cdots, \tag{3.55}$$

where for the translation symmetry (3.50) only $n = 1$ and $n = 2$ are presented in (3.53) and (3.54). Despite of the fact that each of these groups appear to be almost trivial, since they are simple translational groups in the dependent coordinates, they exhibit an increasing complexity with increasing tensor order if written in the **R**-**P**-formulation.

3.3.2 Symmetries of the LMN hierarchy

Symmetries of the LMN hierarchy implied by Euler and Navier-Stokes symmetries Similarly as the MPC hierarchy, the LMN equations are invariant under the classical symmetries of Navier-Stokes equations. This issue was investigated in [32], where the form of classical symmetries written in terms of PDF's was derived. Below, we shortly recall the findings.

The invariance under time and space translations, cf. (3.48), can be very easily inspected. Similar to the scaling of the Burgers equation (2.25), which leads to the scaling of the PDF in (2.27), the PDF equation (3.37) for $\nu = 0$ is invariant under two scaling groups

$$\bar{T}_2 : t^* = t\,,\ \ \boldsymbol{x}^*_{(l)} = \mathrm{e}^{k_2}\boldsymbol{x}_{(l)}\,,\ \ \boldsymbol{v}^*_{(l)} = \mathrm{e}^{k_2}\boldsymbol{v}_{(l)},\ \ f^*_{n+1} = \mathrm{e}^{-3(n+1)k_2} f_{n+1}, \tag{3.56}$$
$$\bar{T}_3 : t^* = \mathrm{e}^{k_3}t\,,\ \ \boldsymbol{x}_{(l)} = \boldsymbol{x}_{(l)}\,,\ \ \boldsymbol{v}^*_{(l)} = \mathrm{e}^{-k_3}\boldsymbol{v}_{(l)}\,,\ \ f^*_{n+1} = \mathrm{e}^{3(n+1)k_3} f_{n+1}, \tag{3.57}$$

which can be compared to the analogous symmetry of the MPCE, cf. (3.48). The scaling of the PDFs f_{n+1} assures that the normalization property (3.30) is satisfied for the transformed function f^*_{n+1}. In the viscous case the two above symmetries reduce to one scaling group with $k_3 = 2k_2$.

It was shown in [32] that the LMN equations are invariant with respect to the Galilean transformations $t^* = t$, $\boldsymbol{x}^*_i = \boldsymbol{x}_i + \boldsymbol{v}_0\,t$, $\boldsymbol{v}^*_i = \boldsymbol{v}_i + \boldsymbol{v}_0$, where $\boldsymbol{v}_0$ is a constant vector.

Statistical symmetries of the LMN hierarchy The statistical symmetries of the LMN hiererchy were first identified in [32] and their derivation will be shortly recalled below. We discuss here a set of transformations of the PDFs under which the LMN equations (3.37) turn out to be invariant and which corresponds to the set of statistical symmetries (3.50) and (3.51) found for the MPCE, where the moments **H** are, respectively shifted by a constant and scaled. For simplicity of notation, it is convenient to first derive symmetries of equation (3.36), where only the one- and two-point PDF are present. Finally, a generalisation to the n-point PDF will be given.

Statistical scaling symmetry As it was shown in [32] scaling of moments according to (3.51) transforms a PDF of a turbulent signal into a non-continuous PDF with delta function at the origin. Such function would correspond to a PDF of a turbulent signal interuppted by intervals where $\boldsymbol{U} = 0$. In reference [32] we first presented a one-point PDF as a Fourier transform of the characteristic function Φ_1 (see e.g. [24])

$$f_1(\boldsymbol{v}; \boldsymbol{x}, t) = \frac{1}{(2\pi)^3} \int \mathrm{e}^{-\mathrm{i}\boldsymbol{v}\cdot\boldsymbol{y}} \Phi_1(\boldsymbol{y}; \boldsymbol{x}, t)\, \mathrm{d}\boldsymbol{s} \;, \tag{3.58}$$

where the index (0) has been skipped. The one-point velocity statistics can be calculated as the n-th order derivative of Φ_1 at the origin

$$\left.\frac{\partial \Phi_1}{\partial y_{i_{(0)}}}\right|_{\boldsymbol{y}=0} = \mathrm{i}\overline{U_{i_{(0)}}(\boldsymbol{x}, t)},$$
$$\left.\frac{\partial^2 \Phi_1}{\partial y_{i_{(0)}} \partial y_{j_{(0)}}}\right|_{\boldsymbol{y}=0} = (\mathrm{i}^2)\overline{U_{i_{(0)}}(\boldsymbol{x}, t) U_{j_{(0)}}(\boldsymbol{x}, t)}, \quad \ldots . \tag{3.59}$$

Hence, the Taylor-series expansion of the characteristic function in (3.45) can be written as

$$\Phi_1 = 1 + \mathrm{i}\overline{U_{i_{(0)}}(\boldsymbol{x}, t)} y_{i_{(0)}} - \frac{1}{2!}\overline{U_{i_{(0)}}(\boldsymbol{x}, t) U_{j_{(0)}}(\boldsymbol{x}, t)} y_{i_{(0)}} y_{j_{(0)}} + \cdots , \tag{3.60}$$

with summation over repeating indices $i_{(0)}, j_{(0)}, \cdots = 1, \ldots, 3$.

If we substitute the transformation of moments (3.51) into (3.60) we obtain

$$\Phi_1^* = 1 + \mathrm{i}\mathrm{e}^{k_s}\overline{U_{i_{(0)}}(\boldsymbol{x}, t)} y_{i_{(0)}} - \frac{1}{2!}\mathrm{e}^{k_s}\overline{U_{i_{(0)}}(\boldsymbol{x}, t) U_{j_{(0)}}(\boldsymbol{x}, t)} y_{i_{(0)}} y_{j_{(0)}} + \cdots . \tag{3.61}$$

We note that the symmetry (3.51) transforms moments of the velocity, starting from the first-order moment, whereas the first term in the above Taylor series expansion is in fact the normalisation of the PDF ($\Phi_n(0) = 1$, hence $\int f_{n+1}\, \mathrm{d}\boldsymbol{v}_{(0)} \ldots \mathrm{d}\boldsymbol{v}_{(n)} = 1$). This term cannot be scaled, in order not to violate the properties of the PDF. Substituting (3.61) into equation (3.58) we obtain the transformed PDF, which can be written in the following form

$$f_1^*(\boldsymbol{v}; \boldsymbol{x}, t) = \delta(\boldsymbol{v}) + \mathrm{e}^{k_s}(f_1 - \delta(\boldsymbol{v}_{(0)})) . \tag{3.62}$$

As it was shown in [32], the symmetry (3.62) may be extended to the $n+1$-point PDF according to

$$f_{n+1}^* = \delta(\boldsymbol{v}_{(0)}) \cdots \delta(\boldsymbol{v}_{(n)}) + \mathrm{e}^{k_s}\left[f_{n+1} - \delta(\boldsymbol{v}_{(0)}) \cdots \delta(\boldsymbol{v}_{(n)})\right] . \tag{3.63}$$

This statistical symmetry has an absolutely remarkable property, which has been described by turbulence researchers for decades: intermittency. Here laminar and turbulent signals alternate. This property is exactly represented by this symmetry, because the delta function corresponds to a laminar flow, whereas turbulence is described by a broadband PDF. In symmetry (3.63), exactly these two parts are linked and hence it was named intermittency symmetry.

Being a PDF, the function f^*_{n+1} must satisy all properties of a PDF.

I. The positivity constraint for PDF implies

$$\delta(\boldsymbol{v}_{(0)})\cdots\delta(\boldsymbol{v}_{(n)})+\mathrm{e}^{k_s}(f_{n+1}-\delta(\boldsymbol{v}_{(0)})\cdots\delta(\boldsymbol{v}_{(n)}))\geq 0,\quad \forall\boldsymbol{v}\in\mathbb{R}^3,\forall\boldsymbol{x}\in\mathbb{R}^3, \tag{3.64}$$

which, for a continuous function f_{n+1} implies that $\mathrm{e}^{k_s}\leq 1$, hence $k_s\leq 0$. Apparently, such restrictions for the scaling parameter k_s means that some group axioms are not satisfied by the transformation (3.63).

II. The normalization condition (3.30) is satisfied.

III. The coincidence property (3.32) is satisfied as $\delta(\boldsymbol{v}_{(0)})\cdots\delta(\boldsymbol{v}_{(n)})=\delta(\boldsymbol{v}_{(0)})\cdots\delta(\boldsymbol{v}_{(n-1)}-\boldsymbol{v}_{(n)})$.

IV. The divergence or continuity condition (3.31) is satisfied, since $\delta(\boldsymbol{v}_{(0)})\cdots\delta(\boldsymbol{v}_{(n)})$ does not depend on the space variable.

V. The separation property (3.33) is not satisfied unless the PDF f_{n+1} itself is a delta function.

If we note that $\delta(\boldsymbol{v}_{(0)})\cdots\delta(\boldsymbol{v}_{(n)})$ does not depend on time and space variables, we see that the LMN equations (3.37) are in fact invariant under the transformation (3.63). The moments calculated from the transformed PDFs read

$$\begin{aligned}
&\overline{U_{i_{(0)}}(\boldsymbol{x}_{(0)},t)\ldots U_{i_{(n)}}(\boldsymbol{x}_{(n)},t)}^{*}\\
&=\int\left[f_{n+1}\mathrm{e}^{k_s}+(1-\mathrm{e}^{k_s})\delta(\boldsymbol{v}_{(0)})\cdots\delta(\boldsymbol{v}_{(n)})\right]v_{i_{(0)}}\ldots v_{i_{(n)}}\,\mathrm{d}\boldsymbol{v}_{(0)}\ldots\mathrm{d}\boldsymbol{v}_{(n)}=\\
&=\mathrm{e}^{k_s}\overline{U_{i_{(0)}}(\boldsymbol{x}_{(0)},t)\ldots U_{i_{(n)}}(\boldsymbol{x}_{(n)},t)}=\mathrm{e}^{k_s}\mathbf{H}_{\{n+1\}},
\end{aligned} \tag{3.65}$$

which is identical to equation (3.51).

Translation symmetry of moments If we substitute the translation symmetry of moments (3.50) into the Taylor-series expansion (3.60), we obtain

$$\begin{aligned}
&\Phi_1^*(\boldsymbol{s};\boldsymbol{x},t)=\Phi_1(\boldsymbol{s};\boldsymbol{x},t)\\
&\quad\underbrace{+\,\mathrm{i}C_{i_{(0)}}y_{i_{(0)}}-\frac{1}{2!}C_{i_{(0)}j_{(0)}}y_{i_{(0)}}y_{j_{(0)}}-\frac{1}{3!}\mathrm{i}C_{i_{(0)}j_{(0)}k_{(0)}}y_{i_{(0)}}y_{j_{(0)}}y_{k_{(0)}}\cdots}_{\phi(\boldsymbol{y})}.
\end{aligned} \tag{3.66}$$

The underbraced sum is the Taylor series expansion of a function $\phi(\boldsymbol{y})$ which equals 0 at the origin and its derivatives at the origin equals, respectively $\mathrm{i}C_{i_{(0)}}$, $-C_{i_{(0)}j_{(0)}}$,

$-\mathrm{i}C_{i_{(0)}j_{(0)}k_{(0)}}$ etc. If equation (3.66) is substituted into equation (3.58), the transformed PDF reads

$$f_1^*(\boldsymbol{v};\boldsymbol{x},t) = f(\boldsymbol{v};\boldsymbol{x},t) + \frac{1}{(2\pi)^3}\int \mathrm{e}^{-\mathrm{i}\boldsymbol{v}\cdot\boldsymbol{y}}\phi(\boldsymbol{y})\,\mathrm{d}\boldsymbol{y} = f_1(\boldsymbol{v};\boldsymbol{x},t) + \psi(\boldsymbol{v}) \ , \quad (3.67)$$

where $\psi(\boldsymbol{v})$ is an inverse Fourier transform of $\phi(\boldsymbol{y})$ and

$$\int \psi(\boldsymbol{v})\,\mathrm{d}\boldsymbol{v} = 0\,, \quad (3.68)$$

which follows from the fact that $\phi(0) = 0$. Note that neither $\phi(\boldsymbol{y})$ nor $\psi(\boldsymbol{v})$ depend on $\boldsymbol{x}$ or time t.

In [32] the considerations were also generalised to the case of the $n+1$-point PDF leading to the followig form of the transformed PDF

$$f_{n+1}^* = f_{n+1} + \psi(\boldsymbol{v}_{(0)})\delta(\boldsymbol{v}_{(0)} - \boldsymbol{v}_{(1)})\ldots\delta(\boldsymbol{v}_{(0)} - \boldsymbol{v}_{(n)}). \quad (3.69)$$

It can be readily verified that the transformations (3.69) correspond to the statistical symmetry (3.50) where the MPC $\mathbf{H}$ tensors are translated. Indeed, under the transformations (3.69), the moments of the PDFs are mapped to

$$\begin{aligned}
&\overline{U_{i_{(0)}}(\boldsymbol{x}_{(0)},t)\ldots U_{i_{(n)}}(\boldsymbol{x}_{(n)},t)}^* \equiv \left(\int f_{n+1}v_{i_{(0)}}\cdots v_{i_{(n)}}\,\mathrm{d}\boldsymbol{v}_{(0)}\cdots\,\mathrm{d}\boldsymbol{v}_{(n)}\right)^*\\
&= \overline{U_{i_{(0)}}(\boldsymbol{x}_{(0)},t)\ldots U_{i_{(n)}}(\boldsymbol{x}_{(n)},t)}\\
&+ \int \psi(\boldsymbol{v}_{(0)})\,\delta(\boldsymbol{v}_{(0)} - \boldsymbol{v}_{(1)})\ldots\delta(\boldsymbol{v}_{(n)} - \boldsymbol{v}_{(n-1)})\,v_{i_{(0)}}\cdots v_{i_{(n)}}\,\mathrm{d}\boldsymbol{v}_{(0)}\cdots\,\mathrm{d}\boldsymbol{v}_{(n)}\\
&= \overline{U_{i_{(0)}}(\boldsymbol{x}_{(0)},t)\ldots U_{i_{(n)}}(\boldsymbol{x}_{(n)},t)}\\
&+ \int \psi(\boldsymbol{v}_{(0)})\,v_{i_{(0)}}\cdots v_{i_{(n)}}\,\mathrm{d}\boldsymbol{v}_{(0)} = \mathbf{H}_{\{n+1\}} + \mathbf{C}_{\{n+1\}}.
\end{aligned} \quad (3.70)$$

In [32] the PDF conditions, which of the above transformed PDF functions have to satisfy, were analysed

I. The coincidence property (3.32) is satisfied by the transformed PDFs (3.69).

II. Because of the probabilistic interpretation of the PDFs, it is locally required that

$$f_{n+1} + \psi(\boldsymbol{v}_{(0)})\delta(\boldsymbol{v}_{(0)} - \boldsymbol{v}_{(1)})\ldots\delta(\boldsymbol{v}_{(0)} - \boldsymbol{v}_{(n)}) \geq 0,\ \forall \boldsymbol{v} \in \mathbb{R}^3, \forall \boldsymbol{x} \in \mathbb{R}^3, \quad (3.71)$$

while the normalization condition globally imposes

$$\int \psi(\boldsymbol{v})\,\mathrm{d}\boldsymbol{v} = 0, \quad (3.72)$$

which agrees with equation (3.68).

III. The divergence or continuity condition (3.31) is satisfied, since ψ does not depend on the space variable.

IV. The separation property (3.33) is not satisfied, as the transformation of the PDF is independent of the spatial variable and thus cannot satisfy this limiting behavior. On the other hand, let us note that, while it is reasonable to require this property, this is never used in the derivation of the equations of the LMN hierarchy. Moreover, this property is not satisfied by the corresponding symmetries of the MPC equations, either.

It was noted in [32] that the function f^*_{n+1} transformed according to (3.69) is always a solution of equation (3.37). However, such f^*_{n+1} may not be a PDF any more, as it may have negative values. Then, (3.69) may not necessarily be a Lie group but still is a symmetry of the LMN equation equation (3.37). This implies that the constants $C_{\{n+1\}}$ obtained in equation (3.70) are not arbitrary, but due to the condition (3.71) on the functions f^*_{n+1} and ψ we expect that they might be limited to a certain range. We conclude that considering the symmetries of the LMN hierarchy provides additional restrictions on the group parameters, which were not observed in the MPC approach.

3.3.3 Symmetries of the Hopf functional equation

Symmetries of the Hopf equation implied by Euler and Navier-Stokes symmetries In order to verify the invariance of the Hopf functional equation under the scaling groups T_2 and T_3 in (3.2) and T_{NS} in (3.4) we first consider transformations of $n+1$-point characteristic functions. From the relation

$$\Phi^*_{n+1} = \int \mathrm{e}^{i\boldsymbol{v}^*_{(0)}\cdot\boldsymbol{y}^*_{(0)}} \dots \mathrm{e}^{i\boldsymbol{v}^*_{(n)}\cdot\boldsymbol{y}^*_{(n)}} f^*_{n+1} \mathrm{d}\boldsymbol{v}^*_{(0)} \dots \mathrm{d}\boldsymbol{v}^*_{(n)} \tag{3.73}$$

we find, that the scaling symmetries (3.56) and (3.57) will hold, if $\boldsymbol{y}^*_{(i)} = \mathrm{e}^{-k_2}\boldsymbol{y}_{(i)}$ and $\boldsymbol{y}^*_{(i)} = \mathrm{e}^{k_3}\boldsymbol{y}_{(i)}$ for each i, as in such a case the exponent $\boldsymbol{v}^*_{(i)} \cdot \boldsymbol{y}^*_{(i)} = \boldsymbol{v}_{(i)} \cdot \boldsymbol{y}_{(i)}$ remains unchanged and using (3.56) we obtain

$$\begin{aligned} \Phi^*_{n+1} &= \int \mathrm{e}^{3(n+1)k_2} \mathrm{e}^{i\boldsymbol{v}^*_{(0)}\cdot\boldsymbol{y}^*_{(0)}} \dots \mathrm{e}^{i\boldsymbol{v}^*_{(n)}\cdot\boldsymbol{y}^*_{(n)}} f_{n+1} \mathrm{e}^{-3(n+1)k_2} \mathrm{d}\boldsymbol{v}_{(0)} \dots \mathrm{d}\boldsymbol{v}_{(n)} \\ &= \Phi_{n+1} \ . \end{aligned} \tag{3.74}$$

The same holds for the second scaling group (3.57), i.e. the $n+1$-point characteristic function (3.73) transforms as $\Phi^*_{n+1} = \Phi_{n+1}$. We expect that the same should hold in the limit $n \to \infty$ i.e. for the characteristic functional.

In this case the first, space scaling group might be problematic at first sight as in the definition of the characteristic functional

$$\Phi^* = \overline{\mathrm{e}^{i\int \boldsymbol{U}^*(\boldsymbol{x}^*,t)\cdot\boldsymbol{y}^*(\boldsymbol{x})\mathrm{d}\boldsymbol{x}^*}} \ , \tag{3.75}$$

the integral over space $\boldsymbol{x}$ is present in the exponent and as we remember from formula (3.56), the space scales as $\boldsymbol{x}^* = \boldsymbol{x}\mathrm{e}^{k_2}$. The answer is that in the continuum case instead of the discrete k-th variable $y_{i_{(k)}}$ we deal with $y_i(\boldsymbol{x})\mathrm{d}\boldsymbol{x}$. The sums are replaced by integrals in the continuum limit and hence $y_i\mathrm{d}\boldsymbol{x}$ should scale as $y_{i_{(k)}}$ in the discrete case, i.e. $y_i^*\mathrm{d}\boldsymbol{x}^* = \mathrm{e}^{-k_2}y_i\mathrm{d}\boldsymbol{x}$. Because $\mathrm{d}\boldsymbol{x}$ scales as $\mathrm{d}\boldsymbol{x}^* = \mathrm{d}\boldsymbol{x}\mathrm{e}^{3k_2}$ it follows that $y_i^* = \mathrm{e}^{-4k_2}y_i$.

To sum up, it can be shown that the Hopf functional equation (3.44) for $\nu = 0$ is invariant under the following scaling transformation of variables

$$T_2 : \Phi^* = \Phi\,,\ \boldsymbol{x}^* = \mathrm{e}^{k_2}\boldsymbol{x}\,,\ t^* = t\,,\ y_i^*\mathrm{d}\boldsymbol{x}^* = \mathrm{e}^{-k_2}y_i\mathrm{d}\boldsymbol{x}\,,\ \boldsymbol{y}^* = \mathrm{e}^{-4k_2}\boldsymbol{y}, \tag{3.76}$$

$$T_3 : \Phi^* = \Phi\,,\ \boldsymbol{x}^* = \boldsymbol{x}\,,\ t^* = \mathrm{e}^{k_3}t\,,\ y_i^*\mathrm{d}\boldsymbol{x}^* = \mathrm{e}^{k_3}y_i\mathrm{d}\boldsymbol{x}\,,\ \boldsymbol{y}^* = \mathrm{e}^{k_3}\boldsymbol{y}. \tag{3.77}$$

For non-vanishing viscosity $\nu \neq 0$, instead of two scaling groups we obtain one scaling according to (3.4).

The generalized Galilean invariance $T_7 - T_9$ is given in (3.2). A proof for the admittance of the classical Galilean group of equation (3.43) under the assumption $\frac{\mathrm{d}\boldsymbol{f}}{dt} = \boldsymbol{U}_0$ may be taken from [23].

Statistical symmetries of the Hopf equation As it was discussed in Section 3.2.5, E. Hopf proposed to present the solution of the Hopf functional as the infinite series expansion (3.45), where

$$C_{n+1} = \int K_{i_{(0)}\cdots i_{(n)}}(\boldsymbol{x}_{(0)},\ldots,\boldsymbol{x}_{(n)},t)y_{i_{(0)}}(\boldsymbol{x}_{(0)})\cdots y_{i_{(n)}}(\boldsymbol{x}_{(n)})\mathrm{d}\boldsymbol{x}_{(0)}\cdots\mathrm{d}\boldsymbol{x}_{(n)} \tag{3.78}$$

with the kernel function given by equation (3.47).

If we substitute the statistical symmetries of moments given by (3.51) from section 3.3.1 into the equations (3.46) and (3.47) we find that it transforms the kernel functions as $K^*_{i_{(0)}\cdots i_{(n)}} = \mathrm{e}^{k_s}K_{i_{(0)}\cdots i_{(n)}}$, and hence, equation (3.45) reads

$$\Phi^* = 1 + \mathrm{e}^{k_s}\left(C_1 + C_2 + \cdots\right), \tag{3.79}$$

or

$$\Phi^* = 1 + \mathrm{e}^{k_s}\left(\Phi - 1\right). \tag{3.80}$$

Similarly, the statistical symmetry (3.50) translates the kernel functions (3.46) by a constant which leads to the following translation of the $n+1$th term in the Taylor series expansion

$$\Phi^*_{n+1} = \Phi_{n+1} + \int C_{i_{(0)}\cdots i_{(n)}}y_{i_{(0)}}(\boldsymbol{x}_{(0)})\cdots y_{i_{(n)}}(\boldsymbol{x}_{(n)})\mathrm{d}\boldsymbol{x}_{(0)}\cdots\mathrm{d}\boldsymbol{x}_{(n)}. \tag{3.81}$$

Hence, the translation symmetry of the characteristic functional Φ, corresponding to (3.50) can be written in the following form

$$\Phi^* = \Phi + \Psi([\boldsymbol{y}(\boldsymbol{x})]), \tag{3.82}$$

where Ψ is a functional such that its nth functional derivative at the origin equals $C_{i_{(0)}\cdots i_{(n)}}$.

To sum up the content of the preceding sections, it was shown that all approaches for the full statistical description of turbulence, namely MPC hierarchy, LMN hierarchy for PDF's and the Hopf characteristic functional equations are invariant under classical symmetries of Navier-Stokes equations and additionally under the set of statistical symmetries: translation of moments and statistical scaling. Through the analysis of the PDF equations the statistical translation and statistical scaling symmetries (3.63) and (3.69) were identified as been closely connected to intermittency. Both symmetries acting simultaneously transforming the PDF e.g. from a turbulent to an intermittent (laminar/turbulent) or laminar PDF. Hence, the statistical symmetries indicate the fact that solutions of Navier-Stokes equations may have physically very different character. Such transformations have only be observed in the statistical approach; hence the statistical symmetries were not found in the Lie group analysis of the Navier-Stokes equations in its classical form (3.1).

4 Summary and outlook

Within the present article it was shown that the admitted set of symmetry groups of the infinite set of multi-point correlation equations is considerably extended by three classes of groups compared to those originally stemming from the Euler and the Navier-Stokes equations. Moreover, it was shown that the corresponding extended set of symmetries is also admitted by the LMN equations for velocity PDF and the Hopf functional approach. However, so far completeness of all admitted symmetries of the MPC equation has not been shown. This appears to be essential not only from a theoretical point of view but rather essential to understand scaling properties of higher moments.

Implicitly, symmetries have been used in engineering-type semi-empirical turbulence modeling for several decades since essentially all symmetries of Euler and the Navier-Stokes equations have been made part of modern turbulence models. Still, this is only partially true for the new statistical symmetries. In fact, some of them have been employed even in very early turbulence models. However, most of the statistical symmetries have never been made use of and, in fact, it will be difficult to make turbulence models consistent with some of the symmetries such as the new scaling symmetry (3.51). Completely new turbulence modeling concepts will be needed to account for these symmetries, which, apparently, imply important turbulence properties such as intermittency.

The present approach is intentionally limited to incompressible flows, whereby turbulence is also of great importance in many other areas. An apparent extension of the presented theoretical concepts to turbulent heat transfer problems can be achieved by adding the scalar heat transfer equation to the Navier-Stokes equations (3.1). This might be either in the most simple case of a passive temperature equation or, if buoyancy is considered, the temperature equation again, while the momentum equation (3.1) is extended by the Boussinesq approximation. In both cases, the multi-point system (3.15) is extended by an additional tensor composed of the temperature correlated with m velocities. Note, that because of the linearity of the temperature equation and the linearity of the temperature in the momentum equation due the Boussinesq approximation, it is not necessary to derive correlations and equations which contain the temperature more than once.

Other extensions such as to Navier-Stokes equations for compressible fluids, i.e. the gas dynamics equations prolonged by viscous terms, are less straight forward. The reason is that due to the variable density and energy a consistent multi-point concept has not been derived yet.

Finally, depending on the Mach number Ma, the gas-dynamics equations may change type and hence, for $Ma > 1$ correlations may only be uniquely defined within the Mach cone. This, apparently, posses an additionally non-trivial constraint on the problem which is unsolved to date.

References

[1] Andreev V K, Rodionov A A, Group analysis of the equations of planar flows of an ideal fluid in Lagrange coordintes, *Dokl. Akad. Nauk SSSR* **298** (1988), 1358–1361.

[2] Avsarkisov V, Oberlack M, Hoyas S, New scaling laws for turbulent Poiseuille flow with wall transpiration, *J. Fluid Mech.* **746** (2014), 99–122.

[3] Bluman G, Cheviakov A, Anco S, *Applications of Symmetry Methods to Partial Differential Equations*, (Appl. Math. Sci.) **168**, Springer-Verlag New York, 2010.

[4] Cantwell B J, Similarity transformations for the two-dimensional, unsteady, stream-function equation, *J. Fluid Mech.* **85** (1978), 257–271.

[5] Friedrich R, Daitche A, Kamps O, Lüff J, Voßkuhle M, Wilczek M, The Lundgren-Monin-Novikov hierarchy: Kinetic equations for turbulence, *Com. Ren. Phy.* **13** (2012), 929.

[6] Gaeta G, Symmetry of stochastic non-variational differential equations, *Physics Reports* **686** (2017), 1–62.

[7] Grigoriev Y N, Ibragimov N H, Kovalev V F, Meleshko S V, *Symmetries of Integro-differential Equations: with Applications in Mechanics and Plasma Physics*, Springer Netherlands, 2010.

[8] Hopf E, Statistical hydromechanics and functional calculus, *J. Rational Mech. Anal.* **1** (1952), 87–122.

[9] Hosokawa I, Monin-Lundgren hierarchy versus the Hopf equation in the statistical theory of turbulence, *Phys. Rev. E* **73** (2006), 067301.

[10] Keller L, Friedmann A, Differentialgleichungen für die turbulente Bewegung einer kompressiblen Flüssigkeit, *Proc. First. Int. Congr. Appl. Mech.* (1924), 395–405.

[11] Khujadze J, Oberlack M, DNS and scaling laws from new symmetry groups of ZPG turbulent boundary layer flow, *Theoret. Comput. Fluid Dyn.* **18** (2004), 391–411.

[12] Kozlov R, On symmetries of stochastic differential equations, *Communications in Nonlinear Science and Numerical Simulation* **17** (2012), 4947 – 4951.

[13] Kozlov R, On symmetries of the Fokker-Planck equation, *Journal of Engineering Mathematics* **82** (2013), 39–57.

[14] Lundgren T S, Distribution functions in the statistical theory of turbulence, *Phys. Fluids* **10** (1967), 969–975.

[15] Mehdizadeh A, *Direct Numerical Simulation, Lie Group Analysis and Modeling of a Turbulent Channel Flow with Wall-normal Rotation*, PhD thesis, TU Darmstadt, 2010.

[16] Mehdizadeh A, Oberlack M, Analytical and numerical investigations of laminar and turbulent Poiseuille-Ekman flow at different rotation rates, *Phys. of Fluids* **22** (2010), 105104.

[17] Monin A, Yaglom A, *Statistical Fluid Mechanics. Chap. II Statistical description of turbulence*, MIT Press Cambridge, 1971.

[18] Monin A S, Equations of turbulent motion, *Prikl. Mat. Mekh* **31** (1967), 1057–1967.

[19] Novikov E A, Kinetic equations for a vortex field, *Soviet Physics-Doklady* **12** (1968), 1006–1008.

[20] Oberlack M, *Symmetrie, Invarianz und Selbstähnlichkeit in der Turbulenz*, Habilitation thesis, RWTH Aachen, 2000.

[21] Oberlack M, Rosteck A, New statistical symmetries of the multi-point equations and its importance for turbulent scaling laws, *Discrete Contin. Dyn. Sys., Ser. S* **3** (2010), 451–471.

[22] Oberlack M, Wacławczyk M, On the extension of Lie group analysis to functional differential equations, *Arch. Mech.* **58** (2006), 597.

[23] Oberlack M, Wacławczyk M, Rosteck A, Avsarkisov V, Symmetries and its importance for statistical turbulence theory, *Bulletin of the JSME: Mechanical Engineering Reviews* **2** (2015), 1–72.

[24] Pope S B, *Turbulent Flows*, Cambridge University Press Cambridge, 2000.

[25] Rosteck A, Oberlack M, Lie algebra of the symmetries of the multi-point equations in statistical turbulence theory, *J Nonlinear Math. Phys.* **18** (2011), 251–264.

[26] Rosteck A M, *Scaling Laws in Turbulence - A Theoretical Approach Using Lie-Point Symmetries*, PhD thesis, TU Darmstadt, 2013.

[27] Shen H, Wray A A, Stationary turbulent closure via the Hopf functional equation, *Journal of Statistical Physics* **65** (1991), 33–52.

[28] Ünal G, Symmetries of Itô and Stratonovich dynamical systems and their conserved quantities, *Nonlinear Dynamics* **32** (2003), 417–426.

[29] Vishik M I, Analytical solutions of Hopf equation which corresponds to parabolic equations or the Navier-Stokes system in *Problems of Mechanics and Mathematical Physics*, Nauka Moscow, 69–97, 1976.

[30] Wacławczyk M, Grebenev V N, Oberlack M, Lie symmetry analysis of the Lundgren-Monin-Novikov equations for multi-point probability density functions of turbulent flow, *Journal of Physics A: Mathematical and Theoretical* **50** (2017), 175501.

[31] Wacławczyk M, Oberlack M, Application of the extended Lie group analysis to the Hopf functional formulation of the Burgers equation, *J. Math. Phys.* **54** (2013), 072901.

[32] Wacławczyk M, Staffolani N, Oberlack M, Rosteck A, Wilczek M, Friedrich R, Statistical symmetries of the Lundgren-Monin-Novikov hierarchy, *Phys. Rev. E* **90** (2014), 013022.

[33] Wilczek M, Daitche A, Friedrich R, On the velocity distribution in homogeneous isotropic turbulence: correlations and deviations from gaussianity, *J. Fluid Mech.* **676** (2011), 191.

[34] Wilczek M, Daitche A, Friedrich R, Theory for the single-point velocity statistics of fully developed turbulence, *EPL* **93** (2011), 34003.

D1. Integral transforms and ordinary differential equations of infinite order

Alan Chávez [a], *Humberto Prado* [b] *and Enrique G Reyes* [b]

[a] *Departamento de Matemáticas,*
Universidad Nacional de Trujillo,
Av. Juan Pablo II s/n. Trujillo-Perú
alancallayuc@gmail.com

[b] *Departamento de Matemática y Ciencia de la Computación,*
Universidad de Santiago de Chile,
Casilla 307 Correo 2, Santiago, Chile.
e_g_reyes@yahoo.ca ; enrique.reyes@usach.cl ; humberto.prado@usach.cl

Abstract

We review our work on ordinary differential equations of infinite order and we apply our theory to equations defined with the help of the Riemann zeta function which are of interest for modern theoretical physics. We interpret these equations with the help of the Laplace and Borel transforms, we study existence, uniqueness and regularity of solutions to the equation

$$f(\partial_t)\phi = J(t)\ , \quad t \geq 0\ ,$$

and we analyze the delicate issue of the initial value problem.

Dedicated to the memory of Prof. Braulio Flores Timble,
formerly at the Faculty of Engineering, USACH.

1 Introduction

We first learned about equations in an infinite number of derivatives from the papers [12, 13] on non-local cosmology. In these works the authors consider the equation

$$\Box e^{-c\Box}\phi = U(x, \phi)\ , \quad c > 0\ , \tag{1.1}$$

in which $\Box$ is the d'Alambert operator and U is some nonlinear function of two variables, and they explain how to obtain it from the highly sophisticated non-commutative action functional first proposed by Witten in [55]. We have reported our attempts at understanding *Euclidean* versions of nonlinear equations such as (1.1) in several papers. A summary of our work appears in [38]: basically, we study nonlinear equations of the form

$$f(\Delta)\phi = U(\cdot, \phi)$$

using fixed point theorems applied to operators defined on spaces connected with generalized versions of Sobolev spaces. On the other hand, the Lorentzian equation (1.1) is still ahead of us. What we have developed up to now is a theory for handling linear "1 + 0" Lorentzian equations, this is, linear ordinary differential equations of the form

$$f(\partial_t)\phi = J(t) \,. \tag{1.2}$$

While this is of course just a first step in the study of Lorentzian equations "in infinitely many derivatives", we have found that this particular case is already very rich: Laplace and Borel transforms (see [24, 9]) are used in essential ways; there exist sophisticated examples of equations (1.2); the formulation of initial value problems is quite subtle and, we can *prove* that the left hand side of (1.2) can indeed be understood as a power series in the operator ∂_t, a fact we have found has been simply assumed in the classical theory on the subject (see for instance [18]) and in the physical literature (see for instance [30, 44]).

In this chapter we summarize our present understanding of (1.2). First of all we present some differential operators of infinite order appearing in some branches of mathematics and physics. In actual fact, these equations were first considered in the XIXth Century! The reader is referred to [10] for a very interesting and inspiring older paper on this subject. The reference [18] mentioned above summarizes these earlier researches. It appears to us that the theory lost some of its appeal due to the lack of physical applications (see however references to mid-twentieth century physical theories in [20]). This climate truly changed with the advent of string theory and modern cosmology; see for instance [55, 12, 13, 44, 30, 36] and references therein. Rather than extracting equations from these references, we mention in the second section two recent equations worthy of investigation, and we point out the existence of papers in other areas of mathematics and physics in which these equations also appear.

The third section of this work is on the mathematical tools we need: Laplace transform and Borel transform; the latter can be understood as a wide extension of the former to the complex domain. Then in Section 4 we define the operator $f(\partial_t)$ for appropriate analytic functions f, and we use our definition to solve (1.2). Our proposal for an understanding of the initial value problem for (1.2) is summarized in Section 5. Regretfully, our approach in Section 4 leaves out a most important equation, the linearized version of the Riemann zeta function model of string theory introduced by Dragovich in [28, 27], and one of the equations we present in Section 2. With this motivation at hand, in Section 6 we generalize our work through the use of Borel transform, first used in the context of equations such as (1.2) in [17]. Finally, in Section 7 we explicitly apply our theory to an equation of infinite order defined with the help of the Riemann zeta function.

This chapter is based on [37, 36, 17, 20, 21]. Since we are interested here in presenting our research rather than in introducing new results, we omit most of the proofs. They are to be found in the papers just cited. A final remark: in previous papers on the subject we have used "nonlocal equation" or "equation with infinitely many derivatives" indistinctly. In this work (and in [20, 21]) we are more careful:

it can be proven (see Sections 4 and 6 below and references [17, 20, 21]) that these two expressions can be understood rigorously and that they do coincide *locally* (in a sense to be specified in the text) but they are different in full generality. Below we use "nonlocal equation" and "nonlocal operator" unless we are carrying out an informal discussion or, we have proved that we can think of the expression $f(\partial_t)$ as a power series.

2 Differential operators of infinite order in mathematics and physics

We mention two appearences of nonlocal operators in mathematics, and we also discuss briefly its interest for string theory. We remark that these operators appear elsewhere as well; see for instance [34, 32, 33] for differential operators of infinite order in Lie groups, Lie algebras and representation theory, and [8, 14, 31] for equations of infinite order in non-local theories of gravity.

2.1 Does the Riemann zeta function satisfy a differential equation?

"Does the Riemann zeta function satisfy a differential equation?" is the title of a work in Analytic Number Theory by R.A. van Gorder published in 2015 in the Journal of Number Theory, see [50]. The aim of his work is to show that the Riemann zeta function satisfies, formally, a differential equation of infinite order.

Let $a \in \mathbb{Z}$ and set $D := \dfrac{d}{dz}$ (the complex differentiation operator). We define the following infinite order differential operator

$$exp(aD) = 1 + \sum_{k=1}^{\infty} \frac{a^k}{k!} D^k \, .$$

Here we just assume that this operator is well behaved and we do not worry about convergence issues. Let us define the following family of functions:

$$p_n(z) = \begin{cases} 1 & if \quad n = 0 \\ \dfrac{1}{(n+1)!} \prod_{j=0}^{n-1} (z+j) & if \quad n \geq 1. \end{cases} \tag{2.1}$$

Using these functions we define (again, formally) $L_n := p_n(z) exp(aD)$, and we consider the infinite sum of the operators L_n,

$$T = \sum_{n=0}^{\infty} L_n \, . \tag{2.2}$$

The operator T is again an infinite order differential operator. We have [50, Theorem 3.1]:

Theorem 1. *The Riemann zeta function ζ satisfies formally the non-homogeneous infinite order differential equation:*

$$T[\zeta(z) - 1] = \frac{1}{z-1} .$$

The importance of this theorem stems from the fact that it is known that there is no algebraic differential equation of finite order which is satisfied by the Riemann zeta function, see [5].

2.2 Distribution of zeros of entire functions

The study of the distribution of zeros of entire functions is very interesting by its implications for the analysis of the non-trivial zeros of the Riemann zeta function; see for example the comments appearing in [16, p. 1725–1726] and references therein. There exist results on the distribution of zeros of real entire functions in which nonlocal operators play a role.

The Laguerre-Pólya class, denoted by $\mathcal{LP}$, is defined as the collection of entire functions f having only real zeros, and such that f has the following factorization [9, sections 2.6 and 2.7]:

$$f(z) = cz^m e^{\alpha z - \beta z^2} \prod_k \left(1 - \frac{z}{\alpha_k}\right) e^{z/\alpha_k} ,$$

where $c, \alpha, \beta, \alpha_k$ are real numbers, $\beta \geq 0$, $\alpha_k \neq 0$, m is a non-negative integer, and $\sum_{k=1}^{\infty} \alpha_k^{-2} < \infty$.

Let D be the differentiation operator and $\phi \in \mathcal{LP}$; the next lemma says that the operator $\phi(D)$ (defined via power series, see [42]) is well defined in $\mathcal{LP}$, see [42, Theorem 8, p. 360].

Lemma 1. *Let $\phi, f \in \mathcal{LP}$ such that $\phi(z) = e^{-\alpha z^2}\phi_1(z)$ and $f(z) = e^{-\beta z^2} f_1(z)$, where ϕ_1, f_1 have genus 0 or 1 and $\alpha, \beta \geq 0$. If $\alpha\beta < 1/4$, then $\phi(D)f \in \mathcal{LP}$.*

The notion of the genus of a function is explained in [9, p. 22].

The following result, stated in [15, Theorem 1], is our promised application of infinite order differential operators to the problem of distribution of zeros of some real entire functions.

Theorem 2. *Let $\phi, f \in \mathcal{LP}$ such that $\phi(z) = e^{-\alpha z^2}\phi_1(z)$ and $f(z) = e^{-\beta z^2} f_1(z)$, where ϕ_1, f_1 have genus 0 or 1 and $\alpha, \beta \geq 0$. If $\alpha\beta < 1/4$ and ϕ has infinitely many zeros, then $\phi(D)f$ has only simple and real zeros.*

Further information on the connection between differential operators of infinite order and the distribution of zeros of entire functions is in [15, 42] and references therein.

2.3 Differential equations of infinite order in string theory

Physicists have investigated the inner consistency of string theory by using p-adic strings, see [1, 51, 52, 53, 11] and also the more recent review [25].

Let p be a prime number. We consider the following Lagrangian formulation of the open p-adic scalar tachyon field after [3, 26, 44, 51, 52, 11]:

$$L_p = \frac{m_p^D}{g_p^2} \frac{p^2}{p-1} \Big(-\frac{1}{2} \phi p^{-\Box/(2m_p^2)} \phi + \frac{1}{p+1} \phi^{p+1} \Big) , \tag{2.3}$$

where $\Box$ is the d'Alembert operator defined by a Lorentzian metric, and m_p^D, g_p^2 are physical constants which we do not discuss. This Lagrangian is defined only formally; it is well defined in "1 + 0" case of Lorentzian metrics and in the cosmological case (for instance in de Sitter cosmology) for which $\Box$ becomes an ordinary differential operator, see [45, 36] and references therein. The equation of motion for (2.3) is

$$p^{\Box/(2m_p^2)} \phi = \phi^p . \tag{2.4}$$

We remark that the $1+0$ dimensional version of this equation, $p^{a\,\partial_t^2} \phi = \phi^p$, $a > 0$, has been studied rigorously by Vladimirov and Volovich, see [51, 52].

Taking the Lagrangian (2.3) as a starting point, Dragovich, see [26], considers the following model

$$L = \sum_{n=1}^{\infty} C_n L_n = \sum_{n=1}^{\infty} C_n \frac{m_n^D}{g_n^2} \frac{n^2}{n-1} \Big(-\frac{1}{2} \phi n^{-\Box/(2m_n^2)} \phi + \frac{1}{n+1} \phi^{n+1} \Big), \tag{2.5}$$

in which all the Lagrangians L_n given by (2.3) are considered. In this situation, explicit Lagrangians occur for different choices of the coefficients C_n in (2.5); for example, let $h \in \mathbb{R}$ and consider the coefficients

$$C_n = \frac{n-1}{n^{2+h}} .$$

Then Dragovich's Lagrangian becomes

$$L_h = \frac{m^D}{g^2} \Big(-\frac{1}{2} \phi \sum_{n=1}^{\infty} n^{-\Box/(2m_n^2)-h} \phi + \sum_{n=1}^{\infty} \frac{n^{-h}}{n+1} \phi^{n+1} \Big) .$$

Now, using the Euler product for the Riemann zeta function

$$\zeta(s) := \sum_{n=1}^{\infty} \frac{1}{n^s} , \qquad Re(s) > 1 ,$$

see [41], we see that Dragovich's Lagrangian becomes

$$L_h = \frac{m^D}{g^2} \Big(-\frac{1}{2} \phi \zeta\big(\frac{\Box}{2m^2} + h\big) \phi + \sum_{n=1}^{\infty} \frac{n^{-h}}{n+1} \phi^{n+1} \Big) .$$

In this case, the equation of motion is

$$\zeta\left(\frac{\Box}{2m^2}+h\right)\phi=U(\phi)\,. \tag{2.6}$$

Motivated by (2.6), we consider below a linear version of Equation (2.6) in the $1+0$ dimensional case, that is,

$$\zeta\left(\partial_t^2+h\right)\phi=J\,,\qquad t\geq 0\,. \tag{2.7}$$

We take the opportunity of stressing that the restriction $t\geq 0$ is natural on physical grounds. For example, it is motivated in [3] by pointing out that classical versions of non-local cosmological models contain singularities at the beginning of time.

3 Mathematical theory for nonlocal equations

3.1 The appearance of integral transforms

We consider that our discussion in the previous section is more than enough motivation for the study of, at least, general linear nonlocal equations of the form

$$f(\partial_t)\phi=g(t),\ t\geq 0\,, \tag{3.1}$$

and for studying initial value problems (IVP) associated with Equation (3.1).

3.2 Laplace transform and Hardy space

The Laplace transform, see for instance [24, 4], is a useful tool for the study of initial value problems in ordinary differential equations. It has been used already in the context of nonlocal equations —using the fact that it determines an isometric isomorphism between locally integrable functions on the positive semi-axis of exponential decay and Widder space, see [4]— in [37, 36]. Here we use the Laplace transform as a tool to obtain correspondences between $L^p(\mathbb{R}_+)$ spaces and Hardy spaces $H^p(\mathbb{C}_+)$.

Let $\mathbb{R}_+$ be the infinite interval $[0,+\infty)$ and $\mathbb{C}_+$ the right half-plane $\{s\in\mathbb{C}\ :\ Re(s)>0\}$. The space $L^p(\mathbb{R}_+)$, $0<p<\infty$, is the Lebesgue space of measurable functions ϕ on $[0,\infty)$ such that

$$||\phi||_{L^p(\mathbb{R}_+)}:=\left(\int_0^\infty|\phi(x)|^p dx\right)^{\frac{1}{p}}<\infty,$$

and the qth Hardy space $H^q(\mathbb{C}_+)$ is the space of all functions Φ which are analytic on $\mathbb{C}_+$ and such that the integral $\mu_q(\Phi,x)$ given by

$$\mu_q(\Phi,x):=\left(\frac{1}{2\pi}\int_{-\infty}^{\infty}|\Phi(x+iy)|^q dy\right)^{\frac{1}{q}},$$

is uniformly bounded for $x > 0$. It is known that $H^q(\mathbb{C}_+)$ becomes a Banach space with the norm $||\Phi||_{H^q(\mathbb{C}_+)} := \sup_{x>0} \mu_q(\Phi, x)$.

The simplest context in which the Laplace transform (and its inverse!) can be defined is the following:

Definition 1. A function $g : \mathbb{R} \to \mathbb{C}$ belongs to the class $\mathcal{T}_a$ if and only if $g(t)$ is identically zero for $t < 0$ and the integral

$$\int_0^\infty e^{-st} g(t)\, dt$$

converges absolutely for $Re(s) > a$.

The *Laplace transform* of a function $g \in \mathcal{T}_a$ is the integral

$$\mathcal{L}(g)(s) = \int_0^\infty e^{-st} g(t)\, dt\,, \qquad Re(s) > a\,.$$

As proven in [24, Theorem 3.1], if $g \in \mathcal{T}_a$, then $\mathcal{L}(g)(s)$ converges absolutely and uniformly for $Re(s) \geq x_0 > a$ (a finer statement is in [24, Theorem 23.1]). It is also known (see [24, Theorem 6.1]) that the Laplace transform $\mathcal{L}(g)$ is an analytic function for $Re(s) > a$. The following results can be also proven in this general framework, see [24]:

Proposition 1.

1. *Let us fix a function $g \in \mathcal{T}_a$. If $D^n g(t)$ belongs to $\mathcal{T}_a$, then*

$$\mathcal{L}(D^n g)(s) = s^n\, \mathcal{L}(g)(s) - \sum_{j=1}^{n} (D^{j-1} g)(0) s^{n-j}\,.$$

2. *If $g \in \mathcal{T}_a$, then for any $\sigma > a$ we have the identity*

$$g(t) = \frac{1}{2\pi i} \int_{\sigma - i\infty}^{\sigma + i\infty} e^{st}\, \mathcal{L}(g)(s)\, ds\,. \tag{3.2}$$

The inverse Laplace transform (determined by the inversion formula (3.2)) is denoted by $\mathcal{L}^{-1}$. Regretfully, as we explain in [17], the class $\mathcal{T}_a$ is not adequate for defining and solving initial value problems, and we must look for more refined contexts. A better alternative is to consider the Laplace transform as a correspondence from Lebesgue spaces $L^p(\mathbb{R}_+)$ into Hardy spaces $H^q(\mathbb{C}_+)$. The following classic Representation theorem was first presented by Doetsch in [23]:

Theorem 3. *(Doetsch's Representation Theorem)*

(i) If $\phi \in L^p(\mathbb{R}_+)$*, where* $1 < p \leq 2$*, and* $\Phi = \mathcal{L}(\phi)$*, then* $\Phi \in H^{p'}(\mathbb{C}_+)$ *with* $\frac{1}{p} + \frac{1}{p'} = 1$*. Moreover if* $x > 0$ *there exists a positive constant* $C(p)$ *such that:*

$$\mu_{p'}(\Phi, x) \leq C(p) \left(\int_0^\infty e^{-pxt} |\phi(t)|^p dt \right)^{\frac{1}{p}} .$$

(ii) If $\Phi \in H^p(\mathbb{C}_+)$*, where* $1 < p \leq 2$*, then there exists* $\phi \in L^{p'}(0, \infty)$ *with* $\frac{1}{p} + \frac{1}{p'} = 1$ *such that* $\Phi = \mathcal{L}(\phi)$*. The function* ϕ *is given by the inversion formula*

$$\phi(t) := \lim_{v \to \infty} \frac{1}{2\pi} \int_{-v}^{v} e^{(\sigma + i\eta)t} \Phi(\sigma + i\eta) d\eta , \qquad \sigma \geq 0 ,$$

and for $x > 0$ *there exists a positive constant* $K(p)$ *such that*

$$\left(\int_0^\infty e^{-p'xt} |\phi(t)|^{p'} dt \right)^{\frac{1}{p'}} \leq K(p) \mu_p(\Phi, x).$$

We remark that the restrictions on p appearing in both parts of the theorem imply that neither (i) is the converse of (ii) nor (ii) is the converse of (i), except in the case $p = p' = 2$. In this particular case, Doetsch's representation theorem becomes the important Paley-Wiener theorem, which states that the Laplace transform is a unitary isomorphism between $L^2(\mathbb{R}_+)$ and $H^2(\mathbb{C}_+)$. For more general versions of Theorem 3 (*e.g.* correspondences between appropriate weighted Lebesgue and Hardy spaces) see [7, 47] and references therein.

Theorem 3 allows us to *define* the operator $f(\partial_t)$ for large families of analytic functions f, roughly speaking by mimicking the definition of classical pseudo-differential operators, see [40]. We point out immediately that there are interesting cases of analytic functions for which the definition of $f(\partial_t)$ via Laplace transform breaks down: one such is the function $s \to \zeta(s^2 + h)$, $h \in \mathbb{R}$, appearing in the Dragovich model discussed in Section 2. Thus, we must look for further generalizations of our theory. Fortunately, there exists a very flexible generalization of Laplace transform to which we now turn.

3.3 Entire functions of exponential type and the Borel transform

The Borel transform has been used to study convolution equations and infinite order differential equations in [18, 22, 21]. We begin with the definition of entire functions of exponential type after [9].

Definition 2. An entire function $\phi : \mathbb{C} \to \mathbb{C}$ is said to be of finite exponential type τ_ϕ and finite order ρ_ϕ, if τ_ϕ and ρ_ϕ are the infimum of the positive numbers τ, ρ such that the following inequality holds:

$$|\phi(z)| \leq C e^{\tau |z|^\rho}, \quad \forall z \in \mathbb{C} \,, \; \textit{and some } C > 0.$$

When $\rho_\phi = 1$, the function ϕ is said to be of *exponential type*, or of *exponential type* τ_ϕ when we need to specify its type. Explicit formulas for calculating ρ_ϕ and τ_ϕ appear in [9, Theorem 2.2.2] and in [9, Formula 2.2.12, p. 11]. The space of functions of exponential type will be denoted by $Exp(\mathbb{C})$.

Definition 3. Let ϕ be an entire function of exponential type τ_ϕ. If $\phi(z) = \sum_{n=0}^{\infty} a_n z^n$, the Borel transform of ϕ is defined by

$$B(\phi)(z) := \sum_{n=0}^{\infty} \frac{a_n n!}{z^{n+1}} \, .$$

The series $B(\phi)(z)$ converges uniformly for $|z| > \tau_\phi$, see [49, p. 106] and [9], and therefore it defines an analytic function on $\{z \in \mathbb{C} : |z| > \tau_\phi\}$. Further details on functions of exponential type and Borel transform appear in [29, 49]

We can calculate the Borel transform of an entire function ϕ of exponential type τ_ϕ (see [9]) using the complex Laplace transform: if z is such that $|z| = r > \tau_\phi$ and z is in the direction θ, then

$$B(\phi)(re^{i\theta}) = e^{i\theta} \int_0^\infty \phi(te^{i\theta}) e^{-rt} dt \, . \tag{3.3}$$

In particular, if $z \in \mathbb{R}$ is large enough, then $B(\phi)$ can be viewed as the analytic continuation of its real Laplace transform,

$$\mathcal{L}(\phi)(z) = \int_0^\infty \phi(t) e^{-zt} dt \, . \tag{3.4}$$

For $\phi \in Exp(\mathbb{C})$, we let $s(B(\phi))$ denote the set of singularities of the Borel transform of ϕ, and by $S(\phi)$ the *conjugate diagram* of $B(\phi)$, this is, the closed convex hull of the set of singularities $s(B(\phi))$. The set $S(\phi)$ is a convex compact subset of $\mathbb{C}$, and $B(\phi)$ is an analytic function in $\mathbb{S} \setminus S(\phi)$, where $\mathbb{S}$ is the extended complex plane $\mathbb{C} \cup \{\infty\}$ and we have set $B(\phi)(\infty) = 0$.

For a domain $\Omega \subset \mathbb{C}$, the set Ω^c indicates the complement of Ω in the extended complex plane $\mathbb{S}$. We define:

Definition 4. Let Ω be a simple connected domain; we define the space $Exp(\Omega)$ as the set of all entire functions ϕ of exponential type such that its Borel transform $B(\phi)$ has all its singularities in Ω and that it has an analytic continuation, called $\mathcal{B}(\phi)$, to Ω^c.

Remark 1. Since Ω^c is closed, the fact that $\mathcal{B}(\phi)$ is analytic in Ω^c means that there exist an open set $U \subset \mathbb{S}$ such that $\mathcal{B}(\phi)$ is analytic in U and $\Omega^c \subset U$. Therefore, using the alternative definition of Borel transform (3.3), we understand $\mathcal{B}(\phi)$ as the analytic continuation of its real Laplace transform (3.4).

In what follows, by *the Borel transform of $\phi \in Exp(\Omega)$*, we mean $\mathcal{B}(\phi)$ jointly with an open set U in the extended plane $\mathbb{S}$, such that $\mathcal{B}(\phi)$ is analytic in U and $\Omega^c \subset U$.

Definition 5. For a function $\phi \in Exp(\Omega)$, we define the set $H_1(\phi)$ as the class of closed rectifiable and simple curves in $\mathbb{C}$ which are homologous and such that the set $s(B(\phi))$ is contained in their bounded region.

The following theorem is a result about the representation of entire functions of exponential type; its proof is in [9, Theorem 5.3.5].

Theorem 4. *(Polya's Representation Theorem). Let ϕ be a function of exponential type and let $\gamma \in H_1(\phi)$. Then,*

$$\phi(z) = \frac{1}{2\pi i} \int_{\gamma} e^{sz} \mathcal{B}(\phi)(s) ds.$$

In particular, if ϕ is of type τ and C_R is the circle of radius $R > \tau$ centered at the origin, then

$$\phi(z) = \frac{1}{2\pi i} \int_{C_R} e^{sz} B(\phi)(s) ds.$$

This theorem allows us to invert the Borel transform. We make the following definition:

Definition 6. If d is a distribution with compact support in $\mathbb{C}$, the $\mathcal{P}$-transform of d is

$$\mathcal{P}(d)(z) := < e^{sz}, d > , \quad z \in \mathbb{C}.$$

In particular, if $d = \mu$ is a complex measure with compact support, then its $\mathcal{P}$-transform is

$$\mathcal{P}(\mu)(z) = \int_{\mathbb{C}} e^{sz} d\mu(s) \text{ , } z \in \mathbb{C} \text{ .}$$

The $\mathcal{P}$-transform is called the Fourier-Laplace transform in [49] and the Fourier-Borel transform in Martineau's classical paper [43]. This definition is in [17]. We reproduce it here in full generality although we use only the case in which d is a complex measure with compact support. The following proposition says that the $\mathcal{P}$-transform "respects" the class of functions of exponential type. It is proven in [17, 21].

Proposition 2. *Let $\mathcal{O} \subset \mathbb{C}$ be a simply connected domain; if μ is a complex measure with compact support contained in $\mathcal{O}$, then $\mathcal{P}(\mu) \in Exp(\mathcal{O})$. Conversely, given any function $\phi \in Exp(\mathcal{O})$ we can choose a measure μ_ϕ such that $\mathcal{P}(\mu_\phi)(z) = \phi(z)$. The measure μ_ϕ is not unique; on the contrary, it can be chosen to have support on any given curve $\gamma \in H_1(\phi)$.*

The proof of the direct implication uses a calculation of the Borel transform of $\mathcal{P}(\mu)$ as the analytic continuation of its real Laplace transform.

To prove the converse implication, it is enough to take $\gamma \in H_1(\phi)$, so that Polya's theorem (Theorem 4) means that ϕ can be represented as $\phi(z) = \mathcal{P}(\mu_\phi)(z)$, where we use the complex measure μ_ϕ defined by

$$d\mu_\phi(s) := \mathcal{B}(\phi)(s)\frac{ds}{2\pi i}\,, \quad s \in \gamma\,. \tag{3.5}$$

4 The operator $f(\partial_t) : L^p(\mathbb{R}_+) \longrightarrow H^q(\mathbb{C}_+)$

4.1 The main definition

It is standard to arrive at a good definition of $f(\partial_t)$ by calculating $f(\partial_t)\phi$ formally: we take a function f analytic around zero, and suppose that $\phi \in L^{p'}(0,\infty)$ is smooth. Let us write

$$f(\partial_t)\phi = \sum_{n=0}^{\infty} \frac{f^{(n)}(0)}{n!}\,\partial_t^n \phi\,.$$

Standard properties of the Laplace transform (see Proposition 1 or [24]) yield, formally,

$$\begin{aligned}
\mathcal{L}(f(\partial_t)\phi)(s) &= \sum_{n=0}^{\infty} \frac{f^{(n)}(0)}{n!}\,\mathcal{L}(\partial_t^n\phi) \\
&= f(s)\mathcal{L}(\phi)(s) - \sum_{n=1}^{\infty}\sum_{j=1}^{n} \frac{f^{(n)}(0)}{n!} s^{n-j}\phi^{(j-1)}(0)\,.
\end{aligned} \tag{4.1}$$

If we define the formal series

$$r(s) = \sum_{n=1}^{\infty}\sum_{j=1}^{n} \frac{f^{(n)}(0)}{n!}\,d_{j-1}\,s^{n-j}\,, \tag{4.2}$$

in which $d = \{d_j : j \geq 0\}$ is a sequence of complex numbers, we can write (4.1) as

$$\mathcal{L}(f(\partial_t)\phi)(s) = f(s)\mathcal{L}(\phi)(s) - r(s)\,, \tag{4.3}$$

in which $r(s)$ is given by (4.2) with $d_j = \phi^{(j)}(0)$.

For appropriated choices of sequences $\{d_{j-1}\}_{j\geq 1}$, the series defined in (4.2) is in fact an analytic function. This is important, since (4.2) encodes a set of "initial conditions". We will expand on this statement in the next section. The following Lemma is in [17].

Lemma 2. *Let $R_1 > 1$ be the maximum radius of convergence of the Taylor series $f_T(s) = \sum_{n=0}^{\infty} \frac{f^{(n)}(0)}{n!}\, s^n$. Set $0 < R < 1$ and suppose that the series*

$$\sum_{j=1}^{\infty} d_{j-1}\frac{1}{s^j} \tag{4.4}$$

is uniformly convergent on compact sets for $|s| > R$. *Then,* (4.2) *is analytic on the disk* $|s| < R_1$.

In particular, see [37, lemma 2.1], if f is an entire function, $R < 1$ and the series (4.4) is convergent for $|s| > R$, then the series (4.2) is an entire function.

Remark 2. There exists a large class of series satisfying conditions of Lemma 2. Indeed, let $r > 0$ and denote by $Exp_r(\mathbb{C})$ the space of entire functions of exponential type $\tau < r$. It is well know that if $\phi \in Exp_r(\mathbb{C})$ and it is of exponential type $\tau < r$, then its Borel transform converges uniformly on $|s| > \tau$ (see again [17, 49]). In our case, if $\phi \in Exp_1(\mathbb{C})$, its Borel transform $\mathcal{B}(\phi)(s)$ is precisely (4.4) for $d_j = \phi^j(0)$.

Motivated by the previous computations and Doetsch's representation theorem, we make the following definition generalizing [37, 36]:

Definition 7. Let f be an analytic function on a region which contains the half-plane $\{s \in \mathbb{C} : Re(s) > 0\}$, and let $\mathcal{H}$ be the space of all $\mathbb{C}$-valued functions on $\mathbb{C}$ which are analytic on (regions of) $\mathbb{C}$. We fix p and p' such that $1 < p \leq 2$ and $\frac{1}{p} + \frac{1}{p'} = 1$, and we consider the subspace D_f of $L^{p'}(0, \infty) \times \mathcal{H}$ consisting of all the pairs (ϕ, r) such that

$$\widehat{(\phi, r)} = f\,\mathcal{L}(\phi) - r \tag{4.5}$$

belongs to the class $H^p(\mathbb{C}_+)$. The domain of $f(\partial_t)$ as a linear operator from $L^{p'}(0, \infty) \times \mathcal{H}$ into $L^{p'}(0, \infty)$ is D_f. If $(\phi, r) \in D_f$ then we define

$$f(\partial_t)\,(\phi, r) = \mathcal{L}^{-1}(\;\widehat{(\phi, r)}\;) = \mathcal{L}^{-1}(f\,\mathcal{L}(\phi) - r)\,. \tag{4.6}$$

This definition presents $f(\partial_t)$ as a *nonlocal operator*, not as a series in infinitely many derivatives! However, in the $p = p' = 2$ case we recover this intuitive description.

4.2 The $L^2(\mathbb{R}_+)$-theory

As just stated, in the $p = p' = 2$ case of the foregoing theory we can justify rigorously the interpretation of $f(\partial_t)$ as an operator in infinitely many derivatives on an appropriated domain. We use a notion of analytic vectors, motivated by Nelson's classical paper [46].

Definition 8. Let A be a linear operator from a Banach space $\mathbb{V}$ to itself, and let $f : \mathbb{R} \to \mathbb{C}$ be a complex valued function, such that $f^{(n)}(0)$ exist for all $n \geq 0$. We say that $v \in \mathbb{V}$ is a f-analytic vector for A if v is in the domain of A^n for all $n \geq 0$ and the series

$$\sum_{n=0} \frac{f^{(n)}(0)}{n!} A^n v\,,$$

defines a vector in $\mathbb{V}$.

As stated in Section 2, the Paley-Wiener theorem (see [39, 54]) is the following special case of Doetsch's representation theorem:

Theorem 5. *The following assertions hold:*

1) *If $g \in L^2(\mathbb{R}_+)$, then $\mathcal{L}(g) \in H^2(\mathbb{C}_+)$.*

2) *Let $G \in H^2(\mathbb{C}_+)$. Then the function*

$$g(t) = \frac{1}{2\pi i} \int_{\sigma - i\infty}^{\sigma + i\infty} e^{st} G(s) ds \,, \sigma > 0 \,,$$

is independent on σ, it belongs to $L^2(\mathbb{R}_+)$ and, it satisfies $G = \mathcal{L}(g)$.

Moreover the Laplace transform $\mathcal{L} : L^2(\mathbb{R}_+) \to H^2(\mathbb{C}_+)$ is a unitary operator.

Analytic vectors for the operator ∂_t on $L^2(\mathbb{R}_+)$ exist. For example, the following lemma holds; see [17]:

Lemma 3. *Let f be an analytic function on a region containing zero, and let R_1 be the maximum radius of convergence of the Taylor series $f_T(s) := \sum_{n=0}^{\infty} \frac{f^{(n)}(0)}{n!} s^n$.*

a) *If p is a polynomial on $\mathbb{R}_+$, and I a finite interval on $\mathbb{R}_+$, then the function $\psi := p \cdot \chi_I$ is an f-analytic vector for ∂_t on $L^2(\mathbb{R}_+)$.*

b) *Let $R_1 > 1$. If $\psi \in C^\infty(\mathbb{R}_+) \cap L^2(\mathbb{R}_+)$ such that for all $n \geq 1$ and some $h \in L^2(\mathcal{I})$ for $\mathcal{I}$ equal to either $\mathbb{R}$ or $\mathbb{R}_+$, we have $||\psi^{(n)}||_{L^2(\mathbb{R}_+)} \leq c(n) ||h||_{L^2(\mathcal{I})}$, with $\{c(n)\}_{n \in \mathbb{N}} =: c \in l^1(\mathbb{N})$. Then ψ is an f-analytic vector for ∂_t on $L^2(\mathbb{R}_+)$.*

Families of functions satisfying the conditions of this lemma are also presented in [17].

The operator $f(\partial_t)$ is defined in the present L^2-context as in Definition 7. The domain D_f of $f(\partial_t)$is quite large; see [17]:

Proposition 3. *Let f be a function which is analytic on a region containing $\mathbb{C}_+$, and let $R_1 > 1$ be the maximum radius of convergence of the Taylor series $f_T(s) := \sum_{n=0}^{\infty} \frac{f^{(n)}(0)}{n!} s^n$. Let ϕ be a f-analytic vector for ∂_t in $L^2(\mathbb{R}_+)$ and suppose that the sequence $\{d_j = \phi^{(j)}(0)\}$ satisfies the condition of Lemma 2. Then, there exists an analytic function r_e on $\mathbb{C}_+$ such that $f\mathcal{L}(\phi) - r_e$ is in the domain D_f of $f(\partial_t)$.*

The proof of this proposition depends essentially on the fact that the Laplace transform is an unitary operator from $L^2(\mathbb{R}_+)$ onto $H^2(\mathbb{C}_+)$. An easy corollary is the following:

Corollary 1. Let f be an entire function and let ϕ be a smooth f-analytic vector for ∂_t in $L^2(\mathbb{R}_+)$. Suppose that the sequence $\{d_j = \phi^{(j)}(0)\}$ satisfies $d_j \leq CR^j$ for $0 < R < 1$. Then $(\phi, r) \in D_f$.

Now let us assume that f is entire. Then, *the operator* $f(\partial_t)$ *is an operator in infinitely many derivatives on the space of* f*-analytic vectors.* In fact, let ϕ be a f-analytic vector for ∂_t in $L^2(\mathbb{R}_+)$, and let

$$r(s) = \sum_{n=1}^{\infty}\sum_{j=1}^{n} \frac{f^{(n)}(0)}{n!}\, \phi^{(j-1)}(0)\, s^{n-j} .$$

If conditions of corollary 1 hold, then

$$\mathcal{L}\left(\sum_{n=0}^{\infty} \frac{f^{(n)}(0)}{n!} \partial_t^n(\phi)\right) = f\mathcal{L}(\phi) - r = \mathcal{L}(f(\partial_t)\phi) ,$$

and therefore

$$f(\partial_t)\phi = \sum_{n=0}^{\infty} \frac{f^{(n)}(0)}{n!} \partial_t^n(\phi) .$$

4.3 Linear nonlocal equations

Definition 7 allows us to solve the nonlocal equation

$$f(\partial_t)(\phi, r) = J , \qquad J \in L^{p'}(0,\infty) . \tag{4.7}$$

We assume that a suitable function $r \in \mathcal{H}$ has been fixed; consequently, we understand Equation (4.7) as an equation for $\phi \in L^{p'}(0,\infty)$ such that $(\phi, r) \in D_f$. We simply write $f(\partial_t)\phi = J$ instead of (4.7). First of all, we formalize what we mean by a solution:

Definition 9. Let us fix a function $r \in \mathcal{H}$. We say that $\phi \in L^{p'}(0,\infty)$ is a solution to the equation $f(\partial_t)\phi = J$ if and only if

1. $\widehat{\phi} = f\,\mathcal{L}(\phi) - r \in H^p(\mathbb{C}_+)$; (i.e., $(\phi, r) \in D_f$);

2. $f(\partial_t)(\phi) = \mathcal{L}^{-1}(\widehat{(\phi, r)}) = \mathcal{L}^{-1}(f\,\mathcal{L}(\phi) - r) = J$.

Our main theorem on existence and uniqueness of the solution to the linear problem (4.7) is the abstract result below. In attention to its importance, we reproduce its proof, although it is modeled after [37, 36] and it appears in [17]:

Theorem 6. *Let us fix a function* f *which is analytic in a region* D *which contains the half-plane* $\{s \in \mathbb{C} : Re(s) > 0\}$. *We also fix* p *and* p' *such that* $1 < p \le 2$ *and* $\frac{1}{p} + \frac{1}{p'} = 1$, *and we consider a function* $J \in L^{p'}(0,\infty)$ *such that* $\mathcal{L}(J) \in H^p(\mathbb{C}_+)$. *We assume that the function* $(\mathcal{L}(J)+r)/f$ *is in the space* $H^p(\mathbb{C}_+)$. *Then, the linear equation*

$$f(\partial_t)\phi = J \tag{4.8}$$

can be uniquely solved on $L^{p'}(0,\infty)$. *Moreover, the solution is given by the explicit formula*

$$\phi = \mathcal{L}^{-1}\left(\frac{\mathcal{L}(J) + r}{f}\right) . \tag{4.9}$$

Proof. We set $\phi = \mathcal{L}^{-1}\left((\mathcal{L}(J)+r)/f\right)$. Since $\mathcal{L}(J) \in H^p(\mathbb{C}_+)$, it follows that the pair (ϕ, r) is in the domain D_f of the operator $f(\partial_t)$: indeed, an easy calculation using Theorem 3 shows that $\widehat{\phi} = \mathcal{L}(J)$, which is an element of $H^p(\mathbb{C}_+)$ by hypothesis. We can then check (using Theorem 3 again) that ϕ defined by (4.9) is a solution of (4.8).

We prove uniqueness using Definition 9: let us assume that ϕ and ψ are solutions to Equation (4.8). Then, item 2 of Definition 9 implies $f\,\mathcal{L}(\phi-\psi) = 0$ on $\{s \in \mathbb{C} : Re(s) > a\}$. Set $h = \mathcal{L}(\phi - \psi)$ and suppose that $h(s_0) \neq 0$ for s_0 in the half-plane just defined. By analyticity, $h(s) \neq 0$ in a suitable neighborhood U of s_0. But then $f = 0$ in U, so that (again by analyticity) f is identically zero. ■

In the following section we impose further conditions on J and f so that we can assure that (4.9) is smooth at $t = 0$, and we use these conditions to study the initial value problem for (4.8). Hereafter we assume the natural decay condition (see [36, 20] and references therein)

$$\left|\frac{r(s)}{f(s)}\right| \leq \frac{C}{|s|^q} \tag{4.10}$$

for $|s|$ sufficiently large and some real number $q > 0$, and we examine three special cases of Theorem 6: (a) the function r/f has no poles; (b) the function r/f has a finite number of poles; (c) the function r/f has an infinite number of poles. As usual, we refer to the original papers [37, 36, 20] for full proofs.

Corollary 2. Assume that the hypotheses of Theorem 6 hold, that $\mathcal{L}(J)/f$ is in $H^p(\mathbb{C}_+)$, and that r/f is an entire function such that (4.10) holds. Then, solution (4.9) to Equation (4.8) is simply $\phi = \mathcal{L}^{-1}\left(\frac{\mathcal{L}(J)}{f}\right)$.

Corollary 3. Assume that the hypotheses of Theorem 6 hold, and that r/f has a finite number of poles ω_i $(i = 1, \ldots, N)$ of order r_i to the left of $Re(s) = 0$. Suppose also that $\mathcal{L}(J)/f$ is in $H^p(\mathbb{C}_+)$, and that the growth condition (4.10) holds. Then, the solution (4.9) can be written in the form

$$\phi(t) = \frac{1}{2\pi i}\int_{\sigma-i\infty}^{\sigma+i\infty} e^{st}\left(\frac{\mathcal{L}(J)}{f}\right)(s)\,ds + \sum_{i=1}^{N} P_i(t)\,e^{\omega_i t}\ , \ \sigma > 0\,, \tag{4.11}$$

in which $P_i(t)$ are polynomials of degree $r_i - 1$.

Formula (4.11) for the solution $\phi(t)$ appears already in Carmichael's paper [18]; see also [6].

Corollary 4. Assume that the hypotheses of Theorem 6 hold, and that the quotient r/f has an infinite number of poles ω_n of order r_n to the left of $Re(s) = 0$ satisfying $|\omega_n| \leq |\omega_{n+1}|$ for $n \geq 1$. We let σ_n be curves in the half-plane $Re(s) \leq 0$ connecting the points $+ib_n$ and $-ib_n$, in which the real numbers b_n are chosen so that σ_n together with the segment of the line $Re(s) =$ between the points $+ib_n$

and $-ib_n$ encloses exactly the first n poles of $r(s)/f(s)$. Suppose that $\mathcal{L}(J)/f$ is in $H^p(\mathbb{C}_+)$, that the curves σ_n are chosen so that b_n tends to infinity as n tends to infinity, and that

$$\lim_{n\to\infty} \int_{\sigma_n} e^{st} \left(\frac{r}{f}\right)(s)ds = 0 .$$

Then, the solution (4.9) to the linear equation (4.8) can be written in the form

$$\phi(t) = \frac{1}{2\pi i} \int_{\sigma-i\infty}^{\sigma+i\infty} e^{s\,t} \left(\frac{\mathcal{L}(J)}{f}\right)(s)\,ds + \sum_{n=1}^{\infty} P_n(t)\,e^{\omega_n t} \,, \sigma > 0 \,; \tag{4.12}$$

in which $P_n(t)$ are polynomials of degree $r_n - 1$.

Remark 3. The solution appearing in Corollary 4 is not necessarily differentiable; see for instance the explicit example appearing in [36]. This means in particular, that in complete generality we cannot even formulate an initial value problem for equations of the form $f(\partial_t)\phi = J$!

5 The initial value problem

5.1 Generalized initial conditions

We note that our abstract formula (4.9) for the solution ϕ to the equation

$$f(\partial_t)\phi = J \,, \qquad t \geq 0 \,, \tag{5.1}$$

tells us that –considering $r = r_d$, where r_d is given by (4.2)– ϕ depends in principle on an infinite number of arbitrary constants, the values $\phi^{(n)}(0)$. This fact has caused serious interpretative difficulties; see for instance [30]. It seems better to think of r itself as a "generalized initial condition":

Definition 10. A generalized initial condition for the equation is an analytic function r_0 such that $(\phi, r_0) \in D_f$ for some $\phi \in L^{p'}(0,\infty)$. A generalized initial value problem is an equation such as (5.1) together with a generalized initial condition r_0. A solution to a given generalized initial value problem $\{(5.1), r_0\}$ is a function ϕ satisfying the conditions of Definition 9 with $r = r_0$.

Thus, given a generalized initial condition, we find a unique solution for (5.1) using (4.9), much in the same way as given *one* initial condition we find a unique solution to a first order linear ODE. As remarked in the previous section, there is no reason to believe that (for a given r) the unique solution (4.9) to (5.1) will be analytic: we can only conclude that the solution belongs to the class $L^{p'}(0,\infty)$ for some $p' > 0$. *On the other hand*, we can show that –provided f and J satisfy some technical conditions– we *can* define initial value problems subject to a finite number of a priori *local* data. We now summarize this development following [36, 20].

5.2 Classical initial value problems

First we need to ensure differentiability of solutions; we use the following lemma, proved in [20]:

Lemma 4. *Let J be a function such that $\mathcal{L}(J)$ exists and let f be an analytic function. Suppose that there exist an integer $M \geq 0$ and a real number $\sigma > 0$ such that*

$$y \mapsto y^n \left(\frac{\mathcal{L}(J)(\sigma + iy)}{f(\sigma + iy)} \right) \text{ belongs to } L^1(\mathbb{R}) \tag{5.2}$$

for each $n = 0, \ldots, M$; then the function

$$t \mapsto \frac{1}{2\pi i} \int_{\sigma - i\infty}^{\sigma + i\infty} e^{st} \left(\frac{\mathcal{L}(J)}{f} \right)(s)\, ds \tag{5.3}$$

is of class C^M.

Doetsch's representation theorem (Theorem 3) implies that it is enough to assume condition (5.2) for *some* $\sigma > 0$ instead of *for every* $\sigma > 0$. Lemma 4 implies the following:

Lemma 5. *Let the functions f and J satisfy the conditions of Corollary 3, and also assume that they satisfy (5.2) for some $\sigma > 0$. Then, the solution (4.11) to the nonlocal equation (5.1) is of class C^M for $t \geq 0$, and it satisfies the identities*

$$\phi^{(n)}(0) = L_n + \sum_{i=1}^{N} \sum_{k=0}^{n} \binom{n}{k} \omega_i^k \left. \frac{d^{n-k}}{dt^{n-k}} \right|_{t=0} P_i(t)\,, \quad n = 0, \ldots, M\,, \tag{5.4}$$

for some numbers L_n.

This result is proven in [37, 36]. The numbers L_n, for $n = 0, \ldots, M$, are

$$L_n = \left. \frac{d^n}{dt^n} \right|_{t=0} \left(\frac{1}{2\pi i} \int_{\sigma - i\infty}^{\sigma + i\infty} e^{st} \left(\frac{\mathcal{L}(J)}{f} \right)(s)\, ds \right) . \tag{5.5}$$

Our main result on initial value problems is the following.

Theorem 7. *We fix $1 < p \leq 2$ and $p' > 0$ such that $1/p + 1/p' = 1$, and also an integer $N \geq 0$. Let f be a function which is analytic in a region D which contains $\{s \in \mathbb{C} : Re(s) > 0\}$, and let J be a function in $L^{p'}(\mathbb{R}_+)$ satisfying $\mathcal{L}(J) \in H^p(\mathbb{C}_+)$ and $\mathcal{L}(J)/f \in H^p(\mathbb{C}_+)$. We choose points ω_i, $i = 1, \cdots, N$, to the left of $Re(s) = 0$, and positive integers r_i, $i = 1, ..., N$. Set $K = \sum_{i=1}^{N} r_i$ and assume that for some $\sigma > 0$ condition (5.2) holds for each $n = 0, \ldots, M$, $M \geq K$. Then, generically, given K values $\phi_0, \ldots, \phi_{K-1}$, there exists a unique analytic function r_0 such that*

(α) *$\frac{r_0}{f} \in H^p(\mathbb{C}_+)$ and it has a finite number of poles ω_i of order r_i, $i = 1, \ldots, N$;*

(β) $\dfrac{\mathcal{L}(J)+r_0}{f} \in H^p(\mathbb{C}_+)$;

(γ) $\left|\dfrac{r_0}{f}(s)\right| \leq \dfrac{M}{|s|^q}$ *for some* $q \geq 1$ *and* $|s|$ *sufficiently large.*

Moreover, the unique solution $\phi = \mathcal{L}^{-1}\left(\frac{\mathcal{L}(J)+r_0}{f}\right)$ *to Equation* (5.1) *is of class* C^K *and it satisfies* $\phi(0) = \phi_0, \ldots, \phi^{(K-1)}(0) = \phi_{K-1}$.

The proof is not entirely straightforward. A first version of it appeared in [37] and it appears in full detail in [20]. Here we mention only its main points:

We consider the K arbitrary numbers ϕ_n, $n = 0, 1, \ldots, K-1$. Recalling (5.4), Lemma 4, and Lemma 5, we set up the linear system

$$\phi_n = L_n + \sum_{i=1}^{N}\sum_{k=0}^{n}\binom{n}{k}\omega_i^k \left.\frac{d^{n-k}}{dt^{n-k}}\right|_{t=0} P_i(t)\,, \quad n = 0\ldots K-1\,, \tag{5.6}$$

for the coefficients of polynomials

$$P_i(t) = a_{1,i} + a_{2,i}\frac{t}{1!} + \cdots + a_{r_i,i}\frac{t^{r_i-1}}{(r_i-1)!}\,,$$

in which L_n are given by (5.5).

System (5.6) can be solved (generically, depending on the points ω_i) uniquely in terms of the data ϕ_n. We define r_0 on the half-plane $\{s \in \mathbb{C} : Re(s) > 0\}$ as follows:

$$r_0(s) = f(s)\,\mathcal{L}\left(\sum_{i=1}^{N} P_i(t)e^{\omega_i t}\right)(s)\,. \tag{5.7}$$

Then on this half-plane we have the identity

$$\mathcal{L}^{-1}(r_0/f)(t) = \sum_{i=1}^{N} P_i(t)e^{\omega_i t}\,. \tag{5.8}$$

Now we can prove (α), that is, that r_0/f belongs to $H^p(\mathbb{C}_+)$ and that its poles are precisely ω_i. This is done in [20]. We can also check that r_0/f satisfies conditions (β), and (γ) appearing in the enunciate of the theorem. Next, we *define* the function

$$\phi(t) = \frac{1}{2\pi i}\int_{\sigma-i\infty}^{\sigma+i\infty} e^{st}\left(\frac{\mathcal{L}(J)}{f}\right)(s)\,ds + \sum_{i=1}^{N} P_i(t)\,e^{\omega_i t}\,, \tag{5.9}$$

and we show that ϕ solves Equation (5.1) and it satisfies the conditions of the theorem. This check (appearing in [20]) finishes the proof.

Theorem 7 tells us that we can freely choose the first K derivatives $\phi^{(n)}(0)$, $n = 0, \ldots, K-1$ of the solution ϕ to Equation (5.1), and we can also show that from $n = K$ onward, if the derivative $\phi^{(n)}(0)$ exists, it is completely determined by the previous ones. Thus, it does not make sense to formulate an initial value problem with more than K arbitrary initial conditions.

We also note that the above proof shows that the *a priori* given points ω_i become the poles of the quotient r_0/f, that the *a priori* given numbers r_i are their respective orders, and that *it is essential* to give this information in order to have meaningful initial value problems. If no points ω_i are present, the solution to the nonlocal Equation (5.1) is simply $\phi = \mathcal{L}^{-1}(\mathcal{L}(J)/f)$, a formula which fixes completely (for f and J satisfying (5.2)) the values of the derivatives of ϕ at zero. This discussion motivates the following definition:

Definition 11. A classical initial value problem for nonlocal equations is a triplet formed by a nonlocal equation

$$f(\partial_t)\phi = J\ , \tag{5.10}$$

a finite set of data:

$$\left\{N \geq 0\ ;\ \{\omega_i \in \mathbb{C}\}_{1\leq i\leq N}\ ;\ \{r_i \in \mathbb{Z}, r_i > 0\}_{1\leq i\leq N}\ ;\ \{\{\phi_n\}_{0\leq n\leq K-1}, K = \textstyle\sum_{i=1}^N r_i\}\right\} \tag{5.11}$$

and the conditions

$$\phi(0) = \phi_0\ ,\quad \phi'(0) = \phi_1\ ,\quad \cdots .\phi^{(K-1)}(0) = \phi_{K-1}\ . \tag{5.12}$$

A solution to a classical initial value problem given by (5.10), (5.11) and (5.12) is a pair $(\phi, r_0) \in D_f$ satisfying the conditions of Definition 9 with $r = r_0$ such that ϕ is differentiable at zero and (5.12) holds.

5.3 Example: Zeta-nonlocal equations

Motivated by the physical models appearing in [2, 26, 28, 27] and briefly discussed in Section 2, let us apply our constructions to equations of the form

$$\zeta_h(\partial_t)\phi(t) = J(t)\ ,\qquad t \geq 0\ , \tag{5.13}$$

in which h is a real parameter and the symbol ζ_h is the shifted Riemann zeta function

$$\zeta_h(s) := \zeta(h+s) = \sum_{n=1}^{\infty} \frac{1}{n^{s+h}}\ .$$

We need to recall some properties of the Riemann zeta function $\zeta(s) := \sum_{n=1}^{\infty} \dfrac{1}{n^s}$, $Re(s) > 1$, see [41]. It is analytic on its domain of definition and, it has an analytic continuation to the whole complex plane with the exception of the point $s = 1$, at which it has a simple pole with residue 1. The analytic continuation of the Riemann zeta function will be also denoted by ζ, and we will refer to it also as the Riemann zeta function.

The shifted Riemann zeta function ζ_h is analytic for $Re(s) > 1-h$, and uniformly and absolutely convergent for $Re(s) \geq \sigma_0 > 1-h$. Also, see [41, Chp. I.6], ζ_h has infinite "trivial zeros" at the points $\{-2n-h : n \in \mathbb{N}\}$, and "nontrivial" zeros in the region $-h < Re(s) < 1-h$, the "critical region" of ζ_h.

The Euler product expansion for the shifted Riemann zeta function is

$$\zeta_h(s) = \prod_{p\in\mathcal{P}} \left(1 - \frac{1}{p^{s+h}}\right)^{-1},$$

where $\mathcal{P}$ is the set of the prime numbers. Therefore, for $Re(s) = \sigma > 1-h$, we have

$$\begin{aligned} \left|\frac{1}{\zeta_h(s)}\right| &= \left|\prod_{p\in\mathcal{P}} \left(1 - \frac{1}{p^{s+h}}\right)\right| = \left|\sum_{n=1}^{\infty} \frac{\mu(n)}{n^{s+h}}\right| \leq \sum_{n=1}^{\infty} \frac{1}{n^{\sigma+h}} \\ &\leq 1 + \int_1^{\infty} \frac{dx}{x^{\sigma+h}} = \frac{\sigma+h}{\sigma+h-1}, \end{aligned} \tag{5.14}$$

where $\mu(\cdot)$ is the Möebius function: $\mu(1) = 1$, $\mu(n) = 0$ if n is divisible by the square of a prime, and $\mu(n) = (-1)^k$ if n is the product of k distinct prime numbers; see [41, Chp. II.2].

We study Equation (5.13) for values of h in the region $(1, \infty)$, since in this case ζ_h is analytic for $Re(s) > 0$ and the theory developed in the previous sections apply. We start with the following lemma:

Lemma 6. *We fix $1 < p \leq 2$ and $p' > 0$ such that $1/p + 1/p' = 1$. Let us assume that $J \in L^{p'}(0,\infty)$ and that $\mathcal{L}(J)$ is in the space $H^p(\mathbb{C}_+)$. Then, $F = \mathcal{L}(J)/\zeta_h$ belongs to $H^p(\mathbb{C}_+)$.*

The proof follows essentially from inequality (5.14).

Now we note that we can replace the general condition (5.2) for the following assumption on the function J:

(H) For some $M \geq 0$, for each $n = 1, 2, 3, \cdots, M$ and for some $\sigma > 0$ we have,

$$y \to y^n \mathcal{L}(J)(\sigma + iy) \in L^1(\mathbb{R}),$$

since **(H)** is enough to ensure that the hypotheses of Lemma 4 hold. We obtain the following theorem on classical initial value problems for the zeta non-local Equation (5.13):

Theorem 8. *We fix $1 < p \leq 2$ and $p' > 0$ such that $1/p + 1/p' = 1$. Let ζ_h be the shifted Riemann zeta function, and assume that $J \in L^{p'}(0,\infty)$ with $\mathcal{L}(J) \in H^p(\mathbb{C}_+)$. Fix also a number $N \geq 0$, a finite number of points ω_i to the left of Re$(s) = 0$, and (if $N > 0$) a finite number of positive integers r_i, $i = 1, ..., N$. Set $K = \sum_{i=1}^N r_i$ and assume that condition* **(H)** *holds for all $n = 0, \ldots, M$, $M \geq K$. Then, generically, given K initial conditions, $\phi_0, \ldots, \phi_{K-1}$, there exists a unique analytic function r_0 such that* (α) $\dfrac{r_0}{\zeta_h} \in H^p(\mathbb{C}_+)$ *and it has a finite number of poles* ω_i *of order* r_i, $i = 1, \ldots, N$ *to the left of* $Re(s) = 0$*;*

(β) $\dfrac{\mathcal{L}(J) + r_0}{\zeta_h} \in H^p(\mathbb{C}_+)$*;*

(γ) $\left|\dfrac{r_0}{\zeta_h}(s)\right| \leq \dfrac{M}{|s|^q}$ *for some* $q \geq 1$ *and* $|s|$ *sufficiently large.*

Moreover, the unique solution ϕ *to Equation* (5.13) *given by* (4.9) *with* $r = r_0$ *is of class* C^K *and it satisfies* $\phi(0) = \phi_0, \ldots, \phi^{(K-1)}(0) = \phi_{K-1}$.

Proof. The proof consists in checking that the hypotheses of Theorem 7 hold. In fact, ζ_h is analytic on $\{s \in \mathbb{C} : Re(s) > 0\}$, and Lemma 6 tells us that $\mathcal{L}(J)/\zeta_h$ belongs to $H^p(\mathbb{C}_+)$. ■

6 From the Laplace to the Borel transform

Looking back at Dragovich's equation (2.7), we now study the problem

$$\zeta(\partial_t^2 + h)\phi = J\,, \quad t \geq 0\,, \tag{6.1}$$

for appropriate functions J. The behavior of the function $s \mapsto \zeta(s^2 + h)$ requires a generalization of the theory developed so far. Indeed, the properties of the Riemann zeta function imply that the symbol

$$\zeta(s^2 + h) = \sum_{n=0}^{\infty} \frac{1}{n^{s^2+h}} \tag{6.2}$$

is analytic in the region $\Gamma := \{s \in \mathbb{C} : Re(s)^2 - Im(s)^2 > 1 - h\}$, which is not a half-plane; also, its poles are the vertices of the hyperbolas $Re(s)^2 - Im(s)^2 = 1 - h$ and its critical region is the set $\{s \in \mathbb{C} : -h < Re(s)^2 - Im(s)^2 < 1 - h\}$. For example, for $h > 1$, Γ is the region limited by the interior of the dark hyperbola $Re(s)^2 - Im(s)^2 = 1 - h$ containing the real axis:

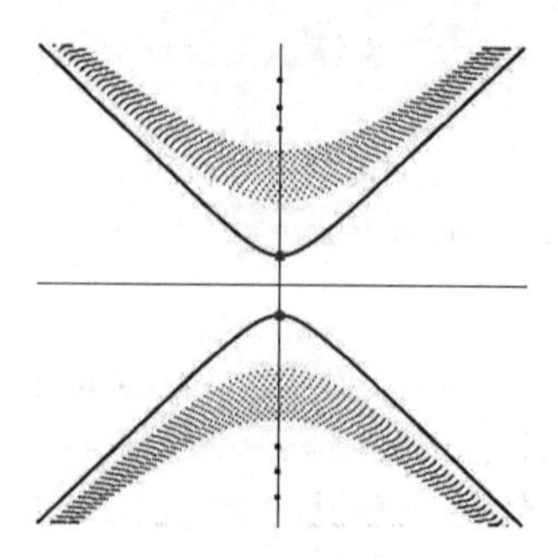

The poles of $\zeta(s^2 + h)$ are the vertices of dark hyperbola, indicated by two thick dots. The trivial zeros of $\zeta(s^2 + h)$ are indicated by thin dots on the imaginary axis; the non-trivial zeros are located on the darker painted region (critical region).

Now, since the Riemann zeta function has an infinite number of zeros on the critical strip (as famously proven by Hadamard and Hardy, see [41] for original references), so does the function $\zeta(s^2+h)$. If we denote by $\mathcal{Z}$ the set of all such zeros we have that $\sup_{z\in\mathcal{Z}} |Re(z)| = +\infty$. Thus, the expression $\mathcal{L}^{-1}\left(\mathcal{L}(J)(s)/\zeta(s^2+h)\right)$ for the solution to Equation (6.1) does not always make sense, since the function $\mathcal{L}(J)(s)/\zeta(s^2+h)$ does not necessarily belongs to $H^p(\mathbb{C}_+)$!

6.1 Functions of $\frac{d}{dt}$ via Borel transform

Let Ω be a simply connected domain (equivalently, let Ω be a Runge domain, see [48, Prop. 17.2]). In what follows we denote by $Hol(\Omega)$ the set of holomorphic functions on Ω.

Definition 12. Let $f \in Hol(\Omega)$, $\phi \in Exp(\Omega)$, and assume that μ_ϕ is the complex measure defined in (3.5) with compact support on the curve $\gamma \in H_1(\phi)$, so that $\mathcal{P}(\mu_\phi) = \phi$. We define

$$f(\partial_t)\phi := \mathcal{P}(f\mu_\phi)\,.$$

In this definition we assume that the curve $\gamma \in H_1(\phi)$ which determines the measure μ_ϕ is in the region Ω. By Cauchy's Theorem, the operator $f(\partial_t)$ is independent of such a γ. Equation

$$f(\partial_t)\phi = J \tag{6.3}$$

is understood as the following integral equation

$$\int_\gamma e^{st} f(s)\mathcal{B}(\phi)(s)\frac{ds}{2\pi i} = g(t), \quad \gamma \in H_1(\phi)\,. \tag{6.4}$$

The assumption that Ω is simply connected is really necessary. An example showing this is in [21]. Now we prove, after [21], that the integral operator $f(\partial_t)$ is locally (*i.e.*, whenever f can be expanded as a power series in an adequate ball contained in Ω) a differential operator of infinite order. Since this is a key result for us, we present its full proof.

Proposition 4. *Let $R > 0$ and assume that $B_R(0) \subset \Omega$. Suppose that $\phi \in Exp(\Omega)$ is such that $s(\mathcal{B}(\phi)) \subset B_R(0)$ and take $f \in Hol(\Omega)$ with $f(z) = \sum_{k=0}^{\infty} a_k z^k$, $|z| < R$. Then, there exist a measure μ_ϕ supported on a curve $\gamma \in H_1(\phi)$ contained in $B_R(0)$ such that $\phi = \mathcal{P}(\mu_\phi)$, and moreover*

$$f(\partial_t)\phi(t) = \mathcal{P}(f\mu_\phi)(t) = \lim_{l\to\infty}\sum_{k=0}^{l} a_k(\partial_t^k\phi)(t),$$

uniformly on compact sets.

Proof. Since $s(\mathcal{B}(\phi))$ is a discrete set, there exist $\delta > 0$ such that $dist(s(\mathcal{B}(\phi)), \{s : |s| = R\}) < \delta$. From [9][Theorem 5.3.12] we have that $\tau_\phi = \sup_{\omega\in s(\mathcal{B}(\phi))} |\omega|$ is the type of ϕ, and so there exists a curve $\gamma \subset (B_{\tau_\phi}(0))^c \cap B_R(0)$ such that $\gamma \in H_1(f)$. Let μ_ϕ be the measure described by Theorem 2 supported on γ. We have $\phi = \mathcal{P}(\mu_\phi)$.

Now, using this measure, we compute:

$$\frac{d}{dz}\phi(z) = \frac{d}{dz}\mathcal{P}(\mu_\phi)(z) = \frac{d}{dz}\int_\gamma e^{sz}d\mu_\phi(s) = \int_\gamma se^{sz}d\mu_\phi(s) = \mathcal{P}(s\mu_\phi).$$

From this, we obtain

$$\sum_{k=0}^{l} a_k \frac{d^k}{dz^k}\phi(z) - \mathcal{P}(f\mu_\phi)(z) = \mathcal{P}(\{\sum_{k=0}^{l} a_k s^k - f(s)\}\mu_\phi)(z) ,$$

and therefore

$$\begin{aligned}\left|\sum_{k=0}^{l} a_k \frac{d^k}{dz^k}\phi(z) - \mathcal{P}(f\mu_{\phi,R})(z)\right| &= \left|\int_\gamma e^{sz}\{\sum_{k=0}^{l} a_k s^k - f(s)\}d\mu_\phi(s)\right| \\ &\leq \int_\gamma e^{|z||s|}\left|\sum_{k=0}^{l} a_k s^k - f(s)\right| |d\mu_\phi|(s) .\end{aligned}$$

Now we take limits as $l \to \infty$. The result follows from the Lebesgue dominated convergence theorem. We note that convergence is uniform over compact subsets of $\mathbb{C}$. ■

Thus, under the hypothesis of this proposition, Equation (6.3) becomes "locally" the infinite order differential equation

$$\sum_{k=0}^{\infty} a_k \frac{d^k}{dt^k}\phi(t) = g(t). \tag{6.5}$$

Proposition 4 also shows that $f(\partial_t)$ is linear on the space of functions ϕ satisfying the hypothesis appearing therein. In fact, we can show the following, after [21]:

Lemma 7. *The operator $f(\partial_t) : Exp(\Omega) \to Exp(\Omega)$ is linear and onto. In other words, the equation*

$$f(\partial_t)\phi = g, \qquad g \in Exp(\Omega) \tag{6.6}$$

always has a solution in $Exp(\Omega)$.

Let us show how the above equation is solved. Let $\mathcal{Z}(f)$ be the set of zeros of f. This is a set of isolated points and so, since $g \in Exp(\Omega)$, there is a curve $\gamma \in H_1(g)$ such that $\mathcal{Z}(f) \cap \gamma = \emptyset$. Also, there is a measure μ_g supported on γ such that $g = \mathcal{P}(\mu_g)$. Set

$$\phi(z) = \mathcal{P}(\frac{\mu_g}{f})(z) = \frac{1}{2\pi i}\int_\gamma \frac{e^{z\eta}\mathcal{B}(g)(\eta)}{f(\eta)}d\eta .$$

It is not difficult to see that it satisfies $f(\partial_t)\phi = g$. Also, it is evident that $\phi \in Exp(\mathbb{C})$ and, we can check (calculating the Borel Transform of ϕ as the analytic continuation of its real Laplace transform) that $s(\mathcal{B}(\phi)) \subset \Omega$.

The following proposition is technically complicated, but interesting as it relates our present approach to Carmichael's point of view; see [18] and Theorem 9 below. A first version of this result is in [17]. The version below is in [21].

Proposition 5. *Let $f \in Hol(\Omega)$ and denote $\mathcal{Z}(f)$ for the set of its zeros. A function $\phi \in Exp(\Omega)$ of exponential type τ_ϕ is a solution of the homogeneous equation $f(\partial_t)\phi = 0$ if and only if there exist polynomials p_k of degree less than the multiplicity of the root $s_k \in \mathcal{Z}(f) \cap B_{\tau_\phi}(0)$, such that*

$$\phi(t) = \sum_{\substack{s_k \in \mathcal{Z}(f) \\ |s_k| < \tau_\phi}} p_k(t) e^{t s_k} .$$

Proof. We present a proof of the *if* part and refer to [17, 21] for the *only if* part. In order to prove that the function

$$\phi(t) = \sum_{\substack{s_k \in \mathcal{Z}(f) \\ |s_k| < \tau_\phi}} p_k(t) e^{t s_k} ,$$

is a solution of the homogeneous equation $f(\partial_t)\phi = 0$, it is enough to see that for a given k

$$f(\partial_t)(p_k(t)e^{ts_k}) = 0 .$$

First, it is evident from (6.7) below that $\phi \in Exp(\Omega)$. Now, we note that for a natural number d and a complex number a_d, the Borel transform of $a_d s^d e^{\lambda s}$ is

$$\mathcal{B}(a_d s^d e^{\lambda s}) = a_d \frac{d!}{(s-\lambda)^{d+1}} . \tag{6.7}$$

Let s_k be a zero of f of order $d_k + 1$, p_k a polynomial of degree $deg(p_k) \leq d_k$; and also $\gamma_k \in H_1(p_k(z)e^{zs_k})$; then using linearity of the Borel transform and the Cauchy theorem we have

$$f(\partial_t)\left(p_k(t)e^{ts_k}\right) = \frac{1}{2\pi i} \int_{\gamma_k} e^{t\eta} f(\eta) \mathcal{B}(p_k(z)e^{zs_k})(\eta) d\eta = 0 .$$

From these computations, we deduce that

$$f(\partial_t)\left(\sum_{s_k \in \mathcal{Z}(f) : |s_k| < \tau_\phi} p_k(t) e^{ts_k} \right) = 0 .$$

■

Corollary 5. Let $R > 0$ and $f \in Hol(B_R(0))$. Then, a function $\phi \in Exp(B_R(0))$ of exponential type τ_ϕ is a solution of the homogeneous equation $f(\partial_t)\phi = 0$, if and only if there exist polynomials p_k of degree less than the multiplicity of the root $s_k \in \mathcal{Z}(f) \cap B_{\tau_\phi}(0)$, such that

$$\phi(t) = \sum_{\substack{s_k \in \mathcal{Z}(f) \\ |s_k| < \tau_\phi}} p_k(t) e^{t s_k} .$$

Theorem 9. *Let $f \in Hol(\Omega)$ and $g \in Exp(\Omega)$. Then a function $\phi \in Exp(\Omega)$ of exponential type τ_ϕ is solution for the non-homogeneous equation $f(\partial_t)\phi = g$ if and only if there exist polynomials p_k of degree less than the multiplicity of the root $s_k \in \mathcal{Z}(f) \cap B_{\tau_\phi}(0)$, such that*

$$\phi(t) = \mathcal{P}(\frac{\mu_g}{f})(t) + \sum_{\substack{s_k \in \mathcal{Z}(f) \\ |s_k| < \tau_\phi}} p_k(t) e^{t s_k} .$$

7 Linear zeta-nonlocal field equations

In this section we apply the foregoing theory to the linear zeta-nonlocal field equation

$$\zeta(\partial_t^2 + h)\phi = g \, , \tag{7.1}$$

in which h is a real parameter. We consider mainly the case in which g is of exponential type, and we explain briefly how to analyze (7.1) for more general data. We will use notation introduced in Section 6 without further ado.

7.1 Zeta-nonlocal field equation with source function in $Exp(\Omega)$

We consider Equation (7.1) with $h > 1$ and refer the reader to [21] for a full analysis. If $h > 1$, the function $\zeta_h(s) := \zeta(s^2 + h)$ has poles at $s = i\sqrt{h-1}$ and $s = -i\sqrt{h-1}$. We warn the reader that in this section we use ζ_h exclusively to stand for $\zeta(s^2 + h)$; it should not be confused with the translated zeta function used in Subsection 5.3. As we have already seen, in this case the behavior of $\zeta_h(s)$ can be represented in the following picture:

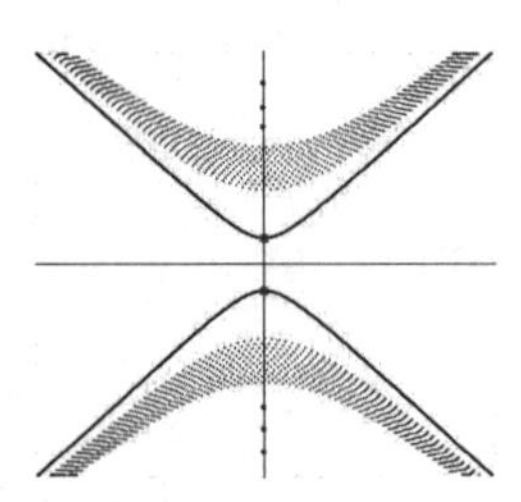

The poles of ζ_h are the vertices of dark hyperbola. The trivial zeros of ζ_h are indicated by thin dots on the imaginary axis; the non-trivial zeros are located on the darker painted region.

We consider the simply connected domain

$$\Omega := \mathbb{C} - \{s \in \mathbb{C} : Re(s) \geq 0, \ \ |Im(s)| = \sqrt{h-1}\} .$$

We see that the function $\zeta_h(s)$ is holomorphic in Ω, and therefore for a source function $g \in Exp(\Omega)$, Equation (7.1) becomes the following integral equation for the measure μ_ϕ:

$$\mathcal{P}(\zeta_h \cdot \mu_\phi)(t) = g(t) . \tag{7.2}$$

Theorem 10. *Let $g \in Exp(\Omega)$. A function $\phi \in Exp(\Omega)$ of exponential type τ_ϕ is a solution to the integral equation (7.2) if and only if there exist polynomials p_k of degree less than the multiplicity of the root $s_k \in \mathcal{Z}(\zeta_h) \cap B_{\tau_\phi}(0)$, such that*

$$\phi(t) = \int_\gamma \frac{e^{ts}}{\zeta(s^2+h)} d\mu_g(s) + \sum_{\substack{s_k \in \mathcal{Z}(\zeta_h) \\ |s_k| < \tau_\phi}} p_k(t) e^{ts_k} ,$$

in which $\gamma \in H_1(g)$ is a curve enclosing the roots $s_k \in \mathcal{Z}(\zeta_h) \cap B_{\tau_\phi}(0)$.

On the other hand, we note that for given $R < \sqrt{h-1}$, the domain Ω contains the ball $B_R(0)$, and since $\zeta_h(s) = \zeta(s^2+h)$ is analytic in this ball, it can be expressed as a power series, say

$$\zeta(s^2+h) = \sum_{k=0}^{\infty} a_k(h) s^k , \qquad |s| < R .$$

Therefore, using Proposition 4, we have that Equation (7.1) can be viewed as the following infinite order differential equation in the space $Exp(B_R(0))$:

$$\sum_{k=0}^{\infty} a_k(h) \frac{d^k}{dt^k} \phi(t) = g(t) . \tag{7.3}$$

In this situation, we have the following result:

Theorem 11. *Let $R < \sqrt{h-1}$ and $g \in Exp(B_R(0))$. Then, a function $\phi \in Exp(B_R(0))$ of exponential type τ_ϕ is a solution of the infinite order zeta-nonlocal field equation (7.3) if and only if there exist polynomials p_k of degree less than the multiplicity of the root $s_k \in \mathcal{Z}(\zeta_h) \cap B_R(0)$, such that*

$$\phi(t) = \int_{|s|=R} \frac{e^{ts}}{\zeta(s^2+h)} d\mu_g(s) + \sum_{\substack{s_k \in \mathcal{Z}(\zeta_h) \\ |s_k| < \tau_\phi}} p_k(t) e^{ts_k} .$$

The initial value problem for (7.3) can be analyzed, in principle, with the help of these two theorems and [17]. We will come back to this problem in a future work.

7.2 Zeta-nonlocal field equation with source function in $\mathcal{L}_>(\mathbb{R}_+)$

If the source function $g(t), t \geq 0$, is an analytic function not necessarily of exponential type, the above approach does not apply. Let us assume that there exists a real number a such that the integral

$$\mathcal{L}(g)(z) = \int_0^\infty e^{-tz} g(t)dt$$

converges absolutely and uniformly on the half-plane $\{z \in \mathbb{C} : Re(z) > a\}$, and for which the function $z \to \mathcal{L}(g)(z)$ is analytic. We also assume that $\mathcal{L}(g)$ has an analytic extension to the left of $Re(z) = a$ until a singularity a_0, and that this new region of analyticity has an angular contour κ_∞ as its boundary.

We denote by $\mathcal{L}_>(\mathbb{R}_+)$ the space of analytic functions satisfying these properties. The problem of interest in this situation is to solve the equation

$$\zeta(\partial_t^2 + h)f = g, \tag{7.4}$$

for a given $g \in \mathcal{L}_>(\mathbb{R}^+)$. The solution of Equation (7.4) if it exists, will not necessarily be an entire function of exponential type. This is a very subtle problem and we only explain our reasoning; we must leave details to our paper [21].

Let $g \in \mathcal{L}_>(\mathbb{R}_+)$ and let the first singularity of the analytic extension of $\mathcal{L}(g)$ up to an angular contour κ_∞ be $a_0 = 0$. Now consider an angle $\frac{\pi}{2} < \psi \leq \pi$, a positive real number $r > 0$ and let κ_r be a finite angular contour contained in κ_∞ as in the following picture:

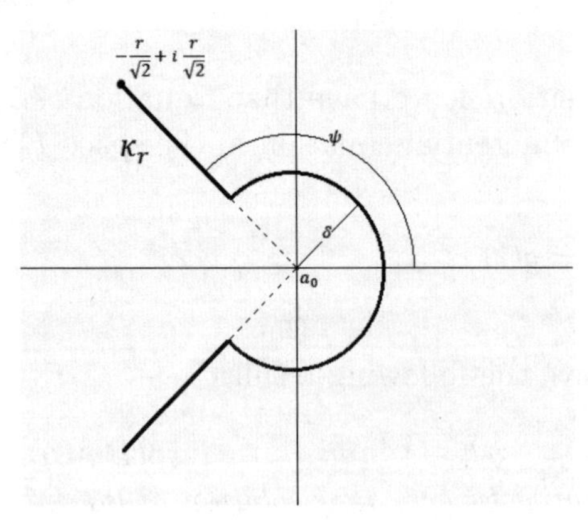

Now we pick the complex measure

$$d\mu_r(s) := \mathcal{X}_{\kappa_r}(s)\mathcal{L}(g)(s)\frac{ds}{2\pi i},$$

where $\mathcal{X}_{\kappa_r}$ denotes the characteristic function of the contour κ_r and we define:

$$g_r(z) := \mathcal{P}(\mu_r)(z) = \int_{\kappa_r} e^{zs}\mathcal{L}(g)(s)\frac{ds}{2\pi i}.$$

We show in [21] that it makes sense to consider the truncated equation

$$\zeta(\partial_t^2 + h)f_r = g_r \ , \quad h > 1. \tag{7.5}$$

We analyze Equation (7.5) in the domain

$$\Omega := \mathbb{C} - \{s \in \mathbb{C} : Re(s) \geq 0, \ |Im(s)| = \sqrt{h-1}\} \ ,$$

and we solve it as in the previous subsection, thereby obtaining "truncated solutions" f_r which are explicitly constructed in [21].

Our crucial final remark is that the functions g_r converge (in the uniform convergence topology) to an analytic extension of the source function g, and that the functions f_r converge (in the uniform convergence topology) to an analytic function f_∞, not necessarily of exponential type. *We can interpret Equation (7.4) in such a way that its solution is precisely* f_∞*!*

8 Future work

There exist interesting equations of infinite order not considered neither in [38] nor in this chapter. For example, the following "infinite order Euler-Arnold equation" was introduced in [35]:

$$\frac{dX}{dt} = -f(-\partial_x^2)^{-1}\left(2X_x f(-\partial_x^2)(X) + X\partial_x f(-\partial_x^2)(X)\right) \ . \tag{8.1}$$

This equation is Hamiltonian and it can be understood in terms of geometry of the loop group $Diff(S^1)$. Moreover, if f satisfies some technical conditions, we can prove that (8.1) is well-posed in a scale of Banach spaces; see [19] and [48]. However, we ignore whether it admits any of the structures commonly associated with integrability or, for instance, peakon solutions. Certainly, progress along these lines would be a considerable advance in the theory of nonlocal equations.

Acknowledgements

The authors thank G. Misiolek (University of Notre Dame) for suggesting the name "Euler-Arnold equation" for Equation (8.1), and for asking whether this equation may be integrable in some reasonable sense. H.P. and E.G.R. have been partially supported by the FONDECYT operating grants # 1170571 and # 1161691 respectively.

References

[1] Aref'eva I Y, Dragovich B, Frampton P H, Volovich I V, The wave function of the universe and p-adic gravity, *Internat. J. Modern Phys. A* **6** (1991), 4341–4258.

[2] Aref'eva I Y, Volovich I V, Quantization of the Riemann zeta-function and cosmology, *Int. J. Geom. Methods Mod. Phys.* **4** (2007), 881-895.

[3] Aref'eva I Y, Volovich I V, Cosmological daemon, *J. High Energ. Phys.* **2011** (2011), 102.

[4] Arendt W, Batty C J K, Hieber M, Neubrander F, *Vector-valued Laplace Transforms and Cauchy Problems*, Birkhäuser Springer Basel AG, Basel, 2011.

[5] Bank S B, Kaufman R P, A note on Hölder's theorem concerning the gamma function, *Math. Ann.* **232** (1978), 115–120.

[6] Barnaby N, Kamran N, Dynamics with infinitely many derivatives: the initial value problem, *J. High Energ. Phys.* **2008** (2008), 40 pp.

[7] Benedetto J J, Heinig H P, Johnson R, Weighted Hardy spaces and the Laplace transform. II, *Math. Nachr.* **132** (1987), 29–55.

[8] Biswas T, Talaganis S, String-inspired infinite-derivative theories of gravity: A brief overview, *Mod. Phys. Lett. A* **30** (2015), 20 pp.

[9] Boas Jr. R P, *Entire Functions*, Academic Press Inc., New York, 1954.

[10] Bourlet C, Sur les opérations en général et les équations différentielles linéaires d'ordre infini, *Annales scientifiques de l'École Normale Supérieure* **14** (1897), 133–190.

[11] Brekke L, Freund P G, Olson M, Witten E, Non-Archimedean string dynamics, *Nuclear Phys. B* **14** (1988), 365–402.

[12] Calcagni G, Montobbio M, Nardelli G, Route to nonlocal cosmology. *Physics Review D* **76** (2007), 126001 (20 pages).

[13] Calcagni G, Montobbio M, Nardelli G, Localization of nonlocal theories. *Physics Letters B* **662** (2008), 285–289.

[14] Calcagni G, *Classical and Quantum Cosmology*, Springer International Publishing, 2017.

[15] Cardon D A, de Gaston S A, Differential operators and entire functions with simple real zeros, *J. Math. Anal. Appl.* **301** (2005), 386–398.

[16] Cardon D A, Convolution operators and zeros of entire functions, *Proc. Amer. Math. Soc.* **130** (2002), 1725–1734.

[17] Carlsson M, Prado H, Reyes E G, Differential equations with infinitely many derivatives and the Borel transform, *Ann. Henri Poincaré* **17** (2016), 2049–2074.

[18] Carmichael R D, Linear differential equations of infinite order, *Bull. AMS* **42** (1936), 193–218.

[19] Chávez A, *Analytical Approach to Some Differential Equations of Infinite Order from Mathematical Physics*, Ph.D. Thesis, Universidad de Chile, 2018.

[20] Chávez A, Prado H, Reyes E G, A Laplace transform approach to linear equations with infinitely many derivatives and zeta-nonlocal field equations, (2017) *arXiv:1705.01525.*

[21] Chávez A, Prado H, Reyes E G, A Borel transform approach to linear zeta-nonlocal field equations, *preprint 2018.*

[22] Dickson D G, Convolution equations and harmonic analysis in spaces of entire functions, *Trans. Amer. Math. Soc.* **184** (1973), 373–385.

[23] Doetsch G, Bedingungen für die Darstellbarkeit einer Funktion als Laplace-integral und eine Umkehrformel für die Laplace-Transformation, *Math. Z.* **42** (1937), 263–286.

[24] Doetsch G, *Introduction to the Theory and Application of the Laplace Transformation*, Springer-Verlag. New York-Heidelberg, 1974.

[25] Dragovich B, Khrennikov A Yu, Kozyrev S V, Volovich I V, Zelenov E I, p-Adic mathematical physics: the first 30 years, *P-Adic Num., Ultrametric Anal. Appl.* **9** (2017), 87–121.

[26] Dragovich B, Zeta-nonlocal scalar fields, *Theor. Math. Phys.* **157** (2008), 1671–1677.

[27] Dragovich B, Towards effective Lagrangians for adelic strings, *Fortschritte der Physik* **57** (2009), 546–551.

[28] Dragovich B, Nonlocal dynamics of p-adic strings, *Theor. Math. Phys.* **164** (2010), 1151–1155.

[29] Dubinskii J A, *Analytic Pseudo-differential Operators and their Applications*, Kluwer Academic Publishers Group, Dordrecht, 1991.

[30] Eliezer D A, Woodard R P, The problem of nonlocality in string theory, *Nuclear Physics B* **325** (1989), 389–469.

[31] Feng L, Woodard R P, Light bending in infinite derivative theories of gravity, *Phys. Rev. D* **95** (2017), 084015.

[32] Goodman R, Differential operators of infinite order on a Lie group. II, *Indiana Univ. Math. J.* **21** (1971/72), 383–409.

[33] Goodman R, Wallach N R, Whittaker vectors and conical vectors, *J. Funct. Anal.* **39** (1980), 199–279.

[34] Goodman R, Differential operators of infinite order on a Lie group. I, *J. Math. Mech.* **19** (1969/70), 879–894.

[35] Górka P, Pons D, Reyes E G, Equations of Camassa-Holm type and the geometry of loop groups, *J. Geometry and Physics* **87** (2015), 190–197.

[36] Górka P, Prado H, Reyes E G, The initial value problem for ordinary differential equations with infinitely many derivatives, *Classical Quantum Gravity* **29** (2012), 065017.

[37] Górka P, Prado H, Reyes E G, Functional calculus via Laplace transform and equations with infinitely many derivatives, *J. Math. Phys.* **51** (2010), 103512.

[38] Górka P, Prado H, Reyes E G, Generalized Euclidean Bosonic String Equations, *Operator Theory: Advances and Applications*, Vol. 224, 147–169, Springer Basel, 2012.

[39] Harper Z, Laplace transform representations and Paley-Wiener theorems for functions on vertical strips, *Doc. Math.* **15** (2010), 235–254.

[40] Hörmander L, *The Analysis of Linear Partial Differential Operators. III*, Springer-Verlag. Berlin, 1985.

[41] Karatsuba A A, Voronin S M, *The Riemann Zeta-Function*, Walter de Gruyter & Co. Berlin, 1992.

[42] Levin B Ja, *Distribution of Zeros of Entire Functions*, American Mathematical Society, Providence, R.I., 1980.

[43] Martineau A, Sur les fonctionnnelles analytiques et la transformation de Fourier-Borel, *J. Analyse Math.* **11** (1963), 1–164.

[44] Moeller N, Zwiebach B, Dynamics with infinitely many time derivatives and rolling tachyons, *J. High Energ. Phys.* **2002** (2002), 034.

[45] Mulryne D J, Nunes N J, Diffusing nonlocal inflation: Solving the field equations as an initial value problem, *Phys. Rev. D* **78** (2008), 063519.

[46] Nelson W, Analytic vectors, *Ann. of Math. (2)* **70** (1959), 572–615.

[47] Rooney P G, Generalized H_p spaces and Laplace transforms, *Abstract Spaces and Approximation (Proc. Conf., Oberwolfach, 1968)* (1969), 258–269.

[48] Trèves F, *Ovcyannikov Theorem and Hyperdifferential Operators*, Instituto de Matemática Pura e Aplicada, Conselho Nacional de Pesquisas, Rio de Janeiro, 1968.

[49] Umarov S, *Introduction to Fractional Pseudo-differential Equations with Singular Symbols*, Springer International Publishing, 2015.

[50] Van Gorder R A, Does the Riemann zeta function satisfy a differential equation?, *J. Number Theory* **147** (2015), 778–788.

[51] Vladimirov V S, On the equation of a p-adic open string for a scalar tachyon field, *Izv. Ross. Akad. Nauk Ser. Mat.* **69** (2005), 55–80.

[52] Vladimirov V S, Volovich Ya I, Nonlinear dynamics equation in p-adic string theory, *Theoret. and Math. Phys.* **138** (2004), 297–309.

[53] Volovich I V, p-adic string, *Classical and Quantum Gravity* **4** (1987), L83–L87.

[54] Yosida K, *Functional Analysis*, Springer-Verlag, Berlin-New York, 1980.

[55] Witten E, Noncommutative geometry and string field theory, *Nucl. Phys. B* **268** (1986), 253–294.

D2. On the rôle of nonlinearity in geostrophic ocean flows on a sphere

A Constantin

Faculty of Mathematics, University of Vienna,
Oskar-Morgenstern Platz 1, 1090 Vienna, Austria
e-mail: adrian.constantin@univie.ac.at

R S Johnson

School of Mathematics, Statistics and Physics, Newcastle University,
Newcastle upon Tyne NE1 7RU, United Kingdom
e-mail: r.s.johnson@ncl.ac.uk

Abstract

We discuss some of the challenges and opportunities presented by the study of geostrophic flows in spherical coordinates.

1 Introduction

The large-scale flow in the ocean is, at leading order, hydrostatically balanced vertically, in the sense that gravitational and pressure gradient forces balance one another, rather than inducing accelerations (see the discussion in [24]). Moreover, at mid- and high latitudes, below a relatively shallow near-surface layer and above an even shallower region near the ocean bed (in which friction effects are relevant for the dynamics), the ocean flow is also close to balance in the horizontal, in the sense that Coriolis forces are balanced by horizontal pressure gradients in what is termed 'geostrophic motion' (from the Greek: '$\gamma\varepsilon\omega$'/'geo' for 'Earth' and '$\sigma\tau\rho o\phi\eta$'/'strophe' for 'turning'); 'ageostrophic flow' refers to the two boundary layers (at the atmosphere-ocean interface and where the ocean moves over a solid bed) in which the flow pattern is modified by friction. In our discussion we will address some basic aspects of geostrophic flow dynamics. In this regard, the advantage of an analytic approach with respect to data-driven studies or numerical simulations is the potential to identify clearly the important processes that control the flow. While numerical approximations undoubtedly play an important rôle in highlighting detailed features of specific cases, and may even provide a useful overall picture, a numerical attempt to encompass all the factors that play a rôle would be, in all likelihood, so excessive as to preclude it as a worthwhile undertaking. It is quite usual nowadays to attempt the identification of flow patterns by using (mainly) statistical methods to detect significant relationships (based on the notion of 'big data'), with numerical simulations used to test the veracity and relevance of the ideas. However, this often increases the risk of finding spurious correlations – statistically robust but unimportant associations – that might hide the underlying structure, and thus obscure, rather than provide, the hoped-for insights.

2 Preliminaries

The Earth is nearly an oblate spheroid (an ellipse rotated about its minor axis), with a small equatorial bulge – the polar radius is about 21 km shorter than the equatorial one (of length 6378 km). The fact that no dynamical consequences of the small deviation from a perfect sphere have been observable in the ocean flows [32] makes a spherical-Earth model adequate for the study of ocean flows. Furthermore, the ocean is a very thin layer at the surface of this sphere, with a mean thickness of about 0.1% of the radius: most oceanic depths are between 3 and 5.5 km.

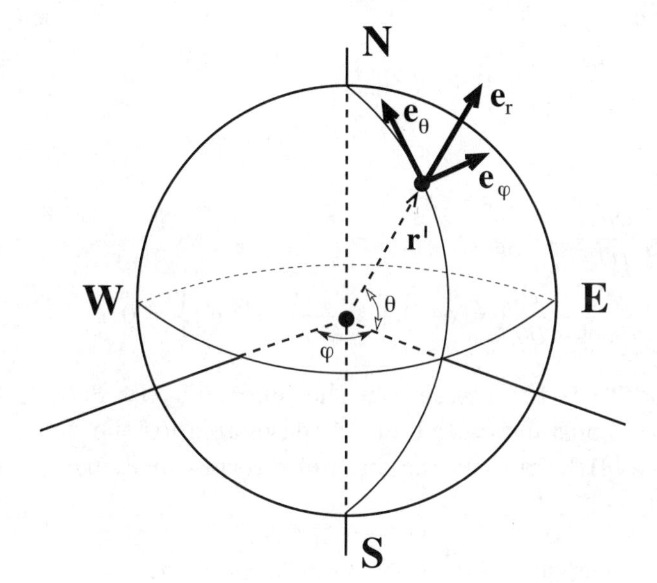

Figure 1. The Earth's rotating spherical coordinate system, where θ is the angle of latitude, φ is the azimuthal angle and $r' = |\mathbf{r}'|$ is the distance from the origin at the Earth's centre. The North/South Poles are at $\theta = \pm\pi/2$ and the Equator is on $\theta = 0$.

In order to study large-scale ocean flows we introduce a set of (right-handed) spherical coordinates, (φ, θ, r'): r' is the distance (radius) from the centre of the sphere, $\theta \in [-\pi/2, \pi/2]$ is the angle of latitude and $\varphi \in [0, 2\pi)$ is the azimuthal angle i.e. the angle of longitude. (We use primes, throughout the formulation of the problem, to denote physical (dimensional) variables; these will be removed in the non-dimensionalisation process.) The unit vectors in this (φ, θ, r')-system, for all points but the two poles, are $(\mathbf{e}_\varphi, \mathbf{e}_\theta, \mathbf{e}_r)$, respectively, and we write the corresponding velocity components as (u', v', w'), with $\mathbf{e}_\varphi$ pointing from West to East, $\mathbf{e}_\theta$ from South to North and $\mathbf{e}_r$ upwards (see Figure 1); we avoid the poles since $\mathbf{e}_\varphi$ and $\mathbf{e}_\theta$ are not well-defined there. The (φ, θ, r')-system is associated with a point fixed on the sphere which is rotating about its polar axis (with an angular speed $\Omega' \approx 7.29 \times 10^{-5}\ \mathrm{rad\,s^{-1}}$).

3 Governing equations

Away from the boundary layers the ocean flow is practically inviscid since friction effects are negligible (see the data in [17, 24]). The reference frame chosen in Section 2 rotates with the Earth, being thus not fixed in space. Because of the spin of the Earth, the surface of the Earth is not an inertial frame and thus, when it is observed from the rotating frame, fictitious forces appear – the Coriolis force and the centrifugal force (see the discussion in [31]). Consequently, the Euler equation and the equation of mass conservation are

$$
\begin{aligned}
&\Big(\frac{\partial}{\partial t'}+\frac{u'}{r'\cos\theta}\frac{\partial}{\partial\varphi}+\frac{v'}{r'}\frac{\partial}{\partial\theta}+w'\frac{\partial}{\partial r'}\Big)(u',v',w')\\
&\quad+\frac{1}{r'}\Big(-u'v'\tan\theta+u'w',\ u'^2\tan\theta+v'w',-u'^2-v'^2\Big)\\
&\quad+2\Omega'\,(-v'\sin\theta+w'\cos\theta,\ u'\sin\theta,\ -u'\cos\theta)\\
&\quad+r'\Omega'^2\,(0,\ \sin\theta\cos\theta,\ -\cos^2\theta)\\
&\quad=-\frac{1}{\rho'}\Big(\frac{1}{r'\cos\theta}\frac{\partial p'}{\partial\varphi},\frac{1}{r'}\frac{\partial p'}{\partial\theta},\frac{\partial p'}{\partial r'}\Big)+(0,0,-g'),
\end{aligned}
\tag{3.1}
$$

and

$$
\begin{aligned}
&\frac{\partial\rho'}{\partial t'}+\frac{u'}{r'\cos\theta}\frac{\partial\rho'}{\partial\varphi}+\frac{v'}{r'}\frac{\partial\rho'}{\partial\theta}+w'\frac{\partial\rho'}{\partial r'}\\
&\quad+\rho'\Big\{\frac{1}{r'\cos\theta}\frac{\partial u'}{\partial\varphi}+\frac{1}{r'\cos\theta}\frac{\partial}{\partial\theta}(v'\cos\theta)+\frac{1}{r'^2}\frac{\partial}{\partial r'}(r'^2w')\Big\}=0,
\end{aligned}
\tag{3.2}
$$

respectively, where $p'(\varphi,\theta,r',t')$ is the pressure in the fluid and $\rho'(\varphi,\theta,r',t')$ the density, with the choice $g'=\text{constant}\approx 9.81\ \mathrm{m\,s^{-2}}$ reasonable for the depths of the oceans on the Earth (see [31]). In (3.1) the quadratic terms involving $1/r'$ are called the *metric terms* and

$$2\mathbf{\Omega}'\wedge\mathbf{V}'=2\Omega'\,(-v'\sin\theta+w'\cos\theta,\ u'\sin\theta,\ -u'\cos\theta),$$

is the Coriolis acceleration, where the wedge notation is used for the vector product of the Earth's angular velocity vector $\mathbf{\Omega}'=\Omega'\,(0,\cos\theta,\sin\theta)$ and the velocity vector $\mathbf{V}'=(u',v',w')$ of the flow. The acceleration associated with pure rotation is

$$\mathbf{\Omega}'\wedge(\mathbf{\Omega}'\wedge\mathbf{X}')=r'\Omega'^2\,(0,\ \sin\theta\cos\theta,\ -\cos^2\theta)$$

with $\mathbf{X}'=r'\,(0,\,0,\,1)$ the position vector of the particle located at longitude φ and latitude θ, a distance r' away from the Earth's centre, since in Cartesian coordinates (denoted below by 'C')

$$\mathbf{X}'=r'\,(\cos\theta\cos\varphi,\ \cos\theta\sin\varphi,\sin\theta)_C$$

while

$$\mathbf{e}_\varphi=(-\sin\varphi,\ \cos\varphi,0)_C,\quad \mathbf{e}_\theta=(-\sin\theta\cos\varphi,\ -\sin\theta\sin\varphi,\cos\theta)_C,\quad \mathbf{e}_r=\frac{1}{r'}\mathbf{X}'.$$

The Coriolis force and the centrifugal force (per unit mass) are the two terms in (3.1) which depend on Ω', usefully moved to the right side of the equation so that they can be combined with the external force of gravity. Since the rate of the work done by a force is the scalar product of the force and the velocity vector of the point of application, the fact that the Coriolis force, $-2\boldsymbol{\Omega}' \wedge \mathbf{V}'$, is orthogonal to $\mathbf{V}'$ ensures that it does no work and therefore it plays no rôle in the energy budget. On the other hand, the centrifugal force (per unit mass), $-\boldsymbol{\Omega}' \wedge (\boldsymbol{\Omega}' \wedge \mathbf{X}')$, is a potential force that may be written as the gradient of the scalar potential $\mathfrak{P}_c(\varphi, \theta, r') = \frac{1}{2}\Omega'^2 r'^2 \cos^2\theta$; note that in the rotating spherical coordinate system the gradient of a scalar function $f(\varphi, \theta, r', t')$ is

$$\nabla f = \frac{1}{r'\cos\theta}\frac{\partial f}{\partial\varphi}\,\mathbf{e}_\varphi + \frac{1}{r'}\frac{\partial f}{\partial\theta}\,\mathbf{e}_\theta + \frac{\partial f}{\partial r'}\,\mathbf{e}_r\,. \tag{3.3}$$

while the material derivative is

$$\frac{Df}{Dt'} = \frac{\partial f}{\partial t} + \frac{u'}{r'\cos\theta}\frac{\partial f}{\partial\varphi} + \frac{v'}{r'}\frac{\partial f}{\partial\theta} + w'\frac{\partial f}{\partial r'}\,. \tag{3.4}$$

Since gravity is also a potential force, we can simplify the governing equations by absorbing it and the centrifugal force into the pressure (see Section 3.1 below). While gravity is directed towards the Earth's centre, the centrifugal force has two components: a meridional component (in the plane perpendicular to the radial direction) and a vertical component (leading to a steady upward acceleration that opposes gravity). Over geological times (i.e. on the scale of millions of years) the equatorward meridional component of the centrifugal force deformed the Earth into an ellipsoid with an equatorial bulge, even though this deformation is very small, corresponding to a difference of about 0.3% between the polar and the equatorial radii. Generally any rotating celestial body will present an equatorial bulge, the size of which depends primarily on its rate of rotation (the faster the spin, the greater the elongation) and on the strength of its gravitational pull compared to the centrifugal effect. In our solar system the Moon, Venus and Mercury all spin so slowly that their equatorial bulges are insignificant, while the rapid rotations of Jupiter and Saturn on their axes explain their pronounced equatorial bulges (being 6.9% and 10.7% thicker around the middle, respectively). On the other hand, while the Sun spins quite rapidly, its gravitational pull is so intense that there is no equatorial bulge worth mentioning. On Earth, for typical mid-latitude ocean flow speeds of $0.1\ \mathrm{m\,s^{-1}}$, the Coriolis acceleration terms are about $1.4 \times 10^{-6}\ \mathrm{m\,s^{-2}}$, with the scale of the centrifugal acceleration (about $3.4 \times 10^{-2}\ \mathrm{m\,s^{-2}}$) several orders of magnitude larger but still only about 0.03% of the gravitational acceleration.

Finally, we note that the density variations in the ocean are quite small compared to the mean density (see the discussion in [31]), with the equatorial regions presenting the greatest ocean stratification with density variations of up to 1% (see the discussion in [11]).

4 Geostrophy and the f- and β-plane approximations

A familiar simplification in oceanographic studies adopts a local 'flat' Cartesian coordinate system. We illustrate these approximations by restricting our discussion to approximations in the constant-density case. In this setting, we absorb gravity and the centrifugal force into a redefined pressure by setting

$$p' = -g'\rho + \frac{1}{2}\rho' r'^2 \Omega'^2 \cos^2\theta + P'(\varphi, \theta, r', t'). \tag{4.1}$$

We can now write (3.1) in the following form:

$$\frac{Du'}{Dt'} + \frac{u'w' - u'v'\tan\theta}{r'} + 2\Omega'(-v'\sin\theta + w'\cos\theta) = -\frac{1}{\rho' r'\cos\theta}\frac{\partial P'}{\partial\varphi}, \tag{4.2}$$

$$\frac{Dv'}{Dt'} + \frac{u'^2\tan\theta + v'w'}{r'} + 2\Omega' u'\sin\theta = -\frac{1}{r'}\frac{\partial P'}{\partial\theta}, \tag{4.3}$$

$$\frac{Dw'}{Dt'} - \frac{u'^2 + v'^2}{r'} - 2\Omega' u'\cos\theta = -\frac{1}{\rho'}\frac{\partial P'}{\partial r'}, \tag{4.4}$$

while (3.2) becomes

$$\frac{1}{r'\cos\theta}\frac{\partial u'}{\partial\varphi} + \frac{1}{r'}\frac{\partial v'}{\partial\theta} - \frac{\tan\theta}{r'}v' + \frac{\partial w'}{\partial r'} + \frac{2}{r'}w' = 0. \tag{4.5}$$

We introduce this discussion by first describing the so-called *traditional approximation* (see [33]). The initial step involves neglecting the metric terms in (4.2)-(4.4), which are curvature effects in spherical coordinates, and the terms in (4.5) which involve velocity components multiplied by $1/r'$. Furthermore, the term $(2\Omega' w'\cos\theta)$ of the Coriolis force is neglected because its effect is small: typically the vertical velocities are much weaker than horizontal velocities (by a factor of 10^{-4}, on average). The last step is to neglect the variation of vertical positions of fluid masses as compared to the Earth's radius R', that is, we replace r' by R' in the denominators of all multiplicative factors in terms of type (3.3) that occur in (4.2)-(4.5). Despite the simplifications made in the traditional approximation, the equations of motion remain complex, due to the persistence of the factor $\cos\theta$ in the denominator of the simplified version of the gradient (3.3) and to the appearance of meridional variations in the two Coriolis parameters

$$f = 2\Omega'\sin\theta, \qquad \tilde{f} = 2\Omega'\cos\theta.$$

The f-plane approximation describes a small region around a latitude and longitude of interest, denoted by the reference values θ_0 and φ_0, respectively. In this setting we replace $\cos\theta$ by $\cos\theta_0$ in the traditional approximation of the gradient terms of type (3.3), with the two Coriolis parameters f and $\tilde{f}$ also replaced by the constant values

$$f_0 = 2\Omega'\sin\theta_0 \quad \text{and} \quad \tilde{f}_0 = 2\Omega'\cos\theta_0,$$

respectively, where f_0 is the familiar 'Coriolis parameter'. After all these simplifications we see by inspection that the equations become

$$\frac{\partial u'}{\partial t} + (\mathbf{v}' \cdot \nabla_{(\bar{x},\bar{y},\bar{z})})u' - f_0 v' = -\frac{1}{\rho'}\frac{\partial P'}{\partial \bar{x}}\,, \tag{4.6}$$

$$\frac{\partial v'}{\partial t} + (\mathbf{v}' \cdot \nabla_{(\bar{x},\bar{y},\bar{z})})v' + f_0 u' = -\frac{1}{\rho'}\frac{\partial P'}{\partial \bar{y}}\,, \tag{4.7}$$

$$\frac{\partial w'}{\partial t} + (\mathbf{v}' \cdot \nabla_{(\bar{x},\bar{y},\bar{z})})w' - \tilde{f}_0 u' = -\frac{1}{\rho'}\frac{\partial P'}{\partial \bar{z}} - g'\,, \tag{4.8}$$

$$\nabla_{(\bar{x},\bar{y},\bar{z})} \cdot \mathbf{v}' = 0\,, \tag{4.9}$$

where $\mathbf{v}' = (u', v', w')$ and

$$(\bar{x}, \bar{y}, \bar{z}) = (R'(\varphi - \varphi_0)\cos\theta_0,\ R'(\theta - \theta_0),\ r')$$

defines a Cartesian coordinate system. Note that $R'(\varphi - \varphi_0)\cos\theta_0$ and $R'(\theta - \theta_0)$ are Cartesian coordinates (west-to-east and south-to-north, i.e. in the directions of $\mathbf{e}_{\varphi_0}$ and $\mathbf{e}_{\theta_0}$, respectively) for motion on the plane that is tangent to the surface of the sphere of radius R', at the reference point of latitude θ_0 and longitude φ_0. Consistently, u', v', and w' are the components of the velocity in this tangent plane.

The β-plane approximation applies to larger regions around a reference latitude θ_0, replacing the exact spherical relation $f = 2\Omega' \sin\theta$ with a truncated Taylor expansion

$$f = 2\Omega' \sin\Big(\theta_0 + (\theta - \theta_0)\Big) \approx 2\Omega' \sin\theta_0 + 2\Omega'(\theta - \theta_0)\cos\theta_0\,; \tag{4.10}$$

so, on the tangent plane, we mimic this by allowing the Coriolis parameter in (4.6)-(4.7) to vary by replacing f_0 by

$$f_0 + \beta\bar{y} \quad \text{where} \quad \beta = \frac{\partial f}{\partial \bar{y}}(0) = \frac{2\Omega' \cos\theta_0}{R'}\,.$$

However, in the β-plane approximation the other Coriolis parameter in (4.8), $\tilde{f}_0$, is kept equal to the constant value used in the f-plane approximation, since allowing it to vary with latitude violates conservation of angular momentum and potential vorticity. This discrepancy between the Taylor truncations of f and $\tilde{f}$ is indicative of the fact that, despite being widely used, the β-plane equations are not a consistent approximation in non-equatorial regions; for a more detailed discussion we refer to [18, 29]. On the other hand, in equatorial regions, the truncated Taylor expansion of $\tilde{f}$ around the reference latitude $\theta_0 = 0$, analogous to (4.10), is

$$\tilde{f} \approx 2\Omega'\,, \tag{4.11}$$

which coincides with that used in the equatorial f-plane approximation. The equatorial f-plane equations, as well as the equatorial β-plane equations, represent a consistent approximation of the governing equations and thus they can be regarded as providing quite accurate insight into equatorial currents [4, 11, 12, 14, 27] and into equatorial wave-current interactions [4, 8, 9, 10, 21, 22, 23].

It is instructive to recall at this point the standard approach to the dynamics of geostrophic flows, based on the f-plane approximation (see [24]). First, we make the hydrostatic approximation for the vertical component: taking advantage of the fact that the vertical speed in ocean flows is very small, we neglect w' in (4.6)-(4.9), and we also ignore the term $\tilde{f}_0 u'$ in (4.8) since $\Omega' u' \ll g'$, thus obtaining the system

$$\frac{\partial u'}{\partial t} + u'\frac{\partial u'}{\partial \bar{x}} + v'\frac{\partial u'}{\partial \bar{y}} - f_0 v' = -\frac{1}{\rho'}\frac{\partial P}{\partial x}, \tag{4.12}$$

$$\frac{\partial v'}{\partial t} + u'\frac{\partial v'}{\partial \bar{x}} + v'\frac{\partial v'}{\partial \bar{y}} + f_0 u' = -\frac{1}{\rho'}\frac{\partial P'}{\partial \bar{y}}, \tag{4.13}$$

$$0 = -\frac{1}{\rho'}\frac{\partial P'}{\partial \bar{z}} - g', \tag{4.14}$$

$$\frac{\partial u'}{\partial \bar{x}} + \frac{\partial v'}{\partial \bar{y}} = 0, \tag{4.15}$$

Supposing now that each of the horizontal flow components u' and v' has a typical magnitude U', and that each varies in time with a characteristic time scale T' and with horizontal position over the characteristic length scale

$$L' = U'T',$$

we non-dimensionalise (4.12)-(4.15) by setting

$$t' = T't, \quad (u', v') = U'(u, v), \quad (\bar{x}, \bar{y}) = L'(x, y), \quad P' = \rho' U'^2 P, \tag{4.16}$$

thus obtaining the nondimensional system

$$u_t + (uu_x + vu_y) - \frac{f_0 L'}{U'} v = -P_x, \tag{4.17}$$

$$v_t + (uv_x + vv_y) + \frac{f_0 L'}{U'} u = -P_y, \tag{4.18}$$

$$u_x + v_y = 0, \tag{4.19}$$

where, due to (4.14), P depends linearly on the vertical spatial variable (within the framework of the hydrostatic approximation). The non-dimensional parameter associated with the Coriolis terms in (4.17)-(4.18) is expressed as the *Rossby number*

$$\mathfrak{R}_o = \frac{U'}{f_0 L'},$$

and $\mathfrak{R}_o \approx 10^{-3}$ for large-scale ocean flows at mid-latitudes, given that typically $U' = 0.1\ \mathrm{m\,s^{-1}}$, $L' = 2000$ km, and for $\theta = 45°$ we have $f_0 = 2\Omega'/\sqrt{2} \approx 1.03 \times 10^{-4}\,\mathrm{s^{-1}}$. However, near the Equator, the small-Rossby-number assumption breaks down since $f_0 \to 0$ as $\theta \to 0$. The smallness of $\mathfrak{R}_o$ in non-equatorial regions means that the acceleration terms in (4.17)-(4.18) are dominated by the Coriolis term, leading to the *geostrophic balance*

$$\begin{cases} v = \mathfrak{R}_o P_x, \\ u = -\mathfrak{R}_o P_y, \end{cases} \tag{4.20}$$

in which the horizontal pressure gradient is balanced by the Coriolis term. Furthermore, (4.19), which expresses the fact that the geostrophic flow is horizontally non-divergent, is merely the compatibiliy condition $P_{xy} = P_{yx}$ for (4.20). Then (4.19) combined with (4.9) yields $w'_{\bar{z}} = 0$, so that the kinematic boundary condition $w' = 0$ on a flat bed yields $w' = 0$ throughout the flow, which is consistent with neglecting the vertical component of the velocity: in the f-plane approximation, the geostrophic flow is, indeed, horizontal.

It is instructive to examine this fundamental geostrophic flow in a little more detail. In particular, let us look at the particle trajectories $(x(t),\, y(t))$ of a smooth and steady geostrophic flow; we see that (4.20) becomes the autonomous Hamiltonian differential system

$$\begin{cases} x'(t) = \mathcal{H}_y\,, \\ y'(t) = -\mathcal{H}_x\,, \end{cases} \tag{4.21}$$

with the smooth Hamiltonian

$$\mathcal{H}(x,y) = -\,\mathfrak{R}_o P(x,y)\,,$$

so that the level sets of $\mathcal{H}$ are invariant sets: if a solution starts in the set, it remains in the set. Consequently a steady geostrophically balanced, horizontal flow is along the contours of constant pressure. The importance of this is the fact that knowledge of the pressure field permits us to draw the phase portrait without solving the system, by simply plotting the level curves $[\mathcal{H}(x,y) = \text{constant}]$ and determining the direction of the particle paths on these level sets.

Near a non-equilibrium point (x_0, y_0), where $\nabla\mathcal{H}(x_0,y_0) \neq (0,0)$, by the Hamiltonian Flow Box Theorem (see [26]), there exist (local) symplectic coordinates (ξ,η) such that the Hamiltonian becomes $\mathcal{H}(\xi,\eta) = \eta$ and the equations of motion become

$$\begin{cases} \xi'(t) = 1\,, \\ \eta'(t) = 0\,. \end{cases} \tag{4.22}$$

Recall that a differentiable change of variables on some open planar set is called symplectic if the Jacobian has determinant 1 at every point, that is, planar symplectic transformations are those that are orientation and area preserving [26]. Furthermore, a symplectic change of variables takes a Hamiltonian system of equations into a Hamiltonian system and conversely, if a change of variables preserves the Hamiltonian form of all Hamiltonian equations, then it must be symplectic. The variables (η,ξ) in (4.22) are called *action-angle variables* and we see that, in appropriate coordinates, the flow of the (nonlinear) dynamical system (4.21) near a regular point is actually a linear flow at constant speed: the system (4.21) is *integrable* since, up to the change of variables, the solutions can be determined by simple quadature.

We now discuss the behaviour of the flow near equilibrium points, which correspond exactly to the critical points of the pressure P. At such an equilibrium point (x_0, y_0), the Jacobian matrix is given by

$$M_J(x_0,y_0) = \begin{pmatrix} \mathcal{H}_{xy}(x_0,y_0) & \mathcal{H}_{yy}(x_0,y_0) \\ -\mathcal{H}_{xx}(x_0,y_0) & -\mathcal{H}_{xy}(x_0,y_0) \end{pmatrix}.$$

The equilibrium (x_0, y_0) is also a critical point of the Hamiltonian $\mathcal{H}:\mathbb{R}^2 \to \mathbb{R}$, to which we associate the Hessian matrix

$$M_H(x_0,y_0) = \begin{pmatrix} \mathcal{H}_{xx}(x_0,y_0) & \mathcal{H}_{xy}(x_0,y_0) \\ \mathcal{H}_{xy}(x_0,y_0) & \mathcal{H}_{yy}(x_0,y_0) \end{pmatrix},$$

which has exactly the same determinant as $M_J(x_0, y_0)$, namely

$$\mathfrak{D}(x_0,y_0) = \mathcal{H}_{xx}(x_0,y_0)\mathcal{H}_{yy}(x_0,y_0) - [\mathcal{H}_{xy}(x_0,y_0)]^2 .$$

A non-degenerate critical point (x_0, y_0) of the Hamiltonian $\mathcal{H}$ is a point where $\mathfrak{D}(x_0, y_0) \neq 0$. Since $M_J(x_0, y_0)$ has zero trace, we deduce that at a non-degenerate critical point, the eigenvalues of $M_J(x_0, y_0)$ are either real and of opposite sign or purely imaginary, depending on the sign of $\mathfrak{D}(x_0, y_0)$. If $\mathfrak{D}(x_0, y_0) < 0$, then $\mathcal{H}$ has a saddle point at (x_0, y_0) and $M_J(x_0, y_0)$ has one positive and one negative real eigenvalue, so that (x_0, y_0) is a saddle point in the sense of planar autonomous dynamical systems. On the other hand, $\mathfrak{D}(x_0, y_0) > 0$ means that $\mathcal{H}$ has a local extremum at the critical point (x_0, y_0), namely a local maximum if $\mathcal{H}_{xx}(x_0, y_0) < 0$ and a local minimum if $\mathcal{H}_{xx}(x_0, y_0) > 0$. Moreover, in this case the equilibrium point will be nonlinearly stable since, for a local minimum, $V(x, y) = \mathcal{H}(x, y) - \mathcal{H}(x_0, y_0)$ is a Lyapunov function, whereas for a local maximum point one can choose $-V$ as the Lyapunov function. In contrast to this, the saddle point is always unstable.

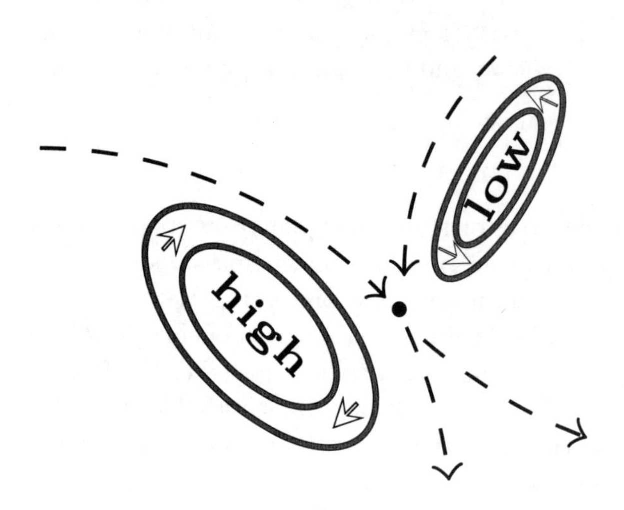

Figure 2. Typical geostrophic ocean flow in the Northern hemisphere: a saddle point (drawn together with its two separatrices that divide its neighbourhood into regions with different behaviour of the trajectories) accommodates the transition between the clockwise flow around a high pressure centre (left) and the anticlockwise flow around a low pressure centre (right). The rotation sense around extrema of the pressure is opposite in the Southern hemisphere, while equatorial flows are quite different since the geostrophic balance breaks down near the Equator.

The Morse-Darboux lemma, a symplectic planar analogue of the Morse lemma (see [6, 19] and the discussion in [3]) allows us to visualise the flow dynamics near a non-degenerate equilibrium point. (We recall that the Morse lemma generalises the Gram-Schmidt diagonalization process for bilinear forms to smooth functions, guaranteeing that there exist coordinates locally transforming the function exactly to the quadratic form generated by the second order partial derivatives evaluated at the critical point.) More precisely, there exists a local change of coordinates mapping the trajectories of (4.21), bijectively and preserving the direction of time, into those of the (normal form) Hamiltonian

$$H(x,y) = \pm(x-x_0)(y-y_0) \quad \text{if} \quad \mathfrak{D}(x_0,y_0) < 0\,,$$

and

$$H(x,y) = \pm\frac{(x-x_0)^2+(y-y_0)^2}{2} \quad \text{if} \quad \mathfrak{D}(x_0,y_0) > 0\,,$$

respectively, with the $\pm$ choice determined by the orientation of the stable/unstable manifolds in the case of the saddle point, and by whether (x_0, y_0) is a local minimum/maximum in the case of the centre. The phase portraits of the normal forms of (4.21) are easily drawn since we deal with linear systems that can be solved explicitly (in the (x,y)-coordinates for the saddle point and by passing to polar coordinates for the centre):

- near the saddle point the trajectories are straight lines parallel to the axes, moving exponentially fast away from the critical point along the unstable direction and towards the point along the stable direction;
- near the centre the trajectories are concentric circles around the centre, orbiting clockwise/counterclockwise depending on whether the critical point is a maximum/minimum of the Hamiltonian function $\mathcal{H}$.

Therefore, near a non-degenerate equilibrium point, the flow represented by (4.21) is topologically conjugate (by means of an area and orientation-preserving bijection) to one of the above dynamical patterns, being merely a distortion of it: closed trajectories will surround a centre while the level set of the saddle consists of two intersecting curves – these are the local stable and the local unstable manifolds. Since $\mathfrak{R}_o P(x,y) = -\mathcal{H}(x,y)$, we infer that the dynamics near non-degenerate local extrema of the pressure depend on whether we are in the northern or the southern hemisphere (see Figure 2).

We conclude our discussion by pointing out briefly how the issue of the dynamics near a general equilibrium point for (4.21) can be addressed by means of Singularity Theory. The goal is to classify the singularities by identifying those which are not equivalent to one another, i.e. which cannot be transformed into one another by smooth transformations; in our planar Hamiltonian setting they are exhausted by the following list of normal forms (see [5]):

- the 1-parameter *parabolic* normal form $H(x,y) = \frac{1}{2}y^2 + \frac{1}{3}x^3 + Ax$, whose phase portraits are depicted in Figure 3 (with no singularities for $A > 0$, with a degenerate singularity at the origin for $A = 0$, while for $A < 0$ the non-degenerate saddle-point $(-\sqrt{A}, 0)$ and the centre $(\sqrt{A}, 0)$ are present);

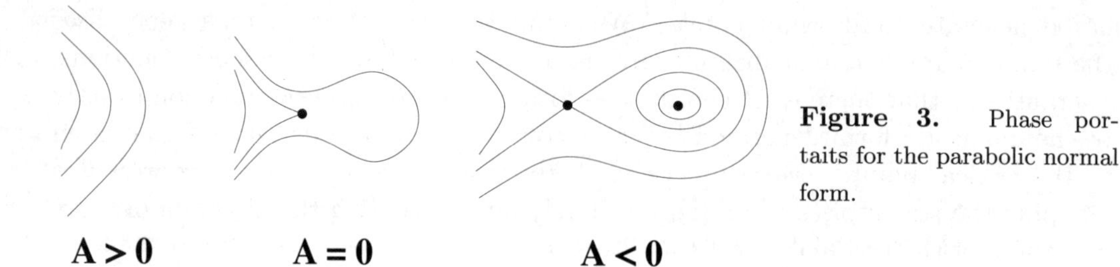

Figure 3. Phase portaits for the parabolic normal form.

- the 3-parameter *umbilic* normal forms $H(x,y) = x^2 y \pm \frac{1}{3} y^3 + A(x^2 \mp y^2) + Bx + Cy$, with the upper/lower signs corresponding to the hyperbolic and elliptic cases, respectively (see [2] for the relevant phase portraits).

5 Geostrophy in spherical coordinates

The discussion in the previous section presents, and expands, the classical description of the geostrophic balance, but its applicability is very limited: we have used the familiar f-plane approximation. In order to develop these ideas for large-scale motions of the oceans, we must either retain higher-order terms (in $\bar{y}$ – very tiresome indeed) or incorporate the spherical geometry. This latter option is often avoided because of the perceived complications but, in this context of (relatively) shallow oceans on a sphere, the development is altogether straightforward. We therefore follow this route by invoking the description in spherical coordinates.

Restricting our discussion to the case $\rho' = \rho'(r')$, redefining the pressure as

$$p' = g' \int_{r'}^{R'} \rho'(r)\,\mathrm{d}r + \frac{1}{2}\rho' r'^2 \Omega'^2 \cos^2\theta + \frac{1}{2}\Omega'^2 \cos^2\theta \int_{r'}^{R'} \hat{r}^2 \frac{\mathrm{d}\rho'}{\mathrm{d}\hat{r}}\,\mathrm{d}\hat{r} + P'(\varphi,\theta,r',t')\,, \tag{5.1}$$

and then writing

$$r' = R' + z'\,, \tag{5.2}$$

we non-dimensionalise the governing equations (3.1)-(3.2) according to

$$t' = T't\,,\ z' = H'z\,, \quad (u',\,v',\,w') = U'\,(u,\,v,\,kw)\,,\ P' = \overline{\rho}' U'^2 P\,,\ \rho' = \overline{\rho}'\rho\,, \tag{5.3}$$

where T' is a typical time scale, H' is the mean depth of the ocean, U' a suitable speed scale and $\overline{\rho}'$ an average density. The scaling factor, k, associated with the vertical component (w) of the velocity, is very small (the typical value is about 10^{-5}) since the vertical motion is so weak that it is almost always inferred rather than measured directly [30]. Defining the shallowness parameter ε and the Rossby number μ by

$$\varepsilon = \frac{H'}{R'}\,, \qquad \mu = \frac{U'}{2\Omega' L'}\,, \tag{5.4}$$

respectively, where $L' = U'T'$ is the horizontal length scale, the Euler equation becomes

$$\mu\frac{\partial}{\partial t}(u,\,v,\,kw)+\frac{1}{2\omega}\Big(\frac{u}{(1+\varepsilon z)\cos\theta}\frac{\partial}{\partial\varphi}+\frac{v}{1+\varepsilon z}\frac{\partial}{\partial\theta}+\frac{k}{\varepsilon}\,w\,\frac{\partial}{\partial z}\Big)(u,\,v,\,kw)$$
$$+\frac{1}{2\omega(1+\varepsilon z)}\Big(-uv\tan\theta+kuw,\;u^2\tan\theta+kvw,\;-u^2-v^2\Big)$$
$$+(-v\sin\theta+kw\,\cos\theta,\;u\sin\theta,\;-u\,\cos\theta)$$
$$=-\frac{1}{2\omega\rho}\Big(\frac{1}{(1+\varepsilon z)\cos\theta}\frac{\partial P}{\partial\varphi},\,\frac{1}{1+\varepsilon z}\frac{\partial}{\partial\theta}\Big[P-\frac{\omega^2\cos^2\theta}{2}\int_1^{1+\varepsilon z}(1+\varepsilon\hat{z})^2\frac{\mathrm{d}\rho}{\mathrm{d}\hat{z}}\mathrm{d}\hat{z}\Big],\;\frac{1}{\varepsilon}\frac{\partial P}{\partial z}\Big),\tag{5.5}$$

where

$$\omega=\frac{\Omega' R'}{U'},\tag{5.6}$$

while the equation of mass conservation becomes

$$\frac{k}{\varepsilon}w\frac{\mathrm{d}\rho}{\mathrm{d}z}+\rho\Big\{\frac{1}{(1+\varepsilon z)\cos\theta}\Big[\frac{\partial u}{\partial\varphi}+\frac{\partial}{\partial\theta}(v\cos\theta)\Big]+\frac{k}{\varepsilon(1+\varepsilon z)^2}\frac{\partial}{\partial z}\Big[(1+\varepsilon z)^2w\Big]\Big\}=0.\tag{5.7}$$

Typically $k = o(\varepsilon)$ (which we assume here) so that, multiplying the third component of (5.5) by ε and subsequently letting $\varepsilon \to 0$ (the shallow-water approximation), we see that the horizontal flow (u, v) is governed by the equations

$$\mu\frac{\partial}{\partial t}(u,\,v)+\frac{1}{2\omega}\Big(\frac{u}{\cos\theta}\frac{\partial}{\partial\varphi}+v\frac{\partial}{\partial\theta}\Big)(u,\,v)+\frac{1}{2\omega}\Big(-uv\tan\theta,\,u^2\tan\theta\Big)$$
$$+(-v\sin\theta,\,u\sin\theta)=-\frac{1}{2\omega\rho}\Big(\frac{1}{\cos\theta}\frac{\partial P}{\partial\varphi},\,\frac{\partial P}{\partial\theta}\Big),\tag{5.8}$$

$$\frac{\partial P}{\partial z}=0,\tag{5.9}$$

and

$$\frac{\partial u}{\partial\varphi}+\frac{\partial}{\partial\theta}(v\cos\theta)=0.\tag{5.10}$$

To explain the relevance of the Rossby number, note that if the time $T' = L'/U'$ it takes for a fluid element moving with speed U' to travel the distance L' is much less than the period of rotation of the Earth, $1/\Omega'$, then the fluid can scarcely sense the Earth's rotation over the time-scale of the motion. Therefore, for rotation to be important, it is necessary that $T' \gg 1/\Omega'$, or equivalently, $\mu \ll 1$. Indeed, we see that $\mu = (R'/2L')/\omega$ and so for reasonably large length scales (e.g. $L' = 2000$ km) with $\omega \approx 4650$ and $\varepsilon \approx 0.002$, given that for large-scale ocean flows in sub-tropical regions typical choices are $U' = 0.1\ \mathrm{m\,s^{-1}}$ and $H' = 4$ km (see [24]), the problem represented by (5.8)-(5.10) can be interpreted by invoking $\omega \to \infty$ with $(R'/2L')$ fixed. Of course, we should adjust the size (scaling) of P so that the flows of interest are driven and maintained, so we impose $\omega \to \infty$ but retain terms P/ω at this stage. Therefore, at leading order in spherical coordinates (with $\varepsilon \to 0$ and $\omega \to \infty$), the large-scale ocean motion is governed by the steady system

$$(-v \sin\theta,\, u \sin\theta) = -\frac{1}{2\omega\rho}\Big(\frac{1}{\cos\theta}\frac{\partial P}{\partial\varphi},\, \frac{\partial P}{\partial\theta}\Big), \tag{5.11}$$

$$\frac{\partial P}{\partial z} = 0\,, \tag{5.12}$$

$$\frac{\partial u}{\partial\varphi} + \frac{\partial}{\partial\theta}(v\cos\theta) = 0\,. \tag{5.13}$$

Equation (5.13) ensures the existence of a stream function $\psi(\varphi,\theta,z)$ for the horizontal components (u,v) of the velocity, with

$$u = -\frac{\partial\psi}{\partial\theta}\,, \qquad v = \frac{1}{\cos\theta}\frac{\partial\psi}{\partial\varphi}\,, \tag{5.14}$$

so that, at every fixed z (and observe that this formulation admits $\rho(z)$), the particle paths of the steady flow are along level sets of the stream function ψ (see [16]): if $(\varphi(t),\,\theta(t))$ are the spherical coordinates (longitude and latitude, respectively) of the particle at time t, then

$$u = \varphi'(t)\cos[\theta(t)]\,, \qquad v = \theta'(t)\,, \tag{5.15}$$

are its horizontal velocity components. Since (5.11) and (5.15) yield

$$\frac{\mathrm{d}}{\mathrm{dt}}\,P(\varphi(t),\,\theta(t),\,z) = 0\,,$$

we see that the flow is along contours of constant pressure (that is, any particle path is confined to a level set of the redefined pressure P). We described in Section 4 how the classical approach reaches this conclusion by relying on the f-plane approximation. The above considerations, at first sight, appear to establish the validity of this result within the framework of shallow-water, small Rossby number, spherical geometry. However, there is a compatibility issue with the system (5.11)-(5.13): invoking (5.13), the equality of the two expressions obtained for $\frac{\partial^2 P}{\partial\varphi\partial\theta}$ from (5.11) leads to $v\cos^2\theta = 0$, that is

$$v \equiv 0\,. \tag{5.16}$$

Returning to the system (5.11)-(5.13), we see that (5.16) (with (5.12)) forces P to be solely a function of θ, with $u(\theta,z)$ determined from the equation

$$u\sin\theta = -\frac{1}{2\omega\rho}\frac{\partial P}{\partial\theta}\,. \tag{5.17}$$

While this type of zonal flow is not without interest, being of some relevance for equatorial flows and for the Antarctic Circumpolar Current (see [12, 13]), it is too simple to give insight into the general large-scale ocean circulation. We saw in

Section 4 that, in plane geometry, we may set $w = 0$ throughout the depth of the layer, but the above considerations show that the Earth's curvature alters this picture quite fundamentally. There are two options: either w is retained (as an O(1) contribution to (5.11)-(5.13)) or nonlinear corrections are needed if neglecting w remains the working hypothesis. Indeed, the observed structure of these large-scale flows on the sphere is that they are nearly purely rotational, i.e. the vertical motion is very small compared with the horizontal motion (so we do have $k = o(\varepsilon)$). Furthermore, the considerations in [15] show that nonlinearity can ameliorate the situation, when spherical geometry is used. We see that the steady-state version of (5.8)-(5.10) is

$$\Big(\frac{u}{\cos\theta}\frac{\partial}{\partial\varphi}+v\frac{\partial}{\partial\theta}\Big)u - uv\tan\theta - 2\omega\, v\sin\theta = -\frac{1}{\rho\cos\theta}\frac{\partial P}{\partial\varphi}, \tag{5.18}$$

$$\Big(\frac{u}{\cos\theta}\frac{\partial}{\partial\varphi}+v\frac{\partial}{\partial\theta}\Big)v + u^2\tan\theta + 2\omega\, u\sin\theta = -\frac{1}{\rho}\frac{\partial P}{\partial\theta}, \tag{5.19}$$

$$0 = \frac{\partial P}{\partial z}, \tag{5.20}$$

$$\frac{\partial u}{\partial\varphi} + \frac{\partial}{\partial\theta}(v\cos\theta) = 0\,. \tag{5.21}$$

The existence of a stream function, $\psi(\varphi,\theta,z)$, satisfying (5.14), is ensured by (5.21) and the elimination of the pressure between equations (5.18) and (5.19) gives the vorticity equation

$$\Big(\psi_\varphi\frac{\partial}{\partial\theta} - \psi_\theta\frac{\partial}{\partial\varphi}\Big)\Big(\frac{1}{\cos^2\theta}\,\psi_{\varphi\varphi} - \psi_\theta\tan\theta + \psi_{\theta\theta} + 2\omega\,\sin\theta\Big) = 0\,, \tag{5.22}$$

in which

$$\nabla^2_\Sigma\psi = \frac{1}{\cos^2\theta}\,\psi_{\varphi\varphi} - \psi_\theta\tan\theta + \psi_{\theta\theta}\,. \tag{5.23}$$

is the Laplace-Beltrami expression of the flow vorticity in spherical coordinates, at leading order. Defining

$$\Psi(\varphi,\theta,z) = \psi + \omega\,\sin\theta\,, \tag{5.24}$$

we may write equation (5.22) in the form

$$\begin{aligned}(\Psi - \omega\,\sin\theta)_\varphi\Big(\frac{1}{\cos^2\theta}\,\Psi_{\varphi\varphi} - \Psi_\theta\tan\theta + \Psi_{\theta\theta}\Big)_\theta \\ -(\Psi - \omega\,\sin\theta)_\theta\Big(\frac{1}{\cos^2\theta}\,\Psi_{\varphi\varphi} - \Psi_\theta\tan\theta + \Psi_{\theta\theta}\Big)_\varphi = 0\,.\end{aligned} \tag{5.25}$$

Throughout regions where

$$\nabla_{(\varphi,\theta)}(\Psi - \omega\,\sin\theta) \neq 0\,,$$

the rank theorem (see [28]) permits us to express the solution of (5.25) in the form

$$\frac{1}{\cos^2\theta}\,\Psi_{\varphi\varphi} - \Psi_\theta\tan\theta + \Psi_{\theta\theta} = F(\Psi - \omega\,\sin\theta)\,, \tag{5.26}$$

where F is an arbitrary function. Note, however, that variant (5.26) of the vorticity equation (5.22) does not capture all possible solutions: for example,

$$\Psi = \omega \sin\theta + A$$

corresponds to the stationary-flow solution $\psi = A$ of (5.22), where A is an arbitrary constant, but fails to solve (5.26) if $F(A) \neq 0$. The left side of (5.26) can be written as

$$\nabla_\Sigma^2(\psi - \omega \sin\theta) = \nabla_\Sigma^2 \psi + 2\omega \sin\theta$$

and represents the total vorticity of the flow, comprising two components: the vorticity solely due to the rotation of the Earth ($2\omega \sin\theta$: 'spin vorticity') and that due to the underlying motion of the ocean and not driven by the rotation of the Earth ($\nabla_\Sigma^2\psi$: 'oceanic' or 'relative' vorticity). One of these contributions (the spin vorticity) is completely prescribed, but that associated with the movement of the ocean is specific to the particular flow conditions. If we ignore the planetary (spin) vorticity by setting $\omega = 0$, equation (5.26) becomes the equation describing stationary vortex structures in an ideal fluid; see Chapter 5 in [1] for a systematic analysis of its exact solutions and for graphical illustrations. The presence of planetary vorticity in equation (5.26) alters considerably the underlying mathematical structure of the problem due to the intricate coupling between the oceanic and the planetary vorticity components. This new equation offers some exciting prospects for future investigations. For example, some explicit solutions are available (see [15]):

1. In the case $F \equiv 0$ (zero oceanic vorticity), the solution

$$\Psi = \alpha \ln\left\{\varphi^2 + \left[A + \ln\left(\frac{\cos\theta}{1 - \sin\theta}\right)\right]^2\right\},$$

   where α and A are arbitrary constants, corresponds to the classical solution for irrotational flows in two-dimensional planar geometry.

2. For $F(\xi) = 2\omega \tanh(\frac{2\xi}{\omega})$, the solution

$$\Psi = \omega \sin\theta - \frac{\omega}{2} \tanh^{-1}(\sin\theta)$$

   of the equation $F(\Psi - \omega \sin\theta) = -2\omega \sin\theta$ describes a flow with zero total vorticity since $\nabla_\Sigma^2 = -2\omega \sin\theta$ in this case.

3. For constant vorticity ($F = \gamma$) a solution is

$$\Psi = \frac{\gamma}{\beta}\left\{\varphi^2 - \left[A + \ln\left(\frac{\cos\theta}{1 - \sin\theta}\right)\right]^2 - \beta \ln\cos\theta\right\},$$

   to within an additive constant; the strength of the velocity field is proportional to γ/β, and the choice of β and of A (for a given constant vorticity γ) controls the type of solutions available. The streamlines for the choice $A = -0.4$ and $\beta = 4$ are depicted in Figure 4; note that γ simply measures the magnitude of the velocity field, and is therefore irrelevant in these plots.

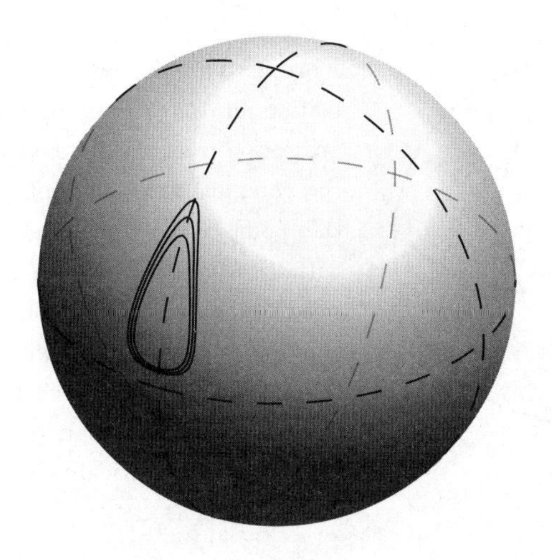

Figure 4. Depiction of three closed streamlines for a geostrophic flow with constant non-zero relative vorticity at mid-latitudes in the Northern hemisphere; the closed streamlines describe a gyre (a large-scale, rotating ocean current).

4. For $F(\xi) = \lambda\xi$, the vorticity equation (5.26) becomes

$$\frac{1}{\cos^2\theta}\,\Psi_{\varphi\varphi} - \Psi_\theta \tan\theta + \Psi_{\theta\theta} = \lambda(\Psi - \omega\,\sin\theta)$$

and explicit solutions are available using spherical harmonic functions; in particular, for $\lambda \neq -2$,

$$\Psi = \frac{\lambda\omega}{\lambda+2}\sin\theta$$

is a solution.

Furthermore, performing the stereographic projection of the unit sphere centred at the origin, from the North Pole to the equatorial plane:

$$\xi = r\,e^{i\varphi} \quad \text{with} \quad r = \frac{\cos\theta}{1-\sin\theta},$$

where (r, φ) are the polar coordinates in the equatorial plane, one can transform (5.26) to the semilinear elliptic equation

$$\Delta\psi + 8\omega\,\frac{1-(x^2+y^2)}{(1+x^2+y^2)^3} - \frac{4F(\psi)}{(1+x^2+y^2)^2} = 0, \tag{5.27}$$

where $\Delta = \partial_x^2 + \partial_y^2$ is the Laplace operator and (x, y) are the Cartesian coordinates in the complex ξ-plane. In spherical geometry, it is natural to investigate the flow described by (5.26) in a region bounded by a level set of the stream function, say $\psi = 0$. This corresponds to solving (5.27) inside the planar region enclosed by the stereographic projection of the level set, with homogeneous Dirichlet boundary conditions.

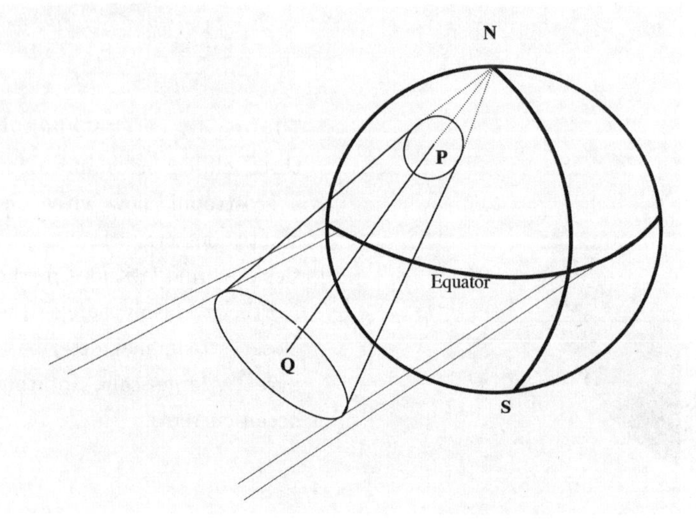

Figure 5. Schematic illustration of the stereographic projection from the sphere to the complex plane, mapping the point P on the sphere with the North Pole N excised to the intersection point Q of the equatorial plane with the ray from N to P. The stereographic projection distorts areas dramatically but preserves angles.

For recent results regarding the study of the solutions to (5.27), using methods from the theory of elliptic partial differential equations and dynamical systems, we refer to [7, 20, 25].

6 Discussion

We have presented an overview of the classical description of geostrophic flows, based on a careful formulation of the problem, and including some additional important material showing how they can be interpreted as an autonomous Hamiltonian system. This approach enables these flows to be viewed in a more general light and suggests that this might open the way to other, and more extensive, investigations. However, the main thrust of the work has been to move beyond the familiar, local description of these flows by showing how the extension to spherical coordinates (not restricted in any way on the sphere) is readily accessible. We have seen that the most direct formulation, which mimics the classical (tangent plane) approach, i.e. linear and for small Rossby number, necessarily leads to an inconsistent system. We have then shown that, although on a strictly theoretical basis we could invoke the existence of a significant vertical structure, the observations of a weak vertical motion suggest an alternative approach: the inclusion of nonlinearity, leading to a consistent development. This has proved not only to be possible and readily accessible, but also produces new and extensive descriptions (of gyres, for example) and indicates that more and various applications might be attainable; in addition, we have encountered a number of intriguing mathematical issues (most notably the new vorticity equation). The conclusion, we submit, is that these flows can, and should, be described using spherical coordinates; indeed, there are many avenues – both applications and mathematical issues – that have opened up for future exploration.

Acknowledgements

This research was supported by the WWTF research grant MA16-009.

References

[1] Andreev V K, Kaptsov O V, Pukhnachov V V, Rodionov A A, *Applications of group-theoretical methods in hydrodynamics*, Kluwer Academic Publishers, Dordrecht, 1988.

[2] Andronov A A, Leontovich E A, Gordon I I, Maier A G, *Theory of bifurcations of dynamic systems on a plane*, Halsted Press, Jerusalem-London, 1973.

[3] Bolsinov A V, Fomenko A T, *Integrable Hamiltonian systems: Geometry, topology, classification*, Chapman & Hall/CRC, Boca Raton, Florida, 2004.

[4] Boyd J P, *Dynamics of the equatorial ocean*, Springer, Berlin, 2018.

[5] Broer H W, Chow S N, Kim Y, Vegter G, A normally elliptic Hamiltonian bifurcation, *Z. Angew. Math. Phys.* **44** (1993), 389–432.

[6] Colin de Verdière Y, Vey J, Le lemme de Morse isochore, *Topology* **18** (1979), 283–293.

[7] Chu J, On a nonlinear model for arctic gyres, *Ann. Mat. Pura Appl.* **197** (2018), 651–659.

[8] Constantin A, An exact solution for equatorially trapped waves, *J. Geophys. Res.: Oceans* **117** (2012), Art. C05029.

[9] Constantin A, Germain P, Instability of some equatorially trapped waves, *J. Geophys. Res.: Oceans, 118* (2013), 2802–2810.

[10] Constantin A, Some nonlinear, equatorially trapped, nonhydrostatic internal geophysical waves, *J. Phys. Oceanogr. 44* (2014), 781–789.

[11] Constantin A, Johnson R S, The dynamics of waves interacting with the Equatorial Undercurrent, *Geophys. Astrophys. Fluid Dyn.* **109** (2015), 311–358.

[12] Constantin A, Johnson R S, An exact, steady, purely azimuthal equatorial flow with a free surface, *J. Phys. Oceanogr. 46* (2016), 1935–1945.

[13] Constantin A, Johnson R S, An exact, steady, purely azimuthal flow as a model for the Antarctic Circumpolar Current, *J. Phys. Oceanogr. 46* (2016), 3585–3594.

[14] Constantin A, Johnson R S, A nonlinear, three-dimensional model for ocean flows, motivated by some observations of the Pacific Equatorial Undercurrent and thermocline, *Physics of Fluids* **29** (2017), Art. 056604.

[15] Constantin A, Johnson R S, Large gyres as a shallow-water asymptotic solution of Euler's equation in spherical coordinates, *Proc. Roy. Soc. A* **473** (2017), Art. 20170063.

[16] Constantin A, Johnson R S, Steady large-scale ocean flows in spherical coordinates, *Oceanography* **31** (2018), 22–30.

[17] Constantin A, Johnson R S, Ekman-type solutions for shallow-water flows on a rotating sphere: a new perspective on a classical problem, (submitted).

[18] Dellar P J, Variations on a beta-plane: derivation of non-traditional beta-plane equations from Hamilton's principle on a sphere, *J. Fluid Mech.* **674** (2011), 174–195.

[19] Eliasson L H, Normal forms for Hamiltonian systems with Poisson commuting integrals – elliptic case, *Comment. Math. Helv.* **65** (1990), 4–35.

[20] Haziot S, Explicit two-dimensional solutions for the ocean flow in arctic gyres, *Monatsh. Math.* (to appear).

[21] Henry D, An exact solution for equatorial geophysical water waves with an underlying current, *Eur. J. Mech. B (Fluids)* **38** (2013), 18–21.

[22] Henry D, Equatorially trapped nonlinear water waves in a β-plane approximation with centripetal forces, *J. Fluid Mech.* **804** (2016), R1, 11 pp.

[23] Ionescu-Kruse D, Matioc A V, Small-amplitude equatorial water waves with constant vorticity: dispersion relations and particle trajectories, *Discrete Contin. Dyn. Syst.* **34** (2014), 3045–3060.

[24] Marshall J, Plumb R A, *Atmosphere, ocean and climate dynamics: an introductory text*, Academic Press, London, 2016.

[25] Marynets K, A nonlinear two-point boundary-value problem in geophysics, *Monatsh. Math.* (to appear).

[26] Meyer K R, Hall G R, *Introduction to Hamiltonian dynamical systems and the n-body problem*, Springer, New York, 1992.

[27] Martin C I, On the existence of free-surface azimuthal equatorial flows, *Appl. Anal.* **96** (2017), 1207–1214.

[28] Newns W F, Functional dependence, *Amer. Math. Monthly* **74** (1967), 911–920.

[29] Paldor N, *Shallow water waves on the rotating Earth*, Springer, Cham, 2015.

[30] Talley L D, Pickard G L, Emery W J, Swift J H, *Descriptive physical oceanography: an introduction*, Elsevier, London, 2011.

[31] Vallis G K, *Atmosphere and ocean fluid dynamics*, Cambridge University Press, Cambridge, 2006.

[32] Wunsch C, *Modern observational physical oceanography*, Princeton University Press, Princeton, 2015.

[33] Zeitlin V, *Geophysical fluid dynamics: understanding (almost) everything with rotating shallow water models*, Oxford University Press, Oxford, 2018.

D3. Review of results on a system of the type many predators - one prey

A V Osipov [a] *and G Söderbacka* [b]

[a] *Faculty of Matematics and Mechanics, Saint Peterburg University, Saint Peterburg, Russia*

[b] *Department of Mathematics, Åbo Akademi, Finland*

Abstract

We here give a review of results on a class of systems of the type n predators - one prey. The case with two predators is considered more carefully. The behaviour of such systems can be very rich. We discuss extinction and different simple and complicated coexistence of the predators.

1 Introduction

In this work we review some of the behaviour of n competing predators feeding on the same prey in a system of the type

$$X_i' = p_i\,\varphi_i(S)\,X_i - d_i\,X_i, \quad i = 1, ..., n, \tag{1.1a}$$

$$S' = H(S) - \sum_{i=1}^{n} q_i\,\varphi_i(S)\,X_i, \tag{1.1b}$$

where the variable S represents the prey and the variables X_i represent the predators. They are, of course, non-negative. The function φ_i is assumed non-decreasing. The function H describes the behaviour of the prey without predators and is usually of logistic type. For general surveys of such models we refer to [3, 27, 26, 30, 20]. As an introductory textbook at student level we refer to [6]. In [14] one can find a little further going introduction. The order of equations with the prey equation as the last one is different from most other publications, but we think it is easier to have the prey on a vertical axis and let the predators be presented on horizontal axes. The behaviour in the case of two predators is discussed more in detail. We mainly consider the case where

$$H(S) = r\,S\left(1 - \frac{S}{K}\right), \; \varphi_i(S) = \frac{S}{S + A_i}, \tag{1.2}$$

where the parameters r, K and A_i are positive.

In order to give a good background for comparison, we begin with a review of the behaviour of the known Lotka-Volterra [18, 29] and Rosenzweig-McArthur systems [23], where only one predator is present. We also give some elementary mathematical background, used for proving the most elementary properties of our systems.

We give general conditions for dissipativity and extinction of one predator for any number of predators. The type of coexistence is reviewed in more detail in the case of two predators. The coexistence can be periodic or chaotic. There can be different types of chaos. We consider one case where the chaos can be modelled by a one-dimensional Poincaré map and another case where we conjecture it is connected to spiral chaos.

The case of more than two predators is discussed only briefly.

At the end we discuss the problem of populations becoming unacceptably small in the models and how to modify the model to be more realistic. Suggested by biologists we introduce modifications, where the predator feeds on another prey (where it anyhow cannot survive in the long run) when the main prey population is too small. Thus in some way we have here two preys, but only one is consumed at the same time. This is more realistic in arctic regions than southern regions, where the predator switches are not so abrupt.

2 Lotka-Volterra equations

Introducing the Lotka-Volerra equations [18, 28] we use the unusual notations x and s for the predator and the prey respectively in order not to create confusion with notations of more general models with many predators. But the order of the equations, with the prey equation as the first is the same as in most publications. Thus in our notations the equations take the form

$$s' = as - bxs,\ x' = -cx + dxs, \tag{2.1}$$

where s is the prey and x the predator, derivatives are taken with respect to time.

The following assumptions are usually cited.

1. The prey population finds ample food at all times.
2. The food supply of the predator population depends entirely on the size of the prey population.
3. The rate of change of population is proportional to its size.
4. During the process, the environment does not change in favour of one species and genetic adaptation is inconsequential.
5. Predators have limitless appetite.

The Lotka-Volterra equation is integrable. It is easy to check that $V' = \frac{\partial V}{\partial s} s' + \frac{\partial V}{\partial x} x' = 0$ and the integrals are

$$V(s,x) = ds - c \ln s + bx - a \ln x. \tag{2.2}$$

These integrals play an important role as Lyapunov functions for estimating basins of attraction in more complicated models.

The solutions are closed curves $V(s,x) = const.$ In Figure 1 we see solution curves in the sx-space representing the values $(s(t), x(t))$ of the variables on a solution for a given initial value $x(0) = x_0$, $s(0) = s_0$. Such a curve is also called a *trajectory* and such types of plots are called *phase portraits.* The trajectories of this system are periodic, except for $s = 0$ or $x = 0$ and at the equilibrium $\left(\frac{c}{d}, \frac{a}{b}\right)$, which is a center.

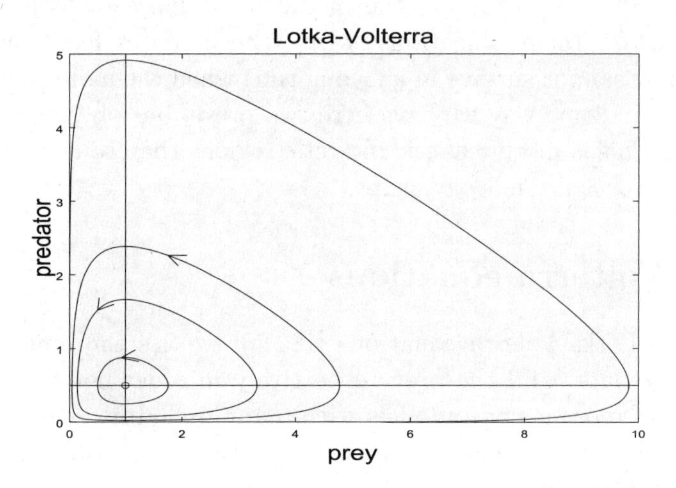

Figure 1. Solution curves for Lotka-Volterra equations for parameters $a = 1$, $b = 2$, $c = 1$, $d = 1$ and intial values $s_0 = x_0 = 0.5$, $s_0 = x_0 = 0.2$ and $s_0 = x_0 = 0.1$. The horizontal and the vertical lines are the null isoclines for which $s' = 0$ and $x' = 0$ respectively.

Some solutions of the Lotka-Volterra equations depending on time are plotted in Figure 2. We notice that here we can see the development in time not seen in the phase portrait figure. Especially, we can see a situation, where the populations are so small that they in practice are hard to observe, but for some short periods they are very big. This corresponds to the big cycles in the phase portrait. Here the big cycle is created for initial values $s_0 = x_0 = 0.01$. For the other initial values the populations do not change in size so much and have a nicer wave form.

We notice that by scaling variables the Lotka-Volterra equations can be written with only two new parameters a and λ. The equations then take the form $s' = (a - x)\, s,\ x' = (s - \lambda)\, x$.

We also notice that the Lotka-Volterra equations are very sensitive to changes in the system, any small change in the system can destroy the cycles seen in the phase portrait. This is because an equilibrium which is a center can with arbitrarily small changes in the system change its type to another one, mostly an attractor or repellor.

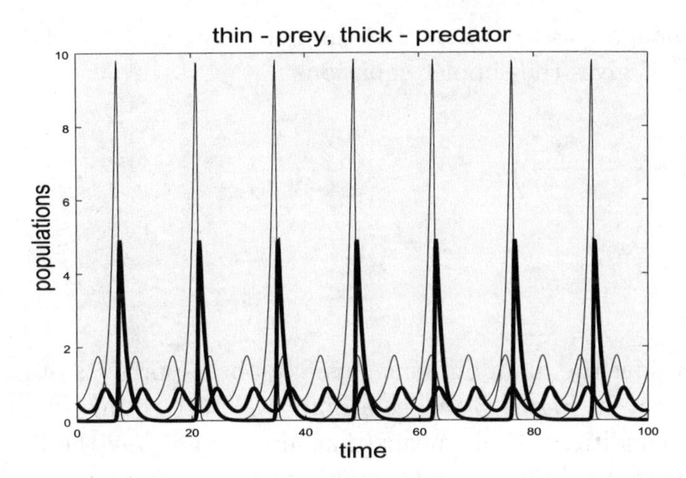

Figure 2. The populations in the Lotka-Volterra model for parameters $a = 1$, $b = 2$, $c = 1$, $d = 1$ with initial conditions $s_0 = x_0 = 0.5$ and $s_0 = x_0 = 0.01$.

3 Rosenzweig-McArthur equations

The sensitivity of the behaviour with respect to changes in the parameters implies that the Lotka-Volterra model needs modifications from a mathematical point of view. But also from biological points of view the assumptions need to be modified to more realistic ones. This is done in the following model created by Rosenzweig and McArthur [23]. The equations for the model are given by using the functions 1.2 in the equations 1.1. We write them in that form, where we still keep the prey equation as the first one.

$$S' = r\,S\left(1 - \frac{S}{K}\right) - \frac{q\,X\,S}{S + A}, \tag{3.1a}$$

$$X' = -d\,X + \frac{p\,X\,S}{S + A}. \tag{3.1b}$$

We formulate some results about the behaviour of this system. These can be found in many places, for example, in [3, 20]. Let $a = \frac{H}{K}$.

1. The predator cannot survive if $d \geq \frac{p}{1+a}$.

2. If $\frac{p\,(1-a)}{1+a} < d < \frac{p}{1+a}$ the system has a stable equilibrium as global attractor where the species coexist.

3. If $d < \frac{p\,(1-a)}{1+a}$ the system has a stable unique cycle as global attractor and the predator and the prey coexist in a cyclic manner.

The time change $\tau = r\,t$, where τ is the new time, and the change of variables $s = \frac{S}{K}$, $x = \frac{q}{rK}\,X$ gives the simpler equations

$$\dot{s} = s\left(1 - s - \frac{x}{s+a}\right), \tag{3.2a}$$

$$\dot{x} = m\,\frac{s-\lambda}{s+a}x, \tag{3.2b}$$

where $\lambda = \frac{dA}{(p-d)K}$, $m = \frac{p-d}{r}$.

We now consider the main different possible phase portraits of the Rosenzweig-McArthur system. In Figure 3 we see a limit cycle attracting all trajectories with positive initial conditions. This means that after some time the behaviour will be periodic and there is only one possible stable periodic behaviour for the coexistence of the predator and the prey.

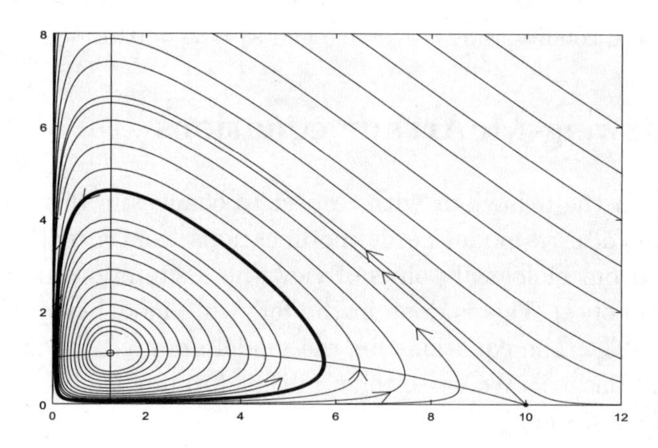

Figure 3. Phase portrait for Rosenzweig-McArthur equations for r=1, p=5, q=5, d=1, K=10, A=5. The vertical line is the zero-isocline $x' = 0$ and the curve is the zero-isocline $s' = 0$.

In Figure 4 we find a phase portrait where all trajectories with positive initial conditions tend to a stable equilibrium. The predators and the prey coexist with constant populations.

In Figure 5 the predator cannot survive and all trajectories with positive initial values tend to the equilibrium where $x = 0$.

Figure 6 gives one more example, where the predator goes extinct.

4 Mathematical tools

Here we review the main mathematical results to be used in the analysis of the models. For more detail we refer to textbooks [2, 15]. They consist mainly of a linearization tool for examining local behaviour and Lyapunov functions for examining

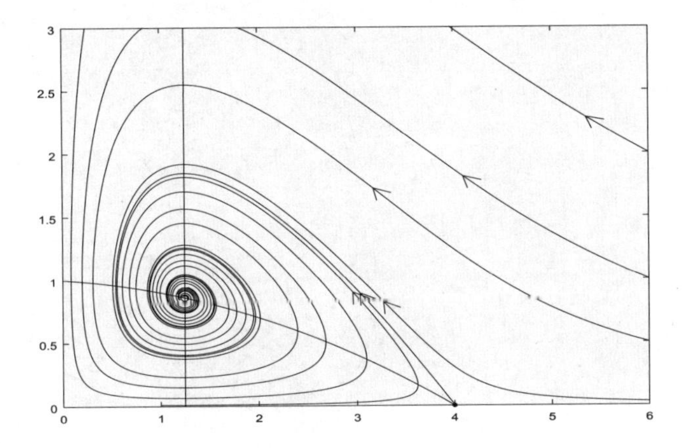

Figure 4. Phase portrait for Rosenzweig-McArthur equations for r=1, p=5, q=5, d=1, K=4, A=5. The vertical line is the zero-isocline $x' = 0$ and the curve is the zero-isocline $s' = 0$.

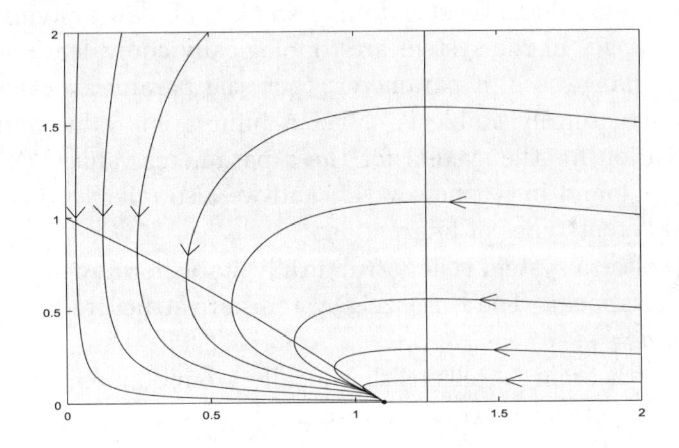

Figure 5. Phase portrait for Rosenzweig-McArthur equations for r=1, p=5, q=5, d=1, K=1.1, A=5. The vertical line is the zero-isocline $x' = 0$ and the curve is the zero-isocline $s' = 0$.

simple non-chaotic global behaviour. Thc tools for examining chaotic behaviour are mainly out of the scope of this review and we refer to other sources for that.

In order to find out whether a system has changed its structure or not when parameters are changed we have to define what we mean when we say that the systems have the same structure. This is given by an equivalence relation.

Definition 1. Two dynamical systems are **topologically equivalent**, if there is a homeomorphism h mapping orbits of system 1 to orbits of system 2 homeomorphically, and preserving the orientation of the orbits.

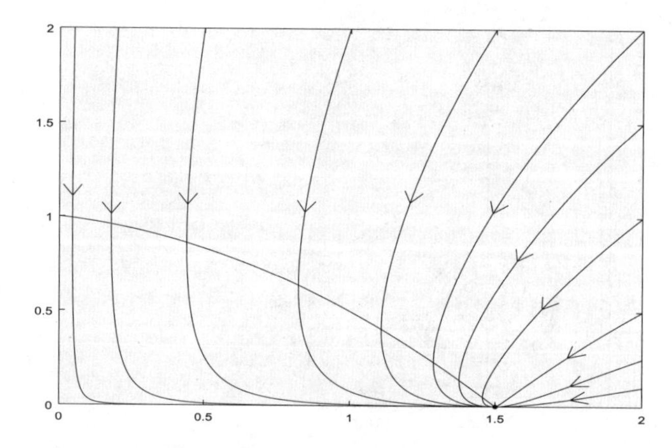

Figure 6. Phase portrait for Rosenzweig-McArthur equations for r=1, p=2, q=2, d=3, K=1,5, A=2. The added curve is the zero-isocline $s' = 0$.

A given system is said to be *structurally stable* if all new systems obtained from small enough changes in the system are topologically equivalent with this original system. If the change is in a parameter, then the parameter value at which the system is not structurally stable is called a bifurcation value and we say there occurs a bifurcation for the system for these parameter values. A general survey of bifurcations is found in Kuznetsov [17] and we also refer to this book for exact definitions of different types of bifurcations.

The Lotka-Volterra system is not structurally stable for any parameters, because the center can disappear. The Rosenzeig systems are structurally stable, except for $d = \frac{p(1-a)}{1+a}$ and $d = \frac{p}{1+a}$.

The following theorem [9] tells us that locally around an equilibrium the system is in most cases topologically equivalent to its linearisation.

Theorem 1. Grobman-Hartman Theorem. *If J is the Jacobian matrix at an equilibrium and the real parts of the eigenvalues of J are non-zero, then there is a neighbourhood of the equilibrium in which the system is topologically equivalent to the linear system $x' = Jx$ given by the Jacobian matrix at the equilibrium.*

We distinguish between three types of equilibria:

- *sinks* (stable), eigenvalues are real and less than zero (nodes), or complex with negative real parts (focus)

- *sources*, eigenvalues are real and greater than zero (nodes), or complex with positive real parts (focus)

- *saddles*, real parts of eigenvalues have different signs, the equilibrium attracts in some directions and is repelling in some directions

All sinks are topologically equivalent. Also all sources and all saddles with the same number of eigenvalues with positive real part are topologically equivalent.

We use the theorem to find the local behaviour of the equilibria of the Rosenzweig-McArthur system using the simplified form of the equations. The standard system 3.2, after a change in time and with changed order of the equations, takes the form

$$x' = m\,(s-\lambda)\,x,\; s' = (h(s)-x)\,s,\; h(s) = (1-s)(s+a). \tag{4.1}$$

Possible equilibria are: $(0,0)$, $(0,1)$, $(h(\lambda),\lambda)$.

The general Jacobian matrix is

$$J = \begin{pmatrix} m(s-\lambda) & m\,x \\ -s & h'(s)s + h(s) - x \end{pmatrix}.$$

The eigenvalues can be seen from the following three calculations.

$$J(0,0) = \begin{pmatrix} -m\,\lambda & 0 \\ 0 & a \end{pmatrix}$$

from which follows that $(0,0)$ is always a saddle.

$$J(0,1) = \begin{pmatrix} m\,(1-\lambda) & 0 \\ -1 & -a \end{pmatrix}$$

from which follows that $(0,1)$ is stable for $\lambda > 1$, and a saddle for $\lambda < 1$.

$$J(h(\lambda),\lambda) = \begin{pmatrix} 0 & m\,h(\lambda) \\ -\lambda & 1-a-2\lambda \end{pmatrix}$$

from which follows that $(h(\lambda),\lambda)$ is stable for $\frac{1-a}{2} < \lambda < 1$, and a source for $\lambda < \frac{1-a}{2}$.

The equilibrium $(0,1)$ changes type for $\lambda = 1$ and the equilibrium $(h(\lambda),\lambda)$ for $\lambda = \frac{1-a}{2}$. These values correspond to the parameter relations $d = \frac{p}{1+a}$ and $d = \frac{p\,(1-a)}{1+a}$ respectively in the original Rosenzweig system. The type of bifurcation at the first parameter value is called *saddle-node* and at the second *Andronov-Hopf bifurcation*. These are the most common bifurcations in a two-dimensional system.

For practical applications it is very important to know what initial conditions of a trajectory leads to what final behaviour. For this we use the definition

Definition 2. The **basin of attraction of** a stable equilibrium is the set of initial conditions for which the trajectory tends to the equilibrium.

To estimate basins of attraction we use *Lyapunov functions.*

We give a known theorem.

Theorem 2. . *Let V be a function defined in a neighbourhood N of the equilibrium. Suppose in the neighbourhood N*

- $V(x) \geq 0$
- $V'(x) < 0$ *except at the equilibrium*
- *if $V(x) < c$, then $x \in N$*

Then the region M defined by $V(x) < c$ is in the basin of attraction of the equilibrium.

There is a modified version of the theorem telling that the conclusion still holds if $V' \leq 0$ in N and the set defined by $V' = 0$ does not contain whole trajectories except for the equilibrium. For more details and proofs we again refer to the known textbooks [2, 15].

Using Lotka-Volterra integrals for system 4.1 as Lyapunov function it is possible to prove that $(h(\lambda), \lambda)$ is globally stable for $\frac{1-a}{2} < \lambda < 1$. We show how to do it in the case where $a > 1$. As Lyapunov function we try

$$V = x - h(\lambda) \ln x - m\,(\lambda \ln s - s). \tag{4.2}$$

Differentiation with respect to time gives $V' = m\,(\lambda - s)(h(\lambda) - h(s))$. From $a > 1$ follows $h(s) > h(\lambda)$ for $s < \lambda$ and $h(s) < h(\lambda)$ for $s > \lambda$ and thus $V' < 0$ for $s \neq \lambda$. The trajectories intersect $s = \lambda$ transversally without tangency and thus the set $V' = 0$ does not contain whole trajectories except the equilibrium. The integral curves $V =$ *constant* fill up the whole region where x and s are positive and V has a minimum at $(h(\lambda), \lambda)$. Then it follows from the modification of the theorem that the equilibrium is a global attractor if we as a Lyapunov function choose $V(x, s) - V(h(\lambda), \lambda)$.

In case $\lambda < \frac{1-a}{2}$ there is a unique globally attracting limit cycle. This was first proved by Cheng [5]. The uniqueness of limit cycles for this and similar systems can also be proved using the known Zhang Zhi-fen theorems [31].

Estimates for the size of the cycle for critically small a and λ are given in [13, 19]. The size of the cycle we determine by the maximal and minimal populations on the cycle. The cycle is called big if at least one population sometimes gets small.

5 Systems with more predators

Here we consider the general system 1.1 for n predators feeding on the same prey. We assume $H(0) = H(K) = 0$ for some $K > 0$, $H'(K) < 0$, $H''(s) < 0$ and $\varphi_i(0) = 0$, $\varphi_i'(s) > 0$.

The functions φ_i and H are of the class $C^2[0, \infty)$ and the variables x_i and s are non-negative: $x_i \geq 0$, $s \geq 0$.

The change of variables $s = \frac{S}{K}$ and $x_i = \frac{q_i}{K} X_i$ gives the system

$$x_i' = \phi_i(s)\, x_i,\; i = 1 ... n, \tag{5.1a}$$

$$s' = h(s) - \sum_{i=1}^{n} \psi_i(s)\, x_i, \tag{5.1b}$$

where

$$h(s) = \frac{1}{K} H(sK),\; \psi_i(s) = \varphi_i(sK), \phi_i(s) = p_i \psi_i(s) - d_i. \tag{5.2}$$

The system satisfies the conditions $A_1 - A_4$ in [21]:

A_1 : All the considered functions are of the class $C^2[0,\infty)$ and the variables x_i and s are non-negative: $x_i \geq 0,\, s \geq 0$.

A_2 : $\psi_i(0) = 0, \quad \psi_i'(s) > 0 \;\text{ for }\; s > 0.$

Here and further we will assume that i takes values from the set $\{1,\, 2, \ldots, n\}$.

A_3 : $\phi_i'(s) > 0$ for $s > 0$ and there exists $\lambda_i > 0$ such that $\phi_i(\lambda_i) = 0$.

A_4 : $h(0) = h(1) = 0, \quad h'(1) < 0$ and $h''(s) < 0$ for $s > 0$.

Further here we will also assume:

A_5 : $0 < \lambda_n < \cdots < \lambda_2 < \lambda_1 < 1$.

We observe that if $\lambda_i \geq 1$ for some i, then the corresponding predator cannot survive.

We make some local analysis of system 5.1 in the case of two predators. Generalization to many predators can be done analogously.

The system then has four equilibria:

$$O_0 = (0,0,0),\; O_1 = (0,0,1),\; P_1 = (h(\lambda_1), 0, \lambda_1),\; P_2 = (0, h(\lambda_2), \lambda_2).$$

The Jacobian matrix is

$$J = \begin{pmatrix} \phi_1(s) & 0 & \phi_1'(s)\, x \\ 0 & \phi_2(s) & \phi_2'(s)\, x \\ -\psi_1(s) & -\psi_2(s) & h'(s) - \psi_1'(s)\, x - \psi_1'(s)\, y \end{pmatrix}.$$

The eigenvalues at O_0 are the negative $\phi_1(0)$ and $\phi_2(0)$ and the positive $h'(0)$ and thus the equilibrium is a saddle. The eigenvalues at O_1 are the positive $\phi_1(1)$ and $\phi_2(1)$ and the negative $h'(1)$ and thus the equilibrium is a saddle. The equilibrium P_1 has one positive eigenvalue $\phi_2(\lambda_1)$ and the other eigenvalues depend on the sign of $h'(\lambda_1) - \psi_1'(\lambda_1)\, h(\lambda_1)$. Thus P_1 is a saddle if $h'(\lambda_1) - \psi_1'(\lambda_1)\, h(\lambda_1) < 0$ and a source if $h'(\lambda_1) - \psi_1'(\lambda_1)\, h(\lambda_1) > 0$. The equilibrium P_2 has one negative eigenvalue $\phi_1(\lambda_2)$ and the other eigenvalues depend on the sign of $h'(\lambda_2) - \psi_2'(\lambda_2)\, h(\lambda_2)$. Thus P_2 is stable if $h'(\lambda_2) - \psi_2'(\lambda_2)\, h(\lambda_2) < 0$ and a saddle if $h'(\lambda_2) - \psi_2'(\lambda_2)\, h(\lambda_2) > 0$.

We notice there is no equilibrium where the predators coexist. This is also true for more than two predators. But it is well known that they can coexist in a cyclic way and thus the system is a known counterexample to the exclusion principle in Biology [20]. Further we are going to look at the general behaviour of these systems.

We examine the global behaviour mostly for some special kinds of these systems and we mostly concentrate on two predators. Some mostly numerical results can be found for three predators in [7].

The general behaviour of such systems has been studied in [11, 10] and parameter conditions for different behaviour are given in [12]. The possibility for coexistence of predators arising from bifurcations was examined in [24, 4, 16]. Many works have considered the situation where the function h is linear and thus condition A_4 is violated. A review of these is found in [26]. More general systems are considered in [25].

The most standard example of system 5.1, which we will consider now, is obtained from the functions in (1.2).

We assume $p_i > d_i$. If not, the corresponding predator will not survive. Using the time change $\tau = r\,t$, where τ is the new time, and the variable changes $s = \frac{S}{K}$, $x_i = \frac{q_i}{r\,K} X_i$ we get the simplified equations

$$x_i' = m_i \frac{s - \lambda_i}{s + a_i} x_i, \tag{5.3a}$$

$$s' = \left(1 - s - \sum_{i=1}^{n} \frac{x_i}{s + a_i}\right) s, \tag{5.3b}$$

where

$$a_i = \frac{A_i}{K},\ m_i = \frac{p_i - d_i}{r},\ \lambda_i = \frac{d_i\,A_i}{K\,(p_i - d_i)}. \tag{5.4}$$

Complicated behaviour of the system in the case $m_i = 1$ and where $n = 2$ was examined in [22, 8, 7].

To return to the original system we can use the formulas

$$S = K s,\ X_i = \frac{rK}{q_i} x,\ t = \frac{\tau}{r} \tag{5.5}$$

and

$$A_i = K\,a_i,\ d_i = \frac{r\,m_i\,\lambda_i}{a_i},\ p_i = \frac{r\,m_i}{a_i}\,(a_i + \lambda_i). \tag{5.6}$$

5.1 Dissipativity

We consider system 5.1. We find a positively invariant set for the system. More results on dissipativity are found in [21]. Let

$$w_i = \sup_{0<s<1} \tilde{w}_i(s),\ \tilde{w}_i(s) = \frac{\phi_i(s) - \frac{h(s)}{1-s}}{\psi_i(s)}. \tag{5.7}$$

The value of w_i is finite and positive, because $\tilde{w}_i(s) \to -\infty$ for $s \to 0+$ and $\tilde{w}_i(s) \to \frac{\phi_i(1)-h'(1)}{\psi_i(1)} = Q_i > 0$ for $s \to 1-$.

We define V by

$$\frac{x_1}{w_1} + \frac{x_2}{w_2} + \ldots + \frac{x_n}{w_n} + s = V. \tag{5.8}$$

We now claim the following:

Statement 1. . *The set formed by the inequalities s, $x_i \geq 0$ and $V \leq 1$ is positively invariant for system 5.1 satisfying conditions $A_1 - A_5$ and all trajectories of the system with positive initial values enter the set in finite time.*

Proof. It can be checked directly, that if $s = 1$ then $V' < 0$ except at $(0, .., 0, 1)$ where $V' = 0$. In other points we get

$$V' = h(s) + \sum_{i=1}^{n} \left[\frac{\phi_i(s)}{w_i} - \psi_i(s) \right] x_i = \tag{5.9}$$

$$= \frac{h(s)}{1-s}(1-V) + \frac{h(s)}{1-s}(V-s) + \sum_{i=1}^{n} \frac{x_i}{w_i} \left[\phi_i(s) - w_i\psi_i(s)\right] = \tag{5.10}$$

$$= \frac{h(s)}{1-s}(1-V) + \sum_{i=1}^{n} \frac{x_i}{w_i} \left[\frac{h(s)}{1-s} + \phi_i(s) - w_i\psi_i(s) \right] < 0. \tag{5.11}$$

From here the statment follows.

Let us examine the case where

$$h(s) = (1-s)s,\ \phi_i(s) = \frac{s^{b_i} - \lambda_i}{s^{b_i} + a_i},\ \psi_i(s) = \frac{s^{b_i}}{s^{b_i} + a_i}. \tag{5.12}$$

If $0 < b_i \leq 1$ then $\tilde{w}_i(s)$ increases and $w_i = Q_i$. If $b_i > 1$ then it is possible that $w_i > Q_i$. For example $b_i = 2$, $a_i = 1$, $\lambda_i = 0.1$ gives $\tilde{w}_i(0.5) = 3.1$ and $Q_i = 2.9$.

5.2 Extinction

We here look for the competition between predators i and j for system 5.3. We begin with finding sufficient conditions for the extinction of one of them. More general results on extinction are found in [21].

We assume $\lambda_j < \lambda_i$ and look at the function η defined by

$$\eta(s)=\frac{\phi_j(s)}{\phi_i(s)}=\eta_1(s)\,\eta_2(s), \tag{5.13}$$

where

$$\eta_1(s)=\frac{s-\lambda_j}{s-\lambda_i},\ \eta_2(s)=\frac{s+a_i}{s+a_j}. \tag{5.14}$$

We use notations $\gamma=\eta(0)$, $\alpha=\eta(1)$. We introduce two numbers

$$\kappa_0=\max_{s\in[0,\lambda_j]}\eta(s),\ \kappa_1=\min_{s\in[\lambda_i,1]}\eta(s). \tag{5.15}$$

They exist and are positive

We assume $\kappa_0<\kappa_1$. We observe that if κ is a number such that $\kappa_0<\kappa<\kappa_1$ then

$$\kappa\phi_i(s)-\phi_j(s)<0 \tag{5.16}$$

for all $s\in(0,1)$.

Really, for $s\in(0,\lambda_j]$, $\phi_i(s)$ is negative and $\phi_j(s)$ is non-positive and from $\eta(s)<\kappa_0$ we get $\phi_j(s)>\kappa_0\phi_i(s)>\kappa\phi_i(s)$ implying (5.16). For $s\in(\lambda_j\,\lambda_i]$, $\phi_i(s)\leq 0<\phi_j(s)$ implying (5.16). Finally for $s\in(\lambda_i,\,1)$ both $\phi_i(s)$ and $\phi_j(s)$ are positive and from $\eta(s)>\kappa_1$ we get $\phi_j(s)>\kappa_1\phi_i(s)>\kappa\phi_i(s)$ implying (5.16).

We consider the case where $a_i\leq a_j$ and show that then always $\kappa_0<\kappa_1$.

From $\lambda_j<\lambda_i$ follows $\eta_1(s)<1$ for $0<s<\lambda_j$ and $\eta_1(s)>1$ for $s>\lambda_i$. Thus we get $\eta(s)<\eta_2(s)$ for $0<s<\lambda_j$ and $\eta(s)>\eta_2(s)$ for $s>\lambda_i$. Because η_2 is increasing or constant for $a_1\leq a_2$ we conclude that $\eta(s_1)>\eta_2(s_1)\geq\eta_2(s_0)>\eta(s_0)$ for $s_0<\lambda_j$ and $s_1>\lambda_i$ from which follows $\kappa_0<\kappa_1$ and (5.16).

Consider now the function U defined by $U(x,y)=\ln\left(\frac{x_i^\kappa}{x_j}\right)$. For the time derivative we get $U'=\kappa\phi_i(s)-\phi_j(s)<0$ and predator x_i goes extinct.

We now consider the case where $a_j\leq a_i$.

We observe that in this case both η_1 and η_2 are decreasing and thus also η. Then $\kappa_0=\eta(0)=\gamma$ and $\kappa_1=\eta(1)=\alpha$.

As in the previous case predator x_i goes extinct if $\gamma<\alpha$.

We can resume the results in a statement

Statement 2. *Let* $L=\frac{\lambda_i(1-\lambda_j)}{\lambda_j(1-\lambda_i)}$ *and* $\lambda_i>\lambda_j$. *If* $a_j>\frac{a_i}{a_i\,L+L-1}$ *the predator* j *goes extinct.*

Noticing that the inequality in the statement is equivalent to $\gamma<\alpha$ we conclude the statement from the discussion above.

Because $L>1$ the predator j always goes extinct when $a_j>1$ and there is some i such that $\lambda_i<\lambda_j$.

These regions give only sufficient conditions for the extinction of x_i. In practice the predators go extinct more often which can be seen from the numerical results we describe here. In Figures 7 and 8 we have analyzed the extinction or coexistence of predators in the case of two predators. In these figures we have fixed λ_1 and λ_2 and a_1 takes values from 0.005 to 0.1 on the horizontal axis and a_2 takes values from 0.005 to 0.02 on the vertical axis in both figures. From our experiments we cannot see that the extinction depends strongly on m_i. So the order of dependence will be an open question. But the other behaviour does change as seen from the figures.

Red region is for x_1 and green for x_2 going extinct. Dark blue is for possible complicated coexistence and cyan for simple periodic. Magenta is for two periodic and orange for three periodic.

If we instead fix a_1 and a_2 the picture looks a little bit different even if showing the same result. We give an example in Figure 9. In region 1 predator x_2 goes extinct and in region 4 predator x_1 goes extinct and region 2 and 3 are for simple and complicated coexistence correspondingly. The conclusion about extinction is made if the population becomes less than e^{-100} before the iterate 5110 and the conclusion of the type of coexistence is made from the last iterates of those 5110 if all populations survive. One iterate gives the next intersection with $s = \lambda_2$ and the intial point of the trajectory is always taken as $(0.5, 0.5, 0.5)$. The picture should be taken as approximate and can differ from reality for some small number of parameter values. Some of the non-smooth bifurcation boundaries are due to slow convergence and the accuracy of the decision for periodicity or intersecting $s = \lambda_2$, but often not. Hence we have a number of open problems to describe these phenomena.

Figure 7. Regions of extinction and coexistence for $\lambda_1 = 0.2, \lambda_2 = 0.1$, $m_1 = m_2 = 1$. Red region for x_1 and green for x_2 going extinct, other colours for different types of coexistence.

5.3 Chaotic behaviour

In this subsection we give more results for two predators in the case of complicated coexistence. Even if mostly the coexistence is simply cyclic (either a simple cycle or

Figure 8. Regions of extinction and coexistence for $\lambda_1 = 0.2, \lambda_2 = 0.1$, $m_1 = 1, m_2 = 2$. Red region for x_1 and green for x_2 going extinct, other colours for different types of coexistence.

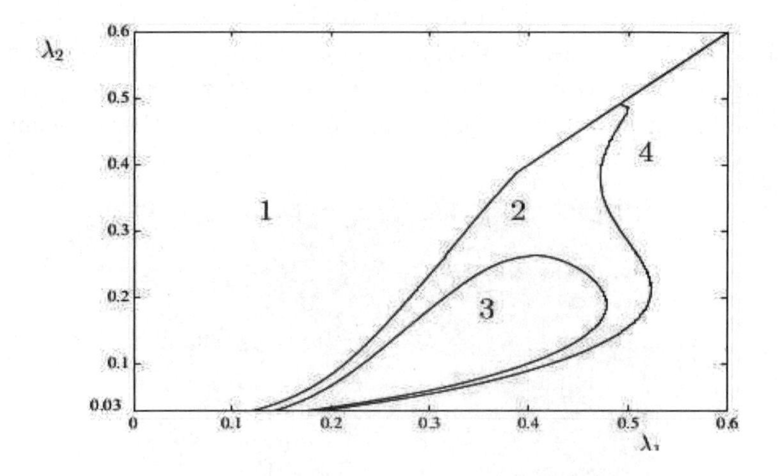

Figure 9. Regions of extinction and coexistence for $a_1 = 0.2, a_2 = 0.02$, $m_1 = m_2 = 1$. The figure is copied from [21] with permission.

a two periodic cycle), for many parameters it can be much more complicated. It is a hard open problem to classify all kinds of complicated behaviour in this system. A general technique is to discretize the dynamics. An effective tool is the Poincaré map introduced below, because when it is well defined it saves most of the dynamic properties. For two predators the map reduces the dimension from three to two. It is also a hard open problem to classify all kinds of Poincaré maps for this system and to find out when it can be well defined. In the cases where the parameters λ_i are small some simplifications can often be done.

In [22] a one-dimensional model map was introduced for the chaotic case in some

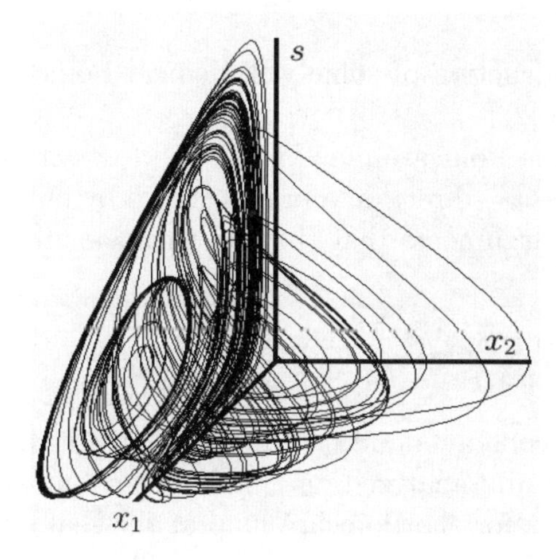

Figure 10. Chaotic attractor where $a_1 = 0.5$, $a_2 = 0.002$, $\lambda_1 = 0.33, \lambda_2 = 0.2$, $m_1 = 1, m_2 = 2$.

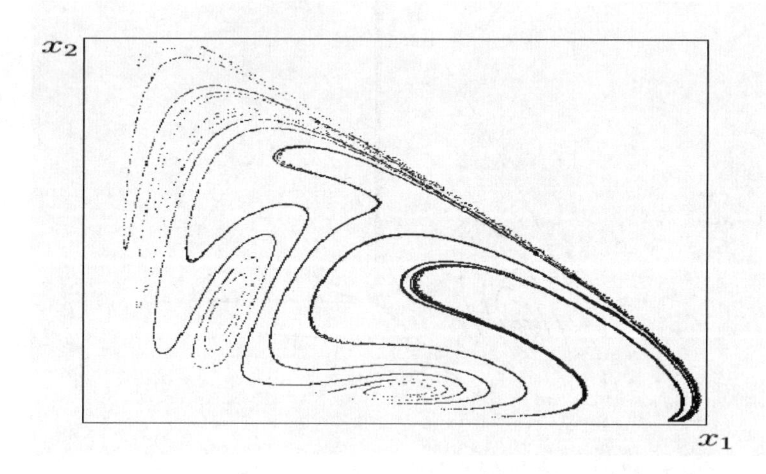

Figure 11. Intersection with $s = \lambda_2$ of chaotic attractor where $a_1 = 0.5$, $a_2 = 0.002$, $\lambda_1 = 0.33, \lambda_2 = 0.2$, $m_1 = 1, m_2 = 2$.

cases. In Figure 10 we see an attractor where the model does not work and in Figure 11 the intersections of that attractor with $s = \lambda_2$ for $s' < 0$.

We give the definition of the Poincaré map as an important tool in examining complicated behaviour.

Definition 3. Poincaré map in R^n. The Poincaré map P is defined on a transversal $(n-1)$-dimensional hypersurface Q (without tangency with trajectories). The image $P(x)$ is defined as the next intersection with Q of a trajectory with initial condition x.

The map gives possibilities to study the iterates of points under a map instead of whole trajectories.

Conditions for construction of some well defined Poincaré maps on $s = const$, $s' < 0$ are obtained in [22]

In the case where the Poincaré map is well-defined, very often there is a strong contraction in the $(x_1 + x_2)$-direction and it is shown by numerical experiments and theoretical estimating arguments that the one dimensional model map given by

$$f(v) = \beta + v - \frac{k_1 + k_2 e^v}{1 + e^v} u, \tag{5.17}$$

where β, u and k_i are constants and $v = \ln(x_2/x_1)$ gives a good approximation.

This map is derived and analyzed for simple behaviour in [8, 7].

Bifurcation diagrams for the Poincaré map of the real system in the variable v are produced in [8, 7, 22].

When our construction of the Poincaré map does not work the chaotic attractor can be more complicated.

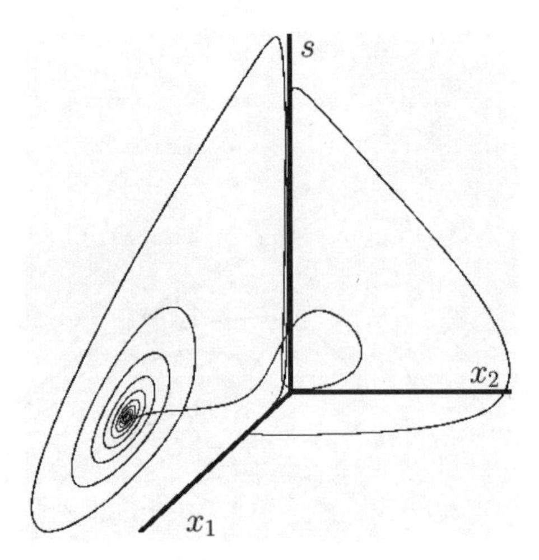

Figure 12. Inner attractor for $a_2 = 0.002$, $\lambda_2 = 0.2$, $a_1 = 0.5$, $\lambda_1 = 0.33$, $m_1 = 1, m_2 = 1$.

We also invite the reader to examine the behaviour from a bifurcation diagram in Figure 13. The horizontal axis represents the value of the bifurcation parameter a_1. The vertical axis represents the values of $\ln\left(\frac{x_2}{x_1}\right)$ on points on an attractor for the given value of the bifurcation parameter. The initial value for all parameters was $x_1 = x_2 = s = 0.5$.

Open problem. Can we have more than one attractor where all species coexist?

From our experiments the reader should be convinced that the answer is most likely to be positive.

We conjecture that attractors like those in Figure 12 are connected with the spiral chaos studied in [1].

We have thus examined cases when the complicated behaviour can be studied using simple Poincaré maps and, especially when the Poincaré map can be well

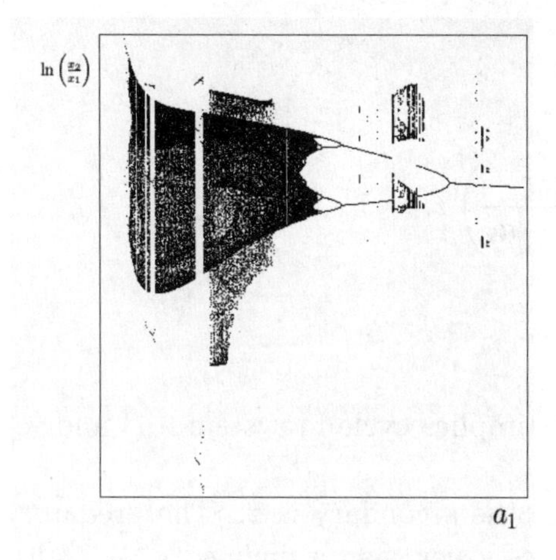

Figure 13. Bifurcation diagram for $a_2 = 0.02$, $\lambda_2 = 0.2$, $\lambda_1 = 0.35$, $m_1 = 1, m_2 = 1$. The range of a_1 is from 0.1 to 2.

approximated by a simple formula. We have also found some properties for some complicated behaviour indicating it can be connected to spiral chaos. What happens when our simple Poincaré map construction does not work and whether there is something else than spiral chaos is an open hard problem.

In most of the chaotic behaviour studied here one population can get very low, and there is the question whether this can be realistic. Some modifications of the system, where low populations are avoided is given in the next subsection. Anyhow there is also chaos in systems, where the populations are not getting too low. Such an example we can see for $a_1 = 0.3433$, $a_2 = 0.006$, $\lambda_1 = 0.45, \lambda_2 = 0.28$, $m_1 = 0.7$, $m_2 = 0.7$.

6 Modified standard system

The standard system has cycles with very low populations for small a and λ. In nature this does not occur because the predator changes behaviour to feeding on other preys, where however it cannot survive for ever. Because this change is sudden in Arctic regions (stochastic in Middle EU) we get a system with switches.

We consider two 3-dimensional systems

$$s' = \left(1 - s - \frac{x}{s + a_1}\right) s, \tag{6.1a}$$

$$z' = (1 - z)z, \tag{6.1b}$$

$$x' = \frac{s - \lambda_1}{s + a_1} x, \tag{6.1c}$$

$$s' = (1-s)s, \tag{6.2a}$$

$$z' = \left(1 - z - \frac{x}{z+a_2}\right) z, \tag{6.2b}$$

$$x' = \frac{z-\lambda_2}{z+a_2} x. \tag{6.2c}$$

Suppose $\lambda_1 < \frac{1-a_1}{2}$ (implies cycle in system 6.1) and $\lambda_2 > 1$ (implies extinction in system 6.2).

The main prey is s, the secondary is z. The predator feeds on the main prey until its density becomes lower than a limit $\epsilon_- < \lambda_1$ and then feeds on z until the main prey reaches a density of ϵ_+, $\epsilon_- < \epsilon_+ \leq \lambda_1$.

For this switched system in the case that λ_2 is not too great, the cycle becomes smaller and chaos can arise.

For example, for $a_1 = a_2 = 0.1$, $\lambda_2 = 1.1$, $\lambda_1 = 0.1, \epsilon_+ = \lambda_1$, $\epsilon_- = 0.1\,\lambda_1$, there is chaos.

Open problem. Examine the dynamics of systems of many predators - one prey with this behaviour of feeding on the prey.

Acknowledgements

We are in debt to N Lundström and T Lindström for serious comments. We are very thankful to N Euler and L Maligranda for technical advices and to Ph Chanfreau and G Högnäs for language consultation.

References

[1] Afraimovich V S, Gonchenko S V, Lerman L M, Shilnikov A L, Turaev D V, Scientific Heritage of L.P. Shilnikov, *Regular and Chaotic Dynamics*, **19**, 4, (2014), 435–460.

[2] Arrowsmith D K, Place C M, *Dynamical Systems, Differential Equations, Maps and Chaotic Behaviour*, Chapman Hall, 1992.

[3] Bazykin A D, *Mathematical Biophysics in Interacting Populations*, Nauka, 1985.

[4] Butler G J, Waltman P, Bifurcation from a limit cycle in a two predator-one prey ecosystem modeled on a chemostat, *J. Math. Biol.*, **12**, 3, (1981), 295–310.

[5] Cheng K S, Uniqueness of limit cycle for a predator-prey system, *SIAM J. Appl. Anal.*, **12**, 4, (1981), 541–548.

[6] Chou C S, Friedman A. *Introduction to Mathematical Biology, Modeling, Analysis and Simulations*, Springer, 2016.

[7] Eirola T, Osipov A V, Söderbacka G, *Chaotic regimes in a dynamical system of the type many predators one prey.* Research reports A 386, Helsinki University of Technology, 1996.

[8] Eirola T, Osipov A V, Söderbacka G, On the appearance of chaotic regimes in one dynamical system of type two perdators one prey, *Actual Problems of Modern Mathematics*, Boxitogorsk, **1**, (1996), 39–70

[9] Hartman P, *Ordinary Differential Equations*, SIAM, 2002.

[10] Hsu S B, Hubell S P, Waltman P, Competing predators, *SIAM J. Appl. Math.*, bf 35, 4, (1978), 617–625.

[11] Hsu S B, Limiting behaviour for competing species, *SIAM J. Appl. Math.*, **34**, 4, (1978), 760–763.

[12] Hsu S B, Hubell S P, P. Waltman P, A contribution to the theory of competing predators, *Ecological Monographs*, **48**, 3, (1978) 337–349.

[13] Hsu S-B, Shi J. Relaxation oscillation profile of limit cycle in predator-prey system, *Discrete and Continuous Dynamical Systems Series B*, **11**, 4, (2009), 893–911.

[14] Iannelli M, Pugliese A, *An Introduction to Mathematical Population Dynamics*, Springer, 2014

[15] Jordan D W, Smith P, *Nonlinear Ordinary Differential Equations*, Oxford Univ Press, 2007.

[16] Keener J P, Oscillatory coexistence in the chemostat: a codimension two unfolding, *SIAM J. Appl. Math.*, **43**, 5, (1983), 1005–1018.

[17] Kuznetsov J A, *Elements of Applied Bifurcation Theory*, Applied Mathematical Series, **112**, Springer, New York, 1998.

[18] Lotka A J, Contribution to the theory of periodic reaction, *J. Phys. Chem.*, **14**,3, (1910), 271-274.

[19] Lundström N L P, Söderbacka G, Estimates of size of cycle in a predator-prey system, *Differential Equations and Dynamical Systems*, 2018, DOI: 10.1007/s12591-018-0422-x

[20] Murray J D. Mathematical Biology I, Third Ed, *Interdisciplinary Appl Math* **17**, 2002.

[21] Osipov A V, Söderbacka G, Extinction and coexistence of predators, *Dinamicheskie Sistemy*, **6**, 1, (2016), 55–64.

[22] Osipov A V, Söderbacka G, Poincaré map construction for some classic two predators - one prey systems, *Internat. J. Bifur. Chaos Appl. Sci. Engrg*, **27**, (2017), no 8, 1750116, 9 pp.

[23] Rosenzweig M, MacArthur R. Graphical representation and stability conditions of predator-prey interaction, *American Naturalist*,**97**, (1963), 209-223.

[24] Smith H L, The interaction of steady state and Hopf bifurcations in a two-predator-one-prey competition model, *SIAM J. Appl. Math.*, **42**, 1 , (1982), 27–43.

[25] Smith H L, Thieme H R, Dynamical systems and population persistence, *Graduate Studies in Mathematics*, (AMS, Providence, RI) **118**, 2011.

[26] Smith H L, Waltman P, *The Theory of the Chemostat: Dynamics of Microbial Competition.* Cambridge University Press, 1995.

[27] Turchin P, *Complex Population Dynamics*, Princeton University Press, 2003.

[28] Volterra V, Variazioni e fluttuazioni del numero d'individui in specie animali conviventi, *Mem. Acad. Lincei Roma.*, **2**, (1926), 31–113.

[29] Volterra V, *Leçons sur la Theorie Mathematique de la Lutte pour la Vie.* Gauthier-Villars, Paris, 1931.

[30] Yodzis P, *Introduction to Theoretical Ecology.* Harper and Row, New York, 1989, p. 384.

[31] Zhang Z-f, Ding T-r, Huang W-z, Dong Z-x, *Qualitative Theory of Differential Equations*, AMS, Providence, Rhode Island, 1991.

D4. Ermakov-type systems in nonlinear physics and continuum mechanics

Colin Rogers and Wolfgang K Schief

School of Mathematics and Statistics, The University of New South Wales, Sydney, NSW 2052, Australia

Abstract

Nonlinear coupled systems of Ermakov-type have extensive applications in both physics and continuum mechanics. Here, applications are described in such diverse areas as rotating shallow water theory, magnetogasdynamics, multi-layer hydrodynamics and many-body theory. Recent work on hybrid Ermakov-Painlevé systems is reviewed.

1 Overview

Nonlinear coupled systems of Ermakov-Ray-Reid type have their origin in classical work [30] of Ermakov (Figure 1) and were originally introduced by Ray and Reid in [59, 60]. These systems, which adopt the form

$$\begin{aligned} \ddot{\alpha} + \omega(t)\alpha &= \frac{1}{\alpha^2\beta}\Phi(\beta/\alpha), \\ \ddot{\beta} + \omega(t)\beta &= \frac{1}{\alpha\beta^2}\Psi(\alpha/\beta), \end{aligned} \tag{1.1}$$

admit a distinctive integral of motion, namely, the invariant

$$\mathcal{I} = \frac{1}{2}(\alpha\dot{\beta} - \beta\dot{\alpha})^2 + \int^{\beta/\alpha} \Phi(z)dz + \int^{\alpha/\beta} \Psi(w)dw \tag{1.2}$$

together with concomitant nonlinear superposition principles. Here, a dot indicates a derivative with respect to the independent variable t. Later in [82], 2+1-dimensional Ermakov-Ray-Reid systems were constructed while extensions of Ermakov-type systems to arbitrary order and dimension and which admit its characteristic invariant were presented in [100]. Therein, alignment of a 2+1-dimensional Ermakov system and an integrable Ernst system of general relativity was shown to produce a novel integrable hybrid of the 2+1-dimensional sinh-Gordon system of [47, 49] and a generalised Ermakov system. In [7], it was established that, like soliton systems, Ermakov-Ray-Reid systems admit underlying linear structure, albeit of another kind. Multi-component Ermakov-Ray-Reid systems were introduced in [89] via symmetry reduction of a classical 2+1-dimensional multi-layer hydrodynamic model. Novel algebraic structure underlying these multi-component systems was uncovered in [6].

Figure 1. V.P. Ermakov (1845 – 1922)

The remarkable occurrence of integrable Ermakov-Ray-Reid systems in nonlinear optics is well-documented. Thus, in a pioneering paper, Wagner *et al* [106], starting from Maxwell's equations, derived via the paraxial approximation, dynamical equations for the envelope of the self-trapping field for optical beams. A coupled pair of nonlinear dynamical equations for the evolution of the transverse radii of elliptical beams in a polarized medium was set down. This pair turns out to be a Ermakov-Ray-Reid system [85]. Such systems arise in both self-trapping and self-focusing nonlinear optics contexts [38, 24, 37, 36, 35]. In particular, they model the evolution of the size and shape of a light spot and wave front in an elliptical Gaussian beam [24, 37]. Modulated versions of nonlinear Schrödinger (NLS) models arise, in particular, in the context of soliton management in periodic nonlinear optics systems [52]. The variational approximation method introduced in [5] and reviewed in [51] has been applied in [84] to isolate integrable Hamiltonian Ermakov-Ray-Reid reductions of modulated 3+1-dimensional NLS models arising in nonlinear optics. In [26], the contribution of orbital angular momentum to the suppression of the collapse of spiralling elliptic solitons in Kerr media has been analysed via a variational approach. An analogous system modelled by a 2+1-dimensional NLS equation incorporating a harmonic trap was subsequently derived in [1] by a variational approximation in the context of elliptic cloud evolution in a Bose-Einstein condensate. In [84], these coupled nonlinear reductions derived in [26] and [1] were both be shown to be Ermakov-Ray-Reid systems.

In [74], a novel Hamiltonian Ermakov-Ray-Reid system was derived in the context of a classical 2+1-dimensional rotating shallow water system with underlying rigid circular paraboloidal bottom topography. Interest in the latter system originated in a study of tidal oscillations in an elliptic basin by Goldsbrough [34]. This work, in return, is related to that of Kirchhoff [46] on vortex structures in the classical (2+1)-dimensional Euler system. Ball [9, 10] obtained key results on the time evolution of moments of inertia and invariance properties of the rotating shallow water system. Invariance theorems in this context were subsequently derived by Lie group methods in [66, 50]. The Ermakov-Ray-Reid system derived in [74] was shown therein to describe the time evolution of the semi-axes of the elliptical moving shoreline on the underlying paraboloidal basin.

The procedure adopted in [74] had its genesis in that used in [67] to construct the general solution of an eight dimensional nonlinear dynamical system descriptive of the time evolution of upper ocean warm-core elliptical eddies as earlier described in [25]. The introduction of physical variables corresponding to modulated versions of the divergence, spin, shear and normal deformation rates rendered the warm-core system analytically tractable. In particular, a novel class of exact solutions with a Ermakov connection termed pulsrodons which pulse periodically and rotate was isolated. Lyapunov stability of pulsrodons and their duals was subsequently addressed by Holm [40] in an elegant Lagrangian treatment using Hamiltonian dynamics of the shallow water system with plane bottom topography. The general class of exact solutions of the 2+1-dimensional shallow water system with circular paraboloidal bottom geometry as generated via the Ermakov-Ray-Reid reduction of [74] has a complicated structure with underlying nonlinear superposition principle. However, as in [67], pulsrodon solutions may be shown to exist associated with the classical Ermakov reduction. In [91], an elliptic vortex-type ansatz was introduced into a 2+1-dimensional system governing rotating homentropic magneto-gasdynamics with a parabolic gas law and a nonlinear dynamical system analogous to that of [67] derived but with an additional algebraic constraint. It is remarked that such 2+1-dimensional magneto-gasdynamic systems but with pressure-density relation $p \sim \rho$ had been earlier investigated in another manner by Neukirch and Cheung in [55]. Hamiltonian structure and integrability aspects of the dynamical system were investigated in [91] and, in particular, magneto-gasdynamic versions of pulsrodons isolated via a classical Ermakov connection. In later work [99], a class of confined magneto-gasdynamic flows was delimited with an elliptic cylindrical boundary with time-dependent semi-axes determined by an integrable Ermakov-Ray-Reid system.

Calogero-Moser many-body systems and their integrability properties have an extensive literature and have importance both in classical mechanics and quantum physics (see e.g. [15, 53, 58, 17]). In recent work, it has been established that classes of known many-body systems may be embedded in solvable non-autonomous multi-component Ermakov systems [80].

The established importance of Ermakov systems in nonlinear optics has been detailed in [85, 84]. In the present work, the occurrence of Ermakov-Ray-Reid systems in both 2+1-dimensional rotating shallow water theory and magneto-gasdynamics is described. Multi-component Ermakov systems and their connections, in turn, with N-layer hydrodynamics and classical many-body problems are reviewed. Hybrid multi-component Ermakov-Painlevé systems are likewise detailed.

2 A rotating shallow water system. Ermakov-Ray-Reid reduction

Here, the concern is with the 2+1-dimensional version of the rotating shallow water system

$$\frac{\partial h}{\partial t} + \operatorname{div}(h\mathbf{q}) = 0, \quad \frac{\partial \mathbf{q}}{\partial t} + \mathbf{q}\cdot\nabla\mathbf{q} + f(\mathbf{k}\times\mathbf{q}) + \nabla(h+Z) = 0, \tag{2.1}$$

so that

$$z = h(x,y,t) + Z(x,y) \tag{2.2}$$

determines the upper free surface and $z = Z(x,y)$ designates the rigid bottom topography: f denotes the usual Coriolis parameter. This classical nonlinear system is readily derived *ab initio* via a hydrostatic approximation and incorporates kinematic and boundary conditions in a manner reviewed in [50]. It was established therein that the hydrodynamic system (2.1), (2.2) admits privileged Lie group symmetries when the underlying basin is either a circular or elliptical paraboloid. In the sequel, attention is restricted to circular paraboloidal geometries so that the underlying rigid basin is given by

$$z = A(x^2+y^2). \tag{2.3}$$

2.1 Elliptic vortex ansatz

An elliptic vortex ansatz is now introduced with

$$\mathbf{q} = L(t)\mathbf{x} + M(t), \quad h = \mathbf{x}^T E(t)\mathbf{x} + h_0(t), \quad \mathbf{x} = \begin{pmatrix} x - q(t) \\ y - p(t) \end{pmatrix} \tag{2.4}$$

where

$$L = \begin{pmatrix} u_1(t) & u_2(t) \\ v_1(t) & v_2(t) \end{pmatrix}, \quad M = \begin{pmatrix} \dot{q}(t) \\ \dot{p}(t) \end{pmatrix}, \quad E = \begin{pmatrix} a(t) & b(t) \\ b(t) & c(t) \end{pmatrix}. \tag{2.5}$$

Insertion of the above ansatz into the continuity and momentum equations, in turn, yields

$$\begin{pmatrix} \dot{a} \\ \dot{b} \\ \dot{c} \end{pmatrix} + \begin{pmatrix} 3u_1 + v_2 & 2v_1 & 0 \\ u_2 & 2(u_1+v_2) & v_1 \\ 0 & 2u_2 & u_1 + 3v_2 \end{pmatrix} \begin{pmatrix} a \\ b \\ c \end{pmatrix} = \mathbf{0} \tag{2.6}$$

together with

$$\dot{h}_0 + (u_1 + v_2)h_0 = 0 \tag{2.7}$$

and

$$\begin{pmatrix} \dot{u_1} \\ \dot{u_2} \\ \dot{v_1} \\ \dot{v_2} \end{pmatrix} + \begin{pmatrix} L^T & -fI \\ fI & L^T \end{pmatrix} \begin{pmatrix} u_1 \\ u_2 \\ v_1 \\ v_2 \end{pmatrix} + 2\begin{pmatrix} a \\ b \\ b \\ c \end{pmatrix} + 2A\begin{pmatrix} 1 \\ 0 \\ 0 \\ 1 \end{pmatrix} = \mathbf{0}, \tag{2.8}$$

where I denotes the 2×2 identity matrix, augmented by

$$\ddot{p} + f\dot{q} + 2Ap = 0, \quad \ddot{q} + f\dot{p} + 2Aq = 0. \tag{2.9}$$

It will be required that the moving shoreline $h = 0$ be closed so that

$$\Delta = ac - b^2 > 0. \tag{2.10}$$

It proves convenient to proceed in terms of new variables (cf. [67])

$$\begin{aligned} &G = u_1 + v_2, \quad G_R = \frac{1}{2}(v_1 - u_2), \quad G_S = \frac{1}{2}(v_1 + u_2), \quad G_N = \frac{1}{2}(u_1 - v_2), \\ &B = a + c, \quad B_S = b, \quad B_N = \frac{1}{2}(a - c). \end{aligned} \tag{2.11}$$

In the preceding, G and G_R represent, in turn, the divergence and spin of the velocity field while G_S and G_N denote shear and normal deformation rates. In terms of the expressions (2.11), the system (2.6)–(2.8) yields

$$\begin{aligned} &\dot{B} + 2[BG + 2(B_N G_N + B_S G_S)] = 0, \\ &\dot{B}_S + 2B_S G + G_S B - 2B_N G_R = 0\,, \quad \dot{B}_N + 2B_N G + G_N B + 2B_S G_R = 0, \\ &\dot{G} + G^2/2 + 2(G_N^2 + G_S^2 - G_R^2) - 2fG_R + 2B + 4A = 0, \\ &\dot{G}_R + GG_R + fG/2 = 0, \\ &\dot{G}_N + GG_N - fG_S + 2B_N = 0\,, \quad \dot{G}_S + GG_S + fG_N + 2B_S = 0 \end{aligned} \tag{2.12}$$

together with

$$\dot{h}_0 + h_0 G = 0. \tag{2.13}$$

If we now set

$$\Omega = \left(G_R + \frac{f}{2}\right)^{-1/2} \tag{2.14}$$

for $G_R \neq -f/2$ then

$$G = 2\dot{\Omega}/\Omega, \tag{2.15}$$

while (2.13) yields

$$h_0 = c_I/\Omega^2, \tag{2.16}$$

where c_I is an arbitrary constant of integration.

New modulated variables are now introduced according to

$$\bar{B} = \Omega^4 B, \quad \bar{B}_S = \Omega^4 B_S, \quad \bar{B}_N = \Omega^4 B_N, \quad \bar{G}_S = \Omega^2 G_S, \quad \bar{G}_N = \Omega^2 G_N \tag{2.17}$$

so that the system (2.12) yields

$$\begin{aligned}
&\dot{\bar{B}} + 4(\bar{B}_N\bar{G}_N + \bar{B}_S\bar{G}_S)/\Omega^2 = 0, \\
&\dot{\bar{B}}_S + (\bar{B}\bar{G}_S - 2\bar{B}_N)/\Omega^2 + f\bar{B}_N = 0, \\
&\dot{\bar{B}}_N + (\bar{B}\bar{G}_N + 2\bar{B}_S)/\Omega^2 - f\bar{B}_S = 0, \\
&\dot{\bar{G}}_N - f\bar{G}_S + 2\bar{B}_N/\Omega^2 = 0, \\
&\dot{\bar{G}}_S + f\bar{G}_N + 2\bar{B}_S/\Omega^2 = 0
\end{aligned} \tag{2.18}$$

together with

$$\Omega^3\ddot{\Omega} + \frac{1}{4}f^2\Omega^4 - 1 + \bar{G}_N^2 + \bar{G}_S^2 + \bar{B} + 2A\Omega^4 = 0. \tag{2.19}$$

Combination of $(2.18)_2$ and $(2.18)_3$ together with use of $(2.18)_1$ produces the integral of motion

$$\bar{B}_S^2 + \bar{B}_N^2 - \frac{\bar{B}^2}{4} = c_{\text{II}}, \tag{2.20}$$

while $(2.18)_{4,5}$ combine to produce the integral of motion

$$\bar{G}_S^2 + \bar{G}_N^2 - \bar{B} = c_{\text{III}}, \tag{2.21}$$

where c_{II}, c_{III} are arbitrary constants. It is noted that (2.20) shows that

$$\Delta = -c_{\text{II}}/\Omega^8, \tag{2.22}$$

whence the closed moving shoreline condition (2.10) requires that $c_{\text{II}} < 0$.

In [74], a parametrisation of the integrals of motion (2.20) and (2.21) which was compatible with the construction of pulsrodon solutions was introduced, namely

$$\begin{aligned}
&\bar{B}_S = -\sqrt{c_{\text{II}} + \frac{1}{4}\bar{B}^2}\cos\phi(t), \quad \bar{B}_N = -\sqrt{c_{\text{II}} + \frac{1}{4}\bar{B}^2}\sin\phi(t), \\
&\bar{G}_S = -\sqrt{c_{\text{III}} + \bar{B}}\sin\theta(t), \quad \bar{G}_N = -\sqrt{c_{\text{III}} + \bar{B}}\cos\theta(t).
\end{aligned} \tag{2.23}$$

It was subsequently shown that

$$\begin{aligned}
&\dot{\phi} = f + \frac{1}{\Omega^2}\left[-2 + \frac{\bar{B}}{2}(\bar{B} + c_{\text{IV}}) \Big/ \left(\frac{\bar{B}^2}{4} + c_{\text{II}}\right)\right], \\
&\dot{\theta} = f - \frac{1}{\Omega^2}(\bar{B} + c_{\text{IV}})/(\bar{B} + c_{\text{III}})
\end{aligned} \tag{2.24}$$

and

$$(\Omega^2\bar{B})^{\cdot\cdot} + (f^2 + 8A)\,\Omega^2 B = -2(fc_{\text{IV}} + c_{\text{V}}), \tag{2.25}$$

where c_{IV}, c_V are additional arbitrary constants. It was observed in [74] that the latter relation constitutes a generalisation of a result obtained in [25] in the case of plane bottom topography, wherein

$$\mathcal{I}_{\mathcal{E}} = \iint (x^2+y^2)h\, dxdy \sim \Omega^2 \bar{B} \tag{2.26}$$

with the double integral being taken over the eddy $\mathcal{E}$. In the more general present circular paraboloidal basin geometry, the moment of inertia evolution equation embodied in (2.25) corresponds to that originally obtained by Ball [9] and subsequently shown to be derivable by Lie group methods in [50]. Here, (2.25) yields

$$\Omega^2 \bar{B} = c_{VI}\cos(\sqrt{f^2+8A}\,t) + c_{VII}\sin(\sqrt{f^2+8A}\,t) - 2(c_V + f c_{IV})/(f^2+8A), \tag{2.27}$$

as long as $f^2+8A \neq 0$, where c_{VI} and c_{VII} are integration constants. Moreover, if $\bar{B}$ is non-constant, it is shown in [74] to be determined via an elliptic integral relation resulting in an additional constant of integration $\bar{B}|_{t=0} = c_{VIII}$.

The complement of 8 arbitrary constants, namely $c_I, c_{II}, \ldots, c_{VIII}$ necessary for the general solution of the 8-dimensional nonlinear dynamical system (2.12)–(2.13) is now at hand. The general solution of the coupled linear system (2.9) produces, corresponding to a Lie group invariance, an additional 4 parameters into the class of solutions of the original shallow water system. In summary, the shallow water velocity components are given by

$$\begin{aligned} u_1 &= \frac{\dot{\Omega}}{\Omega} + \frac{1}{\Omega^2}\sqrt{c_{III}+\bar{B}}\cos\theta(t), \quad v_1 = -\frac{1}{\Omega^2}\sqrt{c_{III}+\bar{B}}\sin\theta(t) + \frac{1}{\Omega^2} - \frac{f}{2}, \\ u_2 &= -\frac{1}{\Omega^2}\sqrt{c_{III}+\bar{B}}\sin\theta(t) - \frac{1}{\Omega^2} + \frac{f}{2}, \quad v_2 = \frac{\dot{\Omega}}{\Omega} - \frac{1}{\Omega^2}\sqrt{c_{III}+\bar{B}}\cos\theta(t), \end{aligned} \tag{2.28}$$

while the elliptic moving shoreline parameters are

$$\begin{aligned} a &= \frac{1}{\Omega^4}\left[\frac{\bar{B}}{2} - \sqrt{c_{II} + \frac{\bar{B}^2}{4}}\sin\phi(t)\right], \quad b = -\frac{1}{\Omega^4}\sqrt{c_{II} + \frac{\bar{B}^2}{4}}\cos\phi(t), \\ c &= \frac{1}{\Omega^4}\left[\frac{\bar{B}}{2} + \sqrt{c_{II} + \frac{\bar{B}^2}{4}}\sin\phi(t)\right], \end{aligned} \tag{2.29}$$

augmented by the relation (2.16) to determine h_0.

2.2 Ermakov-Ray-Reid connection

The semi axes of the moving shoreline ellipse are given by

$$\Phi = \sqrt{\frac{2h_0}{\sqrt{(a-c)^2+4b^2}-(a+c)}} = \sqrt{\frac{h_0}{(B_N^2+B_S^2)^{1/2}-B/2}} \tag{2.30}$$

and

$$\Psi = \sqrt{\frac{2h_0}{-\sqrt{(a-c)^2+4b^2}-(a+c)}} = \sqrt{\frac{h_0}{-(B_N^2+B_S^2)^{1/2}-B/2}}. \tag{2.31}$$

On use of the integral of motion (2.20), these relations become

$$\Phi = \Omega\sqrt{c_{\rm I}}\Big/\sqrt{\left(c_{\rm II} + \frac{\bar{B}^2}{4}\right)^{1/2} - \bar{B}/2}, \quad \Psi = \Omega\sqrt{c_{\rm I}}\Big/\sqrt{-\left(c_{\rm II} + \frac{\bar{B}^2}{4}\right)^{1/2} - \bar{B}/2} \tag{2.32}$$

so that

$$\bar{B} = -\sqrt{-c_{\rm II}}\left[\frac{\Phi}{\Psi} + \frac{\Psi}{\Phi}\right] < 0, \tag{2.33}$$

where it is required that $c_{\rm II} < 0$ and $c_{\rm II} + \bar{B}^2/4 \geq 0$. In [74], it was established that $\dot{\Psi}\Phi - \Psi\dot{\Phi}$ depends only on Φ/Ψ and that the semi-axes Φ, Ψ of the elliptical moving shoreline on the paraboloidal basin are governed by a Ermakov-Ray-Reid system

$$\begin{aligned}
\ddot{\Phi} + \left(\frac{f^2}{4} + 2A\right)\Phi &= \frac{1}{\Phi^2\Psi}\left[ZZ'/[1 + (\Psi/\Phi)^2] - \frac{\Psi}{\Phi}\left(Z^2 + \frac{k}{4}\right)/[1 + (\Psi/\Phi)^2]^2\right], \\
\ddot{\Psi} + \left(\frac{f^2}{4} + 2A\right)\Psi &= \frac{1}{\Psi^2\Phi}\left[-ZZ'/[1 + (\Psi/\Phi)^2] - \frac{\Phi}{\Psi}\left(Z^2 + \frac{k}{4}\right)/[1 + (\Phi/\Psi)^2]^2\right],
\end{aligned} \tag{2.34}$$

where $\dot{\Psi}\Phi - \Psi\dot{\Phi} := Z(\Phi/\Psi)$. The above result holds *a fortiori* both for the elliptical warm-core model of [25] and for the elliptical vortex analysis of [40] corresponding to the plane bottom topography.

It is seen that the Ermakov-Ray-Reid system is Hamiltonian with

$$\ddot{\Phi} = -\frac{\partial V}{\partial \Phi}, \quad \ddot{\Psi} = -\frac{\partial V}{\partial \Psi}, \tag{2.35}$$

where

$$V = \frac{1}{2}\left(\frac{f^2}{4} + 2A\right)(\Phi^2 + \Psi^2) - \frac{1}{2}\frac{Z^2 + k/4}{\Phi^2 + \Psi^2}, \tag{2.36}$$

and the Hamiltonian is given by

$$\mathcal{H} = \frac{1}{2}\left[\dot{\Phi}^2 + \dot{\Psi}^2 + \left(\frac{f^2}{4} + 2A\right)(\Phi^2 + \Psi^2) - \frac{Z^2 + k/4}{\Phi^2 + \Psi^2}\right]. \tag{2.37}$$

It is remarked that Hamiltonian Ermakov-Ray-Reid systems may be solved in an algorithmic manner as described in [74].

3 Hamiltonian Ermakov-Ray-Reid reduction in magneto-gasdynamics. The pulsrodon

Neukirch *et al* [55, 54, 56], in a series of papers on 2+1-dimensional magneto-gasdynamics, investigated restricted situations wherein the nonlinear acceleration terms in the momentum equation either vanish or are conservative. Here, an alternative procedure is adopted. Thus, a generalisation of the elliptic vortex approach of [67] is introduced in the context of the 2+1-dimensional magneto-gasdynamic Lundquist system. The resultant nonlinear matrix dynamical system is shown to admit a remarkable universal property along with underlying integrable Hamiltonian Ermakov-Ray-Reid structure.

In the sequel, our concern is with a 2+1-dimensional magneto-gasdynamic system incorporating rotation, namely

$$\begin{aligned}
&\rho_t + \operatorname{div}(\rho \mathbf{q}) = 0, \\
&\rho[\mathbf{q}_t + \mathbf{q}\cdot\nabla\mathbf{q} + f\mathbf{k}\times\mathbf{q}] - \mu \operatorname{curl}\mathbf{H}\times\mathbf{H} + \nabla p = \mathbf{0}, \\
&\operatorname{div}\mathbf{H} = 0, \\
&\mathbf{H}_t = \operatorname{curl}(\mathbf{q}\times\mathbf{H}), \\
&S_t + \mathbf{q}\cdot\nabla S = 0
\end{aligned} \tag{3.1}$$

together with the equation of state

$$p = p(\rho, S), \quad \left.\frac{\partial p}{\partial \rho}\right|_S > 0 \tag{3.2}$$

and with

$$\mathbf{q}\cdot\mathbf{k} = 0, \quad \mathbf{H} = \nabla A \times \mathbf{k} + h\mathbf{k}. \tag{3.3}$$

In the above, the magneto-gas density $\rho(\mathbf{x},t)$, pressure $p(\mathbf{x},t)$, gas velocity $\mathbf{q}(\mathbf{x},t)$, magnetic induction $\mathbf{H}(\mathbf{x},t)$, magnetic flux $A(\mathbf{x},t)$ and entropy $S(\mathbf{x},t)$ are all assumed to be dependent on the planar spatial variable $\mathbf{x} = x\mathbf{i} + y\mathbf{j}$ and time t, while f denotes the Coriolis constant.

3.1 An Euler-type representation

It was established in [99] that the nonlinear Lundquist system (3.1) augmented by the constraints (3.3) may be encapsulated in a compact Euler-type system

$$\rho_t + \operatorname{div}(\rho\mathbf{q}) = 0, \quad \mathbf{q}_t + \mathbf{q}\cdot\nabla\mathbf{q} + f\mathbf{k}\times\mathbf{q} + \nabla\Pi = \mathbf{0}, \tag{3.4}$$

wherein Π adopts the form

$$\Pi = \tau(t)\mathcal{R}(\rho/w(t)) + \kappa(t), \tag{3.5}$$

and

$$\nabla(\nabla^2 A) = 0. \tag{3.6}$$

The structure of the above nonlinear system turns out to be privileged in that it admits a generalised elliptic vortex representation with

$$\mathbf{q} = L(t)\bar{\mathbf{x}} + \mathbf{m}(t), \tag{3.7}$$

and

$$\rho = w(t)\mathcal{R}^{-1}(\bar{\mathbf{x}}^T E(t)\bar{\mathbf{x}} + e) \tag{3.8}$$

where $L(t)$ and $E(t) = E^T(t)$ are 2×2 time-dependent matrices and

$$\bar{\mathbf{x}} = \begin{pmatrix} x - \bar{q}(t) \\ y - \bar{p}(t) \end{pmatrix}, \quad \mathbf{m} = \begin{pmatrix} \dot{\bar{q}}(t) \\ \dot{\bar{p}}(t) \end{pmatrix}. \tag{3.9}$$

Thus, insertion into the Euler system (3.4) produces the 2×2 matrix system

$$\dot{E} + EL + L^T E = 0, \quad \dot{L} + L^2 + fML + 2\tau E = 0, \quad M = \begin{pmatrix} 0 & -1 \\ 1 & 0 \end{pmatrix} \tag{3.10}$$

together with

$$\dot{w} + (\operatorname{tr} L)w = 0 \tag{3.11}$$

and the adjoined linear system

$$\dot{\mathbf{m}} + fM\mathbf{m} = 0, \quad \mathbf{m} = \dot{\bar{\mathbf{q}}}. \tag{3.12}$$

It is observed that, remarkably, the nonlinear system (3.10) has the universal property that it is independent of $\mathcal{R}$.

In [91], it has been demonstrated that the magneto-gasdynamic system (3.10), (3.11) with a parabolic pressure-density relation

$$p = p_0 + \delta\rho + \epsilon\rho^2, \quad \frac{\partial p}{\partial \rho} > 0 \tag{3.13}$$

admits an elliptic vortex ansatz. The underlying nonlinear dynamical system was solved to isolate novel classes of confined magneto-gasdynamic flows. Here, it will be established that the matrix system (3.10) is Hamiltonian if $\tau = \tau(w)$. Moreover, if $\tau \sim w$ it is reduced to a Ermakov-Ray-Reid system. To render the system (3.10), (3.11) into canonical form, a change of variables, namely

$$\begin{gathered} \bar{L} = w^{-1}\left[DLD^{-1} + \frac{f}{2}M\right], \quad D = \exp\left(\frac{f}{2}Mt\right), \\ \bar{L}^* = \bar{L} - \frac{1}{2}(\operatorname{tr}\bar{L})\,I, \quad \bar{Q} = M\bar{E} = \frac{MDED^{-1}}{w}, \end{gathered} \tag{3.14}$$

is used. The matrix system then adopts the compact form

$$\dot{\bar{Q}} + w[\bar{Q}, \bar{L}^*] = 0, \quad \dot{\bar{L}}^* + \tau[\bar{Q}, M] = 0 \tag{3.15}$$

together with

$$(\operatorname{tr}\bar{L})^{\cdot} - 2w\det\bar{L}^* - \frac{w}{2}(\operatorname{tr}\bar{L})^2 + \frac{f^2}{2w} - 2\tau\operatorname{tr}(M\bar{Q}) = 0. \tag{3.16}$$

In the preceding, $[A,B]=AB-BA$ denotes the usual commutator of the matrices A,B.

On introduction of the parametrization

$$\bar{L}^*=\begin{pmatrix}\psi & \phi-\bar{G}\\ \phi+\bar{G} & -\psi\end{pmatrix},\quad \bar{Q}=M\begin{pmatrix}\bar{B}/2+\bar{\tau} & \bar{\sigma}\\ \bar{\sigma} & \bar{B}/2-\bar{\tau}\end{pmatrix},\quad w=\Omega^{-2},\tag{3.17}$$

the coupled matrix system (3.15) yields

$$\begin{aligned}\dot{\bar{\sigma}}&=-\frac{\bar{B}\phi-2\bar{G}\bar{\tau}}{\Omega^2}, & \dot{\phi}&=-2\tau\bar{\sigma},\\ \dot{\bar{\tau}}&=-\frac{\bar{B}\psi+2\bar{G}\bar{\sigma}}{\Omega^2}, & \dot{\psi}&=-2\tau\bar{\tau},\\ \dot{\bar{B}}&=-\frac{4\bar{\sigma}\phi+\bar{\tau}\psi}{\Omega^2}, & \dot{\bar{G}}&=0,\end{aligned}\tag{3.18}$$

while (3.16) reduces to

$$\Omega^3\ddot{\Omega}+\frac{f^2}{4}\Omega^4-\bar{G}^2+\phi^2+\psi^2+\tau\Omega^2\bar{B}=0.\tag{3.19}$$

The system (3.18), in turn, is seen to admit the 1st integrals

$$\bar{G}=c_{\rm I},\quad \bar{\sigma}^2+\bar{\tau}^2-\frac{\bar{B}^2}{4}=c_{\rm II},\quad 2(\bar{\tau}\phi-\bar{\sigma}\psi)-\bar{G}\bar{B}=c_{\rm III},\tag{3.20}$$

while the identity

$$(\bar{\sigma}\phi+\bar{\tau}\psi)^2+(\bar{\tau}\phi-\bar{\sigma}\psi)^2\equiv(\bar{\sigma}^2+\bar{\tau}^2)(\phi^2+\psi^2)\tag{3.21}$$

shows that

$$\phi^2+\psi^2=[\Omega^4\dot{\bar{B}}^2+4(c_{\rm I}\bar{B}+c_{\rm III})^2]/4(\bar{B}^2+4c_{\rm II}).\tag{3.22}$$

In addition, it may be verified that

$$\frac{\dot{\phi}\psi-\dot{\psi}\phi}{\phi^2+\psi^2}=\frac{\tau(c_{\rm I}\bar{B}+c_{\rm III})}{\phi^2+\psi^2},\tag{3.23}$$

whence

$$\arctan\left(\frac{\phi}{\psi}\right)=\int\tau\frac{c_{\rm I}\bar{B}+c_{\rm III}}{\phi^2+\psi^2}dt.\tag{3.24}$$

Thus, on use of (3.22), it is seen that the original matrix dynamical system is encoded in the coupled pair of nonlinear equations

$$\begin{aligned}&\ddot{\Omega}+\left[\frac{f^2}{4}+\frac{\dot{\bar{B}}^2}{4(\bar{B}^2+4c_{\rm II})}\right]\Omega=\frac{1}{\Omega^3}\left[c_{\rm I}^2-\tau\Omega^2\bar{B}-\frac{(c_{\rm I}\bar{B}+c_{\rm III})^2}{\bar{B}^2+4c_{\rm II}}\right],\\ &\ddot{\bar{B}}+2\left(\frac{\dot{\Omega}}{\Omega}\right)\dot{\bar{B}}-\frac{\dot{\bar{B}}^2}{\bar{B}^2+4c_{\rm II}}\bar{B}=2\tau\frac{\bar{B}^2+4c_{\rm II}}{\Omega^2}+\frac{4(c_{\rm I}\bar{B}+c_{\rm III})(c_{\rm III}\bar{B}-4c_{\rm I}c_{\rm II})}{\Omega^4(\bar{B}^2+4c_{\rm II})},\end{aligned}\tag{3.25}$$

provided that $\dot{\bar{B}}\neq 0$. Once Ω and $\bar{B}$ have been determined, the residual quantities $\bar{\sigma},\bar{\tau}$ and ϕ,ψ are obtained via (3.22) and (3.24).

3.2 Hamiltonian structure and integrability

The semi-axes of the elliptic curves of constant density ρ are, up to a multiplicative constant, given by

$$\mathcal{A} = \Omega\sqrt{-\bar{B} + \sqrt{\bar{B}^2 + 4c_{\mathrm{II}}}}\,, \quad \mathcal{B} = \Omega\sqrt{-\bar{B} - \sqrt{\bar{B}^2 + 4c_{\mathrm{II}}}} \tag{3.26}$$

where $\bar{B}, c_{\mathrm{II}} < 0$ and $\bar{B}^2 + 4c_{\mathrm{II}} \geq 0$. In terms of $\mathcal{A}, \mathcal{B}$, the system (3.25) becomes

$$\begin{aligned} \ddot{\mathcal{A}} + \omega^2\mathcal{A} &= \frac{1}{\mathcal{A}^2\mathcal{B}}\left[\frac{d}{d(\mathcal{B}/\mathcal{A})}\left(\frac{\mathcal{B}}{\mathcal{A}}\mathcal{J}(\mathcal{B}/\mathcal{A})\right) - 4c_{\mathrm{II}}\tau(t)\mathcal{A}\mathcal{B}\right], \\ \ddot{\mathcal{B}} + \omega^2\mathcal{B} &= \frac{1}{\mathcal{B}^2\mathcal{A}}\left[\frac{d}{d(\mathcal{A}/\mathcal{B})}\left(\frac{\mathcal{A}}{\mathcal{B}}\mathcal{J}(\mathcal{A}/\mathcal{B})\right) - 4c_{\mathrm{II}}\tau(t)\mathcal{A}\mathcal{B}\right], \end{aligned} \tag{3.27}$$

where $\omega = f/2$ and

$$\mathcal{J}(\xi) = \mathcal{J}(\xi^{-1}) = (c_{\mathrm{III}} + 2c_{\mathrm{I}}\sqrt{-c_{\mathrm{II}}})\frac{\xi}{(\xi+1)^2} + (c_{\mathrm{III}} - 2c_{\mathrm{I}}\sqrt{-c_{\mathrm{II}}})\frac{\xi}{(\xi-1)^2}. \tag{3.28}$$

If $\tau = \tau(w)$ so that τ depends only on $\mathcal{A}\mathcal{B} = 2\Omega^2\sqrt{-c_{\mathrm{II}}}$ then the system admits the 1st integral

$$\mathcal{H} = \frac{1}{2}(\dot{\mathcal{A}}^2 + \dot{\mathcal{B}}^2) + \frac{1}{2}\omega^2(\mathcal{A}^2 + \mathcal{B}^2) + \frac{\mathcal{J}(\mathcal{A}/\mathcal{B})}{\mathcal{A}\mathcal{B}} + 4c_{\mathrm{II}}\int\frac{\tau}{\mathcal{A}\mathcal{B}}d(\mathcal{A}\mathcal{B}). \tag{3.29}$$

In addition, the nonlinear coupled system (3.27) is of Ermakov-Ray-Reid type when

$$\tau \sim (\mathcal{A}\mathcal{B})^{-1} \sim w \tag{3.30}$$

and it then admits a second integral of motion, namely the invariant

$$\mathcal{I} = \frac{1}{2}(\mathcal{A}\dot{\mathcal{B}} - \mathcal{B}\dot{\mathcal{A}})^2 + \int^{\mathcal{B}/\mathcal{A}}\Phi(\zeta)d\zeta + \int^{\mathcal{A}/\mathcal{B}}\Psi(\eta)d\eta \tag{3.31}$$

so that it is completely integrable. Importantly, it was shown that the constraint (3.30) is admissible with gas law $p = \rho^\gamma$ when the adiabatic index $\gamma = 2$. It is remarked that in [92], a non-isothermal gasdynamic system with origin in work of Dyson [29] on spinning gas clouds was likewise shown when $p = \rho^2$ to admit a Hamiltonian reduction of Ermakov-Ray-Reid type. A Lax pair was constructed in that integrable case and, remarkably, the matrices in the elliptic vortex ansatz may be interpreted as Lax matrices.

The general class of exact solutions of the present 2+1-dimensional magneto-gasdynamic system as generated via the Ermakov connection has a complicated structure associated with its inherent nonlinear superposition principles. Accordingly, it is important to isolate subclasses of physically relevant solutions which admit ready interpretation. In this connection, in [91], magneto-gasdynamic analogues were obtained of the pulsrodon as originally derived in the context of the

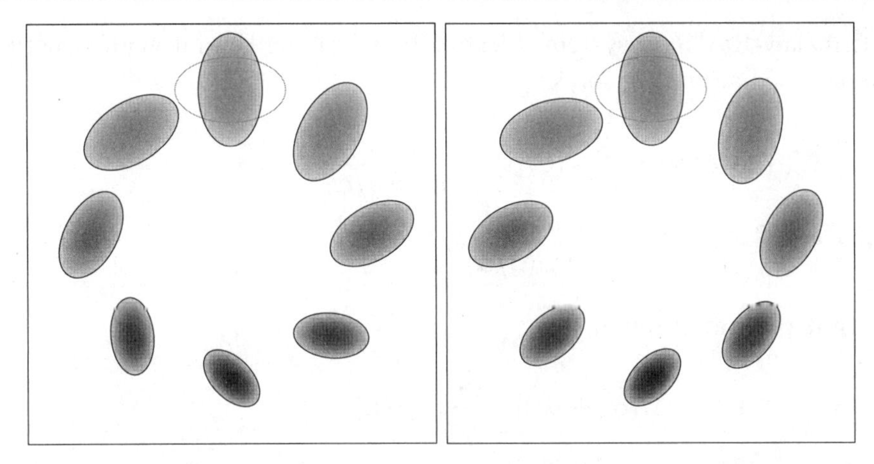

Figure 2. Density plots of pulsrodon solutions at different times. The black ellipses represent the boundary of the flow, that is, density $\rho = 0$. In the interior, darker grey corresponds to higher density. The grey ellipse indicates the position of the pulsrodon after one revolution about the origin. In the corresponding period, the pulsrodon rotates about its centre by 270° (left) and 90° (right) respectively.

evolution of oceanographic warm core eddies in [67]. The pulsating rotating phenomena as described by pulsrodons likewise arise in 2+1-dimensional shallow water hydrodynamics [74, 3] and in the context of rotating gas cloud theory [92]. Pulsrodons have also been shown in [4] to occur in two-layer hydrodynamics as modelled by a Ermakov system. Both the pulsrodon and associated dual solutions have been shown by Holm [40] to be orbitally Lyapunov stable within the class of elliptical vortex solutions. Density plots of the elliptical pulsrodon evolution in the present magneto-gasdynamic context of [91] are displayed in Figure 2.

4 Hamiltonian Ermakov-Ray-Reid systems. Parametrisation and integration

It was established in [74] that the Ermakov-Ray-Reid system

$$\begin{aligned} \ddot{\alpha} + \Omega'(\alpha^2+\beta^2)\alpha &= \frac{1}{\alpha^2\beta}\Phi(\beta/\alpha), \\ \ddot{\beta} + \Omega'(\alpha^2+\beta^2)\beta &= \frac{1}{\alpha\beta^2}\Psi(\alpha/\beta) \end{aligned} \tag{4.1}$$

is Hamiltonian with

$$\ddot{\alpha} = -\frac{\partial V}{\partial \alpha}, \quad \ddot{\beta} = -\frac{\partial V}{\partial \beta} \tag{4.2}$$

when

$$V = \frac{1}{\alpha^2}J(\beta/\alpha) + \frac{1}{2}\Omega(\alpha^2+\beta^2). \tag{4.3}$$

Thus, Ermakov-Ray-Reid systems (4.1) with underlying Hamiltonian structure may be parametrised according to

$$\begin{aligned}\ddot{\alpha} + \Omega'(\alpha^2+\beta^2)\alpha &= \frac{2}{\alpha^3}J(\beta/\alpha) + \frac{\beta}{\alpha^4}\frac{dJ(\beta/\alpha)}{d(\beta/\alpha)},\\ \ddot{\beta} + \Omega'(\alpha^2+\beta^2)\beta &= -\frac{1}{\alpha^3}\frac{dJ(\beta/\alpha)}{d(\beta/\alpha)}.\end{aligned} \tag{4.4}$$

They admit the Hamiltonian

$$\mathcal{H} = \frac{1}{2}[\dot{\alpha}^2 + \dot{\beta}^2 + \Omega(\alpha^2+\beta^2)] + \frac{1}{\alpha^2}J(\beta/\alpha) \tag{4.5}$$

together with the Ermakov invariant

$$\mathcal{I} = \frac{1}{2}(\alpha\dot{\beta} - \dot{\alpha}\beta)^2 + \left(\frac{\alpha^2+\beta^2}{\alpha^2}\right)J(\beta/\alpha). \tag{4.6}$$

Combination of the pair of integrals of motion (4.5) and (4.6) yields

$$\begin{aligned}(\alpha^2+\beta^2)\mathcal{H} - \mathcal{I} &= \frac{1}{2}[\dot{\alpha}^2 + \dot{\beta}^2 - (\alpha\dot{\beta} - \dot{\alpha}\beta)^2 + (\alpha^2+\beta^2)\Omega(\alpha^2+\beta^2)]\\ &= \frac{1}{2}[(\alpha\dot{\alpha} + \beta\dot{\beta})^2 + (\alpha^2+\beta^2)\Omega(\alpha^2+\beta^2)],\end{aligned} \tag{4.7}$$

so that $\Sigma = \alpha^2 + \beta^2$ is determined via

$$\frac{1}{8}\dot{\Sigma}^2 + \frac{1}{2}\Omega(\Sigma)\Sigma - \mathcal{H}\Sigma + \mathcal{I} = 0. \tag{4.8}$$

Moreover, on introduction of Λ according to

$$\Lambda = \frac{2\alpha\beta}{\alpha^2+\beta^2}, \tag{4.9}$$

and new independent variable $\bar{t}$ by

$$d\bar{t} = \Sigma^{-1}dt, \tag{4.10}$$

the invariant (4.6) shows that

$$\pm\frac{1}{2\sqrt{2}}\sqrt{\frac{\Lambda}{(1-\Lambda^2)[\Lambda\mathcal{I} - \mathcal{L}(\Lambda)]}}\,d\Lambda = d\bar{t}, \tag{4.11}$$

where $\mathcal{L}(\Lambda) = (\beta/\alpha)J(\beta/\alpha)$. The original Ermakov variables α, β are given in terms of Σ together with Λ as obtained by integration of (4.11), via the relations

$$\alpha + \beta = \pm\sqrt{\Sigma(1+\Lambda)}, \quad \alpha - \beta = \pm\sqrt{\Sigma(1-\Lambda)}. \tag{4.12}$$

4.1 Autonomisation of the Ermakov-Ray-Reid system

Here, new dependent and independent variables a, b and z are introduced into the non-autonomous system (1.1) according to (see [7])

$$\alpha = a(z)\phi, \quad \beta = b(z)\phi, \quad z = \psi/\phi, \tag{4.13}$$

where ϕ and ψ are linearly independent solutions with unit Wronskian $\phi\dot{\psi} - \psi\dot{\phi}$ of the linear base equation

$$\ddot{\phi} + \omega(t)\phi = 0. \tag{4.14}$$

The Ermakov-Ray-Reid system is then reduced to the autonomous system

$$a_{zz} = \frac{1}{a^2 b}\Phi(b/a), \quad b_{zz} = \frac{1}{b^2 a}\Psi(a/b). \tag{4.15}$$

In particular, the non-autonomous Hamiltonian system

$$\begin{aligned} \ddot{\alpha} + \omega(t)\alpha &= \frac{2}{\alpha^3}J(\beta/\alpha) + \frac{\beta}{\alpha^4}\frac{dJ(\beta/\alpha)}{d(\beta/\alpha)}, \\ \ddot{\beta} + \omega(t)\beta &= -\frac{1}{\alpha^3}\frac{dJ(\beta/\alpha)}{d(\beta/\alpha)} \end{aligned} \tag{4.16}$$

reduces to the integrable autonomous Ermakov-Ray-Reid system

$$\begin{aligned} a_{zz} &= \frac{2}{a^3}J(b/a) + \frac{b}{a^4}\frac{dJ(b/a)}{d(b/a)}, \\ b_{zz} &= -\frac{1}{a^3}\frac{dJ(b/a)}{d(b/a)} \end{aligned} \tag{4.17}$$

with Hamiltonian

$$\mathcal{H} = \frac{1}{2}(a_z^2 + b_z^2) + \frac{1}{a^2}J(b/a) \tag{4.18}$$

and Ermakov invariant

$$\mathcal{I} - \frac{1}{2}(ab_z - a_z b)^2 + \left(\frac{a^2 + b^2}{a^2}\right) J(b/a). \tag{4.19}$$

5 Multi-component Ermakov systems. Genesis in N-layer hydrodynamics

In [81], a symmetry reduction of a classical two-layer nonlinear shallow water system as derived in Stoker [101] was shown to lead to a standard two-component Ermakov-Ray-Reid system. This work was subsequently extended in [89] to derive a novel multi-component Ermakov system via the 2+1-dimensional N-layer hydrodynamic system

$$\begin{aligned}
&\frac{\partial u_i}{\partial t}+u_i\frac{\partial u_i}{\partial x}+v_i\frac{\partial u_i}{\partial y}-fv_i+\frac{\partial p_i}{\partial x}=0,\\
&\frac{\partial v_i}{\partial t}+u_i\frac{\partial v_i}{\partial x}+v_i\frac{\partial v_i}{\partial y}+fu_i+\frac{\partial p_i}{\partial y}=0, \qquad i=1,\dots,N\\
&\frac{\partial}{\partial t}(\eta_{i-1}-\eta_i)+\frac{\partial}{\partial x}[u_i(\eta_{i-1}-\eta_i)]+\frac{\partial}{\partial y}[v_i(\eta_{i-1}-\eta_i)]=0.
\end{aligned} \tag{5.1}$$

In the above, the quantities $\{u_i, v_i, p_i\}$ represent leading order terms in shallow water approximations to velocity components and pressure while $\eta_{i-1}(x, y, t)$ denotes the surface of the upper boundary of the ith layer. In the hydrostatic approximation adopted here, the pressure terms p_i are given by

$$p_i=\sum_{j=0}^{i-1} g_j\eta_j, \tag{5.2}$$

where g_j represents the buoyancy jump at the interface $\eta = \eta_j(x, y, t)$. In the sequel, the Coriolis parameter $f \neq 0$ is scaled to be unity under $t \to ft$, $x \to fx$, $y \to fy$.

In [89], symmetry reduction of the 2+1-dimensional N-layer system (5.1) via the ansatz

$$u_i=x\bar{u}_i(t)+y\bar{v}_i(t), \quad v_i=-x\bar{v}_i(t)+y\bar{u}_i(t), \quad \eta_i=(x^2+y^2)\bar{H}_i(t) \tag{5.3}$$

was shown to lead to the system

$$\begin{aligned}
&\dot{\bar{v}}_i+2\bar{u}_i\bar{v}_i-\bar{u}_i=0,\\
&\dot{\bar{u}}_i-\bar{v}_i^2+\bar{u}_i^2+\bar{v}_i+2\sum_{j=0}^{i-1} g_j\bar{H}_j=0,\\
&\dot{\bar{H}}_{i-1}-\dot{\bar{H}}_i+4(\bar{H}_{i-1}-\bar{H}_i)\bar{u}_i=0,
\end{aligned} \tag{5.4}$$

with $\dot{\bar{H}}_N = 0$ in the case of a rigid bottom topography. Here, we set $\bar{H}_N = 0$. Importantly, the nonlinear system (5.4) admits a parametrisation in terms of a set of dependent variables $\{\Omega_1, \dots, \Omega_N\}$, namely

$$\ddot{\Omega}_i+\frac{1}{4}\Omega_i+\left(\sum_{j=1}^{N}\frac{a_{ij}}{\Omega_j^4}\right)\Omega_i=0, \tag{5.5}$$

where the a_{ij} are constants and

$$\bar{u}_i=\dot{\Omega}_i/\Omega_i, \quad \bar{v}_i=\frac{1}{2}-\Omega_i^{-2}, \quad \bar{H}_{i-1}-\bar{H}_i=c_i\Omega_i^{-4}, \tag{5.6}$$

so that

$$\bar{H}_j=\sum_{k=j+1}^{N} c_k\Omega_k^{-4}. \tag{5.7}$$

The system (5.5) motivated the construction of the N-component Ermakov system as introduced in [89]. In the 3-component case, it adopts the form

$$\begin{aligned}\ddot{\Omega}_1 + \omega(t)\Omega_1 &= \frac{\Phi_1(\Omega_1/\Omega_3, \Omega_2/\Omega_3)}{\Omega_3^3},\\ \ddot{\Omega}_2 + \omega(t)\Omega_2 &= \frac{\Phi_2(\Omega_1/\Omega_3, \Omega_2/\Omega_3)}{\Omega_3^3},\\ \ddot{\Omega}_3 + \omega(t)\Omega_3 &= \frac{\Phi_3(\Omega_1/\Omega_3, \Omega_2/\Omega_3)}{\Omega_3^3},\end{aligned} \tag{5.8}$$

where the Φ_i are dependent on their indicated arguments. Introduction into (5.8) of the new variables

$$\Omega_i = \bar{\Omega}_i(z)\phi, \quad z = \psi/\phi, \tag{5.9}$$

where again ϕ, ψ are linearly independent solutions with unit Wronskian of (4.14), results in the autonomous 3-component Ermakov system

$$\bar{\Omega}_{izz} = \frac{\Phi_i(\bar{\Omega}_1/\bar{\Omega}_3, \bar{\Omega}_2/\bar{\Omega}_3)}{\bar{\Omega}_3^3}, \quad i = 1, 2, 3. \tag{5.10}$$

The latter, in turn, shows that

$$\begin{aligned}\bar{\Omega}_3\bar{\Omega}_{1zz} - \bar{\Omega}_1\bar{\Omega}_{3zz} &= \frac{1}{\bar{\Omega}_3^2}\Phi_1 - \frac{\bar{\Omega}_1}{\bar{\Omega}_3^3}\Phi_3,\\ \bar{\Omega}_3\bar{\Omega}_{2zz} - \bar{\Omega}_2\bar{\Omega}_{3zz} &= \frac{1}{\bar{\Omega}_3^2}\Phi_2 - \frac{\bar{\Omega}_2}{\bar{\Omega}_3^3}\Phi_3,\end{aligned} \tag{5.11}$$

whence if we set

$$U = \bar{\Omega}_1/\bar{\Omega}_3, \quad V = \bar{\Omega}_2/\bar{\Omega}_3, \tag{5.12}$$

then (5.11) yields

$$U_{\bar{t}\bar{t}} = \Phi_1(U, V) - U\Phi_3(U, V), \quad V_{\bar{t}\bar{t}} = \Phi_2(U, V) - V\Phi_3(U, V), \tag{5.13}$$

where

$$d\bar{t} = \bar{\Omega}_3^{-2}dz. \tag{5.14}$$

If the system (5.13) is now set in correspondence with the canonical two-component Ermakov-Ray-Reid system

$$U_{\bar{t}\bar{t}} + \omega_{\text{II}}U = \frac{U}{V^4}f(U/V), \quad V_{\bar{t}\bar{t}} + \omega_{\text{II}}V = \frac{V}{U^4}g(U/V) \tag{5.15}$$

with constant ω_{II}, then the original 3-component system (5.8) becomes

$$\begin{aligned}\ddot{\Omega}_1 + \omega(t)\Omega_1 &= \frac{\Omega_1}{\Omega_2^4}f(\Omega_1/\Omega_2) + \frac{\Omega_1}{\Omega_3^4}[\Phi_3(\Omega_1/\Omega_3, \Omega_2/\Omega_3) - \omega_{\text{II}}],\\ \ddot{\Omega}_2 + \omega(t)\Omega_2 &= \frac{\Omega_2}{\Omega_1^4}g(\Omega_1/\Omega_2) + \frac{\Omega_2}{\Omega_3^4}[\Phi_3(\Omega_1/\Omega_3, \Omega_2/\Omega_3) - \omega_{\text{II}}],\\ \ddot{\Omega}_3 + \omega(t)\Omega_3 &= \frac{\Omega_1}{\Omega_3^3}\Phi_3(\Omega_1/\Omega_3, \Omega_2/\Omega_3),\end{aligned} \tag{5.16}$$

wherein Φ_3 remains an arbitrary function of its indicated arguments. The system (5.16) introduced in [89] was termed a 3-component Ermakov system. Application of the sequence of transformations (5.9), (5.12) and (5.14) reduces it to the two-component system (5.15) augmented by the base Ω_3-equation in (5.16). The latter, remarkably, reduces to the linear Schrödinger equation

$$(1/\bar{\Omega}_3)_{zz} + \Phi_3(\bar{\Omega}_1/\bar{\Omega}_3, \bar{\Omega}_2/\bar{\Omega}_3)(1/\bar{\Omega}_3) = 0 \tag{5.17}$$

in $1/\bar{\Omega}_3$. Thus, once the Ermakov-Ray-Reid system (5.15) has been solved for $U(\bar{t}) = \bar{\Omega}_1/\bar{\Omega}_3$ and $V(\bar{t}) = \bar{\Omega}_2/\bar{\Omega}_3$, the linear equation (5.17) determines $\bar{\Omega}_3 = \bar{\Omega}_3(\bar{t})$. Integration of the relation (5.14) then determines $z = z(\bar{t})$ and hence $t = t(\bar{t})$. The solution of the 3-component Ermakov system (5.16) is then given parametrically in terms of $\bar{t}$ via the relations

$$\Omega_i = \bar{\Omega}_i(\bar{t})\phi(t), \quad t = t(\bar{t}). \tag{5.18}$$

In [89], this integration procedure was illustrated in detail for a 3-component Ermakov system (5.16) for which the 2-component Ermakov-Ray-Reid system (5.15) corresponds to that derived in a two-layer hydrodynamic context in [81].

The N-component Ermakov systems $N > 3$ are now generated in an iterative manner. Thus, if we proceed to the 4-component system

$$\ddot{\Omega}_i + \omega(t)\Omega_i = \Phi_i(\Omega_1/\Omega_4, \Omega_2/\Omega_4, \Omega_3/\Omega_4), \quad i = 1, \ldots, 4 \tag{5.19}$$

then the requirement that it be reducible to a 3-component Ermakov system of the type (5.16) leads to a 4-component Ermakov system which, *in extenso*, adopts the form

$$\begin{aligned}
\ddot{\Omega}_1 + \omega(t)\Omega_1 &= \frac{\Omega_1}{\Omega_2^4} f(\Omega_1/\Omega_2) + \frac{\Omega_1}{\Omega_3^4}[h(\Omega_1/\Omega_3, \Omega_2/\Omega_3) - \omega_{\mathrm{I}}] \\
&\quad + \frac{\Omega_1}{\Omega_4^4}[\Phi(\Omega_1/\Omega_4, \Omega_2/\Omega_4, \Omega_3/\Omega_4) - \omega_{\mathrm{II}}], \\
\ddot{\Omega}_2 + \omega(t)\Omega_2 &= \frac{\Omega_2}{\Omega_1^4} g(\Omega_1/\Omega_2) + \frac{\Omega_2}{\Omega_3^4}[h(\Omega_1/\Omega_3, \Omega_2/\Omega_3) - \omega_{\mathrm{I}}] \\
&\quad + \frac{\Omega_2}{\Omega_4^4}[\Phi(\Omega_1/\Omega_4, \Omega_2/\Omega_4, \Omega_3/\Omega_4) - \omega_{\mathrm{II}}], \\
\ddot{\Omega}_3 + \omega(t)\Omega_3 &= \frac{1}{\Omega_3^3} h(\Omega_1/\Omega_3, \Omega_2/\Omega_3) + \frac{\Omega_3}{\Omega_4^4}[\Phi(\Omega_1/\Omega_4, \Omega_2/\Omega_4, \Omega_3/\Omega_4) - \omega_{\mathrm{II}}], \\
\ddot{\Omega}_4 + \omega(t)\Omega_4 &= \frac{1}{\Omega_4^3}\Phi(\Omega_1/\Omega_4, \Omega_2/\Omega_4, \Omega_3/\Omega_4).
\end{aligned} \tag{5.20}$$

The general N-component Ermakov system as introduced in [89] is given by

$$\ddot{\Omega}_i + \omega(t)\Omega_i = \frac{\Phi_i^{(N)}(\Omega_1/\Omega_N, \ldots, \Omega_{N-1}/\Omega_N)}{\Omega_N^3}, \quad i = 1, \ldots, N, \tag{5.21}$$

wherein the functions $\Phi_i^{(N)}$ are defined recursively by

$$
\begin{aligned}
&\frac{\Phi_k^{(n)}}{\Omega_n^3} = \frac{\Phi_k^{(n-1)}}{\Omega_{n-1}^3} + \frac{\Omega_k}{\Omega_n^4}[\Phi_n^{(n)}(\Omega_1/\Omega_n, \ldots, \Omega_{n-1}/\Omega_n) - \omega_{n-1}], \\
&\Phi_1^{(1)} = \Phi_1^{(1)}(\Omega_1/\Omega_2), \qquad n = 2, \ldots, N, \quad k = 1, \ldots, n-1.
\end{aligned}
\tag{5.22}
$$

The cases $N = 1, 2$ correspond to the classical one-component Ermakov equation and two-component Ermakov-Ray-Reid system respectively. In general, the N-component Ermakov system as determined by (5.21), (5.22) is shown in [89] to be sequentially reducible to a canonical Ermakov-Ray-Reid system of the type (1.1) augmented by $N - 2$ linear Schrödinger equations.

In the next section, it is established that, remarkably, classes of known autonomous 3-body and 4-body systems may be embedded in solvable, non-autonomous multi-component Ermakov systems of the type (5.16) and (5.20) respectively. The Ermakov connection as established in [73] is exploited to construct new classes of integrable non-autonomous many-body type systems parametrised in terms of an arbitrary function $J(y/x)$ where x and y are Jacobi variables.

6 Multi-component Ermakov and many-body system connections

Calogero-Moser many-body systems and their integrability properties have an extensive literature with regard to their importance in both classical and quantum contexts (see, e.g., [15, 53, 58, 17] and literature cited therein). In [44], a link was established between a particular two component autonomous Ermakov system and a 3-body problem while subsequently, in [39], a Hamiltonian which turns out to be associated with a Ermakov-Ray-Reid system was related to that of the Calogero 3-body system of [15]. In [73], it has recently been established that a range of established 3-body and 4-body systems may be embedded in solvable non-autonomous Ermakov systems. This work has been extended in [80]. Here, these Ermakov connections are described, in turn, for a generalised version of the Calogero-Marchiori 3-body system [18] and of a 4-body system as introduced in [8].

6.1 A generalised Calogero-Marchiori 3-body system

Here, we consider a generalised non-autonomous version of a class of 3-body problems investigated by Calogero and Marchiori in [15], namely [73]

$$
\ddot{x}_i + \omega(t) x_i = \frac{\partial V}{\partial x_i}, \quad i = 1, 2, 3,
\tag{6.1}
$$

where

$$\begin{aligned} V = & \frac{\lambda_1}{(x_1 - x_2)^2} + \frac{\lambda_2}{(x_2 - x_3)^2} + \frac{\lambda_3}{(x_3 - x_1)^2} \\ & + \frac{\nu_1}{(x_1 + x_2 - 2x_3)^2} + \frac{\nu_2}{(x_2 + x_3 - 2x_1)^2} + \frac{\nu_3}{(x_3 + x_1 - 2x_2)^2} \\ & + \frac{\mu}{x_1^2 + x_2^2 + x_3^2} - \frac{\delta}{2(x_1 + x_2 + x_3)^2}. \end{aligned} \tag{6.2}$$

The autonomous 3-body problem with $\lambda_1 = \lambda_2 = \lambda_3 = \lambda$ and $\delta = 0$ which incorporates a term $\sim (\sum x_i^2)^{-1}$ was investigated by Bachkhaznadji [8]. The original 3-body problem of Calogero [15] is the autonomous case with $\mu = 0$ and the $\nu_i = 0,\ i = 1, 2, 3$. On introduction of the centre of mass co-ordinate R and Jacobi co-ordinates x, y according to

$$R = \frac{1}{3}(x_1 + x_2 + x_3), \quad x = \frac{1}{\sqrt{2}}(x_1 - x_2), \quad y = \frac{1}{\sqrt{6}}(x_1 + x_2 - 2x_3) \tag{6.3}$$

so that

$$x_1 = R + \frac{1}{\sqrt{2}}x + \frac{1}{\sqrt{6}}y, \quad x_2 = R - \frac{1}{\sqrt{2}}x + \frac{1}{\sqrt{6}}y, \quad x_3 = R - \sqrt{\frac{2}{3}}y, \tag{6.4}$$

the 3-body system determined by (6.1), (6.2) becomes

$$\begin{aligned} \ddot{x} + \omega(t)x = & -\frac{\lambda_1}{x^3} + \frac{4\lambda_2}{(\sqrt{3}y - x)^3} - \frac{4\lambda_3}{(x + \sqrt{3}y)^3} \\ & - \frac{12\nu_2}{(x + \sqrt{3}y)^3} - \frac{12\nu_3}{(3x - \sqrt{3}y)^3} - \frac{2\mu x}{[3R^2 + x^2 + y^2]^2}, \\ \ddot{y} + \omega(t)y = & -\frac{4\sqrt{3}\lambda_2}{(\sqrt{3}y - x)^3} - \frac{4\sqrt{3}\lambda_3}{(x + \sqrt{3}y)^3} \\ & - \frac{\nu_1}{3y^3} - \frac{4\sqrt{3}\nu_2}{(x + \sqrt{3}y)^3} - \frac{4\sqrt{3}\nu_3}{(3x - \sqrt{3}y)^3} - \frac{2\mu y}{[3R^2 + x^2 + y^2]^2}, \\ \ddot{R} + \omega(t)R = & -\frac{2\mu R}{[3R^2 + x^2 + y^2]^2} + \frac{\delta}{27R^3}. \end{aligned} \tag{6.5}$$

The latter constitutes a specialisation of a 3-component Ermakov system of the type (5.16) with $\omega_{\text{II}} = \delta/27$ and

$$\Phi_3(\Omega_1/\Omega_3, \Omega_2/\Omega_3) = \frac{-2\mu}{[3 + (\Omega_1/\Omega_3)^2 + (\Omega_2/\Omega_3)^2]^2} + \frac{\delta}{27} \tag{6.6}$$

with $\Omega_1 = x$, $\Omega_2 = y$ and $\Omega_3 = R$.

6.2 A 4-body system

A multi-component Ermakov connection is established here with a non-autonomous version of a 4-body system as recently investigated in [8]. This adopts the form

$$x_i + \omega(t)x_i = \frac{\partial W}{\partial x_i}, \quad i = 1, \ldots, 4, \tag{6.7}$$

where

$$W = \frac{\lambda_1}{(x_1 - x_2)^2} + \frac{\lambda_2}{(x_2 - x_3)^2} + \frac{\lambda_3}{(x_3 - x_1)^2} + \frac{\mu}{x_1^2 + x_2^2 + x_3^2 + x_4^2} - \frac{\sigma}{(x_1 + x_2 + x_3 - 3x_4)^2} - \frac{\delta}{2(x_1 + x_2 + x_3 + x_4)^2}. \tag{6.8}$$

The change of variables employed in the autonomous 4-body problem of [8] is now adopted, namely

$$x = \frac{1}{\sqrt{2}}(x_1 - x_2), \quad y = \frac{1}{\sqrt{6}}(x_1 + x_2 - 2x_3), \quad z = \frac{1}{2\sqrt{3}}(x_1 + x_2 + x_3 - 3x_4),$$
$$R = \frac{1}{2}(x_1 + x_2 + x_3 + x_4) \tag{6.9}$$

so that

$$x_1 = \frac{R}{2} + \frac{1}{\sqrt{2}}\left(x + \frac{y}{\sqrt{3}} + \frac{z}{\sqrt{6}}\right), \quad x_2 = \frac{R}{2} + \frac{1}{\sqrt{2}}\left(-x + \frac{y}{\sqrt{3}} + \frac{z}{\sqrt{6}}\right),$$
$$x_3 = \frac{R}{2} - \sqrt{\frac{2}{3}}y + \frac{z}{2\sqrt{3}}, \quad x_4 = \frac{R}{2} - \frac{\sqrt{3}}{2}z. \tag{6.10}$$

Insertion of the latter relations into the 4-body system (6.7), (6.8), on reduction, yields

$$\ddot{x} + \omega(t)x = -\frac{\lambda_1}{x^3} + \frac{4\lambda_2}{(\sqrt{3}y - x)^3} - \frac{4\lambda_3}{(x + \sqrt{3}y)^3} - \frac{2\mu x}{(R^2 + x^2 + y^2 + z^2)^2},$$
$$\ddot{y} + \omega(t)y = -\frac{4\sqrt{3}\lambda_2}{(\sqrt{3}y - x)^3} - \frac{4\sqrt{3}\lambda_3}{(x + \sqrt{3}y)^3} - \frac{2\mu y}{(R^2 + x^2 + y^2 + z^2)^2},$$
$$\ddot{z} + \omega(t)z = \frac{1}{6}\frac{\sigma}{z^3} - \frac{2\mu z}{(R^2 + x^2 + y^2 + z^2)^2},$$
$$\ddot{R} + \omega(t)R = -\frac{2\mu R}{(R^2 + x^2 + y^2 + z^2)^2} + \frac{\delta}{4R^3}. \tag{6.11}$$

This constitutes the specialisation of the 4-component Ermakov system (5.20) with

$$\Phi(x/R, y/R, z/R) = -\frac{2\mu}{[1 + (x/R)^2 + (y/R)^2 + (z/R)^2]^2} + \frac{\delta}{4},$$
$$h = \omega_{\mathrm{I}} = \frac{\sigma}{6}, \quad \omega_{\mathrm{II}} = \frac{\delta}{4} \tag{6.12}$$

under the correspondence

$$(\Omega_1, \Omega_2, \Omega_3, \Omega_4) := (x, y, z, R). \tag{6.13}$$

It was established in [73] that the above class of 4-body systems given by (6.7), (6.8) may be embedded via the Hamiltonian parametrisation (4.4) in a broad class of integrable non-autonomous 4-component Ermakov systems parametrised in terms of $J(y/x)$ where x, y are Jacobi co-ordinates, namely

$$
\begin{aligned}
\ddot{x}_1 + \omega(t)x_1 &= \frac{1}{\sqrt{2}}J_1 + \frac{1}{\sqrt{6}}J_2 + \frac{2\sigma}{(x_1+x_2+x_3-3x_4)^3} + \frac{\delta}{(x_1+x_2+x_3+x_4)^3}, \\
\ddot{x}_2 + \omega(t)x_2 &= \frac{1}{\sqrt{6}}J_2 - \frac{1}{\sqrt{2}}J_1 + \frac{2\sigma}{(x_1+x_2+x_3-3x_4)^3} + \frac{\delta}{(x_1+x_2+x_3+x_4)^3}, \\
\ddot{x}_3 + \omega(t)x_3 &= -\frac{2}{\sqrt{6}}J_2 + \frac{2\sigma}{(x_1+x_2+x_3-3x_4)^3} + \frac{\delta}{(x_1+x_2+x_3+x_4)^3}, \\
\ddot{x}_4 + \omega(t)x_4 &= -\frac{6\sigma}{(x_1+x_2+x_3-3x_4)^3} + \frac{\delta}{(x_1+x_2+x_3+x_4)^3},
\end{aligned}
\tag{6.14}
$$

where J_1 and J_2 are given in terms of $J(y/x)$ and $J'(y/x)$ by the relations

$$
J_1 = \frac{2}{x^3}J(y/x) + \frac{y}{x^4}J'(y/x), \quad J_2 = -\frac{1}{x^3}J'(y/x) \tag{6.15}
$$

and x, y are the Jacobi variables as given in (6.3).

In [73], non-autonomous integrable extensions were also presented for a 3-body system of Khare and Bhaduri [45] and the Smorodinski-Winternitz Hamiltonian system of [107]. It remains to determine the complete extent to which many-body problems in the literature may be interpreted as multi-component Ermakov systems and thereby, in particular, extended to integrable non-autonomous counterparts. This is under current investigation.

7 Multi-component Ermakov-Painlevé systems

Physical systems which incorporate spatial modulation arise in a wide range of important contexts. In classical continuum mechanics, they occur *inter alia* in elastodynamics, visco-elastodynamics and in crack and loading boundary value problems in the elastostatics of inhomogeneous media (see, e.g., [43, 11, 21, 22] and work cited therein). Nonlinear Schrödinger (NLS) equations incorporating both temporal and spatial modulation have, in recent years, been shown to arise in the theory of Bose-Einstein condensates and in nonlinear optics [13, 14, 102, 108, 110, 109]. In [88], a class of model NLS equations has been investigated wherein the spatial modulation of the nonlinear term is driven by an integrable Ermakov-Ray-Reid system. Application was made to a modulated NLS equation relevant to the analysis of Bloch wave propagation in optical lattices as investigated in [108].

Novel hybrid two-component Ermakov-Painlevé and Ermakov-NLS systems have recently been introduced in a series of papers [68, 69, 2, 94, 78, 71, 79, 77, 75]. Here, multi-component modulated Ermakov-Painlevé II systems are presented in a natural development of the work of [89] on multi-component Ermakov systems but with a nonlinear integrable base equation instead of the classical linear canonical equation

(4.14). Thus, n-component modulated systems are introduced of the type

$$\begin{aligned}
\ddot{\phi}_1 + \omega(\rho,t)\phi_1 &= \frac{1}{\rho^3}\Phi_1(\phi_1/\rho,\ldots,\phi_n/\rho),\\
\ddot{\phi}_2 + \omega(\rho,t)\phi_2 &= \frac{1}{\rho^3}\Phi_2(\phi_1/\rho,\ldots,\phi_n/\rho),\\
&\vdots\\
\ddot{\phi}_n + \omega(\rho,t)\phi_n &= \frac{1}{\rho^3}\Phi_n(\phi_1/\rho,\ldots,\phi_n/\rho),
\end{aligned} \tag{7.1}$$

where, here, in general, the modulation ρ is determined by an integrable nonlinear base equation

$$\ddot{\rho} + \omega(\rho,t)\rho = 0. \tag{7.2}$$

If one sets

$$\psi_i = \frac{\phi_i}{\rho}, \quad i = 1,\ldots,n \tag{7.3}$$

together with

$$s = \int \frac{dt}{\rho^2(t)}, \tag{7.4}$$

it is seen that the system (7.1), (7.2) is equivalent to the de-coupled autonomous system

$$\psi_1'' = \Phi_1(\psi_1,\ldots,\psi_n), \quad \psi_2'' = \Phi_2(\psi_1,\ldots,\psi_n), \quad \ldots, \quad \psi_n'' = \Phi_n(\psi_1,\ldots,\psi_n), \tag{7.5}$$

augmented by (7.2). In the system (7.5), a prime denotes a derivative with respect to s. The above multi-component system (7.1), (7.2) is analytically tractable if its autonomous reduction (7.5) and the base equation (7.2) are either C-integrable or S-integrable in the sense of Calogero [16] and provided that, in addition, the integral in (7.4) may be evaluated to determine s. The procedure may be applied, in particular, if the canonical system (7.5) constitutes a multi-component Ermakov system of [89]. It may also be implemented if (7.5) is taken as, for instance, the S-integrable n-component Toda lattice system.

7.1 The Ermakov base equation

With regard to the nonlinear base equation (7.2), if ρ is governed by the classical Ermakov equation

$$\ddot{\rho} + \omega(t)\rho = \frac{\zeta}{\rho^3} \tag{7.6}$$

then the integral (7.4) to determine s is readily evaluated. Thus,

$$\begin{aligned} s &= \int \frac{dt}{\rho^2(t)} = \int \frac{dt}{c_1\alpha^2(t) + 2c_2\alpha(t)\beta(t) + c_3\beta^2(t)} \\ &= \frac{1}{W} \int^{\beta(t)/\alpha(t)} \frac{dz}{c_1 + 2c_2 z + c_3 z^2} = \mathcal{S}(\beta/\alpha), \end{aligned} \tag{7.7}$$

where $\alpha(t), \beta(t)$ are two linearly independent solutions of (4.14) with corresponding constant Wronskian $W = \alpha\dot{\beta} - \beta\dot{\alpha}$ and where the constants c_i are such that

$$c_1 c_3 - c_2^2 = \frac{\zeta}{W^2}. \tag{7.8}$$

In the above, use has been made of the nonlinear superposition principle

$$\rho = \sqrt{c_1\alpha^2(t) + 2c_2\alpha(t)\beta(t) + c_3\beta^2(t)} \tag{7.9}$$

for the Ermakov equation (7.6). The latter result is readily retrieved via Lie group methods [86, 95].

7.2 The Ermakov-Painlevé II base equation

In the case of the multi-component Ermakov-Painlevé II system, the base equation (7.2) is taken as the prototype single-component Ermakov-Painlevé II equation

$$\ddot{\rho} = \rho^3 - \frac{t\rho}{2} - \frac{(\alpha + 1/2)^2}{4\rho^3} \tag{7.10}$$

as obtained by insertion of

$$\Sigma = \rho^2 \tag{7.11}$$

in the integrable Painlevé XXXIV equation

$$\ddot{\Sigma} = \dot{\Sigma}^2/2\Sigma - t\Sigma + 2\Sigma^2 - (\alpha + 1/2)^2/2\Sigma. \tag{7.12}$$

It is recalled [23] that the latter is linked to the canonical Painlevé II equation

$$\ddot{Y} = 2Y^3 + tY + \alpha \tag{7.13}$$

via the Hamiltonian system

$$\dot{Y} = -Y^2 - \frac{t}{2} + \Sigma, \quad \dot{\Sigma} = 2YZ + \alpha + \frac{1}{2}. \tag{7.14}$$

Accordingly, the 'Painlevé II-modulated' multi-component Ermakov system adopts the form

$$\begin{aligned} &\ddot{\phi}_1 + \left[-\rho^2 + \frac{t}{2} + \frac{(\alpha + 1/2)^2}{4\rho^4}\right]\phi_1 = \frac{1}{\rho^3}\Phi_1(\phi_1/\rho, \ldots, \phi_n/\rho), \\ &\ddot{\phi}_2 + \left[-\rho^2 + \frac{t}{2} + \frac{(\alpha + 1/2)^2}{4\rho^4}\right]\phi_2 = \frac{1}{\rho^3}\Phi_2(\phi_1/\rho, \ldots, \phi_n/\rho), \\ &\qquad\vdots \\ &\ddot{\phi}_n + \left[-\rho^2 + \frac{t}{2} + \frac{(\alpha + 1/2)^2}{4\rho^4}\right]\phi_n = \frac{1}{\rho^3}\Phi_n(\phi_1/\rho, \ldots, \phi_n/\rho), \end{aligned} \tag{7.15}$$

wherein the Φ_i are determined by the recurrence relations (5.22). In the reduction to the corresponding autonomous system (7.5), it is required here to evaluate the integral (7.4), wherein ρ is a solution of the hybrid Ermakov-Painlevé II equation (7.10). This involves evaluation of

$$s = \int \frac{dt}{\Sigma(t)} \tag{7.16}$$

for positive solutions Σ of the Painlevé XXXIV equation (7.12). It turns out, remarkably, that a celebrated Bäcklund transformation for the Painlevé II equation may be used to generate an infinite sequence of solutions of Painlevé XXXIV for which the integral (7.16) may be evaluated.

7.2.1 Application of a Bäcklund transformation

Let Y_α be a solution of the Painlevé II equation (7.13) corresponding to the parameter α and Σ_α be an associated solution of the Painlevé XXXIV equation. Then, another pair of solutions $(Y_{\alpha+1}, \Sigma_{\alpha+1})$ corresponding to the parameter $\alpha+1$ is given by (see, e.g., [23])

$$Y_{\alpha+1} = -Y_\alpha - \frac{\alpha+1/2}{\Sigma_\alpha}, \quad \Sigma_{\alpha+1} = -\Sigma_\alpha + 2Y_{\alpha+1}^2 + t. \tag{7.17}$$

Moreover, the action of this Bäcklund transformation on

$$s_\alpha = \int \frac{dt}{\Sigma_\alpha(t)} \tag{7.18}$$

may be shown to be given by [94]

$$s_{\alpha+1} = \frac{(\alpha+1/2)s_\alpha + \ln(\Sigma_{\alpha+1}\Sigma_\alpha)}{\alpha+3/2}. \tag{7.19}$$

Accordingly, if we start with the solution $Y_0 = 0$, $\Sigma_0 = t/2$ of the Hamiltonian system (7.14) corresponding to the parameter $\alpha = 0$ then $s_0 = 2\ln|t|$ and the Bäcklund transforms Y_1, Σ_1 and s_1 are determined by (7.17) and (7.19). Iterative action of the Bäcklund transformation leads to a sequence of rational solutions of Painlevé II and Painlevé XXXIV which can be expressed in terms of Yablonski-Vorob'ev polynomials [20]. This, in turn, corresponds to a sequence of solutions of the base Ermakov-Painlevé II equation (7.10) in ρ together with associated integrals s_α, provided Σ_α is positive. The importance of positive solutions of Painlevé XXXIV arises naturally in the physical setting of two-ion electrodiffusion. Thus, in the electrolytic context of [12], the scaled electric field was shown to be governed by the Painlevé II equation and the associated ion concentrations by Painlevé XXXIV. The constraint that the ion concentrations be positive was examined in detail in [12] for the sequence of exact solutions of Painlevé XXXIV in terms of Yablonskii-Vorob'ev polynomials as generated via the iterated action of the Bäcklund transformation for Painlevé II.

A sequence of explicit solutions (Y_α, Σ_α, s_α) corresponding to parameters $\alpha = n + 1/2$, $n \in \mathbb{N}$ expressed in terms of classical Airy functions may be likewise generated iteratively by means of the Bäcklund transformation. Positivity in certain regions may be guaranteed as demonstrated in [12] so that these solutions are applicable in the current procedure when the base equation (7.2) in ρ is of Ermakov-Painlevé II type (7.10).

7.3 Iterative reduction of Ermakov-Painlevé II systems

In the case when (7.5) constitutes an n-component Ermakov system for the ψ_i, $i = 1, \ldots, n$ as set down in [89] then for a given solution ρ of (7.10), the corresponding solutions of the hybrid n-component Ermakov-Painlevé II system (7.15) are determined by the relations (7.3) and (7.4). The n-component Ermakov system is iteratively reducible to a standard Ermakov-Ray-Reid system of the type (1.1) augmented by $n-2$ linear equations [89]. To illustrate the procedure, we consider a 4-component Ermakov-type system

$$\begin{aligned}
\ddot{\phi}_1 + \omega\phi_1 &= \frac{\phi_1}{\phi_2^4}\, f(\phi_1/\phi_2) + \frac{\phi_1}{\phi_3^4}\, [h(\phi_1/\phi_3, \phi_2/\phi_3) - \omega_{\mathrm{I}}] \\
&\quad + \frac{\phi_1}{\phi_4^4}\, [\Phi(\phi_1/\phi_4, \phi_2/\phi_4, \phi_3/\phi_4) - \omega_{\mathrm{II}}], \\
\ddot{\phi}_2 + \omega\phi_2 &= \frac{\phi_2}{\phi_1^4}\, g(\phi_1/\phi_2) + \frac{\phi_2}{\phi_3^4}\, [h(\phi_1/\phi_3, \phi_2/\phi_3) - \omega_{\mathrm{I}}] \\
&\quad + \frac{\phi_2}{\phi_4^4}\, [\Phi(\phi_1/\phi_4, \phi_2/\phi_4, \phi_3/\phi_4) - \omega_{\mathrm{II}}], \\
\ddot{\phi}_3 + \omega\phi_3 &= \frac{1}{\phi_3^3}\, h(\phi_1/\phi_3, \phi_2/\phi_3) + \frac{\phi_3}{\phi_4^4}\, [\Phi(\phi_1/\phi_4, \phi_2/\phi_4, \phi_3/\phi_4) - \omega_{\mathrm{II}}], \\
\ddot{\phi}_4 + \omega\phi_4 &= \frac{1}{\phi_4^4}\, \Phi(\phi_1/\phi_4, \phi_2/\phi_4, \phi_3/\phi_4).
\end{aligned} \tag{7.20}$$

In the specialisation of the above system, wherein $\Phi = 0$ and $\phi_4 := \rho$ is governed by the single component Ermakov-Painlevé II equation (7.10), a hybrid Ermakov-Painlevé II system results, namely

$$\begin{aligned}
&\ddot{\phi}_1 + \left[-\rho^2 + \frac{t}{2} + \frac{(\alpha+\frac{1}{2})^2}{4\rho^4}\right]\phi_1 = \frac{\phi_1}{\phi_2^4}\, f(\phi_1/\phi_2) + \frac{\phi_1}{\phi_3^4}\, [h(\phi_1/\phi_3, \phi_2/\phi_3) - \omega_{\mathrm{I}}] - \omega_{\mathrm{II}}\frac{\phi_1}{\rho^4}, \\
&\ddot{\phi}_2 + \left[-\rho^2 + \frac{t}{2} + \frac{(\alpha+\frac{1}{2})^2}{2\rho^4}\right]\phi_2 = \frac{\phi_2}{\phi_1^4}\, g(\phi_1/\phi_2) + \frac{\phi_2}{\phi_3^4}\, [h(\phi_1/\phi_3, \phi_2/\phi_3) - \omega_{\mathrm{I}}] - \omega_{\mathrm{II}}\frac{\phi_2}{\rho^4}, \\
&\ddot{\phi}_3 + \left[-\rho^2 + \frac{t}{2} + \frac{(\alpha+\frac{1}{2})^2}{2\rho^4}\right]\phi_3 = \frac{1}{\phi_3^3}\, h(\phi_1/\phi_3, \phi_2/\phi_3) - \omega_{\mathrm{II}}\frac{\phi_3}{\rho^4}, \\
&\ddot{\rho} + \left[-\rho^2 + \frac{t}{2} + \frac{(\alpha+\frac{1}{2})^2}{2\rho^4}\right]\rho = 0.
\end{aligned} \tag{7.21}$$

On introduction of the change of variables (7.3), (7.4) it is seen that the above 4-component Ermakov-Painlevé II system (7.3) is reduced to a 3-component Ermakov system of a type set down in [89], namely

$$\begin{aligned}
\psi_1'' + \omega_{\mathrm{II}}\psi_1 &= \frac{\psi_1}{\psi_2^4}\, f(\psi_1/\psi_2) + \frac{\psi_1}{\psi_3^4}\, [h(\psi_1/\psi_3,\ \psi_2/\psi_3) - \omega_{\mathrm{I}}],\\
\psi_2'' + \omega_{\mathrm{II}}\psi_2 &= \frac{\psi_2}{\psi_1^4}\, g(\psi_1/\psi_2) + \frac{\psi_2}{\psi_3^4}\, [h(\psi_1/\psi_3,\ \psi_2/\psi_3) - \omega_{\mathrm{I}}],\\
\psi_3'' + \omega_{\mathrm{II}}\psi_3 &= \frac{1}{\psi_3^3}\, h(\psi_1/\psi_3,\ \psi_2/\psi_3)
\end{aligned} \tag{7.22}$$

together with the single component Ermakov-Painlevé II equation in ρ. With

$$\left.\psi_i^* = \psi_i/\psi_3, \quad i=1,2, \quad \psi_3^* = 1/\psi_3, \quad ds^* = \psi_3^{-2} ds \quad \right\}\mathcal{R} \tag{7.23}$$

the system (7.22), in turn, reduces to a standard Ermakov-Ray-Reid system

$$\psi_{1,s^*s^*}^* + \omega_{\mathrm{I}}\psi_1^* = \frac{\psi_1^*}{\psi_2^{*4}}\, f(\psi_1^*/\psi_2^*), \quad \psi_{2,s^*s^*}^* + \omega_{\mathrm{I}}\psi_2^* = \frac{\psi_2^*}{\psi_1^{*4}}\, g(\psi_1^*/\psi_2^*) \tag{7.24}$$

augmented by a classical Ermakov equation

$$\psi_{3,s^*s^*}^* + h(\psi_1^*,\ \psi_2^*)\psi_3^* = \frac{\omega_{\mathrm{II}}}{\psi_3^{*3}}. \tag{7.25}$$

Thus, once the canonical system (7.24) has been solved for $\psi_1^*(s^*) = \psi_1/\psi_3$ and $\psi_2^*(s^*) = \psi_2/\psi_3$, the Ermakov equation (7.25) determines $\psi_3^*(s^*)$ and hence $s^*(s)$ via integration of the relation in (7.23). The ψ_i, $i=1,2,3$ are then given parametrically in terms of s through s^*. The ϕ_i, $i=1,2,3$ of the original hybrid Ermakov-Painlevé II system (7.3) are then obtained by means of the relations (7.3) and (7.4).

It is interesting to observe that (7.23) constitutes a reciprocal-type transformation with $\mathcal{R}^2 = I$. Reciprocal-type transformations which leave invariant 1+1-dimensional nonlinear governing equations in gasdynamics and magneto-gasdynamics were set down in [61, 62]. They have subsequently been shown to have diverse physical applications such as to the solution of nonlinear moving boundary problems [63, 64, 70, 72, 32], the analysis of oil/water migration through a porous medium [97] and in Cattaneo-type hyperbolic heat conduction [87]. Reciprocal transformations have also been applied in the analysis of discontinuity wave propagation [27] and most recently Korteweg capillarity theory [93]. In soliton theory, the composition of reciprocal and gauge transformations has been used in both 1+1 and 2+1 dimensions to link integrable equations and the inverse scattering schemes in which they are embedded [48, 57, 83, 76, 65, 98, 90].

The above procedure is readily adduced to construct integrable multi-component Ermakov-Painlevé III and Ermakov-Painlevé IV systems with nonlinear base equation in ρ taken as hybrid single-component Ermakov-Painlevé III and Ermakov-Painlevé IV equations, namely

$$\ddot{\rho} - \left[\frac{\dot{\rho}^2}{\rho} - \frac{\dot{\rho}}{\rho t} + \frac{1}{2\rho^2 t}(\alpha\rho^4 + \beta) + \frac{\alpha\rho^4}{2} + \frac{\delta}{2\rho^4}\right]\rho = 0 \tag{7.26}$$

and

$$\ddot{\rho} - \left[\frac{3}{4}\rho^4 + 2t\rho^2 + t^2 - \alpha - \frac{\beta}{2\rho^4}\right]\rho = 0 \tag{7.27}$$

respectively.

To conclude, it is remarked parenthetically that the novel autonomisation procedure encapsulated in (7.2)–(7.4) and originally introduced in a Ermakov context may be likewise applied to other important systems of nonlinear physics wherein the autonomous system (7.5) is analytically tractable. For instance, the S-integrable Toda lattice system

$$\psi_n'' = -e^{\psi_{n+1}} + 2e^{\psi_n} - e^{\psi_{n-1}}, \quad n \in \mathbb{Z}, \tag{7.28}$$

descriptive of the one-dimensional motion of a chain of particles interconnected by nonlinear springs with an exponential potential [103, 104], constitutes a rich class of such systems. This leads to a novel non-autonomous modulated Toda lattice system of the form

$$\ddot{\phi}_n + \omega(\rho,t)\phi_n = \frac{1}{\rho^3}[-e^{\phi_{n+1}/\rho} + 2e^{\phi_n/\rho} - e^{\phi_{n-1}/\rho}], \quad \ddot{\rho} + \omega(\rho,t)\rho = 0, \quad n \in \mathbb{Z} \tag{7.29}$$

which inherits the integrability properties of (7.28) such as invariance under a Bäcklund transformation with concomitant nonlinear superposition principle whereby, in particular, multi-soliton type solutions may be generated iteratively in an algebraic manner [105, 19, 96].

The rich Lie group structure underlying the Toda lattice system (7.28) gives rise to a variety of reductions of interest (see, e.g., [90] and references cited therein). For instance, a reduction of period p corresponds to the Kac-Moody Lie algebra $A_{p-1}^{(1)}$ [42]. In particular, a modulated two-component Toda system may be obtained corresponding to imposing period 3 on the associated canonical system, so that one may set $\psi_1 + \psi_2 + \psi_3 = 0$ without loss of generality. The Toda lattice system (7.28) then adopts the form of the stationary Fordy-Gibbons system [33] with associated modulated system

$$\begin{aligned}
&\ddot{\phi}_1 + \omega(\rho,t)\phi_1 = \frac{1}{\rho^3}[-e^{\phi_2/\rho} + 2e^{\phi_1/\rho} - e^{-\phi_1/\rho - \phi_2/\rho}],\\
&\ddot{\phi}_2 + \omega(\rho,t)\phi_2 = \frac{1}{\rho^3}[e^{-\phi_1/\rho - \phi_2/\rho} + 2e^{\phi_2/\rho} - e^{\phi_1/\rho}],\\
&\ddot{\rho} + \omega(\rho,t)\rho = 0.
\end{aligned} \tag{7.30}$$

A deeper reduction represented by $\psi_1 = \psi_2$, corresponding to the twisted Kac-Moody Lie algebra $A_2^{(2)} \subset A_2^{(1)}$ leads to the modulated integrable nonlinear system

$$\ddot{\phi}_1 + \omega(\rho,t)\phi_1 = \frac{1}{\rho^3}[e^{\phi_1/\rho} - e^{-2\phi_1/\rho}], \quad \ddot{\rho} + \omega(\rho,t)\rho = 0. \tag{7.31}$$

This is a modulated version of the travelling wave reduction of the classical Tzitzéica equation. The latter has applications both in affine differential geometry and 1+1-dimensional anisentropic gasdynamics (see, e.g., [90] and literature cited therein).

It is remarked that a modulated version of the classical Tzitzéica equation arises in affine differential geometry [28]. Travelling wave solutions of the unmodulated Tzitzéica equation may be obtained in terms of Jacobi elliptic functions [41].

The Kac-Moody Lie algebra $D_3^{(2)} \subset A_3^{(1)}$ representing a particular period 6 reduction of the Toda lattice system gives rise to a stationary reduction of the classical Demoulin system with corresponding integrable modulated version

$$\begin{aligned} &\ddot{\phi}_1 + \omega(\rho,t)\phi_1 = \frac{1}{\rho^3}[e^{\phi_1/\rho} - e^{-\phi_1/\rho-\phi_2/\rho}], \\ &\ddot{\phi}_2 + \omega(\rho,t)\phi_2 = \frac{1}{\rho^3}[e^{\phi_2/\rho} - e^{-\phi_1/\rho-\phi_2/\rho}], \\ &\ddot{\rho} + \omega(\rho,t)\rho = 0. \end{aligned} \tag{7.32}$$

The associated autonomous system

$$\psi_1'' = e^{\psi_1} - e^{-\psi_1-\psi_2}, \quad \psi_2'' = e^{\psi_2} - e^{-\psi_1-\psi_2} \tag{7.33}$$

represents a travelling wave reduction of the classical Demoulin system which arises, in particular, in the classification of minimal surfaces in projective differential geometry [31]. It is noted that the reduction $\phi_1 = \phi_2$ leads to the modulated system (7.31).

References

[1] Abdullaev J, Desyatnikov A S, Ostravoskaya E A, Suppression of collapse for matter waves with orbital angular momentum, *J. Opt.* **13** (2011) 064023.

[2] Amster P, Rogers C, On a Ermakov-Painlevé II reduction in three-ion electrodiffusion. A Dirichlet boundary value problem, *Discrete and Continuous Dynamical Systems* **35** (2015) 3277–3292.

[3] An H, Numerical pulsrodons of the (2+1)-dimensional rotating shallow water system, *Phys. Lett. A* **375** (2011) 1921–1925.

[4] An H, Kwong M K, Zhu H, On multi-component Ermakov systems in a two-layer fluid: integrable Hamiltonian structure and exact vortex solutions, *Stud. Appl. Math.* **136** (2016) 139–152.

[5] Anderson D, Bonnedal M, Variational approach to nonlinear focussing of Gaussian laser beams, *Phys. Fluids* **22** (1979), 105–109.

[6] Athorne C, Projective lifts and generalised Ermakov and Bernoulli systems, *J. Math. Anal* **233** (1999) 552–563.

[7] Athorne C, Rogers C, Ramgulam U, Osbaldestin A, On linearisation of the Ermakov system, *Phys. Lett. A* **143** (1990) 207–212.

[8] Bachkhaznadji A, Lassaut M,, Solvable few-body quantum problems, *Few-Body Systems* **56** (2015) 1–17.

[9] Ball F K, Some general theorems concerning the finite motion of a shallow liquid lying on a paraboloid, *J. Fluid Mech.* **17** (1963) 240–256.

[10] Ball F K, The effect of rotation on the simpler modes of motion in an elliptic paraboloid, *J. Fluid Mech.* **22** (1965) 529–545.

[11] Barclay D W, Moodie T B, Rogers C, Cylindrical impact waves in inhomogeneous Maxwellian visco-elastic media, *Acta Mechanica* **29** (1978) 93–117.

[12] Bass L, Nimmo J J C, Rogers C, Schief W K, Electrical structures of interfaces: a Painlevé II model, *Proc. Roy. Soc. London A* **466** (2010) 2117–2136.

[13] Belmonte-Beita J, Pérez-Garcia V M, Vekslechnik V, Lie symmetries and solitons in nonlinear systems with spatially inhomogeneous nonlinearities, *Physical Review Letters* **98** (2007) 064102-1–064102-4.

[14] Belmonte-Beita J, Pérez-Garcia V M, Vekslechnik V, Konotop V V, Localised nonlinear waves in systems with time and space-modulated nonlinearities, *Physical Review Letters* **100** (2008) 164102-1–164102-4.

[15] Calogero F, Solution of a three-body problem in one dimension, *J. Math. Phys.* **10** (1969) 2191–2196.

[16] Calogero F, Universal integrable nonlinear pdes, in *Applications of Analytic and Geometric Methods to Nonlinear Differential Equations* Ed. P A Clarkson, Nato ASI Series, Mathematical and Physical Sciences, Kluwer Academic Publishers (1993) 109–114.

[17] Calogero F, *Classical Many-Body Problems Amenable to Exact Treatments*, Lecture notes in Physics, Springer, Berlin, 2001.

[18] Calogero F, Marchioro C, Exact solution of one-dimensional three-body scattering problems with two-body and/or three-body inverse square potentials, *J. Math. Phys.* **15** (1974) 1425–1430.

[19] Chen H H, Liu C S, Bäcklund transformation solutions of the Toda lattice equation, *J. Math. Phys.* **16** (1975) 1428–1430.

[20] Clarkson P A, Remarks on the Yablonskii-Vorob'ev polynomials, *Phys. Lett. A* **319** (2003) 137–144.

[21] Clements D L, Atkinson C, Rogers C, Antiplane crack problems for an inhomogeneous elastic material, *Acta Mechanica* **29** (1978) 199–211.

[22] Clements D L, Rogers C, On the Bergman operator method and anti-plane contact problems involving an inhomogeneous half-space, *SIAM J. Appl. Math.* **34** (1978) 764–773.

[23] Conte R, Ed *The Painlevé Property: One Century Later*, Springer-Verlag, New York, 1999.

[24] Cornolti F, Lucchesi M, Zambon B, Elliptic Gaussian beam self-focussing in nonlinear media, *Opt. Commun.* **75** (1990) 129–135.

[25] Cushman-Roisin B, Heil W H, Nov D, Oscillation and rotations of elliptical warm-core rings, *J. Geophys. Res.* **90** (1985) 11756–11764.

[26] Desyatnikov A S, Buccoliero D, Dennis M R, Kivshar Y S, Suppression of collapse for spiralling elliptic solitons, *Phys. Rev. Lett.* **104** (2010) 053902-1–4.

[27] Donato A, Ramgulam U, Rogers C, The 3+1-dimensional Monge-Ampère equation in discontinuity wave theory: application of a reciprocal transformation, *Meccanica* **27** (1992) 257–262.

[28] Dunajski M, Plansanglate P, Strominger-Yau-Zaslov geometry, affine spheres and Painlevé III, *Commun. Math. Phys.* **290** (2009) 997–1024.

[29] Dyson F J, Dynamics of a spinning gas cloud, *J. Math. Mech.* **18** (1968) 91–101.

[30] Ermakov V P, Second-order differential equations: conditions of complete integrability, Univ. Izy. Kiev **20** (1880) 1–25.

[31] Ferapontov E V, Schief W K, Surfaces of Demoulin: differential geometry, Bäcklund transformation and integrability, *J. Geom. Phys.* **30** (1999) 343–363.

[32] Fokas A S, Rogers C, Schief W K, Evolution of methacrylate distribution during wood saturation. A nonlinear moving boundary problem, *Appl. Math. Lett.* **18** (2005) 321–328.

[33] Fordy A P, Gibbons J, Integrable nonlinear Klein-Gordon equations, *Commun. Math. Phys.* **77** (1980) 21–30.

[34] Goldsbrough G R, The tidal oscillations in an elliptic basin of variable depth, *Proc. Roy. Soc. London A* **130** (1930) 157–167.

[35] Goncharenko A M, Logvin Y A, Samson A M, Self-focussing of two orthogonally polarized light beams in a nonlinear medium, *Opt. Quantum Electron* **25** (1993) 97–104.

[36] Goncharenko A M, Logvin Y A, Samson A M, Shapovalov P S, Rotating ellipsoidal gaussian beams in nonlinear media, *Opt. Commun.* **81** (1991) 225–230.

[37] Goncharenko A M, Logvin Y A, Samson A M, Shapovalev P S, Turovets S I, Ermakov Hamiltonian systems in nonlinear optics of elliptic Gaussian beams, *Phys. Lett. A* **160** (1991) 138–142.

[38] Guiliano C R, Marburger J H, Yariv A, Enhancement of self-focussing threshold in sapphire with elliptical beams, *Appl. Phys. Lett.* **21** (1972) 58–60.

[39] Haas F, Goedert J, On the Hamiltonian structure of Ermakov systems, *J. Phys. A: Math. Gen.* **29** (1996) 4083–4092.

[40] Holm D D, Elliptical vortices and integrable Hamiltonian dynamics of the rotating shallow-water equations, *J. Fluid Mech.* **227** (1991) 393–406.

[41] Huber A, A note on a class of solitary-like solutions of the Tzitzeica equation generated by a similarity reduction, *Physica D: Nonlinear Phenomena* **237** (2008) 1079–1087.

[42] Kac V G, *Infinite Dimensional Lie Algebras*, Cambridge University Press, 1985.

[43] Karal F C, Keller J B, Elastic wave propagation in homogeneous and inhomogeneous media, *J. Acoust. Soc. America* **31** (1959) 694–705.

[44] Katayama N, On an extension of Perelomov transformation, *Bul. Osaka Dref. Col. Tech.* **25** (1991) 15–17.

[45] Khare A, Bhaduri R K, Some algebraically solvable three-body problems in one-dimension, *J. Phys. A: Math. Gen.* **27** (1994) 2213–2223.

[46] Kirchhoff G, *Vorlesungen über mathematische Physik*, Vol I, Leipzig: Teubner, 1876.

[47] Konopelchenko B, Rogers C, On 2+1-dimensional nonlinear systems of Loewner-type, *Phys. Lett. A* **158** (1991) 391–397.

[48] Konopelchenko B, Rogers C, Bäcklund and reciprocal transformations: gauge connections, in Ames W F, ed, *Nonlinear Equations in the Applied Sciences*, Academic Press, New York (1991) 317–362.

[49] Konopelchenko B, Rogers C, On generalised Loewner systems: novel integrable equations in 2+1-dimensions, *J. Math. Phys.* **34** (1993) 214–242.

[50] Levi D, Nucci M C, Rogers C, Winternitz P, Group theoretical analysis of a rotating shallow liquid in a rigid container, *J. Phys. A: Mathematical and General* **22** (1989) 4743–4767.

[51] Malomed B A, Variational methods in nonlinear optics and related fields, in *Progress in Optics*, Vol **43**, E. Wolf Ed, Elsevier Science, North Holland, Amsterdam (2002), 171–193.

[52] Malomed B A, *Soliton Management in Periodic Systems*, Springer, New York, 2006.

[53] Moser J, Three integrable Hamiltonian systems connected with isospectral deformations, *Adv. Math.* **16** (1975) 197–220.

[54] Neukirch T, Quasi-equilibria: a special class of time-dependent solutions for two-dimensional magnetohydrodynamics, *Phys. Plasmas* **2** (1995) 4389–4399.

[55] Neukirch T, Cheung D L G, A class of accelerated solutions of the two-dimensional ideal magnetohydrodynamic equations, *Proc. Roy. Soc. Lond. A* **457** (2001) 2547–2566.

[56] Neukirch T, Priest E R, Generalization of a special class of time-dependent solutions of the two-dimensional magnetohydrodynamics equations to arbitrary pressure profiles, *Phys. Plasmas* **7** (2000) 3105–3107.

[57] Oevel W, Rogers C, Gauge transformations and reciprocal links in 2+1 dimensions, *Rev. Math. Phys.* **5** (1993) 299–330.

[58] Perelomov A M, *Integrable Systems of Classical Mechanics and Lie Algebras*, Birkhäuser, Basel, 1990.

[59] Ray J R, Nonlinear superposition law for generalised Ermakov systems, *Phys. Lett. A* **78** (1980) 4–6.

[60] Reid J L, Ray J R, Ermakov systems, nonlinear superposition and solution of nonlinear equations of motion, *J. Math. Phys.* **21** (1980), 1583–1587.

[61] Rogers C, Reciprocal relations in non-steady one-dimensional gasdynamics, *Zeit. angew. Math. Phys.* **19** (1968) 58–63.

[62] Rogers C, Invariant transformations in non-steady gasdynamics and magneto-gasdynamics, *Zeit. angew. Math. Phys.* **20** (1969) 370–382.

[63] Rogers C, Application of a reciprocal transformation to a two-phase Stefan problem, *J. Phys. A: Math. Gen.* **18** (1985) L105–L109.

[64] Rogers C, On a class of moving boundary problems in nonlinear heat conduction. Application of a Bäcklund transformation, *Int. J. Nonlinear Mech.* **21** (1986) 249–256.

[65] Rogers C, The Harry Dym equation in 2+1 dimensions: a reciprocal link with the Kadomtsev-Petvishvili equation, *Phys. Lett. A* **120** (1987) 15-18.

[66] Rogers C, Generation of invariance theorems for nonlinear boundary-value problems in shallow water theory: application of MACSYMA, *Numerical & Applied Mathematics*, IMACS Meeting Proceedings, Paris (1989) 69–74.

[67] Rogers C, Elliptic warm-core theory: the pulsrodon, *Phys. Lett. A* **138** (1989) 267–273.

[68] Rogers C, A novel Ermakov-Painlevé II system: $N+1$-dimensional coupled NLS and elastodynamic reductions, *Stud. Appl. Math.* **133** (2014) 214–231.

[69] Rogers C, Hybrid Ermakov-Painlevé IV systems, *J. Nonlinear Mathematical Physics* **21** (2014) 628–642.

[70] Rogers C, On a class of reciprocal Stefan moving boundary problems, *Zeit. angew. Math. Phys.* **66** (2015) 2069–2079.

[71] Rogers C, On hybrid Ermakov-Painlevé systems. Integrable reduction, *J. Nonlinear Mathematical Physics* **24** (2017) 239–249.

[72] Rogers C, Moving boundary problems for an extended Dym equation. Reciprocal connections, *Meccanica* **52** (2017) 3531-3540.

[73] Rogers C, Multi-component Ermakov and non-autonomous many-body systems connections, to be published *Ricerche di Matematica* (2018).

[74] Rogers C, An H, Ermakov-Ray-Reid systems in 2+1-dimensional rotating shallow water theory, *Stud. Appl. Math.* **125** (2010) 275–299.

[75] Rogers C, Bassom A P, Clarkson P A, Ermakov-Painlevé IV systems: canonical reduction, to be published *J. Math. Anal. Appl.* (2018).

[76] Rogers C, Carillo S, On reciprocal properties of the Caudrey-Dodd-Gibbon and Kaup-Kupershmidt hierarchies, *Physica Scripta* **36** (1987) 865–869.

[77] Rogers C, Chow K, On modulated NLS-Ermakov systems, *J. Nonlinear Mathematical Physics* **24** Supplement 1 (2017) 61–74.

[78] Rogers C, Clarkson P A, Ermakov-Painlevé II symmetry reduction of a Korteweg capillarity system, *Symmetry, Integrability and Geometry: Methods and Applications* **13** (2017) 018.

[79] Rogers C, Clarkson P A, Ermakov-Painlevé reduction in cold plasma physics. Application of a Bäcklund transformation, to be published *J. Nonlinear Mathematical Physics* (2018).

[80] Rogers C, Clarkson P A, On hybrid integrable many-body-Painlevé systems. Ermakov connections, to be submitted (2018).

[81] Rogers C, Hoenselaers C, Ramgulam U, Ermakov structure in 2+1-dimensional systems. Canonical reduction, in *Modern Group Analysis: Advanced Analytical and Computational Methods in Mathematical Physics* (N.H. Ibragimov et al Eds) Kluwer Academic Publishers, (1993) 317–328.

[82] Rogers C, Hoenselaers C, Ray J R, On 2+1-dimensional Ermakov systems, *J. Phys. A: Mathematical & General* **26** (1993) 2625–2633.

[83] Rogers C, Nucci M C, On reciprocal Bäcklund transformations and the Korteweg-de Vries hierarchy, *Physica Scripta* **33** (1986) 289–292.

[84] Rogers C, Malomed B, An H, Ermakov-Ray-Reid reductions of variational approximations in nonlinear optics, *Stud. Appl. Math.* **129** (2012) 389–413.

[85] Rogers C, Malomed B, Chow K, An H, Ermakov-Ray-Reid systems in nonlinear optics, *J. Phys. A: Math. Theor.* **43** (2010) 455214 (15pp).

[86] Rogers C, Ramgulam U, A nonlinear superposition principle and Lie group invariance: application in rotating shallow water theory, *Int. J. Nonlinear Mech.* **24** (1989) 229–236.

[87] Rogers C, Ruggeri T, A reciprocal Bäcklund transformation: application to a nonlinear hyperbolic model in heat conduction, *Lett. Il Nuovo Cimento* **44** (1985) 289–296.

[88] Rogers C, Saccomandi G, Vergori L, Ermakov-modulated nonlinear Schrödinger models. Integrable reduction, *J. Nonlinear Mathematical Physics* **23** (2016) 108–126.

[89] Rogers C, Schief W K, Multi-component Ermakov systems: structure and linearization, *J. Math. Anal. Appl.* **198** (1996) 194–220.

[90] Rogers C, Schief W K, *Bäcklund and Darboux Transformations. Geometry and Modern Applications in Soliton Theory*, Cambridge Texts in Applied Mathematics, Cambridge University Press, 2002.

[91] Rogers C, Schief W K, The pulsrodon in 2+1-dimensional magnetogasdynamics. Hamiltonian structure and integrability, *J. Math. Phys.* **52** (2011) 083701 (20pp).

[92] Rogers C, Schief W K, On the integrability of a Hamiltonian reduction of a 2+1-dimensional non-isothermal rotating gas cloud system, *Nonlinearity* **24** (2011) 3165–3178.

[93] Rogers C, Schief W K, The classical Korteweg capillarity system: geometry and invariant transformations, *J. Phys. A: Math. Theor.* **47** (2014) 345201 (20pp).

[94] Rogers C, Schief W K, On Ermakov-Painlevé II systems. Integrable reduction, *Meccanica* **51** (2016) 2967–2974.

[95] Rogers C, Schief W K, Winternitz P, Lie theoretical generalization and discretization of the Pinney equation, *J. Math. Anal. Appl.* **216** (1997) 246–264.

[96] Rogers C, Shadwick W F, *Bäcklund transformations and Their Applications*, Academic Press, New York, 1982.

[97] Rogers C, Stallybrass M P, Clements D L, On two-phase filtration under gravity and with boundary infiltration. Application of a Bäcklund transformation, *J. Nonlinear Analysis, Theory, Methods and Applications* **7** (1983) 785–799.

[98] Rogers C, Wong P, On reciprocal Bäcklund transformations of inverse scattering schemes, *Physica Scripta* **30** (1983) 10–14.

[99] Schief W K, An H, Rogers C, Universal and integrable aspects of an elliptic vortex representation in 2+1-dimensional magnetogasdynamics, *Stud. Appl. Math.* **130** (2013) 49–79.

[100] Schief W K, Rogers C, Bassom A, Ermakov systems with arbitrary order and dimension. Structure and linearization, *J. Phys. A: Math. Gen.* **29** (1996) 903–911.

[101] Stoker J J, *Water Waves*, Interscience Publ., New York, 1957.

[102] Tang X Y, Shukla P K, Solution of the one-dimensional spatially inhomogeneous cubic-quintic nonlinear Schrödinger equation with external potential, *Phys. Rev. A* **76** (2007) 013612.

[103] Toda M, Vibration of a chain with nonlinear interaction, *J. Phys. Soc. Japan* **22** (1967) 431–436.

[104] Toda M, *Theory of Nonlinear Lattices*, Springer Verlag, Berlin, 1989.

[105] Wadati M, Toda M, Bäcklund transformation for the exponential lattice, it J. Phys. Soc. Japan **39** (1975) 1196–1203.

[106] Wagner W G, Haus H A, Marburger J H, Large-scale self-trapping of optical beams in the paraxial ray approximation, *Phys. Rev.* **175** (1968) 256–266.

[107] Winternitz P, Smorodinski Y A, Uhlir M, Fris I, Symmetry groups in classical and quantum mechanics, *Sov. J. Nucl. Phys.* **4** (1967) 444–450.

[108] Zhang J F, Li Y S, Wo L, Malomed B A, Matter-wave solitons and finite amplitude Bloch waves in optical lattices with a spatially modulated nonlinearity, *Phys. Rev. A* **82** (2010) 033614.

[109] Zhong W P, Belic M R, Huang T, Solitary waves in the nonlinear Schrödinger equation with spatially modulated Bessel nonlinearity, *J. Opt. Soc. Am. B* **30** (2013) 1276–1283.

[110] Zhong W P, Belic M R, Malomed B A, Huang T, Solitary waves in the nonlinear Schrödinger equation with Hermite-Gaussian modulation of the local nonlinearity, *Phys. Rev. E* **84** (2011) 046611.

Subject Index

A

action-angle variables 507
adjoint
  operator 319
  symmetry 319
Adler–Kostant–Symes theory
  (AKS) 189
Airy function 253
algebraic invariant surface 297
algebraic traveling wave
  solution 297
algebraically solvable 2, 3, 11
algebrogeometric approach 64
analytic vector 479
Askey-Wilson polynomials 147
  zeros 150
asymptotically isochronous 1, 2, 6
asymptotics .. 278, 279, 282, 283, 289
attractor 534, 536

B

Bäcklund transformation
  bilinear 234
  discrete 81
  KdV 76
  sine-Gordon 77
Benjamin-Ono equation 241
Bernoulli equation 331
β-plane approximation 505
Bianchi permutability theorem 77
bifurcation 537
bilinear Bäcklund
  transformations 234
bilinear D-operator 234
bilinear-fractional integral 98
birational mapping 64, 95
Borel transform 475, 476
boundary value problems 16
Boussinesq equation 238, 248
breather 281, 283, 284
Buhl problem 188
Burgers' equation 392

C

Calogero-Degasperis-Ibragimov-
  Shabat equation 325
chaotic 533
classical initial value
  problem 486
Clebsch system 99
Clunie's lemma 165
Cole-Hopf transformation 52
commuting discrete flows 78
complex Ginzburg-Landau
  equation 173
conservation law 318
  conserved current Φ^t 318
  conserved flux Φ^x 318
Consistency-Around-a-Cube
  (CAC) 78–80
contact geometry linearization
  covering scheme 205
contiguity 5
continuity 5
Coriolis force 502
cylindrical KdV equation 253

D

deautonomisation 48, 65
degenerate soliton
  solutions 279, 282
degree growth 49, 69
  bounded 69

exponential..........50, 59, 69, 70
linear....................62, 69, 70
quadratic...............50, 58, 69
determining equations...........411
discrete Painlevé
equations............46, 49, 51, 67
diophantine...................11, 12
discrete soliton equations......74–93
discrete soliton solutions......81, 82
discrete time..................11, 12
discretization of continuous
equations...................85–88
dissipativity.....................531
Doetsch's representation
theorem.......................474
Dragovich's linear
equation.............488, 492, 494
dynamical degree..........49, 66, 69
dynamical systems..............352
dynamical systems solvable by
algebraic operations.............1

E

elliptic solutions.......161, 163, 180
elliptic vortex ansatz............544
entire functions...................10
equilibrium.................524, 529
Ermakov invariant...............541
Ermakov-Ray-Reid
systems..............541, 548, 555
Euler equation...................502
Euler operator...................318
Euler top.........................119
Euler-Chasles correspondence.....47
exponential type.................475
express method....................58

F

f-plane approximation...........504
first integral.....................411
Fokas method.....................15
Fokker-Planck equation..........430
Fourier transform.................17
full deautonomisation.....53, 67, 69

G

Gambier mapping................52
game of musical chairs.............9
gauge theories..................359
generalized basic hypergeometric
function.......................126
polynomials...................127
zeros..........................139
generalized hypergeometric
function.......................125
polynomials...................125
zeros..........................136
generalized initial condition......483
generic t-dependent polynomial
of arbitrary degree...............1
geostrophic
balance.......................506
ocean flows...................500
geostrophy.......................503
global
relation........................30
solution......................306
goldfish......................7, 10
gradient property...............368
Grassmann variable..............391

H

Halburd's method.................57
Hamiltonian.....................2, 9
Ermakov systems....548, 552–554
system....................97, 507
Hardy space......................473
heavenly
dispersionless systems.........188
equations and systems.........189
integrable...................199
Hermite decomposition.....166, 178
Hirota's bilinear method

discrete 86–91

Hirota-Kimura
- basis 96
- method 94

Hopf
- functional 454, 463
- functional equation 455

I

identities 2–4, 12

indeterminacy
- point of 46, 47, 50, 64

infinite hierarchies 11

initial-values problem 8

integrable heavenly
- dispersionless equations 199
- superflows 217

integrable
- map 95
- PDEs 15
- system 95, 507

integral transforms 16

integrating factor 318

integration/integrability by
- quadratures 422, 427

invariant measure 97

invariant relation 47

inverse scattering transform 22

ISM 263, 286, 288

isochronous 1, 2, 6, 13

isospectral matrices 122

isotropy stratification 359

Ito equation 243, 250

Itô formula 410

J

Jacobi polynomials 133
- zeros 133

K

k-equation 330
- 7th-order 342
- hierarchy of 339
- recursion operator for 348

Kadomtsev-Petviashvili
- equation 237, 249

Kahan's method 94

Kirchhoff equations 99

Kirchhoff case 113

Korteweg-de Vries
- discrete equation 75
- equation (continuous) ... 234, 248, 398
- modified 239

Korteweg-de-Vries-Burgers
- equation 295
- generalized 296

Krichever-Novikov equation 325

Kupershmidt equation 327
- 1st potential 327
- 2nd potential 327
- 3rd potential 328
- 4th potential 328
- 5th potential 329
- 7th-order equations 340–342
- hierarchies of 338–342
- nonlocal invariance of 333–337
- recursion operators for ... 343–348

Kuramoto-Sivashinsky
- equation 163, 295

L

Lagrange top 116

Lagrange–d'Alembert principle .. 188

Laplace transform 474

lattice equations 74–93
- Bianchi permutability theorem . 77
- continuum limits 82–85
- defining evolution 76
- potential KdV 75
- potential modified KdV 75
- soliton equations 74–93

Laurent series 164

Lax pair 24
Lax pairs, discrete 80–82
Lax–Sato
  commutator 188
  compatibility 188
  equations and relationships 193
  integrability 189
  linearization 210
  representations 189
Lebesgue space 473
Lie group classification 420
Lie module 357
Lie series 370
Lie-algebraic integrability
  scheme 188
Lie-Bäcklund symmetries 317–318
Lie-Poisson
  bracket 99
  structure 189
limit cycle 524
linearisable boundary conditions .. 32
linearisable mappings 52, 62
linearization covering
  method 207
linearization of an
  evolutionary system 393
linearly connected symmetries ... 414
Lundgren-Monin-Novikov
  (LMN) equations 451, 458
Lundquist magneto-gasdynamic
  system 549
Lyapunov function 508, 522, 528

M

many-body
  problems 2
  systems 559–562
matrix NLS 273, 277
McMillan mapping 46, 50, 58
modulated Toda-lattice
  systems 568
Molien function 368
Morse-Darboux lemma 508
multi-component
  Ermakov systems 557–559
  Ermakov-Painlevé
    systems 562–568
multidimensional consistency
  (MDC) 78–80
multiple pole solutions 278, 282
multi-point correlation
  (MPC) equations 447, 456
multipotentialisation 321

N

N-layer hydrodynamics 556
Navier-Stokes equations 444
nc AKNS system 264, 271, 274
nc KdV-type equations 260, 273
Nevanlinna theory 165
Newtonian equations
  of motion 1, 2, 7, 8
Nizhnik-Novikov-Veselov
  equation 246
non-confining mappings 61
nonintegrable
  mappings 48, 51, 54, 60, 69, 70
nonlinearly coupled systems
  of ODEs 1, 2, 4–8, 11
nonlinear evolution equation 295
nonlocal equation 481
nonlocal invariance 333
nonlocal operator
  and Borel transform 489
  and Laplace transform 479
nonlocal variables 399
nonlocalities 399
normal forms 352, 354

O

Olmedilla's conjecture 283
one-lump solution 250

operator approach 259
orbit space 358

P

$\mathcal{P}$-transform 477
Painlevé 44
Painlevé test 320
Paley-Wiener theorem 479
perturbation theory 352
phase transition 355
Picard lattice 66, 69
Plemelj formula................... 21
Poincaré
  domain 366
  map........................... 536
polarization...................... 94
projection techniques 261, 268
projective mapping 52
pulsrodons 553
Polya's representation theorem .. 477
potentialisation 321, 348
  multipotentialisation 321
predator 520
prey 520
probability density function
  (PDF) 441

Q

q-derivative operator 127
q-Racah polynomials 151
  zeros 153
QRT mapping 47
quadratic vector field 94
quadratic-fractional integral 97

R

R-structure 189
Racah polynomials 144
  zeros 146
Ramani equation 251
recursion operator 318, 396
  integro-differential 318, 396
relative equilibrium 353
relative (oceanic) vorticity 514
Riemann-Hilbert problem 16
Riemann zeta function 486
ripplons 254
Rossby number 506
rotating shallow water system ... 544

S

Salem number 69
Sawada-Kotera equation 245, 329
  7th-order 342
  hierarchy 339
  recursion operator for 348
Schur Lemma 364, 366
Schwarzian derivative 320
Schwarzian Korteweg-de Vries
  equation (SKdV) 319
  hierarchy 321
semi-discrete potential KdV 83
shallow water waves
  a model equation for 245
simple Lie groups 364
sine-Gordon equation 240
singularity 44, 46
singularity confinement 46, 47
  anticonfinement 63
  late confinement 50, 56, 68, 69
singularity pattern 48
  anticonfined 48, 53, 63
  confined 48, 51, 53, 64
  cyclic 48, 49, 53, 59, 67
  unconfined .. 48, 52, 53, 56, 61, 62
soliton solutions 235, 237
  of the matrix mKdV 288
  of the vector mKdV 287, 289
solvable by
  algebraic operations 1–5
space of initial

conditions 64, 65, 69
spin vorticity 514
spontaneous linearization 369
steady solutions 296
stereographic projection 515
statistical
scaling symmetries . . . 444, 459, 462
symmetries 437
translation
symmetries 444, 458, 460
stochastic differential
equations 410
stratified manifold 352, 358
subequation method 169
super-derivative 391
superequation
by Tian and Liu 403
superfield 391
superposition formula 236
Sylvester equation 262, 274, 276
symmetries
adjoint 319
condition for 318
of an evolutionary system 393
of the Hopf equation 462
of the LMN hierarchy 458
of the MPC equations 456
statistical 437
symmetric criticality principle . . . 359
symmetry-integrable
equations 317
hierarchy 318
system of ODEs 1, 4

T

third kind, mapping 53, 62
three-wave system 119
Toda lattice
2d-Toda lattice 283
traveling wave 295
trivializing a conservation law . . . 399

U

Unified method 15

V

vorticity 513

W

weight technique 392
Whitney stratification 359
Wilson polynomials 141
zeros 143
Wronskian relations 98

◎

编辑手记

世界著名数学家 W. W. 索耶(W. W. Sawyer)曾说过:

> 20 世纪初的数学家的目标是弄清 19 世纪数学家没有搞清楚的那些逻辑要点,解决 19 世纪数学家没有解出的那些问题,对回答得笨拙的问题给出干净的答案,对讨论得浮浅的问题深入识破其原委,统一原来分离开的东西,推广原来作为特殊情况处理的东西. 一件 20 世纪的发现会被认为是极有意义的,是因为它照亮了 19 世纪的大部分问题. 一点儿不谈论上个世纪而只介绍本世纪的数学,就好像不解释前两幕剧中发生了什么,单单表演第三幕一样.

本书就是这样一部纵横 19 世纪和 20 世纪数学与物理交互作用的系列英文著作中的一部,中文书名或可译为《非线性系统及其绝妙的数学结构:第 1 卷》.

本书的主编为诺伯特·欧拉(Norbert Euler),墨西哥人,目前是国际科学中心 A. C. (墨西哥库埃纳瓦卡)的客座教授. 25 年来,他一直在全球多所大学教授各种本科和研究生水平的数学课程. 他是一位活跃的研究人员,迄今为止已发表了 80 多篇关于非线性系统主题的同行评审研究文章,还是多本书的合著者,也参与了一些国际期刊的编辑工作.

正如本书前言中所述:

> 本书的目的是提供一种对非线性系统数学描述的艺术状态的全面说明. 本书包含了 20 篇由非线性系统的不同方面的权威专家撰写的受邀文章,其中包括常微分方程与偏微分方程、微分方程和 q 微分方程、离散或格点方程、非交换方程与矩阵方程、随机方程与超对称方程. 本书内容分为

四个主要部分:Part A:可积系统. Part B:求解方案与求解结构. Part C:非线性系统的对称方法. Part D:应用中的非线性系统. 下面我们对每部分给出一个简短的描述.

Part A 由五篇文章构成,A1 到 A5. 这部分的作者主要提出了如何发现可积系统的基础问题. 在 A1 中,作者 F. 卡洛杰罗(F. Calogero)描述了一种用于识别通过代数运算求解的微分方程的非线性系统的新技术. 在 A2 中,A. S. 福卡斯(A. S. Fokas)和 B. 佩利翁(B. Pelloni)对所谓的福卡斯方法进行了全面的回顾,以解决半直线上应用于非线性偏微分方程的初始边值问题. 在 A3 中,作者 B. 戈拉马迪科(B. Grammaticos),A. 拉曼(A. Ramani),R. 威洛(R. Willox)和 T. 马赛(T. Mase)描述了如何通过奇点的方法去识别离散可积性,其奇点产生于二阶有理映射. 在 A4 中,作者 J. 伊塔林塔(J. Hietarinta)对离散孤子方程或格点孤子方程进行了一个基本的介绍,其中详细讨论了多维设置. 在这部分最后的一篇文章 A5 中,作者 Yu. B. 苏里斯(Yu. B. Suris)和 M. 彼得雷亚(M. Petrera)讨论了海德-广太-木村(Hahan-Hirota-Kimura)离散化的新结果,并在其他事情中提供了运动的积分.

Part B 占据了本书最多的篇幅,由六篇文章组成,B1 到 B6. 本部分的作者描述了获得非线性系统显式解和/或描述系统的解结构的不同方法. 在 B1 中,作者 O. 比弘(O. bihun)讨论了构建等谱矩阵的一个一般方法,该矩阵是根据确定多项式的零点来定义的. 该篇文章显示了该矩阵是如何导致可解非线性一阶常微分系统出现的. 在 B2 中,作者 R. 孔特(R. Conte)、T. W. 额(T. W. Ng)和 C. F. 吴(C. F. Wu)提供了一个介绍作者开发出来的方法的教程,该方法是为了找到不可积的、自治的、任意阶的代数常微分方程的所有亚纯特解. 还给出了物理学的若干例子. 在 B3 中,作者 O. E. 亨托什(O. E. Hentosh)、Ya. A. 普雷卡尔帕茨基(Ya. A. Prykarpatsky)、D. 布莱克默(D. Blackmore)和 A. 普雷卡尔帕茨(A. Prykarpatski)通过强调普法伊弗(Pfeiffer)和拉克斯-佐藤(Lax-Sato)型解来回顾布尔(Buhl)兼容向量场方程问题. 他们进一步分析了相关的李(Lie)代数结构和天文方程的可积性. 一个有意思的关于拉格朗日-达朗贝尔(Lagrange-d' Alembert)原则的相关内容在本部分中也被讨论了. 在 B4 中,作者 X. B. 胡(X. B. Hu)关心的是与双线性型的非线性积分方程的双线性贝可隆(Bäcklund)变换相关的叠加公式和比安基(Bianchi)恒等式. 作者利用这个生成了孤子解、有理解,以及某些非线性积分方程的其他特解. 还提供了一些例子. 在 B5 中,作者 C. 席博尔德(C. Schiebold)构建了 AKNS 系统的 $m\times n$-矩阵值解. 作者给出了一个关于多级解的完整的渐近描述,包括弱束缚呼吸子的波包. 还研究了矢量孤子的碰撞. 在 B6 中,作者 C. 瓦尔斯使(C. Valls)用了一种新方法,该方法是由 A. 高尔(A. Gasull)和 H. 嘉科米尼(H. Giacomimi)引入的,表征广义科尔泰沃赫-德弗里斯-伯格斯(Korteweg-de Vries-Burgers)方程以及仓本-西瓦辛斯(Kuramoto-Sivashinsky)方程的所有行波解.

Part C 中作者集中讨论了非线性系统的对称方法. 本部分由 C1 到 C5 五篇文章组成. 在 C1 中,作者 M. 欧拉(M. Euler)和 N. 欧拉(N. Euler)描述了在 1+1 维中可以构造某些非线性对称可积演化方程的非局部不变性的多电位化过程. 还给出了五阶库珀施密特(Kupershmidt)方程及其层次结构的完整说明. 在 C2 中,作者 G. 加耶塔(G. Gaeta)讨论了庞加莱-迪拉克(Poincaré-Dulac)标准形式中矢量场几何的不同方面. 这在很大程度上依赖于米歇尔(Michel)理论. 还详细讨论了具有先验对称性的系统在物理学中很常见的情况. 在 C3 中,作者 A. V. 基谢廖夫(A. V. Kiselev)和 T. 沃尔夫(T. Wolf)提供了针对使用 SsTools 环境逐步计算非线性耦合玻色子-费米子 $N=1$ 超对称系统的对称代数的非局部递归的一个非正式讨论. 在 C4 中,作者 R. 科兹洛夫(R. Kozlov)讨论了伊藤(Itô)随机微分方程的李点对称,其对应于自变量(时间)和因变量的变换,这些变换保留了布朗(Brown)运动方程的微分形式和性质. 在 C5 中,作者 M. 奥波拉克(M. Oberlack)、M. 瓦克劳奇克(M. Waclawczyk)和 V. 格列别涅夫(V. Grebenev)描述了湍流的随机对称,并讨论了它们在理解湍流统计(比如间歇性和非高斯(Gauss)性)中的重要性.

Part D 提供了非线性系统的某些应用. 此部分包含四篇文章,D1 到 D4. 在 D1 中,作者 A. 查韦斯(A. Chávez)、H. 普拉多(H. Prado)和 E. G. 雷耶斯(E. G. Reyes)回顾了他们关于无限阶的常微分方程的积分变换的工作,并将该理论应用到了方程之中,该方程是在现代理论物理学特别关注的黎曼(Riemann)zeta 函数的帮助下定义的. 在 D2 中,作者 A. 江诗丹顿(A. Constantin)和 R. S. 约翰逊(R. S. Johnson)讨论了地转海洋流在球体上的非线性作用,并指出了其中的一些挑战. 在 D3 中,作者 A. V. 奥西波夫(A. V. Osipov)和 G. 索德巴卡(G. Söderbacka)回顾了他们对一类 n 捕食者--猎物型系统的研究结果,还讨论了追捕的灭绝和各种不同的简单与复杂的共存. 在 D4 中,作者 C. 罗杰斯(C. Rogers)和 W. K. 席费尔(W. K. Schief)讨论了非线性物理和连续介质力学中的叶尔马科夫(Ermakov)型系统. 他们描述了在旋转浅水理论、磁气体动力学、多层流体动力学和多体理论中的应用. 作者还回顾了他们最近关于叶尔马科夫-潘勒韦(Ermakov-Painlevé)系统的工作.

本书的目录为:

前言
作者
Part A:非线性可积系统
A1. 通过代数运算求解非线性耦合微分方程系统
A2. 半线上的可积非线性偏微分方程
A3. 检测离散可积性:奇点方法

A4. 离散孤子方程的基本介绍

A5. 海德-广太-木村离散化可积性的新结果

Part B:解法与解法的结构

B1. 由特殊多项式和根据其零点定义的等谱矩阵满足的动力系统

B2. 微分方程的亚纯解的奇点方法

B3. 布尔问题的法伊弗-佐藤解和天文方程的拉格朗日-达朗贝尔原则

B4. 双线性型中非线性可积方程的叠加公式

B5. AKNS 系统方程的矩阵解

B6. 广义 KdV-伯格斯方程和仓木-西瓦辛斯方程的代数行波

Part C:非线性系统的对称方法

C1. 库珀施密特方程的多能化非局部不变性及其高阶分类

C2. 动力系统范式的几何学

C3. 使用 SsTools 环境的进化超系统的计算对称性与递归算子

C4. 伊藤随机微分方程的对称性及其应用

C5. 湍流的统计对称性

Part D:应用中的非线性系统

D1. 积分变换与无限阶的常微分方程

D2. 地转海洋流在球体上的非线性作用

D3. 对多捕食者--猎物型系统的结果的回顾

D4. 非线性物理学与连续力学中的叶尔马科夫型系统

主题索引

至于为什么我们要不惜重金从国外引进这些学术著作是被许多关心我们的读者经常问到的.这里借《经济观察报》访谈北京万圣书园老板刘苏里的一段话作为回应.

经济观察报:开放的问题,其实是涉及与西方交流的问题,通过书籍的相互了解肯定是一个重要的渠道了.从这个方面说,现有的书籍能满足读者需求吗?

刘苏里:跟西方如何交流,我感觉这个问题现在不但没解决,还有回头的迹象.我们最直接的感受是引进版图书数量上的萎缩.

从 20 世纪 80 年代开放初期,我们经历了一个对世界从不了解到了解的过程,图书引进持续了 30 多年.引进版图书占到万圣销售量差不多六七成吧.同时,也要看到问题的另一个侧面,就是在全世界的知识生产体系里,我们在知识、思想、观念这方面的生产能力仍然很弱,中国需要引进,需要交流,否则怎么追赶?引进版图书的数量不能萎缩,更不能停下脚步,广大读者的需要其实反映了人们的心态.

要知道,中国开放的几十年,受益于与外部世界的交流,也受益于外部世界的思想、观念、理论、方法、科学、教育、文化和技术.外部世界不只开阔了中国人的眼界,还增加了中国人的财富,特别是提高了中国人的自我检省能力和归属感.这个进程不能中断.

对此笔者深以为然!

刘培杰

2023 年 12 月 30 日

于哈工大

刘培杰数学工作室
已出版(即将出版)图书目录——原版影印

书　名	出版时间	定　价	编号
数学物理大百科全书.第 1 卷(英文)	2016—01	418.00	508
数学物理大百科全书.第 2 卷(英文)	2016—01	408.00	509
数学物理大百科全书.第 3 卷(英文)	2016—01	396.00	510
数学物理大百科全书.第 4 卷(英文)	2016—01	408.00	511
数学物理大百科全书.第 5 卷(英文)	2016—01	368.00	512
zeta 函数,q-zeta 函数,相伴级数与积分(英文)	2015—08	88.00	513
微分形式:理论与练习(英文)	2015—08	58.00	514
离散与微分包含的逼近和优化(英文)	2015—08	58.00	515
艾伦・图灵:他的工作与影响(英文)	2016—01	98.00	560
测度理论概率导论,第 2 版(英文)	2016—01	88.00	561
带有潜在故障恢复系统的半马尔柯夫模型控制(英文)	2016—01	98.00	562
数学分析原理(英文)	2016—01	88.00	563
随机偏微分方程的有效动力学(英文)	2016—01	88.00	564
图的谱半径(英文)	2016—01	58.00	565
量子机器学习中数据挖掘的量子计算方法(英文)	2016—01	98.00	566
量子物理的非常规方法(英文)	2016—01	118.00	567
运输过程的统一非局部理论:广义波尔兹曼物理动力学,第 2 版(英文)	2016—01	198.00	568
量子力学与经典力学之间的联系在原子、分子及电动力学系统建模中的应用(英文)	2016—01	58.00	569
算术域(英文)	2018—01	158.00	821
高等数学竞赛:1962—1991 年的米洛克斯・史怀哲竞赛(英文)	2018—01	128.00	822
用数学奥林匹克精神解决数论问题(英文)	2018—01	108.00	823
代数几何(德文)	2018—04	68.00	824
丢番图逼近论(英文)	2018—01	78.00	825
代数几何学基础教程(英文)	2018—01	98.00	826
解析数论入门课程(英文)	2018—01	78.00	827
数论中的丢番图问题(英文)	2018—01	78.00	829
数论(梦幻之旅):第五届中日数论研讨会演讲集(英文)	2018—01	68.00	830
数论新应用(英文)	2018—01	68.00	831
数论(英文)	2018—01	78.00	832

刘培杰数学工作室
已出版(即将出版)图书目录——原版影印

书　　名	出版时间	定　价	编号
湍流十讲(英文)	2018－04	108.00	886
无穷维李代数:第 3 版(英文)	2018－04	98.00	887
等值、不变量和对称性(英文)	2018－04	78.00	888
解析数论(英文)	2018－09	78.00	889
《数学原理》的演化:伯特兰·罗素撰写第二版时的手稿与笔记(英文)	2018－04	108.00	890
哈密尔顿数学论文集(第 4 卷):几何学、分析学、天文学、概率和有限差分等(英文)	2019－05	108.00	891
偏微分方程全局吸引子的特性(英文)	2018－09	108.00	979
整函数与下调和函数(英文)	2018－09	118.00	980
幂等分析(英文)	2018－09	118.00	981
李群,离散子群与不变量理论(英文)	2018－09	108.00	982
动力系统与统计力学(英文)	2018－09	118.00	983
表示论与动力系统(英文)	2018－09	118.00	984
分析学练习. 第 1 部分(英文)	2021－01	88.00	1247
分析学练习. 第 2 部分,非线性分析(英文)	2021－01	88.00	1248
初级统计学:循序渐进的方法:第 10 版(英文)	2019－05	68.00	1067
工程师与科学家微分方程用书:第 4 版(英文)	2019－07	58.00	1068
大学代数与三角学(英文)	2019－06	78.00	1069
培养数学能力的途径(英文)	2019－07	38.00	1070
工程师与科学家统计学:第 4 版(英文)	2019－06	58.00	1071
贸易与经济中的应用统计学:第 6 版(英文)	2019－06	58.00	1072
傅立叶级数和边值问题:第 8 版(英文)	2019－05	48.00	1073
通往天文学的途径:第 5 版(英文)	2019－05	58.00	1074
拉马努金笔记. 第 1 卷(英文)	2019－06	165.00	1078
拉马努金笔记. 第 2 卷(英文)	2019－06	165.00	1079
拉马努金笔记. 第 3 卷(英文)	2019－06	165.00	1080
拉马努金笔记. 第 4 卷(英文)	2019－06	165.00	1081
拉马努金笔记. 第 5 卷(英文)	2019－06	165.00	1082
拉马努金遗失笔记. 第 1 卷(英文)	2019－06	109.00	1083
拉马努金遗失笔记. 第 2 卷(英文)	2019－06	109.00	1084
拉马努金遗失笔记. 第 3 卷(英文)	2019－06	109.00	1085
拉马努金遗失笔记. 第 4 卷(英文)	2019－06	109.00	1086
数论:1976 年纽约洛克菲勒大学数论会议记录(英文)	2020－06	68.00	1145
数论:卡本代尔 1979:1979 年在南伊利诺伊卡本代尔大学举行的数论会议记录(英文)	2020－06	78.00	1146
数论:诺德韦克豪特 1983:1983 年在诺德韦克豪特举行的 Journees Arithmetiques 数论大会会议记录(英文)	2020－06	68.00	1147
数论:1985－1988 年在纽约城市大学研究生院和大学中心举办的研讨会(英文)	2020－06	68.00	1148

刘培杰数学工作室
已出版(即将出版)图书目录——原版影印

书　　名	出版时间	定　价	编号
数论:1987年在乌尔姆举行的 Journees Arithmetiques 数论大会会议记录(英文)	2020—06	68.00	1149
数论:马德拉斯 1987:1987 年在马德拉斯安娜大学举行的国际拉马努金百年纪念大会会议记录(英文)	2020—06	68.00	1150
解析数论:1988年在东京举行的日法研讨会会议记录(英文)	2020—06	68.00	1151
解析数论:2002 年在意大利切特拉罗举行的 C. I. M. E. 暑期班演讲集(英文)	2020—06	68.00	1152
量子世界中的蝴蝶:最迷人的量子分形故事(英文)	2020—06	118.00	1157
走进量子力学(英文)	2020—06	118.00	1158
计算物理学概论(英文)	2020—06	48.00	1159
物质,空间和时间的理论:量子理论(英文)	2020—10	48.00	1160
物质,空间和时间的理论:经典理论(英文)	2020—10	48.00	1161
量子场理论:解释世界的神秘背景(英文)	2020—07	38.00	1162
计算物理学概论(英文)	2020—06	48.00	1163
行星状星云(英文)	2020—10	38.00	1164
基本宇宙学:从亚里士多德的宇宙到大爆炸(英文)	2020—08	58.00	1165
数学磁流体力学(英文)	2020—07	58.00	1166
计算科学:第1卷,计算的科学(日文)	2020—07	88.00	1167
计算科学:第2卷,计算与宇宙(日文)	2020—07	88.00	1168
计算科学:第3卷,计算与物质(日文)	2020—07	88.00	1169
计算科学:第4卷,计算与生命(日文)	2020—07	88.00	1170
计算科学:第5卷,计算与地球环境(日文)	2020—07	88.00	1171
计算科学:第6卷,计算与社会(日文)	2020—07	88.00	1172
计算科学.别卷,超级计算机(日文)	2020—07	88.00	1173
多复变函数论(日文)	2022—06	78.00	1518
复变函数入门(日文)	2022—06	78.00	1523
代数与数论:综合方法(英文)	2020—10	78.00	1185
复分析:现代函数理论第一课(英文)	2020—07	58.00	1186
斐波那契数列和卡特兰数:导论(英文)	2020—10	68.00	1187
组合推理:计数艺术介绍(英文)	2020—07	88.00	1188
二次互反律的傅里叶分析证明(英文)	2020—07	48.00	1189
旋瓦兹分布的希尔伯特变换与应用(英文)	2020—07	58.00	1190
泛函分析:巴拿赫空间理论入门(英文)	2020—07	48.00	1191
卡塔兰数入门(英文)	2019—05	68.00	1060
测度与积分(英文)	2019—04	68.00	1059
组合学手册.第一卷(英文)	2020—06	128.00	1153
*一代数、局部紧群和巴拿赫*一代数丛的表示.第一卷,群和代数的基本表示理论(英文)	2020—05	148.00	1154
电磁理论(英文)	2020—08	48.00	1193
连续介质力学中的非线性问题(英文)	2020—09	78.00	1195
多变量数学入门(英文)	2021—05	68.00	1317
偏微分方程入门(英文)	2021—05	88.00	1318
若尔当典范性:理论与实践(英文)	2021—07	68.00	1366
伽罗瓦理论.第4版(英文)	2021—08	88.00	1408
R 统计学概论	2023—03	88.00	1614
基于不确定静态和动态问题解的仿射算术(英文)	2023—03	38.00	1618

刘培杰数学工作室
已出版(即将出版)图书目录——原版影印

书　　名	出版时间	定　价	编号
典型群,错排与素数(英文)	2020—11	58.00	1204
李代数的表示:通过 gln 进行介绍(英文)	2020—10	38.00	1205
实分析演讲集(英文)	2020—10	38.00	1206
现代分析及其应用的课程(英文)	2020—10	58.00	1207
运动中的抛射物数学(英文)	2020—10	38.00	1208
2—纽结与它们的群(英文)	2020—10	38.00	1209
概率,策略和选择:博弈与选举中的数学(英文)	2020—11	58.00	1210
分析学引论(英文)	2020—11	58.00	1211
量子群:通往流代数的路径(英文)	2020—11	38.00	1212
集合论入门(英文)	2020—10	48.00	1213
酉反射群(英文)	2020—11	58.00	1214
探索数学:吸引人的证明方式(英文)	2020—11	58.00	1215
微分拓扑短期课程(英文)	2020—10	48.00	1216
抽象凸分析(英文)	2020—11	68.00	1222
费马大定理笔记(英文)	2021—03	48.00	1223
高斯与雅可比和(英文)	2021—03	78.00	1224
π 与算术几何平均:关于解析数论和计算复杂性的研究(英文)	2021—01	58.00	1225
复分析入门(英文)	2021—03	48.00	1226
爱德华·卢卡斯与素性测定(英文)	2021—03	78.00	1227
通往凸分析及其应用的简单路径(英文)	2021—01	68.00	1229
微分几何的各个方面.第一卷(英文)	2021—01	58.00	1230
微分几何的各个方面.第二卷(英文)	2020—12	58.00	1231
微分几何的各个方面.第三卷(英文)	2020—12	58.00	1232
沃克流形几何学(英文)	2020—11	58.00	1233
彷射和韦尔几何应用(英文)	2020—12	58.00	1234
双曲几何学的旋转向量空间方法(英文)	2021—02	58.00	1235
积分:分析学的关键(英文)	2020—12	48.00	1236
为有天分的新生准备的分析学基础教材(英文)	2020—11	48.00	1237
数学不等式.第一卷.对称多项式不等式(英文)	2021—03	108.00	1273
数学不等式.第二卷.对称有理不等式与对称无理不等式(英文)	2021—03	108.00	1274
数学不等式.第三卷.循环不等式与非循环不等式(英文)	2021—03	108.00	1275
数学不等式.第四卷.Jensen 不等式的扩展与加细(英文)	2021—03	108.00	1276
数学不等式.第五卷.创建不等式与解不等式的其他方法(英文)	2021—04	108.00	1277

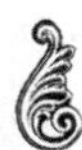

刘培杰数学工作室
已出版(即将出版)图书目录——原版影印

书　　名	出版时间	定　价	编号
冯·诺依曼代数中的谱位移函数:半有限冯·诺依曼代数中的谱位移函数与谱流(英文)	2021—06	98.00	1308
链接结构:关于嵌入完全图的直线中链接单形的组合结构(英文)	2021—05	58.00	1309
代数几何方法.第1卷(英文)	2021—06	68.00	1310
代数几何方法.第2卷(英文)	2021—06	68.00	1311
代数几何方法.第3卷(英文)	2021—06	58.00	1312
代数、生物信息和机器人技术的算法问题.第四卷,独立恒等式系统(俄文)	2020—08	118.00	1199
代数、生物信息和机器人技术的算法问题.第五卷,相对覆盖性和独立可拆分恒等式系统(俄文)	2020—08	118.00	1200
代数、生物信息和机器人技术的算法问题.第六卷,恒等式和准恒等式的相等 问题、可推导性和可实现性(俄文)	2020—08	128.00	1201
分数阶微积分的应用:非局部动态过程,分数阶导热系数(俄文)	2021—01	68.00	1241
泛函分析问题与练习:第2版(俄文)	2021—01	98.00	1242
集合论、数学逻辑和算法论问题:第5版(俄文)	2021—01	98.00	1243
微分几何和拓扑短期课程(俄文)	2021—01	98.00	1244
素数规律(俄文)	2021—01	88.00	1245
无穷边值问题解的递减:无界域中的拟线性椭圆和抛物方程(俄文)	2021—01	48.00	1246
微分几何讲义(俄文)	2020—12	98.00	1253
二次型和矩阵(俄文)	2021—01	98.00	1255
积分和级数.第2卷,特殊函数(俄文)	2021—01	168.00	1258
积分和级数.第3卷,特殊函数补充:第2版(俄文)	2021—01	178.00	1264
几何图上的微分方程(俄文)	2021—01	138.00	1259
数论教程:第2版(俄文)	2021—01	98.00	1260
非阿基米德分析及其应用(俄文)	2021—03	98.00	1261
古典群和量子群的压缩(俄文)	2021—03	98.00	1263
数学分析习题集.第3卷,多元函数:第3版(俄文)	2021—03	98.00	1266
数学习题:乌拉尔国立大学数学力学系大学生奥林匹克(俄文)	2021—03	98.00	1267
柯西定理和微分方程的特解(俄文)	2021—03	98.00	1268
组合极值问题及其应用:第3版(俄文)	2021—03	98.00	1269
数学词典(俄文)	2021—01	98.00	1271
确定性混沌分析模型(俄文)	2021—06	168.00	1307
精选初等数学习题和定理.立体几何.第3版(俄文)	2021—03	68.00	1316
微分几何习题:第3版(俄文)	2021—05	98.00	1336
精选初等数学习题和定理.平面几何.第4版(俄文)	2021—05	68.00	1335
曲面理论在欧氏空间 En 中的直接表示(俄文)	2022 01	68.00	1444
维纳—霍普夫离散算子和托普利兹算子:某些可数赋范空间中的诺特性和可逆性(俄文)	2022—03	108.00	1496
Maple 中的数论:数论中的计算机计算(俄文)	2022—03	88.00	1497
贝尔曼和克努特问题及其概括:加法运算的复杂性(俄文)	2022—03	138.00	1498

刘培杰数学工作室
已出版(即将出版)图书目录——原版影印

书　　名	出版时间	定　价	编号
复分析:共形映射(俄文)	2022－07	48.00	1542
微积分代数样条和多项式及其在数值方法中的应用(俄文)	2022－08	128.00	1543
蒙特卡罗方法中的随机过程和场模型:算法和应用(俄文)	2022－08	88.00	1544
线性椭圆型方程组:论二阶椭圆型方程的迪利克雷问题(俄文)	2022－08	98.00	1561
动态系统解的增长特性:估值、稳定性、应用(俄文)	2022－08	118.00	1565
群的自由积分解:建立和应用(俄文)	2022－08	78.00	1570
混合方程和偏差自变数方程问题:解的存在和唯一性(俄文)	2023－01	78.00	1582
拟度量空间分析:存在和逼近定理(俄文)	2023－01	108.00	1583
二维和三维流形上函数的拓扑性质:函数的拓扑分类(俄文)	2023－03	68.00	1584
齐次马尔科夫过程建模的矩阵方法:此类方法能够用于不同目上的的复杂系统研究、设计和完善(俄文)	2023－03	68.00	1594
周期函数的近似方法和特性:特殊课程(俄文)	2023－04	158.00	1622
扩散方程解的矩函数:变分法(俄文)	2023－03	58.00	1623
多赋范空间和广义函数:理论及应用(俄文)	2023－03	98.00	1632
分析中的多值映射:部分应用(俄文)	2023－06	98.00	1634
数学物理问题(俄文)	2023－03	78.00	1636
函数的幂级数与三角级数分解(俄文)	2024－01	58.00	1695
星体理论的数学基础:原子三元组(俄文)	2024－01	98.00	1696
素数规律:专著(俄文)	2024－01	118.00	1697
狭义相对论与广义相对论:时空与引力导论(英文)	2021－07	88.00	1319
束流物理学和粒子加速器的实践介绍:第2版(英文)	2021－07	88.00	1320
凝聚态物理中的拓扑和微分几何简介(英文)	2021－05	88.00	1321
混沌映射:动力学、分形学和快速涨落(英文)	2021－05	128.00	1322
广义相对论:黑洞、引力波和宇宙学介绍(英文)	2021－06	68.00	1323
现代分析电磁均质化(英文)	2021－06	68.00	1324
为科学家提供的基本流体动力学(英文)	2021－06	88.00	1325
视觉天文学:理解夜空的指南(英文)	2021－06	68.00	1326
物理学中的计算方法(英文)	2021－06	68.00	1327
单星的结构与演化:导论(英文)	2021－06	108.00	1328
超越居里:1903年至1963年物理界四位女性及其著名发现(英文)	2021－06	68.00	1329
范德瓦尔斯流体热力学的进展(英文)	2021－06	68.00	1330
先进的托卡马克稳定性理论(英文)	2021－06	88.00	1331
经典场论导论:基本相互作用的过程(英文)	2021－07	88.00	1332
光致电离量子动力学方法原理(英文)	2021－07	108.00	1333
经典域论和应力:能量张量(英文)	2021－05	88.00	1334
非线性太赫兹光谱的概念与应用(英文)	2021－06	68.00	1337
电磁学中的无穷空间并矢格林函数(英文)	2021－06	88.00	1338
物理科学基础数学.第1卷,齐次边值问题、傅里叶方法和特殊函数(英文)	2021－07	108.00	1339
离散量子力学(英文)	2021－07	68.00	1340
核磁共振的物理学和数学(英文)	2021－07	108.00	1341
分子水平的静电学(英文)	2021－08	68.00	1342
非线性波:理论、计算机模拟、实验(英文)	2021－06	108.00	1343
石墨烯光学:经典问题的电解解决方案(英文)	2021－06	68.00	1344
超材料多元宇宙(英文)	2021－07	68.00	1345
银河系外的天体物理学(英文)	2021－07	68.00	1346
原子物理学(英文)	2021－07	68.00	1347
将光打结:将拓扑学应用于光学(英文)	2021－07	68.00	1348
电磁学:问题与解法(英文)	2021－07	88.00	1364
海浪的原理:介绍量子力学的技巧与应用(英文)	2021－07	108.00	1365

刘培杰数学工作室
已出版(即将出版)图书目录——原版影印

书　　名	出版时间	定　价	编号
多孔介质中的流体:输运与相变(英文)	2021—07	68.00	1372
洛伦兹群的物理学(英文)	2021—08	68.00	1373
物理导论的数学方法和解决方法手册(英文)	2021—08	68.00	1374
非线性波数学物理学入门(英文)	2021—08	88.00	1376
波:基本原理和动力学(英文)	2021—07	68.00	1377
光电子量子计量学.第1卷,基础(英文)	2021—07	88.00	1383
光电子量子计量学.第2卷,应用与进展(英文)	2021—07	68.00	1384
复杂流的格子玻尔兹曼建模的工程应用(英文)	2021—08	68.00	1393
电偶极矩挑战(英文)	2021—08	108.00	1394
电动力学:问题与解法(英文)	2021—09	68.00	1395
自由电子激光的经典理论(英文)	2021—08	68.00	1397
曼哈顿计划——核武器物理学简介(英文)	2021—09	68.00	1401
粒子物理学(英文)	2021—09	68.00	1402
引力场中的量子信息(英文)	2021—09	128.00	1403
器件物理学的基本经典力学(英文)	2021—09	68.00	1404
等离子体物理及其空间应用导论.第1卷,基本原理和初步过程(英文)	2021—09	68.00	1405
磁约束聚变等离子体物理:理想MHD理论(英文)	2023—03	68.00	1613
相对论量子场论.第1卷,典范形式体系(英文)	2023—03	38.00	1615
相对论量子场论.第2卷,路径积分形式(英文)	2023—06	38.00	1616
相对论量子场论.第3卷,量子场论的应用(英文)	2023—06	38.00	1617
涌现的物理学(英文)	2023—05	58.00	1619
量子化旋涡:一本拓扑激发手册(英文)	2023—04	68.00	1620
非线性动力学:实践的介绍性调查(英文)	2023—05	68.00	1621
静电加速器:一个多功能工具(英文)	2023—06	58.00	1625
相对论多体理论与统计力学(英文)	2023—06	58.00	1626
经典力学.第1卷,工具与向量(英文)	2023—04	38.00	1627
经典力学.第2卷,运动学和匀加速运动(英文)	2023—04	58.00	1628
经典力学.第3卷,牛顿定律和匀速圆周运动(英文)	2023—04	58.00	1629
经典力学.第4卷,万有引力定律(英文)	2023—04	38.00	1630
经典力学.第5卷,守恒定律与旋转运动(英文)	2023—04	38.00	1631
对称问题:纳维尔—斯托克斯问题(英文)	2023—04	38.00	1638
摄影的物理和艺术.第1卷,几何与光的本质(英文)	2023—04	78.00	1639
摄影的物理和艺术.第2卷,能量与色彩(英文)	2023—04	78.00	1640
摄影的物理和艺术.第3卷,探测器与数码的意义(英文)	2023—04	78.00	1641

书　　名	出版时间	定　价	编号
拓扑与超弦理论焦点问题(英文)	2021—07	58.00	1349
应用数学:理论、方法与实践(英文)	2021—07	78.00	1350
非线性特征值问题:牛顿型方法与非线性瑞利函数(英文)	2021—07	58.00	1351
广义膨胀和齐性:利用齐性构造齐次系统的李雅普诺夫函数和控制律(英文)	2021—06	48.00	1352
解析数论焦点问题(英文)	2021—07	58.00	1353
随机微分方程:动态系统方法(英文)	2021—07	58.00	1354
经典力学与微分几何(英文)	2021—07	58.00	1355
负定相交形式流形上的瞬子模空间几何(英文)	2021—07	68.00	1356

书　　名	出版时间	定　价	编号
广义卡塔兰轨道分析:广义卡塔兰轨道计算数字的方法(英文)	2021—07	48.00	1367
洛伦兹方法的变分:二维与三维洛伦兹方法(英文)	2021—08	38.00	1378
几何、分析和数论精编(英文)	2021—08	68.00	1380
从一个新角度看数论:通过遗传方法引入现实的概念(英文)	2021—07	58.00	1387
动力系统:短期课程(英文)	2021—08	68.00	1382
几何路径:理论与实践(英文)	2021—08	48.00	1385

刘培杰数学工作室
已出版(即将出版)图书目录——原版影印

书　　名	出版时间	定　价	编号
论天体力学中某些问题的不可积性(英文)	2021－07	88.00	1396
广义斐波那契数列及其性质(英文)	2021－08	38.00	1386
对称函数和麦克唐纳多项式:余代数结构与 Kawanaka 恒等式(英文)	2021－09	38.00	1400
杰弗里·英格拉姆·泰勒科学论文集:第 1 卷. 固体力学(英文)	2021－05	78.00	1360
杰弗里·英格拉姆·泰勒科学论文集:第 2 卷. 气象学、海洋学和湍流(英文)	2021－05	68.00	1361
杰弗里·英格拉姆·泰勒科学论文集:第 3 卷. 空气动力学以及落弹数和爆炸的力学(英文)	2021－05	68.00	1362
杰弗里·英格拉姆·泰勒科学论文集:第 4 卷. 有关流体力学(英文)	2021－05	58.00	1363
非局域泛函演化方程:积分与分数阶(英文)	2021－08	48.00	1390
理论工作者的高等微分几何:纤维丛、射流流形和拉格朗日理论(英文)	2021－08	68.00	1391
半线性退化椭圆微分方程:局部定理与整体定理(英文)	2021－07	48.00	1392
非交换几何、规范理论和重整化:一般简介与非交换量子场论的重整化(英文)	2021－09	78.00	1406
数论论文集:拉普拉斯变换和带有数论系数的幂级数(俄文)	2021－09	48.00	1407
挠理论专题:相对极大值,单射与扩充模(英文)	2021－09	88.00	1410
强正则图与欧几里得若尔当代数:非通常关系中的启示(英文)	2021－10	48.00	1411
拉格朗日几何和哈密顿几何:力学的应用(英文)	2021－10	48.00	1412
时滞微分方程与差分方程的振动理论:二阶与三阶(英文)	2021－10	98.00	1417
卷积结构与几何函数理论:用以研究特定几何函数理论方向的分数阶微积分算子与卷积结构(英文)	2021－10	48.00	1418
经典数学物理的历史发展(英文)	2021－10	78.00	1419
扩展线性丢番图问题(英文)	2021－10	38.00	1420
一类混沌动力系统的分歧分析与控制:分歧分析与控制(英文)	2021－11	38.00	1421
伽利略空间和伪伽利略空间中一些特殊曲线的几何性质(英文)	2022－01	68.00	1422
一阶偏微分方程:哈密尔顿—雅可比理论(英文)	2021－11	48.00	1424
各向异性黎曼多面体的反问题:分段光滑的各向异性黎曼多面体反边界谱问题:唯一性(英文)	2021－11	38.00	1425
项目反应理论手册. 第一卷,模型(英文)	2021－11	138.00	1431
项目反应理论手册. 第二卷,统计工具(英文)	2021－11	118.00	1432
项目反应理论手册. 第三卷,应用(英文)	2021－11	138.00	1433
二次无理数:经典数论入门(英文)	2022－05	138.00	1434

刘培杰数学工作室
已出版(即将出版)图书目录——原版影印

书　　名	出版时间	定　价	编号
数,形与对称性:数论,几何和群论导论(英文)	2022—05	128.00	1435
有限域手册(英文)	2021—11	178.00	1436
计算数论(英文)	2021—11	148.00	1437
拟群与其表示简介(英文)	2021—11	88.00	1438
数论与密码学导论:第二版(英文)	2022—01	148.00	1423
几何分析中的柯西变换与黎兹变换:解析调和容量和李普希兹调和容量、变化和振荡以及一致可求长性(英文)	2021—12	38.00	1465
近似不动点定理及其应用(英文)	2022—05	28.00	1466
局部域的相关内容解析:对局部域的扩展及其伽罗瓦群的研究(英文)	2022—01	38.00	1467
反问题的二进制恢复方法(英文)	2022—03	28.00	1468
对几何函数中某些类的各个方面的研究:复变量理论(英文)	2022—01	38.00	1469
覆盖、对应和非交换几何(英文)	2022—01	28.00	1470
最优控制理论中的随机线性调节器问题:随机最优线性调节器问题(英文)	2022—01	38.00	1473
正交分解法:涡流流体动力学应用的正交分解法(英文)	2022—01	38.00	1475
芬斯勒几何的某些问题(英文)	2022—03	38.00	1476
受限三体问题(英文)	2022—05	38.00	1477
利用马利亚万微积分进行 Greeks 的计算:连续过程、跳跃过程中的马利亚万微积分和金融领域中的 Greeks(英文)	2022—05	48.00	1478
经典分析和泛函分析的应用:分析学的应用(英文)	2022—03	38.00	1479
特殊芬斯勒空间的探究(英文)	2022—03	48.00	1480
某些图形的施泰纳距离的细谷多项式:细谷多项式与图的维纳指数(英文)	2022—05	38.00	1481
图论问题的遗传算法:在新鲜与模糊的环境中(英文)	2022—05	48.00	1482
多项式映射的渐近簇(英文)	2022—05	38.00	1483
一维系统中的混沌:符号动力学,映射序列,一致收敛和沙可夫斯基定理(英文)	2022—05	38.00	1509
多维边界层流动与传热分析:粘性流体流动的数学建模与分析(英文)	2022—05	38.00	1510
演绎理论物理学的原理:一种基于量子力学波函数的逐次置信估计的一般理论的提议(英文)	2022—05	38.00	1511
R^2 和 R^3 中的仿射弹性曲线:概念和方法(英文)	2022—08	38.00	1512
算术数列中除数函数的分布:基本内容、调查、方法、第二矩、新结果(英文)	2022—05	28.00	1513
抛物型狄拉克算子和薛定谔方程:不定常薛定谔方程的抛物型狄拉克算子及其应用(英文)	2022—07	28.00	1514
黎曼-希尔伯特问题与量子场论:可积重正化、戴森-施温格方程(英文)	2022—08	38.00	1515
代数结构和几何结构的形变理论(英文)	2022—08	48.00	1516
概率结构和模糊结构上的不动点:概率结构和直觉模糊度量空间的不动点定理(英文)	2022—08	38.00	1517

刘培杰数学工作室
已出版(即将出版)图书目录——原版影印

书　　名	出版时间	定　价	编号
反若尔当对:简单反若尔当对的自同构(英文)	2022—07	28.00	1533
对某些黎曼—芬斯勒空间变换的研究:芬斯勒几何中的某些变换(英文)	2022—07	38.00	1534
内诣零流形映射的尼尔森数的阿诺索夫关系(英文)	2023—01	38.00	1535
与广义积分变换有关的分数次演算:对分数次演算的研究(英文)	2023—01	48.00	1536
强子的芬斯勒几何和吕拉几何(宇宙学方面):强子结构的芬斯勒几何和吕拉几何(拓扑缺陷)(英文)	2022—08	38.00	1537
一种基于混沌的非线性最优化问题:作业调度问题(英文)	2023—03	38.00	1538
广义概率论发展前景:关于趣味数学与置信函数实际应用的一些原创观点(英文)	2023—03	48.00	1539
纽结与物理学:第二版(英文)	2022—09	118.00	1547
正交多项式和q—级数的前沿(英文)	2022—09	98.00	1548
算子理论问题集(英文)	2022—09	108.00	1549
抽象代数:群、环与域的应用导论:第二版(英文)	2023—01	98.00	1550
菲尔兹奖得主演讲集:第三版(英文)	2023—01	138.00	1551
多元实函数教程(英文)	2022—09	118.00	1552
球面空间形式群的几何学:第二版(英文)	2022—09	98.00	1566
对称群的表示论(英文)	2023—01	98.00	1585
纽结理论:第二版(英文)	2023—01	88.00	1586
拟群理论的基础与应用(英文)	2023—01	88.00	1587
组合学:第二版(英文)	2023—01	98.00	1588
加性组合学:研究问题手册(英文)	2023—01	68.00	1589
扭曲、平铺与镶嵌:几何折纸中的数学方法(英文)	2023—01	98.00	1590
离散与计算几何手册:第三版(英文)	2023—01	248.00	1591
离散与组合数学手册:第二版(英文)	2023—01	248.00	1592
分析学教程.第1卷,一元实变量函数的微积分分析学介绍(英文)	2023—01	118.00	1595
分析学教程.第2卷,多元函数的微分和积分,向量微积分(英文)	2023—01	118.00	1596
分析学教程.第3卷,测度与积分理论,复变量的复值函数(英文)	2023—01	118.00	1597
分析学教程.第4卷,傅里叶分析,常微分方程,变分法(英文)	2023—01	118.00	1598

刘培杰数学工作室
已出版(即将出版)图书目录——原版影印

书　　名	出版时间	定　价	编号
共形映射及其应用手册(英文)	2024—01	158.00	1674
广义三角函数与双曲函数(英文)	2024—01	78.00	1675
振动与波:概论:第二版(英文)	2024—01	88.00	1676
几何约束系统原理手册(英文)	2024—01	120.00	1677
微分方程与包含的拓扑方法(英文)	2024—01	98.00	1678
数学分析中的前沿话题(英文)	2024—01	198.00	1679
流体力学建模:不稳定性与湍流(英文)	2024—03	88.00	1680
动力系统:理论与应用(英文)	2024—03	108.00	1711
空间统计学理论:概述(英文)	2024—03	68.00	1712
梅林变换手册(英文)	2024—03	128.00	1713
非线性系统及其绝妙的数学结构.第1卷(英文)	2024—03	88.00	1714
非线性系统及其绝妙的数学结构.第2卷(英文)	2024—03	108.00	1715
Chip-firing 中的数学(英文)	2024—04	88.00	1716
阿贝尔群的可确定性:问题、研究、概述(俄文)	2024—05	716.00(全7册)	1727
素数规律:专著(俄文)	2024—05	716.00(全7册)	1728
函数的幂级数与三角级数分解(俄文)	2024—05	716.00(全7册)	1729
星体理论的数学基础:原子三元组(俄文)	2024—05	716.00(全7册)	1730
技术问题中的数学物理微分方程(俄文)	2024—05	716.00(全7册)	1731
概率论边界问题:随机过程边界穿越问题(俄文)	2024—05	716.00(全7册)	1732
代数和幂等配置的正交分解:不可交换组合(俄文)	2024—05	716.00(全7册)	1733

联系地址:哈尔滨市南岗区复华四道街10号　哈尔滨工业大学出版社刘培杰数学工作室
邮　　编:150006
联系电话:0451—86281378　　13904613167
E-mail:lpj1378@163.com